T0189901

Lecture Notes in Computer Science 3152

Commenced Publication in 1973
Founding and Former Series Editors:
Gerhard Goos, Juris Hartmanis, and Jan van Leeuwen

Lecture Notes in Computer Science 3122

Matt Franklin (Ed.)

Advances in Cryptology – CRYPTO 2004

24th Annual International Cryptology Conference
Santa Barbara, California, USA, August 15-19, 2004
Proceedings

 Springer

Volume Editor

Matt Franklin
University of California, Department of Computer Science
One Shields Avenue, Davis, CA 95616, USA
E-mail: franklin@cs.ucdavis.edu

Library of Congress Control Number: Applied for

CR Subject Classification (1998): E.3, G.2.1, F.2.1-2, D.4.6, K.6.5, C.2, J.1

ISSN 0302-9743
ISBN 978-3-540-22668-0   ISBN 978-3-540-28628-8 (eBook)
DOI 10.1007/978-3-540-28628-8

Springer is a part of Springer Science+Business Media

springeronline.com

© International Association for Cryptologic Research

Typesetting: Camera-ready by author, data conversion by Olgun Computergrafik
Printed on acid-free paper   SPIN: 11302322   06/3142   5 4 3 2 1 0

Preface

Crypto 2004, the 24th Annual Crypto Conference, was sponsored by the International Association for Cryptologic Research (IACR) in cooperation with the IEEE Computer Society Technical Committee on Security and Privacy and the Computer Science Department of the University of California at Santa Barbara.

The program committee accepted 33 papers for presentation at the conference. These were selected from a total of 211 submissions. Each paper received at least three independent reviews. The selection process included a Web-based discussion phase, and a one-day program committee meeting at New York University.

These proceedings include updated versions of the 33 accepted papers. The authors had a few weeks to revise them, aided by comments from the reviewers. However, the revisions were not subjected to any editorial review.

The conference program included two invited lectures. Victor Shoup's invited talk was a survey on chosen ciphertext security in public-key encryption. Susan Landau's invited talk was entitled "Security, Liberty, and Electronic Communications". Her extended abstract is included in these proceedings.

We continued the tradition of a Rump Session, chaired by Stuart Haber. Those presentations (always short, often serious) are not included here.

I would like to thank everyone who contributed to the success of this conference. First and foremost, the global cryptographic community submitted their scientific work for our consideration. The members of the Program Committee worked hard throughout, and did an excellent job. Many external reviewers contributed their time and expertise to aid our decision-making. James Hughes, the General Chair, was supportive in a number of ways. Dan Boneh and Victor Shoup gave valuable advice. Yevgeniy Dodis hosted the PC meeting at NYU.

It would have been hard to manage this task without the Web-based submission server (developed by Chanathip Namprempre, under the guidance of Mihir Bellare) and review server (developed by Wim Moreau and Joris Claessens, under the guidance of Bart Preneel). Terri Knight kept these servers running smoothly, and helped with the preparation of these proceedings.

June 2004   Matt Franklin

CRYPTO 2004

August 15–19, 2004, Santa Barbara, California, USA

Sponsored by the
International Association for Cryptologic Research (IACR)

in cooperation with
IEEE Computer Society Technical Committee on Security and Privacy,
Computer Science Department, University of California, Santa Barbara

General Chair
James Hughes, StorageTek

Program Chair
Matt Franklin, U.C. Davis, USA

Program Committee

Bill Aiello .... AT&T Labs, USA
Jee Hea An .... SoftMax, USA
Eli Biham .... Technion, Israel
John Black .... University of Colorado at Boulder, USA
Anne Canteaut .... INRIA, France
Ronald Cramer .... University of Aarhus, Denmark
Yevgeniy Dodis .... New York University, USA
Yuval Ishai .... Technion, Israel
Lars Knudsen .... Technical University of Denmark, Denmark
Hugo Krawczyk .... Technion/IBM, Israel/USA
Pil Joong Lee .... POSTECH/KT, Korea
Phil MacKenzie .... Bell Labs, USA
Tal Malkin .... Columbia University, USA
Willi Meier .... Fachhochschule Aargau, Switzerland
Daniele Micciancio .... U.C. San Diego, USA
Ilya Mironov .... Microsoft Research, USA
Tatsuaki Okamoto .... NTT, Japan
Rafail Ostrovsky .... U.C.L.A., USA
Torben Pedersen .... Cryptomathic, Denmark
Benny Pinkas .... HP Labs, USA
Bart Preneel .... Katholieke Universiteit Leuven, Belgium
Alice Silverberg .... Ohio State University, USA
Nigel Smart .... Bristol University, UK
David Wagner .... U.C. Berkeley, USA
Stefan Wolf .... University of Montreal, Canada

Advisory Members

Dan Boneh (Crypto 2003 Program Chair) .... Stanford University, USA
Victor Shoup (Crypto 2005 Program Chair) ...... New York University, USA

External Reviewers

Masayuki Abe	Pierrick Gaudry	Marine Minier
Siddhartha Annapuredy	Rosario Gennaro	Bodo Moeller
Frederik Armknecht	Craig Gentry	Håvard Molland
Daniel Augot	Shafi Goldwasser	David Molnar
Boaz Barak	Jovan Golic	Tal Mor
Elad Barkan	Rob Granger	Sara Miner More
Amos Beimel	Jens Groth	François Morain
Mihir Bellare	Stuart Haber	Waka Nagao
Daniel Bleichenbacher	Shai Halevi	Phong Nguyen
Dan Boneh	Helena Handschuh	Antonio Nicolosi
Carl Bosley	Danny Harnik	Jesper Nielsen
Ernie Brickell	Johan Haståd	Miyako Ohkubo
Ran Canetti	Alejandro Hevia	Kazuo Ohta
Jung Hee Cheon	Jim Hughes	Roberto Oliveira
Don Coppersmith	Yong Ho Hwang	Seong-Hun Paeng
Jean-Sébastien Coron	Oleg Izmerly	Dan Page
Nicolas Courtois	Markus Jakobsson	Dong Jin Park
Christophe De Cannière	Stanislaw Jarecki	Jae Hwan Park
Anand Desai	Rob Johnson	Joonhah Park
Simon-Pierre Desrosiers	Yael Tauman Kalai	Matthew Parker
Irit Dinur	Jonathan Katz	Rafael Pass
Mario di Raimondo	Dan Kenigsberg	Kenny Paterson
Orr Dunkelman	Dmitriy Kharchenko	Erez Petrank
Glenn Durfee	Aggelos Kiayias	David Pointcheval
Iwan Duursma	Eike Kiltz	Prashant Puniya
Stefan Dziembowski	Kihyun Kim	Tal Rabin
Andreas Enge	Ted Krovetz	Haavard Raddum
Nelly Fazio	Klaus Kursawe	Zulfikar Ramzan
Serge Fehr	Eyal Kushilevitz	Oded Regev
Marc Fischlin	Joseph Lano	Omer Reingold
Matthias Fitzi	In-Sok Lee	Renato Renner
Caroline Fontaine	Arjen Lenstra	Leonid Reyzin
Michael J. Freedman	Yehuda Lindell	Vincent Rijmen
Atsushi Fujioka	Hoi-Kwong Lo	Phillip Rogaway
Eiichiro Fujisaki	Pierre Loidreau	Pankaj Rohatgi
Martin Gagne	Anna Lysyanskaya	Adi Rosen
Steven Galbraith	John Malone-Lee	Karl Rubin
Juan Garay	Dominic Mayers	Alex Russell

Amit Sahai
Gorm Salomonsen
Louis Salvail
Tomas Sander
Hovav Shacham
Ronen Shaltiel
Jonghoon Shin
Victor Shoup
Thomas Shrimpton
Berit Skjernaa
Adam Smith
Jerome A. Solinas
Jessica Staddon

Martijn Stam
Jacques Stern
Douglas Stinson
Koutarou Suzuki
Keisuke Tanaka
Edlyn Teske
Christian Tobias
Yuuki Tokunaga
Vinod Vaikuntanathan
Brigitte Vallee
R. Venkatesan
Frederik Vercauteren
Felipe Voloch

Luis von Ahn
Jason Waddle
Shabsi Walfish
Andreas Winter
Christopher Wolf
Juerg Wullschleger
Go Yamamoto
Yeon Hyeong Yang
Sung Ho Yoo
Young Tae Youn
Dae Hyun Yum
Moti Yung

Table of Contents

Public Key Cryptanalysis

Zero-Knowledge

Hash Collisions

Secure Computation

Invited Talk

Stream Cipher Cryptanalysis

Public Key Encryption

Bounded Storage Model

Key Management

Computationally Unbounded Adversaries

Author Index

On Multiple Linear Approximations*

Alex Biryukov**, Christophe De Cannière***, and Michaël Quisquater***

Katholieke Universiteit Leuven, Dept. ESAT/SCD-COSIC,
Kasteelpark Arenberg 10,
B–3001 Leuven-Heverlee, Belgium
{abiryuko,cdecanni,mquisqua}@esat.kuleuven.ac.be

Abstract. In this paper we study the long standing problem of information extraction from multiple linear approximations. We develop a formal statistical framework for block cipher attacks based on this technique and derive explicit and compact gain formulas for generalized versions of Matsui's Algorithm 1 and Algorithm 2. The theoretical framework allows both approaches to be treated in a unified way, and predicts significantly improved attack complexities compared to current linear attacks using a single approximation. In order to substantiate the theoretical claims, we benchmarked the attacks against reduced-round versions of DES and observed a clear reduction of the data and time complexities, in almost perfect correspondence with the predictions. The complexities are reduced by several orders of magnitude for Algorithm 1, and the significant improvement in the case of Algorithm 2 suggests that this approach may outperform the currently best attacks on the full DES algorithm.

Keywords: Linear cryptanalysis, multiple linear approximations, stochastic systems of linear equations, maximum likelihood decoding, key-ranking, DES, AES.

1 Introduction

Linear cryptanalysis [8] is one of the most powerful attacks against modern cryptosystems. In 1994, Kaliski and Robshaw [5] proposed the idea of generalizing this attack using multiple linear approximations (the previous approach considered only the best linear approximation). However, their technique was mostly limited to cases where all approximations derive the same parity bit of the key. Unfortunately, this approach imposes a very strong restriction on the approximations, and the additional information gained by the few surviving approximations is often negligible.

In this paper we start by developing a theoretical framework for dealing with multiple linear approximations. We first generalize Matsui's Algorithm 1 based

---

* This work was supported in part by the Concerted Research Action (GOA) Mefisto-2000/06 of the Flemish Government.
** F.W.O. Researcher, Fund for Scientific Research – Flanders (Belgium).
*** F.W.O. Research Assistant, Fund for Scientific Research – Flanders (Belgium).

on this framework, and then reuse these results to generalize Matsui's Algorithm 2. Our approach allows to derive compact expressions for the performance of the attacks in terms of the biases of the approximations and the amount of data available to the attacker. The contribution of these theoretical expressions is twofold. Not only do they clearly demonstrate that the use of multiple approximations can significantly improve classical linear attacks, they also shed a new light on the relations between Algorithm 1 and Algorithm 2.

The main purpose of this paper is to provide a new generally applicable cryptanalytic tool, which performs strictly better than standard linear cryptanalysis. In order to illustrate the potential of this new approach, we implemented two attacks against reduced-round versions of DES, using this cipher as a well established benchmark for linear cryptanalysis. The experimental results, discussed in the second part of this paper, are in almost perfect correspondence with our theoretical predictions and show that the latter are well justified.

This paper is organized as follows: Sect. 2 describes a very general maximum likelihood framework, which we will use in the rest of the paper; in Sect. 3 this framework is applied to derive and analyze an optimal attack algorithm based on multiple linear approximations. In the last part of this section, we provide a more detailed theoretical analysis of the assumptions made in order to derive the performance expressions. Sect. 4 presents experimental results on DES as an example. Finally, Sect. 5 discusses possible further improvements and open questions. A more detailed discussion of the practical aspects of the attacks and an overview of previous work can be found in the appendices.

2   General Framework

In this section we discuss the main principles of statistical cryptanalysis and set up a generalized framework for analyzing block ciphers based on maximum likelihood. This framework can be seen as an adaptation or extension of earlier frameworks for statistical attacks proposed by Murphy et al. [11], Junod and Vaudenay [3, 4, 14] and Selçuk [12].

2.1   Attack Model

We consider a block cipher E_k which maps a plaintext $P \in \mathcal{P}$ to a ciphertext $C = E_k(P) \in \mathcal{C}$. The mapping is invertible and depends on a secret key $k \in \mathcal{K}$. We now assume that an adversary is given N different plaintext–ciphertext pairs (P_i, C_i) encrypted with a particular secret key k^* (a known plaintext scenario), and his task is to recover the key from this data. A general statistical approach — also followed by Matsui's original linear cryptanalysis — consists in performing the following three steps:

Distillation phase. In a typical statistical attack, only a fraction of the information contained in the N plaintext–ciphertext pairs is exploited. A first step therefore consists in extracting the relevant parts of the data, and discarding all information which is not used by the attack. In our framework, the distillation operation is denoted by a function $\psi : \mathcal{P} \times \mathcal{C} \to \mathcal{X}$ which is applied to each plaintext–ciphertext pair. The result is a vector $\mathbf{x} = (x_1, \ldots, x_N)$ with $x_i = \psi(P_i, C_i)$, which contains all relevant information. If $|\mathcal{X}| \ll N$, which is usually the case, we can further reduce the data by counting the occurrence of each element of $\mathcal{X}$ and only storing a vector of counters $\mathbf{t} = (t_0, \ldots, t_{|\mathcal{X}|-1})$. In this paper we will not restrict ourselves to a single function ψ, but consider m separate functions ψ_j, each of which maps the text pairs into different sets $\mathcal{X}_j$ and generates a separate vector of counters $\mathbf{t}_j$.

Analysis phase. This phase is the core of the attack and consists in generating a list of key candidates from the information extracted in the previous step. Usually, candidates can only be determined up to a set of equivalent keys, *i.e.*, typically, a majority of the key bits is transparent to the attack. In general, the attack defines a function $\sigma : \mathcal{K} \to \mathcal{Z}$ which maps each key k onto an equivalent key class $z = \sigma(k)$. The purpose of the analysis phase is to determine which of these classes are the most likely to contain the true key k^* given the particular values of the counters $\mathbf{t}_j$.

Search phase. In the last stage of the attack, the attacker exhaustively tries all keys in the classes suggested by the previous step, until the correct key is found. Note that the analysis and the searching phase may be intermixed: the attacker might first generate a short list of candidates, try them out, and then dynamically extend the list as long as none of the candidates turns out to be correct.

2.2   Attack Complexities

When evaluating the performance of the general attack described above, we need to consider both the data complexity and the computational complexity. The data complexity is directly determined by N, the number of plaintext–ciphertext pairs required by the attack. The computational complexity depends on the total number of operations performed in the three phases of the attack. In order to compare different types of attacks, we define a measure called the *gain* of the attack:

Definition 1 (Gain). *If an attack is used to recover an n-bit key and is expected to return the correct key after having checked on the average M candidates, then the* gain *of the attack, expressed in bits, is defined as:*

$$\gamma = -\log_2 \frac{2 \cdot M - 1}{2^n} \qquad (1)$$

Let us illustrate this with an example where an attacker wants to recover an n-bit key. If he does an exhaustive search, the number of trials before hitting the correct key can be anywhere from 1 to 2^n. The average number M is $(2^n + 1)/2$, and the gain according to the definition is 0. On the other hand, if the attack immediately derives the correct candidate, M equals 1 and the gain is $\gamma = n$. There is an important caveat, however. Let us consider two attacks which both require a single plaintext–ciphertext pair. The first deterministically recovers one bit of the key, while the second recovers the complete key, but with a probability of 1/2. In this second attack, if the key is wrong and only one plaintext–ciphertext pair is available, the attacker is forced to perform an exhaustive search. According to the definition, both attacks have a gain of 1 bit in this case. Of course, by repeating the second attack for different pairs, the gain can be made arbitrary close to n bits, while this is not the case for the first attack.

2.3   Maximum Likelihood Approach

The design of a statistical attack consists of two important parts. First, we need to decide on how to process the N plaintext–ciphertext pairs in the distillation phase. We want the counters t_j to be constructed in such a way that they concentrate as much information as possible about a specific part of the secret key in a minimal amount of data. Once this decision has been made, we can proceed to the next stage and try to design an algorithm which efficiently transforms this information into a list of key candidates. In this section, we discuss a general technique to optimize this second step. Notice that throughout this paper, we will denote random variables by capital letters.

In order to minimize the amount of trials in the search phase, we want the candidate classes which have the largest probability of being correct to be tried first. If we consider the correct key class as a random variable Z and denote the complete set of counters extracted from the observed data by $\mathbf{t}$, then the ideal output of the analysis phase would consist of a list of classes $\{z\}$, sorted according to the conditional probability $\Pr[Z = z \mid \mathbf{t}]$. Taking the Bayesian approach, we express this probability as follows:

$$\Pr[Z = z \mid \mathbf{t}] = \frac{\Pr[\mathbf{T} = \mathbf{t} \mid z] \cdot \Pr[Z = z]}{\Pr[\mathbf{T} = \mathbf{t}]}. \tag{2}$$

The factor $\Pr[Z = z]$ denotes the a priori probability that the class z contains the correct key k^*, and is equal to the constant $1/|\mathcal{Z}|$, with $|\mathcal{Z}|$ the total number of classes, provided that the key was chosen at random. The denominator is determined by the probability that the specific set of counters $\mathbf{t}$ is observed, taken over all possible keys and plaintexts. The only expression in (2) that depends on z, and thus affects the sorting, is the factor $\Pr[\mathbf{T} = \mathbf{t} \mid z]$, compactly written as $P_z(\mathbf{t})$. This quantity denotes the probability, taken over all possible plaintexts, that a key from a given class z produces a set of counters $\mathbf{t}$. When viewed as a function of z for a fixed set $\mathbf{t}$, the expression $\Pr[\mathbf{T} = \mathbf{t} \mid z]$ is also called the *likelihood* of z given $\mathbf{t}$, and denoted by $L_{\mathbf{t}}(z)$, i.e.,

$$L_{\mathbf{t}}(z) = P_z(\mathbf{t}) = \Pr[\mathbf{T} = \mathbf{t} \mid z].$$

This likelihood and the actual probability $\Pr[Z = z \mid \mathbf{t}]$ have distinct values, but they are proportional for a fixed $\mathbf{t}$, as follows from (2). Typically, the likelihood expression is simplified by applying a logarithmic transformation. The result is denoted by

$$\mathcal{L}_\mathbf{t}(z) = \log L_\mathbf{t}(z)$$

and called the *log-likelihood*. Note that this transformation does not affect the sorting, since the logarithm is a monotonously increasing function.

Assuming that we can construct an efficient algorithm that accurately estimates the likelihood of the key classes and returns a list sorted accordingly, we are now ready to derive a general expression for the gain of the attack.

Let us assume that the plaintexts are encrypted with an n-bit secret key k^*, contained in the equivalence class z^*, and let $\mathcal{Z}^* = \mathcal{Z} \setminus \{z^*\}$ be the set of classes *different* from z^*. The average number of classes checked during the searching phase before the correct key is found, is given by the expression

$$1 + \sum_{z \in \mathcal{Z}^*} \Pr\left[\mathcal{L}_\mathbf{T}(z) \geq \mathcal{L}_\mathbf{T}(z^*) \mid z^*\right],$$

where the random variable $\mathbf{T}$ represents the set of counters generated by a key from the class z^*, given N random plaintexts. Note that this number includes the correct key class, but since this class will be treated differently later on, we do not include it in the sum. In order to compute the probabilities in this expression, we define the sets $\mathcal{T}_z = \{\mathbf{t} \mid \mathcal{L}_\mathbf{t}(z) \geq \mathcal{L}_\mathbf{t}(z^*)\}$. Using this notation, we can write

$$\Pr\left[\mathcal{L}_\mathbf{T}(z) \geq \mathcal{L}_\mathbf{T}(z^*) \mid z^*\right] = \sum_{\mathbf{t} \in \mathcal{T}_z} P_{z^*}(\mathbf{t}).$$

Knowing that each class z contains $2^n/|\mathcal{Z}|$ different keys, we can now derive the expected number of trials M^*, given a secret key k^*. Note that the number of keys that need to be checked in the correct equivalence class z^* is only $(2^n/|\mathcal{Z}|+1)/2$ on the average, yielding

$$M^* = \frac{2^n}{|\mathcal{Z}|} \cdot \left[\frac{1}{2} + \sum_{z \in \mathcal{Z}^*} \sum_{\mathbf{t} \in \mathcal{T}_z} P_{z^*}(\mathbf{t})\right] + \frac{1}{2}. \tag{3}$$

This expression needs to be averaged over all possible secret keys k^* in order to find the expected value M, but in many cases[1] we will find that M^* does not depend on the actual value of k^*, such that $M = M^*$. Finally, the gain of the attack is computed by substituting this value of M into (1).

3   Application to Multiple Approximations

In this section, we apply the ideas discussed above to construct a general framework for analyzing block ciphers using multiple linear approximations.

---

[1] In some cases the variance of the gain over different keys would be very significant. In these cases it might be worth to exploit this phenomenon in a weak-key attack scenario, like in the case of the IDEA cipher.

The starting point in linear cryptanalysis is the existence of unbalanced linear expressions involving plaintext bits, ciphertext bits, and key bits. In this paper we assume that we can use m such expressions (a method to find them is presented in an extended version of this paper [1]):

$$\Pr\left[P[\chi_P^j] \oplus C[\chi_C^j] \oplus K[\chi_K^j] = 0\right] = \frac{1}{2} + \epsilon_j, \quad j = 1, \ldots, m, \tag{4}$$

with (P, C) a random plaintext–ciphertext pair encrypted with a random key K. The notation $X[\chi]$ stands for $X_{l_1} \oplus X_{l_2} \oplus \ldots \oplus X_{l_a}$, where $X_{l_1}, \ldots, X_{l_a}$ represent particular bits of X. The deviation ϵ_j is called the *bias* of the linear expression.

We now use the framework of Sect. 2.1 to design an attack which exploits the information contained in (4). The first phase of the cryptanalysis consists in extracting the relevant parts from the N plaintext–ciphertext pairs. The linear expressions in (4) immediately suggest the following functions ψ_j:

$$x_{i,j} = \psi_j(P_i, C_i) = P_i[\chi_P^j] \oplus C_i[\chi_C^j], \quad i = 1, \ldots, N,$$

with $x_{i,j} \in \mathcal{X}_j = \{0, 1\}$. These values are then used to construct m counter vectors $\mathbf{t_j} = (t_j, N - t_j)$, where t_j and $N - t_j$ reflect the number of plaintext–ciphertext pairs for which $x_{i,j}$ equals 0 and 1, respectively[2].

In the second step of the framework, a list of candidate key classes needs to be generated. We represent the equivalent key classes induced by the m linear expressions in (4) by an m-bit word $z = (z_1, \ldots, z_m)$ with $z_j = k[\chi_K^j]$. Note that m might possibly be much larger than n, the length of the key k. In this case, only a subspace of all possible m-bit words corresponds to a valid key class. The exact number of classes $|\mathcal{Z}|$ depends on the number of *independent* linear approximations (*i.e.*, the rank of the corresponding linear system).

3.1   Computing the Likelihoods of the Key Classes

We will for now assume that the linear expressions in (4) are statistically independent for different plaintext–ciphertext pairs and for different values of j (in the next section we will discuss this important point in more details). This allows us to apply the maximum likelihood approach described earlier in a very straightforward way. In order to simplify notations, we define the probabilities p_j and q_j, and the *imbalances*[3] c_j of the linear expressions as

$$p_j = 1 - q_j = \frac{1 + c_j}{2} = \frac{1}{2} + \epsilon_j.$$

We start by deriving a convenient expression for the probability $P_z(\mathbf{t})$. To simplify the calculation, we first give a derivation for the special key class

---

[2] The vectors $\mathbf{t_j}$ are only constructed to be consistent with the framework described earlier. In practice of course, the attacker will only calculate t_j (this is a minimal sufficient statistic).

[3] Also known in the literature as "correlations".

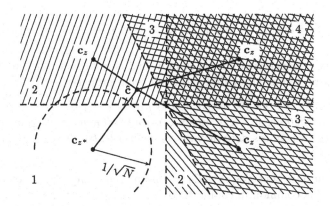

Fig. 1. Geometrical interpretation for $m = 2$. The correct key class z^* has the second largest likelihood in this example. The numbers in the picture represent the number of trials M^* when $\hat{c}$ falls in the associated area.

$z' = (0, \dots, 0)$. Assuming independence of different approximations and of different (P_i, C_i) pairs, the probability that this key generates the counters t_j is given by the product

$$P_{z'}(\mathbf{t}) = \prod_{j=1}^{m} \binom{N}{t_j} \cdot p_j^{t_j} \cdot q_j^{N-t_j} . \tag{5}$$

In practice, p_j and q_j will be very close to $1/2$, and N very large. Taking this into account, we approximate the m-dimensional binomial distribution above by an m-dimensional Gaussian distribution:

$$P_{z'}(\mathbf{t}) \approx \prod_{j=1}^{m} \frac{e^{-\frac{(t_j - p_j \cdot N)^2}{N/2}}}{\sqrt{\pi \cdot N/2}} = \prod_{j=1}^{m} \frac{e^{-\frac{N}{2}(\hat{c}_j - c_j)^2}}{\sqrt{\pi \cdot N/2}} = \frac{e^{-\frac{N}{2}\sum(\hat{c}_j - c_j)^2}}{\left(\sqrt{\pi \cdot N/2}\right)^m} .$$

The variable $\hat{c}_j$ is called the *estimated imbalance* and is derived from the counters t_j according to the relation $N \cdot (1 + \hat{c}_j)/2 = t_j$. For any key class z, we can repeat the reasoning above, yielding the following general expression:

$$P_z(\mathbf{t}) \approx \frac{e^{-\frac{N}{2}\sum(\hat{c}_j - (-1)^{z_j} \cdot c_j)^2}}{\left(\sqrt{\pi \cdot N/2}\right)^m} \tag{6}$$

This formula has a useful geometrical interpretation: if we take a key from a fixed key class z^* and construct an m-dimensional vector $\hat{\mathbf{c}} = (\hat{c}_1, \dots, \hat{c}_m)$ by encrypting N random plaintexts, then $\hat{\mathbf{c}}$ will be distributed around the vector $\mathbf{c}_{\mathbf{z}^*} = ((-1)^{z_1^*} c_1, \dots, (-1)^{z_m^*} c_m)$ according to a Gaussian distribution with a diagonal variance-covariance matrix $1/\sqrt{N} \cdot I_m$, where I_m is an $m \times m$ identity matrix. This is illustrated in Fig. 1. From (6) we can now directly compute the log-likelihood:

$$\mathcal{L}_{\mathbf{t}}(z) = \log L_{\mathbf{t}}(z) = \log P_z(\mathbf{t}) \approx C - \frac{N}{2} \sum_{j=1}^{m} (\hat{c}_j - (-1)^{z_j} \cdot c_j)^2 . \qquad (7)$$

The constant C depends on m and N only, and is irrelevant to the attack. From this formula we immediately derive the following property.

Lemma 1. *The relative likelihood of a key class z is completely determined by the Euclidean distance $|\hat{\mathbf{c}} - \mathbf{c}_{\mathbf{z}}|$, where $\hat{\mathbf{c}}$ is an m-dimensional vector containing the estimated imbalances derived from the known texts, and $\mathbf{c}_{\mathbf{z}} = ((-1)^{z_1} c_1, \ldots, (-1)^{z_m} c_m)$.*

The lemma implies that $\mathcal{L}_{\mathbf{T}}(z) > \mathcal{L}_{\mathbf{T}}(z^*)$ if and only if $|\hat{\mathbf{c}} - \mathbf{c}_{\mathbf{z}}| < |\hat{\mathbf{c}} - \mathbf{c}_{\mathbf{z}^*}|$. This type of result is common in coding theory.

3.2 Estimating the Gain of the Attack

Based on the geometrical interpretation given above, and using the results from Sect. 2.3, we can now easily derive the gain of the attack.

Theorem 1. *Given m approximations and N independent pairs (P_i, C_i), an adversary can mount a linear attack with a gain equal to:*

$$\gamma = -\log_2 \left[2 \cdot \frac{1}{|\mathcal{Z}|} \sum_{z \in \mathcal{Z}^*} \Phi \left(-\sqrt{N} \cdot \frac{|\mathbf{c}_z - \mathbf{c}_{z^*}|}{2} \right) + \frac{1}{|\mathcal{Z}|} \right] , \qquad (8)$$

where $\Phi(\cdot)$ is the cumulative normal distribution function, $\mathbf{c}_{\mathbf{z}} = ((-1)^{z_1} c_1, \ldots, (-1)^{z_m} c_m)$, and $|\mathcal{Z}|$ is the number of key classes induced by the approximations.

Proof. The probability that the likelihood of a key class z exceeds the likelihood of the correct key class z^* is given by the probability that the vector $\hat{\mathbf{c}}$ falls into the half plane $\mathcal{T}_c = \{\mathbf{c} \mid |\hat{\mathbf{c}} - \mathbf{c}_{\mathbf{z}}| \leq |\hat{\mathbf{c}} - \mathbf{c}_{\mathbf{z}^*}|\}$. Considering the fact that $\hat{\mathbf{c}}$ describes a Gaussian distribution around $\mathbf{c}_{z^*}$ with a variance-covariance matrix $1/\sqrt{N} \cdot I_m$, we need to integrate this Gaussian over the half plane $\mathcal{T}_c$ and due to the zero covariances, we immediately find:

$$\Pr\left[\mathcal{L}_{\mathbf{T}}(z) \geq \mathcal{L}_{\mathbf{T}}(z^*) \mid z^* \right] = \Phi \left(-\sqrt{N} \cdot \frac{|\mathbf{c}_z - \mathbf{c}_{z^*}|}{2} \right) .$$

By summing these probabilities as in (3) we find the expected number of trials:

$$M^* = \frac{2^n}{|\mathcal{Z}|} \cdot \left[\frac{1}{2} + \sum_{z \in \mathcal{Z}^*} \Phi \left(-\sqrt{N} \cdot \frac{|\mathbf{c}_z - \mathbf{c}_{z^*}|}{2} \right) \right] + \frac{1}{2} . \qquad (9)$$

The gain is obtained by substituting this expression for M^* in equation (1). $\square$

The formula derived in the previous theorem can easily be evaluated as long as $|\mathcal{Z}|$ is not too large. In order to estimate the gain in the other cases as well, we need to make a few approximations.

Corollary 1. *If* $|\mathcal{Z}|$ *is sufficiently large, the gain derived in Theorem 1 can accurately be approximated by*

$$\gamma \approx -\log_2 \left[2 \cdot \frac{|\mathcal{Z}| - 1}{|\mathcal{Z}|} \cdot \Phi \left(-\sqrt{\frac{N \cdot \bar{c}^2}{2}} \right) + \frac{1}{|\mathcal{Z}|} \right] \triangleq f(N \cdot \bar{c}^2, |\mathcal{Z}|), \qquad (10)$$

where $\bar{c}^2 = \sum_{j=1}^{m} c_j^2$.

Proof. See App. A.

An interesting conclusion that can be drawn from the corollary above is that the gain of the attack is mainly determined by the product $N \cdot \bar{c}^2$. As a result, if we manage to increase $\bar{c}^2$ by using more linear characteristics, then the required number of known plaintext–ciphertext pairs N can be decreased by the same factor, without affecting the gain. Since the quantity $\bar{c}^2$ plays a very important role in the attacks, we give it a name and define it explicitly.

Definition 2. *The* capacity $\bar{c}^2$ *of a system of* m *approximations is defined as*

$$\bar{c}^2 = \sum_{j=1}^{m} c_j^2 = 4 \cdot \sum_{j=1}^{m} \epsilon_j^2.$$

3.3 Extension: Multiple Approximations and Matsui's Algorithm 2

The approach taken in the previous section can be seen as an extension of Matsui's Algorithm 1. Just as in Algorithm 1, the adversary analyses parity bits of the known plaintext–ciphertext pairs and then tries to determine parity bits of internal round keys. An alternative approach, which is called Algorithm 2 and yields much more efficient attacks in practice, consists in guessing parts of the round keys in the first and the last round, and determining the probability that the guess was correct by exploiting linear characteristics over the remaining rounds. In this section we will show that the results derived above can still be applied in this situation, provided that we modify some definitions.

Let us denote by $\mathcal{Z}_O$ the set of possible guesses for the targeted subkeys of the outer rounds (round 1 and round r). For each guess z_O and for all N plaintext–ciphertext pairs, the adversary does a partial encryption and decryption at the top and bottom of the block cipher, and recovers the parity bits of the intermediate data blocks involved in m different $(r-2)$-round linear characteristics. Using this data, he constructs $m' = |\mathcal{Z}_O| \cdot m$ counters t_j, which can be transformed into a m'-dimensional vector $\hat{\mathbf{c}}$ containing the estimated imbalances.

As explained in the previous section, the m linear characteristics involve m parity bits of the key, and thus induce a set of equivalent key classes, which we will here denote by $\mathcal{Z}_I$ (I from *inner*). Although not strictly necessary, we will for simplicity assume that the sets $\mathcal{Z}_O$ and $\mathcal{Z}_I$ are independent, such that each guess $z_O \in \mathcal{Z}_O$ can be combined with any class $z_I \in \mathcal{Z}_I$, thereby determining a subclass of keys $z = (z_O, z_I) \in \mathcal{Z}$ with $|\mathcal{Z}| = |\mathcal{Z}_O| \cdot |\mathcal{Z}_I|$.

At this point, the situation is very similar to the one described in the previous section, the main difference being a higher dimension m'. The only remaining question is how to construct the m'-dimensional vectors $\mathbf{c_z}$ for each key class $z = (z_O, z_I)$. To solve this problem, we will need to make some assumptions. Remember that the coordinates of $\mathbf{c_z}$ are determined by the expected imbalances of the corresponding linear expressions, given that the data is encrypted with a key from class z. For the m counters that are constructed after guessing the correct subkey z_O, the expected imbalances are determined by z_I and equal to $(-1)^{z_{I,1}}c_1, \ldots, (-1)^{z_{I,m}}c_m$. For each of the $m' - m$ other counters, however, we will assume that the wrong guesses result in independent random-looking parity bits, showing no imbalance at all[4]. Accordingly, the vector $\mathbf{c_z}$ has the following form:

$$\mathbf{c_z} = (0, \ldots, 0, (-1)^{z_{I,1}}c_1, \ldots, (-1)^{z_{I,m}}c_m, 0, \ldots, 0)$$

With the modified definitions of $\mathcal{Z}$ and c_z given above, both Theorem 1 and Corollary 1 still hold (the proofs are given in App. A). Notice however that the gain of the Algorithm-2-style linear attack will be significantly larger because it depends on the capacity of linear characteristics over $r - 2$ rounds instead of r rounds.

3.4   Influence of Dependencies

When deriving (5) in Sect. 3, we assumed statistical independence. This assumption is not always fulfilled, however. In this section we discuss different potential sources of dependencies and estimate how they might influence the cryptanalysis.

Dependent plaintext–ciphertext pairs. A first assumption made by equation (5) concerns the dependency of the parity bits $x_{i,j}$ with $1 \le i \le N$, computed with a single linear approximation for different plaintext–ciphertext pairs. The equation assumes that the probability that the approximation holds for a single pair equals $p_j = 1/2 + \epsilon_j$, regardless of what is observed for other pairs. This is a very reasonable assumption if the N plaintexts are chosen randomly, but even if they are picked in a systematic way, we can still safely assume that the corresponding ciphertexts are sufficiently unrelated as to prevent statistical dependencies.

Dependent text mask. The next source of dependencies is more fundamental and is related to dependent text masks. Suppose for example that we want to use three linear approximations with plaintext–ciphertext masks (χ_P^1, χ_C^1), (χ_P^2, χ_C^2), (χ_P^3, χ_C^3), and that $\chi_P^1 \oplus \chi_P^2 \oplus \chi_P^3 = \chi_C^1 \oplus \chi_C^2 \oplus \chi_C^3 = 0$. It is immediately clear that the parity bits computed for these three approximations cannot possibly be independent: for all (P_i, C_i) pairs, the bit computed for the 3rd approximation $x_{i,3}$ is equal to $x_{i,1} \oplus x_{i,2}$.

---

[4] Note that for some ciphers, other assumptions may be more appropriate. The reasoning in this section can be applied to these cases just as well, yielding very similar results.

Even in such cases, however, we believe that the results derived in the previous section are still quite reasonable. In order to show this, we consider the probability that a single random plaintext encrypted with an equivalent key z yields a vector[5] of parity bits $\mathbf{x} = (x_1, \ldots, x_m)$. Let us denote by χ_T^j the concatenation of both text masks χ_P^j and χ_C^j. Without loss of generality, we can assume that the m masks χ_T^j are linearly independent for $1 \leq j \leq l$ and linearly dependent (but different) for $l < j \leq m$. This implies that $\mathbf{x}$ is restricted to a l-dimensional subspace $\mathcal{R}$. We will only consider the key class $z' = (0, \ldots, 0)$ in order to simplify the equations. The probability we want to evaluate is:

$$P_{z'}(\mathbf{x}) = \Pr\left[X_j = x_j \text{ for } 1 \leq j \leq m \mid z'\right]$$

These (unknown) probabilities determine the (known) imbalances c_j of the linear approximations through the following expression:

$$c_j = \sum_{\mathbf{x} \in \mathcal{R}} P_{z'}(\mathbf{x}) \cdot (-1)^{x_j}.$$

We now make the (in many cases reasonable) assumption that all $2^l - m$ masks χ_T, which depend linearly on the masks χ_T^j, but which differ from the ones considered by the attack, have negligible imbalances. In this case, the equation above can be reversed (note the similarity with the Walsh-Hadamard transform), and we find that:

$$P_{z'}(\mathbf{x}) = \frac{1}{2^l} \sum_{j=1}^{m} c_j \cdot (-1)^{x_j}.$$

Assuming that $m \cdot c_j \ll 1$ we can make the following approximation:

$$P_{z'}(\mathbf{x}) \approx \frac{2^m}{2^l} \prod_{j=1}^{m} \frac{1 + c_j \cdot (-1)^{x_j}}{2}.$$

Apart from an irrelevant constant factor $2^m/2^l$, this is exactly what we need: it implies that, even with dependent masks, we can still multiply probabilities as we did in order to derive (5). This is an important conclusion, because it indicates that the capacity of the approximations continues to grow, even when m exceeds twice the block size, in which case the masks are necessarily linearly dependent.

Dependent trails. A third type of dependencies might be caused by merging linear trails. When analyzing the best linear approximations for DES, for example, we notice that most of the good linear approximations follow a very limited number of trails through the inner rounds of the cipher, which might result in dependencies. Although this effect did not appear to have any influence on our experiments (with up to 100 different approximations), we cannot exclude at this point that they will affect attacks using much more approximations.

---

[5] Note a small abuse of notation here: the definition of $\mathbf{x}$ differs from the one used in Sect. 2.1.

Table 1. Attack Algorithm MK 1 and its complexity.

Distillation phase. Obtain N plaintext–ciphertext pairs (p_i, c_i). For $1 \leq j \leq m$, count the number t_j of pairs satisfying $p_i[\chi_P^j] \oplus c_i[\chi_C^j] = 0$ and compute the estimated imbalance $\hat{c}_j = 2 \cdot t_j / N - 1$.

Analysis phase. For each equivalent key class $z \in \mathcal{Z}$, determine the distance

$$|\hat{\mathbf{c}} - \mathbf{c_z}|^2 = \sum_{j=1}^{m} (\hat{c}_j - (-1)^{z_j} \cdot c_j)^2$$

and use these values to construct a sorted list, starting with the class with the smallest distance.

Search phase. Run through the sorted list and exhaustively try all n-bit keys contained in the equivalence classes until the correct key is found.

	Data compl.	Time compl.	Memory compl.
Distillation:	$O(1/\bar{c}^2)$	$O(m/\bar{c}^2)$	$O(m)$
Analysis:	-	$O(m \cdot |\mathcal{Z}|)$	$O(|\mathcal{Z}|)$
Search:	-	$O(2^{n-\gamma})$	$O(|\mathcal{Z}|)$

Dependent key masks. We finally note that we did not make any assumption about the dependency of key masks in the previous sections. This implies that all results derived above remain valid for dependent key masks.

4   Experimental Results

In Sect. 3 we derived an optimal approach for cryptanalyzing block ciphers using multiple linear approximations. In this section, we implement practical attack algorithms based on this approach and evaluate their performance when applied to DES, the standard benchmark for linear cryptanalysis. Our experiments show that the attack complexities are in perfect correspondence with the theoretical results derived in the previous sections.

4.1   Attack Algorithm MK 1

Table 1 summarizes the attack algorithm presented in Sect. 2 (we call this algorithm *Attack Algorithm MK 1*). In order to verify the theoretical results, we applied the attack algorithm to 8 rounds of DES. We picked 86 linear approximations with a total capacity $\bar{c}^2 = 2^{-15.6}$ (see Definition 2). In order to speed up the simulation, the approximations were picked to contain 10 linearly independent key masks, such that $|\mathcal{Z}| = 1024$. Fig. 2 shows the simulated gain for Algorithm MK 1 using these 86 approximations, and compares it to the gain of Matsui's Algorithm 1, which uses the best one only ($\bar{c}^2 = 2^{-19.4}$). We clearly see a significant improvement. While Matsui's algorithm requires about 2^{21} pairs to attain a gain close to 1 bit, only 2^{16} pairs suffice for Algorithm MK 1. The theoretical curves shown in the figure were plotted by computing the gain using

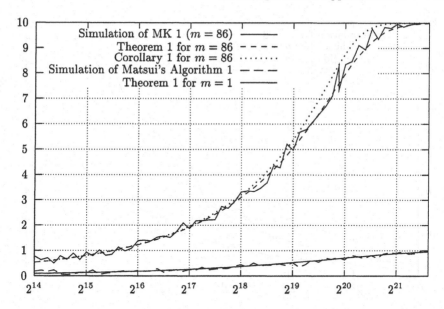

Fig. 2. Gain (in bits) as a function of data (known plaintext) for 8-round DES.

the exact expression for M^* derived in Theorem 1 and using the approximation from Corollary 1. Both fit nicely with the experimental results.

Note, that the attack presented in this section is just a proof of concept, even higher gains would be possible with more optimized attacks. For a more detailed discussion of the technical aspects playing a role in the implementation of Algorithm MK 1, we refer to App. B.

4.2   Attack Algorithm MK 2

In this section, we discuss the experimental results for the generalization of Matsui's Algorithm 2 using multiple linear approximations (called *Attack Algorithm MK 2*). We simulated the attack algorithm on 8 rounds of DES and compared the results to the gain of the corresponding Algorithm 2 attack described in Matsui's paper [9].

Our attack uses eight linear approximations spanning six rounds with a total capacity $\bar{c}^2 = 2^{-11.9}$. In order to compute the parity bits of these equations, eight 6-bit subkeys need to be guessed in the first and the last rounds (how this is done in practice is explained in App. B). Fig. 3 compares the gain of the attack to Matsui's Algorithm 2, which uses the two best approximations ($\bar{c}^2 = 2^{-13.2}$). For the same amount of data, the multiple linear attack clearly achieves a much higher gain. This reduces the complexity of the search phase by multiple orders of magnitude. On the other hand, for the same gain, the adversary can reduce the amount of data by at least a factor 2. For example, for a gain of 12 bits, the data complexity is reduced from $2^{17.8}$ to $2^{16.6}$. This is in a close correspondence with the ratio between the capacities. Note that both simulations were carried

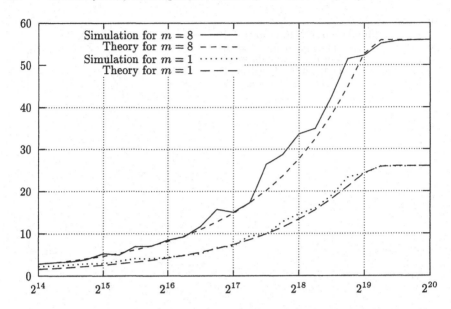

Fig. 3. Gain (in bits) as a function of data (known plaintext) for 8-round DES.

out under the assumption of independent subkeys (this was also the case for the simulations presented in [9]). Without this assumption, the gain will closely follow the graphs on the figure, but stop increasing as soon as the gain equals the number of independent key bits involved in the attack.

As in Sect. 4.1 our goal was not to provide the best attack on 8-round DES, but to show that Algorithm-2 style attacks do gain from the use of multiple linear approximations, with a data reduction proportional to the increase in the joint capacity. We refer to App. B for the technical aspects of the implementation of Algorithm MK 2.

4.3 Capacity – DES Case Study

In Sect. 3 we argued that the minimal amount of data needed to obtain a certain gain compared to exhaustive search is determined by the capacity $\bar{c}^2$ of the linear approximations. In order to get a first estimate of the potential improvement of using multiple approximations, we calculated the total capacity of the best m linear approximations of DES for $1 \leq m \leq 2^{16}$. The capacities were computed using an adapted version of Matsui's algorithm (see [1]). The results, plotted for different number of rounds, are shown in Fig. 4 and 5, both for approximations restricted to a single S-box per round and for the general case. Note that the single best approximation is not visible on these figures due to the scale of the graphs.

Kaliski and Robshaw [5] showed that the first 10 006 approximations with a single active S-box per round have a joint capacity of $4.92 \cdot 10^{-11}$ for 14 rounds

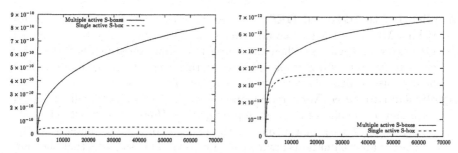

Fig. 4. Capacity (14 rounds).   **Fig. 5.** Capacity (16 rounds).

of DES[6]. Fig. 4 shows that this capacity can be increased to $4 \cdot 10^{-10}$ when multiple S-boxes are allowed. Comparing this to the capacity of Matsui's best approximation ($\bar{c}^2 = 1.29 \cdot 10^{-12}$), the factor 38 gained by Kaliski and Robshaw is increased to 304 in our case. Practical techniques to turn this increased capacity into an effective reduction of the data complexity are presented in this paper, but exploiting the full gain of 10 000 unrestricted approximations will require additional techniques. In theory, however, it would be possible to reduce the data complexity form 2^{43} (in Matsui's case, using two approximations) to about 2^{36} (using 10 000 approximations).

In order to provide a more conservative (and probably rather realistic) estimation of the implications of our new attacks on full DES, we searched for 14-round approximations which only require three 6-bit subkeys to be guessed simultaneously in the first and the last rounds. The capacity of the 108 best approximations satisfying this restriction is $9.83 \cdot 10^{-12}$. This suggests that an MK 2 attack exploiting these 108 approximations might reduce the data complexity by a factor 4 compared to Matsui's Algorithm 2 (*i.e.*, 2^{41} instead of 2^{43}). This is comparable to the Knudsen-Mathiassen reduction [6], but would preserve the advantage of being a known-plaintext attack rather than a chosen-plaintext one.

Using very high numbers of approximations is somewhat easier in practice for MK 1 because we do not have to impose restrictions on the plaintext and ciphertext masks (see App. B). Analyzing the capacity for the 10 000 best 16-round approximations, we now find a capacity of $5 \cdot 10^{-12}$. If we restrict the complexity of the search phase to an average of 2^{43} trials (*i.e.*, a gain of 12 bits), we expect that the attack will require 2^{41} known plaintexts. As expected, this theoretical number is larger than for the MK 2 attack using the same amount of approximations.

5   Future Work

In this paper we proposed a framework which allows to use the information contained in multiple linear approximations in an optimal way. The topics below are possible further improvements and open questions.

---

[6] Note that Kaliski and Robshaw calculated the sum of squared biases: $\sum \epsilon_j^2 = \bar{c}^2/4$.

Application to 16-round DES. The results in this paper suggest that Algorithms MK 1 and MK 2 could reduce the data complexity to 2^{41} known plaintexts, or even less when the number of approximations is further increased. An interesting problem related to this is how to merge multiple lists of key classes (possibly with overlapping key-bits) efficiently.

Application to AES. Many recent ciphers, *e.g.*, AES, are specifically designed to minimize the bias of the best approximation. However, this artificial flattening of the bias profile comes at the expense of a large increase in the number of approximations having the same bias. This suggests that the gain made by using multiple linear approximations could potentially be much higher in this case than for a cipher like DES. Considering this, we expect that one may need to add a few rounds when defining bounds of provable security against linear cryptanalysis, based only on best approximations. Still, since AES has a large security margin against linear cryptanalysis we do not believe that linear attacks enhanced with multiple linear approximations will pose a practical threat to the security of the AES.

Performance of Algorithm MD. Using a very high number of *independent* approximations seems impractical in Algorithms MK 1 and MK 2, but could be feasible with Algorithm MD described in App. B.3. Additionally, this method would allow to replace the multiple linear approximations by multiple linear hulls.

Success rate. In this paper we derived simple formulas for the average number of key candidates checked during the final search phase. Deriving a simple expression for the distribution of this number is still an open problem. This would allow to compute the success rate of the attack as a function of the number of plaintexts and a given maximal number of trials.

6   Conclusions

In this paper, we have studied the problem of generalizing linear cryptanalytic attacks given m multiple linear approximations, which has been stated in 1994 by Kaliski and Robshaw [5]. In order to solve the problem, we have developed a statistical framework based on maximum likelihood decoding. This approach is optimal in the sense that it utilizes all the information that is present in the multiple linear approximations. We have derived explicit and compact gain formulas for the generalized linear attacks and have shown that for a constant gain, the data-complexity N of the attack is proportional to the inverse joint capacity $\bar{c}^2$ of the multiple linear approximations: $N \propto 1/\bar{c}^2$. The gain formulas hold for the generalized versions of both algorithms proposed by Matsui (Algorithm 1 and Algorithm 2).

In the second half of the paper we have proposed several practical methods which deliver the theoretical gains derived in the first part of the paper. We have proposed a key-recovery algorithm MK 1 which has a time complexity $O(m/\bar{c}^2 + m \cdot |\mathcal{Z}|)$ and a data complexity $O(1/\bar{c}^2)$, where $|\mathcal{Z}|$ is the number of solutions of the system of m equations defined by the linear approximations. We have also designed an algorithm MK 2 which is a direct generalization of Matsui's Algorithm 2, as described in [9]. The performances of both algorithms are very close to our theoretical estimations and confirm that the data-complexity of the attack decreases proportionally to the increase in the joint capacity of multiple approximations. We have used 8-round DES as a standard benchmark in our experiments and in all cases our attacks perform significantly better than those given by Matsui. However our goal in this paper was not to produce the most optimal attack on DES, but to construct a new cryptanalytic tool applicable to a variety of ciphers.

References

1. A. Biryukov, C. De Cannière, and M. Quisquater, "On multiple linear approximations (extended version)." Cryptology ePrint Archive: Report 2004/057, http://eprint.iacr.org/2004/057/.

2. J. Daemen and V. Rijmen, *The Design of Rijndael: AES — The Advanced Encryption Standard*. Springer-Verlag, 2002.

3. P. Junod, "On the optimality of linear, differential, and sequential distinguishers," in *Advances in Cryptology - EUROCRYPT 2003* (E. Biham, ed.), Lecture Notes in Computer Science, pp. 17–32, Springer-Verlag, 2003.

4. P. Junod and S. Vaudenay, "Optimal key ranking procedures in a statistical cryptanalysis," in *Fast Software Encryption, FSE 2003* (T. Johansson, ed.), vol. 2887 of *Lecture Notes in Computer Science*, pp. 1–15, Springer-Verlag, 2003.

5. B. S. Kaliski and M. J. Robshaw, "Linear cryptanalysis using multiple approximations," in *Advances in Cryptology - CRYPTO'94* (Y. Desmedt, ed.), vol. 839 of *Lecture Notes in Computer Science*, pp. 26–39, Springer-Verlag, 1994.

6. L. R. Knudsen and J. E. Mathiassen, "A chosen-plaintext linear attack on DES," in *Fast Software Encryption, FSE 2000* (B. Schneier, ed.), vol. 1978 of *Lecture Notes in Computer Science*, pp. 262–272, Springer-Verlag, 2001.

7. L. R. Knudsen and M. J. B. Robshaw, "Non-linear approximations in linear cryptanalysis," in *Proceedings of Eurocrypt'96* (U. Maurer, ed.), no. 1070 in Lecture Notes in Computer Science, pp. 224–236, Springer-Verlag, 1996.

8. M. Matsui, "Linear cryptanalysis method for DES cipher," in *Advances in Cryptology - EUROCRYPT'93* (T. Helleseth, ed.), vol. 765 of *Lecture Notes in Computer Science*, pp. 386–397, Springer-Verlag, 1993.

9. M. Matsui, "The first experimental cryptanalysis of the Data Encryption Standard," in *Advances in Cryptology - CRYPTO'94* (Y. Desmedt, ed.), vol. 839 of *Lecture Notes in Computer Science*, pp. 1–11, Springer-Verlag, 1994.

10. M. Matsui, "Linear cryptanalysis method for DES cipher (I)." (extended paper), unpublished, 1994.

11. S. Murphy, F. Piper, M. Walker, and P. Wild, "Likelihood estimation for block cipher keys," Technical report, Information Security Group, Royal Holloway, University of London, 1995.

12. A. A. Selçuk, "On probability of success in linear and differential cryptanalysis," in *Proceedings of SCN'02* (S. Cimato, C. Galdi, and G. Persiano, eds.), vol. 2576 of *Lecture Notes in Computer Science*, Springer-Verlag, 2002. Also available at https://www.cerias.purdue.edu/papers/archive/2002-02.ps.

13. T. Shimoyama and T. Kaneko, "Quadratic relation of s-box and its application to the linear attack of full round des," in *Advances in Cryptology – CRYPTO'98* (H. Krawczyk, ed.), vol. 1462 of *Lecture Notes in Computer Science*, pp. 200–211, Springer-Verlag, 1998.
14. S. Vaudenay, "An experiment on DES statistical cryptanalysis," in *3rd ACM Conference on Computer and Communications Security, CCS*, pp. 139–147, ACM Press, 1996.

A   Proofs

A.1   Proof of Corollary 1

Corollary 1. *If $|\mathcal{Z}|$ is sufficiently large, the gain derived in Theorem 1 can accurately be approximated by*

$$\gamma \approx -\log_2\left[2\cdot\frac{|\mathcal{Z}|-1}{|\mathcal{Z}|}\cdot\Phi\left(-\sqrt{\frac{N\cdot\bar{c}^2}{2}}\right)+\frac{1}{|\mathcal{Z}|}\right], \qquad (11)$$

where $\bar{c}^2 = \sum_{j=1}^m c_j^2$ is called the total *capacity of the m linear characteristics.*

Proof. In order to show how (11) is derived from (8), we just need to construct an approximation for the expression

$$\frac{1}{|\mathcal{Z}^*|}\sum_{z\in\mathcal{Z}^*}\Phi\left(-\sqrt{N}\cdot\frac{|\mathbf{c}_z-\mathbf{c}_{z^*}|}{2}\right)=\frac{1}{|\mathcal{Z}^*|}\sum_{z\in\mathcal{Z}^*}\Phi\left(-\sqrt{N/4\cdot|\mathbf{c}_z-\mathbf{c}_{z^*}|^2}\right).\quad(12)$$

We first define the function $f(x)=\Phi(-\sqrt{N/4\cdot x})$. Denoting the average value of a set of variables by $E[\cdot]=\widehat{\cdot}$, we can reduce (12) to the compact expression $E[f(x)]$, with $x=|\mathbf{c}_z-\mathbf{c}_{z^*}|^2$. By expanding $f(x)$ into a Taylor series around the average value $\widehat{x}$, we find

$$E[f(x)]=f(\widehat{x})+0+f''(\widehat{x})\cdot E[(x-\widehat{x})^2]+\dots .$$

Provided that the higher order moments of x are sufficiently small, we can use the approximation $E[f(x)]\approx f(\widehat{x})$. Exploiting the fact that the jth coordinate of each vector $\mathbf{c}_z$ is either c_j or $-c_j$, we can easily calculate the average value $\widehat{x}$:

$$\widehat{x}=\frac{1}{|\mathcal{Z}^*|}\sum_{z\in\mathcal{Z}^*}|\mathbf{c}_z-\mathbf{c}_{z^*}|^2=2\cdot\frac{|\mathcal{Z}|}{|\mathcal{Z}^*|}\sum_{j=1}^m c_j^2.$$

When $|\mathcal{Z}|$ is sufficiently large (say $|\mathcal{Z}|>2^8$), the right hand part can be approximated by $2\cdot\sum_{j=1}^m c_j^2=2\cdot\bar{c}^2$ (remember that $\mathcal{Z}^*=\mathcal{Z}\setminus\{z^*\}$, and thus $|\mathcal{Z}^*|=|\mathcal{Z}|-1$). Substituting this into the relation $E[f(x)]\approx f(\widehat{x})$, we find

$$\frac{1}{|\mathcal{Z}^*|}\sum_{z\in\mathcal{Z}^*}\Phi\left(-\sqrt{N}\cdot\frac{|\mathbf{c}_z-\mathbf{c}_{z^*}|}{2}\right)\approx\Phi\left(-\sqrt{\frac{N\cdot\bar{c}^2}{2}}\right).$$

By applying this approximation to the gain formula derived in Theorem 1, we directly obtain expression (11). □

A.2   Gain Formulas for the Algorithm-2-Style Attack

With the modified definitions of $\mathcal{Z}$ and c_z given in Sect. 3.3, Theorem 1 can immediately be applied. This results in the following corollary.

Corollary 2. *Given m approximations and N independent pairs (P_i, C_i), an adversary can mount an Algorithm-2-style linear attack with a gain equal to:*

$$\gamma = -\log_2\left[2 \cdot \frac{1}{|\mathcal{Z}|} \sum_{z \in \mathcal{Z}^*} \Phi\left(-\sqrt{N} \cdot \frac{|\mathbf{c}_z - \mathbf{c}_{z^*}|}{2}\right) + \frac{1}{|\mathcal{Z}|}\right] . \tag{13}$$

The formula above involves a summation over all elements of $\mathcal{Z}^*$. Motivated by the fact that $|\mathcal{Z}^*| = |\mathcal{Z}_O| \cdot |\mathcal{Z}_I| - 1$ is typically very large, we now derive a more convenient approximated expression similar to Corollary 1. In order to do this, we split the sum into two parts. The first part considers only keys $z \in \mathcal{Z}_1^* = \mathcal{Z}_1 \setminus \{z^*\}$ where $\mathcal{Z}_1 = \{z \mid z_O = z_O^*\}$; the second part sums over all remaining keys $z \in \mathcal{Z}_2 = \{z \mid z_O \neq z_O^*\}$. In this second case, we have that $|\mathbf{c}_z - \mathbf{c}_{z^*}|^2 = 2 \cdot \sum_{j=1}^m c_j^2 = 2 \cdot \bar{c}^2$ for all $z \in \mathcal{Z}_2$, such that

$$\sum_{z \in \mathcal{Z}_2} \Phi\left(-\sqrt{N} \cdot \frac{|\mathbf{c}_z - \mathbf{c}_{z^*}|}{2}\right) = |\mathcal{Z}_2| \cdot \Phi\left(-\sqrt{\frac{N \cdot \bar{c}^2}{2}}\right) .$$

For the first part of the sum, we apply the approximation used to derive Corollary 1 and obtain a very similar expression:

$$\sum_{z \in \mathcal{Z}_1^*} \Phi\left(-\sqrt{N} \cdot \frac{|\mathbf{c}_z - \mathbf{c}_{z^*}|}{2}\right) \approx |\mathcal{Z}_1^*| \cdot \Phi\left(-\sqrt{\frac{N \cdot \bar{c}^2}{2}}\right) .$$

Combining both result we find the counterpart of Corollary 1 for an Algorithm-2-style linear attack.

Corollary 3. *If $|\mathcal{Z}|$ is sufficiently large, the gain derived in Theorem 2 can accurately be approximated by*

$$\gamma \approx -\log_2\left[2 \cdot \frac{|\mathcal{Z}| - 1}{|\mathcal{Z}|} \cdot \Phi\left(-\sqrt{\frac{N \cdot \bar{c}^2}{2}}\right) + \frac{1}{|\mathcal{Z}|}\right] , \tag{14}$$

where $\bar{c}^2 = \sum_{j=1}^m c_j^2$ is the total capacity of the m linear characteristics.

Notice that although Corollary 1 and 3 contain identical formulas, the gain of the Algorithm-2-style linear attack will be significantly larger because it depends on the capacity of linear characteristics over $r - 2$ rounds instead of r rounds.

B   Discussion – Practical Aspects

When attempting to calculate the optimal estimators derived in Sect. 3, the attacker might be confronted with some practical limitations, which are often cipher-dependent. In this section we discuss possible problems and propose ways to deal with them.

B.1   Attack Algorithm MK 1

When estimating the potential gain in Sect. 3, we did not impose any restrictions on the number of approximations m. However, while it does reduce the complexity of the search phase (since it increases the gain), having an excessively high number m increases both the time and the space complexity of the distillation and the analysis phase. At some point the latter will dominate, cancelling out any improvement made in the search phase.

Analyzing the complexities in Table 1, we can make a few observations. We first note that the time complexity of the distillation phase should be compared to the time needed to encrypt $N \propto 1/\bar{c}^2$ plaintext–ciphertext pairs. Given that a single counting operation is much faster than an encryption, we expect the complexity of the distillation to remain negligible compared to the encryption time as long as m is only a few orders of magnitude (say $m < 100$).

The second observation is that the number of different key classes $|\mathcal{Z}|$ clearly plays an important role, both for the time and the memory complexities of the algorithm. In a practical situation, the memory is expected to be the strongest limitation. Different approaches can be taken to deal with this problem:

Straightforward, but inefficient approach. Since the number of different key classes $|\mathcal{Z}|$ is bounded by 2^m, the most straightforward solution is to limit the number of approximations. A realistic upper bound would be $m < 32$. The obvious drawback of this approach is that it will not allow to attain very high capacities.

Exploiting dependent key masks. A better approach is to impose a bound on the number l of *linearly independent* key masks χ_K^j. This way, we limit the memory requirements to $|\mathcal{Z}| = 2^l$, but still allow a large number of approximations (for ex. a few thousands). This approach restricts the choice of approximations, however, and thus reduces the maximum attainable capacity. This is the approach taken in Sect. 4.1. Note also that the attack described in [5] can be seen as a special case of this approach, with $l = 1$.

Merging separate lists. A third strategy consists in constructing separate lists and merging them dynamically. Suppose for simplicity that the m key masks χ_K^j considered in the attack are all independent. In this case, we can apply the analysis phase twice, each time using $m/2$ approximations. This will result in two sorted lists of intermediate key classes, both containing $2^{m/2}$ classes. We can then dynamically compute a sorted sequence of final key classes constructed by taking the product of both lists. The ranking of the sequence is determined by the likelihood of these final classes, which is just the sum of the likelihoods of the elements in the separate lists. This approach slightly increases[7] the time complexity of the analysis phase, but will considerably reduce the memory requirements. Note that this approach can be generalized in order to allow some dependencies in the key masks.

---

[7] In cases where the gain of the attack is several bits, this approach will actually decrease the complexity, since we expect that only a fraction of the final sequence will need to be computed.

B.2   Attack Algorithm MK 2

We now briefly discuss some practical aspects of the Algorithm-2-style multiple linear attack, called Attack Algorithm MK 2. As discussed earlier, the ideas of the attack are very similar to Attack Algorithm MK 1, but there are a number of additional issues. In the following paragraphs, we denote the number of rounds of the cipher by r.

Choice of characteristics. In order to limit the amount of guesses in rounds 1 and r, only parts of the subkeys in these rounds will be guessed. This restricts the set of useful $r - 2$-round characteristics to those that only depend on bits which can be derived from the plaintext, the ciphertext, and the partial subkeys. This obviously reduces the maximum attainable capacity.

Efficiency of the distillation phase. During the distillation phase, all N plaintexts need to be analyzed for all $|\mathcal{Z}_O|$ guesses z_O. Since $|\mathcal{Z}_O|$ is rather large in practice, this could be very computational intensive. For example, a naive implementation would require $O(N \cdot |\mathcal{Z}_O|)$ steps and even Matsui's counting trick would use $O(N + |\mathcal{Z}_O|^2)$ steps. However, the distillation can be performed in $O(N + |\mathcal{Z}_O|)$ steps by gradually guessing parts of z_O and re-processing the counters.

Merging Separate lists. The idea of working with separate lists can be applied here just as for MK 1.

Computing distances. In order to compare the likelihoods of different keys, we need to evaluate the distance $|\hat{c} - c_z|^2$ for all classes $z \in \mathcal{Z}$. The vectors $\hat{c}$ and c_z are both $|\mathcal{Z}_O| \cdot m$-dimensional. When calculating this distance as a sum of squares, most terms do not depend on z, however. This allows the distance to be computed very efficiently, by summing only m terms.

B.3   Attack Algorithm MD (distinguishing/key-recovery)

The main limitation of Algorithm MK 1 and MK 2 is the bound on the number of key classes $|\mathcal{Z}|$. In this section, we show that this limitation disappears if our sole purpose is to distinguish an encryption algorithm E_k from a random permutation R. As usual, the distinguisher can be extended into a key-recovery attack by adding rounds at the top and at the bottom.

If we observe N plaintext–ciphertext pairs and assume for simplicity that the a priori probability that they were constructed using the encryption algorithm is $1/2$, we can construct a distinguishing attack using the maximum likelihood approach in a similar way as in Sect. 3. Assuming that all secret keys k are equally probable, one can easily derive the likelihood that the encryption algorithm was used, given the values of the counters $\mathbf{t}$:

$$L_E(\mathbf{t}) \approx \frac{1}{2^m} \prod_{j=1}^{m} \binom{N}{t_j} \cdot \left(p_j^{t_j} \cdot q_j^{N-t_j} + q_j^{t_j} \cdot p_j^{N-t_j} \right).$$

This expression is correct if all text masks and key masks are independent, but is still expected to be a good approximation, if this assumption does not hold (for the reasons discussed in Sect. 3.4). A similar likelihood can be calculated for the random permutation:

$$L_R(\mathbf{t}) = \prod_{j=1}^{m} \binom{N}{t_j} \cdot \left(\frac{1}{2}\right)^N.$$

Contrary to what was found for Algorithm MK 1, both likelihoods can be computed in time proportional to m, *i.e.*, independent of $|\mathcal{Z}|$. The complete distinguishing algorithm, called *Attack Algorithm MD* consists of two steps:

Distillation phase. Obtain N plaintext–ciphertext pairs (P_i, C_i). For $1 \le j \le m$, count the number t_j of pairs satisfying $P_i[\chi_P^j] \oplus C_i[\chi_C^j] = 0$.
Analysis phase. Compute $L_E(\mathbf{t})$ and $L_R(\mathbf{t})$. If $L_E(\mathbf{t}) > L_R(\mathbf{t})$, decide that the plaintexts were encrypted with the algorithm E_k (using some unknown key k).

The analysis of this algorithm is a matter of further research.

C   Previous Work: Linear Cryptanalysis

Since the introduction of linear cryptanalysis by Matsui [8–10], several generalizations of the linear cryptanalysis method have been proposed. Kaliski-Robshaw [5] suggested to use many linear approximations instead of one, but did provide an efficient method for doing so only for the case when all the approximations cover the same parity bit of the key. Realizing that this limited the number of useful approximations, the authors also proposed a simple (but somewhat inefficient) extension to their technique which removes this restriction by guessing a relation between the different key bits. The idea of using non-linear approximations has been suggested by Knudsen-Robshaw [7]. It was used by Shimoyama-Kaneko [13] to marginally improve the linear attack on DES. Knudsen-Mathiassen [6] suggest to convert linear cryptanalysis into a chosen plaintext attack, which would gain the first round of approximation for free. The gain is small, since Matsui's attack gains the first round rather efficiently as well.

A more detailed overview of the history of linear cryptanalysis can be found in the extended version of this paper [1].

Feistel Schemes and Bi-linear Cryptanalysis
(Extended Abstract)

Nicolas T. Courtois

Axalto Smart Cards Crypto Research,
36-38 rue de la Princesse, BP 45, F-78430 Louveciennes Cedex, France
courtois@minrank.org

Abstract. In this paper we introduce the method of bi-linear crypt-analysis (BLC), designed specifically to attack Feistel ciphers. It allows to construct periodic biased characteristics that combine for an arbitrary number of rounds. In particular, we present a practical attack on DES based on a 1-round invariant, the fastest known based on such invariant, and about as fast as the best Matsui's attack. For ciphers similar to DES, based on small S-boxes, we claim that BLC is very closely related to LC, and we do not expect to find a bi-linear attack much faster than by LC. Nevertheless we have found bi-linear characteristics that are strictly better than the best Matsui's result for 3, 7, 11 and more rounds.
For more general Feistel schemes there is no reason whatsoever for BLC to remain only a small improvement over LC. We present a construction of a family of practical ciphers based on a big Rijndael-type S-box that are strongly resistant against linear cryptanalysis (LC) but can be easily broken by BLC, even with 16 or more rounds.

Keywords: Block ciphers, Feistel schemes, S-box design, inverse-based S-box, DES, linear cryptanalysis, generalised linear cryptanalysis, I/O sums, correlation attacks on block ciphers, multivariate quadratic equations.

1 Introduction

In spite of growing importance of AES, Feistel schemes and DES remain widely used in practice, especially in financial/banking sector. The linear cryptanalysis (LC), due to Gilbert and Matsui is the best known plaintext attack on DES, see [4, 25, 27, 16, 21]. (For chosen plaintext attacks, see [21, 2]).

A straightforward way of extending linear attacks is to consider nonlinear multivariate equations. Exact multivariate equations can give a tiny improvement to the last round of a linear attack, as shown at Crypto'98 [18]. A more powerful idea is to use probabilistic multivariate equations, for every round, and replace Matsui's biased linear I/O sums by nonlinear I/O sums as proposed by Harpes, Kramer, and Massey at Eurocrypt'95 [9]. This is known as Generalized Linear Cryptanalysis (GLC). In [10, 11] Harpes introduces partitioning crypt-analysis (PC) and shows that it generalizes both LC and GLC. The correlation cryptanalysis (CC) introduced in Jakobsen's master thesis [13] is claimed even

M. Franklin (Ed.): CRYPTO 2004, LNCS 3152, pp. 23–40, 2004.
© International Association for Cryptologic Research 2004

more general. Moreover, in [12] it is shown that all these attacks, including also
Differential Cryptanalysis are closely related and can be studied in terms of the
Fast Fourier Transform for the cipher round function. Unfortunately, computing
this transform is in general infeasible for a real-life cipher and up till now, non-
linear multivariate I/O sums played a marginal role in attacking real ciphers.
Accordingly, these attacks may be excessively general and there is probably no
substitute to finding and studying in details interesting special cases.

At Eurocrypt'96 Knudsen and Robshaw consider applying GLC to Feistel
schemes [20], and affirm that in this case non-linear characteristics cannot be
joined together. We will demonstrate that GLC can be applied to Feistel ciphers,
which is made possible with our "Bi-Linear Cryptanalysis" (BLC) attack.

2   Feistel Schemes and Bi-linear Functions

Differential [2] and linear attacks on DES [25, 1] have periodic patterns with
invariant equations for some 1, 3 or 8 rounds. In this paper we will present
several new practical attacks with periodic structure for DES, including new
1-round invariants.

2.1   The Principle of the Bi-linear Attack on Feistel Schemes

In one round of a Feistel scheme, one half is unchanged, and one half is linearly
combined with the output of the component connected to the other half. This will
allow bi-linear I/O expressions on the round function to be combined together.
First we will give an example with one product, and extend it to arbitrary bilinear expressions. Then in Section 3 we explain the full method in details (with
linear parts present too) for an arbitrary Feistel schemes. Later we will apply it
to get concrete working attacks for DES and other ciphers.

In this paper we represent Feistel schemes in a completely "untwisted" way,
allowing to see more clearly the part that is not changed in one round. As a
consequence, the orientation changes compared to most of the papers and we
obtain an apparent (but extremely useful) distinction between odd and even
rounds of a Feistel scheme. Otherwise, our notations are very similar to these
used for DES in [23, 18]. For example $L_0[\alpha]$ denotes a sum (XOR) of some subset
α of bits of the left half of the plaintext. Combinations of inputs (or outputs) of
round function number $r = 1, 2, \ldots$ are denoted by $I_r[\alpha]$ (or $O_r[\beta]$). Our exact
notations for DES will be explained in more details when needed, in Section 6.1.
For the time being, we start with a simple rather self-explaining example (cf.
Figure 1) that works for any Feistel cipher.

Proposition 2.1.1 (Combining bi-linear expressions in a Feistel cipher).
For all (even unbalanced) Feistel ciphers operating on $n + n'$ bits with arbitrary
round functions we have: $\forall \alpha \subset \{1, \ldots, n\}, \forall \beta \subset \{1, \ldots, n'\}, \ \forall r \geq 0$:

$$L_r[\alpha]R_r[\beta] \oplus L_0[\alpha]R_0[\beta] = \sum_{i=1}^{\lceil r/2 \rceil} O_{2i-1}[\alpha]I_{2i-1}[\beta] \ \oplus \ \sum_{i=1}^{\lfloor r/2 \rfloor} I_{2i}[\alpha]O_{2i}[\beta] \qquad \square$$

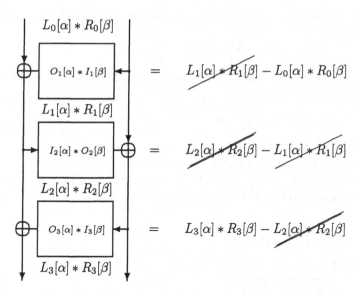

Fig. 1. Fundamental remark: combining bi-linear expressions in a Feistel cipher

From one product this fundamental result extends immediately, by linearity, to arbitrary bi-linear expressions. Moreover, we will see that these bi-linear expressions do not necessarily have to be the same in every round, and that they can be freely combined with linear expressions (BLC contains LC).

3   Bi-linear Characteristics

For simplicity let $n = n'$. In this section we construct a completely general bi-linear characteristic for one round of a Feistel cipher. Then we show how it combines for the next round. Here we study bits locally and denote them by A_i, B_j etc. Later for constructing attacks for many rounds of practical Feistel ciphers we will use (again) the notations $L_i[j_1, \ldots, j_k]$ (cf. Section 6.1).

3.1   Constructing a Bi-linear Characteristic for One Round

Let $\mathcal{S}$ be a homogeneous bi-linear Boolean function $GF(2^n) \times GF(2^n) \to GF(2)$. Let $\mathcal{S}(A_1, \ldots, A_n; B_1, \ldots, B_n) = \sum s_{ij} A_i B_j$.

Let f_K be the round function of a Feistel cipher. We assume that there exist two linear combinations u and v such that the function:

$$(B_1, \ldots, B_n) \mapsto \begin{cases} \sum s_{ij} O_i B_j \oplus \sum u_i O_i \oplus \sum v_i B_i \\ \text{with } (O_1, \ldots, O_n) = f_K(B_1, \ldots, B_n) \end{cases}$$

is biased and equal to 0 with some probability $p \neq 1/2$ with $p = p(K)$ depending in some way on the round key K.

We have $C_i = A_i \oplus O_i$. By bi-linearity (or from Proposition 2.1.1) the following holds:

$$\sum s_{ij}A_iB_j \oplus \sum s_{ij}O_iB_j = \sum s_{ij}C_iB_j$$

From this, for the first round, (could be also any odd-numbered round), we obtain the following characteristic:

$$\left.\begin{array}{l} \sum s_{ij}A_iB_j \oplus \sum u_iA_i \oplus \sum v_iB_i = \\ \sum s_{ij}C_iB_j \oplus \sum u_iC_i \end{array}\right\} \quad \text{with probability } p(K)$$

Finally, we note that, the part linear in the B_i can be arbitrarily split in two parts: $\sum v_iB_i = \sum v_i^{(1)}B_i \oplus \sum v_i^{(2)}B_i$ with $v_i = v_i^{(1)} \oplus v_i^{(2)}$ for all $i = 1, \ldots, n$.
All this is summarized on the following picture:

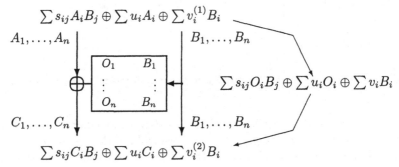

Fig. 2. Constructing a bi-linear characteristic for an odd round of a Feistel cipher

3.2 Application to the Next (Even) Round

The same method can be applied to the next, even, round of a Feistel scheme, with the only difference that the round function is connected in the inverse direction. In this case, to obtain a characteristic true with probability $\neq 1/2$, we need to have a bias in the function:

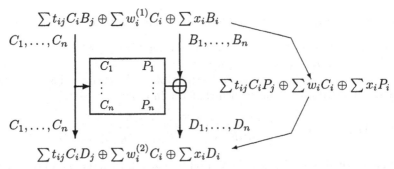

Fig. 3. Constructing a bi-linear characteristic for an even round of a Feistel cipher

$$(C_1, \ldots, C_n) \mapsto \left\{ \begin{array}{l} \sum t_{ij} C_i P_j \oplus \sum w_i C_i \oplus \sum x_i P_i \\ \text{with } (P_1, \ldots, P_n) = f_K(C_1, \ldots, C_n) \end{array} \right.$$

3.3 Combining Approximations to Get a Bi-linear Attack for an Arbitrary Number of Rounds

It is obvious that such I/O sums as specified above can be combined for an arbitrary number of rounds (contradicting [20] page 226). To combine the two characteristics specified above, we require the following three conditions:

1. We need $u = w^{(1)}$.
2. We need $v^{(2)} = x$.
3. We need the homogenous quadratic parts s et t to be correlated (seen as Boolean functions). They do **not** have to be the same (though in many cases they will). In linear cryptanalysis (LC), a correlation between two linear combinations means that these linear combinations have to be the same. In generalized linear cryptanalysis (GLC) [9], and in particular here, for bi-linear I/O sums, it is no longer true. Correlations between quadratic Boolean functions are frequent, and does not imply that $s = t$. For these reasons the number of possible bi-linear attacks is potentially very large.

Summary: We observe that bi-linear characteristics combine exactly as in LC for their linear parts, and that their quadratic parts should be either identical (with orientation that changes in every other round), or correlated.

4 Predicting the Behaviour of Bi-linear Attacks

The behaviour of LC is simple and the heuristic methods of Matsui [25] are known to be able to predict the behaviour of the attacks with good precision (see below). Some attacks work even better than predicted. As already suggested in [9, 20] the study of generalised linear cryptanalysis is **much harder**.

4.1 Computing the Bias of Combined Approximations

A bi-linear attack will use an I/O sum for the whole cipher, being a sum of I/O sums for each round of the cipher such that the terms in the internal variables do cancel. To compute the probability the resulting equation is true, is in general not obvious. Assuming that the I/O sum uses balanced Boolean functions, (otherwise it will be even harder to analyse) one can apply the Matsui's Piling-up Lemma from [25]. This however **can fail**. It is known from [9] that a sum of two very strongly biased characteristics can have a bias much weaker than expected. The resulting bias can even be exactly zero: an explicit example can be found in Section 6.1. of [9]. Such a problem can arise when the connecting characteristics are not independent. This will happen more frequently in BLC than in LC: two linear Boolean functions are perfectly independent unless equal, for non-linear Boolean functions, correlations are frequent. Accordingly, we do not sum independent random variables and the Matsui's lemma may fail.

At this stage there are two approaches: one can try to define a class of attacks that can be proved to work, and restrict oneself only to studying such attacks, or try to explore all possible attacks, including those that do work experimentally without proof. This first approach is adopted in [9]: the Lemma 6 gives a sufficient condition to guarantee that the Piling-up lemma will apply. For this the probability, that the characteristic is true, for a random partial key, should be independent of the input (e.g. the input of the whole round). This explains why Matsui's attacks indeed work well. In [9] it allows to prove that the proposed family of GLC attacks based on homomorphic properties will work as predicted. We will also use this argument in Section 5.

In this paper we frequently adopt rather the second approach: try find as many working attacks as possible, even if current theory does not allow to predict their behaviour with accuracy. A price to pay for this is that each application of Matsui's Lemma will be systematically questioned and confronted to experimental results.

4.2   Key Dependence in Bi-linear Attacks

Another important property of bi-linear cryptanalysis is that the existence of a bias for one characteristic does frequently depend on the key. This does not really happen for LC applied DES, because in DES all key bits are combined linearly and a linear equation will be true with probability either p or $1 - p$ depending on the key. However it will happen for LC and other ciphers, if key bits are involved in a more complex way, for example for ICE [22].

In bi-linear cryptanalysis, the behaviour becomes complex already when the key bits are combined linearly as in DES. Adding a constant (a key bit) to an input of an S-box, does not only modify the constant part in a bi-linear characteristic, but also the linear part. (We note that for DES only the linear part in the output variables will be modified when the key changes). From this, quite frequently two bi-linear characteristics for two parts of a cipher (e.g. for S-boxes) will only connect together for some keys. Such attacks are still very interesting and frequently also do work, with only a slightly weaker bias, for all the other keys. For simplicity, no key bits are displayed in bi-linear characteristics for one or several rounds of a cipher that are studied/displayed in this paper. The values of biases we will present (unless otherwise stated) are given for the reference key being zero. Yet typically we observed that they exist, and slightly vary in value, also for **any** other key (chosen at random). In rare cases, the bias works well only for a fraction of keys (e.g. 25 %): this happens in Appendix B.1.

4.3   Exploring Bi-linear Cryptanalysis

There are different approaches to finding interesting bi-linear attacks to block ciphers. In few cases one can construct attacks that will provably or arguably work (see [9] and later Section 5). Another method is to construct characteristics "by hand" around some particularly strong bias found for one S-box.

We noted the two major difficulties: predicting the bias of combined characteristics, and huge number of possible characteristics (including fragmentation due to the fact they the bias does in general depend on the key). These make it very difficult to have a systematic method (a computer program) that would compute the best bi-linear characteristic for a given cipher. To check if an attack indeed works requires to be able to generate as many plaintexts as for the real attack. To find the best attack is even much harder. It requires to exhaustively search and reject lots of other combinations that should work well but they don't. Each of them has to be tested on an equally large set of plaintexts.

5   The Killer Example for Bi-linear Cryptanalysis

We will construct a practical cipher that is very secure w.r.t. all known attacks for block ciphers, in particular for LC, yet broken by BLC. It mixes two group operations: the XOR and the multiplication in $GF(2^n)$ e.g. $n = 32$ or 64. It uses the inverse in $GF(2^n)$ (cf. Rijndael): let $Inv(X) = X^{-1}$ in $GF(2^n)$ when $X \neq 0$ and 0 otherwise. We build a $2n$-bit Feistel cipher with the i-th round function being:

$$f_i(X) = Inv(X) \cdot (K_i \oplus G(X)) \quad \text{in } GF(2^n), \tag{1}$$

with K_i being the partial key, and G being some function with S-boxes and arbitrary components $\{0,1\}^n \to \{0,1\}^n$. In order to get an insecure cipher, we need to assume that some linear combination of outputs of G is biased. For example, let $Y_1 \oplus Y_5 = 0$ with probability $3/4$. Building a cipher with G alone would be insecure for LC, however here G is composed by a group operation $\cdot$ with $Inv(X)$. The $Inv(X)$ assures global diffusion and very high non-linearity (cf. [3]). Accordingly our round function has very good resistance to linear and differential cryptanalysis for most G, even when $G = 0$. But not against BLC.

First, we can consider a bi-linear attack with bi-linear equations over $GF(2^n)$:
$\forall r \geq 0$:

$$L_r \cdot R_r \oplus L_0 \cdot R_0 = \sum_{i=1}^{\lceil r/2 \rceil} O_{2i-1} \cdot I_{2i-1} \oplus \sum_{i=1}^{\lfloor r/2 \rfloor} I_{2i} \cdot O_{2i} = \sum_{i=1}^{r} I_i \cdot O_i \tag{2}$$

Let $X \cdot Y = (Z_1, \ldots, Z_n)$ with $Z_k = \sum_{ij} M_k^{ij} X_i Y_j$. From (2), or if we prefer, directly from Proposition 2.1.1 and by symmetry $M_k^{ij} = M_k^{ji}$, we get:

$$\forall k \in \{1, \ldots, n\}, \forall r \geq 0 \quad \sum_{ij} M_k^{ij} (L_{ri} R_{rj} \oplus L_{0i} R_{0j}) = \sum_{l=1}^{r} \sum_{ij} M_k^{ij} I_{li} O_{lj} \tag{3}$$

Now, $\forall l \geq 1$, $I_l \cdot O_l = K_l \oplus G(I_l)$ with probability $(1 - 1/2^n)$. We rewrite it:

$$\forall k \in \{1, \ldots, n\}, \forall l \geq 0 \quad \sum_{ij} M_k^{ij} I_{li} O_{lj} = K_{ik} \oplus G_k(I_i) \tag{4}$$

Then we use the linear output bias of G: $G_1 \oplus G_5 = 0$ with probability $3/4$.

$$\forall l \geq 0 \quad \sum_{ij} M_1^{ij} I_{li} O_{lj} \oplus \sum_{ij} M_5^{ij} I_{li} O_{lj} = K_{i1} \oplus G_1(I_i) \oplus K_{i5} \oplus G_5(I_i) \approx C_l \tag{5}$$

The last expression is equal to come constant denoted C_l with probability $3/4$. Finally, we combine with (3) (or equivalently sum these bi-linear expressions over the whole cipher with r rounds).

$$\sum_{ij} \left(M_1^{ij} \oplus M_5^{ij} \right) (L_{ri}R_{rj} \oplus L_{0i}R_{0j}) = \sum_{l=1}^{r} C_l \quad \text{with probability} \quad \frac{1}{2} + \frac{1}{2^{r+1}} \quad (6)$$

What we obtained is a biased bi-linear I/O sum for the whole cipher. We can distinguish this cipher from a random permutation given about 2^{2r+2} plaintexts. For example 16 rounds will be broken on a laptop PC.

Does it work as predicted? In general, as we explain in Section 4.1, it is hard to predict accurately the behaviour of a composed bi-linear attack. However we have little doubt it will work: the $Inv(X)$ should render possible correlation between approximations being combined negligible. In some case we can even prove that this attack works: when $G = 0$, and also when one fixed linear combination of output bits of G is 0, (the other parts can be arbitrary functions). In these cases, dependencies cannot be a problem: we add equations (5) true with probability 1 to get the equation (6) true with probability 1.

Related work: Similar results were previously obtained for some substitution-permutation network (SPN) ciphers. In [9] Harpes, Kramer and Massey give an example of 8-bit SPN that is secure against LC and DC, but insecure for generalised linear cryptanalysis due to a probabilistic homomorphic property of each round relative to quadratic residuosity function modulo 2^8+1. The Jakobsen attack for substitution ciphers that uses probabilistic univariate polynomials from [15] can also be seen as a special case of GLC. However, it is the first time that GLC allows to break a Feistel cipher, which contradicts the impossibility professed by Knudsen and Robshaw [20]. This cipher is built with state-of-art components (inverse in $GF(2^n)$) and can in addition incorporate any additional fashionable component with lots of theory and designer tricks, as a part of G. Due to G it will not have homomorphic properties. Moreover, by adjusting the bias in G, the security of this cipher against BLC will be freely adjusted between (nearly) zero and infinity. It can therefore be arbitrarily weak for BLC, and this even for a very large number of rounds. Yet, the security against the usual attacks (LC, DC) should remain equally good (due to the big Inv S-box).

6   Bi-linear Attacks on DES

6.1   Notation

We ignore the initial and final permutations of DES that have no incidence on the attacks. We use the "untwisted method" of representing DES, as on the right-hand figure, page 254 in [28]. The bit numbering is compatible with the FIPS standard [8], and [23, 18], and differs from Biham, Shamir [2] or Matsui [25, 27]. We denote the bits of the left hand side of the plaintext by $L_0[1] \ldots L_0[n]$. The bits of the right hand side are $R_0[1] \ldots R_0[n]$. Similarly, as in other papers, the

plaintext after i rounds will be L_i, R_i, except that we felt it necessary to have our notations completely "untwisted" which implies that our L_i and R_i for an odd $i = 1, 3, \ldots$ will be inversed compared to [23, 18, 28], Then, we apply the popular convention $X[i_1, \ldots, i_n]$ being $X[i_1] \oplus \ldots \oplus X[i_n]$. For example $L_0[9, 7, 23, 31]$ is the XOR of 4 bits of the left half of the plaintext that are added to the outputs of S1 in the first round. We denote the input bits to the i–th round function by $I_i[1], \ldots, I_i[32]$. Similarly the output bits will be $O_i[1], \ldots, O_i[1]$.
For odd i we have $I_i[j] = R_{i-1}[j] = R_i[j]$ and $O_i[j] = L_{i-1}[j] \oplus L_i[j]$.
For even i we have $I_i[j] = L_{i-1}[j] = L_i[j]$ and $O_i[j] = R_{i-1}[j] \oplus R_i[j]$.

For individual S-boxes, we will denote the inputs/outputs by respectively $O[i]$ and $J[j]$ with i, j being directly the numbers 1..32 in the round function of DES. For example $O[8], O[14], O[25], O[3]$ are the outputs of S-box S5, and $J[16], \ldots, J[21]$ are the inputs of this S-box S5. Depending on the key in round i, we have $I_i[k] = J_i[k]$ or $I_i[k] = J_i[k] + 1$. For better readability, we will avoid naming precisely the key bits involved.

6.2   First Example of Bi-linear Cryptanalysis of DES

Our simulations on DES S-boxes (cf. Appendix A) show that the following two bi-linear characteristics exist for DES S-boxes S1 and S5:

$$O[8, 14, 25, 3] \oplus J[17] \cdot O[3] = 0 \quad \text{for } S5 \text{ with probability } 17/64$$

$$O[17] \oplus J[3] \cdot O[17] = 0 \quad \text{for } S1 \text{ with probability } 47/64$$

From these, acting as if all the key bits were zero ($I_i[k] = J_i[k]$), we deduce the following bi-linear characteristic for two rounds:

$$(*) \quad \left. \begin{array}{l} L_0[3, 8, 14, 25] \oplus L_0[3]R_0[17] \oplus R_0[17] \oplus \\ L_2[3, 8, 14, 25] \oplus L_2[3]R_2[17] \oplus R_2[17] = K[sth] \end{array} \right\} \quad \frac{1}{2} - 1.76 \cdot 2^{-4}$$

The explanation is given on the following picture:

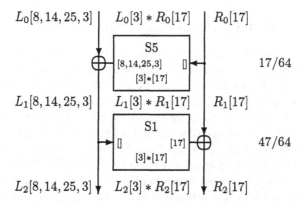

Fig. 4. Our first example - an invariant bi-linear attack on DES (*)

We verified this bias experimentally, and the probability is (we were lucky) equal to the probability that is predicted by Matsui's Piling-Up Lemma.

Key Dependence: Very surprisingly, the above equation (∗) is biased, not only when all key bits are 0, but for every DES key. This can be seen to come from a couple of other (different) bi-linear characteristics from Appendix A.

More rounds: It is easy to see from the picture, and we verified it experimentally, that (∗) is also biased for $1, 2, 3, 4, 5, 6, 7, 8, 9, 10, 11, \ldots$ rounds of DES, and all this happens to work about equally well for an arbitrary key.

Relation to LC: The bias of (∗) is closely related to some prominent equations of Matsui, see the extended version of this paper.

6.3   Invariant Attacks on DES

The equation (∗) is an invariant equation, i.e. the input and the output bi-linear expressions are the same. We have found a simple invariant bi-linear I/O sum for DES that is biased for any key and for any number of rounds. For LC and DES, such simple invariant characteristics do exist, have been found by Biham (page 347 in [1]) in close relation to Davies-Murphy attack. The example (∗) above is one of the best we found for DES, and so far it also **the only known** non-linear 1-round invariant attack on DES that works really well in practice. Our invariant on DES is stronger than Biham's. We recall that Biham uses a bias on a sum of some outputs for two successive DES S-boxes. The best bias obtained by Biham (also exhibited by Matsui in [26] and contained unnoticed in the earlier Davies-Murphy attack [6,7]) is equal to $(35/64 - 1/2)$ for 2 rounds and for S-boxes S7-S8. This gives $1.4 \cdot 2^{-22}$ for 12 rounds. Instead, (∗) gives experimentally only about $1.3 \cdot 2^{-18}$. Accordingly, (∗) is **the strongest known 1-round invariant attack on DES**.

To break full DES requires a bias for 14 rounds (Matsui's 2R method) and the Biham's invariant requires then 2^{50} plaintexts. Our invariant attack requires about 2^{43} plaintexts (the bias of (∗) for 14 rounds is expected to be about 2^{-22}, we did not dispose of a sufficient computing power to compute it exactly).

6.4   How Good Is Our First Example, BLC vs. LC

These new properties of DES give a chosen-plaintext attack on an arbitrary number of rounds of DES, somewhat simpler than Matsui's laborious search for the best linear characteristic. If we try here to predict the resulting bias for 14 rounds by applying the Matsui's Piling-up formula, we would get for 14 rounds the bias of: $1.63 \cdot 2^{-17}$ which means an attack on full DES with only $2^{32.6}$ known plaintexts (!?). Unfortunately, unlike for LC in DES, such predictions are frequently not valid for BLC. Starting from 3 rounds, the bias of our invariant does not follow the prediction at all, yet remains significative. For example if we apply Matsui's Piling-Up Lemma to predict the bias for 4 rounds as 2+2 rounds, we obtain $1.55 \cdot 2^{-6}$, while in practice it is about $1.80 \cdot 2^{-8}$. Our invariant attack seems very bad for 4 rounds, and unfortunately with (∗) we never get a bias better than obtained by Matsui. Yet, it is the best invariant attack on DES known, and for more than 4 rounds the results are again not so bad. Only slightly worse than Matsui. For example for 12 rounds the best result of Matsui from [25] gives $1.19 \cdot 2^{-17}$, while for (*) and a random key our simulation gives $1.3 \cdot 2^{-18}$, To break full DES Matsui requires about 2^{43} plaintexts, and with (*) we also need about 2^{43} (and both are related). In the full version of this paper we give a heuristic argumentation why for DES (but not in general !) the complexity of the best bi-linear attack should be roughly the same than for LC.

For DES and 1-round invariants attacks extended to an arbitrary number of rounds, BLC gives strictly better results than LC. It is also so for more complex periodic constructions and we are going to see that BLC attacks can also be strictly better than any existing linear attack.

6.5 Second Example of Bi-linear Cryptanalysis of DES

In order to exhibit biases really better than Matsui we looked what is the best bi-linear characteristic that exists in DES:

$$J[16, 20] \oplus O[8, 14, 25, 3] \oplus J[16, 17, 20] \cdot O[3] = 0 \quad \text{for } S5 \text{ with probability } 61/64.$$

We note that this equation can be seen as "causing" the existence of the Matsui's best equation (A) for S5: their difference is highly biased. Based mainly on this, we constructed a periodic characteristic for 3, 7, 11 and more rounds that is strictly better than the best results of Matsui for the same number of rounds.

Proposition 6.5.1 (Our Best Attack on 11 Rounds of DES). For all keys, the following equation is biased for 11 rounds of DES:

$$(**)\quad \left. \begin{array}{l} L_0[3, 8, 14, 25] \oplus L_0[3]R_0[16, 17, 20] \oplus R_0[17] \oplus \\ L_{11}[3, 8, 14, 25] \oplus L_{11}[3]R_{11}[16, 17, 20] \oplus R_{11}[17] = \\ K[sth] + K[sth']L_0[3] + K[sth'']L_{11}[3] \end{array} \right\} \quad \frac{1}{2} \pm \text{around } 1.2 \cdot 2^{-15}$$

The exact construction to achieve this is a bit complicated. (cf. Appendix B). The bias of this equation is strictly better than the best linear characteristic for 11 rounds obtained by Matsui (which gives $1.91 \cdot 2^{-16}$ for 11 rounds). It has been verified by computer simulations at every stage. We note also that both are closely related: their difference, is a biased Boolean function.

Our second example allows us to give an attack strictly better than Matsui for 11+2=13 rounds of DES. For the full 16-round DES our results are roughly as good as Matsui (but we hope to improve this soon too). For 17 rounds of DES, as the construction of our second example (**) is periodic, we expect that for 11+4=15 rounds it should also be better than the best bias of Matsui, which would allow to break 15+2=17 rounds of DES faster than by LC. We do not dispose of a sufficient computing power to fully confirm this fact.

7 Conclusion

It was stated that for Feistel ciphers non-linear characteristics cannot be joined together for several rounds, see [20]. In this paper we show that generalised linear cryptanalysis (GLC) is in fact possible for Feistel schemes. To achieve this goal, we introduced bi-linear cryptanalysis (BLC). It gives a new (and the fastest known) 1-round invariant attack on DES. Though more powerful, generalized linear cryptanalysis is unfortunately much harder to study than LC. At present heuristic constructions, to be confirmed (or not) by computer simulations are the only method known to explore it. BLC is related to LC in multiple important ways. It contains LC as a sub-set. LC can be used to construct good bi-linear characteristics and vice-versa. BLC also contains LC as an extension: a combination of biased bi-linear characteristics may extend a concrete combination of biased linear characteristics by adding quadratic polynomials. Yet BLC can be strictly better than any (existing) linear attack. This was demonstrated for 3, 7, 11 and more rounds of DES, and also for s^5DES.

In this paper we only initiate the study of bi-linear cryptanalysis. BLC and GLC extend the role of LC as an essential tool to evaluate the real-life security of many practical ciphers. An interesting contribution of this paper is to point out that, though GLC is excessively general to be systematically explored, the properties of the top-level structure of a cryptographic scheme (e.g. being a Feistel scheme) will determine the type of the attacks (e.g. BLC) that may indeed work. Our new attack can be quite devastating: we constructed a large family of practical ciphers based on big Rijndael-type S-box, that are strongly resistant against LC and all previously known attacks on Feistel ciphers, yet can be broken in practice with BLC for an important number of rounds. Fortunately, for DES, BLC gave only slight improvements over LC and does not cause excessive trouble.

References

1. Eli Biham: *On Matsui's Linear Cryptanalysis*, Eurocrypt'94, LNCS 950, Springer-Verlag pp. 341-355, 1994.
2. Eli Biham and Adi Shamir, *Differential Cryptanalysis of DES-like Cryptosystems.* Journal of Cryptology, vol. 4, pp. 3-72, IACR, 1991.
3. Anne Canteaut, Marion Videau: *Degree of composition of highly nonlinear functions and applications to higher order differential cryptanalysis*, Eurocrypt 2002, LNCS 2332, Springer, 2002.
4. Anne Tardy-Corfdir, Henri Gilbert: *A Known Plaintext Attack of FEAL-4 and FEAL-6*, Crypto'91, LNCS 576, Springer, pp. 172-181, 1992.
5. Nicolas Courtois, Guilhem Castagnos and Louis Goubin: *What do DES S-boxes Say to Each Other ?* Available on eprint.iacr.org/2003/184/.
6. D.W. Davies, *Some Regular Properties of the Data Encryption Standard*, Crypto'82, pp. 89-96, Plenum Press, New-York, 1982.
7. D. Davies and S. Murphy, *Pairs and Triplets of DES S-Boxes*, Journal of Cryptology, vol. 8, Nb. 1, pp. 1-25, 1995.
8. *Data Encryption Standard (DES)*, Federal Information Processing Standards Publication (FIPS PUB) 46-3, National Bureau of Standards, Gaithersburg, MD (1999). http://csrc.nist.gov/publications/fips/fips46-3/fips46-3.pdf
9. C. Harpes, G. Kramer, and J. Massey: *A Generalization of Linear Cryptanalysis and the Applicability of Matsui's Piling-up Lemma*, Eurocrypt'95, LNCS 921, Springer, pp. 24-38. http://www.isi.ee.ethz.ch/ harpes/GLClong.ps

10. Carlo Harpes: *Cryptanalysis of iterated block ciphers*, PhD thesis, No 11625, Swiss Federal Int. of Tech., ETH Series in Information Processing, Ed. J. L. Massey, Hartung-Gorre Verlag Konstanz, 1996, ISBN 3-89649-079-6, ISSN 0942-3044.

11. Carlo Harpes: *Partitioning Cryptanalysis*, Post-Diploma Thesis, Signal and Information Processing Lab., Swiss Federal Institute of Technology, Zurich, March 1995. http://www.isi.ee.ethz.ch/~harpes/pc.ps

12. Thomas Jakobsen, Carlo Harpes: *Non-Uniformity Measures for Generalized Linear Cryptanalysis and Partitioning Cryptanalysis*, Pragocrypt'96, 1996.

13. Thomas Jakobsen: *Correlation Attacks on Block Ciphers*, Master's Thesis, Dept. of Mathematics, Technical University of Denmark, January 1996.

14. Thomas Jakobsen: *Higher-Order Cryptanalysis of Block Ciphers*. Ph.D. thesis, Dept. of Math., Technical University of Denmark, 1999.

15. Thomas Jakobsen: *Cryptanalysis of Block Ciphers with Probabilistic Non-Linear Relations of Low Degree*, Crypto 98, LNCS 1462, Springer, pp. 212-222, 1998.

16. Pascal Junod: *On the complexity of Matsui's attack*, Selected Areas in Cryptography (SAC'01), Toronto, Canada, LNCS 2259, pp. 199-211, Springer, 2001.

17. Burton S. Kaliski Jr, and M.J.B. Robshaw. *Linear Cryptanalysis Using Multiple Approximations*, Crypto'94, LNCS, Springer, pp. 26-39, 1994.

18. Toshinobu Kaneko and Takeshi Shimoyama: *Quadratic Relation of S-box and Its Application to the Linear Attack of Full Round DES*, In Crypto 98, LNCS 1462, p. 200-211, SPringer, 1998.

19. Kwangjo Kim. Sangjin Lee, Sangjoon Park, Daiki Lee: *Securing DES S-boxes against Three Robust Cryptanalysis*, SAC'95, pp.145-157, 1995.

20. Lars R. Knudsen, Matthew J. B. Robshaw: *Non-Linear Characteristics in Linear Cryptoanalysis*. Eurocrypt'96, LNCS 1070, Springer, pp. 224-236, 1996.

21. *Lars R. Knudsen, John Erik Mathiassen: A Chosen-Plaintext Linear Attack on DES. FSE'2000, LNCS 1978, Springer, pp. 262-272, 2001.*

22. Matthew Kwan: *The Design of the ICE Encryption Algorithm*, FSE'97, 4th International Workshop, Haifa, Israel, Springer, LNCS 1267, pp. 69-82, 1997. Available from http://www.darkside.com.au/ice/ice.ps.gz.

23. Susan K. Langford, Martin E. Hellman: *Differential-linear cryptanalysis*, Crypto 94, LNCS 839, pp. 17-25, Springer, 1994.

24. Michael Luby, Charles W. Rackoff, *How to construct pseudorandom permutations from pseudorandom functions*, SIAM Journal on Computing, vol. 17, n. 2, pp. 373-386, April 1988.

25. M. Matsui: *Linear Cryptanalysis Method for DES Cipher*, Eurocrypt'93, LNCS 765, Springer, pp. 386-397, 1993.

26. M. Matsui, *On correlation between the order of S-boxes and the strength of DES*, Eurocrypt'94, LNCS 950, pp. 366-375, Springer, 1995.

27. M.Matsui: *The First Experimental Cryptanalysis of the Data Encryption Standard*, Crypto'94, LNCS 839, Springer, pp. 1-11, 1994.

28. Alfred J. Menezes, Paul C. van Oorschot, Scott A. Vanstone: *Handbook of Applied Cryptography*; CRC Press, 1996.

29. J. Patarin, *How to construct pseudorandom and super pseudorandom permutations from one single pseudorandom function.* Eurocrypt'92,Springer, pp. 256-266, 1992.

30. Adi Shamir: *On the security of DES*, Crypto'85, LNCS 218, Springer, pp. 280-281, 1985.

A   Selected Bi-linear Characteristics of DES S-Boxes

In this section we give some bi-linear characteristics for DES S-boxes. Our results are not exhaustive: the number of possible bi-linear characteristics is huge and we do not have a fast method to find all interesting characteristics. Accordingly we are not certain to have found the best existing characteristics. It is certain that there is no characteristics true with probability 1, as these are easy to check algebraically. Otherwise we explored all cases that use up to two products and we conjecture that the other does not have practical relevance for the security of DES. We give here some interesting results we have found. More will appear in the extended version of this paper.

Table 1. A few selected bi-linear characteristics for DES S-boxes

		equation			remarks and
		input	output	input*output	comments
$S5$	12/64	17	$8, 14, 25, 3$		Matsui's equation A
$S5$	6/64	17	$8, 14, 25, 3$	$[17] * [8, 14, 25, 3]$	gets better
$S5$	58/64			$[17] * [8, 14, 25, 3]$	
$S5$	8/64	17	$8, 14, 25, 3$	$[16, 17, 20] * [8]$	
$S5$	8/64	16, 20	$8, 14, 25$	$[16, 20] * [8, 14, 25]$	
$S5$	**61/64**	16, 20	$8, 14, 25, 3$	$[16, 17, 20] * [3]$	the best in DES
$S5$	47/64		$8, 14, 25$	$17 * 3$	
$S5$	17/64		$8, 14, 25, 3$	$17 * 3$	
$S5$	47/64			$17 * 3$	
$S5$	49/64		3	$17 * 3$	
$S5$	49/64	17		$17 * 3$	
$S5$	17/64	17	3	$17 * 3$	
$S1$	30/64	3	17		Matsui's equation C
$S1$	15/64	3	17	$3 * 17$	gets better
$S1$	47/64		17	$3 * 17$	
$S1$	47/64	3		$3 * 17$	
$S1$	49/64			$3 * 17$	
$S2$	8/64	5	$13, 28, 18$	$8 * 2$	
$S4$	56/64			$[12, 14, 16, 17] * [26, 1]$	(there are many similar)
$S6$	38/64		$11, 19$	$21 * 29$	
$S7$	11/64	25, 28	$32, 12, 7$	$28 * 12, 27 * 22$	
$S8$	40/64		$5, 27, 15$	$29 * 21$	

B   Improved Bi-linear Attacks for DES

The goal of this section is to find or construct examples where bi-linear cryptanalysis gives strictly better bias on DES than the best Matsui's result.

We look at the best Matsui's characteristic on 3 rounds given at the last page of [25]. By itself, it can be considered as very good, even compared to other Matsui's characteristics: it uses twice the best element (A) of Matsui, and nothing between them. Moreover, this element (A) is in itself the best linear characteristic that exist in DES, first described by Shamir in [30]:

(A) $J[17] \oplus O[8, 14, 25, 3] = 0$ for $S5$ with probability $12/64$

From this we get immediately, using Matsui's Piling-Up Lemma from [25], that for 3 rounds, and for any key, the following equation is biased:

$$\left. \begin{array}{l} L_0[8, 14, 25, 3] \oplus R_0[17] \oplus \\ L_3[8, 14, 25, 3] \oplus R_3[17] = K[sth] \end{array} \right\} \quad \frac{1}{2} - 1.56 \cdot 2^{-3}$$

We call Matsui-3 this equation.

B.1   Improving on Matsui-3

We will show that with bi-linear characteristics, there are strictly better equations than Matsui-3. Our simulations looking for the best bi-linear characteristics for DES S-boxes (cf. Appendix A), showed that the best one is the following:

$$J[16, 20] \oplus O[8, 14, 25, 3] \oplus J[16, 17, 20] \cdot O[3] = 0 \quad \text{for } S5 \text{ with probability } 61/64$$

Remark: It is clearly related to, and can be seen as "causing" the existence of the Matsui's equation (A): their difference is naturally biased.

We will use this characteristic. Let KS5 denote the combination of the S-box S5 and the key bits XORed to its inputs. It is easy to see that for KS5, if we denote by $K[sth]$ some constant linear combination of key bits, for any key, one of the following equations is always strongly biased:

$$\left\{ \begin{array}{l} \textbf{(a1)} \ I[16, 20] \oplus O[8, 14, 25, 3] \oplus I[16, 17, 20] \cdot O[3] = K[sth] \\ \qquad\qquad\qquad\qquad \text{or} \\ \textbf{(a2)} \ I[16, 20] \oplus O[8, 14, 25] \oplus I[16, 17, 20] \cdot O[3] = K[sth] \end{array} \right. \quad |\text{bias}| = 1/2 - 3/64$$

In our construction, we will use one of the above, and we will also use another, naturally biased equation, which will be one of the following:

$$\left\{ \begin{array}{l} \textbf{(b)} \ O[16, 17, 20] \oplus I[3] \cdot O[16, 17, 20] = 0 \\ \qquad\qquad\qquad\qquad \text{and} \\ \textbf{(c)} \ I[3] \oplus O[16, 17, 20] \oplus I[3] \cdot O[16, 17, 20] \cdot O[3] = 0 \end{array} \right. \quad |\text{bias}| = 1/2 - 1/4$$

Now we are ready to construct characteristics for 3 rounds of DES.

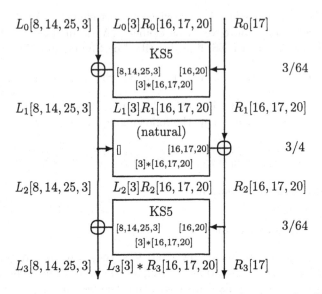

Fig. 5. Combining a1-b-a1 to get a characteristic for 3 rounds of DES

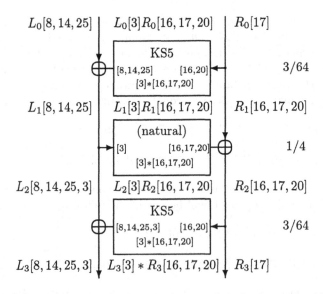

Fig. 6. Combining a2-c-a1 to get a characteristic for 3 rounds of DES

As one should expect, our construction goes as follows:

⋄ In round 1 and 3, depending on the key either a1 or a2 is strongly biased.
⋄ To connect a1 to a1, or a2 with a2, we can use b, as in Figure 5.
⋄ To connect a1 with a2 and the reverse, we use c, as in Figure 6.
⋄ For 3 rounds and for any key, we always have a strong bias on one of the four possibilities: a1-b-a1, a1-c-a2, a2-c-a1, a2-b-a2.

⋄ From Matsui's Piling-Up Lemma, we expect that the whole characteristic will be true with probability $\frac{1}{2} \pm 1.64 \cdot 2^{-3}$. Our simulations show that it is between $\frac{1}{2} \pm 1.65 \cdot 2^{-3}$ and $\frac{1}{2} \pm 1.67 \cdot 2^{-3}$.

⋄ Since, the choice of a1/a2 depends on a linear combination of key bits, We can combine all these into one equation and we get the following result:

Proposition B.1.1 (Our Best Attack on 3 Rounds of DES). For all keys, the following equation is biased for 3 rounds of DES:

$$(**) \quad \left. \begin{array}{l} L_0[3, 8, 14, 25] \oplus L_0[3]R_0[16, 17, 20] \oplus R_0[17] \oplus \\ L_3[3, 8, 14, 25] \oplus L_3[3]R_3[16, 17, 20] \oplus R_3[17] = \\ K[sth] + K[sth']L_0[3] + K[sth'']R_3[3] \end{array} \right\} \quad \frac{1}{2} \pm 1.66 \cdot 2^{-3}$$

In comparison, Matsui-3 gives $\frac{1}{2} - 1.56 \cdot 2^{-3}$. Bi-linear cryptanalysis works better than LC. In the next section we will extend this result (and again beat Matsui) to 7, 11 and more rounds.

Remark: The equation above can be seen as 4 different equations, each of them is highly biased for 1/4 of all keys. We observed that each of the 4 equations is also biased for all DES keys, except that for 3/4 of them the bias is much weaker, we get about $\frac{1}{2} \pm 1.6 \cdot 2^{-7}$.

B.2 Extending the Result for 7, 11 and More Rounds

The idea is to find an element (maybe not very good in itself) that will allow to connect together our (very good) characteristics on 3 rounds. For example, to connect Figure 5 with Figure 6 we use the following element:

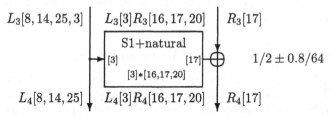

Fig. 7. Connecting the output of a1 to the input of a2

Simulations show that, for any key, this characteristic is true with probability about $1/2 \pm 0.8/64$. The explanation is as follows: the bias is due to to the combination of Matsui's equation (C)

$$(\mathbf{C}) \quad J[3] \oplus O[17] = 0 \quad \text{for } S1 \text{ with probability } 30/64$$

and of the fact that $I[3] \cdot O[16, 17, 20]$ is naturally biased. The same element (Figure 7) does also work to connect a2 to a1.

It remains to be seen how the connection between a1 and a1 or a2 and a2. This is done in a very similar way: we combine (C) with $I[3] \oplus I[3] \cdot O[16, 17, 20]$ that is also naturally biased.

Summary: In every of 4 possible cases, there is a connecting element based on (C). This means that, also for 7 rounds and for any key, again one of the four possibilities is quite biased: a1-b-a1, a1-c-a2, a2-c-a1, a2-b-a2. Again we can recompose it in a single attack:

Proposition B.2.1 (Extension to 7 Rounds of DES). For all keys, the following equation is biased for 7 rounds of DES:

$$\left.\begin{array}{l} L_0[3,8,14,25] \oplus L_0[3]R_0[16,17,20] \oplus R_0[17]\oplus \\ L_7[3,8,14,25] \oplus L_7[3]R_7[16,17,20] \oplus R_3[17] = \\ K[sth] + K[sth']L_0[3] + K[sth'']L_7[3] \end{array}\right\} \quad \frac{1}{2} \pm \text{ about } 2^{-9}$$

This bias is, depending on the key, sometimes better, sometimes worse than Matsui-7 that gives $\frac{1}{2} - 1.95 \cdot 2^{-10}$.

Finally, it is now obvious, that our construction works also for 11, 15, 19 rounds etc. We verified experimentally that for 11 rounds we have:

Proposition B.2.2 (Our Best Attack on 11 Rounds of DES). For all keys, the following equation is biased for 11 rounds of DES:

$$\left.\begin{array}{l} L_0[3,8,14,25] \oplus L_0[3]R_0[16,17,20] \oplus R_0[17]\oplus \\ L_{11}[3,8,14,25] \oplus L_{11}[3]R_{11}[16,17,20] \oplus R_{11}[17] = \\ K[sth] + K[sth']L_0[3] + K[sth'']L_{11}[3] \end{array}\right\} \quad \frac{1}{2} \pm \text{ around } 1.2 \cdot 2^{-15}$$

For a few different keys we have tried (long computation on a PC) the bias was **always strictly better than Matsui-11** that gives $\frac{1}{2} - 1.91 \cdot 2^{-16}$.

Remark: The best characteristics found by Matsui for 3 and 11 rounds [25] are closely related to those presented here: their difference is a biased Boolean function. BLC contains LC not only as a subset, but also as an extension allowing to strictly improve the best linear attacks on DES by adding higher degree monomials.

B.3   Beyond Bi-linear Attacks: Using Cubic Equations

We observed that, for 3 rounds, even better results can be achieved using cubic partially bi-linear characteristics, instead of quadratic bi-linear (**) from Proposition B.1.1. Our simulations show that, for an important fraction of keys:

$$(***) \left.\begin{array}{l} L_0[3,8,14,25] \oplus L_0[3]R_0[16,17,20]R_0[17,18,19,20]\oplus \\ L_3[3,8,14,25] \oplus L_3[3]R_3[16,17,20]R_3[17,18,19,20]\oplus \\ R_0[17] \oplus R_3[17] = K[sth] \end{array}\right\} \quad \frac{1}{2} - 1.82 \cdot 2^{-3}$$

The explanation why this works is quite similar. Though the non-linear part of this equation is not bi-linear, it is well correlated with a truly bi-linear function:

$$L[3]R[16,17,20]R[17,18,19,20] = L[3]R[16,17,20] \text{ with probability } 7/8$$

Unfortunately, the bias of (***) is worse for other keys. On average, the best bias we know for 3 rounds remains (**) from Proposition B.1.1. We also observed that that (***) works for any number of DES rounds and for any key, but again the results are not as good as with (**).

Short Group Signatures

Dan Boneh[1,*], Xavier Boyen[2], and Hovav Shacham[3]

[1] Stanford University
dabo@cs.stanford.edu
[2] Voltage Security
xb@boyen.org
[3] Stanford University
hovav@cs.stanford.edu

Abstract. We construct a short group signature scheme. Signatures in our scheme are approximately the size of a standard RSA signature with the same security. Security of our group signature is based on the Strong Diffie-Hellman assumption and a new assumption in bilinear groups called the Decision Linear assumption. We prove security of our system, in the random oracle model, using a variant of the security definition for group signatures recently given by Bellare, Micciancio, and Warinschi.

1 Introduction

Group signatures, introduced by Chaum and van Heyst [14], provide anonymity for signers. Any member of the group can sign messages, but the resulting signature keeps the identity of the signer secret. In some systems there is a third party that can trace the signature, or undo its anonymity, using a special trapdoor. Some systems support revocation [12, 4, 29, 15] where group membership can be selectively disabled without affecting the signing ability of unrevoked members. Currently, the most efficient constructions [2, 12, 4] are based on the Strong-RSA assumption introduced by Baric and Pfitzman [5].

In the last two years a number of projects have emerged that require the properties of group signatures. The first is the Trusted Computing effort [28] that, among other things, enables a desktop PC to prove to a remote party what software it is running via a process called *attestation*. Group signatures are needed for privacy-preserving attestation [17, Sect. 2.2]. Perhaps an even more relevant project is the Vehicle Safety Communications (VSC) system from the Department of Transportation in the U.S. [18]. The system embeds short-range transmitters in cars; these transmit status information to other cars in close proximity. For example, if a car executes an emergency brake, all cars in its vicinity are alerted. To prevent message spoofing, all messages in the system are signed by a tamper-resistant chip in each car. (MACs were ruled out for this many-to-many broadcast environment.) Since VSC messages reveal the speed and location of the car, there is a strong desire to provide user privacy so that

---

* Supported by NSF and the Packard Foundation.

M. Franklin (Ed.): CRYPTO 2004, LNCS 3152, pp. 41–55, 2004.

the full identity of the car sending each message is kept private. Using group
signatures, where the group is the set of all cars, we can maintain privacy while
still being able to revoke a signing key in case the tamper resistant chip in a car
is compromised. Due to the number of cars transmitting concurrently there is a
hard requirement that the length of each signature be under 250 bytes.

The two examples above illustrate the need for efficient group signatures.
The second example also shows the need for short group signatures. Currently,
group signatures based on Strong-RSA are too long for this application.

We construct short group signatures whose length is under 200 bytes that
offer approximately the same level of security as a regular RSA signature of the
same length. The security of our scheme is based on the Strong Diffie-Hellman
(SDH) assumption [8] in groups with a bilinear map. We also introduce a new assumption in bilinear groups, called the Linear assumption, described in Sect. 3.2.
The SDH assumption was recently used by Boneh and Boyen to construct short
signatures without random oracles [8]. A closely related assumption was used by
Mitsunari et al. [22] to construct a traitor-tracing system. The SDH assumption
has similar properties to the Strong-RSA assumption. We use these properties
to construct our short group signature scheme. Our results suggest that systems
based on SDH are simpler and shorter than their Strong-RSA counterparts.

Our system is based on a new Zero-Knowledge Proof of Knowledge (ZKPK)
of the solution to an SDH problem. We convert this ZKPK to a group signature
via the Fiat-Shamir heuristic [16] and prove security in the random oracle model.
Our security proofs use a variant of the security model for group signatures
proposed by Bellare, Micciancio, and Warinschi [6].

Recently, Camenisch and Lysyanskaya [13] proposed a signature scheme with
efficient protocols for obtaining and proving knowledge of signatures on committed values. They then derive a group signature scheme using these protocols as
building blocks. Their signature scheme is based on the LRSW assumption [21],
which, like SDH, is a discrete-logarithm-type assumption. Their methodology
can also be applied to the SDH assumption, yielding a different SDH-based
group signature.

The SDH group signature we construct is very flexible and we show how to
add a number of features to it. In Sect. 7 we show how to apply the revocation
mechanism of Camenisch and Lysyanskaya [12]. In Sect. 8 we briefly sketch how
to add strong exculpability.

2   Bilinear Groups

We first review a few concepts related to bilinear maps. We follow the notation
of Boneh, Lynn, and Shacham [9]:

1. G_1 and G_2 are two (multiplicative) cyclic groups of prime order p;
2. g_1 is a generator of G_1 and g_2 is a generator of G_2;
3. ψ is a computable isomorphism from G_2 to G_1, with $\psi(g_2) = g_1$; and
4. e is a computable map $e : G_1 \times G_2 \to G_T$ with the following properties:
   - Bilinearity: for all $u \in G_1, v \in G_2$ and $a, b \in \mathbb{Z}$, $e(u^a, v^b) = e(u, v)^{ab}$.
   - Non-degeneracy: $e(g_1, g_2) \neq 1$.

Throughout the paper, we consider bilinear maps $e : G_1 \times G_2 \to G_T$ where all groups G_1, G_2, G_T are multiplicative and of prime order p. One could set $G_1 = G_2$. However, we allow for the more general case where $G_1 \neq G_2$ so that our constructions can make use of certain families of non-supersingular elliptic curves defined by Miyaji et al. [23]. In this paper we only use the fact that G_1 can be of size approximately 2^{170}, elements in G_1 are 171-bit strings, and that discrete log in G_1 is as hard as discrete log in $\mathbb{Z}_q^*$ where q is 1020 bits. We will use these groups to construct short group signatures. We note that the bilinear groups of Rubin and Silverberg [25] can also be used.

We say that two groups (G_1, G_2) as above are a bilinear group pair if the group action in G_1 and G_2, the map ψ, and the bilinear map e are all efficiently computable.

The isomorphism ψ is only needed for the proofs of security. To keep the discussion general, we simply assume that ψ exists and is efficiently computable. (When G_1, G_2 are subgroups of the group of points of an elliptic curve $E/\mathbb{F}_q$, the trace map on the curve can be used as this isomorphism. In this case, $G_1 \subseteq E(\mathbb{F}_q)$ and $G_2 \subseteq E(\mathbb{F}_{q^r})$.)

3   Complexity Assumptions

3.1   The Strong Diffie-Hellman Assumption

Let G_1, G_2 be cyclic groups of prime order p, where possibly $G_1 = G_2$. Let g_1 be a generator of G_1 and g_2 a generator of G_2. Consider the following problem:

q-Strong Diffie-Hellman Problem. The q-SDH problem in (G_1, G_2) is defined as follows: given a $(q + 2)$-tuple $(g_1, g_2, g_2^{\gamma}, g_2^{(\gamma^2)}, \ldots, g_2^{(\gamma^q)})$ as input, output a pair $(g_1^{1/(\gamma+x)}, x)$ where $x \in \mathbb{Z}_p^*$. An algorithm $\mathcal{A}$ has advantage ϵ in solving q-SDH in (G_1, G_2) if

$$\Pr\left[\mathcal{A}(g_1, g_2, g_2^{\gamma}, \ldots, g_2^{(\gamma^q)}) = (g_1^{\frac{1}{\gamma+x}}, x)\right] \geq \epsilon ,$$

where the probability is over the random choice of γ in $\mathbb{Z}_p^*$ and the random bits of $\mathcal{A}$.

Definition 1. *We say that the (q, t, ϵ)-SDH assumption holds in (G_1, G_2) if no t-time algorithm has advantage at least ϵ in solving the q-SDH problem in (G_1, G_2).*

Occasionally we drop the t and ϵ and refer to the q-SDH assumption rather than the (q, t, ϵ)-SDH assumption. The q-SDH assumption was recently used by Boneh and Boyen [8] to construct a short signature scheme without random oracles. To gain confidence in the assumption they prove that it holds in generic groups in the sense of Shoup [27]. The q-SDH assumption has similar properties to the Strong-RSA assumption [5]. We use these properties to construct our short group signature scheme.

3.2 The Linear Diffie-Hellman Assumption

With $g_1 \in G_1$ as above, along with arbitrary generators u, v, and h of G_1, consider the following problem:

Decision Linear Problem in G_1. Given $u, v, h, u^a, v^b, h^c \in G_1$ as input, output **yes** if $a + b = c$ and **no** otherwise.

One can easily show that an algorithm for solving Decision Linear in G_1 gives an algorithm for solving DDH in G_1. The converse is believed to be false. That is, it is believed that Decision Linear is a hard problem even in bilinear groups where DDH is easy. More precisely, we define the advantage of an algorithm $\mathcal{A}$ in deciding the Decision Linear problem in G_1 as

$$
\mathsf{Adv\,Linear}_{\mathcal{A}} \stackrel{\text{def}}{=} \left| \begin{array}{l} \Pr\left[\mathcal{A}(u, v, h, u^a, v^b, h^{a+b}) = \mathsf{yes} : u, v, h \stackrel{\text{R}}{\leftarrow} G_1, a, b \stackrel{\text{R}}{\leftarrow} \mathbb{Z}_p\right] \\ - \Pr\left[\mathcal{A}(u, v, h, u^a, v^b, \eta) = \mathsf{yes} : u, v, h, \eta \stackrel{\text{R}}{\leftarrow} G_1, a, b \stackrel{\text{R}}{\leftarrow} \mathbb{Z}_p\right] \end{array} \right| .
$$

The probability is over the uniform random choice of the parameters to $\mathcal{A}$, and over the coin tosses of $\mathcal{A}$. We say that an algorithm $\mathcal{A}$ (t, ϵ)-decides Decision Linear in G_1 if $\mathcal{A}$ runs in time at most t, and $\mathsf{Adv\,Linear}_{\mathcal{A}}$ is at least ϵ.

Definition 2. *We say that the (t, ϵ)-Decision Linear Assumption (LA) holds in G_1 if no t-time algorithm has advantage at least ϵ in solving the Decision Linear problem in G_1.*

In the full version of the paper we show that the Decision Linear Assumption holds in generic bilinear groups.

Linear Encryption. The Decision Linear problem gives rise to the Linear encryption (LE) scheme, a natural extension of ElGamal encryption. Unlike ElGamal encryption, Linear encryption can be secure even in groups where a DDH-deciding algorithm exists. In this scheme, a user's public key is a triple of generators $u, v, h \in G_1$; her private key is the exponents $x, y \in \mathbb{Z}_p$ such that $u^x = v^y = h$. To encrypt a message $M \in G_1$, choose random values $a, b \in \mathbb{Z}_p$, and output the triple $(u^a, v^b, m \cdot h^{a+b})$. To recover the message from an encryption (T_1, T_2, T_3), the user computes $T_3/(T_1^x \cdot T_2^y)$. By a natural extension of the proof of security of ElGamal, LE is semantically secure against a chosen-plaintext attack, assuming Decision-LA holds.

4   A Zero-Knowledge Protocol for SDH

We are now ready to present the underlying building block for our group signature scheme. We present a protocol for proving possession of a solution to an SDH problem. The public values are $g_1, u, v, h \in G_1$ and $g_2, w \in G_2$. Here $w = g_2^\gamma$ for some (secret) $\gamma \in \mathbb{Z}_p$. The protocol proves possession of a pair (A, x), where $A \in G_1$ and $x \in \mathbb{Z}_p$, such that $A^{x+\gamma} = g_1$. Such a pair satisfies $e(A, wg_2^x) = e(g_1, g_2)$. We use a standard generalization of Schnorr's protocol for proving knowledge of discrete logarithm in a group of prime order [26].

Protocol 1. Alice, the prover, selects exponents $\alpha, \beta \xleftarrow{\text{R}} \mathbb{Z}_p$, and computes a Linear encryption of A:

$$T_1 \leftarrow u^\alpha \qquad T_2 \leftarrow v^\beta \qquad T_3 \leftarrow A h^{\alpha+\beta} \ .$$

She also computes two helper values $\delta_1 \leftarrow x\alpha$ and $\delta_2 \leftarrow x\beta$.

Alice and Bob then undertake a proof of knowledge of values $(\alpha, \beta, x, \delta_1, \delta_2)$ satisfying the following five relations:

$$u^\alpha = T_1 \qquad v^\beta = T_2$$
$$e(T_3, g_2)^x \cdot e(h, w)^{-\alpha-\beta} \cdot e(h, g_2)^{-\delta_1-\delta_2} = e(g_1, g_2)/e(T_3, w)$$
$$T_1^x u^{-\delta_1} = 1 \qquad T_2^x v^{-\delta_2} = 1 \ .$$

This proof proceeds as follows. Alice picks blinding values r_α, r_β, r_x, r_{δ_1}, and r_{δ_2} at random from $\mathbb{Z}_p$. She computes five values based on all these:

$$R_1 \leftarrow u^{r_\alpha} \qquad R_2 \leftarrow v^{r_\beta}$$
$$R_3 \leftarrow e(T_3, g_2)^{r_x} \cdot e(h, w)^{-r_\alpha-r_\beta} \cdot e(h, g_2)^{-r_{\delta_1}-r_{\delta_2}}$$
$$R_4 \leftarrow T_1^{r_x} \cdot u^{-r_{\delta_1}} \qquad R_5 \leftarrow T_2^{r_x} \cdot v^{-r_{\delta_2}} \ .$$

She then sends $(T_1, T_2, T_3, R_1, R_2, R_3, R_4, R_5)$ to the verifier. Bob, the verifier, sends a challenge value c chosen uniformly at random from $\mathbb{Z}_p$. Alice computes and sends back $s_\alpha = r_\alpha + c\alpha$, $s_\beta = r_\beta + c\beta$, $s_x = r_x + cx$, $s_{\delta_1} = r_{\delta_1} + c\delta_1$, and $s_{\delta_2} = r_{\delta_2} + c\delta_2$. Finally, Bob verifies the following five equations:

$$u^{s_\alpha} \stackrel{?}{=} T_1^c \cdot R_1 \tag{1}$$
$$v^{s_\beta} \stackrel{?}{=} T_2^c \cdot R_2 \tag{2}$$
$$e(T_3, g_2)^{s_x} \cdot e(h, w)^{-s_\alpha-s_\beta} \cdot e(h, g_2)^{-s_{\delta_1}-s_{\delta_2}} \stackrel{?}{=} \left(e(g_1, g_2)/e(T_3, w)\right)^c \cdot R_3 \tag{3}$$
$$T_1^{s_x} u^{-s_{\delta_1}} \stackrel{?}{=} R_4 \tag{4}$$
$$T_2^{s_x} v^{-s_{\delta_2}} \stackrel{?}{=} R_5 \ . \tag{5}$$

Bob accepts if all five hold.

Theorem 1. *Protocol 1 is an honest-verifier zero-knowledge proof of knowledge of an SDH pair under the Decision Linear assumption.*

The proof of the theorem follows from the following lemmas that show that the protocol is (1) complete (the verifier always accepts an interaction with an honest prover), (2) zero-knowledge (can be simulated), and (3) a proof of knowledge (has an extractor).

Lemma 1. *Protocol 1 is complete.*

Proof. If Alice is an honest prover in possession of an SDH pair (A, x) she follows the computations specified for her in the protocol. In this case,

$$u^{s_\alpha} = u^{r_\alpha + c\alpha} = (u^\alpha)^c \cdot u^{r_\alpha} = T_1^c \cdot R_1 \ ,$$

so (1) holds. For analogous reasons (2) holds. Further,

$$T_1^{s_x} u^{-s_{\delta_1}} = (u^\alpha)^{r_x + cx} u^{-r_{\delta_1} - cx\alpha} = (u^\alpha)^{r_x} u^{-r_{\delta_1}} = T_1^{r_x} \cdot R_4 \ ,$$

so (4) holds. For analogous reasons (5) holds. Finally,

$$
\begin{aligned}
&e(T_3, g_2)^{s_x} \cdot e(h, w)^{-s_\alpha - s_\beta} \cdot e(h, g_2)^{-s_{\delta_1} - s_{\delta_2}} \\
&= e(T_3, g_2)^{r_x + cx} \cdot e(h, w)^{-r_\alpha - r_\beta - c\alpha - c\beta} \cdot e(h, g_2)^{-r_{\delta_1} - r_{\delta_2} - cx\alpha - cx\beta} \\
&= e(T_3, g_2^x)^c \cdot e(h^{-\alpha-\beta}, wg_2^x)^c \cdot \left(e(T_3, g_2)^{r_x} \cdot e(h, w)^{-r_\alpha - r_\beta} \cdot e(h, g_2)^{-r_{\delta_1} - r_{\delta_2}} \right) \\
&= e(T_3 h^{-\alpha-\beta}, wg_2^x)^c \cdot e(T_3, w)^{-c} \cdot (R_3) \\
&= \left(e(A, wg_2^x)/e(T_3, w) \right)^c \cdot R_3 = \left(e(g_1, g_2)/e(T_3, w) \right)^c \cdot R_3 \ .
\end{aligned}
$$

so (3) holds. □

Lemma 2. *Transcripts of Protocol 1 can be simulated, under the Decision Linear assumption.*

Proof. We describe a simulator that outputs transcripts of Protocol 1.

Pick $A \xleftarrow{R} \mathbb{G}_1$, and $\alpha, \beta \xleftarrow{R} \mathbb{Z}_p$. Set $T_1 \leftarrow u^\alpha$, $T_2 \leftarrow v^\beta$, and $T_3 \leftarrow Ah^{\alpha+\beta}$. Assuming the Decision Linear assumption holds on $\mathbb{G}_1$, the tuples (T_1, T_2, T_3) generated by the simulator are drawn from a distribution that is indistinguishable from the distribution output by any particular prover.

The remainder of this simulator does not assume knowledge of A, x, α, or β, so it can also be used when T_1, T_2, and T_3 are pre-specified. When the pre-specified (T_1, T_2, T_3) are a random Linear encryption of some A, the remainder of the transcript is simulated perfectly.

Now choose a challenge $c \xleftarrow{R} \mathbb{Z}_p$. Select $s_\alpha \xleftarrow{R} \mathbb{Z}_p$, and set $R_1 \leftarrow T_1^c/u^{s_\alpha}$. Then (1) is satisfied. With α and c fixed, a choice for either of r_α or s_α determines the other, and a uniform random choice of one gives a uniform random choice of the other. Therefore s_α and R_1 are distributed as in a real transcript. Choose s_β and R_2 analogously.

Select $s_x, s_{\delta_1}, s_{\delta_2} \xleftarrow{R} \mathbb{Z}_p$ and set $R_4 \leftarrow T_1^{s_x} u^{s_{\delta_1}}$ and $R_5 \leftarrow T_2^{s_x} v^{s_{\delta_2}}$. Again, all the computed values are distributed as in a real transcript. Finally set

$$R_3 \leftarrow e(T_3, g_2)^{s_x} \cdot e(h, w)^{-s_\alpha - s_\beta} \cdot e(h, g_2)^{-s_{\delta_1} - s_{\delta_2}} \cdot \left(e(T_3, w)/e(g_1, g_2) \right)^c \ .$$

This R_3 satisfies (3), and it, too, is properly distributed.

The transcript output is $(T_1, T_2, T_3, R_1, R_2, R_3, R_4, R_5, c, s_\alpha, s_\beta, s_x, s_{\delta_1}, s_{\delta_2})$. As argued above, this transcript is distributed identically to transcripts of Protocol 1, assuming the Decision Linear assumption holds. □

Lemma 3. *There exists an extractor for Protocol 1.*

Proof. Suppose that an extractor can rewind a prover in the protocol above to the point just before the prover is given a challenge c. At the first step of the protocol, the prover sends T_1, T_2, T_3 and R_1, R_2, R_3, R_4, R_5. Then, to challenge value c, the prover responds with s_α, s_β, s_x, s_{δ_1}, and s_{δ_2}. To challenge value $c' \neq c$, the prover responds with s'_α, s'_β, s'_x, s'_{δ_1}, and s'_{δ_2}. If the prover is convincing, all five verification equations (1–5) hold for each set of values.

For brevity, let $\Delta c = c - c'$, $\Delta s_\alpha = s_\alpha - s'_\alpha$, and similarly for Δs_β, Δs_x, Δs_{δ_1}, and Δs_{δ_2}.

Now consider (1) above. Dividing the two instances of this equation, we obtain $u^{\Delta s_\alpha} = T_1^{\Delta c}$. The exponents are in a group of known prime order, so we can take roots; let $\tilde{\alpha} = \Delta s_\alpha / \Delta c$. Then $u^{\tilde{\alpha}} = T_1$. Similarly, from (2), we obtain $\tilde{\beta} = \Delta s_\beta / \Delta c$ such that $v^{\tilde{\beta}} = T_2$.

Consider (4) above. Dividing the two instances gives $T_1^{\Delta s_x} = u^{\Delta s_{\delta_1}}$. Substituting $T_1 = u^{\tilde{\alpha}}$ gives $u^{\tilde{\alpha} \Delta s_x} = u^{\Delta s_{\delta_1}}$, or $\Delta s_{\delta_1} = \tilde{\alpha} \Delta s_x$. Similarly, from (5) we deduce that $\Delta s_{\delta_2} = \tilde{\beta} \Delta s_x$.

Finally, dividing the two instances of (3), we obtain

$$\left(e(g_1, g_2)/e(T_3, w)\right)^{\Delta c} = e(T_3, g_2)^{\Delta s_x} \cdot e(h, w)^{-\Delta s_\alpha - \Delta s_\beta} \cdot e(h, g_2)^{-\Delta s_{\delta_1} - \Delta s_{\delta_2}}$$
$$= e(T_3, g_2)^{\Delta s_x} \cdot e(h, w)^{-\Delta s_\alpha - \Delta s_\beta} \cdot e(h, g_2)^{-\tilde{\alpha} \Delta s_x - \tilde{\beta} \Delta s_x} \ .$$

Taking Δc-th roots, and letting $\tilde{x} = \Delta s_x / \Delta c$, we obtain

$$e(g_1, g_2)/e(T_3, w) = e(T_3, g_2)^{\tilde{x}} \cdot e(h, w)^{-\tilde{\alpha} - \tilde{\beta}} \cdot e(h, g_2)^{-\tilde{x}(\tilde{\alpha} + \tilde{\beta})} \ .$$

This can be rearranged as

$$e(g_1, g_2) = e(T_3 h^{-\tilde{\alpha} - \tilde{\beta}}, w g_2^{\tilde{x}}) \ ,$$

or, letting $\tilde{A} = T_3 h^{-\tilde{\alpha} - \tilde{\beta}}$,

$$e(\tilde{A}, w g_2^{\tilde{x}}) = e(g_1, g_2) \ .$$

Thus the extractor obtains an SDH tuple $(\tilde{A}, \tilde{x})$. Moreover, the $\tilde{A}$ in this SDH tuple is, perforce, the same as that in the Linear encryption (T_1, T_2, T_3). □

5   SDH Signatures of Knowledge

Armed with Theorem 1, we obtain from Protocol 1 a signature scheme secure in the random oracle model by applying the Fiat-Shamir heuristic [16]. Signatures obtained from a proof of knowledge via the Fiat-Shamir heuristic are often called signatures of knowledge. We use a variant of the Fiat-Shamir heuristic, used also by Ateniese et al. [2], where the challenge c rather than the values $R_1, \ldots, R_5$ is transmitted in the signature; the output of the random oracle acts as a checksum for those values not transmitted.

The signature scheme is defined as follows. The public key contains a hash function (viewed as a random oracle) $H : \{0,1\}^* \to \mathbb{Z}_p$, groups G_1 and G_2 with respective generators g_1 and g_2 as in Sect. 2, the random generators u, v, and h of G_1, and $w = g_2^\gamma \in G_2$, where γ is chosen at random in $\mathbb{Z}_p^*$. The private key is an SDH pair (A, x), i.e., a pair such that $A^{x+\gamma} = g_1$. Any such pair is a valid private key.

The signer signs a message $M \in \{0,1\}^*$ using the private key (A, x) as follows. She first undertakes the computation specified in the first round of Protocol 1 to obtain $T_1, T_2, T_3, R_1, R_2, R_3, R_4, R_5$. She obtains the challenge c by giving M and her first-round values to the random oracle:

$$c \leftarrow H(M, T_1, T_2, T_3, R_1, R_2, R_3, R_4, R_5) \in \mathbb{Z}_p \ . \tag{6}$$

She then undertakes the computation specified in the third round of the protocol using the challenge value c to obtain $s_\alpha, s_\beta, s_x, s_{\delta_1}, s_{\delta_2}$. Finally, she outputs the signature σ, computed as

$$\sigma \leftarrow (T_1, T_2, T_3, c, s_\alpha, s_\beta, s_x, s_{\delta_1}, s_{\delta_2}) \ . \tag{7}$$

The verifier uses equations (1–5) to re-derive R_1, R_2, R_3, R_4, and R_5:

$$\tilde{R}_1 \leftarrow u^{s_\alpha}/T_1^c \qquad \tilde{R}_2 \leftarrow v^{s_\beta}/T_2^c \qquad \tilde{R}_4 \leftarrow T_1^{s_x}/u^{s_{\delta_1}} \qquad \tilde{R}_5 \leftarrow T_2^{s_x}/v^{s_{\delta_2}}$$

$$\tilde{R}_3 \leftarrow e(T_3, g_2)^{s_x} \cdot e(h, w)^{-s_\alpha - s_\beta} \cdot e(h, g_2)^{-s_{\delta_1} - s_{\delta_2}} \cdot \bigl(e(T_3, w)/e(g_1, g_2)\bigr)^c \ .$$

He then checks that these, along with the other first-round messages included in σ, give the challenge c, i.e., that

$$c \overset{?}{=} H(M, T_1, T_2, T_3, \tilde{R}_1, \tilde{R}_2, \tilde{R}_3, \tilde{R}_4, \tilde{R}_5) \ . \tag{8}$$

He accepts if this check succeeds.

The Fiat-Shamir heuristic shows that this signature scheme is secure against existential forgery in the random oracle model [1]. Note that a signature comprises three elements of G_1 and six of $\mathbb{Z}_p$.

6   Short Group Signatures from SDH

The signature scheme presented in Sect. 5 is, in fact, also a group signature scheme. In describing the scheme, we follow the definitions given by Bellare et al. [6].

Consider bilinear groups G_1 and G_2 with respective generators g_1 and g_2, as in Sect. 2. Suppose further that the SDH assumption holds on (G_1, G_2), and the Linear assumption holds on G_1. The scheme employs a hash function $H : \{0,1\}^* \to \mathbb{Z}_p$, treated as a random oracle in the proof of security.

KeyGen(n). This randomized algorithm takes as input a parameter n, the number of members of the group, and proceeds as follows. Select $h \overset{\text{R}}{\leftarrow} G_1 \setminus \{1_{G_1}\}$ and $\xi_1, \xi_2 \overset{\text{R}}{\leftarrow} \mathbb{Z}_p^*$, and set $u, v \in G_1$ such that $u^{\xi_1} = v^{\xi_2} = h$. Select $\gamma \overset{\text{R}}{\leftarrow} \mathbb{Z}_p^*$, and set $w = g_2^\gamma$. Using γ, generate for each user i, $1 \le i \le n$, an SDH tuple (A_i, x_i): select $x_i \overset{\text{R}}{\leftarrow} \mathbb{Z}_p^*$, and set $A_i \leftarrow g_1^{1/(\gamma + x_i)}$. The group public key is $gpk = (g_1, g_2, h, u, v, w)$. The private key of the group manager (the party able to trace signatures) is $gmsk = (\xi_1, \xi_2)$. Each user's private key is her tuple $\boldsymbol{gsk}[i] = (A_i, x_i)$. No party is allowed to possess γ; it is only known to the private-key issuer.

Sign(*gpk*, **gsk**[*i*], *M*). Given a group public key *gpk* = (g_1, g_2, h, u, v, w), a user's key **gsk**[*i*] = (A_i, x_i), and a message $M \in \{0,1\}^*$, compute and output a signature of knowledge $\sigma = (T_1, T_2, T_3, c, s_\alpha, s_\beta, s_x, s_{\delta_1}, s_{\delta_2})$ as in the scheme of Sect. 5 (Equation (7)).

Verify(*gpk*, *M*, *σ*). Given a group public key *gpk* = (g_1, g_2, h, u, v, w), a message M, and a group signature σ, verify that σ is a valid signature of knowledge in the scheme of Sect. 5 (Equation (8)).

Open(*gpk*, *gmsk*, *M*, *σ*). This algorithm is used for tracing a signature to a signer. It takes as input a group public key *gpk* = (g_1, g_2, h, u, v, w) and the corresponding group manager's private key *gmsk* = (ξ_1, ξ_2), together with a message M and a signature $\sigma = (T_1, T_2, T_3, c, s_\alpha, s_\beta, s_x, s_{\delta_1}, s_{\delta_2})$ to trace, and proceeds as follows. First, verify that σ is a valid signature on M. Second, consider the first three elements (T_1, T_2, T_3) as a Linear encryption, and recover the user's A as $A \leftarrow T_3/(T_1^{\xi_1} \cdot T_2^{\xi_2})$, following the decryption algorithm given at the end of Sect. 3.2. If the group manager is given the elements $\{A_i\}$ of the users' private keys, he can look up the user index corresponding to the identity A recovered from the signature.

Signature Length. A group signature in the system above comprises three elements of G_1 and six elements of $\mathbb{Z}_p$. Using any of the families of curves described in [9], one can take p to be a 170-bit prime and use a group G_1 where each element is 171 bits. Thus, the total group signature length is 1533 bits or 192 bytes. With these parameters, security is approximately the same as a standard 1024-bit RSA signature, which is 128 bytes.

Performance. The pairings $e(h, w)$, $e(h, g_2)$, and $e(g_1, g_2)$ can be precomputed and cached by both signers and verifiers. The signer can cache $e(A, g_2)$, and, when signing, compute $e(T_3, g_2)$ without evaluating a pairing. Accordingly, creating a group signature requires eight exponentiations (or multi-exponentiations) and no pairing computations. The verifier can derive $\tilde{R}_3$ efficiently by collapsing the $e(T_3, g_2)^{s_x}$ and $e(T_3, w)^c$ pairings into a single $e(T_3, w^c g_2^{s_x})$ term. Thus verifying a group signature requires six multi-exponentiations and one pairing computation. With parameters selected as above, the exponents are in every case 170-bit numbers. For the signer, all bases for exponentiation are fixed, which allows further speedup by precomputation.

6.1   Group Signature Security

We now turn to proving security of the system. Bellare et al. [6] give three properties that a group signature scheme must satisfy:

- correctness, which ensures that honestly-generated signatures verify and trace correctly;
- full-anonymity, which ensures that signatures do not reveal their signer's identity; and
- full-traceability, which ensures that all signatures, even those created by the collusion of multiple users and the group manager, trace to a member of the forging coalition.

For the details of the definitions, see Bellare et al. [6]. We prove the security of our scheme using a variation of these properties. In our proofs, we relax the full-anonymity requirement. As presented [6, Sect. 2], the full-anonymity experiment allows the adversary to query the opening (tracing) oracle before and after receiving the challenge σ. In this respect, the experiment mirrors the indistinguishability experiment against an adaptive CCA2 adversary. We therefore rename this experiment CCA2-full-anonymity. We define a corresponding experiment, CPA-full-anonymity, in which the adversary cannot query the opening oracle. We prove privacy in this slightly weaker model.

Access to the tracing functionality will likely be carefully controlled when group signatures are deployed, so CPA-full-anonymity is a reasonable model to consider. In any case, anonymity and unlinkability, the two traditional group signature security requirements implied by full anonymity [6, Sect. 3], also follow from CPA-full-anonymity. Thus a fully-traceable and CPA-fully-anonymous group signature scheme is still secure in the traditional sense.

In the statements of the theorem, we use big-O notation to elide the specifics of additive terms in time bounds, noting that, for given groups G_1 and G_2, operations such as sampling, exponentiation, and bilinear map evaluation are all constant-time.

Theorem 2. *The SDH group signature scheme is correct.*

Proof. For any group public key $gpk = (g_1, g_2, h, u, v, w)$, and for any user with key $\mathbf{gsk}[i] = (A_i, x_i)$, the key generation algorithm guarantees that $A_i^{\gamma + x_i} = g_1$, so (A_i, x_i) is an SDH tuple for $w = g_2^\gamma$. A correct group signature σ is a proof of knowledge, which is itself a transcript of the SDH protocol given in Sect. 4. Verifying the signature entails verifying that the transcript is correct; thus Lemma 1 shows that σ will always be accepted by the verifier.

Moreover, an honest signer outputs, as the first three components of any signature σ, values $(T_1, T_2, T_3) = (u^\alpha, v^\beta, A_i \cdot h^{\alpha + \beta})$ for some $\alpha, \beta \in \mathbb{Z}_p$. These values form a Linear encryption of A_i under public key (u, v, h), which the group manager, possessing the corresponding private key (ξ_1, ξ_2), can always recover. Therefore any valid signature will always be opened correctly.   □

Theorem 3. *If Linear encryption is (t', ϵ')-semantically secure on G_1 then the SDH group signature scheme is (t, q_H, ϵ)-CPA-fully-anonymous, where $\epsilon = \epsilon'$ and $t = t' - q_H O(1)$. Here q_H is the number of hash function queries made by the adversary and n is the number of members of the group.*

Proof. Suppose $\mathcal{A}$ is an algorithm that (t, q_H, ϵ)-breaks the anonymity of the group signature scheme. We show how to construct a $t + q_H O(1)$-time algorithm $\mathcal{B}$ that breaks the semantic security of Linear encryption (Sect. 3.2) with advantage at least ϵ.

Algorithm $\mathcal{B}$ is given a Linear encryption public key (u, v, h). It generates the remaining components of the group signature public key by following the group signature's key generation algorithm. It then provides to $\mathcal{A}$ the group public key (g_1, g_2, h, u, v, w), and the users' private keys (A_i, x_i).

At any time, $\mathcal{A}$ can query the random oracle H. Algorithm $\mathcal{B}$ responds with elements selected uniformly at random from $\mathbb{Z}_p$, making sure to respond identically to repeated queries.

Algorithm $\mathcal{A}$ requests its full-anonymity challenge by providing two indices, i_0 and i_1, and a message M. Algorithm $\mathcal{B}$, in turn, requests its indistinguishability challenge by providing the two user private keys A_{i_0} and A_{i_1} as the messages whose Linear encryption it must distinguish. It is given a Linear encryption (T_1, T_2, T_3) of A_{i_b}, where bit b is chosen by the Linear encryption challenger.

Algorithm $\mathcal{B}$ generates from this Linear encryption a protocol transcript $(T_1, T_2, T_3, R_1, R_2, R_3, R_4, R_5, c, s_\alpha, s_\beta, s_x, s_{\delta_1}, s_{\delta_2})$ by means of the simulator of Lemma 2. This simulator can generate a trace given (T_1, T_2, T_3), even though $\mathcal{B}$ does not know α, β, or x. Since (T_1, T_2, T_3) is a random Linear encryption of A_{i_b}, the remainder of the transcript is distributed exactly as in a real protocol with a prover whose secret A is A_{i_b}.

Algorithm $\mathcal{B}$ then patches H at $(M, T_1, T_2, T_3, R_1, R_2, R_3, R_4, R_5)$ to equal c. It encounters a collision only with negligible probability. In case of a collision, $\mathcal{B}$ declares failure and exits. Otherwise, it returns the valid group signature $\sigma \leftarrow (T_1, T_2, T_3, c, s_\alpha, s_\beta, s_x, s_{\delta_1}, s_{\delta_2})$ to $\mathcal{A}$.

Finally, $\mathcal{A}$ outputs a bit b'. Algorithm $\mathcal{B}$ returns b' as the answer to its own challenge. Since the encryption of A_{i_b} is turned by $\mathcal{B}$ into a group signature by user i_b, $\mathcal{B}$ answers its challenge correctly whenever $\mathcal{A}$ does.

The keys given to $\mathcal{A}$, and the answers to $\mathcal{A}$'s queries, are all valid and properly distributed. Therefore $\mathcal{A}$ succeeds in breaking the anonymity of the group signature σ with advantage ϵ, and $\mathcal{B}$ succeeds in distinguishing the Linear encryption (T_1, T_2, T_3) with the same advantage.

Algorithm $\mathcal{B}$'s running time exceeds $\mathcal{A}$'s by the amount it takes to answer $\mathcal{A}$'s queries. Each hash query can be answered in constant time, and there are at most q_H of them. Algorithm $\mathcal{B}$ can also create the challenge group signature σ in constant time. If $\mathcal{A}$ runs in time t, $\mathcal{B}$ runs in time $t + q_H O(1)$.   □

The following theorem proves full traceability of our system. The proof is based on the forking lemma [24] and is given in the full version of the paper.

Theorem 4. *If SDH is (q, t', ϵ')-hard on (G_1, G_2), then the SDH group signature scheme is $(t, q_H, q_S, n, \epsilon)$-fully-traceable, where $n = q - 1$, $\epsilon = 4n\sqrt{2\epsilon' q_H} + n/p$, and $t = \Theta(1) \cdot t'$. Here q_H is the number of hash function queries made by the adversary, q_S is the number of signing queries made by the adversary, and n is the number of members of the group.*

7   Revocation

We now discuss how to revoke users in the SDH group signature scheme of Sect. 6. A number of revocation mechanisms for group signatures have been proposed [4, 12]. All these mechanisms can be applied to our system. Here we describe a revocation mechanism along the lines of [12].

Recall that the group's public key in our system is (g_1, g_2, h, u, v, w) where $w = g_2^\gamma \in G_2$ for random $\gamma \in \mathbb{Z}_p^*$ and random $h, u, v \in G_1$. User i's private key is a pair (A_i, x_i) where $A_i = g_1^{1/(\gamma + x_i)} \in G_1$.

Now, suppose we wish to revoke users $1, \ldots, r$ without affecting the signing capability of other users. To do so, the Revocation Authority (RA) publishes a Revocation List (RL) containing the private keys of all revoked users. More precisely, $\mathrm{RL} = \{(A_1^*, x_1), \ldots, (A_r^*, x_r)\}$, where $A_i^* = g_2^{1/(\gamma + x_i)} \in G_2$. Note that $A_i = \psi(A_i^*)$. Here the SDH secret γ is needed to compute the A_i^*'s. In the case where G_1 equals G_2 then $A_i = A_i^*$ and consequently the Revocation List can be derived directly from the private keys of revoked users without having to use γ.

The list RL is given to all signers and verifiers in the system. It is used to update the group public key used to verify signatures. Let $y = \prod_{i=1}^{r}(\gamma + x_i) \in \mathbb{Z}_p^*$. The new public key is $(\bar{g}_1, \bar{g}_2, h, u, v, \bar{w})$ where $\bar{g}_1 = g_1^{1/y}$, $\bar{g}_2 = g_2^{1/y}$, and $\bar{w} = (\bar{g}_2)^\gamma$. We show that, given RL, anyone can compute this new public key, and any unrevoked user can update her private key locally so that it is well formed with respect to this new public key. Revoked users are unable to do so.

We show how to revoke one private key at a time. By repeating the process r times (as the revocation list grows over time) we can revoke all private keys on the Revocation List. We first show how given the public key (g_1, g_2, h, u, v, w) and one revoked private key $(A_1^*, x_1) \in \mathrm{RL}$ anyone can construct the new public key $(\hat{g}_1, \hat{g}_2, h, u, v, \hat{w})$ where $\hat{g}_1 = g_1^{1/(\gamma + x_1)}$, $\hat{g}_2 = g_2^{1/(\gamma + x_1)}$, and $\hat{w} = (\hat{g}_2)^\gamma$. This new public key is constructed simply as:

$$\hat{g}_1 \leftarrow \psi(A_1^*) \qquad \hat{g}_2 \leftarrow A_1^* \qquad \text{and} \qquad \hat{w} \leftarrow g_2 \cdot (A_1^*)^{-x_1} \; ;$$

then $\hat{g}_1 = \psi(A_1)^* = g_1^{1/(\gamma + x_1)}$ and $\hat{w} = g_2 \cdot (A_1^*)^{-x_1} = g_2^{1 - \frac{x_1}{\gamma + x_1}} = (A_1^*)^\gamma = (\hat{g}_2)^\gamma$, as required.

Next, we show how unrevoked users update their own private keys. Consider an unrevoked user whose private key is (A, x). Given a revoked private key, (A_1^*, x_1) the user computes $\hat{A} \leftarrow \psi(A_1^*)^{1/(x - x_1)}/A^{1/(x - x_1)}$ and sets his new private key to be $(\hat{A}, x)$. Then, indeed,

$$(\hat{A})^{\gamma + x} = \psi(A_1^*)^{\frac{\gamma + x}{x - x_1}}/A^{\frac{\gamma + x}{x - x_1}} = \psi(A_1^*)^{\frac{(\gamma + x_1) + (x - x_1)}{x - x_1}}/g_1^{\frac{1}{x - x_1}} = \psi(A_1^*) = \hat{g}_1 \; ,$$

as required. Hence, $(\hat{A}, x)$ is a valid private key with respect to $(\hat{g}_1, \hat{g}_2, h, u, v, \hat{w})$.

By repeating this process r times (once for each revoked key in RL) anyone can compute the updated public key $(\bar{g}_1, \bar{g}_2, h, u, v, \bar{w})$ defined above. Similarly, an unrevoked user with private key (A, x) can compute his updated private key $(\bar{A}, x)$ where $\bar{A} = (\bar{g}_1)^{1/(\gamma + x)}$. We note that it is possible to process the entire RL at once (as opposed to one element at a time) and compute $(\bar{g}_1, \bar{g}_2, h, u, v, \bar{w})$ directly; however this is less efficient when keys are added to RL incrementally.

A revoked user cannot construct a private key for the new public key $(\bar{g}_1, \bar{g}_2, h, u, v, \bar{w})$. In fact, the proof of Theorem 4 shows that, if a revoked user can generate signatures for the new public key $(\bar{g}_1, \bar{g}_2, h, u, v, \bar{w})$, then that user can be used to break the SDH assumption. Very briefly, the reason is that given an SDH challenge one can easily generate a public key tuple $(\bar{g}_1, \bar{g}_2, h, u, v, \bar{w})$ along with the private key for a revoked user $(g_1^{1/(x + \gamma)}, x)$. Then an algorithm that can forge signatures given these two tuples can be used to solve the SDH challenge.

Brickell [11] proposes an alternate mechanism where revocation messages are only sent to signature verifiers, so that there is no need for unrevoked signers to update their keys. Similar mechanisms were also considered by Ateniese et al. [4] and Kiayias et al. [19]. We refer to this as Verifier-Local Revocation (VLR) group signatures. Boneh and Shacham [10] show how to modify our group signature scheme to support this VLR revocation mechanism.

8 Exculpability

In Bellare et al. [6], exculpability (introduced by Ateniese and Tsudik [3]) is informally defined as follows: No member of the group and not even the group manager – the entity that is given the tracing key – can produce signatures on behalf of other users. Thus, no user can be framed for producing a signature he did not produce. They argue that a group signature secure in the sense of full-traceability also has the exculpability property. Thus, in the terminology of Bellare et al. [6], our group signature has the exculpability property.

A stronger notion of exculpability is considered in Ateniese et al. [2], where one requires that even the entity that *issues* user keys cannot forge signatures on behalf of users. Formalizations of strong exculpability have recently been proposed by Kiayias and Yung [20] and by Bellare, Shi, and Zhang [7].

To achieve this stronger property the system of Ateniese et al. [2] uses a protocol (called JOIN) to issue a key to a new user. At the end of the protocol, the key issuer does not know the full private key given to the user and therefore cannot forge signatures under the user's key.

Our group signature scheme can be extended to provide strong exculpability using a similar mechanism. Instead of simply giving user i the private key $(g_1^{1/(\gamma+x_i)}, x_i)$, the user and key issuer engage in a JOIN protocol where at the end of the protocol user i has a triple (A_i, x_i, y_i) such that $A_i^{\gamma+x_i} h_1^{y_i} = g_1$ for some public parameter h_1. The value y_i is chosen by the user and is kept secret from the key issuer. The ZKPK of Sect. 4 can be modified to prove knowledge of such a triple. The resulting system is a short group signature with strong exculpability.

9 Conclusions

We presented a group signature scheme based on the Strong Diffie-Hellman (SDH) and Linear assumptions. The signature makes use of a bilinear map $e : G_1 \times G_2 \to G_T$. When any of the curves described in [9] are used, the group G_1 has a short representation and consequently we get a group signature whose length is under 200 bytes – less than twice the length of an ordinary RSA signature (128 bytes) with comparable security. Signature generation requires no pairing computations, and verification requires a single pairing; both also require a few exponentiations with short exponents.

Acknowledgments

The authors thank the anonymous referees for their valuable feedback.

References

1. M. Abdalla, J. An, M. Bellare, and C. Namprempre. From identification to signatures via the Fiat-Shamir transform: Minimizing assumptions for security and forward-security. In L. Knudsen, editor, *Proceedings of Eurocrypt 2002*, volume 2332 of *LNCS*, pages 418–33. Springer-Verlag, May 2002.
2. G. Ateniese, J. Camenisch, M. Joye, and G. Tsudik. A practical and provably secure coalition-resistant group signature scheme. In M. Bellare, editor, *Proceedings of Crypto 2000*, volume 1880 of *LNCS*, pages 255–70. Springer-Verlag, Aug. 2000.
3. G. Ateniese and G. Tsudik. Some open issues and directions in group signatures. In *Proceedings of Financial Cryptography 1999*, volume 1648, pages 196–211. Springer-Verlag, Feb. 1999.
4. G. Ateniese, G. Tsudik, and D. Song. Quasi-efficient revocation of group signatures. In M. Blaze, editor, *Proceedings of Financial Cryptography 2002*, Mar. 2002.
5. N. Baric and B. Pfitzman. Collision-free accumulators and fail-stop signature schemes without trees. In *Proceedings of Eurocrypt 1997*, pages 480–494. Springer-Verlag, May 1997.
6. M. Bellare, D. Micciancio, and B. Warinschi. Foundations of group signatures: Formal definitions, simplified requirements, and a construction based on general assumptions. In E. Biham, editor, *Proceedings of Eurocrypt 2003*, volume 2656 of *LNCS*, pages 614–29. Springer-Verlag, May 2003.
7. M. Bellare, H. Shi, and C. Zhang. Foundations of group signatures: The case of dynamic groups. Cryptology ePrint Archive, Report 2004/077, 2004. http://eprint.iacr.org/.
8. D. Boneh and X. Boyen. Short signatures without random oracles. In C. Cachin and J. Camenisch, editors, *Proceedings of Eurocrypt 2004*, LNCS, pages 56–73. Springer-Verlag, May 2004.
9. D. Boneh, B. Lynn, and H. Shacham. Short signatures from the Weil pairing. In *Proceedings of Asiacrypt 2001*, volume 2248 of *LNCS*, pages 514–32. Springer-Verlag, Dec. 2001. Full paper: http://crypto.stanford.edu/~dabo/pubs.html.
10. D. Boneh and H. Shacham. Group signatures with verifier-local revocation, 2004. Manuscript.
11. E. Brickell. An efficient protocol for anonymously providing assurance of the container of a private key, Apr. 2003. Submitted to the Trusted Computing Group.
12. J. Camenisch and A. Lysyanskaya. Dynamic accumulators and application to efficient revocation of anonymous credentials. In M. Yung, editor, *Proceedings of Crypto 2002*, volume 2442 of *LNCS*, pages 61–76. Springer-Verlag, Aug. 2002.
13. J. Camenisch and A. Lysyanskaya. Signature schemes and anonymous credentials from bilinear maps. In M. Franklin, editor, *Proceedings of Crypto 2004*, LNCS. Springer-Verlag, Aug. 2004.
14. D. Chaum and E. van Heyst. Group signatures. In D. W. Davies, editor, *Proceedings of Eurocrypt 1991*, volume 547 of *LNCS*, pages 257–65. Springer-Verlag, 1991.
15. X. Ding, G. Tsudik, and S. Xu. Leak-free group signatures with immediate revocation. In T. Lai and K. Okada, editors, *Proceedings of ICDCS 2004*, Mar. 2004.
16. A. Fiat and A. Shamir. How to prove yourself: Practical solutions to identification and signature problems. In A. M. Odlyzko, editor, *Proceedings of Crypto 1986*, volume 263 of *LNCS*, pages 186–194. Springer-Verlag, Aug. 1986.
17. T. Garfinkel, B. Pfaff, J. Chow, M. Rosenblum, and D. Boneh. Terra: A virtual machine-based platform for trusted computing. In *Proceedings of SOSP 2003*, pages 193–206, Oct. 2003.

18. IEEE P1556 Working Group, VSC Project. Dedicated short range communications (DSRC), 2003.
19. A. Kiayias, Y. Tsiounis, and M. Yung. Traceable signatures. In C. Cachin and J. Camenisch, editors, *Proceedings of Eurocrypt 2004*, volume 3027 of *LNCS*, pages 571–89. Springer-Verlag, May 2004.
20. A. Kiayias and M. Yung. Group signatures: Efficient constructions and anonymity from trapdoor-holders. Cryptology ePrint Archive, Report 2004/076, 2004. http://eprint.iacr.org/.
21. A. Lysyanskaya, R. Rivest, A. Sahai, and S. Wolf. Pseudonym systems. In H. Heys and C. Adams, editors, *Proceedings of SAC 1999*, volume 1758 of *LNCS*, pages 184–99. Springer-Verlag, Aug. 1999.
22. S. Mitsunari, R. Sakai, and M. Kasahara. A new traitor tracing. *IEICE Trans. Fundamentals*, E85-A(2):481–4, Feb. 2002.
23. A. Miyaji, M. Nakabayashi, and S. Takano. New explicit conditions of elliptic curve traces for FR-reduction. *IEICE Trans. Fundamentals*, E84-A(5):1234–43, May 2001.
24. D. Pointcheval and J. Stern. Security arguments for digital signatures and blind signatures. *J. Cryptology*, 13(3):361–96, 2000.
25. K. Rubin and A. Silverberg. Supersingular Abelian varieties in cryptology. In M. Yung, editor, *Proceedings of Crypto 2002*, volume 2442 of *LNCS*, pages 336–53. Springer-Verlag, Aug. 2002.
26. C. Schnorr. Efficient signature generation by smart cards. *J. Cryptology*, 4(3):161–174, 1991.
27. V. Shoup. Lower bounds for discrete logarithms and related problems. In W. Fumy, editor, *Proceedings of Eurocrypt 1997*, volume 1233 of *LNCS*, pages 256–66. Springer-Verlag, May 1997.
28. Trusted Computing Group. Trusted Computing Platform Alliance (TCPA) Main Specification, 2003. Online: www.trustedcomputinggroup.org.
29. G. Tsudik and S. Xu. Accumulating composites and improved group signing. In C. S. Laih, editor, *Proceedings of Asiacrypt 2003*, volume 2894 of *LNCS*, pages 269–86. Springer-Verlag, Dec. 2003.

Signature Schemes and Anonymous Credentials from Bilinear Maps

Jan Camenisch[1] and Anna Lysyanskaya[2]

[1] IBM Research, Zurich Research Laboratory, CH–8803 Rüschlikon
jca@zurich.ibm.com
[2] Computer Science Department, Brown University, Providence, RI 02912, USA
anna@cs.brown.edu

Abstract. We propose a new and efficient signature scheme that is provably secure in the plain model. The security of our scheme is based on a discrete-logarithm-based assumption put forth by Lysyanskaya, Rivest, Sahai, and Wolf (LRSW) who also showed that it holds for generic groups and is independent of the decisional Diffie-Hellman assumption. We prove security of our scheme under the LRSW assumption for groups with bilinear maps. We then show how our scheme can be used to construct efficient anonymous credential systems as well as group signature and identity escrow schemes. To this end, we provide efficient protocols that allow one to prove in zero-knowledge the knowledge of a signature on a committed (or encrypted) message and to obtain a signature on a committed message.

1 Introduction

Digital signatures schemes, invented by Diffie and Hellman [20], and formalized by Goldwasser, Micali and Rivest [26], not only provide the electronic equivalent of signing a paper document with a pen but also are an important building block for many cryptographic protocols such as anonymous voting schemes, e-cash, and anonymous credential schemes, to name just a few.

Signature schemes exists if and only if one-way functions exist [32, 35]. However, the efficiency of these general constructions, and also the fact that these signature schemes require the signer's secret key to change between invocations of the signing algorithm, makes these solutions undesirable in practice.

Using an ideal random function (this is the so-called *random-oracle* model), several, much more efficient signature schemes were shown to be secure. Most notably, those are the RSA [34], the Fiat-Shamir [21], and the Schnorr [36] signature schemes. However, ideal random functions cannot be implemented in the plain model [13, 25], and therefore in the plain model, these signature schemes are not provably secure.

Over the years, many researchers have come up with signature schemes that are efficient and at the same time provably secure in the plain model. The most efficient ones provably secure in the standard model are based on the strong RSA assumption [23, 19, 22, 10]. However, no scheme based on an assumption related to the discrete logarithm assumption in the plain (as opposed to random-oracle) model comes close to the efficiency of these schemes.

M. Franklin (Ed.): CRYPTO 2004, LNCS 3152, pp. 56–72, 2004.

In this paper, we propose a new signature scheme that is based on an assumption introduced by Lysyanskaya, Rivest, Sahai, and Wolf [30] and uses bilinear maps. This assumption was shown to hold for generic groups [30], and be independent of the decisional Diffie-Hellman assumption. Our signature scheme's efficiency is comparable to the schemes mentioned above that are based on the Strong RSA assumption.

We further extend our basic signature scheme such that it can be used as a building block for cryptographic protocols. To this end, we provide protocols to prove knowledge of a signature on a committed message and to obtain a signature on a committed message. These protocols yield a group signature scheme [17] or an anonymous credential system [14] (cf. [10]). That is, we obtain the first efficient and secure credential system and group signature/identity escrow schemes [28] that are based solely on discrete-logarithm-related assumptions. We should mention that an anonymous credential system proposed by Verheul [38] is also only based on discrete logarithm related assumptions; however, the scheme is not proven secure. Also note that the recent scheme by Ateniese and de Medeiros [2] requires the strong RSA assumption although no party is required to know an RSA secret key during the operation of the system.

Note that not only are our group signature and anonymous credential schemes interesting because they are based on a different assumption, but also because they are much more efficient than any of the existing schemes. All prior schemes [1, 9, 10, 2] required proofs of knowledge of representations over groups modulo large moduli (for example, modulo an RSA modulus, whose recommended length is about 2K Bits).

Recently, independently from our work, Boneh and Boyen [4] put forth a signature scheme that is also provably secure under a discrete-logarithm-type assumption about groups with bilinear maps. In contrast to their work, our main goal is not just an efficient signature scheme, but a set of efficient protocols to prove knowledge of signatures and to issue signatures on committed (secret) messages. Our end goal is higher-level applications, i.e., group signature and anonymous credential schemes that can be constructed based solely on an assumption related to the discrete logarithm assumption.

In another recent independent work, Boneh, Boyen, and Shacham [5] construct a group signature scheme based on different discrete-logarithm-type assumptions about groups with bilinear pairings. Their scheme yields itself to the design of a signature scheme with efficient protocols as well. In §5 we describe their scheme and its connection to our work in more detail.

Outline of the Paper. In §2 we give our notation and some number-theoretic preliminaries, including bilinear maps and the LRSW assumption. In §3, we give our signature scheme and prove it secure. In §4 we show how our signature yields itself to the design of an anonymous credential system: we give protocols for obtaining a signature on a committed value, and for proving knowledge of a signature on a committed value. In the end of that section, we show how to realize a group signature scheme based on our new signature. Finally, in Section 5, we show that the scheme of Boneh, Boeyn and Shacham can be extended so that a signature scheme with efficient protocols, similar to the one we describe in Sections 3 and 4 can be obtained based on their assumptions as well.

2   Preliminaries

We use notation introduced by Micali [31] (also called the GMR notation), and also notation introduced by Camenisch and Stadler [12]. Here we review it briefly; the complete description can be found in the full version [CL04] of this paper.

If A is an algorithm, and b be a Boolean function, then by $(y \leftarrow A(x) : b(y))$, we denote the event that $b(y) = 1$ after y was generated by running A on input x. By $A^O(\cdot)$, we denote a Turing machine that makes queries to an oracle O. By $Q = Q(A^O(x)) \leftarrow A^O(x)$ we denote the contents of the query tape once A terminates, with oracle O and input x.

A function $\nu(k)$ is *negligible* if for every positive polynomial $p(\cdot)$ and for sufficiently large k, $\nu(k) < \frac{1}{p(k)}$.

Camenisch and Stadler[12] introduced notation for various proofs of knowledge of discrete logarithms and proofs of the validity of statements about discrete logarithms. For instance,

$$PK\{(\alpha, \beta, \gamma) : y = g^\alpha h^\beta \ \wedge \ \tilde{y} = \tilde{g}^\alpha \tilde{h}^\gamma \ \wedge \ (u \le \alpha \le v)\}$$

denotes a *"zero-knowledge Proof of Knowledge of integers α, β, and γ such that $y = g^\alpha h^\beta$ and $\tilde{y} = \tilde{g}^\alpha \tilde{h}^\gamma$ holds, where $u \le \alpha \le v$,"* where $y, g, h, \tilde{y}, \tilde{g}$, and $\tilde{h}$ are elements of some groups $G = \langle g \rangle = \langle h \rangle$ and $\tilde{G} = \langle \tilde{g} \rangle = \langle \tilde{h} \rangle$. The convention is that Greek letters denote quantities the knowledge of which is being proved, while all other parameters are known to the verifier. We will sometimes apply the Fiat-Shamir heuristic to turn such a proof into a signature on a message m, which we will denote as, e.g., $SPK\{(\alpha) : y = g^\alpha\}(m)$.

We also use the standard definition of a digital signature scheme [26].

2.1   Number-Theoretic Preliminaries

We now describe some number-theoretic preliminaries. Suppose that we have a setup algorithm *Setup* that, on input the security parameter 1^k, outputs the setup for $G = \langle g \rangle$ and $\mathsf{G} = \langle \mathsf{g} \rangle$, two groups of prime order $q = \Theta(2^k)$ that have a non-degenerate efficiently computable bilinear map e. More precisely: We assume that associated with each group element, there is a unique binary string that represents it. (For example, if $G = \mathbb{Z}_p^*$, then an element of G can be represented as an integer between 1 and $p - 1$.) Following prior work (for example, Boneh and Franklin [6]), e is a function, $e : G \times G \to \mathsf{G}$, such that

- (Bilinear) For all $P, Q \in G$, for all $a, b \in \mathbb{Z}$, $e(P^a, Q^b) = e(P, Q)^{ab}$.
- (Non-degenerate) There exists some $P, Q \in G$ such that $e(P, Q) \ne 1$, where 1 is the identity of G.
- (Efficient) There exists an efficient algorithm for computing e.

We write: $(q, G, \mathsf{G}, g, \mathbf{g}, e) \leftarrow Setup(1^k)$. It is easy to see, from the first two properties, and from the fact that G and G are both of the same prime order q, that whenever g is a generator of G, $\mathbf{g} = e(g, g)$ is a generator of G.

Such groups, based on the Weil and Tate pairings over elliptic curves (see Silverman [37]), have been extensively relied upon in cryptographic literature over the past few years (cf. [27, 6, 7, 24] to name a few results).

Further, we make the following assumption about the groups G and G.

Assumption 21 (LRSW Assumption) *Suppose that $G = \langle g \rangle$ is a group chosen by the setup algorithm Setup. Let $X, Y \in G$, $X = g^x$, $Y = g^y$. Let $O_{X,Y}(\cdot)$ be an oracle that, on input a value $m \in \mathbb{Z}_q$, outputs a triple $A = (a, a^y, a^{x+mxy})$ for a randomly chosen a. Then for all probabilistic polynomial time adversaries $\mathcal{A}^?$, $\nu(k)$ defined as follows is a negligible function:*

$$\Pr[(q, G, \mathsf{G}, g, \mathbf{g}, e) \leftarrow Setup(1^k); x \leftarrow \mathbb{Z}_q; y \leftarrow \mathbb{Z}_q; X = g^x; Y = g^y;$$
$$(m, a, b, c) \leftarrow \mathcal{A}^{O_{X,Y}}(q, G, \mathsf{G}, g, \mathbf{g}, e, X, Y) \; : \; m \notin Q \; \wedge \; m \in \mathbb{Z}_q \; \wedge$$
$$\wedge \; m \neq 0 \wedge \; a \in G \wedge \; b = a^y \; \wedge \; c = a^{x+mxy}] = \nu(k) \; ,$$

where Q is the set of queries that $\mathcal{A}$ made to $O_{X,Y}(\cdot)$.

This assumption was introduced by Lysyanskaya et al. [30], and considered for groups that are not known to admit an efficient bilinear map. It was also shown, in the same paper, that this assumption holds for generic groups. It is not hard to see that the proof carries over to generic groups G and G with a bilinear map between them.

3   Three Signature Schemes

First, we present a simple signature scheme (Scheme A) and prove it secure under the LRSW assumption. Then, we modify this scheme to get signature schemes that lend themselves more easily to the design of efficient protocols for issuing a signature on a committed value and proving knowledge of a signature on a committed value. The first generalization will allow to sign such that the signature produced is independent of the message (Scheme B), which we generalize further into a scheme that allows to sign blocks of messages (Scheme C).

Schemes A and B are, in fact, special cases of Scheme C. So we really propose just *one* new signature scheme, namely Scheme C. Schemes A and B are just steps that simplify our presentation by making it more modular.

3.1   Scheme A: A Simple Signature Scheme

The signature scheme consists of the following algorithms:

Key generation. The key generation algorithm runs the *Setup* algorithm in order to generate $(q, G, \mathsf{G}, g, \mathbf{g}, e)$. It then chooses $x \leftarrow \mathbb{Z}_q$ and $y \leftarrow \mathbb{Z}_q$, and sets $sk = (x, y)$, $pk = (q, G, \mathsf{G}, g, \mathbf{g}, e, X, Y)$, where $X = g^x$ and $Y = g^y$.

Signature. On input message m, secret key $sk = (x, y)$, and public key $pk = (q, G, \mathsf{G}, g, \mathsf{g}, e, X, Y)$, choose a random $a \in G$, and output the signature $\sigma = (a, a^y, a^{x+mxy})$.

Verification. On input $pk = (q, G, \mathsf{G}, g, \mathsf{g}, e, X, Y)$, message m, and purported signature $\sigma = (a, b, c)$, check that the following verification equations hold.

$$e(a, Y) = e(g, b) \quad \text{and} \quad e(X, a) \cdot e(X, b)^m = e(g, c) . \tag{1}$$

Theorem 1. *Signature Scheme A described above is correct and secure under the LRSW assumption.*

Proof. We first show correctness. The first verification equation holds as $e(a, Y) = e(a, g)^y = e(g, a)^y = e(g, b)$ and the second one holds because $e(X, a) \cdot e(X, b)^m = e(g, a)^x \cdot e(g, a)^{mxy} = e(g, a)^{x+mxy} = e(g, c)$.

We now show security. Without loss of generality, let $\mathsf{g} = e(g, g)$.

Consider the adversary interacting with the signer and outputting a valid signature σ on some message m that he did not query for. It is clear that the signer acts the same way as the oracle $O_{X,Y}$ defined in the LRSW assumption. Therefore, in order to prove security, we must show that the forgery $\sigma = (a, b, c)$ that passes the verification equations, must be of the form (*) $b = a^y$ and (**) $c = a^{x+mxy}$.

Let $a = g^\alpha$, $b = g^\beta$, $c = g^\gamma$. So, we wish to show that $\beta/\alpha = y$, and that $\gamma/\alpha = x + mxy$.

From the first verification equation and the bilinearity of e, we get that

$$\mathsf{g}^{\alpha y} = e(g, g)^{\alpha y} = e(a, Y) = e(g, b) = e(g, g)^\beta = \mathsf{g}^\beta .$$

As g is a generator of G, we can take the logarithm base g on both sides, and obtain $\alpha y = \beta \bmod q$, which gives us (*) as desired.

From the second verification equation, using the above, and, again, the fact that g is a generator:

$$e(X, a) \cdot e(X, b)^m = e(g, c)$$
$$e(g, g)^{x\alpha} e(g, g)^{mx\beta} = e(g, g)^\gamma$$
$$x\alpha + mx\beta = \alpha(x + mxy) = \gamma .$$

3.2   Scheme B: Where Signature Is Independent of the Message

For constructing anonymous credentials, we need a signature scheme where the signature itself is distributed in a way that is information-theoretically independent of the message m being signed. In essence, what is being signed should be an information-theoretically secure commitment (Pedersen commitment) of the message. Thus, we modify Scheme A and obtain Scheme B as follows:

Key generation. Run the *Setup* algorithm to generate $(q, G, \mathsf{G}, g, \mathsf{g}, e)$. Choose $x \leftarrow \mathbb{Z}_q, y \leftarrow \mathbb{Z}_q, z \leftarrow \mathbb{Z}_q$. Let $X = g^x$, $Y = g^y$ and $Z = g^z$. Set $sk = (x, y, z)$, $pk = (q, G, \mathsf{G}, g, \mathsf{g}, e, X, Y, Z)$.

Signature. On input message (m, r), secret key $sk = (x, y, z)$, and public key $pk = (q, G, \mathsf{G}, g, \mathsf{g}, e, X, Y, Z)$ do:
- Choose a random $a \leftarrow G$.
- Let $A = a^z$.
- Let $b = a^y$, $B = A^y$.
- Let $c = a^{x+xym} A^{xyr}$.

Output $\sigma = (a, A, b, B, c)$.

Verification. On input $pk = (q, G, \mathsf{G}, g, \mathsf{g}, e, X, Y, Z)$, message (m, r), and purported signature $\sigma = (a, A, b, B, c)$, check the following:
1. A was formed correctly: $e(a, Z) = e(g, A)$.
2. b and B were formed correctly: $e(a, Y) = e(g, b)$ and $e(A, Y) = e(g, B)$.
3. c was formed correctly: $e(X, a) \cdot e(X, b)^m \cdot e(X, B)^r = e(g, c)$.

Note that the values $(g^m Z^r, a, A, b, B, c)$ are information-theoretically independent of m if r is chosen randomly. This will become crucial when using this signature scheme in the context of an anonymous credential system.

Theorem 2. *Signature Scheme B described above is correct and secure under the LRSW assumption.*

The full proof of this theorem is found in the full version [CL04] of this paper. Here we give a sketch. Correctness follows by inspection. To show security, we consider two types of forgery. Type 1 forgery is on some message (m, r) such that for all previously queried (m_i, r_i) we have $g^m Z^r \neq g^{m_i} Z^{r_i}$. Type 2 forgery is when this is not the case.

The existence of Type-1 forger contradicts the LRSW assumption by reduction from Signature Scheme A. On input a public key $pk = (q, G, \mathsf{G}, g, \mathsf{g}, e, X, Y)$ for Scheme A, our reduction forms a public key $pk' = (q, G, \mathsf{G}, g, \mathsf{g}, e, X, Y, Z)$ for Scheme B by choosing $z \leftarrow \mathbb{Z}_q$ and setting $Z = g^z$. It then runs the forger on input pk', and answers signature queries of the form (m_i, r_i) by transforming them into queries $m_i' = m_i + r_i z \bmod q$ for the signature oracle for Scheme A. It is easy to see that a Type 1 forgery on (m, r) constitutes a successful forgery for the message $m' = m + rz$ in Scheme A.

The existence of Type-2 forger contradicts the discrete logarithm assumption (and therefore the LRSW assumption). The reduction takes as input $(q, G, \mathsf{G}, g, \mathsf{g}, e, Z)$, and sets up the public key for the signature scheme by choosing X and Y. It then runs the forger, answers all the signature queries (since it generated X and Y itself) and obtains a Type-2 forgery, namely (m, r), (m_i, r_i) such that $g^m Z^r = g^{m_i} Z^{r_i}$ for some i. This immediately gives the discrete logarithm of Z to the base g.

3.3   Scheme C: For Blocks of Messages

Scheme B allows us to generate a signature on m in such a way that the signature itself reveals no information about m. Namely, one can choose a random r and sign (m, r) using Scheme B. In general, however, there is no reason that we should limit ourselves to pairs (m, r) when signing. In fact, the construction of Scheme

B can be generalized to obtain Scheme C which can sign tuples $(m^{(0)}, \ldots, m^{(\ell)})$, i.e., blocks of messages.

Scheme C consists of the following algorithms:

Key generation. Run the *Setup* algorithm to generate $(q, G, \mathsf{G}, g, \mathsf{g}, e)$. Choose $x \leftarrow \mathbb{Z}_q, y \leftarrow \mathbb{Z}_q$, and for $1 \leq i \leq \ell, z_i \leftarrow \mathbb{Z}_q$. Let $X = g^x, Y = g^y$ and, for $1 \leq i \leq \ell, Z_i = g^{z_i}$. Set $sk = (x, y, z_1, \ldots, z_\ell)$, $pk = (q, G, \mathsf{G}, g, \mathsf{g}, e, X, Y, \{Z_i\})$.

Signature. On input message $(m^{(0)}, m^{(1)}, \ldots, m^{(\ell)})$, secret key $sk = (x, y, z_1, \ldots, z_\ell)$, and public key $pk = (q, G, \mathsf{G}, g, \mathsf{g}, e, X, Y, \{Z_i\})$ do:
- Choose a random $a \leftarrow G$.
- Let $A_i = a^{z_i}$ for $1 \leq i \leq \ell$.
- Let $b = a^y$, $B_i = (A_i)^y$.
- Let $c = a^{x+xym^{(0)}} \prod_{i=1}^{\ell} A_i^{xym^{(i)}}$.

Output $\sigma = (a, \{A_i\}, b, \{B_i\}, c)$.

Verification. On input $pk = (q, G, \mathsf{G}, g, \mathsf{g}, e, X, Y, \{Z_i\})$, message $(m^{(0)}, \ldots, m^{(\ell)})$, and purported signature $\sigma = (a, \{A_i\}, b, \{B_i\}, c)$, check the following:
1. $\{A_i\}$ were formed correctly: $e(a, Z_i) = e(g, A_i)$.
2. b and $\{B_i\}$ were formed correctly: $e(a, Y) = e(g, b)$ and $e(A_i, Y) = e(g, B_i)$.
3. c was formed correctly: $e(X, a) \cdot e(X, b)^{m^{(0)}} \cdot \prod_{i=1}^{\ell} e(X, B_i)^{m^{(i)}} = e(g, c)$.

The proof that this scheme is secure and correct is deferred to Corollary 1.

4   Anonymous Credential System and Group Signature Scheme

Following Camenisch and Lysyanskaya [10, 29], in order to construct an anonymous credential system, it is sufficient to exhibit a commitment scheme, a signature scheme, and efficient protocols for (1) proving equality of two committed values; (2) getting a signature on a committed value (without revealing this value to the signer); and (3) proving knowledge of a signature on a committed value. We provide all these tools in this section.

Constructing a group signatures scheme or identity escrow scheme additionally requires an encryption scheme that is secure against adaptively chosen ciphertext attacks and a protocol that a committed value is contained in a ciphertext (cf. [12, 3, 11]). Camenisch and Shoup provide an encryption scheme and such a protocol [11]. However, in our case we could also use the Cramer-Shoup encryption scheme [18], provided that the order of the group over which encryption is carried out is the same as the order of the group over which our signature scheme is constructed. This will allow for a more efficient proof that a ciphertext contains information to identify a group member and thus a more efficient group signatures/identity escrow scheme. We will describe the details of this in §4.4.

The reason that our new signature schemes are particularly suitable for the credential scheme application, is the fact that, given one signature on a given message, it is easy to generate another one. Consider Signature Scheme A. From a signature $\sigma = (a, b, c)$ on message m, it is very easy to compute a different signature $\sigma = (\tilde{a}, \tilde{b}, \tilde{c})$ on the same message m: just choose a random $r \in \mathbb{Z}_q$ and let $\tilde{a} = a^r$, $\tilde{b} = b^r$, $\tilde{c} = c^r$. This alone is, of course, not sufficient, but this already shows the way in which the pieces of our credential scheme will fall into place.

4.1 The Relevant Commitment Scheme

Recall the Pedersen commitment scheme [33]: given a group G of prime order q with generators g and h, a commitment to $x \in \mathbb{Z}_q$ is formed by choosing a random $r \leftarrow \mathbb{Z}_q$ and setting the commitment $C = g^x h^r$. This commitment scheme is information-theoretically hiding, and is binding under the discrete logarithm assumption, which is implied by the LRSW assumption. Moreover, there exist in the literature efficient protocols for proving knowledge and equality of committed values (see, for example, [16, 36, 8, 15]).

4.2 Obtaining a Signature on a Committed Value

When Information-Theoretic Hiding Is Not Needed. Consider the signing algorithm for Scheme A. Note that, if the input to the signer is g^m instead of m, the algorithm will still work: on input $M = g^m$, output $a = g^r$, $b = a^y$, and $c = a^x M^{rxy} = a^{x+mxy}$. To maintain security of the signature scheme, however, the user must prove knowledge of m to the signer.

As we will discuss in more detail in §4.4, this leads to a natural application to constructing group signatures: in order to join a group, a new member will choose a secret m, give g^m to the group manager, prove knowledge of m, and obtain the membership certificate (a, b, c) formed as above.

However, note here that the input to the signer, the value g^m, does not unconditionally hide the value m. Thus, if the user wishes to become a member in more than one group using the same secret m (as is the case if we want to build an anonymous credential system), the two group managers can discover that they are talking to the same user. This is easy to see if both group managers use the same generator g for G, because in that case, the user will give g^m to both of them. But this is true even if one group manager uses g, while the other uses $\tilde{g}$: recall that in groups with bilinear pairings, the decisional Diffie-Hellman problem is easy, and so g^m and $\tilde{g}^m$ can be correlated: $e(g^m, \tilde{g}) = e(g, \tilde{g})^m = e(g, \tilde{g}^m)$.

This is why we need Schemes B and C instead of Scheme A. However, we note that for group signatures, Scheme A is sufficient. In the sequel, we will give the description of the protocol for Scheme C, together with a proof of security. Because Scheme A is a special case of Scheme C (in Scheme A, $\ell = 0$), the security of the protocols for A is implied by that for C.

Signing an Information-Theoretically Hidden Message. Signature Schemes B and C are ideally suited for obtaining a signature on a committed value.

Consider Signature Scheme B. Note that to generate a valid signature, the signer need not know (m, r). Instead, it is sufficient that the signer know $M = g^m Z^r$. The values (a, A, b, B) are not a function of (m, r) – so the signer need not know (m, r) to generate them. Suppose that the signer generates them as follows: choose $\alpha \leftarrow \mathbb{Z}_q$, and let $a = g^\alpha$. Choose A, b, and B as prescribed by the signing algorithm. Finally, the signer can compute $c = a^{x+xym} A^{xyr}$ as $c = a^x M^{\alpha xy}$. This will be correct, because:

$$
\begin{aligned}
c &= a^x M^{\alpha xy} \\
&= a^x (g^m Z^r)^{\alpha xy} \\
&= a^x (a^m A^r)^{xy} \qquad \text{because by construction, } A = g^{\alpha z} = Z^\alpha \\
&= a^{x+xym} A^{xyr}
\end{aligned}
$$

More generally, in Signature Scheme C, all the signer needs is the value $M = g^{m^{(0)}} \prod_{i=1}^{\ell} Z_i^{m^{(i)}}$. He can then compute $(a = g^\alpha, \{A_i\}, b, \{B_i\})$ as prescribed, and let $c = a^x M^{\alpha xy}$ as above.

We do not know how to prove such a method for signing secure under the LRSW assumption: the difference from the usual method is that here, the adversary may win by asking a signature query for M for which he does not know the representation in terms of g and Z.

Thus, in order to obtain a signature on a committed value, the protocol needs to be amended by having a recipient of the signature prove that he knows the representation of M in bases g and Z.

Let us give the protocol in detail now. We give the protocol for Signature Scheme C, the ones for Signature Schemes A and B follow from this as they are special cases of Signature Scheme C.

Obtaining a Signature C on a Committed Value. Suppose that $M = g^{m^{(0)}} \prod_{i=1}^{\ell} Z_i^{m^{(i)}}$ is a commitment to a set of messages $(m^{(0)}, \ldots, m^{(\ell)})$ whose signature the user wishes to obtain. Then the user and the signer run the following protocol:

Common Input. The public key $pk = (q, G, \mathbb{G}, g, \mathbf{g}, e, X, Y, \{Z_i\})$, and a commitment M.

User's Input. Values $m^{(0)}, \ldots, m^{(\ell)}$ such that $M = g^{m^{(0)}} \prod_{i=1}^{\ell} Z_i^{m^{(i)}}$.

Signer's Input. Signing key $sk = (x, y, \{z_i\})$.

Protocol. First, the user gives a zero-knowledge proof of knowledge of the opening of the commitment:

$$
PK\{(\mu^{(0)}, \ldots, \mu^{(\ell)}) : M = g^{\mu^{(0)}} \prod_{i=1}^{\ell} Z_i^{\mu^{(i)}}\}
$$

Next, the signer computes $\sigma = (a, \{A_i\}, b, \{B_i\}, c)$ as described above, namely:

- $\alpha \leftarrow \mathbb{Z}_q$, $a = g^\alpha$.
- For $1 \leq i \leq \ell$, let $A_i = a^{z_i}$. Then set $b = a^y$, and for $1 \leq i \leq \ell$, let $B_i = A_i^y$.
- $c = a^x M^{\alpha xy}$.

The user outputs the signature σ.

Theorem 3. *The protocol above is a secure two-party computation of a signature on a discrete-logarithm representation of M under the signer's public key.*

Proof. (Sketch) From the signer's point of view, this protocol is as secure as when the user submits his signature queries in the clear. This is because of the proof of knowledge: there exists an extractor that can discover the value of the message being signed, and ask it of the signer in the clear.

From the user's point of view, as the only place where the user's secret input $(m^{(0)}, \ldots, m^{(\ell)})$ is used is the zero-knowledge proof of knowledge of these values, the only thing that the signer finds out about the message $(m^{(0)}, \ldots, m^{(\ell)})$, is the input value M. Note that if $m^{(\ell)}$ is distributed uniformly at random, then M information-theoretically hides the values $(m^{(0)}, \ldots, m^{(\ell-1)})$.

4.3 Proving Knowledge of a Signature

We first present a protocol to prove knowledge of a signature that works for Scheme A. We then explain why the protocol does not generalize to Scheme B (and thus also Scheme C), show how Scheme C needs to be extended to fix this problem, and obtain Scheme D. We then give a proof of security of Scheme D and a zero-knowledge protocol for proving knowledge of a signature under Scheme D. We note that the protocol to sign a committed (secret) message also works for Scheme D.

The following protocol is a zero-knowledge proof of knowledge of a signed message for Scheme A.

Common input. The public key $pk = (q, G, \mathsf{G}, g, \mathsf{g}, e, X, Y)$.
Prover's input. The message $m \in \mathbb{Z}_q$ and signature $\sigma = (a, b, c)$.
Protocol. The prover does the following:
  1. Compute a blinded version of his signature σ: Choose random $r, r' \in \mathbb{Z}_q$, and blind the signature to form $\tilde{\sigma} := (a^{r'}, b^{r'}, c^{r'r}) = (\tilde{a}, \tilde{b}, \tilde{c}^r) = (\tilde{a}, \tilde{b}, \hat{c})$. Send $(\tilde{a}, \tilde{b}, \hat{c})$ to the verifier.
  2. Let the v_x, v_{xy}, and v_s be as follows:

$$\mathsf{v}_x = e(X, \tilde{a}) \,, \qquad \mathsf{v}_{xy} = e(X, \tilde{b}) \,, \qquad \mathsf{v}_s = e(g, \hat{c}) \,.$$

The Prover and Verifier compute these values (locally) and then carry out the following zero-knowledge proof protocol:

$$PK\{(\mu, \rho) : \mathsf{v}_s^\rho = \mathsf{v}_x \mathsf{v}_{xy}^\mu\} \,.$$

The Verifier accepts if it accepts the proof above and $e(\tilde{a}, Y) = e(g, \tilde{b})$.

Theorem 4. *The protocol above is a zero knowledge proof of knowledge of a signature σ on a message m under Signature Scheme A.*

Proof. First, we prove the zero-knowledge property. The values that the verifier receives from the prover in Step 1 are independent of the actual signature: $\tilde{a}$ and $\tilde{b}$

are just random values satisfying $e(\tilde{a}, Y) = e(g, \tilde{b})$, and $\hat{c}$ is random in G because $\hat{c} = \tilde{c}^{r'}$ for a randomly chosen r'. Therefore, consider the following simulator S: Choose random r and r', and set $\tilde{a} = g^r$, $\tilde{b} = Y^r$, $\hat{c} = g^{r'}$. Then $(\tilde{a}, \tilde{b}, \hat{c})$ is distributed correctly, and so Step 1 is simulated correctly. Then, because in Step 2, the Prover and Verifier execute a zero-knowledge proof, it follows that there exists a simulator S' for this step; just run S'. It is easy to see that S constructed this way is the zero-knowledge simulator for this protocol.

Next, let us prove that this protocol is a proof of knowledge. That is to say, we must exhibit a knowledge extractor algorithm E that, given access to a Prover such that the Verifier's acceptance probability is non-negligible, outputs a value (m, σ), such that σ is a valid signature. Suppose that we are given such a prover. The extractor proceeds as follows: first, it runs the extractor for the proof of knowledge protocol of Step 2. As a result, it obtains the values $r, m \in \mathbb{Z}_q$ such that $\mathsf{v}_s^r = \mathsf{v}_x \mathsf{v}_{xy}^m$. Then:

$$\mathsf{v}_s^r = \mathsf{v}_x \mathsf{v}_{xy}^m$$
$$e(g, \hat{c})^r = e(X, \tilde{a}) e(X, \tilde{b})^m$$
$$e(g, \hat{c}^r) = e(X, \tilde{a}) e(X, \tilde{b})^m$$

And therefore the triple $\sigma = (\tilde{a}, \tilde{b}, \hat{c}^r)$ satisfies the verification equation (1) and hence is a signature on the message m, so our extractor outputs (m, σ).

Let us now try to adapt this protocol for Signature Scheme C. There is one subtlety that arises here: The zero-knowledge simulator needs to be able to come up with something that looks like a blinded signature (let us call it *simulated signature*), even though *the simulator is not given any signature*. In Signature Scheme A this turned out not to be a problem: the simulator simply picked a random r and set $\tilde{a} = g^r$, and $\tilde{b} = Y^r$. Here, this is not going to work, because, in addition to $\tilde{a}$ and $\tilde{b}$, the simulated signature needs to include the values $\{\tilde{A}_i\}$ and $\{\tilde{B}_i\}$. Now, forming $\tilde{A}_i$ is not a problem: $\tilde{A}_i = Z_i^r$. But how do we compute $\tilde{B}_i = \tilde{A}_i^y = g^{rz_i y}$ without knowing z_i or y?

To that end, we may augment the public key for signature scheme C to include a signature on some dummy message, so that the simulator will be given *some* valid signature that includes the correctly formed tuple $(a, \{A_i\}, b, \{B_i\})$, and then, in order to obtain the simulated signature, the simulator will pick a random r, and let $\tilde{a} = a^r$, $\tilde{b} = b^r$, $\tilde{A}_i = A_i^r$, and $\tilde{B}_i = B_i^r$.

An even better solution, in terms of reducing the size of the public key, is actually to include the values $W_i = Y^{z_i}$ in the public key, instead of the signature on the dummy message. It is easy to see that this has no effect on the security of the signature scheme.

Let us now give this new, augmented signature scheme, and prove it secure.

Signature Scheme D. This signature scheme is *the same as Signature Scheme C*, except that the public key also includes the values $\{W_i = Y^{z_i}\}$.

Key generation. Run the *Setup* algorithm to generate $(q, G, \mathsf{G}, g, \mathsf{g}, e)$. Choose $x \leftarrow \mathbb{Z}_q$, $y \leftarrow \mathbb{Z}_q$, and for $1 \le i \le \ell$, $z_i \leftarrow \mathbb{Z}_q$. Let $X = g^x$, $Y = g^y$ and, for $1 \le i \le \ell$, $Z_i = g^{z_i}$ and $W_i = Y^{z_i}$. Set $sk = (x, y, z_1, \ldots, z_\ell)$, $pk = (q, G, \mathsf{G}, g, \mathsf{g}, e, X, Y, \{Z_i\}, \{W_i\})$.

The signature and verification algorithm are identical to the ones of Scheme C.

Theorem 5. *Signature Scheme D is correct and secure under the LRSW assumption.*

The detailed proof of this theorem is given in the full version of this paper. The main idea of the proof of security is that the proof for Scheme B generalizes to the case when we have several Z_i's.

As a forger for Scheme C is also a forger for Scheme D, we have:

Corollary 1. *Signature Scheme C is correct and secure under the LRSW assumption.*

The full description of the protocol and proof of security follow.

Common input. The public key $pk = (q, G, \mathsf{G}, g, \mathsf{g}, e, X, Y, \{Z_i\}, \{W_i\})$.
Prover's input. The block of messages $(m^{(0)}, \ldots, m^{(\ell)})$ and signature $\sigma = (a, \{A_i\}, b, \{B_i\}, c)$.
Protocol. The prover does the following:
1. Compute a blinded version of his signature σ: Choose random $r, r' \in \mathbb{Z}_q$. Form $\tilde{\sigma} = (\tilde{a}, \{\tilde{A}_i\}, \tilde{b}, \{\tilde{B}_i\}, \tilde{c})$ as follows:

$$\tilde{a} = a^r, \quad \tilde{b} = b^r \quad \text{and} \quad \tilde{c} = c^r$$
$$\tilde{A}_i = A_i^r \quad \text{and} \quad \tilde{B}_i = B_i^r \text{ for } 1 \leq i \leq \ell$$

Further, blind $\tilde{c}$ to obtain a value $\hat{c}$ that it is distributed independently of everything else: $\hat{c} = \tilde{c}^{r'}$.
Send $(\tilde{a}, \{\tilde{A}_i\}, \tilde{b}, \{\tilde{B}_i\}, \hat{c})$ to the verifier.
2. Let v_x, v_{xy}, $\mathsf{V}_{(xy,i)}$, $i = 1, \ldots, \ell$, and v_s be as follows:

$$\mathsf{v}_x = e(X, \tilde{a}) \ , \quad \mathsf{v}_{xy} = e(X, \tilde{b}) \ , \quad \mathsf{V}_{(xy,i)} = e(X, \tilde{B}_i) \ , \quad \mathsf{v}_s = e(g, \hat{c})$$

The Prover and Verifier compute these values (locally) and then carry out the following zero-knowledge proof protocol:

$$PK\{(\mu^{(0)}, \ldots, \mu^{(\ell)}, \rho) : (\mathsf{v}_s)^\rho = \mathsf{v}_x(\mathsf{v}_{xy})^{\mu^{(0)}} \prod_{i=1}^{\ell}(\mathsf{V}_{(xy,i)})^{\mu^{(i)}}\}$$

The Verifier accepts if it accepts the proof above and (a) $\{\tilde{A}_i\}$ were formed correctly: $e(\tilde{a}, Z_i) = e(g, \tilde{A}_i)$; and (b) $\tilde{b}$ and $\{\tilde{B}_i\}$ were formed correctly: $e(\tilde{a}, Y) = e(g, \tilde{b})$ and $e(\tilde{A}_i, Y) = e(g, \tilde{B}_i)$.

Theorem 6. *The protocol above is a zero knowledge proof of knowledge of a signature σ on a block of messages $(m^{(0)}, \ldots, m^{(\ell)})$ under Signature Scheme D.*

The proof of this theorem follows the proof of Theorem 4 and is provided in the full version of this paper.

4.4 An Efficient Group Signature Scheme Secure under the LSWR-Assumption

We now present the first efficient group signature (and identity escrow) scheme whose security relies solely on assumptions related to the discrete logarithm problem (in the random oracle model). In contrast, all previous efficient schemes rely on the strong RSA assumption plus the decisional Diffie-Hellman assumption.

Recall that a group signatures scheme allows members of a group to sign anonymously on the group's behalf. In case of disputes, there exists a trusted third party called revocation manager who will be able to open a signature and reveal the identity of the signer. A group signature scheme consists of five procedures: (1) a key generation procedure that produces the public key of the group (and also some keys for the group and revocation manager), (2) a join protocol for a member to get admitted by the group manager, (3) a sign algorithm for an admitted member to sign a message, (4) a verification algorithm to check group signatures for validity with respect to the group's public key, and (5) an opening algorithm that allows the revocation manager to reveal the identity of a signer. A group signature scheme is secure if only the revocation manager can reveal the identity of the signer (anonymity) and if the revocation manager can do this for all valid signatures (traceability) [3].

Our construction follows the approach introduced by Camenisch and Stadler [12]: A member gets a certificate on a membership public key from the group manager when she joins the group. When she wants to sign on behalf of the group, she encrypts her membership public key under the encryption key of the party who will later be able to open group signatures (revocation manager) and then proves that she possesses a certificate on the encrypted membership public key and that she knows its secret key. To make this proof a signature, one usually applies the Fiat-Shamir heuristic to this proof [21].

The public key of the group manager is the public key of our Scheme A, i.e., $pk_M = (q, G, \mathsf{G}, g, \mathsf{g}, e, X, Y)$ and his secret key is $x = \log_g X$ and $y = \log_g Y$. The public key of the revocation manager is the public key of the Cramer-Shoup encryption scheme [18] in the group $\mathsf{G} = \langle \mathsf{g} \rangle$, i.e., $pk_R = (\mathsf{h}, \mathsf{y}_1, \mathsf{y}_2, \mathsf{y}_3)$, with $\mathsf{h} \in_R \mathsf{G}$, $\mathsf{y}_1 = \mathsf{g}^{x_1}\mathsf{h}^{x_2}$, $\mathsf{y}_2 = \mathsf{g}^{x_3}\mathsf{h}^{x_4}$, and $\mathsf{y}_3 = \mathsf{g}^{x_5}$, where $x_1, \ldots, x_5 \in_R \mathbb{Z}_q$ are the revocation manager's secret key[1]. Finally, let $\mathcal{H}() : \{0,1\}^* \to \mathbb{Z}_q$ be a collision resistant hash function (modeled as a random oracle in the proof of security).

The join protocol is as follows. The future group member chooses her membership secret key $k \in_R \mathbb{Z}_q$, sets $P = g^k$, sends P authentically to the group manager, and proves to the group manager the knowledge of $\log_g P$. The group manager replies with a Scheme A signature (a, b, c) on the message committed by P, i.e., computes $a = g^r$, $b = a^y$, and $c = a^x P^{rxy}$, where $r \in_R \mathbb{Z}_q$ (cf. §4.2).

---

[1] The Cramer-Shoup cryptosystem is secure under the decisional Diffie-Hellman (DDH) assumption. Therefore, we cannot use it over group G, because the existence of a bilinear map implies that the DDH problem is tractable. Thus, we use the CS cryptosystem in group G instead.

The group manager stores $\mathsf{P} = e(P, g)$ together with P and the identity of the new group member.

To sign a message m on behalf of the group, the user computes $\mathsf{P} = \mathsf{g}^k = e(P, g)$ and a blinded version of the certificate by choosing random $r, r' \in \mathbb{Z}_q$ and computing $\tilde{\sigma} := (a^{r'}, b^{r'}, c^{r'r}) = (\tilde{a}, \tilde{b}, \tilde{c}^r) = (\tilde{a}, \tilde{b}, \hat{c})$. Next, she encrypts P under the revocation manager's public key pk_R, i.e., she chooses $u \in_R \mathbb{Z}_q$, computes $\mathsf{c}_1 = \mathsf{g}^u$, $\mathsf{c}_2 = \mathsf{h}^u$, $\mathsf{c}_3 = \mathsf{y}_1^u \mathsf{P}$, and $\mathsf{c}_4 = \mathsf{y}_2^u \mathsf{y}_3^{u \mathcal{H}(\mathsf{c}_1 \| \mathsf{c}_2 \| \mathsf{c}_3)}$. Then she computes the following proof-signature (cf. §2):

$$\Sigma = SPK\{(\mu, \rho, \upsilon) : \mathsf{v}_s^\rho = \mathsf{v}_x \mathsf{v}_{xy}^\mu \ \wedge \ \mathsf{c}_1 = \mathsf{g}^\upsilon \ \wedge \ \mathsf{c}_2 = \mathsf{h}^\upsilon \ \wedge$$
$$\wedge \ \mathsf{c}_3 = \mathsf{y}_1^\upsilon \mathsf{g}^\mu \ \wedge \ \mathsf{c}_4 = (\mathsf{y}_2 \mathsf{y}_3^{\mathcal{H}(\mathsf{c}_1 \| \mathsf{c}_2 \| \mathsf{c}_3)})^\upsilon\}(m) \ ,$$

where $\mathsf{v}_x = e(X, \tilde{a})$, $\mathsf{v}_{xy} = e(X, \tilde{b})$, and $\mathsf{v}_s = e(g, \hat{c})$. A group signature consists of $((\tilde{a}, \tilde{b}, \hat{c}), (\mathsf{c}_1, \mathsf{c}_2, \mathsf{c}_3, \mathsf{c}_4), \Sigma)$ and is valid if Σ is a valid SPK as defined above and if $e(\tilde{a}, Y) = e(g, \tilde{b})$ holds.

To open such a group signature, the revocation managers needs to decrypt $(\mathsf{c}_1, \mathsf{c}_2, \mathsf{c}_3, \mathsf{c}_4)$ to obtain P which identifies the group member.

It is not hard to see that, in the random oracle model, this is a secure group signatures scheme under the LRSW and the decisional Diffie-Hellman assumption in G. Let us give a proof sketch for security under the Bellare et al. [3] definition. If an adversary can break anonymity, then one can break the encryption scheme as $(\tilde{a}, \tilde{b}, \hat{c})$ are random values and Σ is derived from an honest-verifier zero-knowledge proof. If an adversary can produce a signature that cannot be opened, i.e., linked to a registered member by the revocation manager, then one can use rewinding to extract a forged signature and break the signature scheme (cf. analysis of the protocol to prove knowledge of a signatures in §4.3). If used as an identity escrow scheme (i.e., if Σ is not a proof-signature but a real protocol between a group member and a verifier), the security proof need not to assume random oracles.

The scheme just described can be extended in several ways. For instance, we could use Scheme D instead of Scheme A and include the user's identity id directly into her membership key P, e.g., $P = g^k Z_1^{\mathrm{id}}$. That is, in the join protocol, the user would send $P' = g^k$ (and prove knowledge of $\log_g P$) and the group manager would then compute P as to ensure that indeed id is contained in P. Then, instead of encrypting P, one could use the Camenisch-Shoup encryption scheme [11] to directly encrypt the identity as one of the discrete logarithms the knowledge of which is proven when proving knowledge of a signature.

5   Constructions Based on the BBS Group Signature

Recently and independently of this work, Boneh, Boyen and Shacham [5] presented a group signature scheme secure under the strong Diffie-Hellman and the Linear assumptions. They showed that, under these assumptions in groups with bilinear pairings, it is hard, on input $(g_1, g_2 = g_1^\gamma)$ to sample tuples of the form

(A, x) where $A = g_1^{1/(\gamma+x)}$ (in other words, $A^{\gamma+x} = g_1$), even given a polynomial number of such samples. In their group signature scheme, such a tuple (A, x) is a user's group membership certificate, while (g_1, g_2) is the public key of the group. At the heart of their construction are (1) a zero-knowledge proof of knowledge of such a tuple; and (2) a scheme for encrypting x. They prove the resulting construction secure under a slightly weaker variant of the Bellare, Micciancio, and Warinschi [3] definition of security.

Boneh, Boyen, and Shacham also modify their main group signature scheme to achieve exculpability, as follows. The public key of the group is augmented by an additional value h; it is now (g_1, g_2, h). The membership certificate of a group member is (A, x, y) such that $A^{\gamma+x} h^y = g_1$. This membership certificate is created via a protocol in which the group manager only learns the value h^y, but not the value y. The unforgeability of membership certificates in this modified scheme can be derived from that of their main scheme. They achieve exculpability because a proof of knowledge of a membership certificate requires the knowledge of the value y.

Note that this latter signature scheme gives rise to the equivalent of our Signature Scheme A, but under a different assumption. Namely, the membership certificate (A, x, y) is a signature on the value y. Just as in our Scheme A, a group member obtains his group membership certificate in such a way that the group manager learns the value h^y but not the value y itself.

Not surprisingly, this signature scheme can be extended to the equivalent of our Schemes B and C using techniques similar to the ones described above. As a result, we can obtain signature schemes with efficient protocols based on the BBS signature. Let us give a sketch for the equivalent for Scheme C. A public key would be $(g_1, g_2, h_0, h_1, \ldots, h_\ell)$. A signature on a block of messages $(m_0, \ldots, m_\ell)$ consists of values (A, x) such that $A^{\gamma+x} \prod_{i=0}^{\ell} h_i^{m_i}$. In order to obtain a signature on a committed block of messages, a user will have to supply the signer with the value $Y = \prod_{i=0}^{\ell} h_i^{m_i}$, and prove knowledge of its representation in the bases $(h_0, \ldots, h_\ell)$. If m_0 is chosen at random, then Y information-theoretically hides $(m_1, \ldots, m_\ell)$. The signer will then generate the signature. A proof of knowledge of a signature on a committed value can be obtained by appropriate modifications to the BBS group signature protocol.

Acknowledgments

We thank Dan Boneh, Xavier Boyen and Hovav Shacham for making their paper [5] available to us as we were preparing the final version of this paper. We also thank the anonymous referees for helpful comments. Anna Lysyanskaya is supported by NSF Career grant CNS-0347661.

References

1. G. Ateniese, J. Camenisch, M. Joye, and G. Tsudik. A practical and provably secure coalition-resistant group signature scheme. In *CRYPTO 2000*, vol. 1880 of *LNCS*, pp. 255–270. Springer Verlag, 2000.

2. G. Ateniese and B. de Medeiros. Efficient group signatures without trapdoors. In *ASIACRYPT 2003*, vol. 2894 of *LNCS*, pp. 246–268. Springer Verlag, 2003.
3. M. Bellare, D. Micciancio, and B. Warinschi. Foundations of group signatures: Formal definition, simplified requirements and a construction based on general assumptions. In *Eurocrypt 2003*, vol. 2656 of *LNCS*, pp. 614–629, 2003.
4. D. Boneh and X. Boyen. Short signatures without random oracles. In *EURO-CRYPT 2004*. Springer Verlag, 2004.
5. D. Boneh, X. Boyen, and H. Shacham. Short group signatures using strong diffie hellman. In *CRYPTO 2004*. Springer Verlag, 2004.
6. D. Boneh and M. Franklin. Identity-based encryption from the Weil pairing. In *CRYPTO 2001*, vol. 2139 of *LNCS*, pp. 213–229. Springer Verlag, 2001.
7. D. Boneh and A. Silverberg. Applications of multilinear forms to cryptography. In *Topics in Algebraic and Noncommutative Geometry, Contemporary Mathematics*, vol. 324, pp. 71–90. American Mathematical Society, 2003.
8. S. Brands. Rapid demonstration of linear relations connected by boolean operators. In *EUROCRYPT '97*, vol. 1233 of *LNCS*, pp. 318–333. Springer Verlag, 1997.
9. J. Camenisch and A. Lysyanskaya. Efficient non-transferable anonymous multi-show credential system with optional anonymity revocation. In *EUROCRYPT 2001*, vol. 2045 of *LNCS*, pp. 93–118. Springer Verlag, 2001.
10. J. Camenisch and A. Lysyanskaya. A signature scheme with efficient protocols. In *Security in communication networks*, vol. 2576 of *LNCS*, pp. 268–289, 2002.
[CL04] Jan Camenisch and Anna Lysyanskaya. Signature schemes and anonymous credentials from bilinear maps. http://eprint.iacr.org, 2004.
11. J. Camenisch and V. Shoup. Practical verifiable encryption and decryption of discrete logarithms. In *CRYPTO 2003*, *LNCS*, pp. 126–144. Springer Verlag, 2003.
12. J. Camenisch and M. Stadler. Efficient group signature schemes for large groups. In *CRYPTO '97*, vol. 1296 of *LNCS*, pp. 410–424. Springer Verlag, 1997.
13. R. Canetti, O. Goldreich, and S. Halevi. The random oracle methodology, revisited. In *Proc. 30th Annual ACM STOC*, pp. 209–218, 1998.
14. D. Chaum. Security without identification: Transaction systems to make big brother obsolete. *Communications of the ACM*, 28(10):1030–1044, Oct. 1985.
15. D. Chaum, J.-H. Evertse, and J. van de Graaf. An improved protocol for demonstrating possession of discrete logarithms and some generalizations. In *EURO-CRYPT '87*, vol. 304 of *LNCS*, pp. 127–141. Springer-Verlag, 1988.
16. D. Chaum and T. P. Pedersen. Wallet databases with observers. In *CRYPTO '92*, vol. 740 of *LNCS*, pp. 89–105. Springer-Verlag, 1993.
17. D. Chaum and E. van Heyst. Group signatures. In *EUROCRYPT '91*, vol. 547 of *LNCS*, pp. 257–265. Springer-Verlag, 1991.
18. R. Cramer and V. Shoup. A practical public key cryptosystem provably secure against adaptive chosen ciphertext attack. In *CRYPTO '98*, vol. 1642 of *LNCS*, pp. 13–25, Berlin, 1998. Springer Verlag.
19. R. Cramer and V. Shoup. Signature schemes based on the strong RSA assumption. In *Proc. 6th ACM CCS*, pp. 46–52. ACM press, nov 1999.
20. W. Diffie and M. E. Hellman. New directions in cryptography. *IEEE Trans. on Information Theory*, IT-22(6):644–654, Nov. 1976.
21. A. Fiat and A. Shamir. How to prove yourself: Practical solution to identification and signature problems. In *CRYPTO '86*, vol. 263 of *LNCS*, pp. 186–194, 1987.
22. M. Fischlin. The Cramer-Shoup strong-RSA signature scheme revisited. In *Public Key Cryptography - PKC 2003*, vol. 2567 of *LNCS*. Springer-Verlag, 2002.
23. R. Gennaro, S. Halevi, and T. Rabin. Secure hash-and-sign signatures without the random oracle. In *EUROCRYPT '99*, vol. 1592 of *LNCS*, pp. 123–139, 1999.

24. C. Gentry and A. Silverberg. Hierarchical ID-based cryptography. In *ASIACRYPT 2002*, vol. 2501 of *LNCS*, pp. 548–566. Springer Verlag, 2002.
25. S. Goldwasser and Y. T. Kalai. On the (in)security of the Fiat-Shamir paradigm. In *Proc. 44th IEEE FOCS*, pp. 102–115. IEEE Computer Society Press, 2003.
26. S. Goldwasser, S. Micali, and R. Rivest. A digital signature scheme secure against adaptive chosen-message attacks. *SIAM Journal on Computing*, 17(2):281–308, Apr. 1988.
27. A. Joux. A one-round protocol for tripartite Diffie-Hellman. In *Proceedings of the ANTS-IV conference*, vol. 1838 of *LNCS*, pp. 385–394. Springer-Verlag, 2000.
28. J. Kilian and E. Petrank. Identity escrow. In *CRYPTO '98*, vol. 1642 of *LNCS*, pp. 169–185, Berlin, 1998. Springer Verlag.
29. A. Lysyanskaya. *Signature Schemes and Applications to Cryptographic Protocol Design*. PhD thesis, Massachusetts Institute of Technology, Cambridge, Massachusetts, Sept. 2002.
30. A. Lysyanskaya, R. Rivest, A. Sahai, and S. Wolf. Pseudonym systems. In *Selected Areas in Cryptography*, vol. 1758 of *LNCS*. Springer Verlag, 1999.
31. S. Micali. 6.875: Introduction to cryptography. MIT course taught in Fall 1997.
32. M. Naor and M. Yung. Universal one-way hash functions and their cryptographic applications. In *Proc. 21st Annual ACM STOC*, pp. 33–43, 1989. ACM.
33. T. P. Pedersen. Non-interactive and information-theoretic secure verifiable secret sharing. In *CRYPTO '91*, vol. 576 of *LNCS*, pp. 129–140. Springer Verlag, 1992.
34. R. L. Rivest, A. Shamir, and L. Adleman. A method for obtaining digital signatures and public-key cryptosystems. *Communications of the ACM*, 21(2):120–126, 1978.
35. J. Rompel. One-way functions are necessary and sufficient for secure signatures. In *Proc. 22nd Annual ACM STOC*, pp. 387–394, Baltimore, Maryland, 1990. ACM.
36. C. P. Schnorr. Efficient signature generation for smart cards. *Journal of Cryptology*, 4(3):239–252, 1991.
37. J. Silverman. *The Arithmetic of Elliptic Curves*. Springer-Verlag, 1986.
38. E. Verheul. Self-blindable credential certificates from the weil pairing. In *ASIACRYPT 2001*, vol. 2248 of *LNCS*, pp. 533–551. Springer Verlag, 2001.

Complete Classification
of Bilinear Hard-Core Functions

Thomas Holenstein, Ueli Maurer, and Johan Sjödin

Department of Computer Science,
Swiss Federal Institute of Technology (ETH),
Zürich, Switzerland
{thomahol,maurer,sjoedin}@inf.ethz.ch

Abstract. Let $f : \{0,1\}^n \to \{0,1\}^l$ be a one-way function. A function $h : \{0,1\}^n \to \{0,1\}^m$ is called a hard-core function for f if, when given $f(x)$ for a (secret) x drawn uniformly from $\{0,1\}^n$, it is computationally infeasible to distinguish $h(x)$ from a uniformly random m-bit string. A (randomized) function $h : \{0,1\}^n \times \{0,1\}^k \to \{0,1\}^m$ is a *general* hard-core function if it is hard-core for every one-way function $f : \{0,1\}^n \to \{0,1\}^l$, where the second input to h is a k-bit uniform random string r. Hard-core functions are a crucial tool in cryptography, in particular for the construction of pseudo-random generators and pseudo-random functions from any one-way function.

The first general hard-core predicate, proposed by Goldreich and Levin, and several subsequently proposed hard-core functions, are bilinear functions in the two arguments x and r. In this paper we introduce a parameter of bilinear functions $h : \{0,1\}^n \times \{0,1\}^k \to \{0,1\}^m$, called exponential rank loss, and prove that it characterizes exactly whether or not h is a general hard-core function. The security proofs for the previously proposed bilinear hard-core functions follow as simple consequences. Our results are obtained by extending the class of list-decodable codes and by generalizing Hast's list-decoding algorithm from the Reed-Muller code to general codes.

Keywords: List-decoding, hard-core functions, Goldreich-Levin predicate.

1 Introduction

Blum and Micali [BM84] showed a hard-core predicate[1] for the exponentiation function modulo a prime, which is widely conjectured to be one-way (except for special primes). They also showed how to construct a pseudo-random generator based on it. Hard-core predicates are also known for some other specific (conjectured) one-way functions.

In a seminal paper [GL89], Goldreich and Levin proved that for any one-way function $f : \{0,1\}^n \to \{0,1\}^l$, the XOR of a random subset of the n bits

---

[1] The term predicate is used throughout to denote a function with range $\{0,1\}$.

M. Franklin (Ed.): CRYPTO 2004, LNCS 3152, pp. 73–91, 2004.

of the input x constitutes a hard-core predicate. This function is randomized (because of the choice of a random subset), and it is easy to see that any general hard-core function must be randomized. An alternative view is to interpret the randomizing input of the hard-core function as an extra input and output of a modified one-way function $f' : \{0,1\}^{2n} \to \{0,1\}^{l+n}$ defined by

$$f'(x,r) = (f(x), r)$$

which now has a deterministic hard-core function $h(x, r)$ [2]. The Goldreich-Levin hard-core function is simply the inner product of x and r, which is a bilinear function $h : \{0,1\}^n \times \{0,1\}^n \to \{0,1\}$.

Any such bilinear map h is characterized by a binary $n \times n$ matrix M, where $h(x,r) = x^T \cdot M \cdot r$. For the Goldreich-Levin predicate, M is simply the identity matrix.

One can show (see [Lub96]) that $m = O(\log n)$ independent Goldreich-Levin predicates are jointly hard-core, i.e., they form a hard-core function $h : \{0,1\}^n \times \{0,1\}^{mn} \to \{0,1\}^m$. An important issue is to reduce the required amount of randomness in a hard-core function. A construction presented in [GL89] (see also [Gol01]) requires only $n+m-1$ instead of mn random bits for an m-bit hard-core function. Goldreich, Rubinfeld, and Sudan [GRS00] reduced the number of random bits down to n, as for the Goldreich-Levin function which produces only one (rather than m) bits. While some of the proofs of these results as they appear in the literature are non-trivial, they will all follow as simple consequences of our main theorem.

More generally, one can consider bilinear functions for vector spaces over any finite field $\mathbb{F}$, i.e., functions $h : \mathbb{F}^n \times \mathbb{F}^k \to \mathbb{F}^m$. We are interested in characterizing which of these functions are general hard-core functions. This characterization turns out to be given by a quite simple parameter of such a bilinear function. The characterization is complete in the sense that when the parameter is below a certain threshold, then the function is hard-core, and otherwise there exist one-way functions (under some reasonable complexity-theoretic assumption) such that h is not a hard-core function for f.

Let us discuss this parameter. For any linear function $\ell : \mathbb{F}^m \to \mathbb{F}$, the function $\ell \circ h$ is a bilinear function $\mathbb{F}^n \times \mathbb{F}^k \to \mathbb{F}$ which can be characterized by an $n \times k$ matrix over $\mathbb{F}$. The parameter of interest, which we call exponential rank loss, is defined as the expected value of the exponentially weighted rank of this matrix, when averaged over all non-zero functions ℓ.

The main technical part of [GL89] consists in showing that an error-correcting code has certain list-decoding properties, i.e., that it is possible to find a list of all codewords in a Hamming ball of a certain size. In this paper we show how to list-decode a larger class of codes. The stated characterization of hard-core functions will then follow.

An application of one-way functions and hard-core predicates are pseudorandom generators. It is easy to obtain a pseudorandom generator from any one-way

---

[2] Yao's method (implicit in [Yao82]) of using several copies of a one-way function and computing the XOR of some of the inputs can also be seen in the same light.

permutation f by iterating f and after each iteration extracting a (the same) hard-core predicate. It is much more complicated and less efficient to use any one-way function (see [HILL99]).

The security of a cryptographic scheme that uses a pseudo-random generator is proven by showing that an algorithm breaking the scheme could distinguish the pseudo-randomness from real randomness. Hast [Has03] showed that in many cryptographic applications, breaking the scheme is actually stronger than just distinguishing the randomness from pseudorandomness with small probability, in the sense that if an algorithm is given a pseudo-random or random input and it breaks the scheme, then it is almost certain that the input was pseudo-random rather than random. Hast then shows that this leads to an improved security analysis for many constructions. The main technical tool is an extension of the list-decoding algorithm to the case where erasures in the codewords are allowed. We use this extension, and furthermore generalize Hast's result by giving list-decoding algorithms that are able to handle erasures for more general codes.

Section 2 introduces the notation and discusses bilinear functions and list-decoding, the main technical tool of the paper. Previous work is also summarized in this section. In Section 3, we analyze a special case of bilinear functions, namely these for which all matrices mentioned above (i.e., for all non-zero linear functions) have full rank. This special case already suffices to prove previous results in the literature. We generalize the algorithm in Section 4 such that it works with *any* bilinear code, where the running time and the produced list will grow linearly with the exponential rank loss of the code. In Section 5 we discuss the application to characterizing hard-core functions.

2   Preliminaries

We use calligraphic letters to denote sets. Capital letters denote random variables over the corresponding sets; and lowercase letters denote specific values of these random variables, i.e., values in the sets.

The notation $f : \mathcal{X} \to \mathcal{Y}$ is used to denote a function f from the domain $\mathcal{X}$ to the range $\mathcal{Y}$. Sometimes, functions take additional randomness (i.e., for every input $x \in \mathcal{X}$ the function only specifies a probability distribution over $\mathcal{Y}$). In this case we write $f : \mathcal{X} \rightsquigarrow \mathcal{Y}$, a notation which also will be used to denote randomized algorithms with domain $\mathcal{X}$ and range $\mathcal{Y}$. If an algorithm has access to a randomized function, we use the term *oracle* for the randomized function.

2.1   Bilinear Functions

Let $\mathbb{F} = \mathrm{GF}(q)$ be the finite field with q elements and let $\mathbb{F}^n$ be the n-dimensional vector space of n-tuples over $\mathbb{F}$. As a special case, we identify $\{0,1\}$ with $\mathrm{GF}(2)$, and the bitstrings $\{0,1\}^n$ of length n with the n-dimensional vector space over $\mathrm{GF}(2)$.

A linear function $\ell : \mathbb{F}^n \to \mathbb{F}$ can be specified by a vector $w \in \mathbb{F}^n$ such that $\ell(v) = \langle w, v \rangle := \sum_i v_i w_i$. We use $\mathscr{L}_n$ to denote the set of all linear functions

$\ell : \mathbb{F}^n \to \mathbb{F}$. Furthermore, $\mathbf{0}$ will denote the zero function $\mathbf{0}(v) \equiv 0$ and we use $\mathscr{L}_n^* := \mathscr{L}_n \backslash \{\mathbf{0}\}$ for the set of all linear functions excluding $\mathbf{0}$.

A *bilinear map* $h : \mathbb{F}^n \times \mathbb{F}^k \to \mathbb{F}$ can be specified by a matrix $M \in \mathbb{F}^{n \times k}$ such that $h(v, w) = v^T M w$. The rank of a bilinear map is just the rank of this matrix. A *bilinear function* $h : \mathbb{F}^n \times \mathbb{F}^k \to \mathbb{F}^m$ is a function where every entry in the output vector is specified by a bilinear map. Note that for any function $\ell \in \mathscr{L}_m$ the concatenation $\ell \circ h$ is a bilinear map. If L is a uniformly chosen random linear function from $\mathscr{L}_m^*$, the *exponential rank loss* $\rho(h)$ is defined as

$$\rho(h) := \mathrm{E}[q^{n - \operatorname{rank}(L \circ h)}].$$

We say that a bilinear function is *full-rank*, if $\operatorname{rank}(\ell \circ h) = n$ for every $\ell \in \mathscr{L}_m^*$ (in which case $\rho(h) = 1$).

2.2   List-Decoding

The main tool in the construction of hard-core functions is the notion of a list-decodable code. Such a code has the property that, given a noisy codeword, it is possible to find a list of *all* codewords which have a certain agreement with the noisy codeword.

Consider a code $\mathcal{C}$ given as a function $\mathcal{C} : \mathcal{X} \to \mathcal{Z}^k$. Note that the input to the function (usually the message) is an element of $\mathcal{X}$ while the output (the codeword) is a k-tuple over $\mathcal{Z}$. The Hamming distance of two words of $\mathcal{Z}^k$ is the number of coordinates in which the words differ. List-decoding is the task of finding for a given $z^k \in \mathcal{Z}^k$ all the values x for which $\mathcal{C}(x)$ has a Hamming distance from z^k that is smaller than some predefined bound. This is in contrast to usual error-correcting, where one aims to find the *one* codeword which is closest to the received word. The most ambitious task is to list-decode close to the noise barrier: given any $\varepsilon > 0$ one wants to find all values x for which $\mathcal{C}(x)$ has a Hamming distance of at most $(1 - \frac{1}{|\mathcal{Z}|} - \varepsilon)k$ from a given word. Since a random word has expected distance $(1 - \frac{1}{|\mathcal{Z}|})k$ from any codeword, this is clearly the best one can expect to achieve.

Instead of considering the function $\mathcal{C}(x)$, one can equivalently consider a function $h : \mathcal{X} \times \{1, \ldots, k\} \to \mathcal{Z}$, such that $h(x, i)$ is the value of $\mathcal{C}(x)$ at the i-th position. More generally we consider functions $h : \mathcal{X} \times \mathcal{Y} \to \mathcal{Z}$ for any domain $\mathcal{Y}$. Analogous, we assume that we have *oracle access* to the noisy word to be decoded: instead of reading the complete word it will be convenient to assume that an oracle $\mathcal{O} : \mathcal{Y} \rightsquigarrow \mathcal{Z}$, on input y, returns the symbol at position y. This allows us to list-decode in sublinear time, i.e., without looking at every position of the word, which in turn allows the codewords to be exponentially large. The oracle is stateless, but may be randomized and is not required to return the same symbol if queried twice with the same input. The agreement of an oracle with a codeword is then expressed as $\Pr[h(x, Y) = \mathcal{O}(Y)]$, where the probability is over the choices of Y and the randomness of the oracle.

Additionally, we allow erasures in the word which will be denoted by $\perp$. Thus, the oracle is a randomized function $\mathcal{O} : \mathcal{Y} \rightsquigarrow \mathcal{Z} \cup \{\perp\}$. The *rate* δ of such an oracle is the probability that a symbol in $\mathcal{Z}$ is returned,

$$\delta := \Pr[\mathcal{O}(Y) \neq \perp].$$

For a fixed word x, the *advantage* ε of $\mathcal{O}$ is defined as

$$\varepsilon := \Pr[\mathcal{O}(Y) = h(x, Y) \mid \mathcal{O}(Y) \neq \perp] - \frac{1}{|\mathcal{Z}|}.$$

This motivates the following definition:

Definition 1 (List-decodable code).[3] *The function* $h : \mathcal{X} \times \mathcal{Y} \to \mathcal{Z}$ *is* (δ, ε)-*list-decodable with* κ *oracle calls and list size* λ *if there exists an oracle algorithm with running time* $\lambda \cdot \text{poly}(\log(|\mathcal{X}|))$ *which, after at most* κ *oracle calls to an oracle* $\mathcal{O} : \mathcal{Y} \rightsquigarrow \mathcal{Z} \cup \{\perp\}$ *with rate at least* δ, *generates a set* Λ *of size at most* λ, *such that for every* x *with* $\Pr[\mathcal{O}(Y) = h(x, Y) \mid \mathcal{O}(Y) \neq \perp] \geq \frac{1}{|\mathcal{Z}|} + \varepsilon$ *the set satisfies* $\Pr[x \in \Lambda] \geq 1/2$.

2.3   Hard-Core Functions

Informally, a one-way function is a function which is easy to evaluate but hard to invert.

Definition 2 (One-way function). *An efficiently computable function family* $f : \{0, 1\}^n \to \{0, 1\}^{p(n)}$ *with* $p(n) \in \text{poly}(n)$ *is a one-way function if for every probabilistic polynomial time (in n) algorithm A the inverting probability* $\Pr[f(A(f(X))) = f(X)]$ *is negligible.*

A hard-core function $h : \{0, 1\}^n \times \{0, 1\}^k \to \{0, 1\}^m$ can intuitively extract bits from the input of a one-way function f such that these bits look random, even given $f(x)$. We can distinguish *(strong) hard-core functions*, where the output is indistinguishable from a random string of length m (which we denote by U^m), and *weak hard-core functions*, where the output of the function is hard to predict.

Definition 3 (Strong hard-core function). *An efficiently computable family* $h : \{0, 1\}^n \times \{0, 1\}^{k(n)} \to \{0, 1\}^{m(n)}$ *of functions, with* $k(n), m(n) \in \text{poly}(n)$ *is a (strong) hard-core function if, for every one way function* $f : \{0, 1\}^n \to \{0, 1\}^{p(n)}$ *and every probabilistic polynomial time algorithm A, the distinguishing advantage given by* $\Pr[A(f(X), R, h(X, R)) = 1] - \Pr[A(f(X), R, U^m) = 1]$, *is negligible in n.*

Definition 4 (Weak hard-core function). *An efficiently computable family* $h : \{0, 1\}^n \times \{0, 1\}^{k(n)} \to \{0, 1\}^{m(n)}$ *with* $k(n), m(n) \in \text{poly}(n)$ *of functions is a weak hard-core function if, for every one-way function* $f : \{0, 1\}^n \to \{0, 1\}^{p(n)}$ *and every probabilistic polynomial time algorithm A, the advantage of A in guessing $h(x, r)$ on input $f(x)$ and r, defined as* $\Pr[A(f(X), R) = h(X, R)] - \frac{1}{2^m}$, *is negligible in n.*

---

[3] We require the list-decoding algorithm to work in time $\lambda \cdot \text{poly}(\log(|\mathcal{X}|))$. Note that in some cases, λ will be superpolynomial in the input size $\log(|\mathcal{X}|)$ and $\log(|\mathcal{Y}|)$.

In general, weak hard-core functions are easier to construct than strong ones. However, we will see that for small outputs the notions are equivalent.

As shown in [Sud00], any list-decodable code $h : \{0,1\}^n \times \{0,1\}^k \to \{0,1\}^m$ as defined above yields a weak hard-core function. To prove this, one assumes for the sake of contradiction that an algorithm B is given which on input $f(x)$ and r predicts $h(x,r)$ with probability higher than $\frac{1}{2^m} + \varepsilon$, for some non-negligible[4] ε. After arguing that B needs to have a reasonable success probability for a significant subset of the possible values for x, one then uses B as the oracle in the list-decoding algorithm. The resulting list, which is small, then contains x with non-negligible probability, and one can find a preimage of $f(x)$ by applying f to all values in the list.

In such a reduction, the running time of the resulting algorithm is dominated by the running time of B. Thus, one is interested in the exact number κ of oracle calls, while the exponent in the running time of the (polynomial) algorithm is of minor importance. In this application, the second input (from $\{0,1\}^k$) corresponds to a random string. As randomness is an expensive resource, one wants k to be as small as possible. We show how to achieve $k = n$ for any n.

2.4   Previous Work

The fundamental result on bilinear list-decodable codes implicitly appears in [GL89], stating that the Reed-Muller code of first order, defined as $h : \{0,1\}^n \times \{0,1\}^n \to \{0,1\}$, $h(x,y) = \langle x, y \rangle = \sum_i x_i y_i$, has an algorithm which efficiently list-decodes it up to an error rate of $1/2 + \varepsilon$, for any $\varepsilon > 0$.

The standard proof used today was found independently by Levin and Rackoff and is given in [Gol01] (see also [Lev87]). In [Has03], Hast introduces the extension of list-decoding algorithms for oracles with erasures. The existence of the resulting algorithm is asserted in the following theorem:

Theorem 5 (Goldreich-Levin, cf. [Has03]). *For any $\varepsilon, \delta > 0$, the function $h : \{0,1\}^n \times \{0,1\}^n \to \{0,1\}$, $h(x,r) = \langle x, r \rangle$ is (δ, ε)-list-decodable with list size $O(\frac{1}{\delta\varepsilon^2})$ and $\Theta(n\frac{1}{\delta\varepsilon^2})$ oracle calls. The list-decoding algorithm needs $\delta\varepsilon^2$ as input.*

This theorem is slightly stronger than the original version in [Has03], where an additional factor n appears in the number of oracle calls and the list size. The version as stated here can be obtained by applying a trick that appears in [Gol01, Section 2.5.2.4][5].

It is natural to generalize this theorem to vector spaces over any finite field. For this, the best known result is given in [GRS00].

Theorem 6. *For any $\delta, \varepsilon > 0$, the function $h : \mathbb{F}^n \times \mathbb{F}^n \to \mathbb{F}$, $h(x,r) = \langle x, r \rangle$ is (δ, ε)-list-decodable with list size $\mathrm{poly}(n, \delta^{-1}\varepsilon^{-1})$ and $\mathrm{poly}(n, \delta^{-1}\varepsilon^{-1})$ oracle calls. The list-decoding algorithm needs $\delta\varepsilon$ as input.*

---

[4] We use *non-negligible* to denote a function which is *not* negligible.

[5] Basically, one uses a linear, asymptotically optimal error-correcting code to find x instead of finding the bits one by one.

The algorithm which is used to prove Theorem 6 is similar to the original algorithm given in [GL89]. The exponents in $\mathrm{poly}(n, \delta^{-1}\varepsilon^{-1})$ are rather high, so we refrain from stating them explicitly.

Näslund shows in [Näs95] that for any one-way function $f(x)$, a hard-core predicate can be obtained if one interprets x as a value in $\mathrm{GF}(2^n)$, and outputs any bit of $ax + b$ for randomly chosen a and b; a result which also follows from the characterization in this paper. Furthermore, he proves that for randomly chosen a, b and prime p the least significant bit of $ax + b \bmod p$ is a hard-core predicate. More generally, in [Näs96] he shows that all bits of $ax + b \bmod p$ are hard-core.

In a different line of research, in [STV01] Sudan et al. give very strong list-decodable codes which are not bilinear, based on Reed-Muller codes. These codes can also be used to obtain hard-core functions for any one-way function.

In [AGS03], Akavia et al. show that list-decoding can also be used to prove specific hard-core results. For example, they give a proof based on list-decodable codes that the least significant bit of RSA is hard-core (which was first shown in [ACGS88]).

3   Full-Rank Bilinear Functions

The main technical goal of this paper is to give a list-decoding procedure for any bilinear function $h : \mathbb{F}^n \times \mathbb{F}^k \to \mathbb{F}^m$. In this section, we will first consider a simple, but very general subset of bilinear functions, namely full-rank bilinear functions h (i.e., $\mathrm{rank}(\ell \circ h) = n$ for every $\ell \neq 0$). We show that these functions have very good list-decoding algorithms.

In a second step we will construct full-rank bilinear functions $h : \mathbb{F}^n \times \mathbb{F}^k \to \mathbb{F}^m$ which are optimal in the sense that for fixed n the dimension k is made as small as possible, while for m every value $0 < m \leq k$ is possible. This allows us to give a very large class of strong hard-core functions.

3.1   List-Decoding of Full-Rank Functions

In this section, we give a list-decoding algorithm for every full-rank bilinear function $h : \mathbb{F}^n \times \mathbb{F}^k \to \mathbb{F}^k$. In particular, for the case $\mathbb{F} = \mathrm{GF}(2)$, we will show that there exists a list-decoding algorithm for h which is as strong as the one guaranteed in Theorem 5.

Theorem 7. *Let $h : \{0,1\}^n \times \{0,1\}^k \to \{0,1\}^m$ be a full rank bilinear function. For any $\delta, \varepsilon > 0$, the function $h(x, y)$ is (δ, ε)-list-decodable with list size $O(\frac{1}{\delta\varepsilon^2})$ and $\Theta(n\frac{1}{\delta\varepsilon^2})$ oracle calls. The list-decoding algorithm needs $\delta\varepsilon^2$ as input.*

For general finite fields, analogously to Theorem 6, the following holds.

Theorem 8. *Let $h : \mathbb{F}^n \times \mathbb{F}^k \to \mathbb{F}^m$ be a full-rank bilinear function. For any $\delta, \varepsilon > 0$, the function $h(x, y)$ is (δ, ε)-list-decodable with list size $\mathrm{poly}(n, \delta^{-1}\varepsilon^{-1})$ and $\mathrm{poly}(n, \delta^{-1}\varepsilon^{-1})$ oracle calls. The list-decoding algorithm needs $\delta\varepsilon$ as input.*

To prove Theorems 7 and 8, we describe an algorithm which, on access to an oracle $\mathcal{O}$ with rate δ, outputs a list of all $x \in \mathbb{F}^n$ which satisfy

$$\Pr[\mathcal{O}(Y) = h(x, Y) \mid \mathcal{O}(Y) \neq \perp] \geq \frac{1}{q^m} + \varepsilon. \tag{1}$$

For this purpose we convert $\mathcal{O}$ to an oracle $\mathcal{O}'$ with the same rate and related advantage, but for a different code. Namely, $\mathcal{O}'$ will have advantage $\varepsilon/2$ on $\langle x, r \rangle$ for any x which satisfies (1), i.e., $\Pr[\mathcal{O}'(R) = \langle x, R \rangle \mid \mathcal{O}'(R) \neq \perp] \geq \frac{1}{q} + \frac{\varepsilon}{2}$. Applying Theorems 5 and 6, respectively, then yields the result.

In the following, let L be a uniform random function from $\mathscr{L}_m^*$, i.e., L is a random variable taking as values functions from $\mathscr{L}_m^*$. We show that if a value z returned by the oracle is better than a random guess for $h(x, y)$, then $L(z)$ is better than a random guess for $L(h(x, y))$ as well. To see why this holds, we first compute the probability that $L(a)$ equals $L(b)$ for two distinct values a and b; this probability is close to $1/q$.

Lemma 9. *For any distinct* $a, b \in \mathbb{F}^m$, $\Pr[L(a) = L(b)] = \dfrac{q^{m-1} - 1}{q^m - 1}$.

Proof. First note that $\Pr[L(a) = L(b)] = \Pr[L(a - b) = 0] = \Pr[L(v) = 0]$ for some $v \neq 0$. If L' is chosen uniformly at random from all functions in $\mathscr{L}_m$ (not excluding $\mathbf{0}$), then $\Pr[L'(v) = 0] = \frac{1}{q}$, and since $\mathbf{0}(v) = 0$ for every v, we can write

$$\frac{1}{q} = \Pr[L'(v) = 0] = \underbrace{\frac{1}{q^m}}_{\Pr[L'=0]} + \underbrace{\frac{q^m - 1}{q^m}}_{\Pr[L'\neq 0]} \Pr[L(v) = 0],$$

which implies the lemma. $\square$

Now we can estimate the probability that $L(Z_1)$ equals $L(Z_2)$ for two random variables Z_1 and Z_2. Later, Z_2 will be $h(x, Y)$ and Z_1 a guess of an oracle for $h(x, Y)$.

Lemma 10. *Let* Z_1 *be a random variable over* $\mathbb{F}^m \cup \{\perp\}$ *and* Z_2 *a random variable over* $\mathbb{F}^m$. *If, for any* $\varepsilon > 0$,

$$\Pr[Z_1 = Z_2 \mid Z_1 \neq \perp] = \frac{1}{q^m} + \varepsilon,$$

then

$$\Pr[L(Z_1) = L(Z_2) \mid Z_1 \neq \perp] \geq \frac{1}{q} + \frac{\varepsilon}{2}.$$

Proof. Obviously, if $Z_1 = Z_2$ we also have $\ell(Z_1) = \ell(Z_2)$ for every $\ell \in \mathscr{L}_m^*$. Using Lemma 9 we obtain

$$\Pr[L(Z_1) = L(Z_2) \mid Z_1 \neq \perp] = \underbrace{\frac{1}{q^m} + \varepsilon}_{\Pr[Z_1 = Z_2 \mid Z_1 \neq \perp]} + \underbrace{\left(\frac{q^m - 1}{q^m} - \varepsilon\right)}_{\Pr[Z_1 \neq Z_2 \mid Z_1 \neq \perp]} \frac{q^{m-1} - 1}{q^m - 1}$$

$$= \frac{1}{q^m} + \varepsilon + \left(\frac{1}{q} - \frac{1}{q^m}\right) - \varepsilon \frac{q^{m-1} - 1}{q^m - 1} \geq \frac{1}{q} + \frac{\varepsilon}{2}. \quad \square$$

Next, we translate a uniform query r into a uniform pair $(\ell, y) \in \mathscr{L}^* \times \{0,1\}^k$, such that $\langle x, r \rangle = \ell(h(x, y))$. We will be able to use this by giving y to the oracle $\mathcal{O}$ which predicts $h(x, y)$ and then apply ℓ to get a prediction for $\langle x, r \rangle$. Since y is uniform we will know the advantage of the oracle in predicting $h(x, y)$, and since ℓ is uniform, we can apply Lemma 10.

Lemma 11. *Let $h : \mathbb{F}^n \times \mathbb{F}^k \to \mathbb{F}^m$ be a full-rank bilinear function. There exists an efficiently computable random mapping $G_h : \mathbb{F}^n \rightsquigarrow \mathbb{F}^k \times \mathscr{L}_m^*$, which, for a uniformly chosen input r outputs a uniform random pair (ℓ, y) such that $\ell(h(x, y)) = \langle x, r \rangle$ for every x.*

Proof. The algorithm implementing G_h first chooses an $\ell \in \mathscr{L}_m^*$ uniformly at random. For a fixed ℓ, let M be the matrix for which $\ell(h(x, y)) = x^T M y$; note that $\mathrm{rank}(M) = n$. As a second step, the algorithm chooses y as a uniform random solution of $My = r$, and returns the pair (ℓ, y). For every fixed ℓ if r is uniformly distributed; the vector y will be uniformly distributed. Furthermore, $\ell(h(x, y)) = x^T M y = x^T r = \langle x, r \rangle$. $\square$

The next lemma proves the claimed conversion; i.e., given an oracle which predicts $h(x, y)$ we implement an oracle which predicts $\langle x, r \rangle$. For this, on input r the algorithm first gets a pair (ℓ, y) using Lemma 11. Then, it queries the given oracle $\mathcal{O}$ with y, applies ℓ to the output and returns the result.

Lemma 12. *Let $h : \mathbb{F}^n \times \mathbb{F}^k \to \mathbb{F}^m$ be a full-rank bilinear function. There is an efficient oracle algorithm A such that for any $\varepsilon > 0$, every $x \in \mathbb{F}^n$ and any oracle $\mathcal{O} : \mathbb{F}^k \rightsquigarrow \mathbb{F}^m$ which satisfies*

$$\Pr[\mathcal{O}(Y) = h(x, Y) \mid \mathcal{O}(Y) \neq \bot] \geq \frac{1}{q^m} + \varepsilon$$

algorithm $A^{\mathcal{O}}$ satisfies

$$\Pr[A^{\mathcal{O}}(R) = \langle x, R \rangle \mid A^{\mathcal{O}}(R) \neq \bot] \geq \frac{1}{q} + \frac{\varepsilon}{2}$$

and $\Pr[A^{\mathcal{O}}(R) \neq \bot] = \Pr[\mathcal{O}(Y) \neq \bot]$. Algorithm A makes one oracle call to $\mathcal{O}$.

Proof. Given a uniformly chosen r, the algorithm first evaluates the function $G_h(r)$ as guaranteed by Lemma 11, to get a uniform pair (ℓ, y) with $\ell(h(x, y)) = \langle x, r \rangle$. It then queries the oracle with y. In case the answer z is not $\bot$ it returns $\ell(z)$; otherwise it returns $\bot$.

Let x be fixed such that

$$\Pr[\mathcal{O}(Y) = h(x, Y) \mid \mathcal{O}(Y) \neq \bot] \geq \frac{1}{q^m} + \varepsilon.$$

Lemma 10 implies that

$$\Pr[L(\mathcal{O}(Y)) = L(h(x, Y)) \mid \mathcal{O}(Y) \neq \bot] \geq \frac{1}{q} + \frac{\varepsilon}{2}.$$

Since (ℓ, y) is uniformly distributed this together with $\ell(h(x, y)) = \langle x, r \rangle$ concludes the proof. $\square$

Lemma 12 can be seen as a reduction of a code to another one, in the sense that given a noisy codeword of one code we can generate a noisy codeword of a related code such that the Hamming distances to codewords are related in some sense. The proofs of Theorems 7 and 8 are now obvious.

Proof (of Theorems 7 and 8). Use Lemma 12, and apply Theorems 5 and 6, respectively. □

3.2   Construction of Full-Rank Functions

As mentioned before, a list-decodable code can be used to obtain a hard-core function, which means that a family of full-rank bilinear functions can be used as a hard-core function. This is stated in the following proposition (a more exact version will be given in Theorem 25, Section 5).

Proposition 13. *Any efficiently computable family of full-rank bilinear functions* $h : \{0,1\}^n \times \{0,1\}^k \to \{0,1\}^m$, *where* $k \in \text{poly}(n)$ *and* $m \in O(\log n)$ *is a strong hard-core function.*

The proposition implies that in order to give a hard-core function it is sufficient to construct a full-rank bilinear function family. In this section, we will present constructions which appear in the literature as hard-core functions, and show that they satisfy $\text{rank}(\ell \circ h) = n$ for every $\ell \neq \mathbf{0}$.

As usual in the context of hard-core functions, we will explain the constructions for vector spaces over $\{0,1\}$. However, all constructions immediately generalize to vector spaces over any finite field.

Recall that any bilinear function $h : \{0,1\}^n \times \{0,1\}^k \to \{0,1\}^m$ can be described by a sequence $M_1, \ldots, M_m$ of $n \times k$ matrices over $\text{GF}(2)$ as $h(x,r) = (x^T M_1 r, \ldots, x^T M_m r)$. It follows that for every ℓ there exists a non-empty subset $I \subseteq \{1, \ldots, m\}$ such that the function $\ell \circ h$ can be written as $\ell(h(x,r)) = x^T (\sum_{i \in I} M_i) r$.

In order to get a full-rank bilinear function it is therefore sufficient to give matrices $M_1, \ldots, M_m$ which satisfy

$$\text{rank}\left(\sum_{i \in I} M_i\right) = n \qquad \text{for every } I \neq \emptyset. \tag{2}$$

Example 14. In [Lub96] it is shown that $O(\log n)$ independent inner product bits give a hard-core function. This function $h : \{0,1\}^n \times \{0,1\}^{nm} \to \{0,1\}^m$ is defined by matrices $M_1, \ldots, M_m$ such that M_i consists of all zeros, except that from column $n(i-1)+1$ to ni it contains a $n \times n$ identity matrix. Here it is obvious that (2) is satisfied.

Example 15. In order to keep the dimension k small, one can obtain a full-rank bilinear function $h : \{0,1\}^n \times \{0,1\}^{n+m-1} \to \{0,1\}^m$ with the construction given in [Gol01] and [GL89]. There, M_i is a matrix of size $n \times (n+m-1)$ which contains only zeros with the exception of an $n \times n$ identity matrix starting at column i. Again, it is obvious that (2) holds.

Note that since rank($\ell \circ h$) cannot be larger than k for any ℓ, it is necessary to have $k \geq n$. If m is small enough this is indeed sufficient:

Theorem 16. *Let vector spaces $\{0,1\}^n$, $\{0,1\}^k$ and $\{0,1\}^m$ over $\{0,1\}$ be given. If $n \leq k$ and $m \leq k$, then there exists a full-rank bilinear function $h : \{0,1\}^n \times \{0,1\}^k \to \{0,1\}^m$.*

Proof. We first note that it is sufficient to give a full-rank bilinear function $\{0,1\}^k \times \{0,1\}^k \to \{0,1\}^k$ for every k, since one can first obtain a bilinear function $\{0,1\}^k \times \{0,1\}^k \to \{0,1\}^m$ by ignoring some of the output coordinates, and in a second step one can get a full-rank bilinear function $\{0,1\}^n \times \{0,1\}^k \to \{0,1\}^m$ by setting some of the inputs to the first arguments to zero.

To construct a full-rank bilinear function $h : \{0,1\}^k \times \{0,1\}^k \to \{0,1\}^k$ we observe that the finite field $\mathrm{GF}(2^k)$ is a vector space over $\{0,1\}$ of dimension k, and for every $x \in \mathrm{GF}(2^k)$ the map $g_x(r) = x \cdot r$ is linear. Let $z_1, \ldots, z_k$ be a basis of $\mathrm{GF}(2^k)$ and let M_i be the matrix which describes the linear mapping g_{z_i} in this basis. Since for any $I \neq \emptyset$ the matrix $\sum_{i \in I} M_i$ describes the linear mapping g_z for some non-zero $z \in \mathrm{GF}(2^k)$, this map is invertible and thus has rank k. $\square$

The bilinear function used in this proof is strongly related to the hard-core function given at the end of [GRS00], and indeed the function given there also satisfies the rank condition needed for Theorem 8[6].

4 General Bilinear Functions

In this section we give a list-decoding algorithm for every (possibly non full-rank) bilinear function. Using the same technique as in Section 3.1 we prove the following analogue of Theorem 7 (recall that $\rho(h) = \mathrm{E}[q^{n-\mathrm{rank}(L \circ h)}]$).

Theorem 17. *Let $h : \{0,1\}^n \times \{0,1\}^k \to \{0,1\}^m$ be any bilinear function. After a preprocessing phase taking time $2^m \cdot \mathrm{poly}(k, n)$, the function $h(x,y)$ is (δ, ε)-list-decodable with list size $O(\frac{\rho(h)}{\delta \varepsilon^2})$ and an expected number $\Theta(n \frac{1}{\delta \varepsilon^2})$ of oracle calls. The algorithm needs $\delta \varepsilon^2$ as input.*

Note that $\Theta(\frac{n}{\delta \varepsilon^2})$ is the *expected* number of queries. For general finite fields Theorem 18 holds.

Theorem 18. *Let $h : \mathbb{F}^n \times \mathbb{F}^k \to \mathbb{F}^m$ be any bilinear function over $\mathbb{F}$. After a preprocessing phase taking time $q^m \cdot \mathrm{poly}(n, k)$, the function $h(x,y)$ is (δ, ε)-list-decodable with list size $\rho(h) \cdot \mathrm{poly}(n, k, \delta^{-1} \varepsilon^{-1})$ and an expected number of $\mathrm{poly}(n, k, \delta^{-1} \varepsilon^{-1})$ oracle calls. The list-decoding algorithm needs $\delta \varepsilon$ as input.*

---

[6] The functions are not identical, but if one considers the "cube" given by stacking the matrices for different linear maps ℓ, then the functions are obtained from each other by a rotation of this cube. It is possible to show that for any two cubes which are obtained by rotation from each other, the corresponding function satisfies the full-rank condition if and only if the same holds for the other cube.

As before we prove these theorems by converting a given oracle $\mathcal{O}$ which on input y predicts $h(x, y)$ to an oracle $\mathcal{O}'$ which on input r predicts $\langle x, r \rangle$. We use Lemma 10 again (and thus Lemma 9), but we modify Lemmas 11 and 12.

A problem is that for some r it may be impossible to choose a pair (ℓ, y) with $\ell(h(x, y)) = \langle x, r \rangle$ for every x. This will force our reduction to return $\perp$ on input r, since there is no way to get a reasonable guess for $\langle x, r \rangle$ from $\mathcal{O}$. Furthermore, the pair (ℓ, y) must be uniformly distributed which makes the conversion return $\perp$ more often. We get the following generalization of Lemma 11:

Lemma 19. *Let $h : \mathbb{F}^n \times \mathbb{F}^k \to \mathbb{F}^m$ be a bilinear function. There exists an efficiently computable mapping $G_h : \mathbb{F}^n \leadsto (\mathbb{F}^k \times \mathscr{L}_m^*) \cup \{\perp\}$ which, on uniformly distributed input r outputs $\perp$ with probability $1 - \frac{1}{\rho(h)}$, and otherwise a uniform random pair (ℓ, y), satisfying $\ell(h(x, y)) = \langle x, r \rangle$ for all x. The algorithm uses a precomputation with time complexity $q^m \cdot \mathrm{poly}(n, k)$.*

Proof. First, as a precomputation, for every $\ell \in \mathscr{L}_m^*$ the algorithm calculates $q^{\mathrm{rank}(\ell \circ h)}$, and stores it in such a way that later it is possible to efficiently draw an element $\ell \in \mathscr{L}_m^*$ with probability $q^{n - \mathrm{rank}(\ell \circ h)} / \hat{\rho}(h)$, where $\hat{\rho}(h) = \sum_{\ell \neq 0} q^{n - \mathrm{rank}(\ell \circ h)} = (q^m - 1) \rho(h)$.

After the precomputation, on input r, the algorithm chooses ℓ according to this probability distribution and obtains the matrix M with $\ell(h(x, y)) = x^T M y$. If the system $My = r$ is solvable, it chooses a solution y uniformly at random and returns (ℓ, y); otherwise it returns $\perp$.

Note that the precomputation can obviously be done in time $q^m \cdot \mathrm{poly}(n, k)$ and every returned pair (ℓ, y) satisfies $\ell(h(x, y)) = \langle x, r \rangle$.

For a fixed ℓ and uniformly chosen r, the probability that there exists a y such that $My = r$ is $q^{\mathrm{rank}(M) - n} = q^{\mathrm{rank}(\ell \circ h) - n}$. Furthermore, conditioned on the event that the system above is solvable, every vector y has the same probability. This implies that the probability that a fixed pair (ℓ, y) is returned is

$$\Pr[G_h(R) = (\ell, y)] = \frac{q^{n - \mathrm{rank}(\ell \circ h)}}{\hat{\rho}(h)} \cdot q^{\mathrm{rank}(\ell \circ h) - n} \cdot \frac{1}{q^k} = \frac{1}{q^k \, \hat{\rho}(h)},$$

which is independent of the pair (ℓ, y). Summing over all possible pairs (ℓ, y) we get $\Pr[G_h(R) \neq \perp] = 1/\rho(h)$. $\square$

We point out that the probability of G_h *not* returning $\perp$ cannot be made any higher. To see why, first note that a pair (ℓ, y) can only be the answer for one specific input r. Furthermore, there are $q^k \, \hat{\rho}(h)$ possible pairs (ℓ, y), which can only be output for $y = 0$; implying that every pair can occur with probability at most $q^{-k} \hat{\rho}^{-1}(h)$.

Along the same line of reasoning as in Section 3, we can now prove the generalized version of Lemma 12.

Lemma 20. *Let $h : \mathbb{F}^n \times \mathbb{F}^k \to \mathbb{F}^m$ be a bilinear function. There is an efficient oracle algorithm A such that for any $\varepsilon > 0$, every $x \in \mathbb{F}^n$ and any oracle $\mathcal{O} : \mathbb{F}^k \leadsto \mathbb{F}^m$ which satisfies*

$$\Pr[\mathcal{O}(Y) = h(x, Y) \mid \mathcal{O}(Y) \neq \perp] \geq \frac{1}{q^m} + \varepsilon,$$

algorithm $A^{\mathcal{O}}$ satisfies

$$\Pr[A^{\mathcal{O}}(R) = \langle x, R \rangle \mid A^{\mathcal{O}}(R) \neq \perp] \geq \frac{1}{q} + \frac{\varepsilon}{2}$$

and $\Pr[A^{\mathcal{O}}(R) \neq \perp] = \frac{1}{\rho(h)} \Pr[\mathcal{O}(R) \neq \perp]$. *The algorithm makes one query to $\mathcal{O}$ with probability $\frac{1}{\rho(h)}$. It uses a preprocessing phase with time complexity $q^m \cdot \text{poly}(n, k)$.*

Proof. The preprocessing is the one needed for G_h of Lemma 19. On input r, the algorithm first uses G_h to obtain either a pair (ℓ, y) or $\perp$. In the second case, the algorithm returns $\perp$ and does not make an oracle query; this happens with probability $1 - \frac{1}{\rho(h)}$. If a pair (ℓ, y) is returned, the algorithm makes one query $z = \mathcal{O}(y)$. If $z \neq \perp$ the algorithm returns $\ell(z)$, otherwise it returns $\perp$.

We fix ε and x such that $\Pr[\mathcal{O}(Y) = h(x, Y) \mid \mathcal{O}(Y) \neq \perp] \geq \frac{1}{q^m} + \varepsilon$. Lemma 10 implies that $\Pr[L(\mathcal{O}(Y)) = L(h(x, Y)) \mid \mathcal{O}(Y) \neq \perp] \geq \frac{1}{q} + \frac{\varepsilon}{2}$. Conditioned on the event that A makes a query to $\mathcal{O}$ the pair (ℓ, y) is uniformly distributed and satisfies $\ell(h(x, y)) = \langle x, r \rangle$. Also, when A does not make a query to $\mathcal{O}$ it returns $\perp$. This implies

$$\Pr[L(\mathcal{O}(Y)) = L(h(x, Y)) \mid \mathcal{O}(Y) \neq \perp] = \Pr[A^{\mathcal{O}}(R) = \langle x, R \rangle \mid A^{\mathcal{O}}(R) \neq \perp].$$

Finally, we see that A does not return $\perp$ if both G_h of Lemma 19 and $\mathcal{O}$ do not return $\perp$, which happens with probability $\frac{1}{\rho(h)} \Pr[\mathcal{O}(Y) \neq \perp]$.   □

Using this conversion, the proofs of Theorems 17 and 18 are now straightforward.

Proof (of Theorems 17 and 18). Use Lemma 20 and apply Theorems 5 and 6, respectively.

5   Implications for Hard-Core Functions

The results of the previous sections have implications in cryptography, namely for one-way functions. In particular, under a reasonable complexity-theoretic assumption the results allow us to classify basically every bilinear function family $h : \{0, 1\}^n \times \{0, 1\}^k \rightarrow \{0, 1\}^m$ according to whether it is a strong hard-core function or not.

We formulate our results in the context of uniform algorithms, but they immediately generalize to a non-uniform context.

5.1   Weak vs. Strong Hard-Core Functions

In general, it is easier to construct weak hard-core functions than to construct strong ones. For example the identity function $h(x) = x$ is a weak hard-core function for any one-way function f (predicting x given $f(x)$ is the same as inverting f), but not a strong hard-core function (given $f(x)$ it is easy to distinguish x from a random value).

For small output values the two notions are equivalent: every weak hard-core function $h : \{0,1\}^n \times \{0,1\}^k \to \{0,1\}^m$ for $m \in O(\log n)$ is also a strong one. This follows from the fact that any distinguisher for such a function can be converted to a predictor. More concretely, assume that an oracle $\mathcal{O}$ has advantage ε in distinguishing $h(x,y)$ from a random value. It is well known that one can get a predictor with advantage $2^{-m}\varepsilon$ from $\mathcal{O}$ (see for example [Lub96]). The following lemma improves this fact by following the idea of Hast that, in cryptographic applications, a distinguisher often comes from an algorithm which tries to break a scheme; if it succeeds then it is almost certain that the input was not random. This can be used to obtain a predictor with lower rate but higher advantage. In the following lemma we use this idea since the probability p_0 that a distinguisher answers 1 on random input can be very small. By replacing $\bot$ with a uniform random output one obtains the well-known version mentioned above.

Lemma 21. *There exists a randomized oracle algorithm A such that for any $z \in \{0,1\}^m$, oracle $\mathcal{O}$ with*

$$p_0 := \Pr[\mathcal{O}(U^m) = 1]$$

and ε defined by

$$p_0(1 + \varepsilon) = \Pr[\mathcal{O}(z) = 1],$$

algorithm A queries $\mathcal{O}$ once and outputs a value from $\{0,1\}^m \cup \{\bot\}$ such that

$$\Pr[A^{\mathcal{O}} \neq \bot] = p_0$$

and

$$\Pr[A^{\mathcal{O}} = z \mid A^{\mathcal{O}} \neq \bot] = \frac{1}{2^m} + 2^{-m}\varepsilon.$$

Proof. Algorithm A chooses a uniform random value $z' \in \{0,1\}^m$. It then queries $\mathcal{O}(z')$ and outputs z' if the oracle outputs 1. Otherwise, it outputs $\bot$.

The probability that A outputs $\bot$ is $1 - p_0$. The probability that A outputs z is $\frac{1}{2^m}(p_0(1 + \varepsilon))$ and thus the probability that A outputs z conditioned on the event that it does not output $\bot$ is $\frac{1+\varepsilon}{2^m}$. $\quad\square$

As a corollary we obtain the following result:

Corollary 22. *Let $h : \{0,1\}^n \times \{0,1\}^k \to \{0,1\}^m$ be a weak hard-core function and $m \in O(\log n)$. Then, h is a strong hard-core function.*

Proof. Assume that h is not a strong hard-core function. Then, there exists an algorithm A which on input $(f(x), r)$ can distinguish $h(x, r)$ from a uniform random string with non-negligible advantage ε. According to Lemma 21 we can use this algorithm to obtain an algorithm which predicts the same string with success probability at least $\frac{1}{2^m} + \frac{\varepsilon}{2^m}$, and thus h is not a weak hard-core function. $\quad\square$

5.2 List-Decodable Codes and Weak Hard-Core Functions

Every list-decodable code can be used as a weak hard-core function. The idea to prove this is to assume that the function h is *not* a weak hard-core function, and to use the algorithm A which predicts $h(x, r)$ given $f(x)$ and r together with the list-decoding algorithm to find a list which contains x with probability at least $1/2$. Applying f to each element of the list and comparing the input we are guaranteed to find a preimage of $f(x)$ with high probability.

In our case, we would like to use the algorithm guaranteed in Theorem 17. This algorithm requires to know the product $\delta \varepsilon^2$, and works as long as the correct value is *at least as large* as the value given to the algorithm.

Note that the value of x is fixed during a run of algorithm A. Consequently such an algorithm can only be successful if the rate δ_x and advantage ε_x for a fixed x is large enough. However, typically only the rate δ and advantage ε averaged over all x is guaranteed to have a certain value. In order to show that this is sufficient we first prove that $\mathrm{E}[\delta_X \varepsilon_X^2] \geq \delta \varepsilon^2$. In the following lemma, it is useful to think of Z as an indicator variable which is 1 if the predictor guesses correctly; 0 on a wrong guess and $\perp$ if the predictor refuses to produce a guess. The random variable X corresponds to the value of x.

Lemma 23. *Let X be a uniformly distributed random variable over $\mathcal{X}$ and let Z be some random variable in $\{0, 1, \perp\}$. Let $\delta := \Pr[Z \neq \perp]$ and $\delta_x := \Pr[Z \neq \perp \mid X = x]$. Fix any constant c and let $\varepsilon := \Pr[Z = 1 \mid Z \neq \perp] - c$ and $\varepsilon_x := \Pr[Z = 1 \mid X = x \land Z \neq \perp] - c$. Then,*

$$\mathrm{E}[\delta_X \varepsilon_X^2] \geq \delta \varepsilon^2.$$

Proof. First we observe that $\delta = \frac{1}{|\mathcal{X}|} \sum_{x \in \mathcal{X}} \delta_x$. Furthermore we have

$$c + \varepsilon = \frac{\Pr[Z = 1]}{\Pr[Z \neq \perp]} = \frac{\frac{1}{|\mathcal{X}|} \sum_{x \in \mathcal{X}} \Pr[Z = 1 \mid X = x]}{\frac{1}{|\mathcal{X}|} \sum_{x \in \mathcal{X}} \delta_x}$$

$$= \frac{\sum_{x \in \mathcal{X}} \delta_x (c + \varepsilon_x)}{\sum_{x \in \mathcal{X}} \delta_x} = c + \frac{\sum_{x \in \mathcal{X}} \delta_x \varepsilon_x}{\sum_{x \in \mathcal{X}} \delta_x}$$

and thus $\varepsilon = \left(\sum_x \delta_x \varepsilon_x \right) / \left(\sum_x \delta_x \right)$. To show that

$$\mathrm{E}[\delta_X \varepsilon_X^2] = \frac{1}{|\mathcal{X}|} \sum_{x \in \mathcal{X}} \delta_x \varepsilon_x^2 \geq \left(\frac{1}{|\mathcal{X}|} \sum_{x \in \mathcal{X}} \delta_x \right) \left(\frac{\sum_{x \in \mathcal{X}} \delta_x \varepsilon_x}{\sum_{x \in \mathcal{X}} \delta_x} \right)^2 = \delta \varepsilon^2,$$

we note that this is equivalent to

$$\left(\sum_{x \in \mathcal{X}} \delta_x \right) \left(\sum_{x \in \mathcal{X}} \delta_x \varepsilon_x^2 \right) \geq \left(\sum_{x \in \mathcal{X}} \delta_x \varepsilon_x \right)^2,$$

which follows directly from the Cauchy-Schwarz inequality.   □

We now show how to use the list-decoding algorithm to invert a function f. The following lemma is usually used when f is a one-way function, $m \in O(\log n)$ and $\rho(h) \in \text{poly}(n)$, in which case it states that h is a weak hard-core function.

Lemma 24. *Let $f : \{0,1\}^n \to \{0,1\}^p$ be any efficiently computable function family. Let $h : \{0,1\}^n \times \{0,1\}^k \to \{0,1\}^m$ be any efficiently computable bilinear function with $k \in \text{poly}(n)$. There exists an oracle algorithm A such that for any $\mathcal{O} : \{0,1\}^p \times \{0,1\}^k \leadsto \{0,1\}^m \cup \{\perp\}$ which satisfies $\Pr[\mathcal{O}(f(X),Y) \neq \perp] = \delta$, and $\Pr\big[\mathcal{O}(f(X),Y) = h(X,Y) \mid \mathcal{O}(f(X),Y) \neq \perp\big] = \frac{1}{2^m} + \varepsilon$, algorithm $A^{\mathcal{O}}$ is running in time $\frac{\rho(h)}{\delta\varepsilon^2} \cdot \text{poly}(n) + 2^m \cdot \text{poly}(n)$ and satisfies*

$$\Pr\big[f(A^{\mathcal{O}}(f(X))) = f(X)\big] \geq \frac{\delta\varepsilon^2}{4},$$

while making an expected number $\Theta(n\frac{1}{\delta\varepsilon^2})$ of oracle calls to $\mathcal{O}$. Algorithm A needs $\delta\varepsilon^2$ as an input.

If h is a full-rank bilinear function, the term $2^m \cdot \text{poly}(n)$ in the running time can be omitted.

Proof. For any fixed $x \in \{0,1\}^n$, let $\delta_x := \Pr[\mathcal{O}(f(x),Y) \neq \perp]$ and $\varepsilon_x := \Pr[\mathcal{O}(f(x),Y) = h(x,Y) \mid \mathcal{O}(f(x),Y) \neq \perp] - \frac{1}{2^m}$. Using Lemma 23 we obtain $\mathrm{E}[\delta_X \varepsilon_X^2] \geq \delta\varepsilon^2$. Since $0 \leq \delta_x \varepsilon_x^2 \leq 1$ for any x, we can apply Markov's inequality to obtain $\Pr[\delta_X \varepsilon_X^2 > \frac{1}{2}\delta\varepsilon^2] > \frac{1}{2}\delta\varepsilon^2$. A run of the algorithm guaranteed in Theorem 17 with input $\delta\varepsilon^2/2$ thus gives a set Λ of size at most $O(\frac{\rho(h)2^n}{\delta\varepsilon^2})$ containing x with probability at least $\frac{1}{4}\delta\varepsilon^2$, while doing an expected number $\Theta(n\frac{1}{\delta\varepsilon^2})$ of oracle calls. Applying f to each $x \in \Lambda$ and testing if it is correct yields the claimed result. $\square$

5.3 Bilinear Hard-Core Functions

Lemma 21 converts a distinguisher to a predictor, while Lemma 24 uses a predictor to invert a function. Combining these two lemmas gives the following theorem:

Theorem 25. *Let $f : \{0,1\}^n \to \{0,1\}^p$ be any efficiently computable function. Let $h : \{0,1\}^n \times \{0,1\}^k \to \{0,1\}^m$ be any efficiently computable bilinear function with $k \in \text{poly}(n)$. There exists an oracle algorithm A such that for $\varepsilon, \delta > 0$ and any $\mathcal{O} : \{0,1\}^p \times \{0,1\}^k \leadsto \{0,1\}$ which satisfies*

$$\Pr\big[\mathcal{O}(f(X),R,h(X,R))\big] = \delta, \quad \text{and} \quad \Pr\big[\mathcal{O}(f(X),R,U^m)\big] = \delta(1+\varepsilon),$$

algorithm A satisfies

$$\Pr[f(A^{\mathcal{O}}(f(X))) = f(X)] \geq \frac{\delta\varepsilon^2}{4 \cdot 2^{2m}}$$

and makes an expected number of

$$\kappa = \Theta\left(n\frac{2^{2m}}{\delta\varepsilon^2}\right)$$

oracle queries to $\mathcal{O}$. *Algorithm A runs in time* $\frac{\rho(h)2^{2m}}{\delta\varepsilon^2}$ poly$(n) + 2^m$ poly(n) *and needs* $\delta\varepsilon^2$ *as input.*

Proof. Combine Lemma 21 with Lemma 24.   □

This theorem implies that any bilinear function $h : \{0,1\}^n \times \{0,1\}^k \to \{0,1\}^m$ with $m \in O(\log n)$ and $\rho(h) \in$ poly(n) can be used as a hard-core function.

Corollary 26. *Let* $h : \{0,1\}^n \times \{0,1\}^k \to \{0,1\}^m$ *be a bilinear function with* $m \in O(\log n)$ *and* $\rho(h) \in$ poly(n). *Then h is a strong hard-core function.*

Proof. Assume otherwise and use Theorem 25 to arrive at a contradiction.   □

5.4   Bilinear Functions not Suitable as Hard-Core Functions

In this section we also consider bilinear functions $h : \{0,1\}^n \times \{0,1\}^k \to \{0,1\}^m$ for which $m \notin O(\log n)$ or $\rho(h) \notin$ poly(n). One can show that $m \notin O(\log n)$ implies the existence of a function $\tilde{m} \in \omega(\log n)$ which is infinitely often smaller than m. Analogously, $\rho(h) \notin$ poly(n) implies the existence of a function $\tilde{\rho}$ which is strictly superpolynomial (i.e., $\log(\tilde{\rho}) \in \omega(\log n)$) and infinitely often smaller than $\rho(h)$. We say that a hard-core function is *regular* if $m \in O(\log n)$ or a polynomial time computable function $\tilde{m}$ as above exists; and $\rho \in$ poly(n) or a polynomial time computable $\tilde{\rho}$ as above exists.

   We show that any regular bilinear function not satisfying the conditions of Corollary 26 is not a hard-core function if some reasonable complexity-theoretic assumption holds, namely the existence of a one-way permutation with exponential security.

Definition 27 (Very strong one-way permutation).[7] *A family of polynomial time computable functions* $f : \{0,1\}^n \to \{0,1\}^n$ *is a very strong one-way permutation if there exists a constant* $c > 0$, *such that for every algorithm A with running time at most* 2^{cn}, *the inverting probability* $\Pr[f(A(f(X))) = f(X)]$ *is at most* 2^{-cn} *for all but finitely many n.*

   Proving that no such functions exist would be a breakthrough in complexity theory. Furthermore, Gennaro and Trevisan show in [GT00] that in relativized worlds such functions exist, and thus our results exclude a relativizing hardcore result for any bilinear function which does not satisfy the conditions of Corollary 26 unconditionally.

   As a first step, we show that it is impossible to use a bilinear function to extract $\omega(\log n)$ hard bits from x. Such a lemma was already hinted at in [GL89].

Lemma 28. *Let* $h : \{0,1\}^n \times \{0,1\}^k \to \{0,1\}^m$ *be a regular bilinear function with* $m \notin O(\log n)$. *If a very strong one-way permutation exists, then h is not a strong hard-core function.*

---

[7] We use permutations for the sake of simplicity. It is easy to see that arbitrary one-way functions with exponential security suffice to prove Theorem 30.

Proof. Since $m \notin O(\log n)$ and h is regular, there exists a polynomial-time computable function $\tilde{m} \in \omega(\log n)$ with $\tilde{m}(n) < m(n)$ for infinitely many n.

We define a one-way function $f : \{0,1\}^n \to \{0,1\}^n$ for which it is easy to give a distinguisher for $h(x,r)$. For this purpose, let $g : \{0,1\}^{\tilde{m}/2} \to \{0,1\}^{\tilde{m}/2}$ be a very strong one-way permutation. On input $x \in \{0,1\}^n$, split the input x into two parts, $x_1 \in \{0,1\}^{\tilde{m}/2}$ and $x_2 \in \{0,1\}^{n-\tilde{m}/2}$. The output of f is then $g(x_1)$ concatenated with x_2. We see that f is a one-way function, since an algorithm A which inverts f in poly(n)-time with non-negligible success probability can be used to invert g in time $2^{o(\tilde{m}(n))}$ with probability $2^{-o(\tilde{m}(n))}$ for infinitely many n.

Furthermore, for any n with $\tilde{m}(n) < m(n)$ it is easy to distinguish $h(x,r)$ from a random string, given $f(x)$ and r. First, we find x_2 from $f(x)$. Since $h(x,r) = x^T M r$ we see that for fixed x_2 and r only a subspace of dimension at most $\tilde{m}/2$ is possible as output value for $h(x,r)$. Also, it is easy to check whether a given value is within this subspace or not. Since a random value will be in the subspace with probability at most $2^{-\tilde{m}/2}$, h cannot be a hard-core function. □

Using basically the same technique, we can now show that only functions with nearly full rank can be used as hard-core functions.

Lemma 29. *Let $h : \{0,1\}^n \times \{0,1\}^p \to \{0,1\}^m$ be a regular bilinear function with $m \in O(\log n)$ and $\rho(h) \notin$ poly(n). If a very strong one-way permutation exists, then h is not a strong hard-core function.*

Proof. Since h is regular and $\rho(h) \notin$ poly(n), there exists a function $\tilde{\rho}$ such that $\log(\tilde{\rho}) \in \omega(\log n)$ and $\tilde{\rho}(n) < \rho(h)(n)$ for infinitely many n.

As in the proof of Lemma 28, we construct a one-way function $f : \{0,1\}^n \to \{0,1\}^n$ by embedding a preimage of size $\{0,1\}^{\log(\tilde{\rho}(n))/2}$ to a very strong one-way permutation g. Consider an n for which $\tilde{\rho}(n) < \rho(h)(n)$. For such an n it is easy to find a linear map to embed the preimage to g such that for some $\ell \in \mathscr{L}_m^*$ the value of $\ell(h(x,y))$ does not depend on the input to g. As in the proof of Lemma 28 it follows immediately that f is a one-way function, and since $\ell(h(x,y))$ only depends on a part of x which can be found by a linear transformation of the output, h cannot be a hard-core function. □

Together, this implies the following theorem.

Theorem 30. *Let $h : \{0,1\}^n \times \{0,1\}^k \to \{0,1\}^m$ be a regular bilinear function, and assume the existence of a very strong one-way permutation. Then h is a strong hard-core function if and only if $\rho(h) \in$ poly(n) and $m \in O(\log n)$.*

Proof. If $\rho(h) \in$ poly(n) and $m \in O(\log n)$, then h is a hard-core function according to Corollary 26. If $m \in O(\log n)$ and $\rho(h) \notin$ poly(n), then h is not a hard-core function according to Lemma 29. If $m \notin O(\log n)$ then h is not a hard-core function according to Lemma 28. □

Acknowledgments

We would like to thank Gustav Hast and Johan Håstad for helpful discussions. This research was supported by the Swiss National Science Foundation, project no. 2000-066716.01/1.

References

[ACGS88] Werner Alexi, Benny Chor, Oded Golreich, and Claus P. Schnorr. RSA and Rabin functions: Certain parts are as hard as the whole. *Siam Journal on Computation*, 17(2):194–209, 1988.

[AGS03]   Adi Akavia, Shafi Goldwasser, and Samuel Safra. Proving hard-core predicates using list decoding. In *The 44th Annual Symposium on Foundations of Computer Science*, pages 146–157, 2003.

[BM84]   Manuel Blum and Silvio Micali. How to generate cryptographically strong sequences of pseudo-random bits. *Siam Journal on Computation*, 13(4):850–864, 1984.

[GL89]   Oded Goldreich and Leonid A. Levin. A hard-core predicate for all one-way functions. In *Proceedings of the Twenty First Annual ACM Symposium on Theory of Computing*, pages 25–32, 1989.

[Gol01]   Oded Goldreich. *Basic Tools*. Foundations of Cryptography. Cambridge University Press, first edition, 2001. ISBN 0-521-79172-3.

[GRS00]   Oded Goldreich, Ronitt Rubinfeld, and Madhu Sudan. Learning polynomials with queries: The highly noisy case. *Siam Journal on Discrete Mathematics*, 13(4):535–570, 2000.

[GT00]   Rosario Gennaro and Luca Trevisan. Lower bounds on the efficiency of generic cryptographic constructions. In *The 41st Annual Symposium on Foundations of Computer Science*, pages 305–313, 2000.

[Has03]   Gustav Hast. Nearly one-sided tests and the Goldreich-Levin predicate. In Eli Biham, editor, *Advances in Cryptology — EUROCRYPT 2003*, volume 2656 of *Lecture Notes in Computer Science*, pages 195–210, 2003. Extended version to appear in *Journal of Cryptology*.

[HILL99] Johan Håstad, Russell Impagliazzo, Leonid A. Levin, and Michael Luby. A pseudorandom generator from any one-way function. *Siam Journal on Computation*, 28(4):1364–1396, 1999.

[Lev87]   Leonid A. Levin. One-way functions and pseudorandom generators. *Combinatorica*, 7(4):357–363, 1987.

[Lub96]   Michael Luby. *Pseudorandomness and Cryptographic Applications*. Princeton University Press, first edition, 1996. ISBN 0-691-02546-0.

[Näs95]   Mats Näslund. Universal hash functions & hard core bits. In Louis C. Guillou and Jean-Jacques Quisquater, editors, *Advances in Cryptology — EUROCRYPT '95*, volume 921 of *Lecture Notes in Computer Science*, pages 356–366, 1995.

[Näs96]   Mats Näslund. All bits in $ax + b$ mod p are hard. In Neal Koblitz, editor, *Advances in Cryptology — CRYPTO '96*, volume 1109 of *Lecture Notes in Computer Science*, pages 114–128, 1996. Extended Abstract.

[STV01]   Madhu Sudan, Luca Trevisan, and Salil Vadhan. Pseudorandom generators without the XOR lemma. *Journal of Computer and System Sciences*, 62(2):236–266, 2001.

[Sud00]   Madhu Sudan. List decoding: Algorithms and applications. *SIGACTN: SIGACT News (ACM Special Interest Group on Automata and Computability Theory)*, 31(1):16–27, 2000.

[Yao82]   Andrew C. Yao. Theory and applications of trapdoor functions (extended abstract). In *The 23rd Annual Symposium on Foundations of Computer Science*, pages 80–91, 1982.

Finding Collisions on a Public Road, or Do Secure Hash Functions Need Secret Coins?

Chun-Yuan Hsiao and Leonid Reyzin

Boston University Computer Science
111 Cummington Street
Boston MA 02215 USA
{cyhsiao,reyzin}@cs.bu.edu

Abstract. Many cryptographic primitives begin with parameter generation, which picks a primitive from a family. Such generation can use public coins (e.g., in the discrete-logarithm-based case) or secret coins (e.g., in the factoring-based case). We study the relationship between public-coin and secret-coin collision-resistant hash function families (CRHFs). Specifically, we demonstrate that:

- there is a lack of attention to the distinction between secret-coin and public-coin definitions in the literature, which has led to some problems in the case of CRHFs;
- in some cases, public-coin CRHFs can be built out of secret-coin CRHFs;
- the distinction between the two notions is meaningful, because in general secret-coin CRHFs are unlikely to imply public-coin CRHFs.

The last statement above is our main result, which states that there is no black-box reduction from public-coin CRHFs to secret-coin CRHFs. Our proof for this result, while employing oracle separations, uses a novel approach, which demonstrates that there is no black-box reduction *without* demonstrating that there is no relativizing reduction.

1 Introduction

1.1 Background

Collision-Resistant Hashing. Collision-resistant (CR) hashing is one of the earliest primitives of modern cryptography, finding its first uses in digital signatures [Rab78,Rab79] and Merkle trees [Mer82,Mer89]. A hash function, of course, maps (potentially long) inputs to short outputs. Informally, a hash function is collision-resistant if it is infeasible to find two inputs that map to the same output.

It is easy to see there is no meaningful way to formalize the notion of collision-resistance for a single fixed-output-length hash function. Indeed, at least half of the 2^{161} possible 161-bit inputs to SHA-1 [NIS95] have collisions (because SHA-1 has 160-bit outputs). Hence, an algorithm finding collisions for SHA-1 is quite simple: it just has, hardwired in it, two 161-bit strings that collide. It exists, even if no one currently knows how to write it down.

M. Franklin (Ed.): CRYPTO 2004, LNCS 3152, pp. 92–105, 2004.

Due to this simple observation, formal definitions of collision-resistant hashing (first given by Damgård [Dam87]) usually speak of collision-resistant function *families* (CRHFs)[1]. A hash function family is collision-resistant if any adversary, given a function chosen randomly from the family, is unable to output a collision for it.

How to Choose from a Family? Most definitions of CRHFs do not dwell on the issue of *how* a hash function is to be chosen from a family. In this paper, we point out that this aspect of the definition is crucial. Indeed, in any application of collision-resistant hashing, some party P must choose a function from the family by flipping some random coins to produce the function description. As we demonstrate, it is important to distinguish between two cases. In the *public-coin* case these random coins can be revealed as part of the function description. In the *secret-coin* case, on the other hand, knowledge of the random coins may allow one to find collisions, and thus P must keep the coins secret after the description is produced. (For examples of both cases, see Section 2.) We note that the original definition of [Dam87] is secret-coin, and that the secret-coin definition is more general: clearly, a public-coin CRHF will also work if one chooses to keep the coins secret.

1.2   Initial Observations

Importance of the Distinction. The distinction between public-coin and secret-coin CRHFs is commonly overlooked. Some works modify the secret-coin definition of [Dam87] to a public-coin definition, without explicitly mentioning the change (e.g., [BR97,Sim98]). Some definitions (e.g., [Mir01]) are ambiguous on this point. This state of affairs leads to confusion and potential problems, as discussed in three examples below.

Example 1. Some applications use the wrong definition of CRHF. For instance, in Zero-Knowledge Sets of Micali, Rabin and Kilian [MRK03], the prover uses a hash function to commit to a set. The hash function is chosen via a shared random string, which is necessary because the prover cannot be trusted to choose his own hash function (since a dishonest prover could benefit from finding collisions), and interaction with the verifier is not allowed at the commit stage (indeed, the prover does not yet know who the verifier(s) will be). In such a setting, one cannot use secret-coin CRHFs (however, in an apparent oversight, [MRK03] defines only secret-coin CRHFs). A clear distinction between public-coin and secret-coin CRHFs would make it easier to precisely state the assumptions needed in such protocols.

Example 2. The result of Simon [Sim98] seems to claim less than the proof implies. Namely, the [Sim98] theorem that one-way permutations are unlikely to imply CRHFs is stated only for public-coin CRHFs, because that is the

---

[1] It is possible to define a single hash function (with variable output-length; cf. previous paragraph) instead of a collection of them. In this case, it can be collision-resistant only against a uniform adversary.

definition [Sim98] uses. It appears to hold also for secret-coin CRHFs, but this requires re-examining the proof. Such re-examination could be avoided had the definitional confusion been resolved.

Example 3. The original result of Goldwasser and Kalai [GK03] on the security of the Fiat-Shamir transform without random oracles has a gap due to the different notions of CRHF (the gap was subsequently closed, see below). Essentially, the work first shows that if no *secret-coin* CRHFs exist, then the Fiat-Shamir transform can never work. It then proceeds to show, in a sophisticated argument, that if *public-coin* CRHFs exist, then it is possible to construct a secure identification scheme for which the Fiat-Shamir transform always results in an insecure signature scheme. This gap in the result would be more apparent with proper definitions.

Let us elaborate on the third example, as it was the motivating example for our work. It is not obvious how to modify the [GK03] proof to cover the case when secret-coin CRHFs exist, but public-coin ones do not. Very recently, Goldwasser and Kalai [GK] closed this gap by modifying the identification scheme of the second case to show that the Fiat-Shamir transform is insecure if secret-coin (rather than public-coin) CRHFs exist. Briefly, the modification is to let the honest prover choose the hash function during key generation (instead of the public-coin Fiat-Shamir verifier choosing it during the interaction, as in the earlier version).

Despite the quick resolution of this particular gap, it and other examples above demonstrate the importance of distinguishing between the two types of collision-resistant hashing. Of course, it is conceivable that the two types are equivalent, and the distinction between them is without a difference. We therefore set out to discover whether the distinction between public-coin and secret-coin hashing is real, i.e., whether it is possible that public-coin CRHFs do not exist, but secret-coin CRHFs do.

1.3 Our Results

Recall that public-coin hashing trivially implies secret-coin hashing. We prove the following results:

1. Dense[2] secret-coin CRHFs imply public-coin CRHFs; but
2. There is no black-box reduction from secret-coin CRHFs to public-coin CRHFs.

The first result is quite simple. The second, which is more involved, is obtained by constructing oracles that separate secret-coin CRHFs from public-coin CRHFs. Our technique for this oracle separation is different from previous separations (such as [IR89,Sim98,GKM+00,GMR01,CHL02]), as explained below. We note that our second result, as most oracle separations, applies only to uniform adversaries (a notable exception to this is [GT00]).

---

[2] A CRHF is *dense* if a noticeable subset of all keys of a particular length is secure; see Section 3.

Our results suggest that a gap between secret-coin and public-coin CRHFs exists, but only if no dense secret-coin CRHFs exist. They highlight the importance of distinguishing between the two definitions of CRHFs.

In addition to these main results, Section 5 addresses secret vs. public coins in other cryptographic primitives.

1.4 On Oracle Separations

Usually when one constructs a cryptographic primitive P (e.g., a pseudorandom generator [BM84]) out of another cryptographic primitive Q (e.g., a one-way permutation), P uses Q as a subroutine, oblivious to how Q implemented. The security proof for P usually constructs an adversary for Q using any adversary for P as a subroutine. This is known as a *"black-box reduction* from P to Q."

Note that to show that no *general* reduction from P to Q exists requires proving that Q does not exist, which is impossible given the current state of knowledge. However, it is often possible to show that no *black-box* reduction from P to Q exists; this is important because most cryptographic reductions are black-box.

The first such statement in cryptography is due to Impagliazzo and Rudich [IR89]. Specifically, they constructed an oracle relative to which key agreement does not exist, but one-way permutations do. This means that any construction of key agreement from one-way permutations does not relativize (i.e., does not hold relative to an oracle). Hence no black-box reduction from key agreement to one-way permutations is possible, because black-box reductions relativize.

The result of [IR89] was followed by other results about "no black-box reduction from P to Q exists," for a variety of primitives P and Q (e.g., [Sim98,GKM+00,GMR01,CHL02]). Most of them, except [GMR01], actually proved the slightly stronger statement that no *relativizing* reduction from P to Q exists, by using the technique of constructing an oracle.

Our proof differs from most others in that it *directly* proves that no black-box reduction exists, without proving that no relativizing reduction exists. We do so by constructing different oracles for the construction of P from Q and for the security reduction from adversary for P to adversary for Q. This proof technique seems more powerful than the one restricted to a single oracle, although it proves a slightly weaker result. The weaker result is still interesting, however, because it still rules out the most common method of cryptographic reduction. Moreover, the stronger proof technique may yield separations that have not been achievable before.

We note that [GMR01] also directly prove that no black-box reduction exists, without proving that no relativizing reduction exists. Our approach is different from [GMR01], whose approach is to show that for every reduction, there is an oracle relative to which this reduction fails.

For a detailed discussion on black-box reductions, see [RTV04]. All reductions in this paper are what they refer to as *fully black-box* reductions.

2   Definitions of Public-Coin and Secret-Coin CRHFs

Examples. Before we define public-coin and secret-coin hashing formally, consider the following two example hash function families. The first one, keyed by a prime p with a large prime $q|(p-1)$, and two elements $g, h \in \mathbb{Z}_p^*$ of order q, computes $H_{p,g,h}(m) = g^{m_1}h^{m_2}$, where m_1 and m_2 are two halves of m (here we think of m as an element of $Z_q \times Z_q$) [3]. The second one, keyed by a product n of two primes $p_1 \equiv 3 \pmod 8$, and $p_2 \equiv 7 \pmod 8$ and a value $r \in \mathbb{Z}_n^*$, computes $H_{n,r}(m) = 4^m r^{2^{|m|}} \bmod n$ [4].

The first hash function family is secure as long as discrete logarithm is hard. Thus, if one publishes the random coins used to generate p, g and h, the hash function remain secure (as long as the generation algorithm doesn't do anything esoteric, such as computing h as a random power of g). On the other hand, the second hash function family is secure based on factoring, and is entirely insecure if the factors of n are known. Thus, publishing the random coins used to generate p_1 and p_2 renders the hash function insecure, and the coins must be kept secret [5].

Definitions. We say that a function is *negligible* if it vanishes faster than any inverse polynomial. We let PPTM stand for a probabilistic polynomial-time Turing machine. We use $M^?$ to denote an oracle Turing machine, and M^A to denote M instantiated with oracle A.

Let k be the security parameter, and let ℓ be a (length) function that does not expand or shrink its input more than a polynomial amount. Below we define two kinds of CRHFs: namely, secret-coin and public-coin. The secret-coin CRHFs definition is originally due to Damgård [Dam87], and the definition here is adapted from [Rus95].

Definition 1. *A* Secret-Coin *Collision Resistant Hash Family is a collection of functions* $\{h_i\}_{i \in I}$ *for some index set* $I \subseteq \{0,1\}^*$, *where* $h_i : \{0,1\}^{|i|+1} \to \{0,1\}^{|i|}$, *and*

1. *There exists a PPTM* GEN, *called the generating algorithm, so that* GEN$(1^k) \in \{0,1\}^{\ell(k)} \cap I$.
2. *There exists a PPTM* EVA, *called the function evaluation algorithm, so that* $\forall i \in I$ *and* $\forall x \in \{0,1\}^{|i|+1}$, EVA$(i,x) = h_i(x)$.
3. *For all PPTM* ADV, *the probability that* ADV(i) *outputs a pair* (x,y) *such that* $h_i(x) = h_i(y)$ *is negligible in* k, *where the probability is taken over the random choices of* GEN *in generating* i *and the random choices of* ADV.

---

[3] This family is derived from Pedersen commitments [Ped91].

[4] This is essentially the construction of [Dam87] based on the claw-free permutations of [GMR88].

[5] It should be noted, of course, whether it is secure to publish the coins depends not only on the family, but also on the key generating algorithm itself: indeed, the first family can be made insecure if the coins are used to generate h as a power of g, rather than pick h directly. Likewise, the second family could be made secure if it were possible to generate n "directly," without revealing p_1 and p_2 (we are not aware of an algorithm to do so, however).

Definition 2. *A* Public-Coin *Collision Resistant Hash Family is a collection of functions* $\{h_i\}_{i \in \{0,1\}^*}$, *where* $h_i : \{0,1\}^{\ell(|i|)+1} \to \{0,1\}^{\ell(|i|)}$, *and*

1. *A PPTM* GEN *on input* 1^k *outputs a uniformly distributed string* i *of length* k.
2. *There exists a PPTM* EVA, *called the function evaluation algorithm, so that* $\forall i \in \{0,1\}^*$ *and* $\forall x \in \{0,1\}^{\ell(|i|)+1}$, EVA$(i,x) = h_i(x)$.
3. *For all PPTM* ADV, *the probability that* ADV(i) *outputs a pair* (x,y) *such that* $h_i(x) = h_i(y)$ *is negligible in* k, *where the probability is taken over the random choices of* GEN *in generating* i *and the random choices of* ADV.

A pair (x,y) such that $h_i(x) = h_i(y)$ is called a *collision* for h_i.

Remarks. The generating algorithm in the public-coin case is trivially satisfied. We keep it here for comparison with the secret-coin case. Note that in both cases, on security parameter k, GEN outputs a function that maps $\{0,1\}^{\ell(k)+1}$ to $\{0,1\}^{\ell(k)}$. This may seem restrictive as the hash functions only compress one bit. However, it is easy to see that h_i can be extended to $\{0,1\}^n$ for any n, and remain collision-resistant with $\ell(k)$-bit outputs, by the following construction: $h_i^*(x) = h_i(\ldots h_i(h_i(h_i(x_1 \circ x_2 \circ \ldots \circ x_{\ell(k)+1}) \circ x_{\ell(k)+2}) \circ x_{\ell(k)+3}) \ldots \circ x_n)$, where x_j denotes the j-th bit of the input string x.

3 Dense Secret-Coin CRHFs Imply Public-Coin CRHFs

The notion of *dense* public-key cryptosystems was introduced by De Santis and Persiano in [DP92]. By "dense" they mean that a uniformly distributed string, with some noticeable probability, is a secure public key. We adapt the notion of denseness in public-key cryptosystems from [DP92] to the context of CRHFs. Informally, a d-dense secret-coin CRHF is a secret-coin CRHF with the following additional property: if we pick a k-bit string at random, then we have probability at least k^{-d} of picking an index i for a *collision-resistant* function[6].

Note that, for example, the factoring-based secret-coin CRHF from Section 2 is dense, because the proportion of k-bit integers that are products of two equal-length primes is $\Theta(k^{-2})$. In fact, we are not aware of any natural examples of secret-coin CRHFs that are not dense (artificial examples, however, are easy to construct).

Given a d-dense secret-coin CRHF, if we pick k^{d+1} strings of length k at random, then with high probability, at least one of them defines a collision-resistant hash function.

Hence, we can build a public-coin CRHF from such dense secret-coin CRHF as follows.

---

[6] Confusingly, sometimes the term dense is used to denote a function family where each function has a dense domain, e.g., [Hai04]. This is unrelated to our use of the term.

1. Generate k^{d+1} random k-bit strings, independently. These strings specify k^{d+1} hash functions $h_1, h_2, \ldots h_{k^{d+1}}$ in the secret-coin CRHF (strictly speaking, some strings may not define functions at all, because they are not produced by GEN; however, simply define $h_i(x) = 0^{\ell(k)}$ if EVA(i, x) does not produce an output of length k in the requisite number of steps).
2. Through the construction described in Section 2, extend the domain of each of these function to binary strings of length $\ell(k)k^{d+1} + 1$. Let the resulting functions be $h_1^*, \ldots, h_{k^{d+1}}^*$.
3. On an input x of length $\ell(k)k^{d+1} + 1$, output concatenation of $h_1^*(x), h_2^*(x), \ldots, h_{k^{d+1}}^*(x)$.

The resulting hash maps binary strings of length $\ell(k)k^{d+1} + 1$ to binary strings of length $\ell(k)k^{d+1}$, and is collision-resistant because at least one of $h_1^*, h_2^*, \ldots, h_{k^{d+1}}^*$ is. (If an adversary could find a collision in the resulting hash function, then the same collision would work for collision-resistant hash function among $h_1^*, h_2^*, \ldots, h_{k^{d+1}}^*$, immediately leading to a contradiction.)

The above discussion yields the following theorem.

Theorem 1. *The existence of dense secret-coin CRHF implies the existence of public-coin CRHF.*

4 Separating Public-Coin CRHFs from Secret-Coin CRHFs

4.1 Black-Box Reductions

Impagliazzo and Rudich [IR89] provided an informal definition of black-box reductions, and Gertner et al. [GKM+00] formalized it. We recall their formalization.

Definition 3. *A black-box reduction from primitive P to primitive Q consists of two oracle PPTMs M and A_Q satisfying the following two conditions:*

If Q can be implemented, so can P: *$\forall N$ (not necessarily PPTM) implementing Q, M^N implements P; and*
If P is broken, so is Q: *$\forall A_P$ (not necessarily PPTM) breaking M^N (as an implementation of P), $A_Q^{A_P, N}$ breaks N (as an implementation of Q).*

The first condition is only a functional requirement; i.e., the term "implement" says nothing about security, but merely says an algorithm satisfies the syntax of the primitive.

4.2 The Main Result

Theorem 2. *There is no black-box reduction from public-coin CRHF to secret-coin CRHF.*

Proof. The following proposition is at the heart of our approach: it shows that it is sufficient to construct different oracles F and G, such that G is used in the *implementations*, while F and G are used for the *adversaries*. This is in contrast to the single-oracle approach usually taken to prove black-box separations.

Proposition 1. *To show that there is no black-box reduction from* public-coin *collision resistant hashing (P) to* secret-coin *collision resistant hashing (Q), it suffices to construct two oracles* F *and* G *such that,*

1. *there is an oracle PPTM L such that $N = L^{\mathsf{G}}$ implements secret-coin hashing;*
2. *for all oracle PPTM M, if M^{G} implements public-coin hashing, then there exists a probabilistic polynomial time adversary A such that $A_P = A^{\mathsf{F}}$ finds a collision for M^{G};*
3. *there is no oracle PPTM B such that $B^{\mathsf{F},\mathsf{G}}$ finds a collision for N.*

Proof. To show that there is no black-box reduction from *public-coin* collision resistant hashing (P) to *secret-coin* collision resistant hashing (Q), we need to negate the definition of black-box reduction from Section 2; i.e., we need to show that for every oracle PPTMs M and A_Q,

Q **can be implemented:** $\exists N$ that implements Q, and if M^N implements P, then

P **can be broken, without breaking** *Q*: $\exists A_P$ that breaks M^N (as an implementation of P), while $A_Q^{A_P, N}$ does not break N (as an implementation of Q).

Recall that "implement" here has only functional meaning.

The first condition clearly implies that Q can be implemented. The second condition also clearly implies that P can be broken: one simply observes that $M^N = M^{L^{\mathsf{G}}}$, and L is a PPTM; hence, writing M^{G} is equivalent to writing M^N. The third condition implies that P can be broken without breaking Q, essentially because Q can never be broken. More precisely, the third condition is actually stronger than what we need: all we need is that for each A_Q, there is A_P that breaks M^N, while $A_Q^{A_P, N}$ does not break N. Instead, we will show that a single A_P essentially works for all A_Q: namely, $A_P = A^{\mathsf{F}}$, for a fixed oracle F and a polynomial-time A. Such A_P breaks M^N; however, as condition 3 in the proposition statement implies, $A_Q^{A_P, N}$ will be unable to break N, because $A_Q^{A_P, N} = A_Q^{A^{\mathsf{F}}, L^{\mathsf{G}}} = B^{\mathsf{F},\mathsf{G}}$ for some oracle PPTM B.

Remarks. Note that if the implementation has access to not only G but also F, it becomes the usual single-oracle separation. The reason why we do not give the implementation access to F is to avoid "self-referencing" when defining F. To see this, note that F is the "collision finder" and is defined according to the oracles that the implementation has access to[7].

---

[7] Similar concern occurs in [Sim98], where constructing the collision-finder requires more careful design.

The rest of this section is devoted to constructing such F and G and proving that they work.

4.3   The Oracles F and G

In constructing F and G, we will use the Borel-Cantelli Lemma (see, e.g., [AG96]), which states that if the sum of the probabilities of a sequence of events converges, then the probability that infinitely many of these events happen is zero. Formally,

Lemma 1 (Borel-Cantelli Lemma). *Let* $B_1, B_2, \ldots$ *be a sequence of events on the same probability space. Then* $\sum_{n=1}^{\infty} \Pr[B_n] < \infty$ *implies* $\Pr[\bigwedge_{k=1}^{\infty} \bigvee_{n \geq k} B_n] = 0$.

We first construct "random" F (collision-finder) and G (secret-coin hash), and then use the above lemma to show that at least one pair of F and G works.

Intuitively, we want F to break any public-coin hashing but not break some secret-coin hashing. More precisely, F will find a collision if it is supplied with the coins of the generating algorithm and will refuse to do so without the coins.

- G consists of two collections of functions $\{g_i\}_{i \in \mathbb{N}}$ and $\{h_\alpha\}_{\alpha \in \{0,1\}^*}$, where each g_i is a random function from $\{0,1\}^i$ to $\{0,1\}^{2i}$. We will call a binary string *valid* if it is in the range of g, and *invalid* if not. Each h_α is a random function from $\{0,1\}^{|\alpha|+1}$ to $\{0,1\}^{|\alpha|}$ if α is valid, and is a constant function $0^{|\alpha|}$ if α is invalid. We will call queries to h_α valid (resp. invalid) if α is valid (resp. invalid).
- F takes a deterministic oracle machine $M^?$ and 1^ℓ as input, and outputs a collision of length $\ell + 1$ for M^G if M^G satisfies the following conditions.
  1. M^G maps $\{0,1\}^{\ell+1}$ to $\{0,1\}^\ell$.
  2. M^G never queries h_α for some α not obtained by previously querying g. I.e., whenever M^G queries h_α, this α is the answer to some g-query that M^G has previously asked.

When both conditions hold, F picks a random x from $\{0,1\}^{\ell+1}$ that has a collision, then a random y ($\neq x$) that collides to x (i.e., $M^G(x) = M^G(y)$), and outputs (x, y). Otherwise F outputs $\bot$.

Observe that when F outputs (x, y), not only x, but also y is uniformly distributed over all points that have a collision. Indeed, let C be the total number of points that have a collision, and suppose y has c collisions $(x_1, x_2, \ldots, x_c)$: then $\Pr[y \text{ is chosen}] = \sum_{i=1}^{c} 1/c \Pr[x_i \text{ is chosen}] = 1/c \cdot (c/C) = 1/C$.

Remarks. The reason for g being length-doubling is to have a "sparse" function family. More specifically, it should be hard to get a value in the range of g without applying it.

As in [Sim98], there are various ways of constructing F (the collision-finding oracle): one can choose a random pair that collides, or a random x then a random y (possibly equal to x) that collides to x. The second construction has the advantage, in analysis, that both x and y are uniformly distributed but does not always give a "correct" collision, like the first one does. Our F has both properties.

4.4   Secret-Coin Collision-Resistant Hash Family Based on G

In this section we construct a secret-coin CRHF. The construction is straight-forward given the oracle G: the generating algorithm uses g and the hashing uses h. More precisely, on input 1^k the generating algorithm picks a random seed $r \in \{0,1\}^k$ and outputs $\alpha = g_k(r)$. The hash function is h_α. Note that the adversary A (who is trying to find a collision) is given only α but not r. We will show that for measure one of oracles F and G, the probability over r and A's coin tosses that A finds a collision for h_α is negligible. Recall that A has access to both F and G.

Define D as the event that A outputs a collision for h_α in the following experiment:

$$r \leftarrow_R \{0,1\}^k, \ \alpha \leftarrow g_k(r), \ (x,y) \leftarrow A^{\mathsf{F},\mathsf{G}}(\alpha).$$

And in the same experiment, define B as the event that during its computation, A queries F on $M^?$, where $M^?$ is some deterministic oracle machine that queries its oracle on a preimage of α under g_k (i.e., intuitively, $M^?$ has r hardwired in it). Suppose A's running time is bounded by k^c for some constant c. The probability that B happens is at most the probability of inverting the random function g_k. If α has a unique preimage, this is at most $k^c/2^k$; the probability that α has two or more preimages is at most $1/2^k$ (because it's the probability that r collides with another value under g_k); hence $\mathbf{Pr}[B] \leq (k^c+1)/2^k$. The probability that D happens conditioned on $\neg B$ is at most the probability of finding a collision for random function h_α, which is bounded by $k^{2c}/2^{2k}$. Recall that A can be randomized. We thus have

$$\begin{aligned}
\mathbf{Pr}_{\mathsf{F},\mathsf{G},r,A}[D] &= \mathbf{Pr}[B] \cdot \mathbf{Pr}[D|B] + \mathbf{Pr}[\neg B] \cdot \mathbf{Pr}[D|\neg B] \\
&\leq \mathbf{Pr}[B] + \mathbf{Pr}[D|\neg B] \\
&\leq (k^c+1)/2^k + k^{2c}/2^{2k} \\
&\leq 2k^c/2^k .
\end{aligned}$$

By the Markov inequality, $\mathbf{Pr}_{\mathsf{F},\mathsf{G}}[\mathbf{Pr}_{r,A}[D] \geq k^2 \cdot 2k^c/2^k] \leq 1/k^2$. Since $\sum_k 1/k^2$ converges, the Borel-Cantelli lemma implies that for only measure zero of F and G, can there be infinitely many k for which event D happens with probability (over r and A's coins) greater than or equal to $k^{c+2}/2^{k-1}$. This implies that for measure one of F and G, event D happens with probability (over r and A's coins) smaller than $k^{c+2}/2^{k-1}$ (a negligible function) for all large enough k. There are only countably many adversaries A, so we have the following lemma.

Lemma 2. *For measure one of F and G, there is a CRHF using G, which is secure against adversaries using G and F.*

4.5   No Public-Coin Collision-Resistant Hash Family Based on G

In this section we show that any implementation of public-coin hashing using oracle G cannot be collision-resistant against adversaries with oracle access to both F and G [8]. More precisely, let $r \in \{0,1\}^k$ be the public randomness used by the generating algorithm for a family of hash functions, and let $M^?$ be the evaluation algorithm. I.e., $M^G(r, \cdot)$ is the hash function specified by r. Assume that $M_r^G(\cdot) \triangleq M^G(r, \cdot)$ maps $\{0,1\}^{\ell(k)+1}$ to $\{0,1\}^{\ell(k)}$, where ℓ is a function that does not expand or shrink the input by more than a polynomial amount. We will show how to find x and y of length $\ell(k) + 1$ such that $M_r^G(x) = M_r^G(y)$.

An immediate attempt is to query $F(M_r^?, 1^{\ell(k)})$, but notice that M_r^G may query h_α for arbitrary α [9], which prevents F from finding a collision for us. However, these α are likely to be invalid, and hence oracle answers to these queries are likely to be $0^{|\alpha|}$. So we can construct a machine $\tilde{M}_r^?$ that behaves "similar" to $M_r^?$ but only after getting α from g does it query h_α. And instead of finding collision for M_r^G, we find collision for $\tilde{M}_r^G$, which can be done by simply querying $F(\tilde{M}_r^?, 1^{\ell(k)})$.

Suppose the running time of M_r^G is bounded by k^c for some constant $c > 1$. Before simulating M_r^G, $\tilde{M}_r^G$ queries g on all inputs of length smaller than or equal to $4c \log k$. This takes $2k^{4c}$ steps. Now $\tilde{M}_r^G$ simulates M_r^G step by step, except for queries to h_α. If α is the answer to one of the queries $\tilde{M}_r^G$ already asked of G (either before the beginning of the simulation or when simulating M_r^G), then $\tilde{M}_r^G$ actually queries h_α. Else it returns $0^{|\alpha|}$ as the answer to M_r^G without querying h_α.

Now fix r and x. For every $M^?$ the probability, over random G, that $\tilde{M}_r^G(x) \neq M_r^G(x)$ is at most the probability, over G, that M_r^G queries h_α for some valid α of length greater than $8c \log k$ without receiving it from g [10]. Consider the very first time that M_r^G makes such a "long" valid query. Let n_g be the number of queries to g on inputs longer than $4c \log k$, and n_h be the number of invalid queries to h prior to this point. Then the probability in question is upper bounded by $k^c \cdot \frac{k^{4c} - n_g - n_h}{k^{8c} - n_g}$, which is at most $1/k^{3c}$. For every fixed G and r, call an x "bad" if $\tilde{M}_r^G(x) \neq M_r^G(x)$. We have

$$\mathbf{Ex}_{G}[\mathbf{Pr}_{x}[x \text{ is bad}]] = \mathbf{Pr}_{G,x}[x \text{ is bad}] \leq 1/k^{3c}.$$

Next, notice that there are at most half of x that have no collisions, and F would pick its answer (x_F, y_F), uniformly, from those points that have a collision. So for a fixed G, the probability over F that x_F is bad is at most twice the probability over random $x \in \{0,1\}^{\ell(k)+1}$ that x is bad. Also recall that the distribution of y_F is the same as x_F. So for every $M^?$,

$$\mathbf{Ex}_{G}[\mathbf{Pr}_{F}[\text{at least one of } (x_F, y_F) \text{ is bad}]] \leq 4 \cdot \mathbf{Ex}_{G}[\mathbf{Pr}_{x}[x \text{ is bad}]].$$

If none of (x_F, y_F) is bad, this pair would be a collision not only for $\tilde{M}_r^G$ but also for M_r^G. We have

$$\mathbf{Pr}_{F,G,r}[(x_F, y_F) \text{ is not a collision of } M_r^G] \leq 4 \mathbf{Pr}_{G,x,r}[x \text{ is bad}] \leq 4/k^{3c},$$

---

[8] In fact, only F is needed to find a collision.

[9] In particular, those α not obtained by previously querying g.

[10] Recall that g is length-doubling.

then

$$\Pr_{\mathsf{F},\mathsf{G}}[\Pr_r[(x_\mathsf{F}, y_\mathsf{F}) \text{ is not a collision of } M_r^\mathsf{G}] \geq 4/k^c] \leq 1/k^{2c}.$$

Since $\sum_k 1/k^{2c}$ converges, the Borel-Cantelli lemma implies that for only measure zero of F and G, can we have $\Pr_r[(x_\mathsf{F}, y_\mathsf{F})$ is not a collision of $M_r^\mathsf{G}] \geq 4/k^c$ for infinitely many k. In other words, for measure one of F and G, $\Pr_r[(x_\mathsf{F}, y_\mathsf{F})$ is a collision of $M_r^\mathsf{G}] \geq 4/k^c$ for all large enough k. There are only countably many oracle machines $M^?$, each of which can be collision resistant for only measure zero of F and G. We conclude the following.

Lemma 3. *For measure one of F and G, any implementation of public-coin hash function families using G cannot be collision-resistant against adversaries using F.*

This concludes the proof of Theorem 2.

5   Public Coins vs. Secret Coins for Other Primitives

Perhaps the lack of attention in the literature to the distinction between secret- and public-coin primitives is due, in part, to the fact that this distinction is often not meaningful.

For example, for one-way function families, these two notions are equivalent, because a secret-coin one-way function family implies a single one-way function (which trivially implies a public-coin one-way function family). Indeed, take the generating algorithm g and evaluation algorithm f and define $F(r, x) \triangleq (g(r), f_{g(r)}(x))$; this is one-way because an adversary who can come up with (r', x') such that $g(r) = g(r')$ and $f_{g(r')}(x') = f_{g(r)}(x)$ can be directly used to invert $f_{g(r)}(x)$, since $f_{g(r)}(x') = f_{g(r')}(x') = f_{g(r)}(x)$.

On the other hand, for trapdoor permutations (and public-key schemes), the notion of public-coin generation is meaningless: indeed the trapdoor (or the secret key) must be kept secret.

However, it seems that this distinction is interesting for some primitives in addition to collision-resistant hash functions. The relationships between public-coin and secret-coin versions of one-way permutation families and claw-free permutation families are unknown[11]. In particular, claw-free permutations are related to collision-resistant hashing [Dam87,Rus95], which suggests that the distinction for claw-free permutations is related to the distinction for CRHFs.

Acknowledgments.   We thank Yael Tauman Kalai for many helpful discussions, and Ron Rivest for assistance with the history of hashing. Thanks also to the anonymous referees for insightful comments. This work was funded in part by the National Science Foundation under Grant No. CCR-0311485.

---

[11] We believe that the same construction of F and G (up to slight modifications) separates public-coin and secret-coin one-way permutation families.

References

[AG96]   Malcolm Adams and Victor Guillemin. *Measure Theory and Probability.* Springer Verlag, 1996.

[BM84]   M. Blum and S. Micali. How to generate cryptographically strong sequences of pseudo-random bits. *SIAM Journal on Computing*, 13(4):850–863, November 1984.

[BR97]   Mihir Bellare and Phillip Rogaway. Collision-resistant hashing: Towards making uowhfs practical. In Burton S. Kaliski, Jr., editor, *Advances in Cryptology—CRYPTO '97*, volume 1294 of *Lecture Notes in Computer Science*, pages 470–484. Springer-Verlag, 17–21 August 1997.

[CHL02]   Yan-Cheng Chang, Chun-Yun Hsiao, and Chi-Jen Lu. On the imposibilities of basing one-way permutations on central cryptographic primitives. In Yuliang Zheng, editor, *Advances in Cryptology—ASIACRYPT 2002*, volume 2501 of *Lecture Notes in Computer Science*, pages 110–124, Queenstown, New Zealand, 1–5 December 2002. Springer-Verlag.

[Dam87]   Ivan Damgård. Collision-free hash functions and public-key signature schemes. In David Chaum and Wyn L. Price, editors, *Advances in Cryptology—EUROCRYPT 87*, volume 304 of *Lecture Notes in Computer Science*. Springer-Verlag, 1988, 13–15 April 1987.

[DP92]   Alfredo De Santis and Giuseppe Persiano. Zero-knowledge proofs of knowledge without interaction. In *33rd Annual Symposium on Foundations of Computer Science*, pages 427–436, Pittsburgh, Pennsylvania, 24–27 October 1992. IEEE.

[GK]   Shafi Goldwasser and Yael Tauman Kalai. On the (in)security of the Fiat-Shamir paradigm. Available From http://www.mit.edu/~tauman/.

[GK03]   Shafi Goldwasser and Yael Tauman Kalai. On the (in)security of the Fiat-Shamir paradigm. In *44th Annual Symposium on Foundations of Computer Science* [IEE03].

[GKM+00]   Yael Gertner, Sampath Kannan, Tal Malkin, Omer Reingold, and Mahesh Viswanathan. The relationship between public key encryption and oblivious transfer. In *41st Annual Symposium on Foundations of Computer Science* [IEE00], pages 325–335.

[GMR88]   Shafi Goldwasser, Silvio Micali, and Ronald L. Rivest. A digital signature scheme secure against adaptive chosen-message attacks. *SIAM Journal on Computing*, 17(2):281–308, April 1988.

[GMR01]   Yael Gertner, Tal Malkin, and Omer Reingold. On the impossibility of basing trapdoor functions on trapdoor predicates. In *42nd Annual Symposium on Foundations of Computer Science*, Las Vegas, Nevada, October 2001.

[GT00]   Rosario Gennaro and Luca Trevisan. Lower bounds on the efficiency of generic cryptographic constructions. In *41st Annual Symposium on Foundations of Computer Science* [IEE00].

[Hai04]   Iftach Haitner. Implementing oblivious transfer using collection of dense trapdoor permutations. In Naor [Nao04], pages 394–409.

[IEE00]   IEEE. *41st Annual Symposium on Foundations of Computer Science*, Redondo Beach, California, November 2000.

[IEE03]   IEEE. *44th Annual Symposium on Foundations of Computer Science*, Cambridge, Massachusetts, October 2003.

[IR89]   Russell Impagliazzo and Steven Rudich. Limits on the provable conse-
             quences of one-way permutations. In *Proceedings of the Twenty First An-
             nual ACM Symposium on Theory of Computing*, pages 44–61, May 1989.
[Mer82]   Ralph C. Merkle. *Secrecy, Authentication, and Public Key Systems*. UMI
             Research Press, 1982.
[Mer89]   Ralph C. Merkle. A certified digital signature. In G. Brassard, editor,
             Advances in Cryptology—CRYPTO '89, volume 435 of *Lecture Notes in
             Computer Science*, pages 218–238. Springer-Verlag, 1990, 20–24 August
             1989.
[Mir01]   Ilya Mironov. Hash functions: From merkle-damgård to shoup. In Joe
             Kilian, editor, *Advances in Cryptology—CRYPTO 2001*, volume 2139 of
             Lecture Notes in Computer Science, pages 166–181. Springer-Verlag, Au-
             gust 2001.
[MRK03]   Silvio Micali, Michael Rabin, and Joe Kilian. Zero-knowledge sets. In *44th
             Annual Symposium on Foundations of Computer Science* [IEE03], pages
             80–91.
[Nao04]   Moni Naor, editor. *First Theory of Cryptography Conference*, volume 2951
             of *Lecture Notes in Computer Science*. Springer-Verlag, February 2004.
[NIS95]   FIPS publication 180-1: Secure hash standard, April 1995. Available from
             http://csrc.nist.gov/fips/.
[Ped91]   Torben Pryds Pedersen. Non-interactive and information-theoretic se-
             cure verifiable secret sharing. In J. Feigenbaum, editor, *Advances in
             Cryptology—CRYPTO '91*, volume 576 of *Lecture Notes in Computer Sci-
             ence*, pages 129–140. Springer-Verlag, 1992, 11–15 August 1991.
[Rab78]   Michael O. Rabin. Digitalized signatures. In Richard A. Demillo, David P.
             Dobkin, Anita K. Jones, and Richard J. Lipton, editors, *Foundations of
             Secure Computation*, pages 155–168. Academic Press, 1978.
[Rab79]   Michael O. Rabin. Digitalized signatures and public-key functions as
             intractable as factorization. Technical Report MIT/LCS/TR-212, Mas-
             sachusetts Institute of Technology, Cambridge, MA, January 1979.
[RTV04]   Omer Reingold, Luca Trevisan, and Salil Vadhan. Notions of reducibility
             between cryptographic primitives. In Naor [Nao04], pages 1–20.
[Rus95]   A. Russell. Necessary and sufficient conditions for collision-free hashing.
             Journal of Cryptology, 8(2):87–100, 1995.
[Sim98]   Daniel R. Simon. Finding collisions on a one-way street: Can secure hash
             functions be based on general assumptions. In Kaisa Nyberg, editor, *Ad-
             vances in Cryptology—EUROCRYPT 98*, volume 1403 of *Lecture Notes in
             Computer Science*. Springer-Verlag, May 31–June 4 1998.

Security of Random Feistel Schemes
with 5 or More Rounds

Jacques Patarin

Université de Versailles
45 avenue des Etats-Unis
78035 Versailles Cedex, France

Abstract. We study cryptographic attacks on random Feistel schemes.
We denote by m the number of plaintext/ciphertext pairs, and by k the
number of rounds. In their famous paper [3], M. Luby and C. Rackoff have
completely solved the cases $m \ll 2^{n/2}$: the schemes are secure against
all adaptive chosen plaintext attacks (CPA-2) when $k \geq 3$ and against
all adaptive chosen plaintext and chosen ciphertext attacks (CPCA-2)
when $k \geq 4$ (for this second result a proof is given in [9]).
In this paper we study the cases $m \ll 2^n$. We will use the "coefficients
H technique" of proof to analyze known plaintext attacks (KPA), adaptive or non-adaptive chosen plaitext attacks (CPA-1 and CPA-2) and
adaptive or non-adaptive chosen plaitext and chosen ciphertext attacks
(CPCA-1 and CPCA-2). In the first part of this paper, we will show
that when $m \ll 2^n$ the schemes are secure against all KPA when $k \geq 4$,
against all CPA-2 when $k \geq 5$ and against all CPCA-2 attacks when
$k \geq 6$. This solves an open problem of [1], [14], and it improves the result
of [14] (where more rounds were needed and $m \ll 2^{n(1-\varepsilon)}$ was obtained
instead of $m \ll 2^n$). The number 5 of rounds is minimal since CPA-2
attacks on 4 rounds are known when $m \geq O(2^{n/2})$ (see [1], [10]). Furthermore, in all these cases we have always obtained an explicit majoration
for the distinguishing probability. In the second part of this paper, we
present some improved generic attacks. For $k = 5$ rounds, we present a
KPA with $m \simeq 2^{3n/2}$ and a non-adaptive chosen plaintext attack (CPA-
1) with $m \simeq 2^n$. For $k \geq 7$ rounds we also show some improved attacks
against random Feistel generators (with more than one permutation to
analyze and $\geq 2^{2n}$ computations).

1 Introduction

A "Luby - Rackoff construction with k rounds", which is also known as a "random Feistel cipher" is a Feistel cipher in which the round functions $f_1, \ldots, f_k$
are independently chosen as truly random functions (see section 2 for precise
definitions).

Since the famous original paper [3] of M. Luby and C. Rackoff, these constructions have inspired a considerable amount of research. In [8] and [14] a
summary of existing works on this topic is given.

We will denote by k the number of rounds and by n the integer such that
the Feistel cipher is a permutation of $2n$ bits $\rightarrow 2n$ bits. In [3] it was proved

M. Franklin (Ed.): CRYPTO 2004, LNCS 3152, pp. 106–122, 2004.
© International Association for Cryptologic Research 2004

that when $k \geq 3$ these Feistel ciphers are secure against all adaptive chosen plaintext attacks (CPA-2) when the number of queries (i.e. plaintext/ciphertext pairs obtained) is $m \ll 2^{n/2}$. Moreover when $k \geq 4$ they are secure against all adaptive chosen plaintext and chosen ciphertext attacks (CPCA-2) when the number of queries is $m \ll 2^{n/2}$ (a proof of this second result is given in [9]).

These results are valid if the adversary has unbounded computing power as long as he does only m queries.

These results can be applied in two different ways: directly using k truly random functions $f_1, \ldots, f_k$ (that requires significant storage), or in a hybrid setting, in which instead of using k truly random functions $f_1, \ldots, f_k$, we use k pseudo-random functions. These two ways are both interesting for cryptography. The first way gives "locally random permutations" where we have proofs of security without any unproven hypothesis (but we need a lot of storage), and the second way gives constructions for block encryption schemes where the security can be relied on a pseudo-random number generator, or on any one-way function.

In this paper, we will study security when $m \ll 2^n$, instead of $m \ll 2^{n/2}$ for the original paper of M. Luby and C. Rackoff. For this we must have $k \geq 5$, since for $k \leq 4$ some CPA-2 attacks when $m \geq O(2^{n/2})$ exist (see [1], [10]). Moreover the bound $m \ll 2^n$ is the larger bound that we can get, since an adversary with unlimited computing power can always distinguish a k-round random Feistel scheme from a random permutation with $O(k \cdot 2^n)$ queries and $O(2^{kn2^n})$ computations by simply guessing all the round functions (it is also possible to do less computing with the same number of queries by using collisions, see [13]).

The bound $m \ll 2^{n/2}$ is called the 'birthday bound', i.e. it is about the square root of the optimal bound against an adversary with unbounded computing power. In [1] W. Aiello and R. Venkatesan have found a construction of locally random functions ('Benes') where the optimal bound ($m \ll 2^n$) is obtained instead of the birthday bound. However here the functions are not permutations. Similarly, in [4], U. Maurer has found some other construction of locally random functions (not permutations) where he can get as close as wanted to the optimal bound (i.e. $m \ll 2^{n(1-\epsilon)}$ and for all $\epsilon > 0$ he has a construction). In [8] the security of unbalanced Feistel schemes is studied and a security proof in $2^{n(1-\epsilon)}$ is obtained, instead of $2^{n/2}$, but for much larger round functions (from $2n$ bits to ϵ bits, instead of n bits to n bits). This bound is basically again the birthday bound for these functions.

In this paper we will show that 5-round random Feistel schemes resist all CPA-2 attacks when $m \ll 2^n$ and that 6-round random Feistel schemes resist all CPCA-2 attacks when $m \ll 2^n$. Here we are very near the optimal bound, and we have permutations. This solves an open problem of [1], [10]. It also significantly improves the results of [6] in which the 2^n security is only obtained when the number of rounds tends to infinity, and the result of [14] where $2^{n(1-\epsilon)}$ security was proved for CPA-2 after 7 rounds (instead of 5 here) and for CPCA-2 after 10 rounds (instead of 6 here). Moreover we will obtain in this paper some explicit and simple majorations for the distinguishing probabilities. We will also present some improved generic attacks. All these results are summarized in appendix A.

2   Notations

General notations

- $I_n = \{0,1\}^n$ denotes the set of the 2^n binary strings of length n. $|I_n| = 2^n$.
- The set of all functions from I_n to I_n is F_n. Thus $|F_n| = 2^{n \cdot 2^n}$.
- For any $f, g \in F_n$, $f \circ g$ denotes the usual composition of functions.
- For any $a, b \in I_n$, $[a, b]$ will be the string of length $2n$ of I_{2n} which is the concatenation of a and b.
- For $a, b \in I_n$, $a \oplus b$ stands for bit by bit exclusive or of a and b.
- Let f_1 be a function of F_n. Let L, R, S and T be four n-bit strings in I_n. Then by definition

$$\Psi(f_1)[L, R] = [S, T] \stackrel{\text{def}}{\Leftrightarrow} \begin{cases} S = R \\ T = L \oplus f_1(R) \end{cases}$$

- Let $f_1, f_2, \ldots, f_k$ be k functions of F_n. Then by definition:

$$\Psi^k(f_1, \ldots, f_k) = \Psi(f_k) \circ \cdots \circ \Psi(f_2) \circ \Psi(f_1).$$

The permutation $\Psi^k(f_1, \ldots, f_k)$ is called a 'Feistel scheme with k rounds' or shortly Ψ^k. When $f_1, \ldots, f_k$ are randomly and independently chosen in F_n, then $\Psi^k(f_1, \ldots, f_k)$ is called a 'random Feistel scheme with k rounds' or a 'Luby-Rackoff construction with k rounds'.

We will first study 4 rounds (with some limitations on the inputs/outputs), then prove our cryptographic results by adding one or two rounds.

Notations for 4 rounds

- We will denote by $[L_i, R_i]$, $1 \le i \le m$, the m cleartexts. These cleartexts can be assumed to be pairwise distinct, i.e. $i \ne j \Rightarrow L_i \ne L_j$ or $R_i \ne R_j$.
- We call "index" any integer between 1 and m.
- $[R_i, X_i]$ is the output after one round, i.e.

$$\forall i, 1 \le i \le m, X_i = L_i \oplus f_1(R_i).$$

- $[X_i, Y_i]$ is the output after two rounds, i.e.

$$\forall i, 1 \le i \le m, Y_i = R_i \oplus f_2(X_i) = R_i \oplus f_2(L_i \oplus f_1(R_i)).$$

- $[Y_i, S_i]$ is the output after three rounds, i.e.

$$\forall i, 1 \le i \le m, S_i = X_i \oplus f_3(Y_i) = L_i \oplus f_1(R_i) \oplus f_3(Y_i).$$

- $[S_i, T_i]$ is the output after 4 rounds, i.e.

$$\forall i, 1 \le i \le m, T_i = Y_i \oplus f_4(S_i).$$

Notations for 5 rounds. We keep the same notations for L_i, R_i, X_i, Y_i. Now $Z_i = X_i \oplus f_3(Y_i)$, and $[S_i, T_i]$ is still the output: $S_i = Y_i \oplus f_4(Z_i)$ and $T_i = Z_i \oplus f_5(S_i)$.

Part I: Security Results

3   The General Proof Strategy

We will first study the properties of 4-round schemes. Our result on 4-round schemes for proving KPA security will be:

Theorem 3.1 (4 rounds) *For random values* $[L_i, R_i]$, $[S_i, T_i]$, $1 \leq i \leq m$, *such that the* $[L_i, R_i]$, $1 \leq i \leq m$, *are pairwise distinct, with probability* $\geq 1 - \beta$ *we have:*

1. *the number H of* $(f_1, f_2, f_3, f_4) \in F_n^4$ *such that* $\forall i$, $1 \leq i \leq m$,

$$\Psi^4(f_1, f_2, f_3, f_4)[L_i, R_i] = [S_i, T_i]$$

   satisfies:

$$H \geq \frac{|F_n|^4}{2^{2nm}}(1 - \alpha).$$

2. α *and* β *can be chosen* $\ll 1$ *when* $m \ll 2^n$.

For 5 rounds, we will have:

Theorem 3.2 (5 rounds) *There are some values* $\alpha > 0$ *and* $\beta > 0$ *and there is a subset* $E \subset I_{2n}^m$ *such that:*

1. *for all pairwise distinct* $[L_i, R_i]$, $1 \leq i \leq m$, *and for all sequences* $[S_i, T_i]$, $1 \leq i \leq m$, *of E the number H of* $(f_1, f_2, f_3, f_4, f_5) \in F_n^5$ *such that* $\forall i$, $1 \leq i \leq m$,

$$\Psi^5(f_1, f_2, f_3, f_4, f_5)[L_i, R_i] = [S_i, T_i]$$

   satisfies:

$$H \geq \frac{|F_n|^5}{2^{2nm}}(1 - \alpha).$$

2. $|E| \geq (1 - \beta) \cdot 2^{2nm}$, *and* α *and* β *can be chosen* $\ll 1$ *when* $m \ll 2^{n(1-\varepsilon)}$, $\forall \varepsilon > 0$.

Remark

1. Here the set E does not depend on the $[L_i, R_i]$, and it will give security against CPA-2. If E depends on the $[L_i, R_i]$, we will obtain security against CPA-1 only.
2. Instead of fixing a set E, as in theorem 3.2, we can formulate a similar theorem in term of expectancy of the deviation of H from the average value (see[15]: there is a formulation for CPA-1 and another for CPA-2). From these formulas we will get security when $m \ll 2^n$.

For 6 rounds, we will have:

Theorem 3.3 (6 rounds) *There are some values* $\alpha > 0$ *and* $\beta > 0$ *and there is a subset* $E \subset I_n^{4m}$ *such that:*

1. *for all* $[L_i, R_i, S_i, T_i]$, $1 \leq i \leq m$, *of* E, *the number* H *of* $(f_1, f_2, f_3, f_4, f_5, f_6) \in F_n^6$ *such that* $\forall i$, $1 \leq i \leq m$,

$$\Psi^6(f_1, f_2, f_3, f_4, f_5, f_6)[L_i, R_i] = [S_i, T_i]$$

   satisfies:

$$H \geq \frac{|F_n|^6}{2^{2nm}}(1 - \alpha).$$

2. *For all super distinguishing circuit* Φ *with* m *oracle gates, the probability that* $[L_i, R_i, S_i, T_i](\Phi)$, $1 \leq i \leq m$, *be in* E *is* $\geq 1 - \beta$, *when* Φ *acts on a random permutation* f *of* $I_{2n} \rightarrow I_{2n}$ *(here* $[L_i, R_i, S_i, T_i](\Phi)$, $1 \leq i \leq m$, *denotes the successive* $[S_i, T_i] = f[L_i, R_i]$ *or* $[L_i, R_i] = f^{-1}[S_i, T_i]$, $1 \leq i \leq m$, *that will appear).*

3. α *and* β *can be chosen* $\ll 1$ *when* $m \ll 2^n$.

Now from these theorems and from the general "coefficients H technique" theorems given in [11], [12], we will get immediately that when $m \ll 2^n$, Ψ^4 is secure against all KPA, Ψ^5 against all CPA-2 and Ψ^6 against all CPCA-2.

4   Circles

One of the terms of the the deviation of Ψ^k from random permutations will be the probability to get "circles" in the variables, as we will explain below.

Definition. We will say that we have 'a circle in R, X, Y' if there are k indices $i_1, \ldots, i_k$ with $k \geq 3$ and such that:

1. $i_1, i_2, \ldots, i_{k-1}$ are pairwise distinct and $i_k = i_1$.
2. $\forall \lambda$, $1 \leq \lambda \leq k - 2$ we have at least one of the three following conditions:
   - $R_{i_\lambda} = R_{i_{\lambda+1}}$ and $(X_{i_{\lambda+1}} = X_{i_{\lambda+2}}$ or $Y_{i_{\lambda+1}} = Y_{i_{\lambda+2}})$

   or • $X_{i_\lambda} = X_{i_{\lambda+1}}$ and $(R_{i_{\lambda+1}} = R_{i_{\lambda+2}}$ or $Y_{i_{\lambda+1}} = Y_{i_{\lambda+2}})$

   or • $Y_{i_\lambda} = Y_{i_{\lambda+1}}$ and $(R_{i_{\lambda+1}} = R_{i_{\lambda+2}}$ or $X_{i_{\lambda+1}} = X_{i_{\lambda+2}})$

Example. If $R_1 = R_2$ and $X_1 = X_2$, then we have a circle in R, X, Y. If $R_1 = R_2$, $X_2 = X_3$, $Y_3 = Y_1$ then we have a circle in R, X, Y.

We will prove the following theorems.

Theorem 4.1 (For 4 rounds) *When* $[L_i, R_i]$, $1 \leq i \leq m$, *are pairwise distinct and randomly chosen, the probability* p *to obtain a circle in* R, X, Y *with at least one equation in* Y *when* f_1, f_2 *are randomly chosen in* F_n *satisfies:*

$$p \leq \frac{3m^2}{2 \cdot 2^{2n}} + \frac{3m^3}{2^{3n}} \cdot \frac{1}{1 - \frac{2m}{2^n}}.$$

Theorem 4.2 (For 5 rounds) *For all pairwise distinct* $[L_i, R_i]$, $1 \leq i \leq m$ *and for all value* λ, *such that* $\lambda > 0$ *and* $2m\sqrt{\lambda} < 2^n$, *we have: the probability* p *to obtain a circle in* X, Y, Z *with at least one equation* $Z_i = Z_j$ *when* f_1, f_2, f_3 *are randomly chosen in* F_n *satisfies:*

$$p \leq \frac{1}{\lambda} + \frac{m(m-1)}{2 \cdot 2^{2n}} + \frac{m(m-1)(m-2)}{2^{3n}} + \frac{4\lambda m^4}{2^{4n}} \cdot \frac{1}{1 - \frac{2m\sqrt{\lambda}}{2^n}}.$$

Corollary 4.1 *From this theorem 4.2 we get immediately that if* $m \ll 2^n$, *then* (λ *can be chosen such that*), p *is very small. So when* $m \ll 2^n$, *the probability to have a circle in* X, Y, Z *with at least one equation* $Z_i = Z_j$ *is negligible.*

Remark. In [15] we show that the condition 'with at least one equation $Z_i = Z_j$' is important: sometime we cannot avoid some circles in X, Y.

With 6 rounds, we can get a simpler formula:

Theorem 4.3 (For 6 rounds) *For all* $[L_i, R_i]$, $1 \leq i \leq m$ *(such that* $i \neq j \Rightarrow L_i \neq L_j$ *or* $R_i \neq R_j$), *the probability* p *to obtain a circle in* X, Y, Z *with at least one equation in* Z *when* f_1, f_2, f_3, f_4 *are randomly chosen in* F_n *satisfies:*

$$p \leq \frac{3m^2}{2^{2n}} + \frac{11m^3}{2^{3n}} \cdot \frac{1}{1 - \frac{2m}{2^n}}.$$

Proof of theorem 4.1, 4.2, 4.3 are given in the extended version of this paper ([15]). A basic tool for these proofs is:

Theorem 4.4 $\forall \lambda > 0$, *for all pairwise distinct* $[L_i, R_i]$, $1 \leq i \leq m$, *when* f_1 *is randomly chosen in* F_n *we have a probability* $\geq 1 - \frac{1}{\lambda}$ *that the number* N *of* (i, j), $i < j / X_i = X_j$ *satisfies:*

$$N \leq \frac{\lambda m(m-1)}{2 \cdot 2^n}.$$

Proof. This result comes immediately from this lemma:

Lemma 4.1 *For all* $[L_i, R_i]$, $1 \leq i \leq m$, *(such that* $i \neq j \Rightarrow L_i \neq L_j$ *or* $R_i \neq R_j$) *the number of* (f_1, i, j) *such that* $X_i = X_j$, $i < j$, *is* $\leq |F_n| \cdot \frac{m(m-1)}{2 \cdot 2^n}$.

Proof of lemma 4.1. $X_i = X_j$ means $L_i \oplus f_1(R_i) = L_j \oplus f_1(R_j)$. This implies $R_i \neq R_j$ (because $L_i = L_j$ and $R_i = R_j \Rightarrow i = j$). Thus, when (i, j) is fixed, the number of f_1 such that $X_i = X_j$ is exactly $\frac{|F_n|}{2^n}$ if $R_i \neq R_j$, and exactly 0 if $R_i = R_j$. Therefore, since we have at most $m(m-1)/2$ values (i, j), $i < j / R_i \neq R_j$, the total number of (f_1, i, j) such that $X_i = X_j$ is $\leq |F_n| \frac{m(m-1)}{2 \cdot 2^n}$ as claimed.

5   Properties of H with 4 Rounds

We give here the main ideas. See the extended version of this paper for more details ([15]). We will first prove that if the $[Y_i, S_i]$ are given, $1 \leq i \leq m$, (i.e. the output after 3 rounds), then the S_i variables will look random as long as $m \ll 2^n$ (but the Y_i variables will not look random in general). Then, with one more round and the same argument, we will obtain that the $[S_i, T_i]$ variables will look random as long as $m \ll 2^n$. We want to evaluate the number H of f_1, f_2, f_3 such that: $\forall i, 1 \leq i \leq m, S_i = L_i \oplus f_1(R_i) \oplus f_3(Y_i)$ with $Y_i = R_i \oplus f_2(L_i \oplus f_1(R_i))$ (1).

Remarks

1. If $Y_i = Y_j$ with $i \neq j$, then $S_i \neq S_j$. So the S_i variables are not perfectly random in I_n when the Y_i are given. However, here we just say that the $[Y_i, S_i]$ must be pairwise distinct, since Ψ^k is a permutation.
2. If S_i is a constant ($\forall i, 1 \leq i \leq m, S_i = 0$ for example), then all the Y_i variables must be pairwise distinct, and in (1) f_3 is then fixed on exactly m points. However the probability for f_1, f_2 to be such that all the Y_i are pairwise distinct is very small. So in this case $H \ll \frac{|F_n|^3}{2^{nm}}$.
3. Let us consider that instead of (1) we had to evaluate the number J of f_1, f_2, f_3 such that $\forall i, 1 \leq i \leq m, S_i = f_3(Y_i)$ with $Y_i = R_i \oplus f_2(L_i \oplus f_1(R_i))$ (i.e. here we do not have the term $L_i \oplus f_1(R_i)$). Then, for random L_i, R_i and for random f_1, f_2, f_3, we will have about 2 times more collisions $S_i = S_j$ compared with a random variable S_i. So if S_i is random, $J \ll \frac{|F_n|^3}{2^{nm}}$ in this case. For (1) we will prove (among other results) that, unlike here for J, when the S_i are random, we always have $H \simeq \frac{|F_n|^3}{2^{nm}}$.

Analysis of (1). (In appendix B an example is given on what we do here) We will consider that all the Y_i are given (as well as the L_i, R_i, S_i), and we want to study how H can depend on the values S_i. If H has almost always the same value for all the S_i, then (by summation on all the Y_i) we will get $H \simeq \frac{|F_n|^3}{2^{nm}}$, and for all $[L_i, R_i]$ the S_i will look random, as wanted, when f_1, f_2, f_3 are randomly chosen in F_n (this is an indirect way to evaluate H).

In (1), when we have a new value Y_i, whatever S_i is, f_3 is exactly fixed on this point Y_i by (1). However if Y_i is not a new value, we have $Y_i = Y_j \Rightarrow L_i \oplus f_1(R_i) = L_j \oplus f_1(R_j) \oplus S_i \oplus S_j$. For each equation $Y_i = Y_j$, we will introduce a value $\lambda_{k(i,j)} = S_i \oplus S_j$. We want to evaluate the number H' of (f_1, f_2) such that: $\forall i, 1 \leq i \leq m, f_2(L_i \oplus f_1(R_i)) = R_i \oplus Y_i$ (2).

We will fix the points (i, j) where $X_i = X_j$, i.e. we look for solutions (f_1, f_2) such that $X_i = X_j$ exactly on these (i, j), and, again, we want to evaluate how the number H' of (f_1, f_2) can depend on the values S_i (i.e. on the values λ_k).

We will group the equations (2) by the same $f_1(R_i)$, i.e. by "blocks in R, X, Y": two indices i and j are in the same block if we can go from i to j by equations $R_k = R_l$, or $X_k = X_l$, or $Y_k = Y_l$ (Since $X_k = X_l \Rightarrow f_1(R_k) = f_1(R_l) \oplus L_k \oplus L_l$ and $Y_k = Y_l \Rightarrow f_1(R_k) = f_1(R_l) \oplus L_k \oplus L_l \oplus \lambda_{k(i,j)}$, from these relations, we can replace the variable $f_1(R_k)$ by the variable $f_1(R_l)$ instead).

Finally, the only dependencies on the λ_k come when we want to evaluate the number H'' of f_1 such that: $\forall i$, $1 \leq i \leq \alpha$, X_i are pairwise distinct, where α is the number of X_i that we want pairwise distinct (if wanted we can assume $\alpha \leq O\left(\frac{m^2}{2^n}\right)$ since variables with no equation in R, X or Y create no problem). Each X_i has an expression like this: $X_i = f_1(R_j) \oplus \lambda_k \oplus L'_l$ (where L'_l is an expression in $\oplus$ of some L_i values), or like this: $X_i = f_1(R_j) \oplus L'_l$. This gives a number of solutions for f_1 that depends only of the fact that some equations of degree one in the λ_k variables are satisfied or not.

(These equations are $X_i \oplus X_j = X_k \oplus X_l$ where i, j are in the same block in R, X, Y and k, l are in the same block in R, X, Y, so these equations can be written only the L_i and λ_k variables).

Example. In the example given in appendix B, $\lambda_1 = L_1 \oplus L_4 \oplus L_5 \oplus L_7$ is one of these equations, that can be true or not when the λ_i values are fixed (here it comes from $X_1 \oplus X_2 \oplus X_5 \oplus X_7$).

Analysis of the dependencies in the λ_k. First, we can notice that if the system has no solution due to an incompatibility (for example if we want $X_1 = f_1(R_1) \oplus L_1$ and $X_2 = f_1(R_1) \oplus \lambda_1$ to be distinct) then we have a circle in R, X, Y with at least one equation in Y. The probability to get such circles has been evaluated in section 4 and is negligible if $m \ll 2^n$. So we will assume that we have no incompatibility in the system that says that the X_i variables considered are pairwise distinct. Let μ be the number of variables λ_k that satisfied at least one of these equations among the $\left(\frac{\alpha^2}{2}\right)$ equations considered for the evaluation of f_1. Each of the μ special λ_k values can have at most α exceptional relations. So for a λ like this, we have: $H \leq H^* \left(1 - \frac{\alpha}{2^n}\right)^{-\mu}$. The value $\left(1 - \frac{\alpha}{2^n}\right)^{-\mu}$ can be $\gg 1$, but since we have μ exceptional relations of degree one on μ variables λ_i, the weight W_λ of these λ values (i.e. the number of f_1, f_2, f_3 that give these values multiplied by the number of these values) satisfies:

$$W_\lambda \leq \frac{1}{2^{n\mu}} C^\mu_{\frac{\alpha^2}{2}} \left(1 - \frac{\alpha}{2^n}\right)^{-\mu} \quad \text{(we denote by } A_\mu \text{ this expression).}$$

(since we have $\leq \frac{\alpha^2}{2}$ possible equations). We have:

$$A_{\mu+1} \geq A_\mu \Leftrightarrow \left(\frac{\alpha^2}{2} - \mu\right)\left(1 - \frac{\alpha}{2^n}\right)^{-1} \geq 2^n(\mu+1) \Leftrightarrow \mu \leq \text{ about } \frac{\alpha^2}{2 \cdot 2^n}.$$

So the weight W_λ becomes negligible as soon as $\mu \gg \frac{\alpha^2}{2 \cdot 2^n}$.

Remark. If these μ variables λ_i generate almost all the possible relations with these variables, then the weight of these variables is even smaller since we just have to choose these μ variables among the α variables and then they are fixed (since almost all the equations are satisfied, many of these equations give equivalent values for the special λ_i). So we will have a C^μ_α instead of $C^\mu_{\frac{\alpha^2}{2}}$.

Finally we have obtain:

Theorem 5.1 *Let $\mathcal{F}$ be the set of values that we fix: i.e. in $\mathcal{F}$ we have the values of the Y_i, and all the indices (i, j) where we have all the equations $X_i = X_j$. Then if S and S' are two sequences of values of I_n^m such that:*

1. *$\forall i, j,\ Y_i = Y_j \Rightarrow S_i \neq S_j$ (and $S_i' \neq S_j'$).*
2. *No circle in R, X, Y can be created from the equalities $Y_i = Y_j \Rightarrow S_i \oplus S_j = X_i \oplus X_j$ and $R_k = R_l \Rightarrow X_k \oplus X_l = L_k \oplus L_l$.*

Then the number $H_{\mathcal{F}}$ of f_1, f_2, f_3 solutions satisfies:

$$|H_{\mathcal{F}}(S) - H_{\mathcal{F}}(S')| \leq H_{\mathcal{F}}(S) \cdot (q + r)$$

where $q = \frac{m^2}{2 \cdot 2^n}$ comes from the λ_i with very few special equalities, and r is a very small term related to the weight of the λ_i with a lot of special equalities (as we have seen r is negligible when $m \ll 2^n$).

We can do the same for $[S_i, T_i]$, as we did for $[Y_i, S_i]$. So, since by summation, we must obtain all the $(f_1, \ldots, f_4)$ with no circles, from theorem 5.1 we will get our results. Here the set E' depends on E, so this works for non-adaptive attacks. For adaptive attacks see [15] (then we have to eliminate some equations by conditions in $[S_i, T_i]$ independently of $[L_i, R_i]$, or to study the expectancy of the deviation of H).

Remark. Another possibility is to use the result of [5]: with 2 times more rounds, security in CPA-1 can be changed in security in CPCA-2. However we would get like this CPCA-2 for 10 rounds (exactly as in [14]) instead of 6 rounds.

6   Comparing [14] and This Paper

Technically the main differences between [14] and this paper are:

1. Here we introduce a condition: no more than $\frac{\lambda m(m-1)}{2^n}$ indices (i, j), $i < j$ such that $X_i = X_j$ (instead of no more than θ pairwise distinct indices such that $X_{i_1} = X_{i_2} = \ldots = X_{i_\theta}$ of [14]). this gives us security when $m \ll 2^n$ (instead of $m \ll 2^{n(1-\varepsilon)}$ or $m \ll \frac{2^n}{n}$ of [14]).
2. In [14], 3 rounds are needed for half the variables to look random, and then 4 more rounds for the $[S_i, T_i]$. Here we show that the S_i will look random after 4 rounds even if the Z_i are public (with a probability near 1 when $m \ll 2^n$). So for the T_i we can use the same result with only one more round. Like this, we need less rounds in this paper compared with [14].
3. In this paper we study λ_k that come for Ψ^4 from $Y_i \oplus Y_j = 0$ (or similarly $Z_i \oplus Z_j = 0$ for Ψ^5) while in [14] all possible λ_k can be fixed.

Part II: Best Found Attacks

7 Generic Attacks on Ψ^5

We will present here the two best generic attacks that we have found on Ψ^5:

1. A CPA-1 attack on Ψ^5 with $m \simeq 2^n$ and $\lambda = O(2^n)$ computations (This is an improvement compared with $m \simeq 2^{3n/2}$ and $\lambda = O(2^{3n/2})$ of [13]).
2. A KPA on Ψ^5 with $m \simeq 2^{3n/2}$ and $\lambda = O(2^{3n/2})$ computations (This is an improvement compared with $m \simeq 2^{7n/4}$ and $\lambda = O(2^{7n/4})$ of [13]).

1. CPA-1 attack on Ψ^5.
   Let us assume that $R_i =$ constant, $\forall i$, $1 \leq i \leq m$, $m \simeq 2^n$. We will simply count the number N of (i,j), $i < j$ such that $S_i = S_j$ and $L_i \oplus T_i = L_j \oplus T_j$. This number N will be about double for Ψ^5 compared with a truly random permutation.
   Proof:
   If $S_i = S_j$,
   $L_i \oplus T_i = L_j \oplus T_j \Leftrightarrow L_i \oplus Z_i = L_j \oplus Z_j \Leftrightarrow f_1(R_1) \oplus f_3(Y_i) = f_1(R_1) \oplus f_3(Y_j)$
   $\Leftrightarrow f_3(R_1 \oplus f_2(L_i \oplus f_1(R_1))) = f_3(R_1 \oplus f_2(L_j \oplus f_1(R_1)))$ (#).
   This will occur if $f_2(L_i \oplus f_1(R_1)) = f_2(L_j \oplus f_1(R_1))$, or if these values are distinct but have the same images by f_3, so the probability is about two times larger.

   Remarks
   (a) By storing the $S_i \| L_i \oplus T_i$ values and looking for collisions, the complexity is in $\lambda \simeq O(2^n)$.
   (b) With a single value for R_i, we will get very few collisions. However this attack becomes significant if we have a few values R_i and for all these values about 2^n values L_i.

2. KPA on Ψ^5.
   The CPA attack can immediately be transformed in a KPA: for random $[L_i, R_i]$, we will simply count the number N of (i,j), $i < j$ such that $R_i = R_j$, $S_i = S_j$, and $L_i \oplus T_i = L_j \oplus T_j$. We will get about $\frac{m(m-1)}{2^{3n}}$ such collisions for Ψ^5, and about $\frac{m(m-1)}{2 \cdot 2^{3n}}$ for a random permutation. This KPA is efficient when m^2 becomes not negligible compared with 2^{3n}, i.e. when $m \geq$ about $2^{3n/2}$.

Remark. These attacks are very similar with the attacks on 5-round Feistel schemes described by Knudsen (cf [2]) in the case where (unlike us) f_2 and f_3 are permutations (therefore, <u>not</u> random functions). Knudsen attacks are based on this theorem:

Theorem 7.1 (Knudsen, see [2]) *Let $[L_1, R_1]$ and $[L_2, R_2]$ be two inputs of a 5-round Feistel scheme, and let $[S_1, T_1]$ and $[S_2, T_2]$ be the outputs. Let us assume that the round functions f_2 and f_3 are permutations (therefore they are <u>not</u> random functions of F_n). Then, if $R_1 = R_2$ and $L_1 \neq L_2$, it is impossible to have simultaneously $S_1 = S_2$ and $L_1 \oplus L_2 = T_1 \oplus T_2$.*

Proof. This comes immediately from (#) above.

8   Generic Attacks on Ψ^k Generators, $k \geq 6$

Ψ^k has always an even signature. This gives an attack in 2^{2n} if we want to distinguish Ψ^k from random permutations (see [13]) and if we have all the possible cleartext/ciphertext. In this appendix, we will present the best attacks that we know when we want to distinguish Ψ^k from random permutations with an even signature, or when we do not have exactly all the possible cleartext/ciphertext.

1. <u>KPA with k even.</u>
   Let (i,j) be two indices, $i \neq j$, such that $R_i = R_j$ and $S_i \oplus S_j = L_i \oplus L_j$. From [10] or [11] p.146, we know the exact value of H in this case, when k is even. We have:

   $$H = H^* \left(1 + \frac{1}{2^{(\frac{k}{2}-2)n}} - \frac{1}{2^{(\frac{k}{2}-1)n}} - \frac{2}{2^{\frac{kn}{2}}} + \frac{1}{2^{(k-1)n}} \right)$$

   where

   $$H^* = \frac{|F_n|^k}{2^{2nm}} \cdot \frac{1}{1 - \frac{1}{2^{2n}}}$$

   i.e. H^* is the average value of H on two cleartext/ciphertext. So there is a small deviation, of about $\frac{1}{2^{(\frac{k}{2}-2)n}}$, from the average value.
   So in a KPA, when the $[L_i, R_i]$ are chosen at random, and if the f_i functions are chosen at random, we will get slightly more (i,j), $i < j$, with $R_i = R_j$ and $S_i \oplus S_j = L_i \oplus L_j$ from a Ψ^k (with k even) than from a truly random permutation. This can be detected if we have enough cleartext/ciphertext pairs from many Ψ^k permutations. In first approximation, these relations will act like independent Bernoulli variables (in reality the equations are not truly independent, but this is expected to create only a modification of second order).
   If we have N possibilities for (i,j), $i < j$, and if X is the number of (i,j), $i < j/R_i = R_j$ and $S_i \oplus S_j = L_i \oplus L_j$, we expect to have:
   $E(X) \simeq \frac{N}{2^{2n}}$
   $V(X) \simeq \frac{N}{2^{2n}}$
   $\sigma(X) \simeq \frac{\sqrt{N}}{2^n}$.
   We want $\sigma(X) \leq \frac{N}{2^{(\frac{k}{2}-2)n}} \cdot \frac{1}{2^{2n}}$ in order to distinguish Ψ^k from a random permutation. So we want $\frac{\sqrt{N}}{2^n} \leq \frac{N}{2^{\frac{k}{2}n}}$ i.e. $N \geq 2^{(k-2)n}$.
   However, if we have μ available permutations, with about 2^{2n} cleartext/ciphertext for each of these permutations, then $N \simeq 2^{4n}\mu$ (here we know these μ permutations almost on every possible cleartext. If not, μ will be larger and we will do more computations). $N \geq 2^{(k-2)n}$ gives $\mu \geq 2^{(k-6)n}$. This is an attack with $2^{(k-6)n}$ permutations and $2^{2n}\mu \simeq 2^{(k-4)n}$ computations.
2. <u>KPA with k odd.</u>
   In [15], a KPA with k odd is given (it has the same properties as the attack above for k even).

9   Conclusion

For a block cipher from $2n$ bits $\rightarrow 2n$ bits, we generally want to have no better attack than attacks with $\geq 2^{2n}$ computations. If this block cipher is a Feistel scheme we then need to have ≥ 6 rounds since (as shown in this paper) there is a generic attack on 5 rounds with 2^n computations in CPA-1 and $2^{3n/2}$ computations in KPA.

In this paper we have also shown that however, in the model where the adversaries have unlimited computing power but have access to only m cleartext/ciphertext pairs, the maximum possible security (i.e. $m \ll 2^n$) is obtained already for 5 rounds for CPA-1 and CPA-2 attacks. This solves an open problem of [1] and [14]. Moreover 6-round Feistel schemes can resist all CPCA-1 and CPCA-2 attacks when $m \ll 2^n$ (For CPCA-1 or CPCA-2 the case $k = 5$ rounds is still unclear: we only know that the security is between $m \ll 2^{n/2}$ and $m \ll 2^n$). When 2^{2n} is small (for example to generate 1000 pseudorandom permutations with an even signature of 30 bits $\rightarrow$ 30 bits) then more than 6 rounds are needed. In this paper we have studied such attacks, and we have extended the "coefficients H technique" to various cryptographic attacks.

We think that our proof strategy is very general and should be also efficient in the future to study different kinds of functions or permutation generators, such as, for example, Feistel schemes with a different group law than $\oplus$, or unbalanced Feistel schemes.

References

1. W. Aiello and R. Venkatesan. *Foiling Birthday Attacks in Length-Doubling Transformations-Benes: A Non-Reversible Alternative to Feistel. EUROCRYPT '96* (Lecture Notes in Computer Science 1070), pp. 307–320, Springer-Verlag.
2. L. R. Knudsen. *DEAL - A 128 bit Block Cipher. Technical Report #151*, University of Bergen, Departement of Informatics, Norway, February 1998.
3. M. Luby and C. Rackoff. *How to construct pseudorandom permutations from pseudorandom functions. SIAM Journal on Computing*, vol. 17, no 2, pp. 373–386, April 1988.
4. U. Maurer. *A simplified and generalized treatment of Luby-Rackoff pseudorandom permutation generators. EUROCRYPT '92*, pp. 239–255, Springer-Verlag.
5. U. Maurer. *Indistinguishability of Random Systems. EUROCRYPT '02* (Lecture Notes in Computer Science 2332), pp. 110–132, Springer-Verlag.
6. U. Maurer and K. Pietrzak. *The security of Many-Round Luby-Rackoff Pseudo-Random Permutations. EUROCRYPT '03*, pp. –, Springer-Verlag.
7. V. Nachev. *Random Feistel schemes for $m = 3$*, available from the author at: Valerie.nachef@math.u-cergy.fr.
8. M. Naor and O. Reingold. *On the Construction of pseudo-random perlutations: Luby-Rackoff revisited. Journal of Cryptology*, vol. 12, 1999, pp. 29–66. Extended abstract was published in Proc. 29th Ann. ACM Symp. on Theory of Computing, 1997, pp. 189–199.
9. J. Patarin. *Pseudorandom Permutations based on the DES Scheme. Eurocode '90*, LNCS 514, pp. 193–204, Springer-Verlag.

10. J. Patarin. *New results on pseudorandom permutation generators based on the DES scheme. Crypto '91*,pp. 301–312, Springer-Verlag.

11. J. Patarin. *Etude des générateurs de permutations basés sur le schéma du DES. Ph. D. Thesis*, Inria, Domaine de Voluceau, Le Chesnay, France, 1991.

12. J. Patarin. *About Feistel Schemes with 6 (or More) Rounds. Fast Software Encryption 1998*, pp. 103–121.

13. J. Patarin. *Generic Attacks on Feistel Schemes. Asiacrypt '01* (Lecture Notes in Computer Science 2248), pp. 222–238, Springer-Verlag.

14. J. Patarin. *Luby-Rackoff: 7 Rounds are Enough for $2^{n(1-\epsilon)}$ Security. Crypto '03* (Lecture Notes in Computer Science 2729), pp.513–529, Springer-Verlag.

15. J. Patarin. *Extended version of this paper*, avaible from the author.

16. B. Schneier and J. Kelsey. *Unbalanced Feistel Networks and Block Cipher Design. FSE '96* (Lecture Notes in Computer Science 1039), pp. 121–144, Springer-Verlag.

Appendices

A   Summary of the Known Results on Random Feistel Schemes

KPA denotes known plaintext attacks. CPA-1 denotes non-adaptive chosen plaintext attacks. CPA-2 denotes adaptive chosen plaintext attacks. CPCA-1 denotes non-adaptive chosen plaintext and ciphertext attacks. CPCA-2 denotes adaptive chosen plaintext and chosen ciphertext attacks. Non-Homogeneous properties are defined in [12].

This figure 1 present the best known results against unbounded adversaries limited by m oracle queries.

	KPA	CPA-1	CPA-2	CPCA-1	CPCA-2	Non-Homogeneous
Ψ	1	1	1	1	1	1
Ψ^2	$2^{n/2}$	2	2	2	2	2
Ψ^3	$2^{n/2}$	$2^{n/2}$	$2^{n/2}$	$2^{n/2}$	3	2
Ψ^4	2^n	$2^{n/2}$	$2^{n/2}$	$2^{n/2}$	$2^{n/2}$	2
Ψ^5	2^n	2^n	2^n	$\geq 2^{n/2}$	$\geq 2^{n/2}$	2
Ψ^6	2^n	2^n	2^n	2^n	2^n	4 *
$\Psi^k, k \geq 6$	2^n	2^n	2^n	2^n	2^n	$\leq \left(\frac{k}{2} - 1\right)^2$ **

Fig. 1. Minimum number m of queries to distinguish Ψ^k from a random permutation of $I_n \rightarrow I_n$. For simplicity we denote 2^α for $O(2^\alpha)$ i.e. when we have security as long as $m \ll 2^\alpha$. $\geq$ means best security proved.

* ≤ 4 comes from [13] and ≥ 4 comes from [7].

** with k even and with $(k-2)(k-4)$ exceptional equations, so if $k \geq 7$ we need more than one permutation for this property.

	KPA	CPA-1	CPA-2	CPCA-1	CPCA-2
Ψ	1	1	1	1	1
Ψ^2	$2^{n/2}$	2	2	2	2
Ψ^3	$2^{n/2}$	$2^{n/2}$	$2^{n/2}$	$2^{n/2}$	3
Ψ^4	2^n	$2^{n/2}$	$2^{n/2}$	$2^{n/2}$	$2^{n/2}$
Ψ^5	$\leq 2^{3n/2}$	2^n	2^n	$\leq 2^n$	$\leq 2^n$
Ψ^6	$\leq 2^{2n}$	$\leq 2^{2n}$	$\leq 2^{2n}$	$\leq 2^{2n}$	$\leq 2^{2n}$
Ψ^7	$\leq 2^{3n}$	$\leq 2^{3n}$	$\leq 2^{3n}$	$\leq 2^{3n}$	$\leq 2^{3n}$
Ψ^8	$\leq 2^{4n}$	$\leq 2^{4n}$	$\leq 2^{4n}$	$\leq 2^{4n}$	$\leq 2^{4n}$
$\Psi^k, k \geq 6$ *	$\leq 2^{(k-4)n}$	$\leq 2^{(k-4)n}$	$\leq 2^{(k-4)n}$	$\leq 2^{(k-4)n}$	$\leq 2^{(k-4)n}$

Fig. 2. Minimum number λ of computations needed to distinguish a generator Ψ^k (with one or many such permutations available) from random permutations with an even signature of $I_n \rightarrow I_n$. For simplicity we denote α for $O(\alpha)$. $\leq$ means best known attack.

* If $k \geq 7$ these attacks analyze about $2^{(k-6)n}$ permutations of the generator and if $k \leq 6$ only one permutation is needed.

History for Ψ^5. For Ψ^5 the best results of security against CPA-2 was:

- In 1988: $m \ll 2^{n/2}$ (cf [3]).
- In 1998: $m \ll 2^{3n/4}$ (cf [12]).
- In 2003: $m \ll 2^{5n/6}$ (cf [13]).
- In 2004: $m \ll 2^n$ (cf this paper).

However CPCA-2 for Ψ^5 is still unclear: so far we only have the original result of Luby and Rackoff $m \ll 2^{n/2}$.

B   Example for Theorem 3.1

We will illustrate here theorem 3.1 on a small toy example. Let $1, 2, 3, 4, 5, 6, 7$ be our indices ($m = 7$). Let us assume that f_1 is fixed such that $R_4 = R_1$, and $R_7 = R_5$, are our only equations $R_i = R_j$ $i > j$. Let us assume that the Y_i are given, and that $Y_4 = Y_2$, and $Y_7 = Y_3$ are the only equations $Y_i = Y_j$, $i > j$. Then we want to show that λ_1 and λ_2 look random, where $\lambda_1 = X_4 \oplus X_2$ and $\lambda_2 = X_7 \oplus X_3$ when f_1, f_2 are randomly chosen. For this, we fix λ_1 and λ_2, $\lambda_1 \neq 0$, $\lambda_2 \neq 0$, and we look for the number H of (f_1, f_2) that give these values. We want to prove that this number H does not depend significantly on λ_1 and λ_2 (except for well detected values of small weight). H is the number of (f_1, f_2) such that (here we put only pairwise distinct R_i variables):

1. $f_1(R_2) = f_1(R_1) \oplus L_2 \oplus L_4 \oplus \lambda_1$ and $f_1(R_5) = f_1(R_3) \oplus L_3 \oplus L_7 \oplus \lambda_2$ (these two equations do not create any problem: they just fix f_1 on two points).
2. **Block R_1Y:**

   $f_2(L_1 \oplus f_1(R_1)) = R_1 \oplus Y_1$
   $f_2(L_4 \oplus \lambda_1 \oplus f_1(R_1)) = R_2 \oplus Y_2$
   $f_2(L_4 \oplus f_1(R_1)) = R_1 \oplus Y_2$.

Block R_3Y:

$$f_2(L_3 \oplus f_1(R_3)) = R_3 \oplus Y_3$$
$$f_2(L_3 \oplus L_5 \oplus L_7 \oplus \lambda_2 \oplus f_1(R_3)) = R_5 \oplus Y_5$$
$$f_2(L_3 \oplus f_1(R_3) \oplus \lambda_2) = R_5 \oplus Y_3.$$

Block R_6Y:

$$f_2(L_6 \oplus f_1(R_6)) = R_6 \oplus Y_6$$

Let us assume that, for example, all the $R_i \oplus Y_i$ are pairwise distinct. Then we want to evaluate the number of functions f_1 such that all the X_i are pairwise distinct. These conditions are more difficult to analyze since here we do not want equalities, but non equalities.

- If $\lambda_1 \in \{0, L_1 \oplus L_4\}$, or $\lambda_2 \in \{0, L_5 \oplus L_7\}$, we have no solution (these values give a circle in R, X, Y).
- For the X_i to be pairwise distinct, we must choose f_1 such that: $f_1(R_1) \oplus f_1(R_3)$ is not in A, where A is a set of 9 values (or less if we have collisions): $A = \{L_1 \oplus L_3, L_4 \oplus \lambda_1 \oplus L_3, L_4 \oplus L_3, L_1 \oplus L_3 \oplus L_5 \oplus L_7 \oplus \lambda_2, L_4 \oplus \lambda_1 \oplus L_3 \oplus L_5 \oplus L_7 \oplus \lambda_2, L_4 \oplus L_3 \oplus L_5 \oplus L_7 \oplus \lambda_2, L_1 \oplus L_3 \oplus \lambda_2, L_4 \oplus \lambda_1 \oplus L_3 \oplus \lambda_2, L_4 \oplus L_3 \oplus \lambda_2\}$. In the proof of theorem 3.1, we analyze the possible dependencies of $|A|$ with the λ_i values.

C Examples of Unusual Values of H for Ψ^5

Example 1: Large value for H
With $m = 2$, when $R_1 = R_2$, $S_1 = S_2$ and $L_1 \oplus L_2 = T_1 \oplus T_2$, then

$$H = \frac{|F_n|^5}{2^{2nm}} \left(2 - \frac{1}{2^n}\right).$$

So here the value of H is about double than average with only $m = 2$.

Remark: $\forall k \in \mathbf{N}^*$, Ψ^k has always such large H with small m ($m \leq \left(\frac{k}{2} - 1\right)^2$ if k is even), we say that "Ψ^k is not homogeneous": see [12]. However, when $k \geq 7$, the probability that such inputs/outputs exist is generally negligible if we study only one single specific permutation.

Example 2: Small value for H
Here our example cannot be with $m \ll 2^{n/2}$ since we know that we always have

$$H \geq \frac{|F_n|^5}{2^{2nm}} \left(1 - \frac{m(m-1)}{2^n}\right)$$

(the proof is the same for Ψ^4 and Ψ^5).

However, we will show that when $m \to 2^{n/2}$, H can be much smaller than average (i.e. $m \to 2^n$ is not necessary, $m \to 2^{n/2}$ is enough). In this example 2, we will assume:

1. $\forall i, j, 1 \leq i \leq j \leq m, R_i = R_j (= R_1)$.
2. $\forall i, j, 1 \leq i \leq j \leq m, S_i = S_j (= S_1)$.
3. $\forall i, j, 1 \leq i \leq j \leq m, i \neq j \Rightarrow L_i \oplus L_j \neq T_i \oplus T_j$ (in example 3 below we will not need this condition 3).

To get condition 3, we may assume, for example, that $\forall i, 1 \leq i \leq m, L_i = i \oplus \varphi(i)$ and $T_i = \varphi(i)$, where φ is well chosen. So $L_i \oplus L_j = T_i \oplus T_j \Leftrightarrow i = j$.
From 1 we have: $\forall i, j, 1 \leq i \leq j \leq m, X_i \oplus X_j = L_i \oplus L_j$.
From 2 we have: $\forall i, j, 1 \leq i \leq j \leq m, Z_i \oplus Z_j = T_i \oplus T_j$.
H is the number of f_1, f_2, f_3, f_4, f_5 such that: $\forall i, 1 \leq i \leq m$,

$$L_i \oplus f_1(R_1) = X_i$$

$$R_1 \oplus f_2(L_i \oplus f_1(R_1)) = Y_i$$

$$X_i \oplus f_3(Y_i) = Z_i$$

$$Y_i \oplus f_4(T_i \oplus f_5(S_1)) = S_1$$

$$Z_i \oplus f_5(S_1) = T_i$$

So H is $|F_n|^2$ times the number of f_2, f_3, f_4 such that: $\forall i, 1 \leq i \leq m$,

$$\begin{cases} Y_i = R_1 \oplus f_2(L_i \oplus f_1(R_1)) = S_1 \oplus f_4(T_i \oplus f_5(S_1)) \\ f_3(Y_i) = L_i \oplus T_i \oplus f_1(R_1) \oplus f_5(S_1) \end{cases}$$

Since all the $L_i \oplus T_i$ are pairwise distinct, all the Y_i must be pairwise distinct. So for $Y_i, 1 \leq i \leq m$, we have exactly: $2^n(2^n-1)(2^n-2)\ldots(2^n-m+1)$ solutions.
Now when $Y_i, 1 \leq i \leq m$, are fixed, f_2, f_3 and f_4 are fixed on exactly m pairwise distinct points. So $H = \frac{|F_n|^5}{2^{3nm}} 2^n(2^n-1)(2^n-2)\ldots(2^n-m+1)$.
Let H^* be the average value of H (when the $[S_i, T_i]$ are pairwise distinct).

$$H^* = \frac{|F_n|^5}{2^{2n}(2^{2n}-1)(2^n-2)\ldots(2^{2n}-m+1)} \geq \frac{|F_n|^5}{2^{2nm}}.$$

So here:

$$\frac{H}{H^*} \leq (1 - \frac{1}{2^n})(1 - \frac{2}{2^n})\ldots(1 - \frac{m-1}{2^n})$$

$$ln\left(\frac{H}{H^*}\right) \simeq -\frac{1+2+\ldots+(m-1)}{2^n} = -\frac{m(m-1)}{2^n}.$$

So when $m(m-1)$ is not negligible compared with 2^n, H will be significatively smaller than H^*, as claimed.

Remark 1. Here $R_i \oplus S_i$ is not random (since $R_i \oplus S_i$ is constant), and $L_i \oplus T_i$ is not random (in example 3 below we will remove this condition on $L_i \oplus T_i$). These hypothesis are generally unrealistic in a cryptographic attack, where $\forall i, 1 \leq i \leq m, L_i$ or T_i, and R_i or S_i, cannot be chosen.

Remark 2. If we start, as here, from $[L_i, R_i]$ values with R_i constant, then the X_i values are pairwise distinct, so the Y_i values are perfectly random (if we define Y_i only from the relation $Y_i = R_i \oplus f_2(X_i)$). However, the Z_i values are not perfectly random (since the probability to have $Z_i \oplus Z_j = L_i \oplus L_j$ is the probability to have $f_3(Y_i) = f_3(Y_j)$ so is about double than average). Similarly, the $[S_i, T_i]$ values are not perfectly random since the probability to have $S_i = S_j$ and $T_i \oplus T_j = L_i \oplus L_j$ is in relation with the probability to have $f_3(Y_i) = f_3(Y_j)$, so is about double than average. We will use again this idea in example 3 below.

Remark 3. Here when $m \to 2^{n/2}$, we can have circles in Y, S, (and circles in R, Y) and this is a way to explain why in this example H can be much smaller than H^*.

Example 3: Small value for H, with random L_i and T_i
    In this example 3, we will assume:

1. $\forall i, j, 1 \le i \le j \le m, R_i = R_j (= R_1)$.
2. $\forall i, j, 1 \le i \le j \le m, S_i = S_j (= S_1)$.
3. Let $A_i = L_i \oplus T_i$. Then $A_i, 1 \le i \le m$, is random. More precisely it will be enough to assume that the number N of collisions $A_i = A_j, i < j$, is $< \frac{m(m-1)}{2 \cdot 2^n ln2}$ to show that H is small compared with the average value H^*. For random values A_i we have $N \simeq \frac{m(m-1)}{2 \cdot 2^n}$, so it is the case ($\frac{1}{ln2} \simeq 1,44$).

As in example 2, H is $|F_n|^2$ times the number of f_2, f_3, f_4 such that: $\forall i$, $1 \le i \le m$,

$$\begin{cases} Y_i = R_1 \oplus f_2(L_i \oplus f_1(R_1)) = S_1 \oplus f_4(T_i \oplus f_5(S_1)) \\ f_3(Y_i) = L_i \oplus T_i \oplus f_1(R_1) \oplus f_5(S_1) \end{cases}.$$

Since all the $L_i \oplus f_1(R_1)$ are pairwise distinct, and all the $T_i \oplus f_5(S_1)$ are pairwise distinct, f_2 and f_4 are fixed on exactly m points when $Y_i, 1 \le i \le m$, is fixed.
    So H is $\frac{|F_n|^4}{2^{2nm}}$ times the number of Y_i, f_3 such that: $\forall i, 1 \le i \le m, f_3(Y_i) = L_i \oplus T_i \oplus f_1(R_1) \oplus f_5(S_1)$.
    Let A_i be a sequence of values of $I_n, 1 \le i \le m$. We want to evaluate the number h of Y_i, f_3 such that: $\forall i, 1 \le i \le m, f_3(Y_i) = A_i$. Let h^* be the average value for h (average on all sequences A_i). We have $h^* = |F_n|$. For random values Y_i, and random functions f_3, A_i will have about 2 times more collisions $A_i = A_j$, $i < j$, than average sequences A_i.
    So h for random values A_i is $\ll h^*$, and h for values A_i with 2 times more collisions than average is $\gg h^*$. This shows that if in this example 3 $L_i \oplus T_i$ is random, then $H \ll H^*$.

Signed Binary Representations Revisited

Katsuyuki Okeya[1], Katja Schmidt-Samoa[2],
Christian Spahn[2], and Tsuyoshi Takagi[2]

[1] Hitachi, Ltd., Systems Development Laboratory,
292, Yoshida-cho, Totsuka-ku, Yokohama, 244-0817, Japan
ka-okeya@sdl.hitachi.co.jp
[2] Technische Universität Darmstadt, Fachbereich Informatik,
Hochschulstr. 10, D-64289 Darmstadt, Germany
{samoa,takagi}@informatik.tu-darmstadt.de

Abstract. The most common method for computing exponentiation of random elements in Abelian groups are sliding window schemes, which enhance the efficiency of the binary method at the expense of some precomputation. In groups where inversion is easy (e.g. elliptic curves), signed representations of the exponent are meaningful because they decrease the amount of required precomputation. The asymptotic best signed method is wNAF, because it minimizes the precomputation effort whilst the non-zero density is nearly optimal. Unfortunately, wNAF can be computed only from the least significant bit, i.e. right-to-left. However, in connection with memory constraint devices left-to-right recoding schemes are by far more valuable.
In this paper we define the MOF (*Mutual Opposite Form*), a new canonical representation of signed binary strings, which can be computed in any order. Therefore we obtain the first left-to-right signed exponent-recoding scheme for general width w by applying the width w sliding window conversion on MOF left-to-right. Moreover, the analogue right-to-left conversion on MOF yields wNAF, which indicates that the new class is the natural left-to-right analogue to the useful wNAF. Indeed, the new class inherits the outstanding properties of wNAF, namely the required precomputation and the achieved non-zero density are exactly the same.

Keywords: *addition-subtraction chains, exponentiation, scalar multiplication, signed binary, elliptic curve cryptosystem, efficient computation, non-adjacent form (NAF), mutual opposite form (MOF), left-to-right*

1 Introduction

In modern cryptosystems one of the most important basic operations is exponentiation g^d, where g is an element of an Abelian group G and d is an integer. A non-zero positive integer d is uniquely represented by a binary string:

$$d = d_{n-1}|d_{n-2}|...|d_1|d_0,$$

where $a|b$ denotes the concatenation of bits a, b, and $d_i \in \{0,1\}$ for $i = 0, 1, ..., n-1$.

M. Franklin (Ed.): CRYPTO 2004, LNCS 3152, pp. 123–139, 2004.

The most common method for performing an exponentiation is the square-and-multiply algorithm, which computes g^d according to the bits d_i (therefore it is often called binary method). The efficiency of this procedure may be enhanced if precomputation is allowed. In this case, we consider more general representations of the exponent, where each non-zero bit d_i is not restricted to be 1, but is an element of a suitable digit set $\mathcal{T}$ of integers. We call $d = \sum_i d_i 2^i$ a $\mathcal{T}$-representation, if $d_i \in \mathcal{T} \cup \{0\}$ holds for each i. In general, $\mathcal{T}$-representations loose the property of uniqueness. The left-to-right square-and-multiply algorithm is easily adjusted to work with a $\mathcal{T}$-representation of the exponent, namely multiplication by the base g is replaced with multiplication by precomputed elements g^{d_i}, where $d_i \in \mathcal{T}$ is the appropriate digit of d. Therefore, the important features of a $\mathcal{T}$-representation are the number of non-zero digits and the cardinality of $\mathcal{T}$, because they determine the required time and memory consumption for computing g^d, respectively. The research problem here is to find optimized representation classes in the sense of trade-off between high non-zero density and low memory consumption.

1.1 New Motivation for Exponentiation Algorithms

As the ubiquitous computing devices are penetrating our daily life, the importance of memory constraint devices (e.g. smart cards) in cryptography is increasing. Smart cards are equipped with several Kbytes RAM only and most of them are reserved for OS and stack. Thus, cryptographic algorithms should be optimized in terms of memory. For this reason we are reluctant to consume memory except the necessary precomputation related to $\mathcal{T}$ for computing exponentiation. Note that in connection with memory constraint devices, the most popular cryptosystems are based on elliptic curves [Kob87,Mil86], because elliptic curve cryptosystems (ECC) provide high security with moderate key-lengths. As elliptic curve groups are written additively, exponentiation has to be understood as scalar multiplication in this context.

Exponent recoding, i.e. the rewriting of the binary exponent to a $\mathcal{T}$-representation, may be performed from the least significant bit (we say "right-to-left") and from the most significant bit ("left-to-right"), respectively. For the purpose of ECC on memory constraint devices we prefer left-to-right to right-to-left recoding methods. The reason is as follows: In the case of elliptic curve scalar multiplication, the left-to-right evaluation stage is the natural choice (see Section 5 for details). If the exponent recoding is done right-to-left, it is necessary to finish the recoding and to store the recoded string before starting the left-to-right evaluation stage. In other words, we require additional n-bit (i.e. exponential size $\mathcal{O}(n)$) RAM for the right-to-left exponent recoding, where n is the bit size of the scalar.

On the contrary, if a left-to-right recoding technique is available, the recoding and evaluation stage may be merged to obtain an efficient exponentiation on the fly, without storing the recoded exponent at all. Therefore it is an important task to construct a left-to-right recoding scheme, even if the size of $\mathcal{T}$ and the non-zero density are not improved.

1.2   Known Solutions

The most established techniques for generating T representations are window methods (see, e.g., the textbooks [Knu81,MOV96] and the survey paper [Gor98]). Loosely speaking, in the window method with width w successively w consecutive bits of the binary exponent are scanned and, if necessary, replaced by a table-entry according to T. We distinguish fixed window methods like the 2^w-ary method, where the window segmentation of the binary string is predetermined and the more advanced sliding window methods, where zero runs are skipped. As an example, let us consider the sliding window method with width $w = 3$. In this case, T equals $\{1, 3, 5, 7\}$. During the recoding stage, the binary exponent is rewritten by performing the following replacements: $1|1 \mapsto 0|3$, $1|0|1 \mapsto 0|0|5$, and $1|1|1 \mapsto 0|0|7$. Note that the sliding window conversion can be performed left-to-right and right-to-left as well. The results may differ syntactically, but the asymptotic non-zero density of both representations is the same, namely $1/(w+1)$. In the unsigned case (i.e. T consists only of positive integers), sliding window techniques are the method of choice.

However, a nice property of elliptic curves is that inversion is computed virtually for free. In this case, it is meaningful to consider digit sets containing negative integers, too. This reduces precomputation effort, because g^{-i} may be computed from g^i on the fly, such that only the elements $g^{|i|}$ for $i \in T$ have to be precomputed. However, the question arises how to construct a signed T representation. In general, there are two strategies. The first one is to construct a $\{-1, +1\}$ representation of d (also called a signed binary representation) and to apply window methods afterwards. Here, the most common signed binary representation is NAF (non-adjacent-form) [Rei60,IEEE], which can be obtained from the binary representation by applying the conversion $*|1|1 \mapsto * + 1|0|\bar{1}$ repeatedly, where $\bar{1}$ denotes -1 and $*$ stands for any binary digit. However, the carry-over $+1$ occurring in the first digit forces the recoding to be performed from the least significant bit, i.e. right-to-left. The second strategy is to generalize the NAF recoding for $w > 2$ in order to obtain wNAF [Sol00,BSS99] (here, the non-adjacent property states that among any w adjacent bits, at most one is non-zero). According to [BSS99], this strategy is the optimal one for $w > 3$. But unfortunately, this strategy suffers from the same drawback as the first one, namely as carry-overs are required, the recoding is restricted to be done right-to-left. Consequently, all exponentiation strategies based on signed T-representations require $\mathcal{O}(n)$ bits of RAM additional memory to store the recoded exponent. Solely in the case of $w = 2$, Joye and Yen proposed a left-to-right binary recoding algorithm [JY00]. But it has been an unsolved problem to generate a left-to-right recoding algorithm for a general width $w > 2$. Note that the asymptotic non-zero density of wNAF is the same as for the unsigned sliding window method on binary, namely $1/(w+1)$. Therefore, wNAF can be seen as its natural signed analogue, and we guess that there could be a carry-free generation method for wNAF. In this paper, the term carry-free refers to an algorithm that transforms the input string in situ, i.e. in each step only the knowledge of a fixed number of consecutive input bits is necessary.

1.3 Our Contributions

The aim of this paper is to solve both problems as follows: (1) we define a new canonical representation class of signed binary. We call it MOF (Mutual Opposite Form) and prove that each integer can be uniquely represented as a MOF. But the outstanding property of MOF is that it can be efficiently developed from a binary string right-to-left or left-to-right, likewise. Consequently, analogue to the unsigned case, sliding window methods may be applied to receive left-to-right and right-to-left recoding schemes for general width w. Surprisingly, applying the right-to-left width w sliding window method on MOF yields wNAF. However, the observation that in the unsigned case right-to-left sliding window yields an unsigned string with non-adjacent property stresses the analogy between unsigned Binary and signed MOF. Therefore we achieve a *carry-free* wNAF generation, a benefit of its own.

(2) Our major aim is to develop a left-to-right recoding algorithm, and this is achieved straightforwardly by applying the width w sliding window method left-to-right on MOF. We call the so-defined class wMOF and prove that each integer can be uniquely represented as a wMOF and that the asymptotic non-zero density of wMOF equals $1/(w+1)$, which is the same as for wNAF. Therefore the classes wNAF and wMOF may be seen as dual to each other. In general our proposed algorithm asymptotically requires additional $\mathcal{O}(w)$ bits of RAM, which is independent from the bit size n and dramatically reduces the required space comparing with previous methods. Consequently, due to its left-to-right nature, the new scheme is by far more convenient with respect to memory consumption than previous schemes. Interestingly, a straight-forward proof shows that for $w = 2$ the proposed method produces the same output as the Joye-Yen recoding, but 2MOF is more efficient in terms of counting the number of basic operations.

We finish this work with some explicit algorithms, proving that the proposed schemes are indeed useful for practical purposes. For example, we develop generating algorithms for wMOF based on efficient table-lookups, and we show how to exploit wMOF for implementing on-the-fly elliptic curve scalar multiplication.

2   Signed Representations

In this section we review some signed representations, which are important in connection with elliptic curve scalar multiplication. For the sake of simplicity, we only deal with non-negative integers d in the following. We call $d = \sum_i d_i 2^i$ a $\mathcal{T}$-representation, if $\mathcal{T}$ is a set of integers and $d_i \in \mathcal{T} \cup \{0\}$ holds for each i. If $\mathcal{T}$ contains negative integers, we speak of *signed* representations, and if $\mathcal{T}$ equals $\{\pm 1\}$, of *signed binary* representations. In general, signed binary representations are redundant. The most established one is NAF (non-adjacent form), introduced by Reitwiesner 1960 [Rei60]. A generalization of Reitwiesner's NAF recoding idea can be found in [Pro00,Avi61]. NAF can be easily defined by the property that at most one out of two consecutive digits is non-zero. Reitwiesner was able to show that ignoring leading zeros each integer has a unique NAF representation. For this reason, some authors call NAF a canonical signed binary representation [EK94]. In addition, as shown among others by Jedwab and Mitchell [JM89], NAF representation provides the minimal Hamming weight. Consequently, the NAF representation of the exponent is the optimal choice if signed methods are meaningful and no precomputation is considered. It was first pointed out by Morain and Olivos that NAF can be used to speed up elliptic curve scalar multiplication [MO90].

However, the situation is less clear if extra memory is available and precomputation is admitted. In this case, signed representations using larger digit sets T should be taken into account. One strategy to construct a signed representation is to apply sliding window methods on signed binary representations. But as signed binary representation is redundant, the question arises which representation is the best for this purpose. Indeed, this is assumed to be an open problem by De Win et al. [WMPW98]. There are several methods to construct signed binary representations as a base for sliding window schemes [KT92,WMPW98], but none of these can be performed left-to-right. In this paper, we will develop a left-to-right recoding scheme, which is of high value in connection with memory constraint devices.

A different approach is wNAF. Instead of applying window techniques to signed binary representations, wNAF is computed directly from binary strings using a generalization of NAF recoding. First we review the definition of wNAF as stated in [Sol00].

Definition 1 (wNAF). *A sequence of signed digits is called wNAF iff the following three properties hold:*

1. *The most significant non-zero bit is positive.*
2. *Among any w consecutive digits, at most one is non-zero.*
3. *Each non-zero digit is odd and less than 2^{w-1} in absolute value.*

Note that 2NAF and NAF are the same. Algorithm 1 describes the generation of wNAF as proposed by Solinas [Sol00].

Algorithm 1 Generation of wNAF [Sol00]

Input: width w, an n-bit integer d
Output: wNAF $\delta_n|\delta_{n-1}|\ldots|\delta_0$ of d
  $i \leftarrow 0$
  while $d \geq 1$ **do**
    if d is even **then**
      $\delta_i \leftarrow 0$
    else
      $\delta_i \leftarrow d$ mods 2^w; $d \leftarrow d - \delta_i$
    $d \leftarrow d/2$; $i \leftarrow i + 1$
  return $(\delta_n, \delta_{n-1}, ..., \delta_0)$.

Here "mods" means the signed modulo, namely a mods b is defined as a mod b and $-b/2 \leq a < b/2$. The algorithm generates wNAF from the least significant bit, that is right-to-left generation again. The average density of non-zero bits is asymptotically $1/(w+1)$ for $n \to \infty$, and the digit set equals $\mathcal{T} = \{\pm 1, \pm 3, \ldots, \pm(2^{w-1} - 1)\}$ which seems to be minimal. Thus wNAF and its variants like modified window NAF [Möl02] are optimal in the sense of the trade-off between speed and memory for $w > 3$ [BSS99,BHLM01]. There are several other algorithms for generating wNAF, for example see [BSS99,MOC97] but each method needs carry-overs. Note that in the worst case all remaining bits are affected by the carry, therefore the previously known wNAF algorithms can not be considered as local methods. By inspecting Algorithm 1 closely, we observe that this generation can be seen as the natural signed analogue to the right-to-left sliding window method on (unsigned) Binary (here, mod instead of mods is computed). Indeed, the latter method produces a representation that fulfills the nonadjacent requirement (see Definition 1, property 3). Consequently, we conjecture that there might be a signed binary representation that produces wNAF when handled with sliding window conversions. The signed binary representation introduced in the next section will also serve for this purpose.

3 MOF: New Canonical Representation for Signed Binary Strings

In this section we present a new signed representation of integers. The proofs of the propositions in this section are in the full version of this paper [OSST04]. In order to achieve a unique representation, we introduce the following special class of signed binary strings, called the mutual opposite form (MOF).

Definition 2 (MOF). *The n-bit mutual opposite form (MOF) is an n-bit signed binary string that satisfies the following properties:*

1. *The signs of adjacent non-zero bits (without considering zero bits) are opposite.*
2. *The most non-zero bit and the least non-zero bit are 1 and $\bar{1}$, respectively, unless all bits are zero.*

Some zero bits are inserted between non-zero bits that have a mutual opposite sign. An example of MOF is $0100\bar{1}01000\bar{1}001\bar{1}0$. An important observation is that each positive integer can be uniquely represented by MOF. Indeed, we have the following theorem.

Theorem 1. *Let n be a positive integer. $(n+1)$-bit MOF has 2^n pair-wise different representations. There is the bijective map between elements of $(n+1)$-bit MOF and n-bit binary strings.*

From this theorem, any n-bit binary string can be uniquely represented by $(n+1)$-bit MOF. We obviously have the following corollary about the non-zero density of MOF.

Corollary 1. *The average non-zero density of n-bit MOF is $1/2$ for $n \to \infty$.*

3.1 Converting Binary String to MOF

We show a simple and flexible conversion from n-bit binary string to $(n+1)$-bit MOF.

The crucial point is the following observation. The n-bit binary string d can be converted to a signed binary string by computing $\mu = 2d \ominus d$, where '$\ominus$' stands for a bitwise subtraction. Indeed, we convert d as follows:

$$
\begin{array}{rcccccccccc}
2d = d_{n-1} & | & d_{n-2} & | & \ldots & | & d_{i-1} & | & \ldots & | & d_1 & | & d_0 & | \\
\ominus\ d = & | & d_{n-1} & | & \ldots & | & d_i & | & \ldots & | & d_2 & | & d_1 & | & d_0 \\
\hline
\mu = d_{n-1} & | & d_{n-2}-d_{n-1} & | & \ldots & | & d_{i-1}-d_i & | & \ldots & | & d_1-d_2 & | & d_0-d_1 & | & -d_0.
\end{array}
$$

Here the i-th signed bit of μ is denoted by μ_i, namely $\mu_i = d_{i-1} - d_i$ for $i = 1, \ldots, n-1$ and $\mu_n = d_{n-1}, \mu_0 = -d_0$. We can prove that the signed representation μ is MOF.

Proposition 1. *The operation $\mu = 2d \ominus d$ converts binary string d to its MOF μ.*

Algorithm 2 provides an explicit conversion from Binary to MOF.

Algorithm 2 Left-to-Right Generation from Binary to MOF

Input: a non-zero n-bit binary string $d = d_{n-1}|d_{n-2}|\ldots|d_1|d_0$
Output: MOF $\mu_n|\ldots|\mu_1|\mu_0$ of d
$\mu_n \leftarrowtail d_{n-1}$
for $i = n - 1$ down to 1 **do**
    $\mu_i \leftarrowtail d_{i-1} - d_i$
$\mu_0 \leftarrowtail -d_0,$
    return $(\mu_n, \mu_{n-1}, \ldots, \mu_1, \mu_0)$.

In order to generate the i-th bit μ_i, Algorithm 2 stores just two consecutive bits d_{i-1} and d_i. This algorithm converts a binary string to MOF from the most significant bit in an efficient way. Note that it is also possible to convert a binary string to MOF right-to-left. Thus MOF representation is highly flexible.

Remark 1. Interestingly, the MOF representation of an integer d equals the recoding performed by the classical Booth algorithm for binary multiplication [Boo51]. The classical Booth algorithm successively scans two consecutive bits of the multiplier A (right-to-left). Depending on these bits, one of the following operations is performed:

No operation,	if $(a_i, a_{i-1}) \in \{(0,0), (1,1)\}$,
Subtract multiplicand B from the partial product,	if $(a_i, a_{i-1}) = (1, 0)$,
Add multiplicand B to the partial product,	if $(a_i, a_{i-1}) = (0, 1)$,

where a_{-1} is defined as 0. Of course, the design goal of this algorithm was to speed up multiplication when there are consecutive ones in the multiplier A, and to provide a multiplication method that works for signed and unsigned numbers as well. To our knowledge, this representation never served as a fundament of theoretical treatment of signed binary strings.

4   Window Methods on MOF

In this section we show how to decrease the non-zero density of MOF by applying window methods on it. First we consider the right-to-left width w sliding window method which surprisingly yields the familiar wNAF. In contrast to previously known generation methods, the new one is carry-free, i.e. in each step the knowledge of at most $w + 1$ consecutive input bits is sufficient.

Then we define the dual new class wMOF as the result of the analogue left-to-right width w sliding window method on MOF. This conversion leads to the first left-to-right signed recoding scheme for general width w.

4.1   Right-to-Left Case: wNAF

In order to describe the proposed scheme, we need the conversion table for width w. First, we define the conversions for MOF windows of length l, such that the first and the last bit is non-zero:

$$\underbrace{0|...|0|2^{l-2}+1}_{l} \hookleftarrow \begin{cases} 1|\bar{1}|0|\ldots|0|0|1 \\ 1|\bar{1}|0|\ldots|0|1|\bar{1} \end{cases} \qquad \underbrace{0|...|0|2^{l-2}+3}_{l} \hookleftarrow \begin{cases} 1|\bar{1}|0|\ldots|0|1|0|\bar{1} \\ 1|\bar{1}|0|\ldots|0|1|\bar{1}|1 \end{cases} \ldots$$

$$\ldots \underbrace{0|...|0|2^{l-1}-3}_{l} \hookleftarrow \begin{cases} 1|0|\ldots|0|\bar{1}|1|\bar{1} \\ 1|0|\ldots|0|\bar{1}|0|1 \end{cases} \qquad \underbrace{0|...|0|2^{l-1}-1}_{l} \hookleftarrow \begin{cases} 1|0|\ldots|0|0|\bar{1} \\ 1|0|\ldots|0|\bar{1}|1 \end{cases}$$

In addition, we have analogue conversions with all signs changed. To generate the complete table for width w, we have to consider all conversions of length $l = 2, 3, \ldots, w$. If $l < w$ holds, the window is filled with leading zeros.

Example: In the case of $w = 3$, we use the following table for the right-to-left sliding window method:

$$\text{Table}_{3\overleftarrow{SW}} : 001 \hookleftarrow \begin{cases} 001 \\ 01\bar{1} \end{cases} \qquad 00\bar{1} \hookleftarrow \begin{cases} 00\bar{1} \\ 0\bar{1}1 \end{cases} \qquad 003 \hookleftarrow \begin{cases} 10\bar{1} \\ 1\bar{1}1 \end{cases} \qquad 00\bar{3} \hookleftarrow \begin{cases} \bar{1}01 \\ \bar{1}1\bar{1} \end{cases}$$

In an analogue way $\text{Table}_{w\overleftarrow{SW}}$ is defined for general w. Based on this table, Algorithm 3 provides a simple carry-free wNAF generation.

Algorithm 3 Right-to-Left Generation from Binary to wNAF

Input: width w, a non-zero n-bit binary string $d = d_{n-1}|d_{n-2}|\ldots|d_1|d_0$
Output: wNAF $\nu_n|\ldots|\nu_1|\nu_0$ of d
$\quad d_{n+w-2} \hookleftarrow 0; \ d_{n+w-3} \hookleftarrow 0; \ldots; \ d_n \hookleftarrow 0; \ d_{-1} \hookleftarrow 0; \ i \hookleftarrow 0$
$\quad$**while** $i \leq n$ **do**
$\quad\quad$**if** $d_{i-1} = d_i$ **then**
$\quad\quad\quad \nu_i \hookleftarrow 0; \ i \hookleftarrow i+1$
$\quad\quad$**else** {The MOF window begins with a non-zero digit righthand}
$\quad\quad\quad (\nu_{i+w-1},\ldots,\nu_i) \hookleftarrow \text{Table}_{w\overleftarrow{SW}}(d_{i+w-2} - d_{i+w-1}, d_{i+w-3} - d_{i+w-2}, \ldots, d_{i-1} - d_i)$
$\quad\quad\quad i \hookleftarrow i+w$
$\quad$**return** $(\nu_n,\ldots,\nu_1,\nu_0)$

Obviously, the output of Algorithm 3 meets the notations of Definition 1, therefore it is wNAF. If we knew that Definition 1 provides a *unique* representation, we could deduce that Algorithm 3 outputs the same as Algorithm 1. This is true, although we could not find a proof in literature. For the sake of completeness, we prove the following theorem in the full version of this paper [OSST04] via exploiting the uniqueness of MOF representation.

Theorem 2. *Every non-negative integer d has a representation as wNAF, which is unique except for the number of leading zeros.*

4.2 Left-to-Right Case: wMOF

In this section we introduce our new proposed scheme. The crucial observation is that as the generation Binary $\mapsto$ MOF can be performed left-to-right, the combination of this generation and left-to-right sliding window method leads to a complete signed left-to-right recoding scheme dual to wNAF.

In order to describe the proposed scheme, we need the conversion table for width w. The conversions for MOF windows of length l, such that the first and the last bit is non-zero, are defined in exactly the same way as in the right-to-left case (see the table in section (4.1) and reflect the assignments). To generate the complete table for width w, we have to consider all conversions of length $l = 2, 3, \ldots, w$ as before. The only difference is that if $l < w$ holds, the window is filled with *closing* zeros instead of leading ones. As an example, we construct the conversion table $\text{Table}_{4\overrightarrow{SW}}$ for width 4:

$$1000\} \mapsto 1000 \qquad 1\bar{1}00\} \mapsto 0100 \qquad \left.\begin{matrix}1\bar{1}10\\10\bar{1}0\end{matrix}\right\} \mapsto 0030 \qquad \left.\begin{matrix}1\bar{1}01\\1\bar{1}1\bar{1}\end{matrix}\right\} \mapsto 0005 \qquad \left.\begin{matrix}100\bar{1}\\10\bar{1}1\end{matrix}\right\} \mapsto 0007$$

$$\bar{1}000\} \mapsto \bar{1}000 \qquad \bar{1}100\} \mapsto 0\bar{1}00 \qquad \left.\begin{matrix}\bar{1}1\bar{1}0\\\bar{1}010\end{matrix}\right\} \mapsto 00\bar{3}0 \qquad \left.\begin{matrix}\bar{1}10\bar{1}\\\bar{1}1\bar{1}1\end{matrix}\right\} \mapsto 000\bar{5} \qquad \left.\begin{matrix}\bar{1}001\\\bar{1}01\bar{1}\end{matrix}\right\} \mapsto 000\bar{7}$$

The table is complete due to the properties of MOF. Note that because of the equalities $*1\bar{1} = *01$, $*\bar{1}1 = *0\bar{1}$ usually two different MOF-strings are converted to the same pattern. In an analogue way, $\text{Table}_{w\overrightarrow{SW}}$ is defined for general width w. In this case the digit set equals $\mathcal{T} = \{\pm 1, \pm 3, \ldots, \pm 2^{w-1} - 1\}$, which is the same as for wNAF. Therefore, the scheme requires only 2^{w-2} precomputed elements. Algorithm 4 makes use of this table to generate wMOF left-to-right.

In order to deepen the duality between wNAF and wMOF, we give a formal definition of wMOF and prove that it leads to a unique representation of non-negative integers.

Definition 3. *A sequence of signed digits is called wMOF iff the following three properties hold:*

1. *The most significant non-zero bit is positive.*
2. *All but the least significant non-zero digit x are adjoint by $w\text{-}1$ zeros as follows:*
   - *in case of $2^{k-1} < |x| < 2^k$ for an integer $2 \leq k \leq w - 1$ the pattern equals $\underbrace{0\ldots0}_{k}x\underbrace{0\ldots0}_{w-k-1},$*

— *in case of* $|x| = 1$ *either the pattern equals* $x\underbrace{0\ldots0}_{w-1}$ *and the next lower non-zero digit has opposite sign from* x *or the pattern equals* $0x\underbrace{0\ldots0}_{w-2}$ *and the next lower non-zero digit has the same sign as* x.
If x *is the least significant non-zero digit, it is possible that the number of right-hand adjacent zeros is smaller than stated above. In addition it is not possible that the last non-zero digit is a 1 following any non-zero digit.*
3. *Each non-zero digit is odd and less than* 2^{w-1} *in absolute value.*

This definition is directly related to the generation of wMOF. Note that the exceptional case corresponding to the least significant bit takes in account that the last window may be shorter than w.

Algorithm 4 Left-to-Right Generation from Binary to wMOF

Input: width w, a non-zero n-bit binary string $d = d_{n-1}|d_{n-2}|...|d_1|d_0$
Output: wMOF $\delta = \delta_n|\delta_{n-1}|...|\delta_1|\delta_0$ of d
  $d_{-1} \leftarrow 0$; $d_n \leftarrow 0$; $i \leftarrow n$
  while $i \geq w - 1$ **do**
    if $d_i = d_{i-1}$ **then**
      $\delta_i \leftarrow 0$; $i \leftarrow i - 1$
    else {The MOF window begins with a non-zero digit lefthand}
      $(\delta_i, \delta_{i-1}\ldots, \delta_{i-w+1}) \leftarrow \text{Table}_{w\overrightarrow{SW}}(d_{i-1} - d_i, d_{i-2} - d_{i-1}, \ldots, d_{i-w} - d_{i-w+1})$
      $i \leftarrow i - w$
  if $i \geq 0$ **then**
    $(\delta_i, \delta_{i-1}\ldots, \delta_0) \leftarrow \text{Table}_{i+1\overrightarrow{SW}}(d_{i-1} - d_i, d_{i-2} - d_{i-1}, \ldots, d_0 - d_1, -d_0)$
  return $(\delta_n, \delta_{n-1}, ..., \delta_1, \delta_0)$.

Regarding the uniqueness and the non-zero density of wMOF, we have the following two theorems, proven in the full version of this paper [OSST04].

Theorem 3. *Every non-negative integer d has a representation as wMOF, which is unique except for the number of leading zeros.*

Theorem 4. *The average non-zero density of wMOF is asymptotically $1/(w+1)$ for $n \mapsto \infty$.*

We finish this section with a detailed example of the conversion from Binary to MOF and the effects of several sliding window methods.

Bin	1 1 1 0 1 0 0 1 1 0 0 1 0 0 0 1 0 1 0 1 1 1 0 1 0 1 0 1 0 1 1 1
MOF	1 0 0 $\bar{1}$ $\bar{1}$ 1 0 1 0 $\bar{1}$ 0 1 $\bar{1}$ 0 0 1 $\bar{1}$ $\bar{1}$ 1 0 0 $\bar{1}$ $\bar{1}$ $\bar{1}$ $\bar{1}$ $\bar{1}$ $\bar{1}$ $\bar{1}$ 1 0 0 $\bar{1}$
2MOF	1 0 0 0 $\bar{1}$ $\bar{1}$ 0 1 0 $\bar{1}$ 0 0 1 0 0 0 1 0 1 1 0 0 0 $\bar{1}$ 0 $\bar{1}$ 0 $\bar{1}$ 0 $\bar{1}$ 0 0 $\bar{1}$
3MOF	1 0 0 0 0 $\bar{3}$ 0 0 0 3 0 0 1 0 0 0 0 3 0 $\bar{1}$ 0 0 0 0 $\bar{3}$ 0 0 3 0 $\bar{1}$ 0 0 $\bar{1}$
4MOF	0 0 0 7 0 0 0 5 0 0 0 0 $\bar{7}$ 0 0 0 0 0 5 0 0 0 7 0 0 0 5 0 0 3 0 0 $\bar{1}$
NAF	1 0 0 $\bar{1}$ 0 1 0 1 0 $\bar{1}$ 0 0 1 0 0 1 0 $\bar{1}$ 0 $\bar{1}$ 0 0 0 $\bar{1}$ 0 $\bar{1}$ 0 $\bar{1}$ 0 $\bar{1}$ 0 0 $\bar{1}$
3NAF	1 0 0 0 0 $\bar{3}$ 0 0 0 3 0 0 1 0 0 0 1 0 0 3 0 0 0 $\bar{1}$ 0 0 $\bar{3}$ 0 0 3 0 0 $\bar{1}$
4NAF	0 0 0 7 0 0 0 5 0 0 0 $\bar{3}$ 0 0 0 7 0 0 0 $\bar{5}$ 0 0 0 0 $\bar{3}$ 0 0 0 5 0 0 0 7

4.3   Left-to-Right Generation of (w)NAF

Although in the preceding section we have presented left-to-right generated signed representations that are at least as useful as (w)NAFs, from a theoretical point of view it is still an interesting question how to generate the (w)NAF from the most significant bit. The reason for the difficulty is a carry caused by the statement $d \leftharpoonup d - \delta_i$ of Algorithm 1. To illustrate the problem, note that the binary strings 101010 and 101011 that only differ in the last digit are converted to the NAFs 101010 and $10\bar{1}0\bar{1}0\bar{1}$, respectively, which differ completely. Intuitively, it is not possible to generate NAF left-to-right without scanning any higher bits. In this section we exploit the MOF representation to discuss how many bits have to be scanned and how many additional storage is required.

Note that we obtain NAF if we apply the conversions $1\bar{1} \mapsto 01$ and $\bar{1}1 \mapsto 0\bar{1}$ right-to-left on MOF. However, performing the same conversions left-to-right may yield a different result. The critical sequence is of the shape

$$0\underbrace{1\bar{1}\ldots\bar{1}}_{\text{odd}}0, \quad \text{or} \quad 0\underbrace{\bar{1}1\ldots\bar{1}}_{\text{odd}}0.$$

Note that this sequence corresponds to the binary string $1010\ldots011$. If the length of the sequence of alternating bits is even, then both of left-to-right and right-to-left conversions uniquely generate the same string, namely $\bar{b}b\ldots\bar{b}b \mapsto 0\bar{b}\ldots0\bar{b}$ for $b \in \{\pm 1\}$. But if the length is odd, left-to-right we obtain $\bar{b}b\ldots b\bar{b} \mapsto 0\bar{b}\ldots0\bar{b}\bar{b}$, whereas right-to-left generates $\bar{b}b\ldots b\bar{b} \mapsto \bar{b}0b0b\ldots0b$. Consequently, if this sequence appears, we have to scan it completely in order to compute the corresponding NAF. However, the first bit and the length of the critical sequence can uniquely determine the corresponding NAF, hence it is not necessary to store the sequence. Thus, the additional required storage in RAM is at most a few bits, namely the bit length of the critical sequence. Therefore, we obtain Algorithm 5.

Algorithm 5 Left-to-Right Generation Binary to NAF

Input: a non-zero n bit binary string $d = d_{n-1}|d_{n-2}|\ldots|d_1|d_0$
Output: NAF $\nu_n|\nu_{n-1}|\ldots|\nu_1|\nu_0$ of d
 $i \leftharpoonup n$; $d_n \leftharpoonup 0$; $d_{-1} \leftharpoonup 0$; $d_{-2} \leftharpoonup 0$
 while $i > -1$ **do**
  $b \leftharpoonup d_{i-1} - d_i$
  if $b = 0$ **then**
   $\nu_i \leftharpoonup 0$; $i \leftharpoonup i - 1$
  else $\{b \neq 0\}$
   find the largest j s.t. $d_{i-j-1} = d_{i-j}$
   if j is odd **then**
    $\nu_i \leftharpoonup b$; $\nu_{i-1} \leftharpoonup 0$; $\nu_{i-2} \leftharpoonup -b$; $\ldots$; $\nu_{i-j+2} \leftharpoonup 0$; $\nu_{i-j+1} \leftharpoonup -b$; $\nu_{i-j} \leftharpoonup 0$
   else $\{j$ is even$\}$
    $\nu_i \leftharpoonup 0$; $\nu_{i-1} \leftharpoonup b$; $\ldots$; $\nu_{i-j+2} \leftharpoonup 0$; $\nu_{i-j+1} \leftharpoonup b$; $\nu_{i-j} \leftharpoonup 0$
  $i \leftharpoonup i - j - 1$
 return $(\nu_n, \nu_{n-1}, \ldots., \nu_1, \nu_0)$

It is also possible to construct a left-to-right generation algorithm of wNAF, $w > 2$. In this case, the critical sequence is of the following shape

$$\underbrace{0\ldots0}_{w-1}a_ia_{i-1}\ldots a_1a_0\underbrace{0\ldots0}_{w-1}, \tag{1}$$

where the most and least $(w-1)$ bits are zero and no zero run of length $w-1$ appears in $a_ia_{i-1}\ldots a_1a_0$. If it is possible to convert the critical sequence (1) left-to-right to wNAF, then we can generate wNAF from any MOF. In order to find the corresponding wNAF of (1), we scan the whole sequence right-to-left and obtain the segmentations that are produced by the right-to-left sliding window conversion MOF $\mapsto w$NAF. Note that there is no need to store the width w windows, but we must detect and store the length of the zero runs between any two windows. In addition, the content of the left-most window, which may be smaller than w, has to be transfered. Afterwards, the sequence (1) can be rewritten as follows:

$$\underbrace{0\ldots0}_{w-1}|r|b_i|t_i|\ldots|b_2|t_2|b_1|t_1|\underbrace{0\ldots0}_{w-1}, \tag{2}$$

where r consists of at most $w-1$ consecutive bits of MOF (and may be the empty word ε), $b_j \in \{\varepsilon, 0, 00, \ldots, \underbrace{0\ldots0}_{w-2}\}$, and each t_j is a length w pattern of MOF, corresponding to an entry of Table$_w\overleftarrow{SW}$. Here we have to store r and the b_j. Based on these informations, the corresponding wNAF is completely determined left-to-right. Thus we need to store at most $(w - 1 + \log_2(w-2))\frac{n}{w})$ bits.

4.4 Comparison with Previous Methods

In this section we clarify the difference to previous schemes for generating signed representations.

In 1992, Koyama and Tsuruoka developed a new recoding technique to convert a binary string to a signed binary string [KT92]. Following this step, a left-to-right sliding window method is applied. The new signed binary representation has the benefit that it reduces the asymptotic non-zero density, but it requires the sub-optimal digit set $\mathcal{T} = \{\pm1, \pm3, \ldots, \pm(2^w - 3)\}$. If the sliding window method is directly applied to NAF, due to the NAF property fewer possible window contents have to be taken into account, resulting in a smaller digit set $\mathcal{T}$. An easy calculation shows that the largest odd NAF consisting of at most w digits equals $\frac{1}{3}(2^{w+1} - 1)$ for odd w (cf. $1010\ldots01$) and $\frac{1}{3}(2^{w+1} + 1) - 2$ for even w (cf. $1010\ldots1001$). For this reason, De Win et al. prefer the latter method for elliptic curve scalar multiplication [WMPW98]. Although there are slightly more point operations needed to evaluate the scalar multiplication if the exponent is represented as wNAF compared to the [WMPW98] representation, the required precomputation is less in the wNAF case because of the smaller digit set. Indeed, Blake et al. proved that wNAF is asymptotically better than sliding window on NAF schemes if $w > 3$ [BSS99]. In the context of memory constraint devices, a small digit set $\mathcal{T}$ is even more valuable, because fewer precomputed elements have to be stored. But as none of the preceding methods is a left-to-right scheme, each one requires *additional* memory $\mathcal{O}(n)$ to store the recoded string before starting the left-to-right evaluation of the scalar product. Note that in the context of sliding window on signed binary schemes like [KT92,WMPW98] the sliding window conversion may be performed left-to-right, but to obtain the signed binary representation we have to proceed right-to-left in either case.

In contrast, wMOF turns out as a complete left-to-right scheme. Consequently, there is no additional memory required for performing the scalar multiplication. In addition, due to the properties of MOF, the digit set of wMOF is the same as for wNAF and therefore minimal.

In order to compare the proposed algorithms with previous ones, we summarize the memory requirements of the new left-to-right schemes in the following theorem.

Theorem 5. *Algorithm 4 requires only $\mathcal{O}(w)$ bits memory for generating wMOF. Algorithm 5 requires at most $(\log_2 n)$ bits memory for generating NAF left-to-right. For general width w, there is a left-to-right algorithm that generates wNAF with at most $(w - 1 + \log_2(w - 2))\frac{n}{w})$ bit memory.*

Next, we compare the characterizing properties for the proposed schemes and some previous ones. In the second column, the value $\#\mathcal{T}/2$ equals the number of elements, that have to be precomputed and stored. In the last column, we describe the amount of memory (in bits) that is required additionally to this storage, e.g. to construct the signed representation or to store the converted string in right-to-left schemes. As usual, n equals the bit-length of the scalar, and SW is an abbreviation for sliding window.

Table 1. Comparison of Memory Requirement and Non-zero Density

Scheme	$\#\mathcal{T}/2$	1/N.-z. Density	Additional Memory
wNAF [Sol00,BSS99,MOC97]	2^{w-2}	$w + 1$	$\mathcal{O}(n)$
[KT92]	$2^{w-1} - 1$	$w + \frac{3}{2}$	$\mathcal{O}(n)$
NAF+SW as [WMPW98]	$\frac{1}{3}(2^w + (-1)^{w+1})$	$w + \frac{4}{3} - \frac{(-1)^w}{3 \cdot 2^{w-2}}$	$\mathcal{O}(n)$
wMOF, Sec. 4.2	2^{w-2}	$w + 1$	$\mathcal{O}(w)$
l-t-r wNAF, Sec. 4.3	2^{w-2}	$w + 1$	$\mathcal{O}(\log n),\ w = 2$ $\mathcal{O}(\frac{\log w}{w}n),\ w > 2$

5   Applications to Elliptic Curve Scalar Multiplication

Let $K = GF(p)$ be a finite field, where $p > 3$ is a prime. Let E be an elliptic curve over K. The elliptic curve E has an Abelian group structure with identity element $\mathcal{O}$ called the point of infinity. A point $P \in E$ is represented as $P = (x, y)$. The inverse of point $P = (x, y)$ is equal to $-P = (x, -y)$, hence it can be computed virtually for free. The elliptic curve additions $P_1 + P_2$ and $2P$ are denoted by ECADD and ECDBL, respectively, where $P_1, P_2, P \in E$.

As elliptic curves are written additively, exponentiation has to be understood as scalar multiplication. The familiar binary algorithms are adopted by computing ECADD instead of multiplying and ECDBL instead of squaring.

In general, we distinguish two main concepts of performing scalar multiplication: left-to-right and right-to-left. Here, d is represented as $d = \sum_{i=0}^{n} d_i 2^i$, $d_i \in \{0,1\}$, $d_{n-1} = 1$.

Algorithm Binary Method, l-t-r	**Algorithm** Binary Method, r-t-l
Input: P; $d = d_{n-1}|\ldots|d_1|d_0$	**Input:** P; $d = d_{n-1}|\ldots|d_1|d_0$
Output: scalar multiplication dP	**Output:** scalar multiplication dP
$Q \leftarrow P$	$Q_1 \leftarrow P$; $Q_2 \leftarrow \mathcal{O}$
for $i = n-2$ down to 0	**for** $i = 0$ to $n-1$
$\quad Q \leftarrow \text{ECDBL}(Q)$	$\quad$ **if** $d_i = 1$
$\quad$ **if** $d_i = 1$	$\quad\quad Q_2 \leftarrow \text{ECADD}(Q_2, Q_1)$
$\quad\quad Q \leftarrow \text{ECADD}(Q, P)$	$\quad Q_1 \leftarrow \text{ECDBL}(Q_1)$
return Q.	**return** Q_2.

Though in general both methods provide the same efficiency, the left-to-right method is preferable due to the following reasons:

1. The left-to-right method can be adjusted for general $\mathcal{T}$-representations of d like wNAF or wMOF in a more efficient way than the right-to-left method.
2. The ECADD step in the left-to-right method has the fixed input tP, $t \in \mathcal{T}$. Therefore it is possible to speed up these steps if tP is expressed in affine coordinates for each $t \in \mathcal{T}$, since some operations are negligible in this case. The improvement for a 160-bit scalar multiplication is about 15% with NAF over right-to-left scheme in the Jacobian coordinates [CMO98].
3. The right-to-left method needs an auxiliary register for storing $2^i P$.

5.1   Explicit Implementation for $w = 2$

In the following we show how the ideas of Section 4.2 lead to an efficient left-to-right scalar multiplication algorithm. For the sake of simplicity, we begin with the special case $w = 2$. The treatment for general width w can be found in the full version of this paper [OSST04].

Let d be a binary string. The MOF and 2MOF representation of d are denoted by μ and δ, respectively. The proposed scheme scans the two bits of μ from the most significant bit, and if the sequences $1\bar{1}$ or $\bar{1}1$ appear, we perform the following conversions: $1\bar{1} \mapsto 01$ and $\bar{1}1 \mapsto 0\bar{1}$. Two consecutive bits of d determine the corresponding bit of MOF μ. Thus, three consecutive bits of d can generate the corresponding bit of the 2MOF δ. In order to find an efficient implementation, we discuss the relationship of bit representation among μ, δ, and d. The i-th bits of μ, δ, d are denoted by μ_i, δ_i, d_i, respectively. Because of the relation $\mu_i = d_{i-1} - d_i$, we know $\mu_i = 0$ if and only if $d_{i-1} = d_i$. The other 3-bit binary strings (d_i, d_{i-1}, d_{i-2}) where $d_{i-1} \neq d_i$ are only $(d_i, d_{i-1}, d_{i-2}) = (0,1,1), (1,0,0), (0,1,0), (1,0,1)$, corresponding to $(\delta_i, \delta_{i-1}) = (1,0), (-1,0), (0,1), (0,-1)$. Thus, there is a one-to-one map between (δ_i, δ_{i-1}) and (d_i, d_{i-1}, d_{i-2}) leading to the explicit Algorithm 6.

Algorithm 6 Explicit Left-to-Right Generation of 2MOF

Input: a non-zero n-bit binary string $d = d_{n-1}|d_{n-2}|...|d_1|d_0$
Output: 2MOF $\delta = \delta_n|\delta_{n-1}|...|\delta_1|\delta_0$ of d

  $d_{-1} \leftarrow 0$
  $i \leftarrow c + 1$ for the largest c with $d_c \neq 0$
  $\delta_n \leftarrow 0$; $\delta_{n-1} \leftarrow 0$;...; $\delta_{i+1} \leftarrow 0$
  while $i \geq 1$ **do**
    if $d_{i-1} = d_i$ **then**
      $\delta_i \leftarrow 0$; $i \leftarrow i - 1$
    else $\{d_{i-1} \neq d_i\}$
      $\delta_i \leftarrow -d_i + d_{i-2}$; $\delta_{i-1} \leftarrow -d_{i-2} + d_{i-1}$; $i \leftarrow i - 2$
    if $i = 0$ **then**
      $\delta_0 \leftarrow -d_0$
  return $\delta_n, \delta_{n-1}, ..., \delta_1, \delta_0$.

Finally, Algorithm 7 merges the recoding stage and evaluation stage of scalar multiplication.

Algorithm 7 Left-to-Right Scalar Multiplication Algorithm (On the Fly), $w = 2$

Input: a point P, a non-zero n-bit binary string $d = d_{n-1}|d_{n-2}|...|d_1|d_0$
Output: product dP

  $d_{-1} \leftarrow 0$; $d_n \leftarrow 0$
  $i \leftarrow c + 1$ for the largest c with $d_c \neq 0$
  if $d_{i-2} = 0$ **then**
    $Q \leftarrow P$; $i \leftarrow i - 2$
  else $\{d_{i-2} = 1\}$
    $Q \leftarrow \text{ECDBL}(P)$; $i \leftarrow i - 2$
  while $i \geq 1$ **do**
    if $d_{i-1} = d_i$ **then**
      $Q \leftarrow \text{ECDBL}(Q)$; $i \leftarrow i - 1$
    else $\{d_{i-1} \neq d_i\}$
      $Q \leftarrow \text{ECDBL}(Q)$
      if $(d_i, d_{i-2}) = (1, 1)$ **then**
        $Q \leftarrow \text{ECDBL}(Q)$; $Q \leftarrow \text{ECADD}(Q, -P)$
      else if $(d_i, d_{i-2}) = (1, 0)$ **then**
        $Q \leftarrow \text{ECADD}(Q, -P)$; $Q \leftarrow \text{ECDBL}(Q)$
      else if $(d_i, d_{i-2}) = (0, 1)$ **then**
        $Q \leftarrow \text{ECADD}(Q, P)$; $Q \leftarrow \text{ECDBL}(Q)$
      else if $(d_i, d_{i-2}) = (0, 0)$ **then**
        $Q \leftarrow \text{ECDBL}(Q)$; $Q \leftarrow \text{ECADD}(Q, P)$
      $i \leftarrow i - 2$
  if $i = 0$ **then**
    $Q \leftarrow \text{ECDBL}(Q)$; $Q \leftarrow \text{ECADD}(Q, -d_0P)$
  return Q.

The advantage of the previous algorithm is that it reduces the memory requirement since it does not store the converted representation of d.

6   Conclusion

It was an unsolved problem to generate a signed representation *left-to-right* for a general width *w*. In this paper we presented a solution of this problem. The proposed scheme inherits the outstanding properties of *w*NAF, namely the set of pre-computed elements and the non-zero density are same as those of *w*NAF. In order to achieve a left-to-right exponent recoding, we defined a new canonical representation of signed binary strings, called the mutual opposite form (MOF). An *n*-bit integer can be uniquely represented by $(n+1)$-bit MOF, and this representation can be constructed efficiently left-to-right. Then the proposed exponent recoding is obtained by applying the width *w* (left-to-right) sliding window conversion to MOF. The proposed scheme is conceptually easy to understand and it is quite simple to implement. Moreover, if we apply the width *w* (right-to-left) sliding window conversion to MOF, we surprisingly obtain the classical *w*NAF. This is the first *carry-free* algorithm for generating *w*NAF. Therefore the proposed scheme has a lot of advantages and it promises to be a good alternative to *w*NAF. We believe that there will be many new applications of this algorithms for cryptography.

References

[Avi61]   Aviziensis, A., *Signed digit number representations for fast parallel arithmetic*, IRE Trans. Electron. Comput., 10:389-400, (1961).

[BSS99]   Blake, I., Seroussi, G., and Smart, N., *Elliptic Curves in Cryptography*, Cambridge University Press, 1999.

[BHLM01]   Brown, M., Hankerson, D., Lopez, J., and Menezes, A., *Software Implementation of the NIST Elliptic Curves Over Prime Fields*, Topics in Cryptology - CT-RSA 2001, LNCS 2020, (2001), 250-265.

[Boo51]   Booth, A., *A signed binary multiplication technique*, Journ. Mech. and Applied Math., 4(2), (1951), 236-240.

[CMO98]   Cohen, H., Miyaji, A., and Ono, T., *Efficient Elliptic Curve Exponentiation Using Mixed Coordinates*, Advances in Cryptology - ASIACRYPT '98, LNCS1514, (1998), 51-65.

[EK94]   Egecioglu, Ö, and Koc, C, *Exponentiation using Canonical Recoding*, Theoretical Computer Science, 129(2), (1994), 407-417.

[Gor98]   Gordon, D., *A survey of fast exponentiation methods*, Journal of Algorithms, vol.27, (1998), 129-146.

[IEEE]   IEEE P1363, Standard Specifications for Public-Key Cryptography. http://groupe.ieee.org/groups/1363/

[JM89]   Jedwab, J., and Mitchell, C.J., *Minimum Weight Modified Signed-digit Representations and Fast Exponentiation*, Electronics Letters 25, (1989), 1171-1172.

[JY00]   Joye, M., and Yen, S.-M., *Optimal Left-to-Right Binary Signed-digit Exponent Recoding*, IEEE Transactions on Computers 49(7), (2000), 740-748.

[Knu81]   Knuth, D. E., *The art of computer programmming, vol. 2, Seminumerical Algorithms*, 2nd ed., Addison-Wesley, Reading, Mass. (1981).

[Kob87]   Koblitz, N., *Elliptic Curve Cryptosystems*, Math. Comp. 48, (1987), 203-209.

[KT92]   Koyama, K. and Tsuruoka, Y., *Speeding Up Elliptic Curve Cryptosystems using a Signed Binary Windows Method*, Advances in Cryptology - CRYPTO '92, LNCS740, (1992), 345-357.

[Mil86]   Miller, V.S., *Use of Elliptic Curves in Cryptography*, Advances in Cryptology - CRYPTO '85, LNCS218, (1986), 417-426.

[MO90]   Morain, F., Olivos, J., *Speeding Up the Computations on an Elliptic Curve using Addition-Subtraction Chains*, Informa. Theor. Appl., 24, (1990), pp.531-543.

[MOC97]   Miyaji, A., Ono, T., and Cohen, H., *Efficient Elliptic Curve Exponentiation*, Information and Communication Security, ICICS 1997, LNCS 1334, (1997), 282-291.

[MOV96]   Menezes, A., van Oorschot, P. and Vanstone, S., *Handbook of Applied Cryptography*, CRC Press, 1996.

[Möl02]   Möller, B., *Improved Techniques for Fast Exponentiation*, The 5th International Conference on Information Security and Cryptology (ICISC 2002), LNCS 2587, (2003), 298-312.

[OCo99]   O'Connor, L., *An Analysis of Exponentiation Based on Formal Languages*, Advances in Cryptology - EUROCRYPT '99, LNCS1592, (1999), 375-388.

[OSST04]   Okeya, K., Schmidt-Samoa, K., Spahn, C., Takagi, T., *Signed Binary Representations Revisited*, Cryptology ePrint Archive (2004).
            http://eprint.iacr.org/

[Pro00]   Prodinger, H., *On Binary Representations of Integers with Digits {-1,0-1}*, Integers: Electronic Journal of Combinatorial Number Theory 0, (2000)

[Rei60]   Reitwiesner, G. W., *Binary arithmetic*, Advances in Computers, vol.1, (1960), 231-308.

[Sol00]   Solinas, J.A., *Efficient Arithmetic on Koblitz Curves*, Designs, Codes and Cryptography, 19, (2000), 195-249.

[WMPW98]   Win, E., Mister, S., Preneel, B., and Wiener, M., *On the Performance of Signature Schemes Based on Elliptic Curves*, Algorithmic Number Theory, ANTS-III, LNCS 1423, (1998), 252-266.

Compressed Pairings

Michael Scott[1],[*] and Paulo S.L.M. Barreto[2]

[1] School of Computing, Dublin City University
Ballymun, Dublin 9, Ireland
mike@computing.dcu.ie
[2] Escola Politécnica, Universidade de São Paulo
Av. Prof. Luciano Gualberto, tr. 3, 158
BR 05508-900, São Paulo(SP), Brazil
pbarreto@larc.usp.br

Abstract. Pairing-based cryptosystems rely on bilinear non-degenerate maps called pairings, such as the Tate and Weil pairings defined over certain elliptic curve groups. In this paper we show how to compress pairing values, how to couple this technique with that of point compression, and how to benefit from the compressed representation to speed up exponentiations involving pairing values, as required in many pairing based protocols.

Keywords: pairing-based cryptosystem, efficient implementation.

1 Introduction

With the discovery of a viable identity-based encryption scheme based on the Weil pairing [5], pairing-based cryptography has become of great interest to cryptographers. Since then, pairing-based protocols – many with novel properties – have been proposed for key exchange [30], digital signature [6], encryption [5], and signcryption [28]. Although the Weil pairing was initially proposed as a suitable construct for the realisation of such protocols, it is now usually accepted that the Tate pairing is preferable for its greater efficiency. Supersingular elliptic curves were originally proposed as a suitable setting for pairing-based schemes; recent work has shown that certain ordinary curves are equally suitable, and offer greater flexibility in the choice of security parameters [3, 26]. Fast computer algorithms for the computation of the Tate pairing on both supersingular and ordinary curves have been suggested in [1, 3, 12].

The Tate pairing calculation involves an application of Miller's algorithm [24] coupled to a final exponentiation to get a unique value. A typical protocol step requires the calculation of a pairing value followed by a further exponentiation of the result.

In this paper we explore the concept of *compressed pairings*, their efficient computation, and the subsequent processing (typically exponentiation) of pairing values. Our main contribution is to show that one can effectively reduce the

* Supported in part by Enterprise Ireland RIF grant IF/2002/0312/N

M. Franklin (Ed.): CRYPTO 2004, LNCS 3152, pp. 140–156, 2004.

bandwidth occupied by pairing values without impairing security nor processing time; in some cases, one even obtains a 30%–40% speed enhancement. Our work gives further motivation for the approach of Galbraith et al. [14], who investigate the bit security of pairing values and show that taking the trace causes no loss of security.

This paper is organized as follows. Section 2 introduces basic mathematical concepts. Section 3 discusses laddering exponentiation of pairing values, and introduces a laddering variant of the BKLS [1] algorithm to compute pairings. Section 4 describes how to compress pairing values to half length, and establishes a connection with the techniques of point compression and point reduction. Section 5 defines a ternary exponentiation ladder for finite fields in characteristic 3. Section 6 describes how to compress pairing values to one third of their length, presents a more efficient and slightly simpler version of the Duursma-Lee algorithm [11] that enables pairing computation in compressed form, and discusses improved variants of point compression and point reduction in characteristic 3. We summarise our work in section 7.

2 Mathematical Preliminaries

The theory behind elliptic curve cryptography is well documented in standard texts. The reader is referred to [23] for more background.

Let p be a prime number, m a positive integer and $\mathbb{F}_{p^m}$ the finite field with p^m elements; p is said to be the *characteristic* of $\mathbb{F}_p$, and m is its *extension degree*. Unless otherwise stated, we assume $p \neq 2$ throughout this paper.

Let $q = p^m$. An *elliptic curve* $E(\mathbb{F}_q)$ is the set of solutions (x, y) over $\mathbb{F}_q$ to an equation of form $E : y^2 + a_1 xy + a_3 y = x^3 + a_2 x^2 + a_4 x + a_6$, where $a_i \in \mathbb{F}_q$, together with an additional *point at infinity*, denoted O. The same equation defines curves over $\mathbb{F}_{q^k}$ for $k > 0$ (although note that the a_i remain in $\mathbb{F}_q$). The number of points on an elliptic curve $E(\mathbb{F}_{q^k})$, denoted $\#E(\mathbb{F}_{q^k})$, is called the *order* of the curve over the field $\mathbb{F}_{q^k}$.

An (additive) Abelian group structure is defined on E by the well known secant-and-tangent method [29]. Let $n = \#E(\mathbb{F}_{q^k})$. The order of a point $P \in E$ is the least nonzero integer r such that $rP = O$, where rP is the sum of r terms equal to P. The order of a point divides the curve order. For a given integer r, the set of all points $P \in E$ such that $rP = O$ is denoted $E[r]$. We say that $E[r]$ has *embedding degree* k if $r \mid q^k - 1$ and $r \nmid q^s - 1$ for any $0 < s < k$. In this paper we assume $k > 1$. It is in fact not difficult to find suitable curves with this property for relatively small values of k as described in [2, 7, 10]. We are interested here in curves where k is even, as this case facilitates fast calculation of the Tate pairing [3].

For our purposes, a *divisor* is a formal sum $\mathcal{A} = \sum_P a_P(P)$ of points on the curve $E(\mathbb{F}_{q^k})$. An Abelian group structure is defined on the set of divisors by the addition of corresponding coefficients in their formal sums; in particular, $n\mathcal{A} = \sum_P (n\, a_P)(P)$. The *degree* of a divisor $\mathcal{A}$ is the sum $\deg(\mathcal{A}) = \sum_P a_P$. Let $f : E(\mathbb{F}_{q^k}) \to \mathbb{F}_{q^k}$ be a function on the curve and let $\deg(\mathcal{A}) = 0$. We define

$f(\mathcal{A}) \equiv \prod_P f(P)^{a_P}$. The divisor of a function f is $(f) \equiv \sum_P \mathrm{ord}_P(f)(P)$. A divisor $\mathcal{A}$ is called *principal* if $\mathcal{A} = (f)$ for some function (f). A divisor $\mathcal{A}$ is principal if and only if $\deg(\mathcal{A}) = 0$ and $\sum_P a_P P = O$ [23, theorem 2.25]. Two divisors $\mathcal{A}$ and $\mathcal{B}$ are *equivalent*, $\mathcal{A} \sim \mathcal{B}$, if their difference $\mathcal{A} - \mathcal{B}$ is a principal divisor. Let $P \in E(\mathbb{F}_q)[r]$ where r is coprime to q, and let $\mathcal{A}_P$ be a divisor equivalent to $(P) - (O)$; under these circumstances the divisor $r\mathcal{A}_P$ is principal, and hence there is a function f_P such that $(f_P) = r\mathcal{A}_P = r(P) - r(O)$. The (reduced) *Tate pairing* of order r is the map $e_r : E(\mathbb{F}_q)[r] \times E(\mathbb{F}_{q^k}) \to \mathbb{F}_{q^k}^*$ given by $e_r(P, Q) = f_P(\mathcal{D})^{(q^k-1)/r}$ for some divisor $\mathcal{D} \sim (Q) - (O)$. The Tate pairing is bilinear and non-degenerate; assuming $k > 1$, one gets $e_r(P, Q) \neq 1$ if Q is chosen from a coset containing a point of order r which is linearly independent from P. The computation of $f_P(\mathcal{D})$ is achieved by an application of Miller's algorithm [24], whose output is only defined up to an r-th power in $\mathbb{F}_{q^k}^*$. The final exponentiation to the power of $(q^k - 1)/r$ is needed to produce a unique result, and it also makes it possible to compute $f_P(Q)$ rather than $f_P(\mathcal{D})$ [1]. Sometimes we will drop the r subscript of the Tate pairing, writing simply $e(P, Q)$.

2.1   Lucas Sequences

Lucas sequences provide a relatively cheap way of implementing $\mathbb{F}_{q^2}$ exponentiation in a subgroup whose order divides $q + 1$. They have been extensively studied in the literature, and a fast "laddering" algorithm for their computation has been developed [18, 19, 32], using ideas originally developed by Lehmer and Montgomery [20, 27]. Lucas sequences have been suggested as a suitable vehicle for certain public-key schemes (see [4]). The laddering algorithm can in fact be used as an alternative to the standard square-and-multiply approach to exponentiation in any Abelian group, but it is particularly well-suited for Lucas sequences and certain parameterisations of elliptic curves [19]. The authors of [19] go on to emphasise that the laddering algorithm requires very little memory, facilitates parallel computing, and has a natural resistance to side-channel attacks when used in a cryptographic context.

The Lucas sequence consists of a pair of functions $U_k, V_k : \mathbb{F}_q \times \mathbb{F}_q \to \mathbb{F}_q$. Commonly one is interested in computing $U_k(P, 1)$ and $V_k(P, 1)$ for some field element P, in which case we write simply $U_k(P)$ and $V_k(P)$ or omit the arguments altogether. For this distinguished case the sequences are defined as

$$U_0 = 0, \ U_1 = 1, \ U_{k+1} = PU_k - U_{k-1}$$
$$V_0 = 2, \ V_1 = P, \ V_{k+1} = PV_k - V_{k-1}$$

Only the V_k sequence needs to be explicitly evaluated, as we also have the relationship

$$U_k = (PV_k - 2V_{k-1})/(P^2 - 4)$$

The fast laddering algorithm is described in Appendix A. Lucas sequences are useful in the exponentiation of certain field elements, as we will see next.

3   Exponentiating Pairing Values

We consider first the case of embedding degree $k = 2$ (although the following discussion also covers the case $k = 2d$ with the substitution $q \to q^d$). Recall that we assume the characteristic to be odd.

We represent an element of the field $\mathbb{F}_{q^2}$ as $x + iy$, where $x, y \in \mathbb{F}_q$, and $i^2 = \delta$ for some quadratic non-residue $\delta \in \mathbb{F}_q$. Assume in what follows that all arithmetic is in the field $\mathbb{F}_q$.

The final exponentiation in this case consists of a raising to the power of $(q - 1)(q + 1)/r$. This can be considered in two parts – exponentiation to the power of $q - 1$ followed by exponentiation to the power of $(q + 1)/r$. Now if the output of Miller's algorithm is $x + iy \in \mathbb{F}_{q^2}$, then

$$(x + iy)^{q-1} = (x + iy)^q/(x + iy) = (x - iy)/(x + iy)$$

which is obviously much quicker than the standard square-and-multiply algorithm. The element $a + ib \equiv (x + iy)^{q-1}$ calculated in this fashion has the property:

$$a^2 - \delta b^2 = 1 \tag{1}$$

where $a^2 - \delta b^2$ is called the *norm* of $a + ib$; this property, easily verified by simple substitution, is maintained under any subsequent exponentiation. An element of this form in $\mathbb{F}_{q^2}$ is called *unitary* [16]. Also observe that $(a + ib)^{-1} = (a - ib)$ for a unitary element. In fact, any element of $\mathbb{F}_{q^2}$ whose order divides $q + 1$ will have this property.

A unitary element can obviously be determined up to the sign of b from a alone, using equation 1. And this is our first observation - the output of the Tate algorithm contains some considerable redundancy. It could be represented by a single element of $\mathbb{F}_q$ and a single bit to represent the sign of b, rather than as a full element of $\mathbb{F}_{q^2}$.

One can efficiently raise a unitary element of $\mathbb{F}_{q^2}$ to a power m by means of Lucas sequences. This is a consequence of the observation that

$$(a + bi)^m = V_m(2a)/2 + U_m(2a)bi,$$

as one can verify by induction. As pointed out above, only $V_m(2a)$ needs to be explicitly calculated.

If M is a multiplication and S a squaring in $\mathbb{F}_q$, then the computational cost of this method to compute $(a + bi)^m$ is therefore $1M + 1S$ per step, where a step involves the processing associated with a single bit of m (see appendix A). The conventional binary exponentiation algorithm in $\mathbb{F}_{q^2}$ takes 1 squaring and about $1/2$ multiplication in $\mathbb{F}_{q^2}$ for an overall cost of roughly $2S + 5M/2$ per step. If $\delta = -1$, then this can be reduced to $2S + 3M/2$ per step[1]. Thus the improved algorithm costs about 60% as much as the basic binary square-and-multiply method. When memory is not an issue the binary algorithm can be

---

[1] If $a + bi$ is unitary and $\delta = -1$, one can compute $(a + bi)^2$ as $(2a^2 - 1) + [(a + b)^2 - 1]i$, and $(a + bi)(c + di)$ as $(u - v) + (w - u - v)i$ where $u = ac$, $v = bd$, $w = (a + b)(c + d)$.

implemented by using windowing techniques, as described in [15]. However the laddering algorithm proposed here for unitary elements will always be faster than a conventional binary algorithm for a general element in $\mathbb{F}_{q^2}$.

Note that this improvement is relevant not only for the second part of the final exponentiation of the Tate pairing, but for any exponentiation directly involving pairing values, as happens in many pairing-based protocols [5, 17, 28].

3.1   A Laddering Pairing Algorithm

For $U, V \in E(\mathbb{F}_q)$, define $g_{U,V}$ to be the line through U and V. For all $a, b \in \mathbb{Z}$, the line function satisfies $(g_{aP,bP}) = (aP) + (bP) + (-[a+b]P) - 3(O)$.

Let $P \in E(\mathbb{F}_q)$, and for $c \in \mathbb{Z}$ let f_c be a function with divisor $(f_c) = c(P) - (cP) - (c-1)(O)$. One can show that $f_{a+b}(\mathcal{D}) = f_a(\mathcal{D}) \cdot f_b(\mathcal{D}) \cdot g_{aP,bP}(\mathcal{D}) / g_{[a+b]P,-[a+b]P}(\mathcal{D})$ up to a constant nonzero factor. This is called *Miller's formula*. In the computation of the Tate pairing $e_r(P, Q)$ for even k and a careful choice of P and Q (see [1, 3]), this formula can be simplified to $f_{a+b}(Q) = f_a(Q) \cdot f_b(Q) \cdot g_{aP,bP}(Q)$.

Let $(r_t, \ldots, r_0)_2$ be the binary representation of r. By coupling Miller's simplified formula with Montgomery's scalar multiplication ladder, we get a laddering version of the BKLS algorithm [1] to compute $e_r(P, Q)$:

Laddering BKLS algorithm to compute $e_r(P, Q)$:

$v_0 \leftarrow 1, \ v_1 \leftarrow 1$
$R_0 \leftarrow P, \ R_1 \leftarrow 2P$
for $i \leftarrow t - 1$ **downto** 0 **do**
    if $r_i = 0$ **then**
        $v_0 \leftarrow v_0^2 \cdot g_{R_0,R_0}(Q), \ R_1 \leftarrow R_0 + R_1$
        $R_0 \leftarrow 2R_0, \ v_1 \leftarrow v_0 \cdot g_{R_0,P}(Q)$
    else
        $v_1 \leftarrow v_1^2 \cdot g_{R_1,R_1}(Q), \ R_0 \leftarrow R_0 + R_1$
        $R_1 \leftarrow 2R_1, \ v_0 \leftarrow v_1 \cdot g_{R_1,-P}(Q)$
    end if
end for
return $v_0^{(q^k-1)/r}$

Although this algorithm has no computational advantage over the original BKLS, it may be useful in the same context of the laddering algorithms described in [19].

4   Compressing Pairings to Half Length

Instead of keeping the full $a + bi$ value of the Tate pairing, it may be possible for cryptographic purposes to discard b altogether, leaving the values defined only up to conjugation, which means one of the pairing arguments will only be defined up to a sign:

$$e(P,Q) = a + bi \Rightarrow a - bi = (a + bi)^{-1} = e(P,Q)^{-1} = e(P,-Q).$$

This is similar to the point reduction technique, whereby instead of keeping $Q = (x,y)$ one only keeps the abscissa x.

Definition 1. *The $\mathbb{F}_q$-trace of an element $u \in \mathbb{F}_{q^2}$ is the sum of the conjugates of u, $\mathrm{tr}(u) = u + u^q$.*

Notice that $\mathrm{tr}(a + ib) = (a + ib) + (a - ib) = 2a$, in effect discarding the imaginary part. We define the *compressed Tate pairing* $\varepsilon(P,Q)$ as $\mathrm{tr}(e(P,Q))$ [2].

4.1   Point Reduction

Point reduction is an optimization technique introduced by Miller in 1985 [25]. It consists of basing cryptographic protocols solely on the x coordinate of the points involved rather than using both coordinates. This setting is possible because the x coordinate of any multiple of a given point P depends only on the x coordinate of P. A related but less efficient technique is that of point compression, which consists of keeping not only the x coordinate but also a single bit β from the y coordinate to choose between the two roots $y_{\pm} = \pm\sqrt{x^3 + ax + b}$.

Some pairing-based cryptosystems have been originally defined to take profit from point reduction. An example is the BLS signature scheme [6], where the signature of a message represented by a curve point M under the signing key s is the x coordinate σ of the point $S = sM$. This means that, implicitly, the actual signature is $\pm S$ rather than S alone. To verify a BLS signature, the verifier checks whether $e(M,V) = e(\pm S, Q)$, where the verification key is $V = sQ$. Incidentally, the verification key itself can be reduced to its x coordinate (say, ξ), even though this possibility does not seem to have been considered by the authors of BLS.

4.2   Coupling Point Reduction with Compressed Pairings

Verifying a BLS signature involves computing a point $V \in \{V, -V\}$ from ξ, a point $S' \in \{S, -S\}$ from σ and checking whether $e_r(M, V') = e(S', Q)$ or $e(M, V') = e(S', Q)^{-1}$. Using the property that any pairing value z is unitary (and hence $z^{-1} = \bar{z}$), one can simply check whether $\mathrm{tr}(e(M, V')) = \mathrm{tr}(e(S', Q))$. This is especially interesting, since a compressed pairing $\varepsilon(P,Q)$ is precisely $\mathrm{tr}(e(\pm P, \pm Q))$.

An important aside is that exponentiation of compressed pairings must take into account the fact that they are actually traces of full pairings. This means one cannot exponentiate a pairing as if it were a simple $\mathbb{F}_{q^{k/2}}$ value; rather, one must always handle it as a Lucas sequence element.

---

[2] Rubin and Silverberg [13] use traces to compress BLS signatures, but in an entirely different manner, and with a compression factor much closer to 1.

5   A Ternary Exponentiation Ladder

Supersingular curves in characteristic 3 are a popular choice of underlying algebraic structure for pairing-based cryptosystems, since many optimisations are possible in such a setting [1, 11, 12]. Pairing compression is possible for those systems, and we now propose a ternary ladder for Lucas sequences in characteristic 3 that keeps the exponentiation cost in $\mathbb{F}_{q^k}$ within about 33% of the exponentiation cost in $\mathbb{F}_{q^{k/2}}$.

Assume the sequence element index is written in *signed* ternary notation, $K = (d_{t-1}, \ldots, d_0)_3$, with $d_{t-1} = 1$. At step j (counting downwards from $t - 1$ to 0), we want to compute V_{K_j} where $K_j = \sum_{i=j}^{t-1} d_i 3^{i-j}$. Thus, by definition, $K_j = 3K_{j+1} + d_j$.

For $d_j = -1$, we write down the formulas to compute $V_{3K_{j+1}-2}$, $V_{3K_{j+1}-1}$, and $V_{3K_{j+1}}$:

$$V_{3K_{j+1}} = V_{K_{j+1}}^3$$
$$V_{3K_{j+1}-1} = PV_{3K_{j+1}-2} - V_{K_{j+1}-1}^3$$
$$V_{3K_{j+1}-2} = PV_{3K_{j+1}-1} - V_{K_{j+1}}^3$$

Similarly, for $d_j = 1$ we write down the formulas to compute $V_{3K_{j+1}}$, $V_{3K_{j+1}+1}$, and $V_{3K_{j+1}+2}$:

$$V_{3K_{j+1}} = V_{K_{j+1}}^3$$
$$V_{3K_{j+1}+1} = PV_{3K_{j+1}+2} - V_{K_{j+1}+1}^3$$
$$V_{3K_{j+1}+2} = PV_{3K_{j+1}+1} - V_{K_{j+1}}^3$$

In each case, the second and third relations constitute a simple linear system. Solving them, we get these expressions for $V_{3K_{j+1}-1}$, $V_{3K_{j+1}}$, and $V_{3K_{j+1}+1}$:

$$
\begin{aligned}
V_{3K_{j+1}-1} &= (P^2 - 1)^{-1}(PV_{K_{j+1}}^3 + V_{K_{j+1}-1}^3) \\
&= (P^2 - 1)^{-1}[PV_{K_{j+1}}^3 + (PV_{K_{j+1}} - V_{K_{j+1}+1})^3] \\
&= (P^2 - 1)^{-1}[(P + P^3)V_{K_{j+1}}^3 - V_{K_{j+1}+1}^3] \\
V_{3K_{j+1}} &= V_{K_{j+1}}^3 \\
V_{3K_{j+1}+1} &= (P^2 - 1)^{-1}(PV_{K_{j+1}}^3 + V_{K_{j+1}+1}^3) \\
&= (P^2 - 1)^{-1}[PV_{K_{j+1}}^3 + (PV_{K_{j+1}} - V_{K_{j+1}-1})^3] \\
&= (P^2 - 1)^{-1}[(P + P^3)V_{K_{j+1}}^3 - V_{K_{j+1}-1}^3]
\end{aligned}
$$

If $(P^2 - 1)^{-1}$ and $P + P^3$ are precomputed, computing $V_{3K_{j+1}}$ and *one* of $V_{3K_{j+1}-1}$ or $V_{3K_{j+1}+1}$ involves two products and two cubes, and the computation can be carried out using only $V_{K_{j+1}}$ and *one* of $V_{K_{j+1}-1}$ or $V_{K_{j+1}+1}$. We can therefore keep track of which value between these two actually accompanies $V_{K_{j+1}}$, and compute V_{K_j} and V_{K_j+1} at the cost of only 2 products and two cubes per step. Besides, since we are working in characteristic 3, the cost of cubing is negligible compared to the cost of multiplying.

The binary ladder computes V_{K_j} and V_{K_j+1} at the cost of one squaring and one product, or about 1.8 product, per step. However, the step count of the ternary ladder is only about $1/\lg(3)$ of its binary counterpart, and hence its total cost is about 70% of the binary ladder. We point out that the ternary ladder can be used for plain exponentiation in characteristic 3 as an independent technique, even in contexts where compressed pairings are not desired or not an option.

A detailed ternary ladder algorithm is described in Appendix A.

6   Compressing Pairings to a Third of Their Length

Definition 2. *The* $\mathbb{F}_{q^2}$-*trace of an element* $f \in \mathbb{F}_{q^6}$ *is the value* $\mathrm{tr}(f) = f + f^{q^2} + f^{q^4} \in \mathbb{F}_{q^2}$.

The trace is $\mathbb{F}_{q^2}$-linear: $\mathrm{tr}(\alpha u) = \alpha \, \mathrm{tr}(u)$ for any $\alpha \in \mathbb{F}_{q^2}$ and $u \in \mathbb{F}_{q^6}$.

When the elliptic curve has an embedding degree $k = 6$, the Tate pairing algorithm outputs an element of $\mathbb{F}_{q^6}$ of order r, where r divides $q^6 - 1$, but not $q^i - 1$ for $0 < i < 6$. Now $q^6 - 1 = \Phi_1(q)\Phi_2(q)\Phi_3(q)\Phi_6(q)$. Therefore the output of the Tate pairing is an element of order r which divides $\Phi_6(q) = q^2 - q + 1$. For $q \equiv 2 \pmod{3}$, these are precisely the type of points considered in the XTR public key scheme [21] (which is based on the ideas of [8]), and all of the time/space optimizations that have been developed for this scheme [21,31] apply here as well. In particular, we note that laddering algorithms again appear to be optimal [31], and the Tate pairing output can be represented by its $\mathbb{F}_{q^2}$-trace, and hence compressed by a factor of 3. Observe that the compressed value, being a trace, must be implicitly exponentiated using the Lenstra-Verheul algorithm [21, Algorithm 2.3.7] – the trace value *per se* is not even a point of order r.

For supersingular curves in characteristic 3 we can do better than merely take the trace – rather, it is possible to do nearly all computations without resorting to arithmetic any more complex than that on $\mathbb{F}_{q^2}$.

6.1   Simpler Arithmetic for Pairing Computation in Characteristic 3

Let $q = 3^m$ for some $m \equiv 1, 5 \pmod{6}$, let $b = \pm 1$, and let $\sigma, \rho \in \mathbb{F}_{q^6}$ be elements satisfying $\sigma^2 + 1 = 0$ and $\rho^3 - \rho - b = 0$. The *modified Tate pairing* on the supersingular curve $E(\mathbb{F}_{3^m}) : y^2 = x^3 - x + b$ is the mapping $\hat{e}_r(P,Q) = f_P(\phi(Q))^{(q^6-1)/r}$ where $\phi : E(\mathbb{F}_q) \to E(\mathbb{F}_{q^6})$ is the distortion map $\phi(x,y) = (\rho - x, \sigma y)$.

Duursma and Lee showed [11, Theorem 5] that the modified Tate pairing for points $P = (\alpha, \beta)$ and $Q = (x, y)$ can be written as a product of factors of form $g = \beta y \bar{\sigma} - (\alpha + x - \rho + b)^2$. This expression can be rewritten as $g = \lambda - \mu\rho - \rho^2$, where $\mu \equiv \alpha + x + b \in \mathbb{F}_q$ and $\lambda \equiv \beta y \bar{\sigma} - \mu^2 \in \mathbb{F}_{q^2}$. Specifically, the Duursma-Lee algorithm to compute $f_P(\phi(Q))$ is as follows (cf. [11, Algorithm 4]):

Duursma-Lee algorithm to compute $f_P(\phi(Q))$:

$f \leftarrow 1$
for $i \leftarrow 1$ **to** m **do**
    $\alpha \leftarrow \alpha^3, \quad \beta \leftarrow \beta^3$
    $\mu \leftarrow \alpha + x + b, \quad \lambda \leftarrow \beta y \bar{\sigma} - \mu^2$
    $g \leftarrow \lambda - \mu\rho - \rho^2, \quad f \leftarrow f \cdot g$
    $x \leftarrow x^{1/3}, \quad y \leftarrow y^{1/3}$
end for
return f

The output is an element $f \in \mathbb{F}_{q^6}$. We now show that this algorithm can be modified to compute $\text{tr}(f)$ instead, by maintaining a ladder of three values $[\text{tr}(f), \text{tr}(f\rho), \text{tr}(f\rho^2)]$. Since f is initialized to 1, the initial ladder can be computed from ρ alone, namely, $[\text{tr}(1), \text{tr}(\rho), \text{tr}(\rho^2)] = [0, 0, 2]$, as one readily deduces from the definition of ρ:

Theorem 1. *Let $q = 3^m$ for some $m \equiv 1, 5 \pmod 6$, and let $\rho \in \mathbb{F}_{q^6}$ satisfy $\rho^3 - \rho - b = 0$. Then $\text{tr}(\rho) = 0$ and $\text{tr}(\rho^2) = 2$.*

Proof. From $\rho^3 = \rho + b$ it follows by induction that $\rho^{3^n} = \rho + nb$, and hence $\rho^{q^2} = \rho^{3^{2m}} = \rho + 2mb$ and $\rho^{q^4} = \rho^{3^{4m}} = \rho + mb$, so that $\text{tr}(\rho) = \rho + \rho^{q^2} + \rho^{q^4} = \rho + \rho + 2mb + \rho + mb = 0$. Moreover, $(\rho^2)^{3^n} = (\rho^{3^n})^2 = (\rho + nb)^2 = \rho^2 - nb\rho + n^2$, so that $\text{tr}(\rho^2) = \rho^2 + (\rho^2)^{q^2} + (\rho^2)^{q^4} = \rho^2 + \rho^2 - 2mb\rho + (2m)^2 + \rho^2 - mb\rho + m^2 = 2$. $\square$

At each step of the loop, we compute $[\text{tr}(fg), \text{tr}(fg\rho), \text{tr}(fg\rho^2)]$ according to the following theorem:

Theorem 2.

$$\begin{bmatrix} \text{tr}(fg) \\ \text{tr}(fg\rho) \\ \text{tr}(fg\rho^2) \end{bmatrix} = A \cdot \begin{bmatrix} \text{tr}(f) \\ \text{tr}(f\rho) \\ \text{tr}(f\rho^2) \end{bmatrix}, \quad where \ A \equiv \begin{bmatrix} \lambda & -\mu & -1 \\ -b & (\lambda - 1) & -\mu \\ -b\mu & -(\mu + b) & (\lambda - 1) \end{bmatrix}.$$

Proof. Using the $\mathbb{F}_{q^2}$-linearity of the trace and the defining property $\rho^3 = \rho + b$, we have $fg = f(\lambda - \mu\rho - \rho^2) \implies \text{tr}(fg) = \lambda\,\text{tr}(f) - \mu\,\text{tr}(f\rho) - \text{tr}(f\rho^2)$. Similarly, $fg\rho = f(\lambda - \mu\rho - \rho^2)\rho = \lambda f\rho - \mu f\rho^2 - f\rho - bf \implies \text{tr}(fg\rho) = -b\,\text{tr}(f) + (\lambda - 1)\,\text{tr}(f\rho) - \mu\,\text{tr}(f\rho^2)$. Finally, $fg\rho^2 = -bf\rho + (\lambda - 1)f\rho^2 - \mu f\rho - \mu bf \implies \text{tr}(fg\rho^2) = -\mu b\,\text{tr}(f) - (\mu + b)\,\text{tr}(f\rho) + (\lambda - 1)\,\text{tr}(f\rho^2)$. $\square$

Therefore, defining $L \equiv [L_0, L_1, L_2]^T = [\text{tr}(f), \text{tr}(f\rho), \text{tr}(f\rho^2)]^T$ and using the matrix A defined above, the modified algorithm to compute pairing traces reads:

A laddering algorithm to compute $\operatorname{tr}(f_P(\phi(Q)))$:

$L \leftarrow [0,0,2] \quad // = [\operatorname{tr}(1), \operatorname{tr}(\rho), \operatorname{tr}(\rho^2)]$
for $i \leftarrow 1$ **to** m **do**
$\quad \alpha \leftarrow \alpha^3, \quad \beta \leftarrow \beta^3$
$\quad \mu \leftarrow \alpha + x + b, \quad \lambda \leftarrow \beta y \bar{\sigma} - \mu^2$
$\quad L \leftarrow A \cdot L$
$\quad x \leftarrow x^{1/3}, \quad y \leftarrow y^{1/3}$
end for
return L_0

However, to obtain a unique pairing value suitable for pairing-based protocols we need $\operatorname{tr}(f_P(\phi(Q)))^{(q^6-1)/r})$ rather than $\operatorname{tr}(f_P(\phi(Q)))$. Let $e = f_P(\phi(Q))$. The simplest (and seemingly the most efficient) way to do it is to recover e from all three components of $L = [\operatorname{tr}(e), \operatorname{tr}(e\rho), \operatorname{tr}(e\rho^2)]$.

We use the $\mathbb{F}_{q^2}$-linearity of the trace and fact that $\{1, \rho, \rho^2\}$ is a basis of $\mathbb{F}_{q^6}$ with respect to $\mathbb{F}_{q^2}$, i.e. any element $e \in \mathbb{F}_{q^6}$ can be written as $e = x + y\rho + z\rho^2$ where $x, y, z \in \mathbb{F}_{q^2}$. The trick is straightforward:

1. $L_0 = \operatorname{tr}(e) = \operatorname{tr}(x + y\rho + z\rho^2) = x\operatorname{tr}(1) + y\operatorname{tr}(\rho) + z\operatorname{tr}(\rho^2) = 2z \implies z = -L_0$.
2. $L_1 = \operatorname{tr}(e\rho) = \operatorname{tr}(x\rho + y\rho^2 + z(\rho + b)) = bz\operatorname{tr}(1) + (x + z)\operatorname{tr}(\rho) + y\operatorname{tr}(\rho^2) = 2y \implies y = -L_1$.
3. $L_2 = \operatorname{tr}(e\rho^2) = \operatorname{tr}(x\rho^2 + y(\rho + b) + z(\rho^2 + b\rho)) = by\operatorname{tr}(1) + (y + bz)\operatorname{tr}(\rho) + (x + z)\operatorname{tr}(\rho^2) = 2(x + z) \implies x = L_0 - L_2$.

Thus we recover e from the pairing ladder essentially for free. Now one must compute $g = e^{(q^6-1)/r}$, and then take the trace of g. This can be efficiently done using the techniques described in [1, Appendix A.2], at a cost roughly equivalent to a few extra steps of the laddering algorithm.

Each step of this laddering algorithm takes 17 $\mathbb{F}_q$ multiplications. This compares well with the original Duursma-Lee algorithm where each step takes 20 $\mathbb{F}_q$ multiplications, and avoids $\mathbb{F}_{q^6}$ arithmetic in the main loop.

6.2   Implicit Exponentiation in Characteristic 3

It is quite commonplace that the pairing value undergoes further exponentiation as dictated by the underlying cryptographic protocol. We are thus confronted with the task of computing $\operatorname{tr}(g^m)$ given the value of $\operatorname{tr}(g)$. The Lenstra-Verheul algorithm [21, Algorithm 2.3.7] performs this task for characteristic $p \equiv 2 \pmod 3$. We now describe a variant tailored for characteristic 3.

Let $c \in \mathbb{F}_{q^2}$, and let $F(c, X) \equiv X^3 - cX^2 + c^q X - 1 \in \mathbb{F}_{q^2}[X]$ with roots $h_0, h_1, h_2 \in \mathbb{F}_{q^6}$. One can show [21, Lemma 2.2.1] that, if $g \in \mathbb{F}_{q^6}$ is an element of order dividing $\Phi_6(q) = q^2 - q + 1$, then the roots of $F(\operatorname{tr}(g), X)$ are the $\mathbb{F}_{q^2}$-conjugates of g. Defining $c_n \equiv h_0^n + h_1^n + h_2^n$, one can further show [21, Lemmas 2.3.2 and 2.3.4] (see also [9]) that $c_{-n} = c_n^q$ and $c_{u+v} = c_u c_v - c_v^q c_{u-v} + c_{u-2v}$. The proofs of these properties are independent of the field characteristic.

From the above properties, one easily deduces the following relations that hold in characteristic 3:

$$c_{2n} = c_n^2 + c_n^q$$
$$c_{3n} = c_n^3$$
$$c_{3n-1} = c_{2n} \cdot c_{n-1} - c_{n-1}^q \cdot c_{n+1} + c_2$$
$$c_{3n-2} = c^{-q} \cdot (c_{n-1} - c_n)^3 + c^{1-q} \cdot c_{3n-1}$$
$$c_{3n+1} = c_{2n} \cdot c_{n+1} - c_{n+1}^q \cdot c_{n-1} + c_2^q$$
$$c_{3n+2} = c^{-1} \cdot (c_{n+1} - c_n)^3 + c^{q-1} \cdot c_{3n+1}$$

Computing c_{2n} takes two $\mathbb{F}_q$ multiplications, $c_{3n\pm1}$ takes four $\mathbb{F}_q$ multiplications, and $c_{3n\pm2}$ takes six $\mathbb{F}_q$ multiplications.

Define $L_n(c) \equiv \langle c_{3n}, c_{3n+1}, c_{3n+2}, c_{3n+3} \rangle \in (\mathbb{F}_{q^2})^3$. Using the above formulas, one can compute any one of $L_{3n}(c)$, $L_{3n+1}(c)$, or $L_{3n+2}(c)$ from $L_n(c)$ at the cost of 12 $\mathbb{F}_q$ multiplications:

$$L_{3n} = \langle c_{9n}, \ c_{9n+1}, c_{9n+2}, c_{9n+3} \rangle = \langle c_{3(3n)}, \ c_{3(3n+1)-2}, c_{3(3n+1)-1}, c_{3(3n+1)} \rangle$$
$$L_{3n+1} = \langle c_{9n+3}, c_{9n+4}, c_{9n+5}, c_{9n+6} \rangle = \langle c_{3(3n+1)}, c_{3(3n+1)+1}, c_{3(3n+2)-1}, c_{3(3n+2)} \rangle$$
$$L_{3n+2} = \langle c_{9n+6}, c_{9n+7}, c_{9n+8}, c_{9n+9} \rangle = \langle c_{3(3n+2)}, c_{3(3n+2)+1}, c_{3(3n+2)+2}, c_{3(3n+3)} \rangle$$

From the definition of c_n, it is clear that $c_n = \mathrm{tr}(g^n)$ if $c = \mathrm{tr}(g)$. Hence, if $L_{\lfloor n/3 \rfloor}(\mathrm{tr}(g)) = \langle S_0, S_1, S_2, S_3 \rangle$, then $\mathrm{tr}(g^n) = S_{n \bmod 3}$. The total cost of this algorithm, about $7.6 \lg n$ $\mathbb{F}_q$ multiplications, matches the complexity of the ternary ladder introduced in section 5 for $\mathbb{F}_{q^3}$-trace exponentiation. Appendix B lists this algorithm in detail. We point out that this ternary ladder can also be the basis of a characteristic 3 variant of the XTR cryptosystem.

6.3   Coupling Pairing Compression with Point Reduction

A nice feature of this algorithm is that it is compatible with a variant of the point reduction technique.

The conventional approach to compress a point $R = (u, v)$ is to keep only u and a single bit of v; point reduction discards v altogether. In characteristic 3, it is more advantageous to discard u instead, keeping v and a trit of u to distinguish among the solutions of the curve equation $u^3 - u + (b - v^2) = 0$; alternatively, one can reduce R by keeping only v and modifying the cryptographic protocols to allow for any of the three points R_0, R_1, and R_2 that share the same v. Thus, we will show that the input to the laddering algorithm of section 6.1 can be only y (or β); the corresponding x (or α) can be easily recovered except for a trit, and the actual choice of this trit does not affect the compressed pairing value.

Let $z \in \mathbb{F}_{q^6}$ where $q = 3^m$ for odd m, and assume the order r of z divides $\Phi_6(q)$, i.e. $r \mid q^2 - q + 1$. The conjugates of z are z, z^{q^2}, and z^{q^4}, or equivalently z, z^{q-1}, and z^{-q}, since $q^2 \equiv q - 1 \pmod{r}$ and $q^4 \equiv -q \pmod{r}$. The trace of z is the sum of the conjugates, $\mathrm{tr}(z) = z + z^{q-1} + z^{-q}$ [21]. Consider the supersingular elliptic curve $E : y^2 = x^3 - x + b$, $b \in \{-1, 1\}$, whose order is [23, section 5.2.2] $n = q + 1 - t = 3^m + 1 \pm 3^{(m+1)/2}$, where $t = \pm 3^{(m+1)/2}$ is the trace of the Frobenius.

Let $P = (x, y) \in E(\mathbb{F}_q)$, and let $Q \in E(\mathbb{F}_{q^6})$ be a linearly independent point. The conjugates of $e(P, Q)$ are $e(P, Q)$, $e(P, Q)^{q-1} = e([q-1]P, Q)$, and $e(P, Q)^{-q} = e(-qP, Q)$. The following property holds:

Lemma 1. *If $P \in E[r]$, points P, $[q-1]P$, and $-qP$ share precisely the same y coordinate.*

Proof. Let $P = (x, y)$. A simple inspection of the group law for characteristic 3 [1] reveals that $3P = (x^9 - b, -y^9)$, and hence $3^j P = (x^{9^j} - jb, (-1)^j y^{9^j})$. Thus $[q-1]P = q^2 P = 3^{2m} P = (x^{9^{2m}} - 2mb, (-1)^{2m} y^{9^{2m}}) = (x^{3^{4m}} + mb, y^{3^{4m}}) = (x + mb, y)$, where we used the fact that $u^{3^m} = u$ for any $u \in \mathbb{F}_{3^m}$. Similarly, $-qP = q^2(q^2 P) = q^2(x + mb, y) = (x - mb, y)$. $\square$

We see that, for $m \not\equiv 0 \pmod 3$, the x coordinates of P, $[q-1]P$, and $-qP$ are the three solutions to $x^3 - x + (1 - y^2) = 0$, which are exactly $\{x, x+1, x+2\}$. Obviously, the traces of the pairings computed from the conjugates of P are all equal, since $\operatorname{tr}(e(P, Q))$ is simply the sum of the conjugates of $e(P, Q)$. Thus, the actual solution x to the curve equation above used to compute $\operatorname{tr}(e(P, Q))$ is irrelevant. Also, computing x from y is very efficient, since it amounts to solving a linear system (see appendix C).

7   Conclusions

We have introduced the notion of *compressed pairings*, and suggested how they can be realised as traces of ordinary Tate pairings. We also described how compressed pairings can be computed and implicitly exponentiated by means of laddering algorithms, with a compression ratio of $1/2$ in characteristic $p > 3$ and $1/3$ in characteristic 3; our algorithms thus reduce bandwidth requirements without impairing performance. Finally, we showed how to couple compressed pairings with the technique of point compression or point reduction. As a side result, we proposed an efficient laddering algorithm for plain exponentitation in characteristic 3, which can be used even in contexts where compressed pairings are not desired.

Our work constitutes evidence that the security of pairing-based cryptosystems is linked to the security of the Lucas/XTR schemes, and gives further motivation for the approach of Galbraith *et al.* regarding the use of traces to prevent security losses.

We leave it as an open problem to find a method to compute pairings directly in compressed form when the compression ratio is $1/3$ or better on ordinary (non-supersingular) curves in characteristic $p > 3$.

Acknowledgements

We are grateful to Steven Galbraith, Robert Granger, and Waldyr Benits Jr. for their valuable comments during the preparation of this work, and to the anonymous referees for their improvement suggestions.

References

1. P. S. L. M. Barreto, H. Y. Kim, B. Lynn, and M. Scott. Efficient algorithms for pairing-based cryptosystems. In *Advances in Cryptology – Crypto'2002*, volume 2442 of *Lecture Notes in Computer Science*, pages 354–368, Santa Barbara, USA, 2002. Springer-Verlag.
2. P. S. L. M. Barreto, B. Lynn, and M. Scott. Constructing elliptic curves with prescribed embedding degrees. In *Security in Communication Networks – SCN'2002*, volume 2576 of *Lecture Notes in Computer Science*, pages 263–273, Amalfi, Italy, 2002. Springer-Verlag.
3. P. S. L. M. Barreto, B. Lynn, and M. Scott. On the selection of pairing-friendly groups. In *Selected Areas in Cryptography – SAC'2003*, Ottawa, Canada, 2003. to appear.
4. D. Bleichenbacher, W. Bosma, and A. K. Lenstra. Some remarks on lucas-based cryptosystems. In *Advances in Cryptology – Crypto'95*, volume 963 of *Lecture Notes in Computer Science*, pages 386–396, Santa Barbara, USA, 1995. Springer-Verlag.
5. D. Boneh and M. Franklin. Identity-based encryption from the Weil pairing. *SIAM Journal of Computing*, 32(3):586–615, 2003.
6. D. Boneh, B. Lynn, and H. Shacham. Short signatures from the Weil pairing. In *Advances in Cryptology – Asiacrypt'2001*, volume 2248 of *Lecture Notes in Computer Science*, pages 514–532, Gold Coast, Australia, 2002. Springer-Verlag.
7. F. Brezing and A. Weng. Elliptic curves suitable for pairing based cryptography. Cryptology ePrint Archive, Report 2003/143, 2003. Available from http://eprint.iacr.org/2003/143.
8. A. E. Brouwer, R. Pellikaan, and E. R. Verheul. Doing more with fewer bits. In *Advances in Cryptology – Asiacrypt'99*, volume 1716 of *Lecture Notes in Computer Science*, pages 321–332, Singapore, 1999. Springer-Verlag.
9. L. Carlitz. Recurrences of the third order and related combinatorial identities. *Fibonacci Quarterly*, 16(1):11–18, 1978.
10. R. Dupont, A. Enge, and F. Morain. Building curves with arbitrary small MOV degree over finite prime fields. Cryptology ePrint Archive, Report 2002/094, 2002. Available from http://eprint.iacr.org/2002/094.
11. I. Duursma and H.-S. Lee. Tate pairing implementation for hyperelliptic curves $y^2 = x^p - x + d$. In *Advances in Cryptology – Asiacrypt'2003*, volume 2894 of *Lecture Notes in Computer Science*, pages 111–123, Taipei, Taiwan, 2003. Springer-Verlag.
12. S. Galbraith, K. Harrison, and D. Soldera. Implementing the Tate pairing. In *Algorithmic Number Theory Symposium – ANTS V*, volume 2369 of *Lecture Notes in Computer Science*, pages 324–337, Sydney, Australia, 2002. Springer-Verlag.
13. S. Galbraith, K. Harrison, and D. Soldera. Using primitive subgroups to do more with fewer bits. In *Algorithmic Number Theory Symposium – ANTS VI*, volume 3076 of *Lecture Notes in Computer Science*, pages 18–41, Annapolis, USA, 2004. Springer-Verlag.

14. S. Galbraith, H. Hopkins, and I. Shparlinski. Secure bilinear diffie-hellman bits. Cryptology ePrint Archive, Report 2002/155, 2002. Available from http://eprint.iacr.org/2002/155.

15. D. Gordon. A survey of fast exponentiation methods. *Journal of Algorithms*, 27:129–146, 2002.

16. K. Hoffman and R. Kunze. *Linear Algebra*. Prentice Hall, New Jersey, USA, 2nd edition, 1971.

17. A. Joux. A one-round protocol for tripartite Diffie-Hellman. In *Algorithmic Number Theory Symposium – ANTS IV*, volume 1838 of *Lecture Notes in Computer Science*, pages 385–394, Leiden, The Netherlands, 2000. Springer-Verlag.

18. M. Joye and J. J. Quisquater. Efficient computation of full Lucas sequences. *Electronics Letters*, 32(6):537–538, 1996.

19. M. Joye and S. Yen. The montgomery powering ladder. In *Cryptographic Hardware and Embedded Systems - CHES'2002*, volume 2523 of *Lecture Notes in Computer Science*, pages 291–302, Berlin, Germany, 2003. Springer-Verlag.

20. D. H. Lehmer. Computer technology applied to the theory of numbers. In W. J. LeVeque, editor, *Studies in Number Theory*, volume 6 of *MAA Studies in Mathematics*, pages 117–151. Math. Assoc. Amer. (distributed by Prentice-Hall, Englewood Cliffs, N.J.), 1969.

21. A. K. Lenstra and E. R. Verheul. The xtr public key system. In *Advances in Cryptology – Crypto'2000*, volume 1880 of *Lecture Notes in Computer Science*, pages 1–19, Santa Barbara, USA, 2000. Springer-Verlag.

22. R. Lidl and H. Niederreiter. *Finite Fields*. Number 20 in Encyclopedia of Mathematics and its Applications. Cambridge University Press, Cambridge, UK, 2nd edition, 1997.

23. A. Menezes. *Elliptic Curve Public Key Cryptosystems*. Kluwer Academic Publishers, 1993.

24. V. S. Miller. Short programs for functions on curves. Unpublished manuscript, 1986. Available from http://crypto.stanford.edu/miller/miller.pdf.

25. V. S. Miller. Use of elliptic curves in cryptography. In *Advances in Cryptology – Crypto'85*, volume 218 of *Lecture Notes in Computer Science*, pages 417–426, Santa Barbara, USA, 1986. Springer-Verlag.

26. A. Miyaji, M. Nakabayashi, and S. Takano. New explicit conditions of elliptic curve traces for FR-reduction. *IEICE Transactions on Fundamentals*, E84-A(5):1234–1243, 2001.

27. P. L. Montgomery. Speeding the pollard and elliptic curve methods of factorization. *Mathematics of Computation*, 48(177):243–264, 1987.

28. D. Nalla and K. C. Reddy. Signcryption scheme for identity-based cryptosystems. Cryptology ePrint Archive, Report 2003/066, 2002. Available from http://eprint.iacr.org/2003/066.

29. J. H. Silverman. *The Arithmetic of Elliptic Curves*. Number 106 in Graduate Texts in Mathematics. Springer-Verlag, Berlin, Germany, 1986.

30. N. P. Smart. An identity based authenticated key agreement protocol based on the weil pairing. *Electronics Letters*, 38:630–632, 2002.

31. M. Stam and A. K. Lenstra. Speeding up XTR. In *Advances in Cryptology – Asiacrypt'2001*, volume 2248 of *Lecture Notes in Computer Science*, pages 125–143, Gold Coast, Australia, 2001. Springer-Verlag.

32. S. M. Yen and C. S. Laih. Fast algorithms for LUC digital signature computation. *IEE Proceedings on Computers and Digital Techniques*, 142(2):165–169, 1995.

A   Computation of Lucas Sequence Elements

The Lucas sequence $V_n(P, 1)$ for some field element P is defined by the following recurrence relations:

$$V_0 = 2, \; V_1 = P, \; V_{n+1} = PV_n - V_{n-1}.$$

Let $n = (n_t \ldots n_0)_2$ be an integer in binary representation, with $n_t = 1$. The Lucas sequence element $V_n(P, 1)$ can be computed as:

$v_0 \leftarrow 2, \; v_1 \leftarrow P$
for $j \leftarrow t$ downto 0 do
  if $n_j = 1$ then
    $v_0 \leftarrow v_0 v_1 - P, \; v_1 \leftarrow v_1^2 - 2$
  else
    $v_1 \leftarrow v_0 v_1 - P, \; v_0 \leftarrow v_0^2 - 2$
  end if
end for
return v_0

Let $n = (n_t \ldots n_0)_3$ be the *signed* ternary representation of $n \geqslant 0$. The Lucas sequence element $V_n(P, 1)$ in characteristic 3 (as needed for the implicit exponentiation of $\mathbb{F}_{q^3}$-traces of $\mathbb{F}_{q^6}$ values) can be computed using the following algorithm:

$\mu \leftarrow (P^2 - 1)^{-1}, \; T \leftarrow P + P^3$
$v_0 \leftarrow 2, \; v_1 \leftarrow P, \; up \leftarrow$ true
for $j \leftarrow t$ downto 0 do
  $w \leftarrow v_0^3$
  if $n_j = -1$ then
    $v_0 \leftarrow$ if up then $\mu(Tw - v_1^3)$ else $\mu(Pw + v_1^3)$
    $v_1 \leftarrow w$
    $up \leftarrow$ true
  else if $n_j = 1$ then
    $v_0 \leftarrow$ if up then $\mu(Pw + v_1^3)$ else $\mu(Tw - v_1^3)$
    $v_1 \leftarrow w$
    $up \leftarrow$ false
  else /* $n_j = 0$ */
    $v_1 \leftarrow$ if up then $\mu(Pw + v_1^3)$ else $\mu(Tw - v_1^3)$
    $v_0 \leftarrow w$
    $up \leftarrow$ true
  end if
end for
return v_0

B   Implicit Exponentiation of $\mathbb{F}_{q^{k/3}}$-Traces

Let $n = (n_t \ldots n_0)_3$ be the plain ternary representation of $n \geqslant 0$. The following algorithm computes the $\mathbb{F}_{q^2}$-trace $c_n \equiv \text{tr}(g^n)$ of an element $g \in \mathbb{F}_{q^6}$ from its $\mathbb{F}_{q^2}$-trace $c \equiv \text{tr}(g)$.

$c^{-1} \leftarrow c^q \cdot (c^q \cdot c)^{-1}$ // N.B. $(c^q \cdot c) \in \mathbb{F}_q$
$c^{q-1} \leftarrow c^q \cdot c^{-1}$, $c^{-q} \leftarrow (c^{-1})^q$, $c^{1-q} \leftarrow (c^{q-1})^q$, $c_2 \leftarrow c^2 + c^q$
$S_0 \leftarrow 0$, $S_1 \leftarrow c$, $S_2 \leftarrow c_2$, $S_3 \leftarrow c^3$
for $j \leftarrow t$ **downto** 0 **do**
    if $n_j = 0$ **then**
        $S_3' \leftarrow S_1^3$
        $S_2' \leftarrow (S_1^2 + S_1^q) \cdot S_0 - S_0^q \cdot S_2 + c_2$
        $S_1' \leftarrow c^{-q} \cdot (S_0 - S_1)^3 + c^{1-q} \cdot S_2'$
        $S_0' \leftarrow S_0^3$
    else if $n_j = 1$ **then**
        $s_1 \leftarrow S_1$
        $s_2 \leftarrow S_2$
        $S_1' \leftarrow (s_1^2 + s_1^q) \cdot s_2 - s_2^q \cdot S_0 + c_2^q$
        $S_0' \leftarrow s_1^3$
        $S_2' \leftarrow (s_2^2 + s_2^q) \cdot s_1 - s_1^q \cdot S_3 + c_2$
        $S_3' \leftarrow s_2^3$
    else /* $n_j = 2$ */
        $S_0' \leftarrow S_2^3$
        $S_1' \leftarrow (S_2^2 + S_2^q) \cdot S_3 - S_3^q \cdot S_1 + c_2^q$
        $S_2' \leftarrow c^{-1} \cdot (S_3 - S_2)^3 + c^{1-q} \cdot S_1'$
        $S_3' \leftarrow S_3^3$
    end if
end for
return $S_{n \bmod 3}$

C   Solving the Curve Equation in Characteristic 3

Definition 3. *The absolute trace of a field element* $a \in \mathbb{F}_{3^m}$ *is the linear form:*

$$\text{tr}(a) = a + a^3 + a^9 + \cdots + a^{3^{m-1}}.$$

The absolute trace will always be in $\mathbb{F}_3$ as one can easily check by noticing from the above definition that $\text{tr}(a)^3 = \text{tr}(a)$, for all $a \in \mathbb{F}_{3^m}$. Being surjective and linear over $\mathbb{F}_3$, it can always be represented as a (usually sparse) dual vector $T \in \mathbb{F}_{3^m}$ in a given basis, so that one can compute $\text{tr}(u) = T \cdot u$ in no more than $O(m)$ time. In a normal basis $\{\theta^{3^i}\}$ with $\text{tr}(\theta) = 1$, computing $\text{tr}(u)$ amounts to summing up all coefficients of u.

The coordinates of a curve point $P = (x, y)$ are constrained by the curve equation to satisfy $y^2 = x^3 + ax + b$. Thus one can represent a point as either (x, β)

where $\beta \in \mathbb{F}_2$ indicates which of the two roots correspond to $y = \pm\sqrt{x^3 + ax + b}$, or else by (τ, y) where $\tau \in \mathbb{F}_3$ indicates which of the three solutions one has to take of the equation $x^3 + ax + (b - y^2) = 0$. In characteristic 3, cubing is a linear operation, which makes the second possibility more advantageous.

Consider the special equation $x^3 - x - u = 0$ for a given $u \in \mathbb{F}_{3^m}$, which is relevant for supersingular curves in characteristic 3. This equation has a solution if, and only if, $\mathrm{tr}(u) = 0$ [22, theorem 2.25]. This is the case for $1/3$ of the elements in $\mathbb{F}_{3^m}$, since the trace function is linear and surjective. The complexity of solving the cubic equation is only $O(m^2)$, as we show now.

Let $\mathcal{C} : \mathbb{F}_{3^m} \to \mathbb{F}_{3^m}$ be defined by $\mathcal{C}(x) = x^3 - x$. The kernel of $\mathcal{C}$ is $\mathbb{F}_3$ [22, chapter 2,section 1], hence the rank of $\mathcal{C}$ is $m - 1$ [16, section 3.1, theorem 2].

Theorem 3. *The equation $x^3 - x - u = 0$ over $\mathbb{F}_{3^m}$ can be solved in $O(m^2)$ steps.*

Proof. If $\mathbb{F}_{3^m}$ is represented in standard polynomial basis, the cubic equation reduces to a system of linear equations with coefficients in $\mathbb{F}_3$, and can be solved in no more than $O(m^2)$ steps. This is achieved by first checking whether the system has solutions, i.e. whether $\mathrm{tr}(u) = 0$. If so, since the rank of $\mathcal{C}$ is $m - 1$ one obtains an invertible $(m - 1) \times (m - 1)$ matrix A by leaving out the one row and correspondingly one column of the matrix representation of $\mathcal{C}$ on the given basis. A solution of the cubic equation is then given by an arbitrary element $x_0 \in \mathbb{F}_3$ and by the solution of system $A\tilde{x} = \tilde{u}$, which is obtained as $\tilde{x} = A^{-1}\tilde{u}$ in $O(m^2)$ time.

Using a normal basis to represent field elements, it is not difficult to see that the cubic equation can be efficiently solved in $O(m)$ time by the following algorithm (the proof is straightforward and left as an exercise):

Cubic equation solving in normal basis:

$x_0 \leftarrow$ root selector (an arbitrary element from $\mathbb{F}_3$)
for $i \leftarrow 1$ **to** $m - 1$ **do** {
    $x_i \leftarrow x_{i-1} - u_i$
}
x is a solution if, and only if, $x_{m-1} = x_0 + u_0$.

$\square$

Asymptotically Optimal Communication
for Torus-Based Cryptography

Marten van Dijk[1,2] and David Woodruff[1,*]

[1] MIT Computer Science and Artificial Intelligence Laboratory, Cambridge, USA
{marten,dpwood}@mit.edu
[2] Philips Research Laboratories, Eindhoven, The Netherlands

Abstract. We introduce a compact and efficient representation of elements of the algebraic torus. This allows us to design a new discrete-log based public-key system achieving the optimal communication rate, partially answering the conjecture in [4]. For n the product of distinct primes, we construct efficient ElGamal signature and encryption schemes in a subgroup of $F_{q^n}^*$ in which the number of bits exchanged is only a $\phi(n)/n$ fraction of that required in traditional schemes, while the security offered remains the same. We also present a Diffie-Hellman key exchange protocol averaging only $\phi(n) \log_2 q$ bits of communication per key. For the cryptographically important cases of $n = 30$ and $n = 210$, we transmit a 4/5 and a 24/35 fraction, respectively, of the number of bits required in XTR [14] and recent CEILIDH [24] cryptosystems.

1 Introduction

In classical Diffie-Hellman key exchange there are two fixed system parameters - a large prime q and a generator g of the multiplicative group F_q^* of the field F_q. In [10], the idea of working in finite extension fields instead of prime fields was proposed, but no computational or communication advantages were implied. In [26] Schnorr proposed working in a relatively small subgroup of F_q^* of prime order, improving the computational complexity of classical DH, but requiring the same amount of communication.

In [4] it is shown how to combine these two ideas so that the number of bits exchanged in DH key exchange is reduced by a factor of 3. Specifically, it is shown that elements of an order r subgroup G of $F_{q^6}^*$ can be efficiently represented using $2 \log_2 q$ bits if r divides $q^2 - q + 1$, which is one third of the $6 \log_2 q$ bits required for elements of $F_{q^6}^*$. Since the smallest field containing G is $F_{q^6}^*$, one can show [13] that with respect to attacks known today, the security of working in G is the same as that of working in $F_{q^6}^*$ for r large enough. In [14, 15] the XTR public key system was developed using the method of [4] together with an efficient arithmetic to achieve both computational and communication savings. These papers also show how to reduce communication in ElGamal encryption and signature schemes in $F_{q^6}^*$.

---

* Supported by an NDSEG fellowship.

In [4] it was conjectured that one can extend this technique to any n by working in the subgroup of $F_{q^n}^*$ of order $\Phi_n(q)$, where $\Phi_n(x)$ denotes the nth cyclotomic polynomial. Since the degree of $\Phi_n(x)$ is $\phi(n)$, where ϕ is the Euler function, one could transmit a $\phi(n)/n$ fraction of the number of bits needed in classical DH, while achieving the same level of security. For n the product of the first k primes, $\phi(n)/n \to 0$ as $k \to \infty$, so the savings get better and better. In [3, 24], evidence that the techniques of [4] cannot generalize to arbitrary n was presented, and in [3, 24], some specific versions of the conjecture in [4] made in [3] were shown to be false. Also in [24, 25, 23] it is shown that the group of order $\Phi_n(q)$ is isomorphic to the well-studied algebraic torus $T_n(F_q)$ [30] and that a positive answer to the conjecture in [4] is possible if one can construct an efficient rational parameterization of $T_n(F_q)$. However, such a construction is only known when n is a prime power or the product of two prime powers, although it is conjectured to exist for all n [24, 30]. In [24] a construction is given for $n = 6$, which is the basis for the CEILIDH public-key cryptosystem. CEILIDH achieves the same communication as XTR with a few computational differences.

In this paper we finally break the "$n \le 6$ barrier" by constructing, for every n, efficient ElGamal encryption and signature schemes in $F_{q^n}^*$, which require transmitting at most a $\phi(n)/n$ fraction of the bits required in their classical counterparts. Further, we present an asymptotical variant of DH key exchange in which the average number of bits exchanged per key approaches $\phi(n) \log_2 q$. The key property that we use is the fact that $T_n(F_q)$ is *stably rational* (see [30], section 5.1). Specifically, our enabling technique is the construction of efficiently computable bijections θ and θ^{-1} with

$$\theta : T_n(F_q) \times \left(\times_{d|n,\ \mu(n/d)=-1} F_{q^d}^* \right) \to \times_{d|n,\ \mu(n/d)=1} F_{q^d}^*,$$

where $\times$ denotes direct product, and μ is the Möbius function[1]. This allows us to bypass the torus conjecture of [24], by relaxing the problem of efficiently representing a single symbol of $T_n(F_q)$, to the problem of efficiently representing a sequence of symbols in $T_n(F_q)$. Our bijections enable us to compactly represent m elements of $T_n(F_q)$ with $(m\phi(n) + \sum_{d|n,\mu(n/d)=-1} d) \log q$ bits, which for large enough m, is roughly $\phi(n) \log q$ bits per element. We stress that while our key exchange protocol achieves the optimal $n/\phi(n)$ reduction factor asymptotically, our encryption and signature schemes achieve this even for the encrypting or signing of a single message.

Note that the domain and range of θ need not be isomorphic. Indeed, letting G_d denote the cyclic group of order d, if $n = 2$ and $q = 3$, then the domain of θ is isomorphic to $G_4 \times G_2$, while the range is isomorphic to G_8. We show, however, that θ can be decomposed into isomorphisms plus a map requiring a table lookup. We show how to choose q so that constructing and querying this table is extremely efficient.

---

[1] For an integer n, $\mu(n) = 1$ if $n = 1$, $\mu(n) = 0$ if n has a repeated factor, and $\mu(n) = (-1)^k$ if n is a product of k distinct primes (see [11], section 16.3).

Our choice of q and r for fixed n will also affect the security of our scheme. We give an efficient heuristic for choosing q and r for the practical cases of $n = 30$ and $n = 210$, where we achieve a communication reduction by factors of $15/4$ and $35/8$, respectively. Further, for any n, we give an efficient algorithm for choosing q and r with a theoretical guarantee on its performance. This latter algorithm is primarily of theoretical interest, showing how to optimally choose q and r when n tends to infinity for a sufficiently large security requirement.

While our main focus and contribution is on the communication complexity, we also calculate the amount of computation necessary to evaluate θ and θ^{-1} for general n, and we attempt to minimize the number of modular exponentiations. We show that our representation enjoys some of the same computational advantages of CEILIDH over XTR, including the ability to multiply elements of $T_n(F_q)$ directly. This allows us to come close to the non-hybrid version of El-Gamal encryption in [24]. Indeed, in addition to constructing a hybrid ElGamal encryption scheme, we construct a scheme in which to encrypt m messages, we form m ElGamal encryptions in $T_n(F_q)$ plus one additional encryption using a symmetric cipher. Unfortunately, the computational complexity of our scheme is not that practical, whereas XTR for instance, permits very efficient computations if just exponentiation is required. For $n = 30$, we hand-optimize the computation of θ and θ^{-1}. Our analysis for general n shows that all of our protocols and algorithms are (theoretically) efficient in n and the sizes of q and r.

Outline: Section 2 discusses the algebraic and number-theoretic tools we use. In section 3 we construct the bijections θ and θ^{-1}. Section 4 shows how to choose system parameters to guarantee security and efficiency, giving both a practical algorithm for $n = 30$ and $n = 210$, and a theoretical algorithm for general n. In section 5 we discuss our cryptographic applications. Section 6 treats the computational complexity of our bijections, and we conclude in section 7.

2   Preliminaries

2.1   Cyclotomic Polynomials and Algebraic Tori

We first state a few facts about the cyclotomic polynomials. See [19] for more background.

Definition 1. *Let n be a positive integer and let $\zeta_n = e^{2\pi i/n}$. The nth cyclotomic polynomial $\Phi_n(x)$ is defined by:*

$$\Phi_n(x) = \prod_{1 \leq k \leq n,\ \gcd(k,n)=1} (x - \zeta_n^k).$$

It is easy to see that the degree of $\Phi_n(x)$ is $\phi(n)$, where ϕ is the Euler-totient function. We also have:

$$x^n - 1 = \prod_{d|n} \Phi_d(x),$$

and using the Möbius function μ,

$$\Phi_n(x) = \prod_{d|n}(x^d - 1)^{\mu(n/d)}.$$

It can be shown that the cyclotomic polynomials are irreducible polynomials over $\mathbb{Q}$ with integer coefficients. For q a prime power, let F_q denote the finite field with q elements. For integers $n > 0$ we define the algebraic torus[2] $T_n(F_q)$:

$$T_n(F_q) = \{\alpha \in F_{q^n}^* \mid \alpha^{\Phi_n(q)} = 1\}.$$

2.2   Number Theory

The following is the celebrated prime number theorem (see [11], chapter 22):

Theorem 1. *For large enough n, the number of primes less than or equal to n is $\frac{n}{\ln n} + o\left(\frac{n}{\ln n}\right)$.*

We also need the fact that for any $n > 6$, $\phi(n) > n/(6 \ln \ln n)$, and for n the product of the first k distinct primes, $\phi(n) = \Theta(n/\log \log n)$. We use the following density theorem in our analysis:

Theorem 2. (Chebotarev [5, 16]) *For any integer n and any $a \in Z_n^*$, the density of primes p (among the set of all primes) with $p = a \mod n$ is $1/\phi(n)$.*

3   The Bijection

Let q be a prime power, n a positive integer, $F_{q^n}^*$ the multiplicative group of the field of order q^n, and $T_n(F_q)$ the $\phi(n)$-dimensional algebraic torus over F_q. For an integer k, let $[k] = \{1, 2, \ldots, k\}$. The goal of this section is to construct efficiently computable bijections θ and θ^{-1}, where

$$\theta : T_n(F_q) \times \left(\times_{d|n, \, \mu(n/d)=-1} F_{q^d}^*\right) \rightarrow \times_{d|n, \, \mu(n/d)=1} F_{q^d}^*.$$

Our strategy is to first find efficient bijections γ and γ^{-1}, where

$$\gamma : F_{q^n}^* \rightarrow \times_{d|n} T_d(F_q).$$

Note that in general $F_{q^n}^*$ and $\times_{d|n} T_d(F_q)$ need not be isomorphic. Let G_m denote the cyclic group of order m. We first need a few lemmas. The following is an immediate consequence of the structure theorem of abelian groups, but for completeness and to exhibit the efficient isomorphisms, we include it:

Lemma 1. *Suppose $n = r_1 \cdot r_2 \cdots r_k$ for pairwise relatively prime positive integers $r_1, \ldots, r_k$. Then there exist efficiently computable isomorphisms $\rho : G_n \rightarrow \times_{i \in [k]} G_{r_i}$ and $\sigma : \times_{i \in [k]} G_{r_i} \rightarrow G_n$.*

---

[2] Technically, $T_n(F_q)$ just refers to the F_q points of the algebraic torus rather than the torus itself (see [24, 30]).

Proof. For $i \in [k]$, put $d_i = n/r_i$. Since the r_i are pairwise relatively prime, $\gcd(d_1, d_2, \ldots, d_k) = 1$, so there exist integers e_i for which $\sum_{i \in [k]} e_i d_i = 1$. For $\alpha \in G_n$, define $\rho(\alpha) = (\alpha^{d_i})_{i \in [k]}$. Since $(\alpha^{d_i})^{r_i} = 1$, ρ maps elements of G_n to elements in the product group $\times_{i \in [k]} G_{r_i}$. For $(\alpha_i)_{i \in [k]} \in \times_{i \in [k]} G_{r_i}$, define $\sigma((\alpha_i)_{i \in [k]}) = \prod_{i \in [k]} \alpha_i^{e_i}$, where multiplication occurs in G_n.

The claim is that ρ and σ are inverse isomorphisms between G_n and $\times_{i \in [k]} G_{r_i}$. For $\alpha \in G_n$, we have $\sigma(\rho(\alpha)) = \sigma((\alpha^{d_i}))_{i \in [k]} = \prod_{i \in [k]} \alpha^{d_i e_i} = \alpha$. Similarly, for $(\alpha_i)_{i \in [k]} \in \times_{i \in [k]} G_{r_i}$, we have

$$\rho(\sigma((\alpha_i)_{i \in [k]})) = \rho\left(\prod_{i \in [k]} \alpha_i^{e_i}\right) = \left(\prod_{j \in [k]} \alpha_j^{e_j d_i}\right)_{i \in [k]}.$$

Now, $r_j \mid d_i$ if $j \neq i$, so in this case $\alpha_j^{e_j d_i} = 1$. Also, $\alpha_i^{e_i d_i} = \alpha_i^{1 - \sum_{j \neq i} e_j d_j} = \alpha_i^{1 - k r_i}$ for an integer k, so $\alpha_i^{e_i d_i} = \alpha_i$. Hence, $\rho(\sigma((\alpha_i)_{i \in [k]})) = (\alpha_i)_{i \in [k]}$, which shows ρ and σ are inverses. Observe that $\rho(\alpha_1 \cdot \alpha_2) = ((\alpha_1 \cdot \alpha_2)^{d_i})_{i \in [k]} = (\alpha_1^{d_i})_{i \in [k]} \cdot (\alpha_2^{d_i})_{i \in [k]} = \rho(\alpha_1) \cdot \rho(\alpha_2)$, and similarly $\sigma((\alpha_i)_{i \in [k]} \cdot (\alpha_i')_{i \in [k]}) = \prod_{i \in [k]} (\alpha_i \cdot \alpha_i')^{e_i} = \prod_{i \in [k]} (\alpha_i)^{e_i} \prod_{i \in [k]} (\alpha_i')^{e_i} = \sigma((\alpha_i)_{i \in [k]}) \cdot \sigma((\alpha_i')_{i \in [k]})$, which shows that the maps are isomorphisms. Computing ρ and σ just requires multiplication and exponentiation, which can be made efficient by repeated squaring.

Let $U = U(n, q)$ be the smallest positive integer for which $\gcd(\Phi_d(q), \Phi_e(q), \frac{q^n - 1}{U}) = 1$ for all $d \neq e$ with $d \mid n$ and $e \mid n$.

Lemma 2. *For $d \mid n$, let $y_d = \gcd(\Phi_d(q), \frac{q^n - 1}{U})$. Then $F_{q^n}^* \cong G_U \times (\times_{d \mid n} G_{y_d})$. Furthermore, the isomorphisms are efficiently computable.*

Proof. By lemma 1 it suffices to show (1) $q^n - 1 = U \prod_{d \mid n} y_d$, (2) for all d, $\gcd(U, y_d) = 1$, and (3) for all $d \neq e$, $\gcd(y_d, y_e) = 1$.

Using the fact that $q^n - 1 = \prod_{d \mid n} \Phi_d(q)$, the following establishes (1):

$$\frac{q^n - 1}{U} = \gcd\left(\prod_{d \mid n} \Phi_d(q), \frac{q^n - 1}{U}\right) = \prod_{d \mid n} \gcd\left(\Phi_d(q), \frac{q^n - 1}{U}\right) = \prod_{d \mid n} y_d,$$

where the second equality follows from the definition of U. For (2), observe that

$$\gcd(U, y_d) = \gcd\left(U, \Phi_d(q), \frac{q^n - 1}{U}\right) \mid \gcd\left(U, \frac{q^n - 1}{U}\right) = 1,$$

since if prime $p \mid U$, by minimality of U there exist $d \neq e$ for which $p \mid \gcd(\Phi_d(q), \Phi_e(q))$, so if $p \mid \frac{q^n - 1}{U}$, then $p \mid \gcd(\Phi_d(q), \Phi_e(q), \frac{q^n - 1}{U})$, a contradiction. To see (3), note that $\gcd(y_d, y_e) = \gcd(\Phi_d(q), \Phi_e(q), \frac{q^n - 1}{U}) = 1$ by the definition of U.

We use the following bijections with complexity proportional to U, which we later show to be negligible for an appropriate choice of q.

Lemma 3. *For* $d \mid n$, *let* $z_d = \gcd(\Phi_d(q), U)$. *There exist bijections between* G_U *and* $\times_{d\mid n} G_{z_d}$ *requiring* $O(\log U + \log n + \log \log q)$ *time to evaluate and* $O(Un^{1+\epsilon} \log q)$ *space for any* $\epsilon > 0$.

Proof. Using the definition of U,

$$\prod_{d\mid n} |G_{z_d}| = \prod_{d\mid n} \gcd(\Phi_d(q), U) = \gcd\left(\prod_{d\mid n} \Phi_d(q), U\right) = \gcd(q^n - 1, U) = U,$$

so there exists a bijection between the two groups. Choose a generator g of G_U and generators g_d of G_{z_d}. For each $i \in [U]$, make a table entry mapping g^i to a unique tuple $(g_d^{i_d})_{d\mid n}$. Since the sum of the divisors of n is less than $O(n^{1+\epsilon})$ for any $\epsilon > 0$ ([11], section 18.3), the table consumes $O(Un^{1+\epsilon} \log q)$ space. We sort the entries in both directions so that both bijections are efficient. Evaluations of either bijection can then be performed with a binary search in $O(\log U + \log n + \log \log q)$ time.

We need another auxiliary map:

Lemma 4. *Let* y_d *and* z_d *be as in the previous two lemmas. Then,* $\times_{d\mid n} T_d(F_q) \cong \left(\times_{d\mid n} G_{y_d}\right) \times \left(\times_{d\mid n} G_{z_d}\right)$. *Furthermore, the isomorphisms are efficiently computable.*

Proof. It suffices to show for any $d \mid n$, $T_d(F_q) \cong G_{y_d} \times G_{z_d}$, and that this isomorphism is efficiently computable. Note that $y_d z_d = \gcd(\Phi_d(q), \frac{q^n-1}{U}) \gcd(\Phi_d(q), U) = \Phi_d(q)$ since $\gcd(U, \frac{q^n-1}{U}) = 1$ by the definition of U. By the same observation, $\gcd(y_d, z_d) = 1$. Lemma 1 establishes the claim.

The following is immediate from the previous 3 lemmas:

Lemma 5. *Assuming the maps of lemma 3 are efficient, there exist efficiently computable bijections* γ *and* γ^{-1}, *where* $\gamma : F_{q^n}^* \rightarrow \times_{d\mid n} T_d(F_q)$.

We now have the bijection claimed at the beginning:

Theorem 3. *Assuming the maps of lemma 3 are efficient, there exist efficiently computable bijections* θ *and* θ^{-1}, *where* $\theta : T_n(F_q) \times \left(\times_{d\mid n,\ \mu(n/d)=-1} F_{q^d}^*\right) \rightarrow \times_{d\mid n,\ \mu(n/d)=1} F_{q^d}^*$.

Proof. Lemma 5 gives efficient bijections between $T_n(F_q) \times \left(\times_{d\mid n,\ \mu(n/d)=-1} F_{q^d}^*\right)$ and $T_n(F_q) \times \left(\times_{d\mid n,\ \mu(n/d)=-1} (\times_{e\mid d} T_e(F_q))\right)$, and also between $\times_{d\mid n,\ \mu(n/d)=1} F_{q^d}^*$ and $\times_{d\mid n,\ \mu(n/d)=1} (\times_{e\mid d} T_e(F_q))$. By permuting coordinates, the theorem will follow if we show the multiset equality

$$\{n\} \cup \bigsqcup_{d\mid n,\ \mu(n/d)=-1} \{e \text{ s.t. } e \mid d\} = \bigsqcup_{d\mid n,\ \mu(n/d)=1} \{e \text{ s.t. } e \mid d\}.$$

From section 2, $\Phi_n(x) \prod_{\mu(n/d)=-1}(x^d - 1) = \prod_{\mu(n/d)=1}(x^d - 1)$ in the polynomial ring $\mathbb{Q}[x]$. Decomposing this equation into irreducible polynomials, we have $\Phi_n(x) \prod_{\mu(n/d)=-1} \prod_{e|d} \Phi_e(x) = \prod_{\mu(n/d)=1} \prod_{e|d} \Phi_e(x)$, and since $\mathbb{Q}[x]$ is a unique factorization domain, the irreducible polynomials on the left must be the same as those on the right. This gives the desired multiset equality.

4   Parameter Selection

The two constraints on choosing q and r for fixed n are security and efficiency constraints, the latter measured by the size $U(n,q)$ of the tables needed in our bijections. We first discuss the role of security in parameter selection:

4.1   Security Measures

Our schemes derive their security from the same assumptions of XTR and CEILIDH. That is, if there is a successful attack against one of our cryptographic primitives, then there is a successful attack against the corresponding primitive in the underlying group we use, which we assume is impossible. Let $\langle g \rangle \subset F_{q^n}^*$ be a multiplicative group of order r with generator g. The security of our applications relies on the hardness of both the Computational Diffie-Hellman problem (CDH) and the Decisional Diffie-Hellman problem (DDH) in $\langle g \rangle$. The former is the problem of computing g^{xy} given g^x and g^y and the latter is that of distinguishing triples of the form (g^a, g^b, g^{ab}) from (g^a, g^b, g^c) for random a, b, and c. The hardness of both of these problems implies the hardness of the discrete logarithm problem (DL) in $\langle g \rangle$: find x given g^x. Due to the Pohlig-Hellman algorithm [21], the DL problem in $\langle g \rangle$ can be reduced to the DL problem in all prime order subgroups of $\langle g \rangle$, so we might as well assume that r is prime.

There are two known approaches to solving the DL problem in $\langle g \rangle$ [1,7,9,13, 20,27,28], one which attacks the full multiplicative group of F_{q^n} itself using the Discrete Logarithm variant of the Number Field Sieve, and one which concentrates directly on the subgroup $\langle g \rangle$ using Pollard's Birthday Paradox based rho method [22]. Let s be the smallest divisor of n for which $\langle g \rangle$ can be embedded in $F_{q^s}^*$. The heuristic expected running time of the first attack is $L[q^s, 1/3, 1.923]$, where $L[n, v, u] = \exp((u+o(1))(\ln n)^v (\ln \ln n)^{1-v})$. If q is small, e.g. $q = 2$, then the constant 1.923 can be replaced with 1.53. The second attack, due to Pollard, takes $O(\sqrt{r})$ operations in $\langle g \rangle$.

Hence we see that the difficulty of solving the DL problem in $\langle g \rangle$ depends on both the size of the minimal surrounding subfield and on the size of its prime order r. If F_{q^n} is itself the minimal surrounding subfield, as is the case if we choose $r \mid \Phi_n(q)$ with $r > n$, then for sufficiently large r the DL, CDH, and DDH problems in $\langle g \rangle$ are widely believed to be just as hard as solving their classical counterparts w.r.t. an element of prime order $\approx r$ in the prime field of cardinality $\approx q^n$ [14]. As mentioned in [14], when $n \log_2 q \approx 1024$ and $\log_2 r \approx 160$, solving the DL problem in $\langle g \rangle$ is generally believed to be harder than factoring an 1024-bit RSA modulus provided q is not too small.

4.2   Practical Algorithm for $n = 30$ and $n = 210$

Based on our security discussion, it is shown in [4] that, assuming an RSA key length between 1024 and 2048 bits gives adequate security, for $n = 30$ we should choose q to be a prime between 35 and 70 bits long, and for $n = 210$ we should choose q to be a prime between 5 and 10 bits long. Note that for the next value of n for which we achieve a communication savings, $n = 2310 = 2 \cdot 3 \cdot 5 \cdot 7 \cdot 11$, the field size will have to be at least 2310 bits, so any setting of q already exceeds the 2048 bits needed for adequate security.

In [13] it is shown how to quickly find a q and an r meeting these requirements for fixed n. The algorithm is heuristic, and involves choosing random q of a certain size and checking if $\Phi_n(q)$ contains a sufficiently large prime factor r by trial division with the primes up to roughly 10^5. On a 166MHz processor, for $n = 30$ it was shown that it takes 12 seconds to find an r of size between 214 and 251 bits for q of size 32 bits. Note that for $n = 30$ we actually need r to be slightly smaller, as claimed in the previous paragraph. This way we can achieve the largest efficiency gain for a fixed security guarantee. Using the algorithm of [13], fixing the size of r to be approximately 161 bits and searching for an appropriate q took three hours instead of the 12 seconds needed previously. However, there are three reasons we do not consider this to be problematic. First, CPU speeds are easily ten times as fast these days. Second, we don't need to fix the size of r to be exactly 161 bits; we just need to find an r of approximately this size. And third, finding the system parameters is a one-time cost and can be done offline, or even by a trusted third party.

From the efficiency analysis in the next section and lemma 6, one can show that the table size $U(n, q)$ resulting from choosing q at random subject to the above constraints is likely to be small with good probability. Hence, this heuristic algorithm is likely to find a q and an r so that both security and efficiency constraints are met in a reasonable amount of time.

4.3   Theoretical Algorithm for General n with Probabilistic Guarantees

In this section we use properties of the density of primes to design a parameter selection algorithm and rigorously analyze its performance. Unfortunately, since the factorization of $\Phi_n(q)$ for random primes q does not seem to be well-understood, we are forced to choose $q > r$, which with respect to attacks known today, doesn't allow for choosing the optimal q and r for $n = 30$ and $n = 210$ if we just want 2048 bit RSA security. A straightforward calculation shows that for $n = 30$, the following algorithm gives us the largest efficiency gain for a fixed security guarantee if and only if q is at least 558 bits. Hence, we should view the algorithm as theoretical in nature, and apply the heuristic of the previous section for small n.

Let k be a positive integer tending to infinity and let n be the product of the first k primes. We want to choose q so that:

1. $n \log q$ is sufficiently large.
2. There exists a large prime factor r of $\Phi_n(q)$.
3. $U = U(n, q)$ is small.

We say an integer is *squarefree* if it contains no repeated factors. The selection algorithm is as follows:

Parameter Selection Algorithm PSA($n = p_1 \cdots p_k, Q, R$):

1. Let S be the subset of the first k primes p for which $p - 1$ is squarefree, and put $T = \{p_1, \ldots, p_k\} \setminus S$.
2. Find an R-bit prime r for which $r = 1 \bmod n$, and find a $z \in Z_r^*$ of order n.
3. Find a Q-bit prime $q = z + kr > n$, for some integer k, such that:
   (a) For all $p \in S$, $q^{pO_p(q)} \neq 1 \bmod p^3$, where $O_p(q)$ denotes the order of q in Z_p^*.
   (b) For all $p \in T$, $O_p(q) = p - 1$.
4. Find a generator g of the subgroup of order r of $F_{q^n}^*$. Output r, q, and g.

We first claim that if the PSA algorithm terminates, then r and q meet the aforementioned properties. By setting Q large enough, the first property holds. We have $\Phi_n(q) = \Phi_n(z + kr) = \Phi_n(z) + sr$ for some integer s, and since $O_r(z) = n$, $\Phi_n(z) + sr = 0 \bmod r$. Hence by choosing R sufficiently large, the second property holds. To show $U = U(n, q)$ is small, we need the following lemma:

Lemma 6. *Let p be a prime and q an integer such that $p \nmid q$. Then $p \mid U$ if and only if $pO_p(q) \mid n$. In case of the latter, $p^i \mid U$ if and only if $p^i \mid (q^{pO_p(q)} - 1)$.*

Proof. By minimality of U, $p \mid U$ if and only if there exist divisors $d < e$ of n for which $p \mid \gcd(\Phi_d(q), \Phi_e(q))$. Fix two such divisors d and e, let $f = \gcd(d, e)$, and suppose $f < d$. Since $f < d$, $p \mid \Phi_d(q) \mid (q^d - 1)/(q^f - 1) = 1 + q^f + q^{2f} + \cdots + q^{(d/f-1)f}$. Since $p \mid \gcd(\Phi_d(q), \Phi_e(q)) \mid \gcd(q^d - 1, q^e - 1) = q^f - 1$, we have $q^f = 1 \bmod p$, so $d/f = 0 \bmod p$, or $p \mid d/f$. Similarly, $p \mid e/f$. But then $p \mid \gcd(d/f, e/f)$, contradicting our choice of f. Hence, $d = f$ which means $d \mid e$ and $p \mid e/d \mid n$.

Suppose there is another divisor $c < d$ of n for which $p \mid \Phi_c(q)$. Then by the above, $c \mid d$ and $p \mid (d/c)$, and since $p \mid (e/d)$, $p^2 \mid e \mid n$, contradicting the fact that n is squarefree. This means that (d, e) is the unique pair of divisors for which $p \mid \gcd(\Phi_d(q), \Phi_e(q))$. Since $p \mid q^n - 1$, $O_p(q) \mid n$, and since $\gcd(O_p(q), p) = 1$, $pO_p(q) \mid n$. Put $d = O_p(q)$ and $e = pO_p(q)$. Then d is the smallest positive integer for which $q^d = 1$, so $p \mid \Phi_d(q)$. Also, $\Phi_e(q) = (q^e - 1)/(q^d - 1) = 1 + q^d + \cdots + q^{(e/d-1)d} = e/d \bmod p = 0 \bmod p$. Hence if $p \mid \gcd(\Phi_d(q), \Phi_e(q))$, then $d = O_p(q)$ and $e = pO_p(q)$. Conversely, if $pO_p(q) \mid n$, then $p \mid U$ for these d, e.

We have shown $p \mid U$ if and only if $pO_p(q) \mid n$. The above shows that if $p^i \mid U$, then $p^i \mid (\Phi_{O_p(q)}(q) \cdot \Phi_{pO_p(q)}(q)) \mid (q^{pO_p(q)} - 1)$, and conversely if $p^i \mid (q^{pO_p(q)} - 1) \mid (q^n - 1)$, then $p^i \mid \gcd(\Phi_d(q), \Phi_e(q)) \mid U$.

Remark 1. Note that $p^2 \mid (q^{pO_p(q)} - 1)$, since on the one hand we have $p \mid (q^{O_p(q)} - 1)$, and on the other hand we have $(q^n - 1)/(q^{O_p(q)} - 1) = 1 + q^{O_p(q)} + q^{2O_p(q)} + \cdots + q^{(p-1)O_p(q)} = 1 + 1 + \cdots + 1 = 0 \mod p$. Hence if $p \mid U$, then $(q^{pO_p(q)} - 1) \mid (q^n - 1)$, so it follows that $p^2 \mid U$.

The following lemma provides tight asymptotic bounds on $U = U(n, q)$:

Lemma 7. *If the PSA algorithm terminates, $U = \Theta(n^{2C})$, where $C \approx .374$ is Artin's constant.*

Proof. By the previous lemma, if $p \mid U$, then $p \mid n$, so $p \in \{p_1, \ldots, p_k\}$. Now if $p \in T$, $p - 1$ is not squarefree, so $O_p(q) \nmid n$ by step 3b, so $p \nmid U$. On the other hand, if $p \in S$, $p - 1$ is a product of distinct primes in $\{p_1, \ldots, p_k\}$, so $O_p(q) \mid n$ and hence $p \mid U$. Combining this with the remark above, step 3a of the PSA algorithm, and the previous lemma, we conclude that U is exactly the square of the product of primes in S and that the PSA algorithm chooses q so that U is minimal.

To obtain the bound on U it suffices to show that the density of primes p for which $p - 1$ is squarefree is C, where C is Artin's constant [8]. The bound will then hold for large enough k. For a prime p, $p - 1$ is not squarefree if and only if $p = 1 \mod q^2$ for a prime q. By the inclusion-exclusion principle, the multiplicativity of $\phi(\cdot)$, and theorem 2, the density of primes p for which $p - 1$ is squarefree is:

$$1 - \sum_{\text{primes } p} \frac{1}{\phi(p^2)} + \sum_{\text{primes } p,q} \frac{1}{\phi(p^2 q^2)} - \cdots = \prod_{\text{primes } p} \left(1 - \frac{1}{\phi(p^2)}\right) = C.$$

By theorem 1, for sufficiently large k, $p_k \approx k \log k$ and $k \approx \frac{\log k}{\log \log k}$, where the approximation is up to low order terms. Hence, $U \leq p_k^{2Ck} \approx (k \log k)^{2Ck} \approx (\log n)^{2C \frac{\log n}{\log \log n}} \approx n^{2C}$.

Finally, we show the PSA algorithm terminates quickly in expectation:

Efficiency Analysis: By theorem 1, $k \approx \frac{\log n}{\log \log n}$ and $p_k \approx \log n$. Determining S and T in step 1 can therefore be done by trial division in $O(\log^2 n)$ time. We can perform step 2 by choosing a random R-bit number r, efficiently checking if r is prime, and checking if $r = 1 \mod n$. This requires an expected $\phi(n)R = O\left(\frac{Rn}{\log \log n}\right)$ samples r. To find z, we choose a random $\alpha \in Z_r^*$, set $\beta = \alpha^{\frac{q-1}{n}}$, and check that $\beta^d \neq 1 \mod r$ for all proper divisors d of n. In expectation, after $O(\log R)$ trials one such α will be a generator of Z_r^*, for which setting $z = \beta = \alpha^{\frac{q-1}{n}}$ gives z with $O_r(z) = n$. Conversely, if for all proper divisors d of n we have $\beta^d \neq 1 \mod r$, then $O_r(\beta) = n$. Since the number of proper divisors of n is $O(n^\epsilon)$ for any $\epsilon > 0$ ([11], section 18.1), the check in step 2 is efficient.

For step 3, for each $p \in T$, we can find an element $a_p \in Z_p^*$ with $O_p(a_p) = p-1$ by simply trying each of the $p - 1 = O(\log n)$ elements of Z_p^* until we succeed.

We then choose a random integer k for which $q = z + kr$ is a Q-bit number and efficiently check if q is prime. If so, then for each $p \in S$, we can compute $O_p(q)$ in $O(\log n)$ time, then check if $q^{pO_p(q)} \neq 1 \bmod p^3$ by repeated squaring. For each $p \in T$ we check if $q = a_p \bmod p$.

The claim is that the number of random samples k needed in step 3 is only $O(Qn^{1-C})$. Using the fact that the density of primes amongst integers of the form $z + kr$ is $O\left(\frac{1}{\log(z+kr)}\right)$, an integer k for which $z + kr$ is prime can be found with $O(Q)$ samples in expectation. By independence, the density of primes q which are $a_p \bmod p$ for every $p \in T$ is $\prod_{p \in T} \frac{1}{\phi(p)} = \Omega\left(\frac{\log \log n}{n^{1-C}}\right)$, where C is Artin's constant. Fix any $p \in S$. By theorem 2, for all but a negligible fraction of primes q, $q = g^i \bmod p^3$ for g a generator of $Z_{p^3}^*$. Since g is a generator, $q^{pO_p(q)} = 1 \bmod p^3$ if and only if i is a multiple of $\frac{\phi(p^3)}{pO_p(q)}$, and there are only $pO_p(q) \leq p(p-1)$ such multiples. By theorem 2, it is equally likely that $q = g^i$ for any $i \in [\phi(p^3)]$, so the density of primes q for which $q^{pO_p(q)} \neq 1 \bmod p^3$ is at least $1 - 1/p$. By independence, the density of q for which $q^{pO_p(q)} \neq 1 \bmod p^3$ for all $p \in S$ is at least $\prod_{p \in S}(1 - 1/p) = \prod_{p \in S} = \Omega\left(\frac{1}{\log \log n}\right)$. Applying independence one last time, we conclude that q can be found with an expected $O(Qn^{1-C})$ samples k.

Finally, step 4 can be implemented by choosing a random $g \in F_{q^n}^*$ and making sure that $(q^n - 1)/r \neq 1$. The number of generators of $F_{q^n}^*$ is $\phi(q^n - 1)$ which is $\Omega\left(\frac{q^n}{\log n + \log Q}\right)$, so the expected number of samples g needed is $O(\log n + \log Q)$.

5   Cryptographic Applications

Let n be the product of the first k primes, and let r, q, and g be public parameters generated as in section 4. Define $\sigma^-(n) = \sum_{d|n,\ \mu(n/d)=-1} d$ and $\sigma^+(n) = \sum_{d|n,\ \mu(n/d)=1} d$, and observe that $\phi(n) + \sigma^-(n) = \sigma^+(n)$. From section 3, we have an efficiently computable bijection θ and its inverse θ^{-1}, with θ :
$$T_n(F_q) \times \left(\times_{d|n,\ \mu(n/d)=-1} F_{q^d}^*\right) \rightarrow \times_{d|n,\ \mu(n/d)=1} F_{q^d}^*.$$
From the proof of theorem 3, we see that there are a number of choices for θ depending on which coordinate permutation is chosen. While this choice does not affect the communication of our protocols or the size of our encryptions/signatures, it can affect the computational costs. In section 6 we choose a specific permutation and analyze the computational requirements for $n = 30$.

We will think of θ and θ^{-1} as efficiently computatble maps between $T_n(F_q) \times F_q^{\sigma^-(n)}$ and $F_q^{\sigma^+(n)}$ by fixing polynomial representations of F_{q^d} with $d \mid n$. An element of $F_q^{\sigma^-(n)}$ is then just a list of $\sigma^-(n)$ q-ary coefficients with respect to these polynomials, and can be treated as an element of $\times_{d|n,\ \mu(n/d)=-1} F_{q^d}^*$. Let $i_d, i_d + 1, \ldots, i_d + d - 1$ denote the coordinates of an element $x \in F_q^{\sigma^-(n)}$ corresponding to the coefficients of x with respect to the irreducible polynomial for F_{q^d}. Our map may not be well-defined because we may have $(x_{i_d}, x_{i_d+1}, \ldots, x_{i_d+d-1}) = 0$.

However, if $y \in F_q^{\sigma^-(n)}$ is chosen randomly, the probability that some coordinate of y is zero is less than $\sigma^-(n)/q = O(n^\epsilon/q)$ for any $\epsilon > 0$, which is negligible. The same is true of a randomly chosen element of $F_q^{\sigma^+(n)}$. Hence, if we apply θ and θ^{-1} to random $(x_1, x_2) \in T_n(F_q) \times F_q^{\sigma^-(n)}$ and $y \in F_q^{\sigma^+(n)}$, $\theta(x_1, x_2)$ and $\theta^{-1}(y)$ are well-defined with overwhelming probability.

It is possible to modify θ and θ^{-1} if one wants more than a probabilistic guarantee. Define $d^-(n) = \sum_{d|n,\ \mu(n/d)=-1} 1$ and $d^+(n) = \sum_{d|n,\ \mu(n/d)=1} 1$. We can efficiently extend θ to the well-defined map $\tilde{\theta}$,

$$\tilde{\theta} : T_n(F_q) \times F_q^{\sigma^-(n)} \ \rightarrow \ \left(\times_{d|n,\ \mu(n/d)=1} F_{q^d}^* \right) \times \{0,1\}^{d^-(n)},$$

where for each $(x,y) \in T_n(F_q) \times F_q^{\sigma^-(n)}$ and for each $d \mid n$ with $\mu(n/d) = -1$, if $(y_{i_d}, \ldots, y_{i_d+d-1}) = 0$, we replace y_{i_d+d-1} with 1, obtaining a new string y', and define $\tilde{\theta}(x,y) = \theta(x,y') \circ b$, where for all $j \in [d^-(n)]$, $b_j = 1$ if and only if $(y_{i_d}, \ldots, y_{i_d+d-1}) = 0$ for the jth divisor d. Note that $\tilde{\theta}^\leftarrow$, the inverse of $\tilde{\theta}$ restricted to the image of $\tilde{\theta}$, is also well-defined. Similarly, letting β denote θ^{-1}, we can extend β to a well-defined map $\tilde{\beta} : F_q^{\sigma^+(n)} \rightarrow T_n(F_q) \times F_q^{\sigma^-(n)} \times \{0,1\}^{d^+(n)}$ and construct $\tilde{\beta}^\leftarrow$.

The next sections describe our cryptographic applications. For simplicity, in our security analyses we assume θ and θ^{-1} are actually bijections between $T_n(F_q) \times F_q^{\sigma^-(n)}$ and $F_q^{\sigma^+(n)}$, although it should be understood that our protocols can be slightly modified so that $\tilde{\theta}$ or $\tilde{\beta}$ can be used without affecting the security. The only application where this is not immediately obvious is the non-hybrid ElGamal encryption, but step 3 of that protocol can be modified to additionally encrypt the "extra bits" from $\tilde{\beta}$ using, say, the same key used in step 3.

5.1   Diffie-Hellman Key Agreement

For Alice and Bob to agree on a sequence of m secret keys K_i, they engage in the following protocol:

1. Alice and Bob choose random S_0 and T_0 in $\times_{d|n,\ \mu(n/d)=-1} F_{q^d}^*$, respectively, and treat them as elements of $F_q^{\sigma^-(n)}$.
2. For $i = 1$ to m,
    (a) Alice selects a random integer x_i with $1 \leq x_i \leq r$, sets $A_i = g^{x_i}$, computes $\theta(A_i, S_{i-1}) = (a_i, S_i) \in F_q^{\phi(n)} \times F_q^{\sigma^-(n)}$ and transmits a_i to Bob.
    (b) Bob selects a random integer y_i with $1 \leq y_i \leq r$, sets $B_i = g^{y_i}$, computes $\theta(B_i, T_{i-1}) = (b_i, T_i) \in F_q^{\phi(n)} \times F_q^{\sigma^-(n)}$ and transmits b_i to Alice.
3. Alice sends S_m to Bob and Bob sends T_m to Alice.
4. For $i = m$ to 1,
    (a) Alice computes $\theta^{-1}(b_i, T_i) = (B_i, T_{i-1})$, and sets $K_i = B_i^{x_i} = g^{x_i y_i}$.
    (b) Bob computes $\theta^{-1}(a_i, S_i) = (A_i, S_{i-1})$, and sets $K_i = A_i^{y_i} = g^{x_i y_i}$.

The number of bits sent from Alice to Bob (and from Bob to Alice) is about $(m\phi(n)+\sigma^-(n))\log q$, so the rate approaches the optimal $\phi(n)\log q$ bits per key as m gets large. This beats all known schemes for $n \geq 30$. In particular, for $n = 30$, our scheme requires only $8\log q$ bits per shared key while generalizing the scheme in section 4.11 of [14] to $n = 30$ gives a scheme requiring $10\log q$ bits per key exchange. The scheme in [24] would also achieve our rate, but needs an unproven conjecture concerning the rationality of $T_{30}(F_q)$.

Observe that (A_1, S_0) and (B_1, T_0) are random, and since θ is a bijection, the last $\sigma^-(n)$ coordinates of $\theta(A_1, S_0)$ are of a random element in $\times_{d|n, \mu(n/d)=1} F^*_{q^d}$. Hence the probability that some coordinate of S_1 is zero is even less than that for a random element in $F_q^{\sigma^+(d)}$, which is negligible. One can then verify that every application of θ or θ^{-1} is on a random element. It follows from the foregoing discussion and the union bound that the probability of either Alice or Bob ever attempting to apply θ or θ^{-1} on an element outside of the domain is negligible. For deterministic guarantees, one can replace θ and θ^{-1} with $\tilde{\theta}$ and $\tilde{\theta}^{\leftarrow}$, negligibly changing the rate to $\phi(n)\log q + O(n^\epsilon)$ for any $\epsilon > 0$. Given the overwhelming probability guarantees for θ and θ^{-1}, this does not seem necessary.

Security: An eavesdropper obtains $a_1,\ldots,a_m, b_1,\ldots,b_m, S_m$, and T_m. Since θ and θ^{-1} are efficient bijections, this is equivalent to obtaining $A_1,\ldots,A_m$, $B_1,\ldots,B_m, S_0$, and T_0. Since S_0 and T_0 are random, determining a shared secret K_i is equivalent to solving the CDH problem in $\langle g \rangle$, given $A_1,\ldots,A_m, B_1,\ldots,$ B_m.

5.2   ElGamal Signature Schemes

Suppose the message M to be signed is at least $\sigma^-(n)\log q - \log r$ bits long. If this is not the case, one can wait until there are $m > 1$ messages M_i to be signed for which $\sum_i |M_i| \geq \sigma^-(n)\log q - \log r$, then define M to be the concatenation $M_1 \circ \cdots \circ M_m$ and sign M. For a random a, $1 \leq a \leq r - 1$, let a be Alice's private key and $A = g^a$ her public key. Let $h : \{0,1\}^* \to \mathbb{Z}_r$ be a cryptographic hash function. We have the following generalized ElGamal signature scheme (see p.458 of [18] for background):

Signature Generation (M):

1. Alice selects a random secret integer k, $1 \leq k \leq r$, and computes $d = g^k$.
2. Alice then computes $e = k^{-1}(h(M) - ah(d)) \bmod r$.
3. Alice expresses $M \circ e$ as $(R, S) \in F_q^{\sigma^-(n)} \times \{0,1\}^*$, computes $\theta(d, R) = T$, and outputs (S, T) as her signature.

Signature Verification (M, S, T):

1. Bob computes $\theta^{-1}(T) = (d, R)$ and constructs M and e from R and S.
2. Bob accepts the signature if and only if $A^{h(d)} d^e = g^{h(M)}$.

The communication of this scheme is at the optimal $|M| + \log r + \phi(n)\log q$ for ElGamal signature schemes, even for one message (as long as M is large enough).

This beats the $|M| + \log r + (n/3)\log q$ communication of the scheme in [4, 17] when $n \geq 30$, in particular for the practical values $n = 30$ and $n = 210$. Our communication is the same as that in [24], but we do not rely on any conjectures.

Note that our map θ may fail since M need not be random. One can avoid this by excluding the negligibly few M for which θ is not defined (as in RSA or the schemes of [24]), or one can replace θ with $\tilde{\theta}$, as defined above, and communicate an additional $O(n^\epsilon)$ bits of overhead. Alternatively Alice can use a pseudorandom generator to randomize M and communicate the small seed used to Bob, requiring even less communication than the already asymptotically negligible $O(n^\epsilon)$ bits.

We note that a simple modification of our protocol, making it similar in spirit to our key exchange protocol, can allow Alice to sign each M_i individually, allowing for incremental verification.

Security: In this scheme the verifier obtains (S, T), which is equivalent to obtaining M, d, and e. Thus, the security of this scheme reduces to the security of the generalized ElGamal signature scheme in $\langle g \rangle$.

5.3 ElGamal Encryption

We present two flavors of ElGamal encryption. The first is a hybrid scheme with shorter encryptions than the one in [14], while the second is essentially a non-hybrid analogue of ElGamal in $T_n(F_q)$. In the second, to encrypt a sequence of m messages, $m+1$ encryptions are created and m of them are performed directly in $T_n(F_q)$. The first scheme achieves optimal communication, while the second is asymptotically optimal.

Hybrid ElGamal. For random b, $1 \leq b \leq r - 1$, let b be Bob's private key and $B = g^b$ his public key. Suppose Alice wants to encrypt the message $M \in F_q^{\sigma^-(n)}$ with Bob's public key. Let E be an agreed upon symmetric encryption scheme with domain $F_q^{\sigma^-(n)}$. We have the following protocol:

Encryption (M):

1. Alice selects a random secret integer k, $1 \leq k \leq r$, and computes $d = g^k$.
2. From B Alice computes $e = B^k = g^{bk}$.
3. From e Alice derives a key Q for E and computes the encryption of M, $E(M)$, under key Q. Alice writes $E(M)$ as $(R, S) \in F_q^{\sigma^-(n)} \times \{0,1\}^*$.
4. Alice computes $\theta(d, R) = T$ and outputs her encryption (S, T).

Decryption (S, T):

1. Bob computes $\theta^{-1}(T) = (d, R)$.
2. From d and b Bob computes $e = g^{bk}$.
3. From e Bob derives Q and decrypts $E(M) = (R, S)$ to obtain and output M.

The communication of this scheme is at the optimal $|E(M)| + \phi(n)\log q$ bits for hybrid ElGamal encryption. As in our protocol for signature schemes, we achieve this rate even for a single message. This beats the $|E(M)| + (n/3)\log q$ bit scheme in [14] for $n \geq 30$.

It is unlikely that θ or θ^{-1} is applied to an element with any zero coordinates since d is random and $E(M)$ is likely to "look random" in practice, so $\theta(d, R)$ is likely to be a random element of $F_q^{\sigma^+(n)}$ for which it is extremely unlikely that any coordinates are zero. An exact analysis, though, depends on one's choice of E. As in our protocol for signature schemes, one can randomize $E(M)$ to decrease the error probability or replace θ with $\bar{\theta}$ for a deterministic guarantee at the cost of a few bits of communication.

Security: An adversary learns (S, T), which is equivalent to learning d and $E(M)$. Assuming the CDH problem is hard in $\langle g \rangle$, the security of this scheme is just that of the symmetric scheme E, assuming the key Q to E is chosen reasonably from e. To derive Q from e, one can extract bits that are hard to compute by an eavesdropper, see [2].

Almost Non-hybrid ElGamal. In the following, Alice will encrypt a sequence of m messages $M_1, \ldots, M_m$, each in $F_q^{\phi(n)}$. She will form $m + 1$ encryptions, m of which are encryptions in $T_n(F_q)$, and one requiring the use of an agreed upon symmetric encryption scheme E.

In the encryption phase of our scheme we will apply θ^{-1} to $(M_i \circ R)$ for some $R \in F_q^{\sigma^-(n)}$. For semantic security, for all i it must hold that $\theta^{-1}(M_i \circ R) \in \langle g \rangle \times F_q^{\sigma^-(n)}$, which in general may be strictly contained in $T_n(F_q) \times F_q^{\sigma^-(n)}$. For this we adopt the technique in section 3.7 of [25]. Namely, by reserving a few bits of each M_i to be "redundancy bits", if $\langle g \rangle$ has small enough index in $T_n(q)$, then for any R we need only try a few random settings of these bits until $\theta^{-1}(M_i \circ R) \in \langle g \rangle \times F_q^{\sigma^-(n)} = (c, d) \in \langle g \rangle \times F_q^{\sigma^-(n)}$, which we can test by checking if $c^r = 1$. In the following protocol description we ignore this issue and assume whenever θ^{-1} is applied, its image is in $\langle g \rangle \times F_q^{\sigma^-(n)}$.

For random $b_1, b_2, 1 \leq b_1, b_2 \leq r - 1$, let b_1, b_2 be Bob's private keys and $B_1 = g^{b_1}, B_2 = g^{b_2}$ be his public keys. We have the following scheme:

Encryption (M):

1. Alice chooses a random $R_0 \in F_q^{\sigma^-(n)}$.
2. For $i = 1$ to m,
   (a) Alice computes $\theta^{-1}(M_i \circ R_{i-1}) = (c_i, R_i) \in \langle g \rangle \times F_q^{\sigma^-(n)}$.
   (b) Alice chooses a random secret integer $k_i, 1 \leq k_i \leq r$, and forms the encryption $(d_i, e_i) = (g^{k_i}, c_i B_1^{k_i})$.
3. Alice uses the hybrid ElGamal encryption scheme with symmetric cipher E and public key B_2 to encrypt R_m as (T_m, S) with $T_m \in F_q^{\sigma^-(n)}$ and $S \in \{0, 1\}^*$.
4. For $i = m$ to 1,
   (a) Alice computes $\theta(d_i, T_i) = (x_i, W_i) \in F_q^{\phi(n)} \times F_q^{\sigma^-(n)}$.
   (b) Alice computes $\theta(e_i, W_i) = (y_i, T_{i-1}) \in F_q^{\phi(n)} \times F_q^{\sigma^-(n)}$.
5. Alice outputs $x_1, \ldots, x_m, y_1, \ldots, y_m, T_0, S$ as her encryption of $M_1, \ldots, M_m$.

Decryption $(x_1, \ldots, x_m, y_1, \ldots, y_m, T_0, S)$:

1. For $i = 1$ to m,
   (a) Bob computes $\theta^{-1}(y_i \circ T_{i-1}) = (e_i, W_i)$.
   (b) Bob computes $\theta^{-1}(x_i \circ W_i) = (d_i, T_i)$.
   (c) Bob computes $c_i = e_i / d_i^{b_1}$.
2. Bob uses T_m and S, together with b_2, in the decryption procedure of the hybrid ElGamal scheme to recover R_m.
3. For $i = m$ to 1, Bob computes $\theta(c_i, R_i) = M_i \circ R_{i-1}$.
4. Bob outputs $M_1, \ldots, M_m$.

The communication of this scheme is $2m\phi(n) \log q + |E(R_m)| + \phi(n) \log q$ bits. Hence, as m grows, the rate of this scheme approaches $2\phi(n) \log q$, which is optimal for ElGamal type encryption.

Note that the M_i's need not be random, and consequently $\theta^{-1}(M_i, R_{i-1})$ may not be well-defined. Choosing random R_0 will increase the chances that $\theta^{-1}(M_i, R_{i-1})$ is always defined. Alternatively, one can use the ideas of section 5.2 to randomize M_i, or one can use β instead of θ^{-1}. Again, since $E(R_m) = (S, T_m)$ needn't be random even if E is semantically secure, one may want to use $\tilde{\theta}$ in place of θ. This adds a negligible amount to the communication, and as stated earlier, encrypting the extra bits of $\tilde{\beta}$ can be done in step 3.

Security: An adversary learns $x_1, \ldots, x_m, y_1, \ldots, y_m, T_0, S$, which is equivalent to learning $E'(R_m), d_1, \ldots, d_m, e_1, \ldots, e_m$, where E' is the semantically secure hybrid encryption scheme. Assuming DDH is hard in $\langle g \rangle$, (d_i, e_i) is a semantically secure encryption $E''(c_i)$ of c_i for all i. The security of the scheme then follows from the fact that the keypairs (b_1, B_1) and (b_2, B_2) of E', E'' are independent.

6   Computational Complexity

In this section we present efficient algorithms for computing θ and θ^{-1}, analyze their complexity, and suggest an alternative way of improving computational costs with slightly more communication. Each of these is described in turn.

6.1   Algorithm

Before describing θ and θ^{-1}, we need some notation:

- For $d \mid n$, let U_d be the smallest integer for which $\gcd(\Phi_e(q), \Phi_f(q), \frac{q^d - 1}{U_d}) = 1$ for all $e \neq f$ with $e \mid d$ and $f \mid d$.
- For $e \mid d \mid n$, we define $y_{d,e} = \gcd(\Phi_e(q), \frac{q^d - 1}{U_d})$ and $z_{d,e} = \gcd(\Phi_e(q), U_d)$. Generalizing section 3, we can find w_d and $w_{d,e}$ s.t. $\frac{q^d - 1}{U_d} w_d + \sum_{e \mid d} \frac{q^d - 1}{y_{d,e}} w_{d,e} = 1$. Further, we can find $u_{d,e}$ and $v_{d,e}$ for which $\frac{\Phi_e(q)}{y_{d,e}} u_{d,e} + \frac{\Phi_e(q)}{z_{d,e}} v_{d,e} = 1$.
- Let $\rho_e(d) : \{d : e \mid d \mid n, \mu(n/d) = -1\} \rightarrow \{d : e \mid d \mid n, \mu(n/d) = +1\}$ for $e \mid n$, $e \neq n$, be a bijective mapping and define $\rho_n(n) = n$.

A naive implementation of θ consists of the following steps:

1. We first use an isomorphism

$$T_n(F_q) \times \times_{\mu(n/d)=-1} F_{q^d} \longrightarrow T_n(F_q) \times \times_{\mu(n/d)=-1} G_{U_d} \times G_{(q^d-1)/U_d}.$$

2. By using a table lookup we map $\times_{\mu(n/d)=-1} G_{U_d} \longrightarrow \times_{\mu(n/d)=-1} \times_{e|d} G_{z_{d,e}}$ and we use an isomorphism $\times_{\mu(n/d)=-1} G_{(q^d-1)/U_d} \longrightarrow \times_{\mu(n/d)=-1} \times_{e|d} G_{y_{d,e}}$. By the structure theorem of Abelian groups there is an isomorphism $G_{z_{d,e}} \times G_{y_{d,e}} \longrightarrow T_e(F_q)$ for each $d|n$ with $\mu(n/d) = -1$ and $e \mid d$.

3. By using a permutation we obtain a mapping

$$T_n(F_q) \times \times_{\mu(n/d)=-1} \times_{e|d} T_e(F_q) \longrightarrow \times_{\mu(n/d)=+1} \times_{e|d} T_e(F_q).$$

4. By the structure theorem of Abelian groups there is, for each $d|n$ with $\mu(n/d) = +1$ and $e \mid d$, an isomorphism $T_e(F_q) \longrightarrow G_{z_{d,e}} \times G_{y_{d,e}}$. By using a table lookup we map $\times_{\mu(n/d)=+1} \times_{e|d} G_{z_{d,e}} \longrightarrow \times_{\mu(n/d)=+1} G_{U_d}$ and we use an isomorphism $\times_{\mu(n/d)=+1} \times_{e|d} G_{y_{d,e}} \longrightarrow \times_{\mu(n/d)=+1} G_{(q^d-1)/U_d}$.

5. In the last step we use an isomorphism

$$\times_{\mu(n/d)=+1} G_{U_d} \times G_{(q^d-1)/U_d} \longrightarrow \times_{\mu(n/d)=+1} F_{q^d}.$$

Each of the isomorphisms are defined by taking simultaneous exponentiations. An improved implementation combines different isomorphisms in a single simultaneous exponentiation. Each table lookup followed by an exponentiation can be implemented as a single table lookup. This reduces the number of exponentiations and multiplications.

Computation of $\theta(x, (x_d)_{d|n, \mu(n/d)=-1})$ for $(x_d)_{d|n, \mu(n/d)=-1} \in \times_{d|n, \mu(n/d)=-1} F_{q^d}^*$ and $x \in T_n(F_q)$:

1. For $d \mid n$, $\mu(n/d) = -1$,
   (a) Compute $x_d^{(q^d-1)/U_d} \in G_{U_d}$ and map it to $(Z_{d,e})_{e|d} \in \times_{e|d} G_{z_{d,e}}$ by using a table look up.
   (b) Compute $(Z_{\rho_e(d),e} = (Z_{d,e}^{v_{d,e}} x_d^{(q^d-1)u_{d,e}/y_{d,e}})^{\Phi_e(q)/z_{\rho_e(d),e}})_{e|d} \in \times_{e|d} G_{z_{\rho_e(d),e}}$.

2. Compute $Z_{n,n} = x^{\Phi_n(q)/z_{n,n}} \in G_{z_{\rho_n(n),n}}$.

3. For $d \mid n$, $\mu(n/d) = +1$,
   (a) Map $(Z_{d,e})_{\rho_e(d')=d, e|d} \in \times_{e|d} G_{z_{d,e}}$ to $Z_d \in G_{U_d}$ by using a table look up.
   (b) Compute $x_d = Z_d^{w_d} \cdot \prod_{\rho_e(d')=d, e|d, e\neq n} (Z_{d',e}^{v_{d',e}} x_d^{(q^{d'}-1)u_{d',e}/y_{d',e}})^{\Phi_e(q)w_{d,e}/y_{d,e}}$ which is in $G_{U_d} \cdot G_{(q^d-1)/U_d} = F_{q^d}^*$.

4. Multiply x_n with $x^{\Phi_n(q)w_{n,n}/y_{n,n}}$.

5. $\theta(x, (x_d)_{d|n, \mu(n/d)=-1}) = (x_d)_{d|n, \mu(n/d)=+1}$.

The ideas in section 3 can be used to show the algorithm above is well-defined. The improved computation of θ^{-1} is similar, where we make sure to use the inverse of the coordinate permutation used in θ.

6.2   Complexity

For background on efficient computations in fields and subgroups, see [6, 12, 29]. Consider the algorithm for θ. In step 1, for $d \mid n$, $\mu(n/d) = -1$, we perform $1 + \sum_{e|d} 1$ exponentiations in F_{q^d}. Notice that, in step 1b we do not need to compute $Z_{d,e}^{v_{d,e}}$ since it can be combined with the table lookup in step 1a (there is an entry in the table corresponding to $Z_{d,e}^v$ for every v). Step 2 costs 1 exponentiation in F_{q^n}.

For $d \mid n$, $\mu(n/d) = -1$ or $d = n$, we precompute $x_d^{2^i}$, $0 \leq i \leq d \log q$. This costs $d \log q$ multiplications in F_{q^d}. By using the results of the precomputation, an exponentiation x_d^t, for some t, in F_{q^d} costs on average $(d \log q)/2$ multiplications in F_{q^d} (the bit length of the exponent t is $(d \log q)$ and roughly half the time a bit is equal to 1). Each multiplication in F_{q^d} costs $f(d) \leq d^2$ multiplications in F_q. Summarizing, steps 1 and 2 cost about

$$C_1 = \left(3f(n)n + \sum_{d|n,\mu(n/d)=-1} (3 + \sum_{e|d} 1)f(d)d \right) \frac{\log q}{2}$$

multiplications in F_q.

In step 3, for $d \mid n$, $\mu(n/d) = +1$, we need to perform, for each $e \mid d$ with $\rho_e(d') = d$, one exponentiation in $F_{q^{d'}}$. We do not need to compute $Z_{d',e}^{v_{d',e}}$ which can be combined with the table lookup in step 1a.

The cost of step 3, measured in multiplications in the base field F_q, is on average approximately $\sum_{d|n,\mu(n/d)=+1} \sum_{e|d} f(\rho_e^{-1}(d))\rho_e^{-1}(d)(\log q)/2$. Since ρ_e defines a permutation, this expression is equal to

$$C_2 = \left(f(n)n + \sum_{d|n,\mu(n/d)=-1} (\sum_{e|d} 1)f(d)d \right) \frac{\log q}{2}.$$

The total cost is $C_1 + C_2$ multiplications in F_q, where we neglect the cost of table lookups, addition, and multiplication modulo an integer. Since $\sum_{e|d} 1 = O(d^\varepsilon)$, we have $\sum_{d|n,\mu(n/d)=-1} (3 + 2\sum_{e|d} 1)f(d)d = O(\sum_{d|n} d^{3+\varepsilon}) = O((\sum_{d|n} d)^{3+\varepsilon}) = O(n^{3+\varepsilon})$, since the sum of divisors of n is $O(n^{1+\varepsilon})$ for any $\epsilon > 0$. This proves $C_1 + C_2 = O(n^{3+\varepsilon} \log q)$.

The same techniques show θ^{-1} requires $O(n^{3+\epsilon} \log q)$ multiplications in F_q.

6.3   Efficiency Improvements

To improve the efficiency we may use exponentiation algorithms for fixed exponents using vector addition chains. Also, we may group several exponentiations of x_d together into one exponentiation by appropriately choosing the bijections ρ_e. If n is not too large, we may use simultaneous exponentiation to speed up the computations. Full simultaneous exponentiations in every step requires a precomputation of 2^n multiplications. We may optimize by using simultaneous exponentiation to compute intermediate results which we multiply together to compute the full exponentiation. Finally, we may combine the exponentiations required in our applications with the evaluation of θ.

Notice that θ is much more efficient if, for $d \mid n$ with $\mu(n/d) = -1$, $x_d \in G_{(q^d-1)/U_d}$. Then, for $e \mid d \mid n$ with $\mu(n/d) = -1$, $Z_{d,e} = 1$ and $Z_d = 1$. Table lookups can be avoided. Therefore each x_d, for $d \mid n$ with $\mu(n/d) = +1$, can be computed by a single simultaneous exponentiation of $x, x_d \in G_{(q^d-1)/U_d}, d \mid n, \mu(n/d) = -1$, with fixed exponents in step 3. To make use of this, we define a new map τ which maps $(x, (x_d)_{d \mid n, \mu(n/d)=-1})$ into $\theta(x, (x_d^{U_d})_{d \mid n, \mu(n/d)=-1})$ and the table entries of $(x_d^{(q^d-1)/U_d})_{d \mid n, \mu(n/d)=-1}$. This increases the communication cost by

$$\sum_{d \mid n, \mu(n/d)=-1} \log_2 U_d$$

bits which in practice is much less than $\log_2 q$. So at the cost of a small increase in communication we improve the computational efficiency.

Computation of $\tau(x, (x_d)_{d \mid n, \mu(n/d)=-1})$ and $\tau^{\leftarrow}$:

1. For $d \mid n$, $\mu(n/d) = -1$, compute $(x_d' = x_d^{(q^d-1)/U_d})_{d \mid n, \mu(n/d)=-1}$.
2. Compute

$$x_d = \prod_{\rho_e(d')=d, e \mid d, e \neq n} (x_{d'}^{U_{d'}(q^{d'}-1)u_{d',e}/y_{d',e}})^{\Phi_e(q)w_{d,e}/y_{d,e}} \in G_{(q^d-1)/U_d} \subseteq F_{q^d}^*,$$

   for $d \mid n$ with $\mu(n/d) = +1$. Multiply x_n with $x^{\Phi_n(q)w_n/z_{n,n}+\Phi_n(q)w_{n,n}/y_{n,n}}$.
3. $\tau(x, (x_d)_{d \mid n, \mu(n/d)=-1}) = ((x_d)_{d \mid n, \mu(n/d)=+1}, (x_d')_{d \mid n, \mu(n/d)=-1})$.
4. Compute $x_n^{(q^n-1)v_{n,n}/U_n+(q^n-1)u_{n,n}/y_{n,n}} = x$.
5. Compute

$$x_{d'}'^{a_{d'}}\left(\prod_{d=\rho_{e'}(d'), e' \mid d'} (x_d^{(q^d-1)u_{d,e'}/y_{d,e'}})^{\Phi_{e'}(q)w_{d',e'}/y_{d',e'}}\right)^{b_{d'}} = x_{d'}'^{a_{d'}} x_{d'}^{U_{d'}b_{d'}} = x_{d'},$$

   for $d' \mid n$ with $\mu(n/d') = -1$, where $\frac{q^{d'}-1}{U_{d'}}a_{d'} + U_{d'}b_{d'} = 1$.
6. $\tau^{\leftarrow}((x_d)_{d \mid n, \mu(n/d)=+1}, (x_d')_{d \mid n, \mu(n/d)=-1}) = (x, (x_d)_{d \mid n, \mu(n/d)=-1})$.

For $n = 30$, $\{d \mid n : \mu(n/d) = -1\} = \{15, 10, 6, 1\}$ and $\{d \mid n : \mu(n/d) = +1\} = \{30, 5, 3, 2\}$. We define $\rho_1(15) = 5, \rho_3(15) = 30, \rho_5(15) = 5, \rho_{15}(15) = 30, \rho_1(10) = 2, \rho_2(10) = 2, \rho_5(10) = 30, \rho_{10}(10) = 30, \rho_1(6) = 3, \rho_2(6) = 30, \rho_3(6) = 3, \rho_6(6) = 30, \rho_1(1) = 30, \rho_{30} = 30$. We use $f(30) = 234$, $f(15) = 78$, $f(10) = 45$, $f(6) = 18$, $f(5) = 15$, $f(3) = 6$, and $f(2) = 3$ [31]. In step 1, we compute x_{15}', x_{10}', x_6', and x_1' using single exponentiations by using the square and multiply method [18, p. 614]. This costs in total $3(78 \cdot 15 + 45 \cdot 10 + 18 \cdot 6 + 1)(\log q)/2 = 2593.5 \log q$ multiplications in F_q.

In step 2, x_{30} is computed as a simultaneous exponentiation [18, p. 618]in $x \in F_{q^{30}}, x_{15} \in F_{q^{15}}, x_{10} \in F_{q^{10}}, x_6 \in F_{q^6}, x_1 \in F_q$. In a precomputation we compute for each of the 2^5 possible sets $S \subset \{x, x_{15}, x_{10}, x_6, x_1\}$ the product $\prod_{w \in S} w$. The whole precomputation costs at most 2^5 multiplications in $F_{q^{30}}$. In the computation of x_{30} the exponents of x, x_{15}, x_{10}, etc., have bit lengths $30 \log q$, $15 \log q$, $10 \log q$, etc. This means that in the second half of the simultaneous exponentiation (the last $30 \log q - 15 \log q$ bits of the exponents) we only need to square or square-and-multiply with $x \in F_{q^{30}}$. So the average costs in the second half of the simultaneous multiplication is equal to $3(15 \log q)/2$ multiplications in $F_{q^{30}}$. The simultaneous exponentiation corresponding to the bits ranging from position $10 \log q$ to $15 \log q$ involves square or square and multiply with x, x_{15}, or $x \cdot x_{15}$. This costs on average $7(5 \log q)/4$ multiplications (5 is the difference between 15 and 10, on average we need 1 multiplication in 1 out of 4 cases and 2 multiplications in 3 out of 4 cases). Notice that we treat squaring as a single multiplication in this excersise. Continuing this argument we need in total $234(2^5 + 15(3/2) + 5(7/4) + 4(15/8) + 5(31/16) + 1(63/32))(\log q) = 19283.1 \log q$ multiplications in F_q (2^5 comes from preprocessing).

The outputs x_5, x_3 and x_2 are single multiplications in x_{15}, x_6, and x_{10}, respectively costing a total of $3(78 \cdot 15 + 18 \cdot 6 + 45 \cdot 10)(\log q)/2 = 2592 \log q$ multiplications. Concluding, the computation of τ costs approximately $24468.6 \log q$ multiplications in F_q. A single exponentiation in $F_{q^{30}}$ costs $234 \cdot 30(\log q)3/2 = 10530 \log q$ multiplications. Hence, τ costs about 2.32 exponentiations in $F_{q^{30}}$.

In the implementation of $\tau^{\leftarrow}$ we compute x as a single exponentiation in x_{30}, costing $234 \cdot 30(\log q)3/2 = 10530 \log q$ multiplications. In step 5, x_{15} is a simultaneous exponentiation in x_{30} and x_5 (and a table look up for the exponentiation in x'_{15}). This costs $78(2^2 + 25(3/2) + 5(7/4))(\log q) = 3919.5 \log q$ multiplications. Similarly, x_{10} costs $45(2^2 + 28(3/2) + 2(7/4))(\log q) = 2227.5 \log q$ and x_6 costs $18(2^2 + 27(3/2) + 3(7/4))(\log q) = 895.5 \log q$ multiplications. We compute x_1 as a single exponentiation in x_{30}, costing $234 \cdot 30(\log q)3/2 = 10530 \log q$ multiplications. Concluding, the computation of $\tau^{\leftarrow}$ costs approximately $28102.5 \log q$ multiplications, which is equivalent to 2.67 exponentiations in $F_{q^{30}}$.

7   Conclusions and Open Problems

Our fundamental contribution is a compact and efficient representation of elements of $T_n(F_q)$, namely, the construction of bijections θ and θ^{-1} of section 3. This allows us to construct ElGamal signature and encryption schemes meeting the optimal rate of communication, as well as a secret key exchange protocol meeting this rate asymptotically. If the torus conjecture of [24] is proven, the schemes in that paper will also achieve this rate, and moreover, their scheme for DH key exchange will meet the optimal rate even for a single key exchanged. Hence, resolving their conjecture is an important problem. Another important question is whether the computational cost of our schemes can be reduced to a more practical level. Finally, our representation of $T_n(F_q)$ may have other applications.

References

1. L. M. Adelman, J. DeMarrais, *A Subexponential Algorithm for Discrete Logarithms over All Finite Fields*, in Advances in Cryptology – Crypto '93, LNCS 773, Springer-Verlag 1994, 147-158.
2. D. Boneh and R. Venkatesan, *Rounding in lattices and its cryptographic applications*, Proc. 8-rd Annual ACM-SIAM Symposium on Discrete Algorithms, ACM, NY, 1997, 675– 681.
3. W. Bosma, J. Hutton, and E. R. Verheul, *Looking Beyond XTR*, in Advances in Cryptology – Asiacrypt '02, LNCS **2501**, Springer, Berlin, 2002, 46 - 63.
4. A. E. Brouwer, R. Pellikaan, and E. R. Verheul, *Doing More with Fewer Bits*, In Advances of Cryptology – Asiacrypt '99, LNCS **1716**, Springer, 321-332.
5. N. G. Chebotarev, Die Bestimmung der Dichtigkeit einer Menge von Primzahlen, welche zu einer gegebenen Substitutionsklasse gehören. Math. Ann. **95**, 191-228 (1926).
6. H. Cohen and A. K. Lenstra, *Supplement to Implementation of a New Primality Test*, Mathematics of Computation, volume 48, number 177, 1987.
7. D. Coppersmith, *Fast Evaluation of Logarithms in Fields of Characteristic Two*, IEEE Trans. Inform. Theory 30 (1984), 587-594.
8. S. R. Finch, *Artin's Constant*, 2.4 in Mathematical Constants, Cambridge, England: Cambridge University Press (2003), 104-110.
9. D. Gordon, *Discrete Logarithms in GF(p) Using the Number Field Sieve*, SIAM J. Discrete Math. 6 (1993), 312-323.
10. T. ElGamal, *A Public Key Cryptosystem and a Signature Scheme Based on Discrete Logarithms*, IEEE Transactions on Information Theory 31(4), 1985, 469-472.
11. G.H. Hardy and E.M. Wright, An Introduction to the Theory of Numbers, 5th edition, Oxford University Press, 1979.
12. A. Karatsuba and Y. Ofman. *Multiplication of Multidigit Numbers on Automata*, Soviet Physics Doklady, volume 7, 1963, 595-596.
13. A. K. Lenstra, *Using Cyclotomic Polynomials to Construct Efficient Discrete Logarithm Cryptosystems over Finite Fields*, Proceedings of ACISP 97, LNCS 1270, Springer-Verlag 1997, 127-138.
14. A. K. Lenstra and E. R. Verheul, *The XTR Public Key System*, In Advances of Cryptology – Crypto 2000, LNCS **1880**, Springer, 1-19.
15. A. K. Lenstra and E. R. Verheul, *An Overview of the XTR Public Key System*, in Public-key cryptography and computational number theory (Warsaw, 2000), de Gruyter, Berlin, 2001, 151-180.
16. H. W. Lenstra, *The Chebotarev Density Theorem*, URL: http://math.berkeley.edu/ jvoight/notes/oberwolfach/Lenstra-Chebotarev.pdf
17. Seongan Lim, Seungjoo Kim, Ikkwon Yie, Jaemoon Kim, Hongsub Lee, *XTR Extended to $GF(p^{6m})$*, Selected Areas in Cryptography, 8th Annual International Workshop, SAC 2001, 301-312, Springer Verlag, 2001.
18. A. J. Menezes, P. C. van Oorschot, S. A. Vanstone, Handbook of Applied Cryptography, CRC Press, Boca Raton, FL, 1997.
19. T. Nagell, "The Cyclotomic Polynomials" and "The Prime Divisors of the Cyclotomic Polynomial", 46 and 48 in Introduction to Number Theory. New York: Wiley, 158-160 and 164-168, 1951.
20. A. Odlyzko, *Discrete Logarithms: The past and the future*, Designs, Codes and Cryptography, 19 (2000), 129-145.

21. S. C. Pohlig, M. E. Hellman, *An Improved Algorithm for Computing Logarithms over GF(p) and its Cryptographic Significance*, IEEE Trans. on IT, 24 (1978), 106-110.
22. J. M. Pollard, *Monte Carlo methods for index computation (mod p)*, Math. Comp., 32 (1978), 918-924.
23. K. Rubin and A. Silverberg, *Algebraic tori in cryptography*, to appear in High Primes and Misdemeanours: lectures in honour of the 60th birthday of Hugh Cowie Williams, Fields Institute Communications Series, American Mathematical Society, Providence, RI (2004).
24. K. Rubin and A. Silverberg, *Torus-Based Cryptography*, In Advances of Cryptology – Crypto 2003, LNCS **2729**, Springer, 349-365.
25. K. Rubin and A. Silverberg, *Using primitive subgroups to do more with fewer bits*, In Algorithmic Number Theory (ANTS VI), Lecture Notes in Computer Science 3076 (2004), Springer, 18-41.
26. C. P. Schnorr, *Efficient Signature Generation by Smart Cards*, Journal of Cryptology, 4 (1991), 161-174.
27. O. Schirokauer, *Discrete Logarithms and Local Units*, Phil. Trans. R. Soc. Lond. A 345, 1993, 409-423.
28. O. Schirokauer, D. Weber, Th. F. Denny, *Discrete Logarithms: the effectiveness of the index calculus method*, Proceedings ANTS II, LNCS 1122, Springer-Verlag 1996.
29. M. Stam, *Speeding up Subgroup Cryptosystems*, PhD Thesis, Eindhoven University of Technology, 2003.
30. V. Voskresenskii, Algebraic Groups and Their Birational Invariants, Translations of Mathematical Monographs **179**, American Mathematical Society, Providence, RI, 1998.
31. A. Weimerskirch and C. Paar, *Generalizations of the Karatsuba Algorithm for Efficient Implementations*, URL: http://www.crypto.ruhr-uni-bochum.de/Publikationen/, 2003.

How to Compress Rabin Ciphertexts and Signatures (and More)

Craig Gentry

DoCoMo USA Labs
cgentry@docomolabs-usa.com

Abstract. Ordinarily, RSA and Rabin ciphertexts and signatures are $\log N$ bits, where N is a composite modulus; here, we describe how to "compress" Rabin ciphertexts and signatures (among other things) down to about $(2/3) \log N$ bits, while maintaining a tight provable reduction from factoring in the random oracle model. The computational overhead of our compression algorithms is small. We also improve upon Coron's results regarding partial-domain-hash signature schemes, reducing by over 300 bits the hash output size necessary to prove adequate security.

1   Introduction

The hardness of factoring is one of the most fundamental and frequently used assumptions of public-key cryptography; yet cryptosystems that rely on the factoring assumption have relatively poor performance in terms of bandwidth. For example, RSA and Rabin ciphertexts and signatures are typically at least as many bits as the composite modulus N, while recent advances in hardware-based approaches to factoring (e.g., [32]) suggest that N must be more than 1024 bits for strong security. So, factoring-based cryptosystems often do not compare favorably with cryptosystems based on alternative hard problems – e.g., ECC for encryption or DSA for signatures.

Bandwidth consumption is important, in part because fundamental limitations of wireless technology put bandwidth at a premium. For example, Barr and K. Asanović [2] note that wireless transmission of a single bit can cost more than 1000 times as much energy as a 32-bit computation. Since battery efficiency is growing relatively slowly, energy consumption (particularly through wireless transmission) may become a significant bottleneck.

Moreover, signal interference places physical limits on how much data can be transmitted wirelessly in a given region. This was not a problem in wired networks. These limitations are compounded by the lossiness of wireless channels, which necessitates *additional* bandwidth in the form of forward error correction (FEC). FEC is particularly important for cryptographic transmissions, where partial recovery of a ciphertext or digital signature is typically useless.

These considerations make compression algorithms very attractive. In fact, in recent years, substantial progress has been made in constructing "compressed" cryptosystems. For example, XTR [22] and CEILIDH [30] both use "compact

M. Franklin (Ed.): CRYPTO 2004, LNCS 3152, pp. 179–200, 2004.
© International Association for Cryptologic Research 2004

representations" of certain elements to achieve a bandwidth savings. There are also a variety of hybrid cryptosystems, such as signcryption and aggregate signature schemes, in which multiple cryptographic functionalities are somehow represented by a single, relatively short string. However, although such hybrid cryptosystems exist for RSA and Rabin, none of them breaks the "($\log N$)-bit barrier."

OUR DESIGN GOALS. In light of these considerations, we would like to construct a compression algorithm that is broadly applicable to factoring-based schemes, such as RSA and Rabin. Ideally, the compression algorithm should allow RSA and Rabin ciphertexts and signatures to be substantially less than $\log N$ bits without sacrificing any security – i.e., while still using (and retaining the security of) a ($\log N$)-bit modulus. Moreover, the compression algorithm should add minimal computational overhead. If the compression algorithm requires additional computation, this computation should not require use of the secret key, so that it can be performed (more quickly) outside of a "secure environment," such as a smart card.

OUR RESULTS. We essentially achieve our design goals, except that our techniques work only for Rabin-type cryptosystems, not for RSA. Along the way, we also substantially improve upon Coron's results on partial-domain-hash Rabin signature schemes (Rabin-PDH).

Coron [16] proved the security of a variant of the Rabin signing scheme (Rabin-PDH) in which the hash function that is used to hash the message outputs strings of length $\left(\frac{2}{3} + \epsilon\right) \log N$ bits. It turns out that this ϵ has a large effect in practice; if the simulator in the security proof wishes to generate a distribution of signatures whose statistical distance from uniform is less than 2^{-80}, Coron's method requires that the hash output length be at least $\frac{2}{3} \log N + 364$ bits. We provide a perfectly uniform drawing algorithm that reduces the necessary hash output length to only $\frac{2}{3} \log N + 3$ bits; moreover, our security proof is tighter.

Our main result, however, is a compression algorithm that allows a 33% reduction in the bit-length of Rabin signatures and ciphertexts, without any sacrifice in security. (Notice that Coron's result is not a compression algorithm; although the hash output length of Coron's Rabin-PDH scheme may be less than $\log N$ bits, the Rabin-PDH signature itself, which is essentially a modular square root of the hash output, is a ($\log N$)-bit value.) For our improved version of Rabin-PDH signatures, the "entropy" of the hash output is just over $\frac{2}{3} \log N$ bits; thus, it is *theoretically* possible that the signature could also be expressed in about $\frac{2}{3} \log N$ bits. In fact, up to the loss of a few bits, this is precisely what we achieve: a ($\frac{2}{3} \log N + 6$)-bit Rabin-PDH signature, with a tight reduction from factoring N.

Our lossless compression algorithm also works for Rabin encryption, but in reverse. A ($\frac{2}{3} \log N$)-bit plaintext is "decompressed" by mapping it to a ($\log N$)-bit number that has a ($\frac{2}{3} \log N + 3$)-bit modular square. This modular square is a "compressed" Rabin ciphertext. Numerous other cryptosystems also involve computing square roots modulo a composite modulus N, including Fiat-Shamir, Cocks's identity-based encryption scheme, as well as various schemes enabling ring signatures, signcryption, and so on. Our techniques enable a similar 33% bandwidth reduction for these schemes.

RELATED WORK. Like Coron's work, our techniques build upon Brigitte Vallée's elegant analysis of the distribution, in $\mathbb{Z}/N\mathbb{Z}$, of integers in $B_{N,h,h'} = \{x \in [1, N) : h \leq x^2(\mathrm{mod}\,N) < h'\}$ for $h' - h \leq 8N^{2/3}$ – i.e., integers with modular squares in a "narrow" interval. We provide a self-contained discussion of her results in section 3.

Some previous work has been done on compressing Rabin and low-exponent RSA signatures – in particular, Bernstein [7] mentions that one can simply remove the $\frac{1}{e}\log_2 N$ least significant bits of any regular Rabin or RSA signature, and the verifier can use Coppersmith's method [17] to recover those bits. Bleichenbacher [8] describes an improvement: the signer can use continued fractions to express the signature s as $a/b(\mathrm{mod}\,N)$, where a is about $\frac{e-1}{e}\log_2 N$ bits and b is about $\frac{1}{e}\log_2 N$ bits, and send a as the signature. The verifier checks that $c = a^e/H(m)(\mathrm{mod}\,N)$ is an eth power (namely b^e) over $\mathbb{Z}$. The drawback of these methods, though they arguably reduce Rabin signature length to $\frac{1}{2}\log_2 N$ bits, is that they do not allow message recovery; the verifier needs m before verifying, which effectively adds to the signature length. These methods also do not appear to be very broadly applicable; e.g., they do not appear to lead to low-bit-length encryption, signcryption and aggregate signature schemes.

As mentioned above, Coron [16] uses a "compressed" output space for the hash function in a Rabin signature scheme, but the partial-domain hash signatures themselves are still $\log N$ bits.

ORGANIZATION OF THE PAPER. This paper is organized as follows. After noting some preliminaries in Section 2, we describe Vallée's distributional observations and her "quasi-uniform" drawing algorithm in section 3. In section 4, we describe our perfectly uniform drawing algorithm, and our improvement upon Coron's results regarding Rabin-PDH. We describe our compression algorithm in section 5, after which we describe compressed Rabin encryption and signature schemes in section 6. Finally, in Section 7, we mention other cryptosystems – such as signcryption, aggregate signature and ring signature schemes – for which our compression algorithm allows a 33% bandwidth reduction.

2   Preliminaries

We gather some mathematical notation here for convenience. Let $\{0,1\}^*$ denote the set of all bit strings, and let $\{0,1\}^n$ denote the set of all bit-strings of length n. For a real number r, $\lceil r \rceil$ denotes the ceiling of r, that is, the smallest integer value greater than or equal to r. Similarly, $\lfloor r \rfloor$ denotes the floor of r, that is, the largest integer value less than or equal to r. Finally, $\lfloor r \rceil$ denotes the closest integer to r. Let the symbol $\|$ denote concatenation.

Throughout, N will denote a suitable integer modulus. To be suitable, N should at least be computationally hard to factor using any modern factoring algorithm. In practice, one often generates N as the product of two large prime numbers p and q – e.g., 512 bits apiece. However, one could choose N differently for our schemes, if desired. For example, setting $N = p^d q$ for $d > 1$ can lead to efficiency advantages, though one should be wary of setting d too large [11].

Let $B_{N,h,h'} = \{x \in [1, N] : h \le x^2 (\mathrm{mod} N) < h'\}$ for integers h and h' and suitable modulus N – i.e., the set of integers with modular squares in $[h, h')$. Let B be shorthand for $B_{N,h,h'}$ when N, h and h' are understood.

A "lattice" consists of the set of all vectors that can be generated as integer linear combinations of a set of basis vectors. For example, if (a, b) and (c, d) are two basis vectors in two-dimensional space, the lattice that they generate is the set of vectors $\{(k_1 a + k_2 c, k_1 b + k_2 d) : k_1, k_2 \in \mathbb{Z}\}$.

3   Distribution of Numbers with Small Modular Squares

Developing a compressed representation of numbers in $B_{N,h,h'}$ that is efficiently computable and invertible requires an understanding of how numbers in $B_{N,h,h'}$ are distributed in $[0, N/2)$. The compression algorithm works, at a high level, by taking this distribution into account.

In [33], Vallée describes the "global" distribution of $B_{N,h,h'}$ in $[0, N/2)$ in terms of its "local" distribution in each of a set of *Farey intervals* that covers $[0, N/2)$. She then describes each local distribution in terms of points of a lattice that lie in the region between two parabolas. For $h' - h \ge 8N^{2/3}$, the distribution of $B_{N,h,h'}$-elements among the Farey intervals is "quasi-independent," allowing her to construct an algorithm that draws integers from $B_{N,h,h'}$ "quasi-uniformly." Since Vallée's analysis forms the basis of our compression algorithm, we review it in detail in this section.

3.1   Farey Sequences

Some properties of Farey sequences are collected in [20]; we recall them below.

Definition 1 (Farey Sequence). *The Farey sequence $\mathcal{F}_k$ of order k is the ascending sequence $(\frac{0}{1}, \frac{1}{k}, \ldots, \frac{1}{1})$ of fractions $\frac{a_i}{b_i}$ with $1 \le a_i \le b_i \le k$ and $\gcd(a_i, b_i) = 1$.*

The characteristic property of Farey sequences is expressed in the following theorem [20]:

Theorem 1. *If $\frac{a_i}{b_i}$ and $\frac{a_{i+1}}{b_{i+1}}$ are consecutive in $\mathcal{F}_k$, then $b_i a_{i+1} - a_i b_{i+1} = 1$.*

Another useful theorem concerning Farey sequences is the following:

Theorem 2. *If $\frac{a_i}{b_i}$ and $\frac{a_{i+1}}{b_{i+1}}$ are consecutive in $\mathcal{F}_k$, then $b_i + b_{i+1} > k$.*

The latter theorem follows from the fact that $(a_i + a_{i+1})/(b_i + b_{i+1})$, the so-called "mediant" of a_i/b_i and a_{i+1}/b_{i+1}, is between a_i/b_i and a_{i+1}/b_{i+1} and would be in $\mathcal{F}_k$ if $b_i + b_{i+1} \le k$. Farey sequences lead naturally to the notion of a Farey partition, in which the set of mediants partition the interval $[0, N/2)$ into subintervals. The formal definition is as follows.

Definition 2 (Farey Partition). *The Farey partition of order k of the interval $[0, N/2)$ is the set of intervals $J(a_i, b_i) = [\frac{(a_{i-1}+a_i)N}{2(b_{i-1}+b_i)}, \frac{(a_i+a_{i+1})N}{2(b_i+b_{i+1})})$, where $\frac{a_i}{b_i}$ is the i-th term in $\mathcal{F}_k$.*

So that each "end" of $[0, N/2)$ is covered by the partition, we set $(a_0, b_0) = (a_1, b_1)$ and $(a_{z+1}, b_{z+1}) = (a_z, b_z)$, where $a_z/b_z = 1/1$ is the final fraction in the Farey sequence.

Vallée found it convenient to use another set of intervals $I(a_i, b_i)$, called "Farey intervals," that are related to $J(a_i, b_i)$.

Definition 3 (Farey Interval). *The Farey interval $I(a_i, b_i)$ of order k is the open interval with center $\frac{a_i N}{2b_i}$ and radius $\frac{N}{2kb_i}$, where $\frac{a_i}{b_i}$ is the i-th term in $\mathcal{F}_k$.*

Using Theorems 1 and 2, one can easily prove that $I(a_i, b_i)$ contains $J(a_i, b_i)$, and that the interval $I(a_i, b_i)$ is no more than twice as wide as the interval $J(a_i, b_i)$ [1]. One can also prove that every number in $[0, N/2)$ is covered by at least one, and at most two, Farey intervals – e.g., by showing that, for every i, $I(a_{i-1}, b_{i-1})$ intersects $I(a_i, b_i)$, but neither $I(a_{i-1}, b_{i-1})$ nor $I(a_{i+1}, b_{i+1})$ contains the center of $I(a_i, b_i)$. Vallée probably favored using the Farey intervals rather than the $J(a_i, b_i)$ in her analysis, because (roughly speaking) the fact that each $I(a_i, b_i)$ is symmetric about $a_i N/2b_i$ makes her analysis cleaner. A "Farey Covering," which is analogous to a Farey partition, is then defined as follows.

Definition 4 (Farey Covering). *The Farey covering of order k of the interval $[0, N/2)$ is the set Farey intervals $I(a_i, b_i)$ of order k.*

3.2   The Connection between Farey Sequences and B's Distribution

Although it is far from obvious, Farey sequences have a close connection with the distribution in $\mathbb{Z}/N\mathbb{Z}$ of integers in $B_{N,h,h'}$. Vallée observed that the gaps between consecutive integers in B vary widely close to the rationals $a_i N/2b_i$ of small denominator b_i. Close to these rationals, the distribution might be called "clumpy," with large gaps separating sequences of small gaps. However, as one considers wider intervals centered at $a_i N/2b_i$, the distribution of B-elements provably "evens out" – i.e., the ratio of the number of B-elements in the interval, versus the number one would expect if the B-elements were distributed uniformly, approaches 1. Roughly speaking, the width of interval needed before the "clumpiness" can be disregarded is inversely proportional to b_i. This is one reason why Farey intervals are useful for analyzing B's distribution; the diameter of $I(a_i, b_i)$ is also inversely proportional to b_i.

Building on the above observations, Vallée ultimately proved that the number of $B_{N,h,h'}$-elements in $I(a_i, b_i)$ is essentially proportional to the width of $I(a_i, b_i)$ (as one would expect), as long as $h' - h$ is large enough. Formally, Vallée proved the following theorem [33].

Theorem 3. *For $-h = h' \geq 4N^{2/3}$ and $k = \frac{N}{h'}$, the subset $B_{N,h,h'}$ and the Farey covering of order k are quasi-independent.*

Vallée defines *quasi-independence* as follows.

Definition 5 (Quasi-Independence). *A subset X and a covering $\mathcal{Y} = \{Y_j\}$ of $\mathbb{Z}_N$ are quasi-independent if, for all j, the sets X and Y_j are (l_1, l_2)-independent for some positive constants l_1 and l_2 – i.e., $l_1 \leq \frac{P(X \cap Y_j)}{P(X)P(Y_j)} \leq l_2$.*

Clearly, this definition is meaningless unless l_1 and l_2 are independent of N. Vallée proves that $l_1 = \frac{1}{5}$ and $l_2 = 4$ suffice when $-h = h' \geq 4N^{2/3}$ and $k = \frac{N}{h'}$. This means that, for these parameters, any given Farey interval has no more than $l_2/l_1 = 20$ times the "density" of $B_{N,h,h'}$-elements than any other Farey interval.

Interestingly, Vallée's proof of Theorem 3 is essentially constructive. To analyze the distribution of $B_{N,h,h'}$-elements in the "local" region $I(a_i, b_i)$, Vallée associates each $B_{N,h,h'}$-element with a point that is in a particular lattice and that lies in the region between two particular parabolas. She then partitions the lattice into a set of parallel lines. The number of lines may be very large – e.g., superpolynomial in $\log N$. Her distribution analysis then becomes "even more local"; she provides upper and lower bounds on how many associated lattice points can occur on each line (except for at most 6 of the lines, for which she only provides upper bounds). These bounds imply similar bounds on the number of $B_{N,h,h'}$-elements in $I(a_i, b_i)$. Her constructive approach results in what one may call a "quasi-enumeration" of $B_{N,h,h'}$-elements in $I(a_i, b_i)$, in which each element is indexed first by the line of its associated lattice point, and then by the lattice point's position on the line. This quasi-enumeration is crucial to Vallée's "quasi-uniform" drawing algorithm (subsection 3.3), to our uniform drawing algorithm (section 4), and to our algorithms for losslessly compressing $B_{N,h,h'}$-elements (section 5).

Before discussing these algorithms, we review the details of Vallée's analysis. Set x_0 to be the closest integer to $\frac{a_i N}{2b_i}$ (the center of the Farey interval). If $x = x_0 + u$ is in $B_{N,h,h'}$, then $h \leq x_0^2 + 2x_0u + u^2 (\bmod N) < h'$. Now, let $L(x_0)$ be the lattice generated by the vectors $(1, 2x_0)$ and $(0, N)$. Then, $x = x_0 + u$ is in $B_{N,h,h'}$ precisely when there is a w such that $(u, w) \in L(x_0)$ and $h \leq x_0^2 + w + u^2 < h'$. The latter requirement implies that (u, w) is in between the two parabolas defined, in variables u' and w', by the formulas $x_0^2 + w' + u'^2 = h$ and $x_0^2 + w' + u'^2 = h'$. Thus, if we set $u_0 = x_0 - \frac{a_i N}{2b_i}$, then each $x \in B_{N,h,h'} \cap I(a_i, b_i)$ corresponds to a lattice point in:

$$P(a_i, b_i) = \{(u, w) \in L(x_0) \ : \ |u + u_0| \leq \frac{h}{2b_i} \text{ and } h \leq x_0^2 + w + u^2 < h'\}. \quad (1)$$

It may seem like a fairly complicated task to approximate how many lattice points in $L(x_0)$ are between the two parabolas defined above[1], but, as Vallée describes, it is possible to find a lattice basis of $L(x_0)$ in which the basis vectors are each short, with one basis vector being "quasi-horizontal" and the other being "quasi-vertical." The basis is (r, s) with:

---

[1] Indeed, finding all of the $L(x_0)$ points on a *single* parabola is equivalent to finding all of a number's modular square roots, which is equivalent to factoring.

$$r = b_i(1, 2x_0) - a_i(0, N) = (b_i, 2b_i u_0) , \tag{2}$$

$$s = b_{i-1}(1, 2x_0) - a_{i-1}(0, N) = (b_{i-1}, \frac{N}{b_i} + 2b_{i-1}u_0) . \tag{3}$$

Recall that $|u_0| \leq \frac{1}{2}$, and $b_i \leq k$ with $k = \frac{1}{4}N^{1/3}$.

Having computed this short lattice basis, Vallée considers the distribution of $P(a_i, b_i)$-points (and hence B-elements) on individual lines parallel to $vecr$. Each point in $P(a_i, b_i)$ lies on a quasi-horizontal line that intersects the vertical axis at ordinate $w_0 - vN/b_i$ for some rational index $v \in [0, (h'-h)^2/16b_iN + (h'-h)b_i/N]$, where $w_0 = h' - x_0^2 + u_0^2$ and where consecutive indices differ by 1. For lines with indices from $v_1 = \lceil 2(h'-h)b_i/N \rceil$ to $v_2 = \lfloor (h'-h)^2/16b_iN \rfloor$, which intersect the region between the two parabolas in an area she dubs the "legs" (which is in between the "chest" and the "feet"), Vallée proves the following theorem:

Theorem 4. *The number $n(v)$ of points in $P(a_i, b_i)$ on the line with index v in the legs satisfies:* $\frac{1}{2}\frac{(h'-h)}{\sqrt{vb_iN}} \leq n(v) \leq \frac{7}{4}\frac{(h'-h)}{\sqrt{vb_iN}}$.

Her bounds on each individual line in the legs imply lower and upper bounds on the total number of lattice points in the legs, using the inequalities:

$$\sum_{v=v_1}^{v_2} \frac{1}{\sqrt{v}} \geq \int_{v_1}^{v_2+1} \frac{dv}{\sqrt{v}} = 2(\sqrt{v_2+1} - \sqrt{v_1}) , \tag{4}$$

$$\sum_{v=v_1}^{v_2} \frac{1}{\sqrt{v}} \leq \frac{1}{\sqrt{v_1}} + \int_{v_1+1}^{v_2} \frac{dv}{\sqrt{v}} = \frac{1}{\sqrt{v_1}} + 2\sqrt{v_2} . \tag{5}$$

For lines with indices in $[0, 2(h'-h)b_i/N]$ or $[(h'-h)^2/16b_iN, (h'-h)^2/16b_iN + (h'-h)b_i/N]$ that intersect the "chest" or "feet," Vallée provides no nontrivial *lower* bounds on the number of $P(a_i, b_i)$-points they may contain, only upper bounds. For $h'-h = 8N^{2/3}$, one can verify Vallée's results that there are at most 4 lines in the chest, each with fewer than $\frac{2}{b_i}\sqrt{(v_1-1)N/b_i} + 1$ points, and that there are at most 2 lines in the feet, each with fewer than 8 points. Ultimately, Vallée proves Theorem 3 using her lower bounds for the legs, and upper bounds for the chest, legs and feet.

3.3 Vallée's Quasi-uniform Drawing Algorithm

Vallée uses the above results, particularly her lower and upper bounds for the legs, to obtain a concrete algorithm for drawing integers from $B_{N,h,h'}$ quasi-uniformly when $h' - h \geq 8N^{2/3}$. For a *quasi-uniform* drawing algorithm, the respective probabilities of any two $B_{N,h,h'}$-elements being drawn are within a constant factor of each other; formally:

Definition 6 (Quasi-Uniform). *A drawing algorithm C, defined over a finite set U and with values in a subset X of $\mathbb{Z}_N$, is said to be (l_1, l_2)-uniform (or quasi-uniform) for constants l_1 and l_2 if, for all $x \in X$, $\frac{l_1}{|X|} \leq \Pr[u \leftarrow U \mid C(u) = x] \leq \frac{l_2}{|X|}$.*

Vallée's algorithm is as follows:

1. *Randomly Select a Starting Point:* Pick random integer $x \in [0, N/2)$ with uniform distribution.

2. *Determine Farey Interval:* Use continued fractions to compute (a_i, b_i) for which $x \in J(a_i, b_i)$.

3. *Evaluate the Number of Points in $P(a_i, b_i)$:* Compute $x_0 = \lfloor \frac{a_i N}{b_i} \rceil$, count exactly the number n_{c+f} of points in the chest and feet, and obtain a lower bound n_l on the number of points in the legs using Vallée's lower bounds (with Equation 4).

4. *Pick a Point from $P(a_i, b_i)$:* Randomly select an integer in $t \in [1, n_{c+f} + n_l]$ with uniform distribution. If $t \leq n_{c+f}$, output the appropriate point from the chest or feet. Else, use Equation 4 to determine which quasi-horizontal line would contain the n_l^{th} point in the legs if each line met Vallée's lower bounds, and randomly choose a point in $P(a_i, b_i)$ on that line with uniform distribution.

5. *Compute x' from the Chosen Point in $P(a_i, b_i)$:* Let (u, w) be the lattice point output by the previous step. Set $x' = x_0 + u$.

Remark 1. In Step 3, one can quickly can get an exact count for how many points are in the chest and the feet by counting the *exact* number of points on each line, using simple geometry. (Recall that there are at most 4 lines in the chest, 2 in the feet.) A line intersects one of the two parabolas in at most 4 locations, possibly cutting the line into two segments that lie in between the parabolas. After finding the first and last lattice points on each segment, extrapolating the total number of points on each segment is easy since the x-coordinates of consecutive lattice points differ by b_i (see Equation 2). Vallée avoids counting the number of points on lines in the legs, since the number of lines in the legs may be super-polynomial in $\log N$.

The drawing algorithm outputs an $x' \in B_{N,h,h'}$ that is in the same $J(a_i, b_i)$ interval as x. A wider interval (recall that $I(a_i, b_i)$ has diameter $\frac{N}{b_i k}$ for $1 \leq b_i \leq k$, and that $J(a_i, b_i)$ is at least half as wide as $I(a_i, b_i)$) has a higher chance of being chosen in the first two steps. However, once an interval is chosen, any given B-element in that interval has a lower probability of being chosen if the interval is wide than if it is narrow. On balance, these factors even out (this is quasi-independence), and the drawing algorithm is quasi-uniform.

In computing l_2/l_1, there are three things to consider. First, different Farey intervals may have different "densities" of $B_{N,h,h'}$-elements; specifically, the ratio may be as much as 20 (see discussion after Theorem 3). Second, in Step 2, we used $J(a_i, b_i)$ rather than $I(a_i, b_i)$; since $I(a_i, b_i)$ is between 1 and 2 times as wide as $J(a_i, b_i)$, this costs us another factor of 2. Finally, within the $J(a_i, b_i)$ interval, different lines may be closer to the lower bounds or closer to the upper bounds, leading to a factor of $\frac{7/4}{1/2} = \frac{7}{2}$. Thus, l_2/l_1 is at most $20 \cdot 2 \cdot \frac{7}{2} = 140$.

4 Improving Vallée's and Coron's Results

In this section, we describe how to modify Vallée's quasi-uniform drawing algorithm to make it perfectly uniform. Our perfectly uniform drawing algorithm gives us an immediate improvement upon Coron's proof of security for Rabin-PDH; in particular, it allows us to reduce the output size of the partial domain hash function (see subsection 4.2). More generally, the fact that a simulator can draw B-elements uniformly in responding to an adversary's hash queries allows us (when combined with the compression schemes of Section 5) to reduce the bandwidth of several signature-related cryptosystems, including aggregate signature schemes, ring signature schemes and signcryption schemes.

4.1 A Perfectly Uniform Drawing Algorithm

Modifying Vallée's quasi-uniform drawing algorithm to make it perfectly uniform is surprisingly simple. Our modification is based on our observation that, for any $B_{N,h,h'}$-element (with $h' - h \geq 8N^{2/3}$, as required by Vallée), anyone can efficiently compute the *exact* probability $P_{x'}$ that Vallée's quasi-uniform drawing algorithm will output x'. For example, a simulator in a security proof can compute this probability (without, of course, needing the factorization of N).

Assume, for now, that we can efficiently compute $P_{x'}$ for any given x'. Let P_{min} be a lower bound on such probabilities over all $B_{N,h,h'}$-elements. Then, the improved drawing algorithm is as follows:

1. Use Vallée's method to pick an $x' \in B_{N,h,h'}$ quasi-uniformly.
2. Compute $P_{x'}$.
3. Goto Step 1 with probability $(P_{x'} - P_{min})/P_{x'}$.
4. Otherwise, output x'.

Since Vallée's drawing algorithm is quasi-uniform, the expected number of "Goto" loops per draw is a small constant; thus, the simulator's estimated time-complexity increases only by a constant factor. The probability that x' is chosen in Step 1 and that it "survives" Step 3 is the same for all x' – namely, $P_{x'} \cdot (1 - \frac{P_{x'} - P_{min}}{P_{x'}}) = P_{min}$; for this reason, and since each run of Vallée's algorithm is independent, the algorithm is perfectly uniform.

Now, given x', how does one (say, a simulator) compute $P_{x'}$? First, the simulator determines the at most two Farey intervals $I(a_i, b_i)$ and $I(a_{i+1}, b_{i+1})$ that contain x'. For $I(a_i, b_i)$, the simulator computes the index v_i of the quasi-horizontal line l_{v_i} that contains the lattice point (u_i, w_i) associated to x', and the exact number $n(v_i)$ of lattice points on l_{v_i}. Similarly, if there is a second Farey interval $I(a_{i+1}, b_{i+1})$ that contains x', the simulator computes v_{i+1}, $l_{v_{i+1}}$, (u_{i+1}, w_{i+1}), and $n(v_{i+1})$. Then, using the variables x and t from Vallée's drawing algorithm, the probability that x' will be chosen is:

$$(Pr[x \in J(a_i, b_i)]) \cdot (Pr[t_i \in l_{v_i} \mid x \in J(a_i, b_i)]) \cdot \left(\frac{1}{n(v_i)}\right) +$$

$$(Pr[x \in J(a_{i+1}, b_{i+1})]) \cdot (Pr[t_{i+1} \in l_{v_{i+1}} \mid x \in J(a_{i+1}, b_{i+1})]) \cdot \left(\frac{1}{n(v_{i+1})}\right) ,$$

where we use $Pr[t_i \in l_{v_i}]$ to denote the probability that the choice of t in Step 4 of Vallée's algorithm will map to the line l_{v_i}.

Remark 2. So that the above terminology works when (u_i, w_i) (or (u_{i+1}, w_{i+1})) lies in the chest or feet, we can pretend that these n_{c+f} points lie on a single "line."

Focusing on the first summand in the expression above, the simulator can compute each of the two probabilities in this term efficiently. First, the simulator computes the number of integers in $J(a_i, b_i)$; denoting this number by j_i, $Pr[x \in J(a_i, b_i)]$ is simply $j_i/\lceil N/2 \rceil$. Next, for the second probability, suppose that $n_{c+f} + n_l$ is the approximation used in Step 4 of Vallée's algorithm derived from her lower bounds (namely, $n_l = \lceil \frac{(h'-h)}{\sqrt{b_i N}}(\sqrt{v_2 + 1} - \sqrt{v_1}) \rceil$) for the legs, and that $n_{v_i} = \lceil \frac{(h'-h)}{\sqrt{b_i N}} \sqrt{v_i + 1} \rceil - \lceil \frac{(h'-h)}{\sqrt{b_i N}} \sqrt{v_i} \rceil$ is her approximation for the number of points on l_{v_i}. (Warning: our v_i notation collides here with Vallée's definition of v_1 and v_2.) Then, $Pr[t \in l_{v_i} \mid x \in J(a_i, b_i)] = n_{v_i}/(n_{c+f} + n_l)$. In a similar fashion, the simulator can compute the necessary probabilities for $I(a_{i+1}, b_{i+1})$, thereby obtaining a perfectly uniform drawing algorithm.

Vallée was presumably content with finding a quasi-uniform drawing algorithm, since a uniform algorithm would not have improved her result of a provable $\exp(\sqrt{(4/3) \log n \log \log n})$-time factoring algorithm by a significant amount. However, as described below, our uniform drawing algorithm has a significant practical impact on Coron's partial-domain hash variant of Rabin's signature scheme.

4.2   Improving Coron's Results for Rabin-PDH

Coron [16] provided a random-oracle security proof for a partial-domain hash Rabin signature scheme (Rabin-PDH), in which the signature x' is a modular square root (up to a fudge factor) of $\gamma \cdot H(m) + f(m)$, where H is a partial-domain hash with output space $[0, N^\beta]$ for $\frac{2}{3} + \epsilon \le \beta < 1$, f is a possibly constant function, and γ is a constant. In Rabin signing, a common fudge factor is to accept the signature if $x'^2 \equiv c(\gamma \cdot H(m) + f(m))(\mathrm{mod} N)$ for any $c \in \{-2, -1, 1, 2\}$, when $N = pq$ for $p \equiv 3(\mathrm{mod} 8)$ and $q \equiv 7(\mathrm{mod} 8)$. In this case, x' is an integer in $B_{N,h,h'}$ for $h = cf(m)$ and $h' = h + c\gamma N^\beta$ if $c\gamma$ is positive, or for $h = h' + c\gamma N^\beta$ and $h' = cf(m)$ if $c\gamma$ is negative. Coron's proof requires that γ be very small in magnitude (e.g., 16 or 256) [16], so that $h' - h = |c\gamma N^\beta|$ is sufficiently small. One reason that Rabin-PDH was an interesting problem for Coron to analyze was that partial-domain hashes were *already being used* by standardized encoding schemes. For example, ISO 9796-2 defined the encoding $\mu(m) = 4A_{16} \| m \| H(m) \| BC_{16}$.

As mentioned above, Coron provides a proof of security for Rabin-PDH when $h' - h$ is at least $(\frac{2}{3} + \epsilon) \log N$ bits, but this "ϵ" can be quite large in practice. Coron's security proof relies completely on his algorithm for drawing integers from $B_{N,h,h'}$ with a distribution whose distance from uniform is at most $16N^{\frac{-3\epsilon}{13}}$. This statistical distance must be very small, so that an adversary cannot distinguish a real attack from a simulated attack, in which the simulator uses Coron's drawing algorithm to respond to hash queries. For the statistical distance to be at most 2^{-k}, we must have that $4 - \frac{3\epsilon}{13}\log N \leq -k$, which implies that $\epsilon \geq \frac{13(k+4)}{3\log N}$. This implies that $h' - h$ is at least $(\frac{2}{3} + \epsilon)\log N = \frac{2}{3}\log N + \frac{13(k+4)}{3}$ bits. When $k = 80$, for example, $h' - h$ must be at least $\frac{2}{3}\log N + 364$ bits. This means that, for $k = 80$, Coron's technique does not reduce the minimum output size of the hash function at all, until N is at least $3 \cdot 364 = 1092$ bits!

We get a better, and much more practical, provable security result by using our perfectly uniform drawing algorithm. In particular, since our algorithm allows us to draw $B_{N,h,h'}$-elements uniformly for $h' - h \geq 8N^{2/3}$, we can prove a reduction from factoring to Rabin-PDH when $h' - h$ is only $\frac{2}{3}\log N + 3$ bits, over 300 bits less than Coron's result for $k = 80$! Moreover, the proof of security is tighter than Coron's proof for two reasons: 1) the adversary cannot possibly distinguish the simulated distribution from uniform; and 2) Coron's proof, which adapts his proof for RSA-FDH [15], does not provide a tight reduction from factoring (cf. Bernstein [6]).

For completeness, we prove the security of a specific variant of our improved Rabin-PDH, though it should be clear that our drawing algorithm can work with essentially any variant. We pick the one (succintly) described below for its simplicity. Other variants may have advantages; e.g., Bernstein's [6] security reduction is tighter by a small constant, and Bellare and Rogaway [3] describe an encoding scheme that allows (at least partial) recovery of the message being signed.

Let N be the public key, with $N = pq$ for $p \equiv 3(\mod 8)$ and $q \equiv 7(\mod 8)$. Let $A_{a,b}$ be the unique number modulo N that satisfies $A_{a,b} \equiv a(\mod p)$ and $A_{a,b} \equiv b(\mod q)$. Let $H_1 : \{0,1\}^* \to [h, h')$ be the partial-domain hash function with $h'' = h' - h(\mod N) \geq 8N^{2/3}$, and $H_2 : \{0,1\}^* \to \{A_{\pm 1,\pm 1}\}$ be a keyed hash function, with the key known only to the signer. To sign M, the signer first computes $m = H_1(M)$, and then:

1. Sets $s' = m^{(n-p-q+5)/8} \mod n$ if $\left(\frac{m}{N}\right) = 1$; else, sets $s' = (m/2)^{(n-p-q+5)/8}$;
2. Sends $s = s' \cdot H_2(m) \mod n$.

To verify, the recipient checks that either $s^2 \equiv \pm H_1(M)(\mod N)$ or $s^2 \equiv \pm 2 \cdot H_1(M)(\mod N)$. This scheme can be easily modified, à la Bernstein [6], to avoid the computation of Jacobi symbols.

In Appendix A, we prove the following theorem.

Theorem 5. *Assume that there is a chosen-message attack adversary $\mathcal{A}$ that breaks our Rabin-PDH scheme for modulus N in time t with probability ϵ. Then, in the random oracle model, there is an algorithm $\mathcal{B}$ that factors N in time t' with probability ϵ', where $\epsilon' \geq \frac{1}{2}\epsilon(1 - \frac{1}{h''})$, and $t' = O(t + q_H \log^2 N)$.*

5 The Compression Algorithms

In the previous section, we reduced the permissible output size of the hash function in Rabin-PDH to about $\frac{2}{3}\log N$ bits, but Rabin-PDH *signatures* are still $\log N$ bits. In this section, we describe compression algorithms that allow us to compress not only Rabin-PDH signatures, but also Rabin ciphertexts (not to mention aggregate signatures, ring signatures, signcryptions, and so on).

A prerequisite of any compression algorithm is to understand the distribution of what is being compressed. Vallée gives a constructive characterization of the distribution, in $\mathbb{Z}/N\mathbb{Z}$, of integers in $B_{N,h,h'}$; we leverage her characterization to construct a lossless compression algorithms. Roughly speaking, we associate $B_{N,h,h'}$-elements to strings of about $\log_2(h' - h)$ bits that specify the $B_{N,h,h'}$-element's Farey interval and its "address" (according to Vallée's rough enumeration) within that interval. For a B-element in a wider Farey interval, we use fewer bits of the bit string to specify the Farey interval and more bits to specify its address; on balance, it evens out.

Our compression algorithms involve two *nondeterministic quasi-bijections*, $\theta : B_{N,h,h'} \times \mathcal{D} \to \{0,1\}^{c_2+\log_2(h'-h)}$ (used in the signature schemes) and $\pi : \{0,1\}^{-c_1+\log_2(h'-h)} \times \mathcal{D} \to B_{N,h,h'}$ (used in the encryption scheme), for small nonnegative constants c_1 and c_2. These mappings are not actual bijections; we call them "nondeterministic quasi-bijections" since the image of an element under each mapping or its inverse has a small constant cardinality; formally:

Definition 7 (Nondeterministic Quasi-bijection). *For sets $(\mathcal{X}, \mathcal{D}, \mathcal{Y})$ and constants (l_1, l_2, l_3, l_4), we say $\pi : \mathcal{X} \times \mathcal{D} \to \mathcal{Y}$ is an (l_1, l_2, l_3, l_4)-nondeterministic-quasi-bijection if:*

1. *For all $x \in \mathcal{X}$, the cardinality of $\{\pi(x,d) : d \in \mathcal{D}\}$ is in $[l_1, l_2]$.*
2. *For all $y \in \mathcal{Y}$, the cardinality of $\{x : \exists d \in \mathcal{D} \text{ with } \pi(x,d) = y\}$ is in $[l_3, l_4]$.*

Above, $\mathcal{D}$ is an auxiliary set – e.g., it may be used as a source of (a small number of) random dummy bits if one wishes to make π randomized. The purpose of $\mathcal{D}$ is simply to make π an actual "mapping," with a single output for a given input (even though for a single $x \in \mathcal{X}$ there may be multiple outputs). Notice that an actual bijection is a $(1,1,1,1)$-quasi-bijection.

Roughly speaking, our signature scheme uses θ to compress, without loss, a Rabin-PDH signature (an element of $B_{N,h,h'}$) to a short bit string. Since the "entropy" of the hash output in Rabin-PDH is about $\frac{2}{3} \log N$ bits, one may hope that a Rabin-PDH signature can also be this short; in fact, within a few bits, this is precisely the case. To verify the compressed signature, it is decompressed to recover the ordinary Rabin-PDH signature, which is then verified in the normal fashion. Our encryption scheme uses π to map encoded bit strings to integers in $B_{N,h,h'}$, which are then squared to create short ciphertexts. Both θ and π are efficiently computable and efficiently invertible – i.e., it is easy to recover x from $\pi(x,d)$ or x' from $\theta(x',d)$ – without any trapdoor information.

Why don't we just replace π with θ^{-1}? Indeed, we could if θ were a bijection, but (unfortunately) θ maps each $B_{N,h,h'}$-element to possibly several short strings; if we used θ^{-1} to map short encoded messages to $B_{N,h,h'}$-elements, multiple plaintexts would correspond to the same ciphertext, which we wish to avoid. Thus, although the only real difference between π and θ^{-1} is that we reduce the size of π's domain to ensure that it is an injection, we find it convenient to keep the notation separate.

5.1   Mapping B-Elements to Short Strings (The θ Quasi-bijection)

Below, we give one approach to the θ quasi-bijection. Roughly speaking, $\theta(x', d)$ re-expresses a $B_{N,h,h'}$-element x' according to its Farey interval and its "address" (using Vallée's lattice) within the Farey interval. For example, a "naive" way to re-express x' is as (a_i, b_i, v, l), where (a_i, b_i) defines x''s Farey interval, v is the index of the quasi-horizontal line that contains the lattice point associated to x', and l represents the lattice point's position on the line. In this format, x' has at most two representations, one corresponding to each Farey interval that contains x'; the only effect of "d" is to pick one of these representations. We describe a different format below that has tighter compression and does not suffer from the parsing problems of the naive approach.

The θ quasi-bijection below maps $x' \in B_{N,h,h'}$ to a short string in $[0, h'']$, where h'' is a parameter whose value will be calibrated later.

Computing $\theta(x', d)$:

1. Determine (a_i, b_i) for which x' is in $J(a_i, b_i)$.
2. Compute x_{left}, the smallest integer in $[0, h'']$ with $(x_{left} + 1) \cdot \frac{N}{h''}$ in $J(a_i, b_i)$, and x_{right}, the largest integer in $[0, h'']$ with $x_{right} \cdot \frac{N}{h''}$ in $J(a_i, b_i)$.
3. Compute n_{c+f}, the number of lattice points in the chest and feet of $P(a_i, b_i)$, and n_l, an upper bound for the number of points in the legs.
4. Using Vallée's enumeration, select one integer in $x_{right} - x_{left}$ (there may be several) that corresponds to the lattice point (u, w) that is associated to x'. More specifically:
   - If (u, w) is the lth point in the chest or feet, set $c = l$.
   - Otherwise, let s_v be Vallée's upper bound for the number of leg lattice points on quasi-horizontal lines with index at most v. Compute the index v of the line containing (u, w). Let n_v be the actual number of lattice points on the line with index v and let $n'_v = s_v - s_{v-1}$ be Vallée's upper-bound estimate. Suppose that x' is the kth lattice point on the line. Pick an integer $c \in (n_{c+f} + s_{v-1} + n'_v \frac{k-1}{n_v}, n_{c+f} + s_{v-1} + n'_v \frac{k}{n_v}]$.
   - Pick an integer $c' \in ((x_{right} - x_{left}) \frac{c-1}{n_{c+f}+n_l}, (x_{right} - x_{left}) \frac{c}{n_{c+f}+n_l}]$. Set $x = x_{left} + c'$.

Although not mentioned explicitly in the algorithm description above, Vallée's quasi-enumeration, and the steps that use this quasi-enumeration, depend on the values of h and h' (which we assume to be public, and which could be most conveniently be set to 0 and $8N^{2/3}$). Shortly, we will calibrate h'' so that $x_{right} - x_{left}$ is larger than (but within a constant of) $n_{c+f} + n_l$. In computing $\theta(x', d)$, d is used – either deterministically or as a source of random bits – to pick the values of c and c'. Given $\theta(x', d)$, one can recover the value of x' as follows:

Computing $\theta^{-1}(x)$:

1. Determine (a_i, b_i) for which $x \cdot \frac{N}{h''}$ is in $J(a_i, b_i)$.
2. Compute x_{left}, the smallest integer in $[0, h'']$ with $(x_{left}+1) \cdot \frac{N}{h''}$ in $J(a_i, b_i)$, and x_{right}, the largest integer in $[0, h'']$ with $x_{right} \cdot \frac{N}{h''}$ in $J(a_i, b_i)$.
3. Compute n_{c+f}, the number of lattice points in the chest and feet of $P(a_i, b_i)$, and n_l, an upper bound for the number of points in the legs.
4. Compute $c' = x - x_{left}$. From c' and $n_{c+f} + n_l$, compute the value of c. If $c \leq n_{c+f}$, let (u, w) be the cth point in the chest or feet. Otherwise, compute the index v such that $c \in (n_{c+f} + s_{v-1}, n_{c+f} + s_v]$, as well as the value of k (defined as above), and let (u, w) be the kth point on the quasi-horizontal line with index v.
5. Set $x' = \theta^{-1}(x) = \lfloor \frac{a_i N}{b_i} \rceil + u$.

Now, we calibrate h'' to be as small as possible while still allowing the property that at least one bit string in $[0, h'']$ is uniquely associated to each $B_{N,h,h'}$-element. We can ensure this property if, for every interval, $x_{right} - x_{left} \geq n_{c+f} + n_l$ – i.e., the number of bit strings associated to $J(a_i, b_i)$ is at least the number of points in $P(a_i, b_i)$.

Since $x_{left}\frac{N}{h''}$ and $(x_{right}+1)\frac{N}{h''}$ are separated by a distance greater than the width of $J(a_i, b_i)$, we get that $(x_{right}-x_{left}+1)\frac{N}{h''} > \frac{h'-h}{4b_i}$, where the latter term is the half of the diameter of $I(a_i, b_i)$; thus, we get $x_{right} - x_{left} + 1 > \frac{h''(h'-h)}{4b_i N}$. To determine an h'' for which $\frac{h''(h'-h)}{4b_i N} \geq n_{c+f} + n_l$, we use an upper bound proven by Vallée [33]: $n_{c+f} + n_l \leq 4\frac{(h'-h)^2}{2b_i N} = \frac{2(h'-h)^2}{b_i N}$. Thus, if $h'' \geq 8(h'-h)$, then $x_{right} - x_{left} + 1 > n_{c+f} + n_l$. As long as the n_l estimate is an integer, this implies that $x_{right} - x_{left} \geq n_{c+f} + n_l$, as desired. So, we can set $h'' = 8(h'-h)$. For this value of h'', the θ mapping compresses $B_{N,h,h'}$-elements to within 3 bits of the theoretical minimum. The reader can verify that θ outputs an answer for every x' (i.e., $l_1 \geq 1$) and that θ^{-1} has exactly one possible output for each x (i.e., $l_3 = l_4 = 1$).

5.2   Mapping Short Strings to B-Elements (The π Quasi-bijection)

Like θ^{-1}, the π quasi-bijection maps short strings to $B_{N,h,h'}$-elements. However, we would like π to map short strings (e.g., plaintext strings) into $B_{N,h,h'}$ *injectively* (e.g., to allow correct decryption); thus, the set of short strings is smaller than the set of $B_{N,h,h'}$-elements (rather than the reverse). For that reason, π uses Vallée's *lower* bounds (unlike θ). Since π is otherwise similar to θ^{-1}, we relegate a precise description of π to Appendix B.

In terms of performance, all steps of the θ and π quasi-bijections and their inverses are $O(\log^2 N)$, except (possibly) the determination of the Farey interval, which uses continued fractions. However, even the continued fraction step can be computed in $O(\log^2 N)$ time – e.g., using adaptations of techniques from [14].

6 Compressed Rabin-PDH Signing and Compressed Rabin-OAEP+ Encryption

In this section, we describe how to use the θ and π quasi-permutations to achieve a 33% reduction in the size of Rabin signatures and Rabin ciphertexts.

The signature case is easy to describe. Recall that, in Section 4.2, we described how to construct a Rabin-PDH signature s that satisfies either $s^2 \equiv \pm H_1(M)(\bmod N)$ or $s^2 \equiv \pm 2 \cdot H_1(M)(\bmod N)$ for $H_1 : \{0,1\}^* \rightarrow [h, h')$, where $h' - h \geq 8N^{2/3}$. For simplicity, let's assume that $s^2 \equiv H_1(M)(\bmod N)$; the other cases can be handled similarly. In this case, we simply set the compressed Rabin-PDH signature to be $\theta_{N,h,h'}(s,d)$ – i.e., the θ quasi-permutation's compression of s for modulus N and parameters h and h'. To verify the compressed Rabin-PDH signature, the verifier simply recovers s from $\theta_{N,h,h'}(s,d)$, and then verifies s in the normal fashion. Note that anybody can create a compressed Rabin-PDH signature from a (non-compressed) Rabin-PDH signature, and vice versa, without needing trapdoor information – i.e., the compression algorithm is completely separate from the signing process.

The proof of security for compressed Rabin-PDH follows easily from the proof of security for (non-compressed) Rabin-PDH. Specifically, let $\mathcal{A}$ be a chosen-message attack adversary against Compressed Rabin-PDH, and let $\mathcal{B}$ be chosen-message attack adversary against Rabin-PDH that interacts both with a "challenger" and with $\mathcal{A}$. To respond to $\mathcal{A}$'s signature query on M, $\mathcal{B}$ queries the challenger regarding M, receives back Rabin-PDH signature x', and sends x to $\mathcal{A}$, where $x = \theta_{N,h,h'}(x',d)$. Eventually, $\mathcal{A}$ aborts or sends $\mathcal{B}$ a forgery x^* on a message M^* that it has never queried. $\mathcal{B}$ aborts or computes $x'^* = \theta_{N,h,h'}^{-1}(x^*)$ and sends x'^* to the challenger as its forgery.

The encryption case is more complicated, because the compression algorithm cannot be separated from the encryption process. Unfortunately, this fact – together with the fact the encryption scheme is not quite a one-way *permutation* as required by OAEP+, but rather a *quasi-bijection* – requires us redo the entire OAEP+ security proof, albeit with relatively minor modifications. At a high level, encryption and decryption proceed as follows:

Encryption:

1. Compute $x \in [1, h'']$, an encoding of M.
2. Compute $x' = \pi_{N,h,h'}(x,d) \in B_{N,h,h'} \cap [0, N/2)$.
3. Compute $y = x'^2(\bmod N)$.
4. Output $c = y - h$ as the ciphertext.

Decryption:

1. Recover y from c and h.
2. Compute each $x' \in B_{N,h,h'} \cap [0, N/2)$ such that $x'^2 \equiv y(\bmod N)$.
3. For each x', compute the values of $x = \pi_{N,h,h'}^{-1}(x',d)$.
4. For each x, undo the message encoding, and confirm that the message M is encoded correctly.
5. If an x is encoded correctly, output the decryption; otherwise, indicate decryption failure.

For Vallée's parameters, for a given x', there are at most two values of x in Step 3 – i.e., $l_4 = 2$ – so the encoding of at most 4 values of x must be checked. As mentioned in section 5 and further discussed in Appendix B, our preferred parameters are $h' - h = 8N^{2/3}$ and $h'' \leq \frac{(h'-h)}{5}$.

Although we could use any of a variety of encoding schemes, we prove that Compressed Rabin-OAEP+ has a tight reduction to factoring. The OAEP+ encoding scheme uses three hash functions:

$$G : \{0,1\}^{k_0} \to \{0,1\}^m, H' : \{0,1\}^{m+k_0} \to \{0,1\}^{k_1}, \text{ and } H : \{0,1\}^{m+k_1} \to \{0,1\}^{k_0}$$

where m, k_0, k_1 are security parameters. The quantities 2^{-k_0} and 2^{-k_1} should be negligible. Let $n = m + k_0 + k_1 = \log h'' < \frac{2}{3} \log N + \frac{8}{5}$. To encode message $M \in \{0,1\}^m$, the sender:

1. Picks a random $r \in \{0,1\}^{k_0}$.
2. Sets $s \leftarrow (G(r) \oplus M) \| H'(r\|M)$ and $t \leftarrow H(s) \oplus r$.
3. Sets $x \leftarrow s\|t$, an n-bit integer.

In Step 4 of Decryption, the recipient decodes by parsing each candidate x into $s_i\|t_i$ for $s_i \in \{0,1\}^{m+k_1}$ and $t_i \in \{0,1\}^{k_0}$, and then parsing s_i into for $s_i'\|s_i''$ for $s_i' \in \{0,1\}^m$ and $s_i'' \in \{0,1\}^{k_1}$. For each i, the recipient computes $r_i \leftarrow t_i \oplus H(s_i)$ and $M_i \leftarrow s_i' \oplus G(r_i)$, and tests whether $s_i'' = H'(r_i\|M_i)$. If there is a unique i for which the condition is satisfied, the recipient outputs M_i as the correct plaintext; otherwise, it indicates a decryption failure. For technical reasons in the security proof, we require that $d = r$ – i.e., that the encrypter use r as the random bits in the computation of $\pi_{N,h,h'}(x,d)$ – and that the decrypter indicate a decryption failure if this is not done. For compressed Rabin-OAEP+, we prove the following theorem in Appendix C.

Theorem 6. *Let $\mathcal{A}$ be an IND-CCA2 adversary that breaks Compressed Rabin-OAEP+ in time t with advantage ϵ for modulus N. Then $\epsilon \leq \frac{l_2}{l_1}\epsilon' + (q_{H'} + q_D)/2^{k_1} + (q_D + 1)q_G/2^{k_0}$, where ϵ' is the success probability that a particular algorithm $\mathcal{B}$ can factor, $t' = O(t + q_G q_H T_f + (q_G + q_{H'} + q_H + q_D) \log N)$, and T_f is the complexity of encryption.*

7   Extensions

In the full version of the paper, we describe compressed signcryption, aggregate signature and ring signature schemes, in which we achieve a 33% bandwidth reduction in comparison to Rabin-variants of the schemes in [24], [23] and [29]. We also note that our compression algorithms can be applied to allow shorter identity-based secret and public keys for the Fiat-Shamir signature scheme and Cocks' identity-based encryption scheme.

References

1. T.M. Apostol, *Modular Functions and Dirichlet Series in Number Theory*, Springer-Verlag (1976).
2. K. Barr and K. Asanović, *Energy Aware Lossless Data Compression*, in Proc. of MobiSys 2003.
3. M. Bellare and P. Rogaway, *The Exact Security of Digital Signatures – How to Sign with RSA and Rabin*, in Proc. of Eurocrypt 1996, LNCS 1070, pages 399–416. Springer-Verlag, 1996.
4. M. Bellare and P. Rogaway, *Optimal Asymmetric Encryption – How to Encrypt with RSA*, in Proc. of Eurocrypt 1994, LNCS 950, pages 92–111. Springer-Verlag, 1994.
5. D.J. Bernstein, *A Secure Public-Key Signature System with Extremely Fast Verification*, 2000. Available at http://cr.yp.to/djb.html.
6. D.J. Bernstein, *Proving Tight Security for Standard Rabin-Williams Signatures*, 2003. Available at http://cr.yp.to/djb.html.
7. D.J. Bernstein, *Reducing Lattice Bases to Find Small-Height Values of Univariate Polynomials*, 2003. Available at http://cr.yp.to/djb.html.
8. D. Bleichenbacher, *Compressed Rabin Signatures*, in Proc. of CT-RSA 2004.
9. D. Boneh, *Simplified OAEP for the RSA and Rabin Functions*, in Proc. of Crypto 2001, LNCS 2139, pages 275–291. Springer-Verlag, 2001.
10. D. Boneh, C. Gentry, B. Lynn, and H. Shacham, *Aggregate and Verifiably Encrypted Signatures from Bilinear Maps*, in Proc. of Eurocrypt 2003, LNCS 2656, pages 416–432. Springer-Verlag, 2003.
11. D. Boneh, G. Durfee, and N. Howgrave-Graham, *Factoring $N = p^r q$ for Large r*, in Proc. of Crypto 1999, LNCS 1666, pages 326–337. Springer-Verlag, 1999.
12. D. Boneh and R. Venkatesan, *Breaking RSA May Not Be Equivalent to Factoring*, in Proc. of Eurocrypt 1998, LNCS 1233, pages 59–71. Springer-Verlag, 1998.
13. C. Cocks, *An Identity Based Encryption Scheme Based on Quadratic Residues*, in Proc. of Cryptography and Coding 2001, LNCS 2260, Springer (2001). Available at http://www.cesg.gov.uk/technology/id-pkc/media/ciren.pdf.
14. H. Cohen, *A Course in Computational Algebraic Number Theory*, 4th ed., Graduate Texts in Mathematics, Springer, 2000.
15. J.S. Coron, *On the Exact Security of Full Domain Hash*, in Proc. of Crypto 2000, LNCS 1880, pages 229-235. Springer-Verlag, 2000.
16. J.-S. Coron, *Security Proof for Partial-Domain Hash Signature Schemes*, In Proc. of Crypto 2002, LNCS 2442, pages 613–626. Springer-Verlag, 2002.
17. D. Coppersmith, *Finding a Small Root of a Univariate Modular Equation*, in Proc. of Eurocrypt 1996, LNCS 1070, pages 155-165. Springer-Verlag, 1996.
18. U. Feige, A. Fiat, A. Shamir, *Zero-Knowledge Proofs of Identity*, in Jour. of Cryptology (1), pp. 77–94 (1988).
19. A. Fiat, A. Shamir, *How to Prove Yourself: Practical Solutions to Identification and Signature Problems*, in Proc. of Crypto 1986, LNCS 263, pp. 186-194. Springer (1986).
20. G.H. Hardy and E.M. Wright, *An Introduction to the Theory of Numbers*, Oxford Science Publications (5th edition).
21. J. Jonsson, *A OAEP Variant with a Tight Security Proof*, 2003. Available at http://www.math.kth.se/~jakobj/crypto.html.
22. A.K. Lenstra and E.R. Verheul, *The XTR Public Key System*, In Proc. of Crypto 2000, LNCS 1880, pages 1–20. Springer-Verlag, 2000.

23. A. Lysyanskaya, S. Micali, L. Reyzin, and H. Shacham, *Sequential Aggregate Signatures from Trapdoor Homomorphic Permutations*, in Proc. of Eurocrypt 2004, LNCS 3027, pages 74–90. Springer-Verlag, 2004.

24. J. Malone-Lee and W. Mao, *Two Birds One Stone: Signcryption Using RSA*, 2002. Available at http://www.hpl.hp.com/techreports/2002/HPL-2002-293.html.

25. A. Menezes, P. van Oorschot and S. Vanstone, *Handbook of Applied Cryptography*, CRC Press, 1996.

26. S. Micali, A. Shamir, *An Improvement of the Fiat-Shamir Identification and Signature Scheme*, in Proc. of Crypto 1988, LNCS 403, pp. 244–247. Springer-Verlag (1990).

27. H. Ong, C.P. Schnorr, *Fast Signature Generation with a Fiat Shamir - Like Scheme*, in Proc. of Eurocrypt 1990, LNCS 473, pp. 432–440. Springer-Verlag (1990).

28. M.O. Rabin, *Digitalized Signatures and Public-Key Functions as Intractable as Factorization*, MIT/LCS/TR-212, MIT Laboratory for Computer Science, 1979.

29. R.L. Rivest, A. Shamir, and Y. Tauman, *How to Leak a Secret*, in Proc. of Asiacrypt 2001, LNCS 2248, pages 552–565. Springer-Verlag, 2001.

30. K. Rubin and A. Silverberg, *Torus-based Cryptography*, in Proc. of Crypto 2003, LNCS 2729, pages 349–365. Springer-Verlag, 2003.

31. V. Shoup, *OAEP Reconsidered*, in Proc. of Crypto 2001, LNCS 2139, pages 239–259. Springer-Verlag, 2001.

32. A.K. Lenstra, A. Shamir, J. Tomlinson and E. Tromer, *Analysis of Bernstein's Factorization Circuit*, in Proc. of Asiacrypt 2002, LNCS 2501, pages 1–26. Springer-Verlag, 2002.

33. B. Vallée, *Provably Fast Integer Factoring with Quasi-Uniform Small Quadratic Residues*, In Proc. of STOC 1989, pages 98–106.

34. B. Vallée, *Generation of Elements with Small Modular Squares and Provably Fast Integer Factoring Algorithms*, Mathematics of Computation, vol. 56, no. 194, pages 823–849, 1991.

A   Security Proof for Improved Rabin-PDH

To prove our scheme secure against existential forgery under chosen-message attacks, we construct the following game:

Setup: B gives A the public key N, retaining H_1 for use as a random oracle.

Hash Queries: A can make a query M_i to the H_1-oracle at any time. If B has received an identical query before, it responds as it did before. Otherwise, B responds by first generating a random value $c_i \in \{-2, -1, 1, 2\}$ with uniform distribution. It then generates a number $s_i \in B_{N,c_ih,c_ih'}$ with uniform distribution and sets $H_1(M_i) = s_i^2/c_i(\mathrm{mod}N)$. It logs (M_i, s_i) into its H_1-list. (When $c_i = \pm 2$, there is a small complication – namely, s_i must be chosen s.t. not only $c_ih(\mathrm{mod}n) \leq s_i^2(\mathrm{mod}n) < c_ih'(\mathrm{mod}n)$, but also $s_i^2(\mathrm{mod}n) \in \{c_ih(\mathrm{mod}n), \ldots, c_i(h'-1)(\mathrm{mod}n)\}$. The simulator can accomplish this easily simply by discarding the sampled s_i's that don't satisfy the latter inequality (50% of them for $|c_i| = 2$).)

Signature Queries: A can make a query M_i to the H_1-oracle at any time. B responds by using M_i to recover s_i from its H_1-list; it then sends s_i.

Forgery: Eventually, the adversary either aborts or outputs a signature s on a message M for which it has not made a signature query.

One can easily confirm that $\mathcal{B}$'s H_1-query responses, as well as its signature responses, are indistinguishable from uniform; in fact, they are perfectly uniform.

Any forgery that $\mathcal{A}$ manages to generate for message M must satisfy $s^2 \equiv \pm H_1(M)(\bmod N)$ or $s^2 \equiv \pm 2 \cdot H_1(M)(\bmod N)$. If $\mathcal{A}$ made no H_1-query at M, then its probability of success is at most $1/h''$. If $\mathcal{A}$ did make an H_1-query at M, then $\mathcal{B}$ recovers the value s' associated to M from its H_1-list. With probability $\frac{1}{2}$, $\gcd(s - s', N)$ gives a nontrivial factor of N. Thus, $\epsilon' \geq \frac{1}{2}\epsilon(1 - \frac{1}{h''})$, and $t' = O(t + q_H \log^2 N)$.

B   Details of the π Quasi-bijection

Let $x \in [0, h'']$, where h'' is a parameter whose value will be calibrated later. The π quasi-permutations sends x to an element of $B_{N,h,h'}$, as follows.

Computing $\pi(x,d)$:

1. Compute $x \cdot \frac{N}{h''}$, and determine (a_i, b_i) for which the result is in $J(a_i, b_i)$.
2. Compute x_{left}, the smallest integer in $[0, h'']$ with $(x_{left}+1) \cdot \frac{N}{h''}$ in $I(a_i, b_i)$, and x_{right}, the largest integer in $[0, h'']$ with $x_{right} \cdot \frac{N}{h''}$ in $I(a_i, b_i)$.
3. Compute n_{c+f}, the number of lattice points in the chest and feet of $P(a_i, b_i)$, and n_l, a lower bound for the number of points in the legs.
4. Using Vallée's enumeration, select one lattice point (u, w) (there may be several) that corresponds to $x - x_{left}$. More specifically:
   - Pick an integer in $c \in ((n_{c+f} + n_l)\frac{x-x_{left}-1}{x_{right}-x_{left}}, (n_{c+f} + n_l)\frac{x-x_{left}}{x_{right}-x_{left}}]$
   - If $c \leq n_c + n_f$, pick the lattice point (u, w) that has enumeration c in the chest or feet.
   - Otherwise, let s_v be Vallée's lower-bound for the number of leg lattice points on quasi-horizontal lines with index at most v. Compute v such that $s_{v-1} < c - n_{c+f} \leq s_v$. Let n_v be the number of lattice points on the line with index v and let n'_v be Vallée's lower-bound estimate. Pick an integer $c' \in (n_v(\frac{c-n_{c+f}-s_{v-1}-1}{n'_v}), n_v(\frac{c-n_{c+f}-s_{v-1}}{n'_v})]$, and set (u, w) to be the c'^{th} point in $P(a_i, b_i)$ on the line.
5. Set $x' = x_0 + u$, where $x_0 = \lfloor \frac{a_i N}{b_i} \rceil$. Output x'.

We omit the description of $\pi^{-1}(x')$, since it should be clear from the above. Now, we mention some of the properties of the π quasi-permutation.

Choosing the parameters such that $0 < x_{right} - x_{left} \leq n_{c+f} + n_l$ – i.e., such that the lower bound on the number of points in $P(a_i, b_i)$ is greater than the number of bit strings associated to $I(a_i, b_i)$ – ensures that l_1 is at least 1, since one can always find a value for c in the computation of π. Notice that $(x_{right} - x_{left} - 1)\frac{N}{h''} < \frac{h'-h}{2b_i}$, where the latter term is the diameter of $I(a_i, b_i)$. This implies that $x_{right} - x_{left} - 1 < \frac{h''(h'-h)}{2b_i N}$. Now, consider the parameters used by Vallée. Vallée considered the case $-h = h' = 4N^{2/3}$, so that $h' - h = 8N^{2/3}$.

For this value of $h' - h$, Vallée proved a lower bound of $n_{c+f} + n_l \geq \frac{(h'-h)^2}{10b_i N}$ (see [33]). Thus, if $h'' \leq \frac{(h'-h)}{5}$, then $x_{right} - x_{left} - 1 < n_{c+f} + n_l$. As long as the n_l estimate is an integer, this implies that $x_{right} - x_{left} \leq n_{c+f} + n_l$, as desired. To ensure that $x_{right} - x_{left}$ is never zero, we want that $\frac{N}{h''} \leq \frac{N}{k^2} \Rightarrow$ $h'' \geq k^2 = N^{2/3}/16 = \frac{(h'-h)}{128}$, where the latter is the diameter of the narrowest Farey interval. So, we can set h'' to be anything between $\frac{(h'-h)}{128}$ and $\frac{(h'-h)}{5}$; values closer to the latter involve less ciphertext expansion.

On the other hand, we would like l_2 and l_4 to be small positive constants. This ensures that picking x (and d) uniformly and outputting $\pi(x, d)$ is a *quasi-uniform drawing algorithm* for $B_{N,h,h'}$ (this helps get a tight security proof for the encryption scheme). The computation of $\pi^{-1}(x')$ outputs up to two values of x, exactly one for each Farey interval that contains x'; thus $l_4 = 2$. We use Vallée's upper bounds to bound l_2. Specifically, Vallée's computations allow $n_{c+f} + n_l$ to be upper bounded by $(1.004 + 0.125 + \frac{4 - \sqrt{5}}{8})\frac{(h'-h)^2}{2b_i N} < \frac{.7(h'-h)^2}{b_i N}$, allowing us to upper bound the number of possible values of c by 4, for h''. Also, there are at most $\lceil \frac{7}{2} \rceil = 4$ (see Vallée's Leg Theorem) possible values of c', so l_2 is at most $4 \times 4 = 16$. Accordingly, for $h' - h = 8N^{2/3}$ and $h'' = \lfloor \frac{(h'-h)}{5} \rfloor$, one gets a $(1, 16, 1, 2)$ quasi-bijection.

C   Security Proof for Compressed Rabin-OAEP+

Recall the standard definition of security against adaptive chosen-ciphertext attack. An algorithm $\mathcal{A}$ "breaks" the encryption scheme if, in the following game, it outputs the correct value of b in the final stage with more than negligible advantage:

Setup: The challenger generates a Rabin modulus N and hash functions G, H' and H, defined as above. It sends (N, G, H', H) to $\mathcal{A}$.

Phase 1: $\mathcal{A}$ requests the challenger to decrypt ciphertexts of $\mathcal{A}$'s choosing.

Challenge: $\mathcal{A}$ chooses two plaintexts M_0 and M_1 and sends them to the challenger. The challenger randomly chooses bit $b \in \{0, 1\}$, encrypts M_b, and sends the ciphertext c to $\mathcal{A}$.

Phase 2: $\mathcal{A}$ again requests the challenger to decrypt ciphertexts of $\mathcal{A}$'s choosing, other than the Challenge ciphertext.

Output: Finally, $\mathcal{A}$ outputs a bit $b' \in \{0, 1\}$.

We define $\mathcal{A}$'s advantage as: $\text{Adv}(\mathcal{A}) = |\Pr[b' = b] - \frac{1}{2}|$.

In the game above, algorithm $\mathcal{B}$ plays the part of the challenger, using its its control over the random oracles G, H' and H to respond to $\mathcal{A}$'s decryption queries. We say that the system is $(t, \epsilon, q_D, q_G, q_{H'}, q_H)$-secure if no attacker limited to time t, to q_D decryption queries, to q_G G-queries, to $q_{H'}$ H'-queries, and to q_H H-queries, has advantage more than ϵ. Now, we define aspects of the game more precisely.

Hash queries: $\mathcal{A}$ can query G, H' or H at any time. In responding to these queries, $\mathcal{B}$ maintains a G-list, H'-list and H-list logging queries and responses. If

$\mathcal{A}$ makes a query that is contained in one of $\mathcal{B}$'s lists, $\mathcal{B}$ responds the same way it did before. Otherwise, for G, it generates a random m-bit string with uniform distribution, sends this to $\mathcal{A}$ as its G-query response, and logs $\mathcal{A}$'s G-query and its response on its G-list. It responds similarly to H'-queries and H-queries. We use the convention that before $\mathcal{A}$ makes an H'-query on (r_i, M_i), it makes a G-query on r_i and an H-query on $s_i = (G(r_i) \oplus M_i) \| H'(r_i \| M_i)$.

Challenge: At some point, $\mathcal{A}$ produces two plaintexts $M_0, M_1 \in \{0,1\}^m$ on which it wishes to be challenged. $\mathcal{B}$ picks a random $b \in \{0,1\}$ and encrypts M_b in the usual way. Let c^* be the resulting ciphertext, and let s'^*, s''^*, t^*, r^*, and M^* denote the values corresponding to c^* that would be obtained through the decryption process.

Decryption Queries and Probability Analysis: $\mathcal{A}$ can make decryption queries at any time, subject to the constraint that it cannot query the Challenge ciphertext in Phase 2. Our treatment of decryption queries closely tracks Shoup's analysis for trapdoor *permutations* encoded using OAEP+. Shoup's analysis consists of a sequence of games G_i for $0 \le i \le 5$, each game a slight modification of the previous one, where G_0 represents the attack on the encryption scheme, and G_5 is a certain attack in which an adversary obviously has no advantage. Shoup bounds $|\Pr[S_{i-1}] - \Pr[S_i]|$ for $1 \le i \le 5$, where $\Pr[S_i]$ is an adversary's probability of success in game G_i, thereby bounding an adversary's advantage in G_0. To reduce space, our proof draws heavily from Shoup's proof.

In game G_1, the decryption oracle decrypts ciphertext c_i as usual, recovering s_i', s_i'', t_i, r_i, and M_i in the process. The decryption oracle is identical to G_0 (e.g., it can find modular square roots) except that the decryption oracle in G_1 rejects whenever r_i is not on its G-list. Let F_1 be the event that a ciphertext rejected in G_1 would not have been rejected in G_0. Consider a ciphertext $c \ne c^*$ submitted to the decryption oracle. If $r = r^*$ and $M = M^*$, then since there is only a single legitimate ciphertext generated from r^* and M^* (recall that we use r as the random bits in the π quasi-bijection), G_0 would also have rejected. Our analysis of the case of $r_i \ne r^*$ or $M_i \ne M^*$ is identical to Shoup's, leading to the conclusion that $|\Pr[S_0] - \Pr[S_1]| \le q_D/2^{k_1}$.

In game G_2, the decryption oracle is identical to that of G_1, except it rejects when s_i is not on its H-list. Let F_2 be the event that a ciphertext rejected in G_2 would not have been rejected in G_1. For ciphertext $c_i \ne c^*$ with s_i not on the H-list, we consider two cases:

Case 1: $s_i = s^*$. Now, $s_i = s^*$ and $c_i \ne c^*$ implies $t_i \ne t^*$ (again because we made π deterministic given r). Shoup's remaining analysis of this case also works for our situation.

Case 2: $s_i \ne s^*$. Our analysis here is again identical.

Like Shoup, we obtain $|\Pr[S_1] - \Pr[S_2]| = \Pr[F_2] \le q_{H'}/2^{k_1} + q_D q_G/2^{k_0}$.

In game G_3 the decryption oracle does not have access to a trapdoor, but instead maintains a ciphertext-list. After receiving an H'-query (r_i, M_i), it computes all possible values of $x_i' = \pi_{N,h,h'}(s_i \| t_i, r_i)$ and $c_i = x_{i,j}'^2 - h(\bmod N)$. It logs these ciphertexts in its ciphertext-list. Shoup's probability analysis applies to our case: $\Pr[S_2] = \Pr[S_3]$. His time-complexity analysis also applies: over the course of G_3, the decryption oracle's complexity is $O(\min(q_{H'}, q_H))T_f + (q_G + q_{H'} + q_H + q_D)\log N)$, where T_f is the complexity of the encryption function.

Game G_4, in which Shoup replaces the original random oracles with different but identically distribute variables, also works in our case. (See [31] for details of G_4.) Note the new encryption oracle in G_4 is identically distributed to the old one, even though "f" is not a permutation in our case, since Shoup's changes only affect f's input, not f itself. $\Pr[S_3] = \Pr[S_4]$.

Game G_5 is the same as G_3 (we skipped describing G_4) except that the encryption oracle chooses random strings $r^+ \in \{0,1\}^{k_0}$ and $g^+ \in \{0,1\}^m$, and it uses these values in the computation of the ciphertext, as described in [31]. Since g^+ is only used to mask M^*, $\Pr[S_5] = \frac{1}{2}$. Like Shoup, we also obtain in our case that $\Pr[S_4] - \Pr[S_5] \leq \Pr[F_5]$, where F_5 is the event that $\mathcal{A}$ queries G at r^*. However, our proofs finally diverge significantly at this point. Shoup describes an auxiliary game G_5' in which the encryption oracle is modified again to simply output a random number c^+ in the ciphertext space (in our case, $B_{N,h,h'} \cap [0, N/2)$), and then he uses the fact that, for a permutation, c^+ comes from a distribution identical to c^*. We cannot do this, since the π quasi-bijection chooses from $B_{N,h,h'} \cap [0, N/2)$ – and thus from the ciphertext space – only quasi-uniformly.

Instead, we define our c^+ as $f(w)$ for $w \in \{0,1\}^n$ chosen randomly with uniform distribution, and (as always) r^* and s^* are defined with respect to this ciphertext. Then, for reasons analogous to those used by Shoup, if we define F_5' to be the event that $\mathcal{A}$ queries G at r^* in game G_5', we have $\Pr[F_5] = \Pr[F_5']$. Letting F_5'' be the event that $\mathcal{A}$ queries H at s^* in game G_5', we have that $\Pr[F_5'] = \Pr[F_5' \wedge F_5''] + \Pr[F_5' \wedge \neg F_5'']$.

Now, we claim that, if π is an (l_1, l_2, l_3, l_4) quasi-bijection, then $\Pr[F_5' \wedge F_5''] \leq \frac{l_2}{l_1}\mathrm{Adv}(\mathcal{B})$. For brevity, denote the probability $\Pr[F_5' \wedge F_5'' | w']$ – i.e., the probability F_5' and F_5'' occur given the value $w' = \pi_{N,h,h'}(w, r)$ for w as chosen above – by $P_{w'}$, where w' will be treated as a random variable. Notice that, for any w', there exists a v' such that $w'^2 \equiv v'^2 \pmod{N}$ and $\gcd(w', v', N)$ is a nontrivial factor of N; in fact, we can "pair off" the numbers in $B_{N,h,h'}$, so that each w' corresponds to exactly one v'. Suppose that $r'' \in \{0,1\}^{k_0}$ and $s'' \in \{0,1\}^{m+k_1}$ correspond to v'. If $\mathcal{A}$ queries $r'' \in \{0,1\}^{k_0}$ and $s'' \in \{0,1\}^{m+k_1}$ (which occurs with probability $P_{v'}$), then $\mathcal{B}$ can use $w' = f(w)$ to find a nontrivial factor of N by taking every pair r_i, s_i queried by $\mathcal{A}$, deriving the corresponding t'', computing $x'' = \pi_{N,h,h'}(s_i \| t_i)$, and checking whether $\gcd(x'', w', N)$ is a nontrivial factor.

Overall, we have that $\Pr[F_5' \wedge F_5''] = \sum_{w'} \Pr[F_5' \wedge F_5'' | w'] \cdot \Pr[w']$. This probabilility is less than $\frac{l_2}{l_1} \sum_{w'} \Pr[F_5' \wedge F_5'' | w'] \cdot \Pr[v']$ by quasi-uniformity, where each w' is paired off with a v' that gives a nontrivial factor. However, the probability $\sum_{w'} \Pr[F_5' \wedge F_5'' | w'] \cdot \Pr[v']$ is less than $\mathcal{B}$'s probability of success, which proves the claim.

For the same reason as in [31], $\Pr[F_5' \wedge \neg F_5''] \leq q_G/2^{k_0}$. Thus, we get $\Pr[F_5'] \leq \frac{l_2}{l_1}\mathrm{Adv}(\mathcal{B}) + q_G/2^{k_0}$. Collecting all of the results, we get the time and complexity stated in the theorem.

On the Bounded Sum-of-Digits Discrete Logarithm Problem in Finite Fields[*]

Qi Cheng

School of Computer Science
The University of Oklahoma
Norman, OK 73019, USA
qcheng@cs.ou.edu

Abstract. In this paper, we study the bounded sum-of-digits discrete logarithm problem in finite fields. Our results concern primarily with fields $\mathbf{F}_{q^n}$ where $n|q-1$. The fields are called Kummer extensions of $\mathbf{F}_q$. It is known that we can efficiently construct an element g with order greater than 2^n in the fields. Let $S_q(\bullet)$ be the function from integers to the sum of digits in their q-ary expansions. We first present an algorithm that given g^e ($0 \le e < q^n$) finds e in random polynomial time, provided that $S_q(e) < n$. We then show that the problem is solvable in random polynomial time for most of the exponent e with $S_q(e) < 1.32n$, by exploring an interesting connection between the discrete logarithm problem and the problem of list decoding of Reed-Solomon codes, and applying the Guruswami-Sudan algorithm. As a side result, we obtain a sharper lower bound on the number of congruent polynomials generated by linear factors than the one based on Stothers-Mason ABC-theorem. We also prove that in the field $\mathbf{F}_{q^{q-1}}$, the bounded sum-of-digits discrete logarithm with respect to g can be computed in random time $O(f(w)\log^4(q^{q-1}))$, where f is a subexponential function and w is the bound on the q-ary sum-of-digits of the exponent, hence the problem is fixed parameter tractable. These results are shown to be generalized to Artin-Schreier extension $\mathbf{F}_{p^p}$ where p is a prime. Since every finite field has an extension of reasonable degree which is a Kummer extension, our result reveals an unexpected property of the discrete logarithm problem, namely, the bounded sum-of-digits discrete logarithm problem in any given finite field becomes polynomial time solvable in certain low degree extensions.

1 Introduction and Motivations

Most of practical public key cryptosystems base their security on the hardness of solving the integer factorization problem or the discrete logarithm problem in finite fields. Both of the problems admit subexponential algorithms, thus we have to use long parameters, which make the encryption/decryption costly if the parameters are randomly chosen. Parameters of low Hamming weight, or more generally, of small sum-of-digits, offer some remedy. Using them speeds

[*] This research is partially supported by NSF Career Award CCR-0237845.

up the system while seeming to keep the security intact. In particular, in the cryptosystem based on the discrete logarithm problem in finite fields of small characteristic, using small sum-of-digits exponents is very attractive, due to the existence of normal bases [1]. It is proposed and implemented for smart cards and mobile devices, where the computing power is severely limited. Although attacks exploring the specialty were proposed [14], none of them have polynomial time complexity.

Let $\mathbf{F}_{q^n}$ be a finite field. For $\beta \in \mathbf{F}_{q^n}$, if $\beta, \beta^q, \beta^{q^2}, \cdots, \beta^{q^{n-1}}$ form a linear basis of $\mathbf{F}_{q^n}$ over $\mathbf{F}_q$, we call them a *normal basis*. It is known that a normal basis exists for every pair of prime power q and a positive integer n [11, Page 29]. Every element α in $\mathbf{F}_{q^n}$ can be represented as

$$\alpha = a_0\beta + a_1\beta^q + \cdots + a_{n-1}\beta^{q^{n-1}}$$

where $a_i \in \mathbf{F}_q$ for $0 \le i \le n-1$. The power of q is a linear operation, thus

$$\alpha^q = a_0\beta^q + \cdots + a_{n-2}\beta^{q^{n-1}} + a_{n-1}\beta.$$

Hence to compute the q-th power, we only need to shift the digits, which can be done very fast, possibly on the hardware level. Let e be an integer with q-ary expansion

$$e = e_0 + e_1 q + e_2 q^2 + \cdots + e_{n-1}q^{n-1} \quad (0 \le e_i < q \text{ for } 0 \le i \le n-1). \quad (1)$$

The sum-of-digits of e in the q-ary expansion is defined as $S_q(e) = \sum_{i=0}^{n-1} e_i$. When $q = 2$, the sum-of-digits becomes the famous Hamming weight. To compute α^e, we only need to do shiftings and at most $S_q(e)$ many of multiplications. Furthermore, the exponentiation algorithm can be parallelized, which is a property not enjoyed by the large characteristic fields. For details, see [16].

1.1   Related Work

The discrete logarithm problem in finite field $\mathbf{F}_{q^n}$, is to compute an integer e such that $g' = g^e$, given a generator g of a subgroup of $\mathbf{F}_{q^n}^*$ and g' in the subgroup. The general purpose algorithms to solve the discrete logarithm problem are the number field sieve and the function field sieve (for a survey see [13]). They have time complexity

$$\exp(c(\log q^n)^{1/3}(\log\log q^n)^{2/3})$$

for some constant c, when q is small, or n is small.

Suppose we want to compute the discrete logarithm of g^e with respect to base g in the finite field $\mathbf{F}_{q^n}$. If we know that the Hamming weight of e is equal to w, there is an algorithm proposed by Coppersmith (described in [14]), which works well if w is very small. It is a clever adaption of the baby-step giant-step idea, and runs in random time $O(\sqrt{w}\binom{\lceil \log q^n/2 \rceil}{\lfloor w/2 \rfloor})$. It is proved in [14] that the average-case complexity achieves only a constant factor speed-up over the worst case. It is not clear how his idea can be generalized when the exponent has small sum-of-digits in the base $q > 2$. However, we can consider the very special case where $e_i \in \{0,1\}$ for $0 \leq i \leq n - 1$ and $\sum_{0 \leq i \leq n-1} e_i = \lfloor \frac{n}{2} \rfloor$. Recall that e_i's are the digits of e in the q-ary expansion. It can be verified that Coppersmith's algorithm can be applied in this case. The time complexity becomes $O(\sqrt{n}\binom{\lfloor n/2 \rfloor}{\lfloor n/4 \rfloor})$. If $q < n^{O(1)}$, it is much worse than the time complexity of the function field sieve on a general exponent.

If the q-ary sum-of-digits of the exponent is bounded by w, is there an algorithm which runs in time $f(w)\log^c(q^n)$ and solves the discrete logarithm problem in $\mathbf{F}_{q^n}$, for some function f and a constant c? A similar problem has been raised from the parametric point of view by Fellows and Koblitz [10], where they consider the prime finite fields and the bounded Hamming weight exponents. Their problem is listed among the most important open problems in the theory of parameterized complexity [9]. From the above discussions, it is certainly more relevant to cryptography to treat the finite fields with small characteristic and exponents with bounded sum-of-digits.

Unlike the case of the integer factorization, where a lot of special purpose algorithms exist, the discrete logarithm problem is considered more intractable in general. As an example, one should not use a RSA modulus of about 1000 bits with one prime factor of 160 bits. It would be vulnerable to the elliptic curve factorization algorithm. However, in the Digital Signature Standard, adopted by the U.S. government, the finite field has cardinality about 2^{1024} or larger, while the encryption/decryption is done in a subgroup of cardinality about 2^{160}. As another example, one should search for a secret prime as random as possible in RSA, while in the case of the discrete logarithm problem, one may use a finite field of small characteristic, hence the group of very special order. It is believed that no trapdoor can be placed in the group order, as long as it has a large prime factor (see the panel report on this issue in the Proceeding of Eurocrypt 1992). In order to have an efficient algorithm to solve the discrete logarithm, we need that every prime factor of the group order is bounded by a polynomial function on the logarithm of the cardinality of the field. Given the current state of analytic number theory, it is very hard, if not impossible, to decide whether there exists *infinitely* many finite fields of even (or constant) characteristic, where the discrete logarithm can be solved in polynomial time.

In summary, there are several common perceptions about the discrete logarithm problem in finite fields:

1. As long as the group order has a big prime factor, the discrete logarithm problem is hard. We may use exponents with small sum-of-digits, since the discrete logarithm problem in that case seems to be fixed parameter intractable. We gain advantage in speed by using bounded sum-of-digits exponents, and at the same time keep the problem as infeasible as using the general exponents.

2. If computing discrete logarithm is difficult, it should be difficult for any generator of the group. The discrete logarithm problem with respect to one generator can be reduced to the discrete logarithm problem with respect to any generator. Even though in the small sum-of-digits case, a reduction is not available, it is not known that changing the generator of the group affects the hardness of the discrete logarithm problem.

1.2   Our Results

In this paper, we show that those assumptions taken in combination are incorrect. We study the discrete logarithm problem in large multiplicative subgroups of the Kummer and Artin-Schreier extensions with a prescribed base, and prove that the bounded sum-of-digits discrete logarithm are easy in those groups. More precisely we prove constructively:

Theorem 1. *(Main) There exists a random algorithm to find the integer e given g and g^e in $\mathbf{F}_{q^n}$ in time polynomial in $\log(q^n)$ under the conditions:*

1. *$n | q - 1$;*
2. *$0 \le e < q^n$, and $S_q(e) \le n$;*
3. *$g = \alpha + b$ where $\mathbf{F}_q(\alpha) = \mathbf{F}_{q^n}$, $b \in \mathbf{F}_q^*$ and $\alpha^n \in \mathbf{F}_q$.*

Moreover, there does not exist an integer $e' \ne e$ satisfying that $0 \le e' < q^n$, $S_q(e') \le n$ and $g^{e'} = g^e$

The theorem leads directly to a parameterized complexity result concerning the bounded sum-of-digits discrete logarithm, which answers an important open question for special, yet non-negligibly many, cases.

Corollary 1. *There exists an element g of order greater than 2^q in $\mathbf{F}_{q^{q-1}}^*$, such that the discrete logarithm problem with respect to the generator g can be solved in time $f(w) \log^4(q^{q-1})$, where f is a subexponential function and w is the bound of the sum-of-digits of the exponent in q-ary expansion.*

A few comments are in order:

- For a finite field $\mathbf{F}_{q^n}$, if $n | q - 1$, then there exists $g \in \mathbf{F}_{q^n}$ satisfying the condition in the theorem, in the other words, there exists an irreducible polynomial of form $x^n - a$ ($a \in \mathbf{F}_q$) over $\mathbf{F}_q$; if there exists α such that $\mathbf{F}_q(\alpha) = \mathbf{F}_{q^n}$ and $\alpha^n \in \mathbf{F}_q$, then $n | q - 1$.
- As a comparison, Coppersmith's algorithm runs in exponential time in the case where $e_i \in \{0, 1\}$ for $0 \le i \le n - 1$, $S_q(e) = \frac{n}{2}$ and $q < n^{O(1)}$, while our algorithm runs in polynomial time in that case. On the other hand, Coppersmith's algorithm works for every finite field, while our algorithm works in Kummer extensions. Our result has an indirect affect on an arbitrary finite field though, since every finite field has extensions of degree close to a given number, which are Kummer extensions. As an example, suppose we want to find such an extension of $\mathbf{F}_q$ with degree about $\log^2 q$. We first pick a random n close to $\log q$ such that $(n, q) = 1$. Let l be the order of q in $\mathbf{Z}/n\mathbf{Z}$. The field $\mathbf{F}_{(q^l)^n}$ is a Kummer extension of $\mathbf{F}_{q^l}$, and an extension of $\mathbf{F}_q$. According to Theorem 1, there is a polynomial time algorithm which computes the discrete logarithm to some element g in $\mathbf{F}_{q^{ln}}$ provided that the sum-of-digits of the exponent in the q^l-ary expansion is less than n. Hence our result reveals an unexpected property of the discrete logarithm problem in finite fields: the difficulty of bounded sum-of-digits discrete logarithm problem drops dramatically if we move up to extensions and increase the base of the exponent accordingly.

– Numerical evidences suggest that the order of g is often equal to the group order $q^n - 1$, and is close to the group order otherwise. However, it seems hard to prove it. In fact, this is one of the main obstacles in improving the efficiency of AKS-style primality testing algorithm [2]. We make the following conjecture.

Conjecture 1. Suppose that a finite field $\mathbf{F}_{q^n}$ and an element g in the field satisfy the conditions in Theorem 1. In addition, $n \geq \log q$. The order of g is greater than $q^{n/c}$ for an absolute constant c.

– Even though we can not prove that the largest prime factor of the order of g is very big, it seems, as supported by numerical evidences, that the order of g, which is a factor of $q^n - 1$ bigger than 2^n, is rarely smooth. For instance, in the $\mathbf{F}_{2^{889}} = \mathbf{F}_{128^{127}}$, any g generates the whole group $\mathbf{F}^*_{2^{889}}$. The order $2^{889} - 1$ contains a prime factor of 749 bits. One should not attempt to apply the Silver-Pohlig-Hellman algorithm here.

A natural question arises: can the restriction on the sum-of-digits in Theorem 1 be relaxed? Clearly if we can solve the problem under condition $S_q(e) \leq (q-1)n$ in polynomial time, then the discrete logarithm problem in subgroup generated by g is broken. If g is a generator of $\mathbf{F}^*_{q^n}$, then the discrete logarithm problem in $\mathbf{F}_{q^n}$ and any of its subfields to any base are broken. We find a surprising relationship between the relaxed problem and the list decoding problem of Reed-Solomon codes. We are able to prove:

Theorem 2. *Suppose e is chosen randomly from the set*

$$\{0 \leq e < q^n - 1 | S_q(e) < 1.32n\}.$$

There exists an algorithm given g and g^e in $\mathbf{F}_{q^n}$, to find e in time polynomial in $\log(q^n)$, with probability greater than $1 - c^{-n}$ for some constant c greater than 1, under the conditions:

1. $n | q - 1$;
2. $g = \alpha + b$ where $\mathbf{F}_q(\alpha) = \mathbf{F}_{q^n}$, $b \in \mathbf{F}^*_q$ and $\alpha^n \in \mathbf{F}_q$.

Given a polynomial ring $\mathbf{F}_q[x]/(h(x))$, it is an important problem to determine the size of multiplicative subgroup generated by $x - s_1, x - s_2, \cdots, x - s_n$ where $(s_1, s_2, \cdots, s_n) = S$ is a list of distinct elements in $\mathbf{F}_q$, and for all i, $h(s_i) \neq 0$. The lower bound of the order directly affects the time complexity of AKS-style primality proving algorithm. In that context, we usually have $\deg h(x) | n$. Assume that $\deg h(x) = n$. For a list of integers $E = (e_1, e_2, \cdots, e_n)$, we denote

$$(x - s_1)^{e_1} (x - s_2)^{e_2} \cdots (x - s_n)^{e_n}$$

by $(x - S)^E$. One can estimate the number of distinct congruent polynomials of form $(x - S)^E$ modulo $h(x)$ for E in certain set. It is obvious that if $E \in \{(e_1, e_2, \cdots, e_n) | \sum e_i < n - 1, e_i \geq 0\}$, then all the polynomials are in different congruent classes. This gives a lower bound of 4^n. Through a clever use of Stothers-Mason ABC-theorem, Voloch [15] and Berstein [5] proved that if $\sum e_i < 1.1n$, then at most 4 such polynomials can fall in the same congruent class, hence obtained a lower bound of 4.27689^n. We improve their result and obtain a lower bound of 5.17736^n.

Theorem 3. *Use the above notations. Let C be*

$$\{(e_1, e_2, \cdots, e_n) | e_i \geq 0 \text{ for } 1 \leq i \leq n, \sum_{i=1}^{n} e_i < 1.5501n, |\{i | e_i \neq 0\}| = \lfloor 0.7416n \rfloor\}.$$

If there exist pairwise different element $E_1, E_2, \cdots, E_m \in C$ such that

$$(x - S)^{E_1} \equiv (x - S)^{E_2} \equiv \cdots \equiv (x - S)^{E_m} \pmod{h(x)},$$

then $m = O(n^2)$. Note that $|C| = 5.17736^n n^{\Theta(1)}$

By allowing negative exponents, Voloch [15] obtained a bound of 5.828^n. Our bound is smaller than his. However, starting from $|S| = 2\deg h(x)$, our method gives better bounds. Details are left in the full paper. A distinct feature of our bound is that it relates to the list decoding algorithm of Reed-Solomon codes. If a better list decoding algorithm is found, then our bound can be improved accordingly.

1.3   Organization of the Paper

The paper is organized as follows. In Section 2, we list some results of counting numbers with small sum-of-digits. In Section 3, we present the basic idea and the algorithm, and prove Theorem 1 and Corollary 1. In Section 4, we prove Theorem 2 and Theorem 3. In Section 5, we extend the results to Artin-Schreier extensions. We conclude our paper with discussions of open problems.

2   Numbers with Small Sum-of-Digits

Suppose that the q-ary expansion of a positive integer e is

$$e = e_0 + e_1 q + e_2 q^2 + \cdots + e_{n-1} q^{n-1},$$

where $0 \leq e_i \leq q - 1$ for all $0 \leq i \leq n - 1$. How many nonnegative integers e less than q^n satisfy $S_q(e) = w$? Denote the number by $N(w, n, q)$. Then $N(w, n, q)$ equals the number of nonnegative integral solutions of

$$\sum_{i=0}^{n-1} e_i = w$$

under the conditions that $0 \leq e_i \leq q - 1$ for all $0 \leq i \leq n - 1$. The generating function for $N(w, n, q)$ is

$$(1 + x + \cdots + x^{q-1})^n = \sum_i N(i, n, q) x^i.$$

If $w \leq q - 1$, then the conditions $e_i \leq q - 1$ can be removed, we have that $N(w, n, q) = \binom{w+n-1}{n-1}$. It is easy to see that if $q = 2$, we have that $N(w, n, 2) = \binom{n}{w}$. In the later section, we will need to estimate $N(w, n, q)$, where w is n times a small constant less than 2. Since

$$(1 + x + \cdots + x^{q-1})^n$$

$$= (\frac{1 - x^q}{1 - x})^n$$

$$= (1 - x^q)^n \sum_{i=0}^{\infty} \binom{i+n-1}{n-1} x^i$$

$$\equiv (1 - nx^q) \sum_{i=0}^{2q-1} \binom{i+n-1}{n-1} x^i \pmod{x^{2q}}$$

$$\equiv \sum_{i=0}^{q-1} \binom{i+n-1}{n-1} x^i + \sum_{i=q}^{2q-1} (\binom{i+n-1}{n-1} - n\binom{i-q+n-1}{n-1}) x^i \pmod{x^{2q}}$$

Hence $N(w, n, q) = \binom{w+n-1}{n-1} - n\binom{w-q+n-1}{n-1}$ if $q \le w < 2q$.

3 The Basic Ideas and the Algorithm

The basic idea of our algorithm is adopted from the index calculus algorithm. Let $\mathbf{F}_{q^n}$ be a Kummer extension of $\mathbf{F}_q$, namely, $n|q-1$. Assume that $q = p^d$ where p is the characteristic. The field $\mathbf{F}_{q^n}$ is usually given as $\mathbf{F}_p[x]/(u(x))$ where $u(x)$ is an irreducible polynomial of degree dn over $\mathbf{F}_p$. If g satisfies the condition in Theorem 1, then $x^n - \alpha^n$ must be an irreducible polynomial over $\mathbf{F}_q$. Denote α^n by a. To implement our algorithm, it is necessary that we work in another model of $\mathbf{F}_{q^n}$, namely, $\mathbf{F}_q[x]/(x^n - a)$. Fortunately the isomorphism

$$\psi : \mathbf{F}_p[y]/(u(y)) \rightarrow \mathbf{F}_{q^n} = \mathbf{F}_q[x]/(x^n - a)$$

can be efficiently computed. To compute $\psi(v(y))$, where $v(y)$ is a polynomial of degree at most $dn - 1$ over $\mathbf{F}_p$, all we have to do is to factor $u(y)$ over $\mathbf{F}_q[x]/(x^n - a)$, and to evaluate $v(y)$ at one of the roots. Factoring polynomials over finite fields is a well-studied problem in computational number theory, we refer to [3] for a complete survey of results. The random algorithm runs in expected time $O(dn(dn + \log q^n)(dn \log q^n)^2)$, and the deterministic algorithm runs in time $O(dn(dn + q)(dn \log q^n)^2)$. From now on we assume the model $\mathbf{F}_q[x]/(x^n - a)$.

Consider the subgroup generated by $g = \alpha + b$ in $(\mathbf{F}_q[x]/(x^n - a))^*$, recall that $b \in \mathbf{F}_q^*$ and $\alpha = x \pmod{x^n - a}$. The generator g has order greater than 2^n [8], and has a very nice property as follows. Denote $a^{\frac{q-1}{n}}$ by h, we have

$$g^q = (\alpha + b)^q = \alpha^q + b = a^{\frac{q-1}{n}}\alpha + b = h\alpha + b,$$

and more generally

$$(\alpha + b)^{q^i} = \alpha^{q^i} + b = h^i \alpha + b.$$

In other words, we obtain a set of relations: $\log_{\alpha+b}(h^i \alpha + b) = q^i$ for $0 \le i \le n - 1$. This corresponds to the precomputation stage of the index calculus.

The difference is that, in our case, the stage finishes in polynomial time, while generally it requires subexponential time. For a general exponent e,

$$(\alpha + b)^e = (\alpha + b)^{e_0 + e_1 q + \cdots + e_{n-1} q^{n-1}}$$
$$= (\alpha + b)^{e_0} (h\alpha + b)^{e_1} \cdots (h^i \alpha + b)^{e_i} \cdots (h^{n-1}\alpha + b)^{e_{n-1}}.$$

If $f(\alpha)$ is an element in $\mathbf{F}_{q^n}$, where $f \in \mathbf{F}_q[x]$ is a polynomial of degree less than n, and $f(\alpha) = (\alpha + b)^e$ and $S_q(e) < n$, then due to unique factorization in $\mathbf{F}_q[x]$, $f(x)$ can be completely split into the product of linear factors over $\mathbf{F}_q$. We can read the discrete logarithm from the factorizations, after the coefficients are normalized. The algorithm is described as follows.

Algorithm 1 *Input: g, g^e in $\mathbf{F}_{q^n} = \mathbf{F}_q[x]/(x^n - a)$ satisfying the conditions in Theorem 1.*

Output: e.

1. *Define an order in $\mathbf{F}_q$ (for example, use the lexicographic order). Compute and sort the list $(1, h, h^2, h^3, \cdots, h^{n-1})$.*
2. *Suppose that g^e is represented by $f(\alpha)$, where $f \in \mathbf{F}_q[x]$ has degree less than n. Factoring $f(x)$ over $\mathbf{F}_q$, let $f(x) = c(x + d_1)^{e_1} \cdots (x + d_k)^{e_k}$ where $c, d_1, \cdots, d_k$ are in $\mathbf{F}_q$.*
3. *(Normalization) Normalize the coefficients and reorder the factors of $f(x)$ such that their constant coefficients are b and $f(x) = (x + b)^{e_1} \cdots (h_{n-1}x + b)^{e_{n-1}}$, where $h_i = h^i$;*
4. *Output $e_0 + e_1 q + \cdots + e_{n-1} q^{n-1}$;*

The step 1 takes time $O(n \log^2 q \log n + n \log n \log q) = O(n \log n \log^2 q)$. The most time-consuming part is to factor a polynomial over $\mathbf{F}_q$ with degree at most n. The random algorithm runs in expected time $O(n(n + \log q)(n \log q)^2)$ and the deterministic algorithm runs in time $O(n(n + q)(n \log q)^2) = O(n^3 q \log^2 q)$. Normalization and reordering can be done in time $O(n \log n \log q)$, since we have a sorted list of $(1, h, h^2, h^3, \cdots, h^{n-1})$. Thus the algorithm can be finished in random time $O(n(n + \log q)(n \log q)^2)$ and in deterministic time $O(n^3 q \log^2 q)$. This concludes the proof of the main theorem.

Now we are ready to prove Corollary 1. Any $f(x)$ where $f(\alpha) = (\alpha + b)^e \in < \alpha + b > \subseteq \mathbf{F}_{q^{q-1}}$ is congruent to a product of at most $w = S_q(e)$ linear factors modulo $x^{q-1} - a$. If $w < q - 1$, we have an algorithm running in time $O(q^4 \log^2 q)$, according to Theorem 1. So we only need to consider the case when $w \geq q - 1$. The general purpose algorithm will run in random time $f(\log q^{q-1})$, where f is a subexponential function. Theorem 1 follows from the fact that $\log q^{q-1} \leq w \log w$.

4   The Application of the List Decoding Algorithm of Reed-Solomon Codes

A natural question arises: can we relax the bound on the sum-of-digits and still get a polynomial time algorithm? Solving the problem under the condition

$S_q(e) \leq (q-1)n$ basically renders the discrete logarithm problems in $\mathbf{F}_{q^n}$ and any of its subfields easy. Suppose that $g^e = f(\alpha)$ where $f(x) \in \mathbf{F}_q[x]$ has degree less than n. Using the same notations as in the previous section, we have

$$f(\alpha) = (\alpha + b)^{e_0}(h\alpha + b)^{e_2} \cdots (h^{n-1}\alpha + b)^{e_{n-1}}.$$

Hence there exists a polynomial $t(x)$ with degree $\sum_{i=0}^{n-1} e_i - n$ such that

$$f(x) + (x^n - a)t(x) = (x + b)^{e_0}(hx + b)^{e_1} \cdots (h^{n-1}x + b)^{e_{n-1}}.$$

If the cardinality of $\{i | e_i \neq 0\}$ is greater than k then the curve $y = t(x)$ will pass at least k points in the set

$$\{(i, -\frac{f(i)}{i^{q-1} - a}) | i \in \{-b, -\frac{b}{h}, \cdots, -\frac{b}{h^{n-1}}\}\}.$$

To find all the polynomials of degree $d = \sum_{i=0}^{n-1} e_i - n$, which pass at least k points in a given set of n points, is an instance of the list decoding problem of Reed-Solomon codes. It turns out that there are only a few such polynomials, and they can be found efficiently as long as $k \geq \sqrt{nd}$.

Proposition 1. (*Guruswami-Sudan [12]) Given n distinct elements $x_0, x_1, \cdots,$ $x_{n-1} \in \mathbf{F}_q$, n values $y_0, y_1, \cdots, y_{n-1} \in \mathbf{F}_q$ and a natural number d, there are at most $O(\sqrt{n^3 d})$ many univariate polynomials $t(x) \in \mathbf{F}_q[x]$ of degree at most d such that $y_i = t(x_i)$ for at least $\sqrt{nd}$ many points. Moreover, these polynomials can be found in random polynomial time.*

For each $t(x)$, we use the Cantor-Zassenhaus algorithm to factor $f(x) + (x^n - a)t(x)$. There must exist a $t(x)$ such that the polynomial $f(x) + (x^n - a)*t(x)$ can be completely factored into a product of linear factors in $\{h^i x + b | 0 \leq i \leq n-1\}$, and e is computed as a consequence.

4.1   The Proof of Theorem 2

In this section, we consider the case when $S_q(e) \leq 1.32n$. If there are at least $0.5657n \geq \sqrt{0.32n \cdot n}$ number of nonzero e_i's, then we can apply the Guruswami-Sudan algorithm to find all the $t(x)$. In order to prove Theorem 2, it remains to show:

Lemma 1. *Define $A_{n,q}$ as*

$$\{(e_1, e_2, \cdots, e_n) \mid e_1 + e_2 + \cdots + e_n \leq 1.32n, e_i \in \mathbf{Z} \text{ and } 0 \leq e_i \leq q-1 \text{ for } 1 \leq i \leq n.\}$$

and B_n as

$$\{(e_1, e_2, \cdots, e_n) \mid |\{i | e_i \neq 0\}| < 0.5657n\}.$$

We have

$$\frac{|A_{n,q} \cap B_n|}{|A_{n,q}|} < c^{-n}$$

for some constant $c > 1$ when n is sufficiently large.

Proof. The cardinality of $A_{n,q}$ is $\sum_{i=0}^{\lfloor 1.32n \rfloor} N(i,n,q) > \binom{2.32n}{n} > 4.883987...^n$.
The cardinality of $A_{n,q} \cap B_n$ is less than $\sum_{v=\lceil 0.5657n \rceil}^{n} \binom{n}{v}\binom{1.32n}{n-v-1}$. The summands
maximize at $v = 0.5657n$ if $v \geq 0.5657n$. Hence we have

$$\sum_{v=\lceil 0.5657n \rceil}^{n} \binom{n}{v}\binom{\lfloor 1.32n \rfloor}{n-v-1}$$

$$< 0.4343n \binom{n}{\lceil 0.5657n \rceil}\binom{\lfloor 1.32n \rfloor}{\lfloor 0.4343n \rfloor}$$

$$< 4.883799...^n$$

This proves the lemma with $c = 4.883987.../4.883799... > 1$.

4.2 The Proof of Theorem 3

Proof. Let τ be a positive real number less than 1. Define

$$C_{n,q,\tau} = \left\{ (e_1, e_2, \cdots, e_n) \mid \begin{array}{l} e_1 + e_2 + \cdots + e_n = \lfloor (1+\tau)n \rfloor, e_i \in \mathbf{Z} \\ \text{and } 0 \leq e_i \leq q-1 \text{ for } 1 \leq i \leq n \\ \text{and } |\{i|e_i \neq 0\}| = \lfloor \sqrt{\tau}n \rfloor \end{array} \right\}$$

Given $f(x) \in \mathbf{F}_q[x]$, if there exists $E \in C_{n,q,\tau}$, such that $(x-S)^E \equiv f(x)$
$\pmod{h(x)}$, there must exist a polynomial $t(x)$ such that $(x-S)^E = t(x)h(x) + f(x)$, and $t(x)$ is a solution for the list decoding problem with input $\{(s, -\frac{f(s)}{h(s)})|s \in S\}$. According to Propostion 1, there are at most $O(n^2)$ solutions. Thus the
number of congruent classes modulo $h(x)$ that $\{(x-S)^E|E \in C_{n,q,\tau}\}$ has is
greater than $\Omega(|C_{n,q,\tau}|/n^2)$. We have

$$|C_{n,q,\tau}| = \binom{n}{\sqrt{\tau}n}\binom{(1+\tau)n}{\sqrt{\tau}n}$$

$$= n^{\Theta(1)}(\frac{(1+\tau)^{1+\tau}}{\tau^{\sqrt{\tau}}(1-\sqrt{\tau})^{1-\sqrt{\tau}}(1+\tau-\sqrt{\tau})^{1+\tau-\sqrt{\tau}}})^n.$$

It takes the maximum value $5.17736...^n$ at $\tau = 0.5501$.

5 Artin-Schreier Extensions

Let p be a prime. The Artin-Schreier extension of a finite field $\mathbf{F}_p$ is $\mathbf{F}_{p^p}$. It is
easy to show that $x^p - x - a = 0$ is an irreducible polynomial in $\mathbf{F}_p$ for any
$a \in \mathbf{F}_p^*$. So we may take $\mathbf{F}_{p^p} = \mathbf{F}_p[x]/(x^p - x - a)$. Let $\alpha = x \pmod{x^p - x - a}$.
For any $b \in \mathbf{F}_p$, we have

$$(\alpha + b)^p = \alpha^p + b = \alpha + b + a,$$

and similarly

$$(\alpha + b)^{p^i} = \alpha^{p^i} + b = \alpha + b + ia.$$

Hence the results for Kummer extensions can be adopted to Artin-Schreier extensions. For the subgroup generated by $\alpha + b$, we have a polynomial algorithm to solve the discrete logarithm if the exponent has p-ary sum-of-digits less than p. Note that b may be 0 in this case.

Theorem 4. *There exists an algorithm to find the integer e given g and g^e in $\mathbf{F}_{p^p}$ in time polynomial in $\log p^p$ under the conditions:*

1. $0 \le e < p^p$, and $S_q(e) \le p - 1$;
2. $g = \alpha + b$ where $\mathbf{F}_p(\alpha) = \mathbf{F}_{p^p}$, $b \in \mathbf{F}_p$ and $\alpha^p + \alpha \in \mathbf{F}_p^*$.

Moreover, there does not exist an integer $e' \ne e$ satisfying that $0 \le e' < p^p$, $S_q(e') \le n$ and $g^{e'} = g^e$.

Theorem 5. *There exists an element g of order greater than 2^p in $\mathbf{F}_{p^p}^*$, such that the discrete logarithm problem with respect to g can be solved in time $O(f(w)(\log p^p)^4)$, where f is a subexponential function and w is the bound of the sum-of-digits of the exponent in the p-ary expansion.*

Theorem 6. *Suppose that $g = \alpha + b$, where $\mathbf{F}_p(\alpha) = \mathbf{F}_{p^p}$, $b \in \mathbf{F}_p$ and $\alpha^p + \alpha \in \mathbf{F}_p^*$. Suppose e is chosen in random from the set*

$$\{0 \le e < q^n - 1 | S_q(e) < 1.32n\}.$$

There exists an algorithm given g and g^e in $\mathbf{F}_{p^p}$, to find e in time polynomial in $\log(p^p)$, with probability greater than $1 - c^{-n}$ for some constant c greater than 1.

6 Concluding Remarks

A novel idea in the celebrated AKS primality testing algorithm, is to construct a subgroup of large cardinality through linear elements in finite fields. The subsequent improvements [6, 7, 4] rely on constructing a single element of large order. It is speculated that these ideas will be useful in attacking the integer factorization problem. In this paper, we show that they do affect the discrete logarithm problem in finite fields. We give an efficient algorithm which computes the bounded sum-of-digits discrete logarithm with respect to prescribed bases in Kummer extensions. We believe that this is more than a result which deals with only special cases, as every finite field has extensions of reasonable degrees which are Kummer extensions. For instance, if we need to compute the discrete logarithm of s in $\mathbf{F}_q$ base g, we can construct a suitable Kummer extention $\mathbf{F}_{q^n}$, and try to solve the discrete logarithms of a and g with respect to a selected base in the extension. This approach is worth studying. Another interesting problems is to further relax the restriction on the sum-of-digits of the exponent. It is also important to prove or disprove Conjecture 1. If that conjecture is true, the AKS-style primality proving can be made compatible or even better than ECPP or the cyclotomic testing in practice.

Acknowledgments

We thank Professor Pedro Berrizbeitia for very helpful discussions.

References

1. G. B. Agnew, R. C. Mullin, I. M. Onyszchuk, and S. A. Vanstone. An implementation for a fast public-key cryptosystem. *Journal of Cryptology*, 3:63–79, 1991.
2. M. Agrawal, N. Kayal, and N. Saxena. Primes is in P. http://www.cse.iitk.ac.in/news/primality.pdf, 2002.
3. Eric Bach and Jeffrey Shallit. *Algorithmic Number theory*, volume I. The MIT Press, 1996.
4. D. J. Bernstein. Proving primality in essentially quartic random time. http://cr.yp.to/papers/quartic.pdf, 2003.
5. D. J. Bernstein. Sharper ABC-based bounds for congruent polynomials. http://cr.yp.to/, 2003.
6. Pedro Berrizbeitia. Sharpening "primes is in p" for a large family of numbers. http://lanl.arxiv.org/abs/math.NT/0211334, 2002.
7. Qi Cheng. Primality proving via one round in ECPP and one iteration in AKS. In Dan Boneh, editor, *Proc. of the 23rd Annual International Cryptology Conference (CRYPTO)*, volume 2729 of *Lecture Notes in Computer Science*, Santa Barbara, 2003. Springer-Verlag.
8. Qi Cheng. Constructing finite field extensions with large order elements. In *ACM-SIAM Symposium on Discrete Algorithms (SODA)*, 2004.
9. R. G. Downey and M. R. Fellows. *Parameterized Complexity*. Springer-Verlag, 1999.
10. M. Fellows and N. Koblitz. Fixed-parameter complexity and cryptography. In *Proceedings of the Tenth International Symposium on Applied Algebra, Algebraic Algorithms and Error-Correcting Codes (AAECC'93)*, volume 673 of *Lecture Notes in Computer Science*. Springer-Verlag, 1993.
11. Shuhong Gao. *Normal Bases over Finite Fields*. PhD thesis, The University of Waterloo, 1993.
12. Venkatesan Guruswami and Madhu Sudan. Improved decoding of Reed-Solomon and algebraic-geometry codes. *IEEE Transactions on Information Theory*, 45(6): 1757–1767, 1999.
13. A. M. Odlyzko. Discrete logarithms: The past and the future. *Designs, Codes, and Cryptography*, 19:129–145, 2000.
14. D. R. Stinson. Some baby-step giant-step algorithms for the low Hamming weight discrete logarithm problem. *Math. Comp.*, 71:379–391, 2002.
15. J. F. Voloch. On some subgroups of the multiplicative group of finite rings. http://www.ma.utexas.edu/users/voloch/preprint.html, 2003.
16. Joachim von zur Gathen. Efficient exponentiation in finite fields. In *Proc. 32nd IEEE Symp. on Foundations of Comp. Science*, 1991.

Computing the RSA Secret Key Is Deterministic Polynomial Time Equivalent to Factoring

Alexander May

Faculty of Computer Science, Electrical Engineering and Mathematics
University of Paderborn
33102 Paderborn, Germany
alexx@uni-paderborn.de

Abstract. We address one of the most fundamental problems concerning the RSA cryptoscheme: Does the knowledge of the RSA public key/ secret key pair (e, d) yield the factorization of $N = pq$ in polynomial time? It is well-known that there is a *probabilistic* polynomial time algorithm that on input (N, e, d) outputs the factors p and q. We present the first *deterministic* polynomial time algorithm that factors N provided that $e, d < \phi(N)$ and that the factors p, q are of the same bit-size. Our approach is an application of Coppersmith's technique for finding small roots of bivariate integer polynomials.

Keywords: RSA, Coppersmith's method

1 Introduction

One of the most important tasks in public key cryptography is to establish the polynomial time equivalence of

- the problem of computing the secret key from the public information to
- a well-known hard problem P that is believed to be computational infeasible.

This reduction establishes the security of the secret key under the assumption that the problem P is computational infeasible. On the other hand, such a reduction does not provide any security for a public key system itself, since there might be ways to break a system without computing the secret key.

Now let us look at the RSA scheme. We briefly define the RSA parameters: Let $N = pq$ be a product of two primes of the same bit-size. Furthermore, let e, d be integers such that $ed = 1 \mod \phi(N)$, where $\phi(N)$ is Euler's totient function.

For the RSA scheme, we know that there exists a *probabilistic* polynomial time equivalence between the secret key computation and the problem of factoring the modulus N. The proof is given in the original RSA paper by Rivest, Shamir and Adleman [9] and is based on a work by Miller [8].

In this paper, we present a *deterministic* polynomial time algorithm that on input (N, e, d) outputs the factors p, q, provided that p and q are of the same bit-size and that

$$ed \leq N^2.$$

M. Franklin (Ed.): CRYPTO 2004, LNCS 3152, pp. 213–219, 2004.

In the normal RSA-case we have $e, d < \phi(N)$, since e, d are defined modulo $\phi(N)$. This implies that $ed \leq N^2$ as required. Thus, our algorithm establishes the *deterministic* polynomial time equivalence between the secret key computation and the factorization problem in the most common RSA case. We reduce the problem of factoring N to the problem of computing d, the reduction in the opposite direction is trivial.

Our approach is an application of Coppersmith's method [4] for finding small roots of bivariate integer polynomials. We want to point out that some cryptanalytic results [1,2] are based on Coppersmith's technique for solving *modular* bivariate polynomial equations. In contrast to these, we make use of Coppersmith's algorithm for bivariate polynomials with a small root over the integers. Therefore, our result does not depend on the usual heuristic for modular multivariate polynomial equations but is rigorous.

To the best of our knowledge, the only known application of Coppersmith's method for bivariate polynomials with a root over the integers is the so-called "factoring with high bits known" [4]: Given half of the most significant bits of p, one can factor N in polynomial time. Howgrave-Graham [6] showed that this problem can be solved alternatively using an univariate modular approach (see also [5]).

Since our approach directly uses Coppersmith's method for bivariate integer polynomials, the proof of our reduction is brief and simple.

The paper is organized as follows. First, we present in Sect. 2 a deterministic polynomial time algorithm that factors N on input (N, e, d) provided that $ed \leq N^{\frac{3}{2}}$. This more restricted result is interesting, since RSA is frequently used with small e in practice. Additionally, we need only elementary arithmetic in order to prove the result. As a consequence, the underlying algorithm has running time $\mathcal{O}(\log^2 N)$.

Second, we show in Sect. 3 how to improve the previous result to the desired bound $ed \leq N^2$ by applying Coppersmith's method for solving bivariate integer polynomials. We conclude by giving experimental results in Sect. 4.

2   An Algorithm for $ed \leq N^{\frac{3}{2}}$

In this work, we always assume that N is a product of two different prime factors p, q of the same bitsize, wlog $p < q$. This implies

$$p < N^{\frac{1}{2}} < q < 2p < 2N^{\frac{1}{2}}.$$

We obtain the following useful estimates:

$$p + q < 3N^{\frac{1}{2}} \quad \text{and} \quad \phi(N) = N + 1 - (p + q) > \frac{1}{2}N.$$

Let us denote by $\lceil k \rceil$ the smallest integer greater or equal to k. Furthermore, we denote by $\mathbb{Z}_{\phi(N)}^*$ the ring of invertible integers modulo $\phi(N)$.

In the following theorem, we present a very efficient algorithm that on input (N, e, d) outputs the factors of N provided that $ed \leq N^{\frac{3}{2}}$.

Theorem 1 *Let $N = pq$ be an RSA-modulus, where p and q are of the same bit-size. Suppose we know integers e, d with $ed > 1$,*

$$ed = 1 \bmod \phi(N) \quad and \quad ed \leq N^{\frac{3}{2}}.$$

Then N can be factored in time $\mathcal{O}(\log^2 N)$.

Proof: Since $ed = 1 \bmod \phi(N)$, we know that

$$ed = 1 + k\phi(N) \quad \text{for some } k \in \mathbb{N}.$$

Next, we show that k can be computed up to a small constant for our choice of e and d. Therefore, let us define $\tilde{k} = \frac{ed-1}{N}$ as an underestimate of k. We observe that

$$
\begin{aligned}
k - \tilde{k} &= \frac{ed - 1}{\phi(N)} - \frac{ed - 1}{N} \\
&= \frac{N(ed - 1) - (N - p - q + 1)(ed - 1)}{\phi(N)N} \\
&= \frac{(p + q - 1)(ed - 1)}{\phi(N)N}
\end{aligned}
$$

Using the inequalities $p + q - 1 < 3N^{\frac{1}{2}}$ and $\phi(N) \geq \frac{1}{2}N$, we conclude that

$$k - \tilde{k} < 6N^{-\frac{3}{2}}(ed - 1). \tag{1}$$

Since $ed \leq N^{\frac{3}{2}}$, we know that $k - \tilde{k} < 6$. Thus, one of the six values $\left\lceil \tilde{k} \right\rceil + i$, $i = 0, 1, \ldots 5$ must be equal to k. We test these six candidates successively. For the right choice k, we can compute

$$N + 1 + \frac{1 - ed}{k} = p + q.$$

From the value $p + q$, we can easily find the factorization of N.

Our approach uses only elementary arithmetic on integers of size $\log(N)$. Thus, the running time is $\mathcal{O}(\log^2 N)$ which concludes the proof of the theorem. □

3   The Main Result

In this section, we present a polynomial time algorithm that on input (N, e, d) outputs the factorization of N provided that $ed \leq N^2$. This improves upon the result of Theorem 1. However, the algorithm is less efficient, especially when we get close to the bound N^2.

Our approach makes use of the following result of Coppersmith [4] for finding small roots of bivariate integer polynomials.

Theorem 2 (Coppersmith) *Let* $f(x,y)$ *be an irreducible polynomial in two variables over* $\mathbb{Z}$, *of maximum degree* δ *in each variable separately. Let* X, Y *be bounds on the desired solution* (x_0, y_0). *Let* W *be the absolute value of the largest entry in the coefficient vector of* $f(xX, yY)$. *If*

$$XY \leq W^{\frac{2}{3\delta}}$$

then in time polynomial in $\log W$ *and* 2^{δ} *we can find all integer pairs* (x_0, y_0) *with* $f(x_0, y_0) = 0$, $|x_0| \leq X$ *and* $|y_0| \leq Y$.

Now, let us prove our main theorem.

Theorem 3 *Let* $N = pq$ *be an RSA-modulus, where* p *and* q *are of the same bit-size. Suppose we know integers* e, d *with* $ed > 1$,

$$ed = 1 \bmod \phi(N) \quad and \quad ed \leq N^2.$$

Then N *can be factored in time polynomial in the bit-size of* N.

Proof: Let us start with the equation

$$ed = 1 + k\phi(N) \quad \text{for some } k \in \mathbb{N}. \tag{2}$$

Analogous to the proof of Theorem 1, we define the underestimate $\tilde{k} = \frac{ed-1}{N}$ of k. Using (1), we know that

$$k - \tilde{k} < 6N^{-\frac{3}{2}}(ed - 1) < 6N^{\frac{1}{2}}.$$

Let us denote $x = k - \tilde{k}$. Therefore, we have an approximation $\tilde{k}$ for the unknown parameter k in (2) up to an additive error of x.

Next, we also want to find an approximation for the second unknown parameter $\phi(N)$ in (2). Note that

$$N - \phi(N) = p + q - 1 < 3N^{\frac{1}{2}}.$$

That is, $\phi(N)$ lies in the interval $[N - 3N^{\frac{1}{2}}, N]$. We can easily guess an estimate of $\phi(N)$ with additive error at most $\frac{1}{4}N^{\frac{1}{2}}$ by doing a brute-force search on the most significant bits of $\phi(N)$.

More precisely, we divide the interval $[N - 3N^{\frac{1}{2}}, N]$ into 6 sub-interval of length $\frac{1}{2}N^{\frac{1}{2}}$ with centers $N - \frac{2i-1}{4}N^{\frac{1}{2}}$, $i = 1, 2, \ldots, 6$. For the correct choice of i we have

$$\left| N - \frac{2i-1}{4}N^{\frac{1}{2}} - \phi(N) \right| \leq \frac{1}{4}N^{\frac{1}{2}}.$$

Let g denote the term $\frac{2i-1}{4}N^{\frac{1}{2}}$ for the right choice of i. That is, we know $\phi(N) = N - g - y$ for some unknown y with $|y| \leq \frac{1}{4}N^{\frac{1}{2}}$.

Plugging our approximations for k and $\phi(N)$ in (2) leads to

$$ed - 1 - (\tilde{k} + x)(N - g - y) = 0.$$

Let us round $\tilde{k}$ and g to the next integers. Here we omit the rounding brackets $\lceil \tilde{k} \rceil$, $\lceil g \rceil$ for ease of simplicity. Notice that the effect of this rounding on the bounds of the estimation errors x and y can be neglected (x becomes even smaller). Thus, we assume in the following that $\tilde{k}$, g are integers. Therefore, we can define the following bivariate integer polynomial

$$f(x,y) = xy - (N-g)x + \tilde{k}y - \tilde{k}(N-g) + ed - 1$$

with a root $(x_0, y_0) = (k - \tilde{k}, p + q - 1 - g)$ over the integers.

In order to apply Coppersmith's theorem (Theorem 2), we have to bound the size of the root (x_0, y_0). We define $X = 6N^{\frac{1}{2}}$ and $Y = \frac{1}{4}N^{\frac{1}{2}}$. Then, $|x_0| \leq X$ and $|y_0| \leq Y$.

Let W denote the ℓ_∞-norm of the coefficient vector of $f(xX, yY)$. We have

$$W \geq (N-g)X \geq 3N^{\frac{3}{2}}.$$

By Coppersmith's theorem, we have to satisfy the condition $XY \leq W^{\frac{2}{3}}$. Using our bounds, we obtain

$$XY = \frac{3}{2}N < \left(3N^{\frac{3}{2}}\right)^{\frac{2}{3}} \leq W^{\frac{2}{3}}.$$

Thus, we can find the root (x_0, y_0) in time polynomial in the bit-size of W using Coppersmith's method. Note that the running time is also polynomial in the bit-size of N since $W \leq NX = 6N^{\frac{3}{2}}$. Finally, the term $y_0 = p + q - 1 - g$ yields the factorization of N. This concludes the proof of the theorem. $\qquad\square$

We want to point out that Theorem 3 can be easily generalized to the case, where $p + q \leq \text{poly}(\log N) \cdot N^{\frac{1}{2}}$. I.e., we do not necessarily need that p and q are of the same bit-size. All that we have to require is that they are balanced up to some polylogarithmic factor in N.

The following theorem is a direct consequence of Theorem 3. It establishes the polynomial time equivalence of computing d and factoring N in the common RSA case, where $e, d \in \mathbb{Z}^*_{\phi(N)}$.

Theorem 4 *Let $N = pq$ be an RSA-modulus, where p and q are of the same bit-size. Furthermore, let $e \in \mathbb{Z}^*_{\phi(N)}$ be an RSA public exponent.*

Suppose we have an algorithm that on input (N, e) outputs in deterministic polynomial time the RSA secret exponent $d \in \mathbb{Z}^*_{\phi(N)}$ satisfying $ed = 1 \mod \phi(N)$. Then N can be factored in deterministic polynomial time.

4   Experiments

We want to provide some experimental results. We implemented the algorithm introduced in the previous section on an 1GHz Linux-PC. Our implementation of Coppersmith's method follows the description given by Coron [4]. L^3-lattice reduction [7] is done using Shoup's NTL library [10].

We choose $e < \phi(N)$ randomly. Therefore, in every experiment the product ed is very close to the bound N^2. Notice that in Theorem 3, we have to do a small brute-force search on the most significant bits of $\phi(N)$ in order to prove the desired bound. The polynomial time algorithm of Coppersmith given by Theorem 2 requires a similar brute-force search on the most significant bits.

In Table 1, we added a column that states the total number c of bits that one has to guess in order to find a sufficiently small lattice vector. Thus, we have to multiply the running time of the lattice reduction algorithm by a factor of 2^c. As the results indicate, the number c heavily depends on the lattice dimension. Coppersmith's technique yields a polynomial time algorithm when the lattice dimension is of size $\theta(\log W)$. However, we only tested our algorithm for lattices of small fixed dimensions 16, 25 and 36.

Table 1. Results for $ed \approx N^2$

N	c	dim	L^3-time
512 bit	55 bit	16	0.5 min
512 bit	43 bit	25	6 min
512 bit	36 bit	36	53 min
768 bit	80 bit	16	1 min
768 bit	63 bit	25	13 min
768 bit	53 bit	36	128 min
1024 bit	105 bit	16	2.5 min
1024 bit	82 bit	25	26 min
1024 bit	67 bit	36	242 min

Our experiments compare well to the experimental results of Coron [3]: One cannot come close to the bounds of Coppersmith's theorem without reducing lattices of large dimension. Notice that we have to guess a large number of bits. In contrast, by the proof of Coppersmith's theorem (see [4]) the number of bits that one has to guess for lattice dimension $\theta(\log W)$ is a small constant. However, it is a non-trivial task to handle lattices of these dimensions in practice.

One might conclude that our method is of purely theoretical interest. But let us point out that we have a worst case for our approach when the product ed is very close to the bound N^2. In Table 2, we provide some more practical results for the case $ed \approx N^{1.75}$.

Table 2. Results for $ed \approx N^{1.75}$

N	c	dim	L^3-time
512 bit	10 bit	25	6 min
768 bit	13 bit	25	13 min
1024 bit	18 bit	25	26 min

References

1. D. Boneh, G. Durfee, "Cryptanalysis of RSA with private key d less than $N^{0.292}$", IEEE Trans. on Information Theory, Vol. 46(4), pp. 1339–1349, 2000
2. J. Blömer, A. May, "New Partial Key Exposure Attacks on RSA", Advances in Cryptology – Crypto 2003, Lecture Notes in Computer Science Vol. 2729, pp. 27–43, Springer-Verlag, 2003
3. Jean-Sébastien Coron, "Finding Small Roots of Bivariate Integer Polynomial Equations Revisited", Advances in Cryptology – Eurocrypt '04, Lecture Notes in Computer Science Vol. 3027, pp. 492–505, Springer-Verlag, 2004
4. D. Coppersmith, "Small solutions to polynomial equations and low exponent vulnerabilities", Journal of Cryptology, Vol. 10(4), pp. 223–260, 1997.
5. D. Coppersmith, "Finding Small Solutions to Small Degree Polynomials", Cryptography and Lattice Conference (CaLC 2001), Lecture Notes in Computer Science Volume 2146, Springer-Verlag, pp. 20–31, 2001.
6. N. Howgrave-Graham, "Finding small roots of univariate modular equations revisited", Proceedings of Cryptography and Coding, Lecture Notes in Computer Science Vol. 1355, Springer-Verlag, pp. 131–142, 1997
7. A. K. Lenstra, H. W. Lenstra, and L. Lovász, "Factoring polynomials with rational coefficients," Mathematische Annalen, Vol. 261, pp. 513–534, 1982
8. G. L. Miller, "Riemann's hypothesis and tests for primality", Seventh Annual ACM Symposium on the Theory of Computing, pp. 234–239, 1975
9. R. Rivest, A. Shamir and L. Adleman, "A Method for Obtaining Digital Signatures and Public-Key Cryptosystems", Communications of the ACM, Vol. 21(2), pp.120–126, 1978
10. V. Shoup, NTL: A Library for doing Number Theory, online available at http://www.shoup.net/ntl/index.html

Multi-trapdoor Commitments and Their Applications to Proofs of Knowledge Secure Under Concurrent Man-in-the-Middle Attacks*

Rosario Gennaro

IBM T.J.Watson Research Center
P.O.Box 704, Yorktown Heights NY 10598
rosario@watson.ibm.com

Abstract. We introduce the notion of *multi-trapdoor* commitments which is a stronger form of trapdoor commitment schemes. We then construct two very efficient instantiations of multi-trapdoor commitment schemes, one based on the Strong RSA Assumption and the other on the Strong Diffie-Hellman Assumption.

The main application of our new notion is the construction of a *compiler* that takes *any* proof of knowledge and transforms it into one which is secure against a concurrent man-in-the-middle attack (in the common reference string model). When using our specific implementations, this compiler is very efficient (requires no more than four exponentiations) and maintains the round complexity of the original proof of knowledge. The main practical applications of our results are concurrently secure identification protocols. For these applications our results are the first simple and efficient solutions based on the Strong RSA or Diffie-Hellman Assumption.

1 Introduction

A proof of knowledge allows a Prover to convince a Verifier that he knows some secret information w (for example a witness for an NP-statement y). Since w must remain secret, one must ensure that the proof does not reveal any information about w to the Verifier (who may not necessarily act honestly and follow the protocol). Proofs of knowledge have several applications, chief among them identification protocols where a party, who is associated with a public key, identifies himself by proving knowledge of the matching secret key.

However when proofs of knowledge are performed on an open network, like the Internet, one has to worry about an active attacker manipulating the conversation between honest parties. In such a network, also, we cannot expect to control the timing of message delivery, thus we should assume that the adversary has control on when messages are delivered to honest parties.

---

* Extended Abstract. The full version of the paper is available at
http://eprint.iacr.org/2003/214/

M. Franklin (Ed.): CRYPTO 2004, LNCS 3152, pp. 220–236, 2004.

The adversary could play the "man-in-the-middle" role, between honest provers and verifiers. In such an attack the adversary will act as a prover with an honest verifier, trying to make her accept a proof, even if the adversary does not know the corresponding secret information. During this attack, the adversary will have access to honest provers proving other statements. In the most powerful attack, the adversary will start several such sessions at the same time, and interleave the messages in any arbitrary way.

Informally, we say that a proof of knowledge is concurrently non-malleable, if such an adversary will never be able to convince a verifier when she does not know the relevant secret information (unless, of course, the adversary simply relays messages unchanged from an honest prover to an honest verifier).

OUR MAIN CONTRIBUTION. We present a general transformation that takes any proof of knowledge and makes it concurrently non-malleable. The transformation preserves the round complexity of the original scheme and it requires a common reference string shared by all parties.

The crucial technical tool to construct such compiler is the notion of *multi-trapdoor commitments* (MTC) which we introduce in this paper. After defining the notion we show specific number-theoretic constructions based on the Strong RSA Assumption and the recently introduced Strong Diffie-Hellman Assumption. These constructions are very efficient, and when applied to the concurrent compiler described above, this is the whole overhead.

MULTI-TRAPDOOR COMMITMENTS. Recall that a commitment scheme consist of two phases, the first one in which a sender commits to a message (think of it as putting it inside a sealed envelope on the table) and a second one in which the sender reveals the committed message (opens the envelope).

A trapdoor commitment scheme allows a sender to commit to a message with information-theoretic privacy. I.e., given the transcript of the commitment phase the receiver, even with infinite computing power, cannot guess the committed message better than at random. On the other hand when it comes to opening the message, the sender is only computationally bound to the committed message. Indeed the scheme admits a *trapdoor* whose knowledge allows to open a commitment in any possible way. This trapdoor should be hard to compute efficiently.

A *multi-trapdoor* commitment scheme consists of a family of trapdoor commitments. Each scheme in the family is information-theoretically private. The family admits a *master* trapdoor whose knowledge allows to open *any* commitment in the family in any way it is desired. Moreover each commitment scheme in the family admits its own specific trapdoor. The crucial property in the definition of multi-trapdoor commitments is that when given the trapdoor of one scheme in the family it is infeasible to compute the trapdoor of another scheme (unless the master trapdoor is known).

CONCURRENT COMPOSITION IN DETAIL. When considering a man-in-the-middle attacker for proofs of knowledge we must be careful to define exactly what kind of concurrent composition we allow.

Above we described the case in which the attacker acts as a verifier in several concurrent executions of the proof, with several provers. We call this *left-concurrency* (as usually the provers are positioned on the left of the picture). On the other hand *right-concurrency* means that the adversary could start several concurrent executions as a prover with several verifiers.

Under these attacks, we need to prove that the protocols are zero-knowledge (i.e. simulatable) and also proofs of knowledge (i.e. one can extract the witness from the adversary). When it comes to extraction one also has to make the distinction between *on-line* and *post-protocol* extraction [27]. In an on-line extraction, the witness is extracted as soon as the prover successfully convinces the verifier. In a post-protocol extraction procedure, the extractor waits for the end of all the concurrent executions to extract the witnesses of the successful executions.

In the common reference string it is well known how to fully (i.e. both left and right) simulate proofs of knowledge efficiently, using the result of Damgård [16]. We use his techniques, so our protocols are fully concurrently zero-knowledge. Extraction is more complicated. Lindell in [30] shows how to do post-protocol extraction for the case of right concurrency. We can use his techniques as well. But for many applications what really matters is on-line extraction. We are able to do that only under left-concurrency[1]. This is however enough to build fully concurrently secure applications like identification protocols.

PRIOR WORK. Zero-knowledge protocols were introduced in [24]. The notion of proof of knowledge (already implicit in [24]) was formalized in [21, 6].

Concurrent zero-knowledge was introduced in [20]. They point out that the typical simulation paradigm to prove that a protocol is zero-knowledge fails to work in a concurrent model. This work sparked a long series of papers culminating in the discovery of non-constant upper and lower bounds on the round complexity of concurrent zero-knowledge in the black-box model [13, 34], unless extra assumptions are used such as a common reference string. Moreover, in a breakthrough result, Barak [2] shows a constant round non-black-box concurrent zero-knowledge protocol, which however is very inefficient in practice.

If one is willing to augment the computational model with a common reference string, Damgård [16] shows how to construct very efficient 3-round protocols which are concurrent (black-box) zero-knowledge.

However all these works focus only on the issue of zero-knowledge, where one has to prove that a verifier who may engage with several provers in a concurrent fashion, does not learn any information. Our work focuses more on the issue of *malleability* in proofs of knowledge, i.e. security against a man-in-the-middle who may start concurrent sessions.

The problem of malleability in cryptographic algorithms, and specifically in zero-knowledge proofs, was formalized by Dolev *et al.* in [19], where a non-malleable ZK proof with a polylogarithmic number of rounds is presented. This protocol, however, is only *sequentially* non-malleable, i.e. the adversary can only

---

[1] However, as we explain later in the Introduction, we could achieve also right-concurrency if we use so-called Ω-protocols

start sessions sequentially (and non concurrently) with the prover. Barak in [3] shows a constant round non-malleable ZK proof in the non-black-box model (and thus very inefficient).

Using the taxonomy introduced by Lindell [29], we can think of concurrent composition as the most general form of composition of a protocol *with itself* (i.e. in a world where only this protocol is run). On the other hand it would be desirable to have protocols that arbitrarily compose, not only with themselves, but with any other "secure" protocol in the environment they run in. This is the notion of *universal composable security* as defined by Canetti [11]. Universally composable zero-knowledge protocols are in particular concurrently non-malleable. In the common reference string model (which is necessary as proven in [11]), a UCZK protocols for Hamiltonian Cycle was presented in [12]. Thus UCZK protocols for any NP problem can be constructed, but they are usually inefficient in practice since they require a reduction to the Hamiltonian Cycle problem.

As it turns out, the common reference string model is necessary also to achieve concurrent non-malleability (see [30]). In this model, the first theoretical solution to our problem was presented in [17]. Following on the ideas presented in [17] more efficient solutions were presented in [27, 22, 31].

Our compiler uses ideas from both the works of Damgård [16] and Katz [27], with the only difference that it uses multi-trapdoor instead of regular trapdoor commitments in order to achieve concurrent non-malleability.

SIMULATION-SOUND TRAPDOOR COMMITMENTS. The notion of Simulation-Sound Trapdoor Commitments (SSTC), introduced in [22] and later refined and improved in [31], is very related to our notion of MTC. The notion was introduced for analogue purposes: to compile (in a way similar to ours) any Σ-protocol into one which is left-concurrently non-malleable. They show generic constructions of SSTC and specific direct constructions based on the Strong RSA Assumption and the security of the DSA signature algorithm.

The concept of SSTC is related to ours, though we define a weaker notion of commitment (we elaborate on the difference in Section 3). The important contribution of our paper with respect to [22, 31] is twofold: (i) we show that this weaker notion is sufficient to construct concurrently non-malleable proofs; (ii) because our notion is weaker, we are able to construct more efficient number theoretic instantiations. Indeed our Strong RSA construction is about a factor of 2 faster than the one presented in [31]. This efficiency improvement is inherited by the concurrently non-malleable proof of knowledge, since in both cases the computation of the commitment is the whole overhead[2].

---

[2] In [22, 31] Ω-protocols are introduced, which dispense of the need for rewinding when extracting and thus can be proven to be left and right-concurrently non-malleable (and with some extra modification even universally composable). It should be noted that if we apply our transformation to the so-called Ω-protocols introduced by [22], then we obtain on-line extraction under both left and right concurrency. However we know how to construct efficient direct constructions of Ω-protocols only for knowledge of discrete logarithms, and even that is not particularly efficient. Since for the applications we had in mind left-concurrency was sufficient, we did not follow this path in this paper.

REMARK. Because of space limitations, all the proofs of the Theorems, and various technical details are omitted and can be found in the full version of the paper.

2   Preliminaries

In the following we say that function $f(n)$ is negligible if for every polynomial $Q(\cdot)$ there exists an index n_Q such that for all $n > n_Q$, $f(n) \leq 1/Q(n)$.

Also if $A(\cdot)$ is a randomized algorithm, with $a \leftarrow A(\cdot)$ we denote the event that A outputs the string a. With $Prob[A_1; \ldots; A_k : B]$ we denote the probability of event B happening after $A_1, \ldots, A_k$.

2.1   One-Time Signatures

Our construction requires a *strong* one-time signature scheme which is secure against chosen message attack. Informally this means that the adversary is given the public key and signatures on any messages of her choice (adaptively chosen after seeing the public key). Then it is infeasible for the adversary to compute a signature of a new message, or a different signature on a message already asked. The following definition is adapted from [25].

Definition 1. (SG, Sig, Ver) *is a* strong *one-time secure signature if for every probabilistic polynomial time forger* $\mathcal{F}$, *the following*

$$
Prob \left[
\begin{array}{c}
(sk, vk) \leftarrow SG(1^n) \; ; \; M \leftarrow \mathcal{F}(vk) \; ; \\
sig \leftarrow Sig(M, sk) \; ; \; \mathcal{F}(M, sig, vk) = (M', sig') \; : \\
Ver(M', sig', vk) = 1 \; and \\
(M \neq M' \; or \; sig \neq sig')
\end{array}
\right]
$$

is negligible in n.

One-time signatures can be constructed more efficiently than general signatures since they do not require public key operations (see [7, 8, 28]). Virtually all the efficient one-time signature schemes are strong.

2.2   The Strong RSA Assumption

Let N be the product of two primes, $N = pq$. With $\phi(N)$ we denote the Euler function of N, i.e. $\phi(N) = (p-1)(q-1)$. With Z_N^* we denote the set of integers between 0 and $N - 1$ and relatively prime to N.

Let e be an integer relatively prime to $\phi(N)$. The RSA Assumption [35] states that it is infeasible to compute e-roots in Z_N^*. I.e. given a random element $s \in_R Z_N^*$ it is hard to find x such that $x^e = s \bmod N$.

The Strong RSA Assumption (introduced in [4]) states that given a random element s in Z_N^* it is hard to find $x, e \neq 1$ such that $x^e = s \bmod N$. The assumption differs from the traditional RSA assumption in that we allow the adversary to freely choose the exponent e for which she will be able to compute e-roots.

We now give formal definitions. Let $RSA(n)$ be the set of integers N, such that N is the product of two $n/2$-bit primes.

Assumption 1 *We say that the Strong RSA Assumption holds, if for all probabilistic polynomial time adversaries $\mathcal{A}$ the following probability*

$$Prob[\ N \leftarrow RSA(n)\ ;\ s \leftarrow Z_N^*\ :\ \mathcal{A}(N,s) = (x,e)\ \ s.t.\ \ x^e = s \bmod N\ ]$$

is negligible in n.

A more efficient variant of our protocol requires that N is selected as the product of two *safe* primes, i.e. $N = pq$ where $p = 2p' + 1$, $q = 2q' + 1$ and both p', q' are primes. We denote with $SRSA(n)$ the set of integers N, such that N is the product of two $n/2$-bit safe primes. In this case the assumptions above must be restated replacing $RSA(n)$ with $SRSA(n)$.

2.3   The Strong Diffie-Hellman Assumption

We now briefly recall the Strong Diffie-Hellman (SDH) Assumption, recently introduced by Boneh and Boyen in [9].

Let G be cyclic group of prime order q, generated by g. The SDH Assumption can be thought as an equivalent of the Strong RSA Assumption over cyclic groups. It basically says that no attacker on input $G, g, g^x, g^{x^2}, g^{x^3}, \ldots$, for some random $x \in Z_q$, should be able to come up with a pair (e, h) such that $h^{x+e} = g$.

Assumption 2 *We say that the ℓ-SDH Assumption holds over a cyclic group G of prime order q generated by g, if for all probabilistic polynomial time adversaries $\mathcal{A}$ the following probability*

$$Prob[\ x \leftarrow Z_q\ :\ \mathcal{A}(g, g^x, g^{x^2}, \ldots, g^{x^\ell}) = (e \in Z_q, h \in G)\ \ s.t.\ \ h^{x+e} = g\ ]$$

is negligible in $n = |q|$.

Notice that, depending on the group G, there may not be an efficient way to determine if $\mathcal{A}$ succeeded in outputting (e, h) as above. Indeed in order to check if $h^{x+e} = g$ when all we have is g^{x^i}, we need to solve the Decisional Diffie-Hellman (DDH) problem on the triple $(g^x g^e, h, g)$. Thus, although Assumption 2 is well defined on any cyclic group G, we are going to use it on the so-called *gap-DDH* groups, i.e. groups in which there is an efficient test to determine (with probability 1) on input (g^a, g^b, g^c) if $c = ab \bmod q$ or not. The *gap-DDH* property will also be required by our construction of multi-trapdoor commitments that uses the SDH Assumption[3].

2.4   Definition of Concurrent Proofs of Knowledge

POLYNOMIAL TIME RELATIONSHIPS. Let $\mathcal{R}$ be a polynomial time computable relationship, i.e. a language of pairs (y, w) such that it can be decided in polynomial time in $|y|$ if $(y, w) \in \mathcal{R}$ or not. With $\mathcal{L_R}$ we denote the language induced by $\mathcal{R}$ i.e. $\mathcal{L_R} = \{y : \exists w : (y, w) \in \mathcal{R}\}$.

---

[3] Gap-DDH groups where Assumption 2 is believed to hold can be constructed using bilinear maps introduced in the cryptographic literature by [10].

More formally an ensemble of polynomial time relationships $\mathcal{PTR}$ consists of a collection of families $\mathcal{PTR} = \cup_n \mathcal{PTR}_n$ where each $\mathcal{PTR}_n$ is a family of polynomial time relationships $\mathcal{R}_n$. To an ensemble $\mathcal{PTR}$ we associate a randomized *instance generator* algorithm IG that on input 1^n outputs the description of a relationship $\mathcal{R}_n$. In the following we will drop the suffix n when obvious from the context.

PROOFS OF KNOWLEDGE. In a proof of knowledge for a relationship $\mathcal{R}$, two parties, Prover P and Verifier V, interact on a common input y. P also holds a secret input w, such that $(y, w) \in \mathcal{R}$. The goal of the protocol is to convince V that P indeed knows such w. Ideally this proof should not reveal any information about w to the verifier, i.e. be zero-knowledge.

The protocol should thus satisfy certain constraints. In particular it must be *complete*: if the Prover knows w then the Verifier should accept. It should be *sound*: for any (possibly dishonest) prover who does not know w, the verifier should almost always reject. Finally it should be *zero-knowledge*: no (poly-time) verifier (no matter what possibly dishonest strategy she follows during the proof) can learn any information about w.

Σ-PROTOCOLS. Many proofs of knowledge belong to a class of protocols called Σ-protocols. These are 3-move protocols for a polynomial time relationship $\mathcal{R}$ in which the prover sends the first message a, the verifier answers with a random challenge c, and the prover answers with a third message z. Then the verifier applies a local decision test on y, a, c, z to accept or not.

Σ-protocols satisfy two special constraints:

Special soundness. A cheating prover can only answer *one* possible challenge c. In other words we can compute the witness w from two accepting conversations of the form (a, c, z) and (a, c', z').

Special zero-knowledge. Given the statement y and a challenge c, we can produce (in polynomial time) an accepting conversation (a, c, z), with the same distribution of real accepting conversations, without knowing the witness w. Special zero-knowledge implies zero-knowledge with respect to the honest verifier.

All the most important proofs of knowledge used in cryptographic applications are Σ-protocols (e.g. [36, 26]).

We will denote with $a \leftarrow \Sigma_1[y, w]$ the process of selecting the first message a according to the protocol Σ. Similarly we denote $c \leftarrow \Sigma_2$ and $z \leftarrow \Sigma_3[y, w, a, c]$.

MAN-IN-THE-MIDDLE ATTACKS. Consider now an adversary $\mathcal{A}$ that engages with a verifier V in a proof of knowledge. At the same time $\mathcal{A}$ acts as the verifier in another proof with a prover P. Even if the protocol is a proof of knowledge according to the definition in [6], it is still possible for $\mathcal{A}$ to make the verifier accept even without knowing the relevant secret information, by using P as an oracle. Of course $\mathcal{A}$ could always copy the messages from P to V, but it is not hard to show (see for example [27]) that she can actually prove even a different statement to V.

In a *concurrent* attack, the adversary $\mathcal{A}$ is activating several sessions with several provers, in any arbitrary interleaving. We call such an adversary a *concurrent man-in-the-middle*. We say that a proof of knowledge is concurrently non-malleable if such an adversary fails to convince the verifier in a proof in which he does not know the secret information. In other words a proof of knowledge is concurrently non-malleable, if for any such adversary that makes the verifier accept with non-negligible probability we can extract a witness.

Since we work in the common reference string model we define a proof system as tuple (crsG,P,V), where crsG is a randomized algorithm that on input the security parameter 1^n outputs the common reference string crs. In our definition we limit the prover to be a probabilistic polynomial time machine, thus technically our protocols are *arguments* and not *proofs*. But for the rest of the paper we will refer to them as proofs.

If $\mathcal{A}$ is a concurrent man-in-the-middle adversary, let $\pi_{\mathcal{A}}(n)$ be the probability that the verifier V accepts. That is

$$\pi_{\mathcal{A}} = Prob[\mathcal{R}_n \leftarrow \mathsf{IG}(1^n)\ ;\ crs \leftarrow \mathsf{crsG}(1^n)\ ;\ [\mathcal{A}^{\mathsf{P}(y_1),\ldots,\mathsf{P}(y_k)}, \mathsf{V}](crs, y) = 1]$$

where the statements $y, y_1, \ldots, y_k$ are adaptively chosen by $\mathcal{A}$. Also we denote with $\mathsf{View}[\mathcal{A}, \mathsf{P}, \mathsf{V}]_{crs}$ the view of $\mathcal{A}$ at the end of the interaction with P and V on common reference string crs.

Definition 2. *We say that* (crsG,P,V) *is a concurrently non-malleable proof of knowledge for a relationship* $(\mathcal{PTR}, \mathsf{IG})$ *if the following properties are satisfied:*

Completeness. *For all* $(y, w) \in \mathcal{R}_n$ *(for all* $\mathcal{R}_n$*) we have that* $[\mathsf{P}(y, w), \mathsf{V}(y)]=1$.
Witness Extraction. *There exist a probabilistic polynomial time* knowledge extractor KE, *a function* $\kappa : \{0,1\}^* \rightarrow [0,1]$ *and a negligible function* ϵ, *such that for all probabilistic polynomial time concurrent man-in-the-middle adversary* $\mathcal{A}$, *if* $\pi_{\mathcal{A}}(n) > \kappa(n)$ *then* KE, *given rewind access to* $\mathcal{A}$, *computes* w *such that* $(y, w) \in \mathcal{R}_n$ *with probability at least* $\pi_{\mathcal{A}}(n) - \kappa(n) - \epsilon(n)$.
Zero-Knowledge. *There exist a probabilistic polynomial time* simulator SIM $=$ $(\mathsf{SIM}_1, \mathsf{SIM}_\mathsf{P}, \mathsf{SIM}_\mathsf{V})$, *such that the two random variables*

$$Real(n) = [\ crs \leftarrow \mathsf{crsG}(1^n)\ ,\ \mathsf{View}[\mathcal{A}, \mathsf{P}, \mathsf{V}]_{crs}\ ]$$
$$Sim(n) = [\ crs \leftarrow \mathsf{SIM}_1(1^n)\ ,\ \mathsf{View}[\mathcal{A}, \mathsf{SIM}_\mathsf{P}, \mathsf{SIM}_\mathsf{V}]_{crs}\ ]$$

are indistinguishable.

Notice that in the definition of zero-knowledge the simulator does *not* have the power to rewind the adversary. This will guarantee that the zero-knowledge property will hold in a concurrent scenario. Notice also that the definition of witness extraction assumes only left-concurrency (i.e. the adversary has access to many provers but only to one verifier).

3   Multi-trapdoor Commitment Schemes

A trapdoor commitment scheme allows a sender to commit to a message with information-theoretic privacy. I.e., given the transcript of the commitment phase the receiver, even with infinite computing power, cannot guess the committed message better than at random. On the other hand when it comes to opening the message, the sender is only computationally bound to the committed message. Indeed the scheme admits a *trapdoor* whose knowledge allows to open a commitment in any possible way (we will refer to this also as *equivocate* the commitment). This trapdoor should be hard to compute efficiently.

A *multi-trapdoor* commitment scheme consists of a family of trapdoor commitments. Each scheme in the family is information-theoretically private. We require the following properties from a multi-trapdoor commitment scheme:

1. The family admits a *master* trapdoor whose knowledge allows to open *any* commitment in the family in any way it is desired.
2. Each commitment scheme in the family admits its own specific trapdoor, which allows to equivocate that specific scheme.
3. For any commitment scheme in the family, it is infeasible to open it in two different ways, unless the trapdoor is known. However we do allow the adversary to equivocate on a few schemes in the family, by giving it access to an oracle that opens a given committed value in any desired way. The adversary must selects this schemes, *before* seeing the definition of the whole family. It should remain infeasible for the adversary to equivocate any other scheme in the family.

The main difference between our definition and the notion of SSTC [22, 31] is that SSTC allow the adversary to choose the schemes in which it wants to equivocate even *after* seeing the definition of the family. Clearly SSTC are a stronger requirement, which is probably why we are able to obtain more efficient constructions.

We now give a formal definition. A (non-interactive) multi-trapdoor commitment scheme consists of five algorithms: CKG, Sel, Tkg, Com, Open with the following properties.

CKG is the master key generation algorithm, on input the security parameter it outputs a pair PK, TK where PK is the master public key associated with the family of commitment schemes, and TK is called the *master trapdoor*.

The algorithm Sel selects a commitment in the family. On input PK it outputs a specific public key pk that identifies one of the schemes.

Tkg is the specific trapdoor generation algorithm. On input PK,TK,pk it outputs the specific trapdoor information tk relative to pk.

Com is the commitment algorithm. On input PK,pk and a message M it outputs $C(M) = \mathsf{Com}(\mathsf{PK}, \mathsf{pk}, M, R)$ where R are the coin tosses. To open a commitment the sender reveals M, R and the receiver recomputes C.

Open is the algorithm that opens a commitment in any possible way given the trapdoor information. It takes as input the keys PK,pk, a commitment $C(M)$ and its opening M, R, a message $M' \neq M$ and a string T. If $T = \mathsf{TK}$ or $T = \mathsf{tk}$ then Open outputs R' such that $C(M) = \mathsf{Com}(\mathsf{PK}, \mathsf{pk}, M', R')$.

We require the following properties. Assume PK and all the pk's are chosen according to the distributions induced by CKG and Tkg.

Information Theoretic Security. For every message pair M, M' the distributions $C(M)$ and $C(M')$ are statistically close.

Secure Binding. Consider the following game. The adversary $\mathcal{A}$ selects k strings $(\mathsf{pk}_1, \ldots, \mathsf{pk}_k)$. It is then given a public key PK for a multi-trapdoor commitment family, generated with the same distribution as the ones generated by CKG. Also, $\mathcal{A}$ is given access to an oracle $\mathcal{EQ}$ (for Equivocator), which is queried on the following string $C = \mathsf{Com}(\mathsf{PK}, \mathsf{pk}, M, R), M, R, \mathsf{pk}$ and a message $M' \neq M$. If $\mathsf{pk} = \mathsf{pk}_i$ for some i, and is a valid public key, then $\mathcal{EQ}$ answers with R' such that $C = \mathsf{Com}(\mathsf{PK}, \mathsf{pk}, M', R')$ otherwise it outputs *nil*. We say that $\mathcal{A}$ wins if it outputs $C, M, R, M', R', \mathsf{pk}$ such that $C = \mathsf{Com}(\mathsf{PK}, \mathsf{pk}, M, R) = \mathsf{Com}(\mathsf{PK}, \mathsf{pk}, M', R')$, $M \neq M'$ and $\mathsf{pk} \neq \mathsf{pk}_i$ for all i. We require that for all efficient algorithms $\mathcal{A}$, the probability that $\mathcal{A}$ wins is negligible in the security parameter.

We can define a stronger version of the **Secure Binding** property by requiring that the adversary $\mathcal{A}$ receives the trapdoors tk_i's matching the public keys pk_i's, instead of access to the equivocator oracle $\mathcal{EQ}$. In this case we say that the multi-trapdoor commitment family is *strong*[4].

3.1 A Scheme Based on the Strong RSA Assumption

The starting point for the our construction of multi-trapdoor commitments based on the Strong RSA Assumption, is a commitment scheme based on the (regular) RSA Assumption, which has been widely used in the literature before (e.g. [14, 15]).

The master public key is a number N product of two large primes p, q, and s a random element of Z_N^*. The master trapdoor is the factorization of N, i.e. the integers p, q. The public key of a scheme in the family is an ℓ-bit prime number e such that $GCD(e, \phi(N)) = 1$. The specific trapdoor of the scheme with public key e is the e-root of s, i.e. a value $\sigma_e \in Z_N^*$ such that $\sigma_e^e = s \bmod N$.

To commit to $a \in [1..2^{\ell-1}]$ the sender chooses $r \in_R Z_N^*$ and computes $A = s^a \cdot r^e \bmod N$. To decommit the sender reveals a, r and the previous equation is verified by the receiver.

Proposition 1. *Under the Strong RSA Assumption the scheme described above is a multi-trapdoor commitment scheme.*

Sketch of Proof: Each scheme in the family is unconditionally secret. Given a value $A = s^a \cdot r^e$ we note that for each value $a' \neq a$ there exists a unique value r' such that $A = s^{a'}(r')^e$. Indeed this value is the e-root of $A \cdot s^{a-a'}$. Observe, moreover that r' can be computed efficiently as $\sigma_e^{a-a'}$, thus knowledge of σ_e allows to open a commitment (for which we know an opening) in any desired way.

---

[4] This was actually our original definition of multi-trapdoor commitments. Phil MacKenzie suggested the possibility of using the weaker approach of giving access to an equivocator oracle (as done in [31]) and we decided to modify our main definition to the weaker one, since it suffices for our application. However the strong definition may also have applications, so we decided to present it as well.

We now argue the **Secure Binding** property under the Strong RSA Assumption. Assume we are given a Strong RSA problem istance N, σ. Let's now run the **Secure Binding** game. The adversary is going to select k public keys which in this case are k primes, $e_1, \ldots, e_k$. We set $s = \sigma^E$ where $E = \prod_{i=1}^{k} e_i$ and return N, s as the public key of the multi-trapdoor commitment family. This will easily allow us to simulate the oracle $\mathcal{EQ}$, as we know the e_i-roots of s, i.e. the trapdoors of the schemes identified by e_i.

Assume now that the adversary equivocates a commitment scheme in the family identified by a prime $e \neq e_i$. The adversary returns a commitment A and two distinct openings of it (a, r) and (a', r'). Thus

$$A = s^a r^e = s^{a'} (r')^e \implies s^{a-a'} = \left(\frac{r'}{r}\right)^e \tag{1}$$

Let $\delta = a - a'$. Since $a, a' < e$ and e and the e_i's are all distinct primes we have that $GCD(\delta E, e) = 1$. We can find integers α, β such that $\alpha \delta E + \beta e = 1$. Now we can compute (using Shamir's GCD trick [37] and Eq.(1))

$$\sigma = \sigma^{\alpha \delta E + \beta e} = (\sigma^E)^{\alpha \delta} \cdot \sigma^{\beta e} = (s^\delta)^\alpha \cdot \sigma^{\beta e} = \left(\frac{r'}{r}\right)^{\alpha e} \sigma^{\beta e} \tag{2}$$

By taking e-roots on both sides we find that $\sigma_e = \left(\frac{r'}{r}\right)^\alpha s^\beta$.   □

Remark: The commitment scheme can be easily extended to any message domain $\mathcal{M}$, by using a collision-resistant hash function H from $\mathcal{M}$ to $[1..2^{\ell-1}]$. In this case the commitment is computed as $s^{H(a)} r^e$. In our application we will use a collision resistant function like SHA-1 that maps inputs to 160-bit integers and then choose e's larger than 2^{160}.

3.2   A Scheme Based on the SDH Assumption

Let G be a cyclic group of prime order q generated by g. We assume that G is a *gap-DDH* group, i.e. a group such that deciding Diffie-Hellman triplets is easy. More formally we assume the existence of an efficient algorithm DDH-Test which on input a triplet (g^a, g^b, g^c) of elements in G outputs 1 if and only if, $c = ab \bmod q$. We also assume that the Assumption 2 holds in G.

The master key generation algorithm selects a random $x \in Z_q$ which will be the master trapdoor. The master public key will be the pair g, h where $h = g^x$ in G. Each commitment in the family will be identified by a specific public key pk which is simply an element $e \in Z_q$. The specific trapdoor tk of this scheme is the value f_e in G, such that $f_e^{x+e} = g$.

To commit to a message $a \in Z_q$ with public key pk $= e$, the sender chooses at random $\phi \in Z_q$ and computes $h_e = (h \cdot g^e)^\phi$. It then runs Pedersen's commitment [33] with bases g, h_e, i.e., it selects a random $r \in Z_q$ and computes $A = g^a h_e^r$. The commitment to a is the value A.

To open a commitment the sender reveals a and $F = g^{\phi \cdot r}$. The receiver accepts the opening if DDH-Test$(F, h \cdot g^e, A \cdot g^{-a}) = 1$.

Proposition 2. *Under the SDH Assumption the scheme described above is a multi-trapdoor commitment scheme.*

Sketch of Proof: Each scheme in the family is easily seen to be unconditionally secret. The proof of the **Secure Binding** property follows from the proof of Lemma 1 in [9], where it is proven that the trapdoors f_e can be considered "weak signatures". In other words the adversary can obtain several $f_{e_1}, \ldots, f_{e_\ell}$ for values $e_1, \ldots, e_\ell$ chosen before seeing the public key g, h, and still will not be able (under the $(\ell + 1)$-SDH) to compute f_e for a new $e \neq e_i$.

The proof is then completed if we can show that opening a commitment in two different ways for a specific e is equivalent to finding f_e.

Assume we can open a committment $A = g^\alpha$ in two ways $a, F = g^\beta$ and $a', F' = g^{\beta'}$ with $a \neq a'$. The **DDH-Test** tells us that $\alpha - a = \beta(x + e)$ and $\alpha - a' = \beta'(x + e)$, thus $a - a' = (\beta' - \beta)(x + e)$ or

$$g^{(a-a')} = \left(\frac{F'}{F}\right)^{(x+e)} \implies f_e = \left(\frac{F'}{F}\right)^{(a-a')^{-1}}$$

By the same reasoning, if we know f_e and we have an opening F, a and we want to open it as a' we need to set $F' = F \cdot f_e^{a-a'}$. $\square$

4   The Protocol

In this section we describe our full solution for non-malleable proofs of knowledge secure under concurrent composition using multi-trapdoor commitments.

INFORMAL DESCRIPTION. We start from a Σ-protocol as described in Section 2. That is the prover P wants to prove to a verifier V that he knows a witness w for some statement y. The prover sends a first message a. The verifier challenges the prover with a random value c and the prover answers with his response z.

We modify this Σ-protocol in the following way. We assume that the parties share a common reference string that contains the master public key PK for a multi-trapdoor commitment scheme. The common reference string also contains a collision-resistant hash function H from the set of verification keys vk of the one-time signature scheme, to the set of public keys pk in the multi-trapdoor commitment scheme determined by the master public key PK.

The prover chooses a key pair (sk, vk) for a one-time strong signature scheme. The prover computes pk$= H(\text{vk})$ and $A = \text{Com}(\text{PK}, \text{pk}, a, r)$ where a is the first message of the Σ-protocol and r is chosen at random (as prescribed by the definition of Com). The prover sends vk, A to the verifier. The crucial trick is that we use the verification key vk to determine the value pk used in the commitment scheme.

The verifier sends the challenge c. The prover sends back a, r as an opening of A and the answer z of the Σ-protocol. It also sends sig a signature over the whole transcript, computed using sk. The verifier checks that a, r is a correct opening of A, that sig is a valid signature over the transcript using vk and also

---

CNM-POK

Common Reference String: PK the master public key for a multi-trapdoor commitment scheme. A collision resistant hash function H which maps inputs to public keys for the multi-trapdoor commitment scheme determined by PK.

Common Input: A string y.

Private Input for the Prover: a witness w for the statement y, i.e. $(y, w) \in \mathcal{R}$.

- The Prover computes $(\mathsf{sk}, \mathsf{vk}) \leftarrow \mathsf{SG}(1^n)$; $\mathsf{pk} = H(\mathsf{vk})$; $a \leftarrow \Sigma_1[y, w]$; $r \in_R Z_N^*$; $A = \mathsf{Com}(\mathsf{PK}, \mathsf{pk}, a, r)$
  The Prover sends A and vk to the Verifier.

$$P \xrightarrow{\quad A, \mathsf{vk} \quad} V$$

- The Verifier selects a random challenge $c \leftarrow \Sigma_2$ and sends it to the Prover.

$$P \xleftarrow{\quad c \quad} V$$

- The Prover computes $z \leftarrow \Sigma_3[y, w, a, c]$ and $\mathsf{sig} = \mathsf{Sig}_{\mathsf{sk}}(y, A, c, a, r, z)$. He sends a, r, z, sig to the Verifier.

$$P \xrightarrow{\quad a, r, z, \mathsf{sig} \quad} V$$

- The Verifier accepts iff $A = \mathsf{Com}(\mathsf{PK}, \mathsf{pk}, a, r)$; $\mathsf{Ver}_{\mathsf{vk}}(y, A, c, a, r, z) = 1$ and $\mathsf{Acc}(y, a, c, z) = 1$.

Fig. 1. A Concurrently Non-malleable Proof of Knowledge

that (a, c, z) is an accepting conversation for the Σ-protocol. The protocol is described in Figure 1.

Theorem 1. *If multi-trapdoor commitments exist, if H is a collision-resistant hash function, and if (SG,Sig,Ver) is a strong one-time signature scheme, then* CNM-POK *is a concurrently non-malleable proof of knowledge (see Definition 2).*

4.1   The Strong RSA Version

In this section we are going to add a few comments on the specific implementations of our protocol, when using the number-theoretic constructions described in Sections 3.1 and 3.2. The main technical question is how to implement the collision resistant hash function H which maps inputs to public keys for the multi-trapdoor commitment scheme.

The SDH implementation is basically ready to use "as is". Indeed the public keys pk of the multi-trapdoor commitment scheme are simply elements of Z_q, thus all is needed is a collision-resistant hash function with output in Z_q.

On the other hand, for the Strong RSA based multi-trapdoor commitment, the public keys are prime numbers of the appropriate length. A prime-outputting collision-resistant hash function is described in [23]. However we can do better than that, by modifying slightly the whole protocol. We describe the modifications (inspired by [32, 15]) in this section.

MODIFYING THE ONE-TIME SIGNATURES. First of all, we require the one-time signature scheme (SG, Sig, Ver) to have an extra property: i.e. that the distribution induced by SG over the verification keys vk is the uniform one[5]. Virtually all the known efficient one-time signature schemes have this property.

Then we assume that the collision resistant hash function used in the protocol is drawn from a family which is *both* a collision-resistant collection and a collection of families of universal hash functions[6].

Assume that we have a randomly chosen hash function H from such a collection mapping n-bit strings (the verification keys) into k-bit strings and a prime $P > 2^{k/2}$.

We modify the key generation of our signature scheme as follows. We run SG repeatedly until we get a verification key vk such that $e = 2P \cdot H(\text{vk}) + 1$ is a prime. Notice that $\ell = |e| > \frac{3}{2}k$. Let us denote with SG' this modified key generation algorithm.

We note the following facts:

- $H(\text{vk})$ follows a distribution over k-bit strings which is statistically close to uniform; thus using results on the density of primes in arithmetic progressions (see [1], the results hold under the Generalized Riemann Hypothesis) we know that this process will stop in polynomial time, i.e. after an expected ℓ iterations.
- Since e is of the form $2PR + 1$, and $P > e^{1/3}$, primality testing of all the e candidates can be done deterministically and very efficiently (see Lemma 2 in [32]).

Thus this is quite an efficient way to associate primes to the verification keys.

Notice that we are not compromising the security of the modified signature scheme. Indeed the keys of the modified scheme are a polynomially large fraction of the original universe of keys. Thus if a forger could forge signature on this modified scheme, then the original scheme is not secure as well.

ON THE LENGTH OF THE PRIMES. In our application we need the prime e to be relatively prime to $\phi(N)$ where N is the RSA modulus used in the protocol. This can be achieved by setting $\ell > n/2$ (i.e. $e > \sqrt{N}$). In typical applications (i.e. $|N| = 1024$) this is about 512 bits (we can obtain this by setting $|P| = 352$ and k, the length of the hash function output, to 160). Since the number of iterations to choose vk depends on the length of e, it would be nice to find a way to shorten it.

---

[5] This requirement can be relaxed to asking that the distribution has enough min-entropy.

[6] This is a reasonable assumption that can be made on families built out of a collision-resistant hash function (such as SHA-1). See also [18] for analysis of this type of function families.

If we use safe RSA moduli, then we can enforce that $GCD(e, \phi(N)) = 1$ by choosing e small enough (for 1024-bit safe moduli we need them to be smaller than 500 bits). In this case the collision-resistant property will become the limiting factor in choosing the length. By today's standards we need k to be at least 160. So the resulting primes will be ≈ 240 bits long.

4.2   Identification Protocols

The main application of our result is the construction of concurrently secure identification protocols. In an identification protocol, a prover, associated with a public key pk, communicates with a verifier and tries to convince her to be the legitimate prover (i.e. the person knowing the matching secret key sk.) An adversary tries to mount an impersonation attack, i.e. tries to make the verifier accept without knowing the secret key sk.

The adversary could be limited to interact with the real prover only before mounting the actual impersonation attack [21]. On the other hand a more realistic approach is to consider the adversary a "man-in-the-middle" possibly in a concurrent fashion [5]. Clearly such an attacker can always relays messages unchanged between the prover and the verifier. In order to make a security definition meaningful, one defines a successful impersonation attack as one with a transcript different from the ones between the attacker and the real prover[7].

It is not hard to see that CNM-POK is indeed a concurrently secure identification protocol. It is important to notice that we achieve full concurrency here, indeed the extraction procedure in the proof of Theorem 1 does not "care" if there are many other executions in which the adversary is acting as a prover. Indeed we do not need to rewind *all* executions, but only one in order to extract the one witness we need. Thus if there are other such executions "nested" inside the one we are rewinding, we just run them as the honest verifier.

Acknowledgments

I would like to thank Hugo Krawczyk for conversations that started the research on this paper. Thanks also to Dario Catalano, Shai Halevi, Jonathan Katz, Dah-Yoh Lim, Yehuda Lindell and especially Phil MacKenzie for helpful conversations and advice.

References

1. E. Bach and J. Shallit. *Algorithmic Number Theory - Volume 1.* MIT Press, 1996.
2. B. Barak. *How to go beyond the black-box simulation barrier.* Proc. of 42nd IEEE Symp. on Foundations of Computer Science (FOCS'01), pp.106–115, 2001.

---

[7] In [5] an even more powerful adversary is considered, one that can even *reset* the internal state of the prover. The resulting notion of security implies security in the concurrent model. We do not consider the resettable scenario, but our protocols are more efficient than the ones proposed in [5].

3. B. Barak. *Constant-round Coin Tossing with a Man in the Middle or Realizing the Shared Random String Model.* Proc. of 43rd IEEE Symp. on Foundations of Computer Science (FOCS'02), pp.345–355, 2001.
4. N. Barić, and B. Pfitzmann. *Collision-free accumulators and Fail-stop signature schemes without trees.* Proc. of EUROCRYPT'97 (LNCS 1233), pp.480–494, Springer 1997.
5. M. Bellare, M. Fischlin, S. Goldwasser and S. Micali. *Identification Protocols Secure against Reset Attacks.* Proc. of EUROCRYPT'01 (LNCS 2045), pp.495–511, Springer 2001.
6. M. Bellare and O. Goldreich. *On defining proofs of knowledge.* Proc. of CRYPTO'92 (LNCS 740), Springer 1993.
7. D. Bleichenbacher and U. Maurer. Optimal Tree-Based One-time Digital Signature Schemes. *STACS'96*, LNCS, Vol. 1046, pp.363–374, Springer-Verlag.
8. D. Bleichenbacher and U. Maurer. *On the efficiency of one-time digital signatures.* Proc. of ASIACRYPT'96 (LNCS 1163), pp.145–158, Springer 1996.
9. D. Boneh and X. Boyen. *Short Signatures without Random Oracles.* Proc. of EUROCRYPT'04 (LNCS 3027), pp.382–400, Springer 2004.
10. D. Boneh and M. Franklin. *Identity-Based Encryption from the Weill Pairing.* SIAM J. Comp. 32(3):586–615, 2003.
11. R. Canetti. *Universally Composable Security: A new paradigm for cryptographic protocols.* Proc. of 42nd IEEE Symp. on Foundations of Computer Science (FOCS'01), pp.136–145, 2001.
12. R. Canetti and M. Fischlin. *Universally Composable Commitments.* Proc. of CRYPTO'01 (LNCS 2139), pp.19–40, Springer 2001.
13. R. Canetti, J. Kilian, E. Petrank and A. Rosen. *Concurrent Zero-Knowledge requires $\tilde{\Omega}(\log n)$ rounds.* Proc. of 33rd ACM Symp. on Theory of Computing (STOC'01), pp.570–579, 2001.
14. R. Cramer and I. Damgård. New Generation of Secure and Practical RSA-based signatures. *Proc. of Crypto '96* LNCS no. 1109, pages 173-185.
15. R. Cramer and V. Shoup. Signature schemes based on the Strong RSA assumption. *Proc. of 6th ACM Conference on Computer and Communication Security 1999.*
16. I. Damgård. *Efficient Concurrent Zero-Knowledge in the Auxiliary String Model.* Proc. of EUROCYPT'00 (LNCS 1807), pp.174–187, Springer 2000.
17. A. De Santis, G. Di Crescenzo, R. Ostrovsky, G. Persiano and A. Sahai. *Robust Non-Interactive Zero Knowledge.* Proc. of CRYPTO'01, (LNCS 2139), pp.566-598, Springer 2001.
18. Y. Dodis, R. Gennaro, J. Håstad, H. krawczyk and T. Rabin. *Randomness Extraction and Key Derivation using the CBC, Cascade and HMAC Modes.* This proceedings.
19. D. Dolev, C. Dwork and M. Naor. *Non-malleable Cryptography.* SIAM J. Comp. 30(2):391–437, 2000.
20. C. Dwork, M. Naor and A. Sahai. *Concurrent Zero-Knowledge.* Proc. of 30th ACM Symp. on Theory of Computing (STOC'98), pp.409–418, 1998.
21. U. Feige, A. Fiat and A. Shamir. *Zero-Knowledge Proofs of Identity.* J. of Crypt. 1(2):77–94, Springer 1988.
22. J. Garay, P. MacKenzie and K. Yang. *Strengthening Zero-Knowledge Protocols Using Signatures.* Proc. of EUROCRYPT'03 (LNCS 2656), pp.177–194, Springer 2003. Final version at eprint.iacr.org
23. R. Gennaro, S. Halevi and T. Rabin. Secure Hash-and-Sign Signatures Without the Random Oracle. *Proc. of Eurocrypt '99* LNCS no. 1592, pages 123-139.

24. S. Goldwasser, S. Micali, and C. Rackoff. The knowledge complexity of interactive proof-systems. *SIAM. J. Computing*, 18(1):186–208, February 1989.

25. S. Goldwasser, S. Micali, and R. Rivest. A digital signature scheme secure against adaptive chosen-message attacks. *SIAM J. Computing*, 17(2):281–308, April 1988.

26. L.C. Guillou and J.J. Quisquater. *A Practical Zero-Knowledge Protocol Fitted to Security Microprocessors Minimizing both Transmission and Memory.* Proc. of EUROCRYPT'88 (LNCS 330), pp.123–128, Springer 1989.

27. J. Katz. *Efficient and Non-Malleable Proofs of Plaintext Knowledge and Applications.* Proc. of EUROCRYPT'03 (LNCS 2656), pp.211–228, Springer 2003.

28. L. Lamport. Constructing Digital Signatures from a One-Way Function. *Technical Report SRI Intl.* CSL 98, 1979.

29. Y. Lindell. *Composition of Secure Multi-Party Protocols.* Lecture Notes in Computer Science vol.2815, Springer 2003.

30. Y. Lindell. *Lower Bounds for Concurrent Self Composition.* Proc of the 1st Theory of Cryptography Conference (TCC'04), LNCS 2951, pp.203–222, Springer 2004.

31. P. MacKenzie and K. Yang. *On Simulation-Sound Trapdoor Commitments.* Proc. of EUROCRYPT'04 (LNCS 3027), pp.382–400, Springer 2004.

32. U. Maurer. *Fast Generation of Prime Numbers and Secure Public-Key Cryptographic Parameters.* J. of Crypt. 8(3):123–156, Springer 1995.

33. T. Pedersen. Non-interactive and information-theoretic secure verifiable secret sharing. In *Crypto '91*, pages 129–140, 1991. LNCS No. 576.

34. M. Prabhakaran, A. Rosen and A. Sahai. *Concurrent Zero-Knowledge with logarithmic round complexity.* Proc. of 43rd IEEE Symp. on Foundations of Computer Science (FOCS'02), pp.366–375, 2002.

35. R. Rivest, A. Shamir and L. Adelman. A Method for Obtaining Digital Signature and Public Key Cryptosystems. *Comm. of ACM*, 21 (1978), pp. 120–126

36. C. P. Schnorr. Efficient signature generation by smart cards. *Journal of Cryptology*, 4:161–174, 1991.

37. A. Shamir. On the generation of cryptographically strong pseudorandom sequences. ACM Trans. on Computer Systems, 1(1), 1983, pages 38-44.

Constant-Round Resettable Zero Knowledge with Concurrent Soundness in the Bare Public-Key Model

Giovanni Di Crescenzo[1], Giuseppe Persiano[2], and Ivan Visconti[3]

[1] Telcordia Technologies, Piscataway, NJ, USA
giovanni@research.telcordia.com
[2] Dip. di Informatica ed Appl., Univ. di Salerno, Baronissi, Italy
giuper@dia.unisa.it
[3] Département d'Informatique, École Normale Supérieure, Paris, France
ivan.visconti@ens.fr

Abstract. In the bare public-key model (BPK in short), each verifier is assumed to have deposited a public key in a file that is accessible by all users at all times. In this model, introduced by Canetti *et al.* [STOC 2000], constant-round black-box concurrent and resettable zero knowledge is possible as opposed to the standard model for zero knowledge. As pointed out by Micali and Reyzin [Crypto 2001], the notion of soundness in this model is more subtle and complex than in the classical model and indeed four distinct notions have been introduced (from weakest to strongest): one-time, sequential, concurrent and resettable soundness.

In this paper we present the first *constant-round* concurrently sound resettable zero-knowledge argument system in the bare public-key model for $\mathcal{NP}$. More specifically, we present a 4-round protocol, which is optimal as far as the number of rounds is concerned. Our result solves the main open problem on resettable zero knowledge in the BPK model and improves the previous works of Micali and Reyzin [EuroCrypt 2001] and Zhao *et al.* [EuroCrypt 2003] since they achieved concurrent soundness in stronger models.

1   Introduction

The classical notion of a zero-knowledge proof has been introduced in [1]. Roughly speaking, in a zero-knowledge proof a prover can prove to a verifier the validity of a statement without releasing any additional information. In order to prove that a zero-knowledge protocol does not leak information it is required to show the existence of a probabilistic polynomial-time algorithm, referred to as *Simulator*, whose output is indistinguishable from the output of the interaction between the prover and the verifier. Since its introduction, the concept of a zero-knowledge proof and the simulation paradigm have been widely used to prove the security of many protocols. More recently, it has been recognized that in several practical settings the original notion of zero knowledge (which in its original formulation

M. Franklin (Ed.): CRYPTO 2004, LNCS 3152, pp. 237–253, 2004.

only considered one prover and one verifier that carried out the proof procedure in isolation) was insufficient. For example, the notion of *concurrent zero knowledge* [2] formalizes security in a scenario in which several verifiers access concurrently the same prover and maliciously coordinate their actions so to extract information from the prover. Motivated by considerations regarding smart cards, the notion of *resettable zero knowledge* (rZK, in short) was introduced in [3]. An rZK proof remains "secure" even if the verifier is allowed to tamper with the prover and to reset the prover in the middle of a proof to any previous state and then asks different questions. It is easy to see that concurrent zero knowledge is a special case of resettable zero knowledge and, currently, rZK is the strongest notion of zero knowledge that has been studied. Unfortunately, if we only consider black-box zero knowledge, constant-round concurrent zero knowledge is only possible for trivial languages (see [4]). Moreover, the existence of a constant-round concurrent zero-knowledge argument in the non-black-box model (see [5] for the main results in the non-black-box model) is currently an open question. Such negative results have motivated the introduction of the *bare public-key model* [3] (BPK, in short). Here each possible verifier deposits a public key pk in a public file and keeps private the associated secret information sk. From then on, all provers interacting with such a verifier will use pk and the verifier cannot change pk from proof to proof. Canetti *et al.* [3] showed that constant-round rZK is possible in the BPK model. However, the fact that the verifier has a public key means that it is vulnerable to an attack by a malicious prover that opens several sessions with the same verifier in order to violate the soundness condition. This is to be contrasted with the standard models for interactive zero knowledge [1] or non-interactive zero knowledge [6] where, as far as soundness is concerned, it does not matter whether a malicious prover is interacting once or multiple times with the same verifier.

Indeed, in [7], Micali and Reyzin pointed out, among other contributions, that the known constant-round rZK arguments in the BPK model did not seem to be sound if a prover was allowed to concurrently interact with several instances of the same verifier. In other words, the known rZK arguments in the BPK were not *concurrently sound*.

Micali and Reyzin gave in [7] a 4-round argument system which is sequentially sound (*i.e.*, the soundness holds if a prover can play only sequential sessions) and probably is not concurrently sound, and they also showed that the same holds for the five-round protocol of Canetti *et al.* [3]. Moreover they proved that resettable soundness cannot be achieved in the black-box model. In [8], Barak *et al.* used non-black-box techniques in order to obtain a constant-round rZK argument of knowledge but their protocol enjoys only sequential soundness.

In order to design a concurrently sound resettable zero-knowledge argument system, Micali and Reyzin proposed (see [9]) the upper bounded public-key (UPK, in short) model in which a honest verifier possesses a counter and uses the same private key no more than a fixed polynomial number of times. A weaker model than the UPK model but still stronger than the BPK model is the weak public-key (WPK, in short) model introduced in [10]. In this model an honest verifier can use the same key no more than a fixed polynomial number of times for each statement to be proved.

Other models were proposed in order to achieve constant-round concurrent zero knowledge. In particular, in [2, 11] a constant-round concurrent zero-knowledge proof system is presented by relaxing the asynchrony of the model or the zero-knowledge property. In [12] a constant-round concurrent zero-knowledge proof system is presented by requiring a pre-processing stage in which both the provers and the verifiers are involved. In [13] a constant-round concurrent zero-knowledge proof is presented assuming the existence of a trusted auxiliary string. All these models are considered stronger than the BPK model.

Our results. In this paper we present the first *constant-round concurrently* sound *resettable* zero-knowledge argument system in the BPK model for $\mathcal{NP}$. In particular we show a 4-round argument that is optimal in light of a lower bound for concurrent soundness proved in [7]. We stress that our result is the best one can hope for in terms of combined security against malicious provers and verifiers if we restrict ourselves to black-box zero knowledge, since in this setting simultaneously achieving resettable soundness and zero knowledge has been shown to be possible only for languages in BPP by [7]. Our construction employs the technique of *complexity leveraging* used in the previous results [3, 7, 10] in order to prove the soundness of their protocols and is based on the existence of a verifiably binding cryptosystem semantically secure against subexponential adversaries. The existence of cryptographic primitives secure against subexponential adversaries is used also in [3, 7, 10] and the existence of a constant-round black-box rZK argument system in the BPK model assuming only cryptographic primitives secure against polynomial-time adversaries is an interesting open question.

Finally, we describe a simple 3-round sequentially sound and sequential zero-knowledge argument system in the BPK model for all $\mathcal{NP}$.

2   Definitions

The BPK model. The Bare Public-Key (BPK, in short) model assumes that:
1. there exists a public file F that is a collection of records, each containing a public key;
2. an (honest) prover is an interactive deterministic polynomial-time algorithm that takes as input a security parameter 1^n, F, an n-bit string x, such that $x \in L$ and L is an NP-language, an auxiliary input y, a reference to an entry of F and a random tape;
3. an (honest) verifier V is an interactive deterministic polynomial-time algorithm that works in the following two stages: 1) in a first stage on input a security parameter 1^n and a random tape, V generates a key pair $(\mathrm{pk}, \mathrm{sk})$ and stores pk in one entry of the file F; 2) in the second stage, V takes as input sk, a statement $x \in L$ and a random string, V performs an interactive protocol with a prover, and outputs "accept" or "reject";
4. the first interaction of each prover starts after that all verifiers have completed their first stage.

Definition 1. *Given an NP-language L and its corresponding relation R_L, we say that a pair $\langle P, V \rangle$ is complete for L, if for all n-bit strings $x \in L$ and any witness y such that $(x, y) \in R_L$, the probability that V interacting with P on input y, outputs "reject" is negligible in n.*

Malicious provers in the BPK model. Let s be a positive polynomial and P^* be a probabilistic polynomial-time algorithm that takes as first input 1^n.

P^* is an *s-sequential malicious* prover if it runs in at most $s(n)$ stages in the following way: in stage 1, P^* receives a public key **pk** and outputs an n-bit string x_1. In every even stage, P^* starts from the final configuration of the previous stage, sends and receives messages of a single interactive protocol on input **pk** and can decide to abort the stage in any moment and to start the next one. In every odd stage $i > 1$, P^* starts from the final configuration of the previous stage and outputs an n-bit string x_i.

P^* is an *s-concurrent malicious* prover if on input a public key **pk** of V, can perform the following $s(n)$ interactive protocols with V: 1) if P^* is already running i protocols $0 \leq i < s(n)$ he can start a new protocol with V choosing the new statement to be proved; 2) he can output a message for any running protocol, receive immediately the response from V and continue.

Attacks in the BPK model. In [7] the following attacks have been defined.

Given an s-sequential malicious prover P^* and an honest verifier V, a *sequential attack* is performed in the following way: 1) the first stage of V is run on input 1^n and a random string so that a pair (**pk**, **sk**) is obtained; 2) the first stage of P^* is run on input 1^n and **pk** and x_1 is obtained; 3) for $1 \leq i \leq s(n)/2$ the $2i$-th stage of P^* is run letting it interact with V that receives as input **sk**, x_i and a random string r_i, while the $(2i + 1)$-th stage of P^* is run to obtain x_i.

Given an s-concurrent malicious prover P^* and an honest verifier V, a *concurrent attack* is performed in the following way: 1) the first stage of V is run on input 1^n and a random string so that a pair (**pk**, **sk**) is obtained; 2) P^* is run on input 1^n and **pk**; 3) whenever P^* starts a new protocol choosing a statement, V is run on inputs the new statement, a new random string and **sk**.

Definition 2. *Given a complete pair $\langle P, V \rangle$ for an NP-language L in the BPK model, then $\langle P, V \rangle$ is a concurrently (resp. sequentially) sound interactive argument system for L if for all positive polynomial s, for all s-concurrent (resp s-sequential) malicious prover P^*, for any false statement "$x \in L$" the probability that in an execution of a concurrent (resp. sequential) attack V outputs "accept" for such a statement is negligible in n.*

The strongest notion of zero knowledge, referred to as resettable zero knowledge, gives to a verifier the ability to rewind the prover to a previous state. This is significantly different from a scenario of multiple interactions between prover and verifier since after a rewinding the prover uses the same random bits.

We now give the formal definition of a black-box resettable zero-knowledge argument system for $\mathcal{NP}$ in the bare public-key model.

Definition 3. *An interactive argument system* $\langle P, V \rangle$ *in the BPK model is black-box resettable zero-knowledge if there exists a probabilistic polynomial-time algorithm S such that for any probabilistic polynomial time V^*, for any polynomials s, t, for any $x_i \in L$, $|x_i| = n$, $i = 1, \ldots, s(n)$, V^* runs in at most t steps and the following two distributions are indistinguishable:*

1. *the output of V^* that generates F with $s(n)$ entries and interacts (even concurrently) a polynomial number of times with each $P(x_i, y_i, j, r_k, F)$ where y_i is a witness for $x_i \in L$, $|x_i| = n$ and r_k is a random tape for $1 \leq i, j, k \leq s(n)$;*
2. *the output of S interacting with V^* on input $x_1, \ldots, x_{s(n)}$.*

Moreover we define such an adversarial verifier V^* as an (s, t)-resetting malicious verifier.

An important tool used this paper is that of a non-interactive zero-knowledge argument system.

Definition 4. *A pair of probabilistic polynomial-time algorithms* (NIPK,NIVK) *is a non-interactive zero-knowledge argument system for an $\mathcal{NP}$ language L if there exists a polynomial $k(\cdot)$,*

1. *(Completeness) for all $x \in L$, with $|x| = n$ and NP-witness y for $x \in L$,*

$$Pr[\sigma \xleftarrow{R} \{0,1\}^{k(n)}; \Pi \leftarrow \text{NIPK}(x, y, \sigma) : \text{NIVK}(x, \Pi, \sigma) = 1] = 1.$$

2. *(Soundness) for all $x \notin L$*

$$Pr[\sigma \xleftarrow{R} \{0,1\}^{k(n)}; \exists \Pi : \text{NIVK}(x, \Pi, \sigma) = 1]$$

   is negligible.
3. *(Simulatability) there exists a probabilistic polynomial-time algorithm S such that the family of distributions*

$$\{(\sigma, \Pi) \xleftarrow{R} S(x) : (\sigma, \Pi)\}_{x \in L} \text{ and } \{\sigma \xleftarrow{R} \{0,1\}^{k(n)}; \Pi \xleftarrow{R} \text{NIPK}(x, y, \sigma) : (\sigma, \Pi)\}_{x \in L}$$

   are computationally indistinguishable.

We assume, without loss of generality, that a random reference string of length n is sufficient for proving theorems of length n (that is, we assume $k(n) = n$).

3 Concurrently Sound rZK Argument System for $\mathcal{NP}$ in the BPK Model

In this section we present a constant-round concurrently sound resettable zero-knowledge argument in the BPK model for all $\mathcal{NP}$ languages.

In our construction we assume the existence of an encryption scheme that is secure with respect to sub-exponential adversaries and that is *verifiably binding*. We next review the notion of semantic security adapted for sub-exponential adversaries and present the notion of a *verifiably binding* cryptosystem.

An encryption scheme is a triple of efficient algorithms $PK = (G, E, D)$. The key generator algorithm G on input a random k-bit string r (the security parameter) outputs a pair (pk, sk) of public and private key. The public key pk is used to encrypt a string m by computing $E(pk, m; r)$ where r is a random string of length $|m|$.

Semantic security [14] is defined by considering the following experiment for encryption scheme $PK = (G, E, D)$ involving a two-part adversary $\mathcal{A} = (\mathcal{A}_0, \mathcal{A}_1)$. The key generator G is run on a random k-bit string and keys (pk, sk) are given in output. Two $POLY(k)$-bit strings ω_0 and ω_1 are returned by $\mathcal{A}_0$ on input pk. Then b is taken at random from $\{0, 1\}$ and an encryption ξ of ω_b is computed. We say that adversary $\mathcal{A}$ is *successful for* PK if the probability that $\mathcal{A}_1$ outputs b on input pk, ω_0, ω_1 and ξ is non-negligibly (in k) greater than $1/2$. We say that PK is η-*secure* if no adversary running in time $o(2^{k^\eta})$ is successful. The classical notion of semantic security is instead obtained by requiring that no polynomial-time adversary is successful.

Roughly speaking, a *verifiably binding* cryptosystem PK is a cryptosystem for which 1) given a string pk and an integer k, it is easy to verify that pk is a legal public key with security parameter k and 2) to each ciphertext corresponds at most one plaintext.

More formally,

Definition 5. *An η-secure encryption scheme* $PK = (G, E, D)$ *is* verifiably binding *iff:*

  1. (binding): for any probabilistic polynomial-time algorithm $\mathcal{A}$ it holds that

$$Pr[(pk, m_0, m_1, r_0, r_1) \leftarrow \mathcal{A}(1^k) : E(pk, m_0; r_0) = E(pk, m_1; r_1)]$$

  is negligible in k;

  2. (verifiability): there exists a probabilistic polynomial-time algorithm VER such that if pk belongs to the output space of G on input a k-bit string then $\mathrm{VER}(pk, 1^k) = 1$; $\mathrm{VER}(pk, 1^k) = 0$ otherwise.

Assumptions. To prove the properties of our protocol we make the following complexity theoretic assumptions:

1. The existence of an η-secure verifiably binding encryption scheme $PK = (G, E, D)$ for some $\eta > 0$.
   We briefly note that the El Gamal encryption scheme [15] is verifiably binding since an exponentiation in Z_q^* is one to one and it can be easily verified that a positive integer q is a prime.
2. The existence of a one-to-one length-preserving one-way function $f : \{0, 1\}^* \rightarrow \{0, 1\}^*$ which, in turn, implies the existence of a pseudo-random family of functions $\mathcal{R} = \{R_s\}$.
3. The existence of a non-interactive zero-knowledge proof system (NIZK, in short) $(\mathrm{NIPM}, \mathrm{NIVM})$ for an $\mathcal{NP}$-complete language.

4. The existence of a 3-round witness indistinguishable argument of knowledge $\mathtt{WI} = (\mathtt{WI}_1, \mathtt{WI}_2, \mathtt{WI}_3)$ for a specific polynomial-time relation that we define in the following way. Let f be a one-to-one length-preserving one-way function and let $\mathtt{PK}$ be an η-secure verifiably binding encryption scheme. Then define the polynomial-time relation $\mathcal{C} = C(\mathtt{PK}, f)$ as consisting of all pairs $((\mathtt{pk}, v), (\mathtt{wit}))$, where $\mathtt{pk}$ is a public key of the output space of $\mathtt{G}$ and v is a string and either $\mathtt{wit} = \mathtt{sk}$ and $(\mathtt{pk}, \mathtt{sk})$ is in the output space of $\mathtt{G}$ or $\mathtt{wit} = u$ and $f(u) = v$.

Before describing our protocol formally, let us try to convey the main idea behind it. Fix an $\mathcal{NP}$ language L and let x be the input statement. The prover generates a puzzle (in our construction, the puzzle consists of a string v and solving the puzzle consists in finding the inverse $f^{-1}(v)$ of the one-to-one length-preserving one-way function f) and sends it to the verifier. The verifier uses $\mathtt{WI}$ to prove knowledge of the private key $\mathtt{sk}_i$ associated to her public key $\mathtt{pk}_i$ or knowledge of the solution of the puzzle given to her by the prover. Moreover, the prover and the verifier play a coin tossing protocol, based on the encryption scheme $\mathtt{PK}$ to generate a reference string for the NIZK proof that $x \in L$.

In our implementation of the FLS-paradigm [16], in the interaction between the prover and the verifier, the verifier will use his knowledge of the private key to run $\mathtt{WI}$. In order to prove concurrent soundness, we show an algorithm $\mathcal{A}$ that interacts with a (possibly) cheating prover P^* and breaks an η-secure encryption scheme in time $o(2^{k^\eta})$. The puzzle helps algorithm $\mathcal{A}$ in simulating the verifier with respect to a challenge public key $\mathtt{pk}$ for which it does not have access to the private key. Indeed, $\mathcal{A}$ instead of proving knowledge of the private key associated to $\mathtt{pk}$ proves knowledge of the solution of the puzzle by performing exhaustive search. By carefully picking the size of the puzzle (and thus the time required to solve it) we can make sure $\mathcal{A}$ runs in time $o(2^{k^\eta})$.

Note that when $\mathcal{A}$ inverts the one-to-one length-preserving one-way function and computes the witness-indistinguishable argument of knowledge, it runs in subexponential time in order to simulate the verifier without performing rewinds. Straight-line quasi-polynomial time simulatable argument systems were studied in detail in [17], where this relaxed simulation notion is used to decrease the round complexity of argument systems. We use a similar technique but for subexponential time simulation of arguments of knowledge.

If the steps described above were executed sequentially, we would have an 8-round protocol (one round for the prover to send the puzzle, three rounds for the coin tossing, three rounds for the witness-indistinguishable argument of knowledge, and one round for the NIZK). However, observe that the coin-tossing protocol and the 3-round witness-indistinguishable argument of knowledge can be performed in parallel thus reducing the the round complexity to 5 rounds. Moreover, we can save one more round, by letting the prover send the puzzle in parallel with the second round of the witness indistinguishable argument of knowledge. To do so, we need a special implementation of this primitive since, when the protocol starts, only the size of the statement is known and the statement itself is part of the second round. Let us now give the details of our construction.

The public file. The public file F contains entries consisting in public keys with security parameter k for the public-key cryptosystem PK.

Private inputs. The private input of the prover consists of a witness y for $x \in L$. The private input of the verifier consists of the secret key $\mathbf{sk}_i$ corresponding to the public key $\mathbf{pk}_i$.

The protocol. Suppose that the prover wants to prove that $x \in L$ and denote by $n = \mathrm{POLY}(k)$ the length of x. We denote by i the index of the verifier in the public file so that the verifier knows the private key $\mathbf{sk}_i$ associated with the i-th public key $\mathbf{pk}_i$ of the public file F.

In the first round V randomly picks an n-bit string σ_v that will be used as V's contribution to the reference string for the non-interactive zero-knowledge protocol. V compute the encryption ξ of σ_v using an n-bit string r_v as randomness and by using public key $\mathbf{pk}_i$. Moreover, V runs $\mathtt{WI}_1$ in order to compute the first message a_1 of the witness-indistinguishable argument of knowledge. Then V sends (ξ, a_1) to P. In the second round P verifies that $\mathbf{pk}_i$ is a legal public key for PK with k as security parameter and then computes its contribution to the random string to be used for the non-interactive argument by picking a random seed s and computing $(u, \sigma_p) = R_s(x \circ y \circ F \circ \xi \circ a_1 \circ i)$ ("$\circ$" denotes concatenation) where $\{R_s\}$ is a family of pseudorandom functions. The string u has length $k' \leq k$ (to be determined later) whereas σ_p has length n and is P's contribution for the reference string. P runs $\mathtt{WI}_2$ to compute the second message a_2 of the witness-indistinguishable argument of knowledge. Moreover P computes $v = f(u)$ where f is a one-to-one length-preserving one-way function and sends (σ_p, a_2, v) to the verifier. In the third round of the protocol V uses his knowledge of the private key to run $\mathtt{WI}_3$ obtaining a_3, so that she proves that she knows either the private key associated with $\mathbf{pk}_i$ or $f^{-1}(v)$. V then sends a_3, σ_v and r_v to P. In the last round of the protocol P verifies that the witness-indistinguishable argument of knowledge is correct and that ξ is an encryption of σ_v. Then P runs algorithm NiPM on input x and using $\sigma = \sigma_p \oplus \sigma_v$ as reference string obtaining a proof Π_p that is sent to V. A more formal description of the protocols is found in Figure 1.

Theorem 1. *If there exists an η-secure verifiably binding encryption scheme, a one-to-one length-preserving one-way function then there exists a* constant-round concurrently sound resettable zero-knowledge *argument for all languages in $\mathcal{NP}$ in the BPK model.*

Proof. Consider the protocol found in Figure 1.

Completeness. If $x \in L$ then P can always compute the proof Π and V accepts it.

Concurrent soundness. Assume by contradiction that the protocol is not concurrently sound. Thus there exists an s-concurrent malicious prover P^* that by,

Common input: the public file F, n-bit string $x \in L$ and index i that specifies the i-th entry of F. Public key $\mathbf{pk}_i$ has security parameter k.
P's private input: a witness y for $x \in L$.
V's private input: private key $\mathbf{sk}_i$.

V-round-1:
1. randomly pick $\sigma_v \leftarrow \{0,1\}^n$ and $r_v \leftarrow \{0,1\}^n$;
2. compute $\xi = \mathrm{E}(\mathbf{pk}_i, \sigma_v; r_v)$ and $a_1 = \mathtt{WI}_1(1^k)$;
3. send (ξ, a_1) to P;

P-round-2:
1. verify that $\mathbf{pk}_i$ is a public key with security parameter k for PK;
2. randomly pick $s \leftarrow \{0,1\}^n$ and compute $R = R_s(x \circ y \circ F \circ \xi \circ a_1 \circ i)$; let u be the string consisting of the first k' bits of R and σ_p the string consisting of the next n bits of R;
3. compute $a_2 = \mathtt{WI}_2(a_1)$;
4. compute $v = f(u)$ where f is a one-to-one length preserving one-way function;
5. send (σ_p, a_2, v) to V;

V-round-3:
1. verify that v is a k'-bit string;
2. set $\sigma = \sigma_p \oplus \sigma_v$;
3. run algorithm $\mathtt{WI}_3$ on input instance $(\mathbf{pk}_i, v)$, messages a_1, a_2 using $\mathbf{sk}_i$ as a witness and obtaining a_3;
4. send (σ_v, a_3, r_v) to P;

P-round-4:
1. verify that $\xi = \mathrm{E}(\mathbf{pk}_i, \sigma_v; r_v)$;
2. set $\sigma = \sigma_p \oplus \sigma_v$;
3. verify that (a_1, a_2, a_3) is the correct transcript of the 3-round witness indistinguishable argument on input instance $(\mathbf{pk}_i, v)$;
4. run NiPM on input instance x, y as a witness and σ as reference string obtaining proof Π;
5. send Π to V;

V-decision: verify that Π is a proof by running algorithm NiVM on input x, Π and σ.

Fig. 1. The 4-round concurrently sound rZK argument system for $\mathcal{NP}$ in the BPK model. The values k and k' are determined as functions of n in the proof of concurrent soundness.

concurrently interacting with V, has non-negligible probability $p(n)$ of making the verifier accept some $x \notin L$ of length n. We assume we know the index of the session j^* in which the prover will succeed in cheating (this assumption will be later removed) and exhibit an algorithm $\mathcal{A}$ that has black-box access to P^* (*i.e.*, $\mathcal{A}$ simulates the work of a verifier V) and breaks the encryption scheme PK in $o(2^{k^n})$ steps, thus reaching a contradiction.

We now describe algorithm $\mathcal{A}$. $\mathcal{A}$ runs in two stages. First, on input the challenge public key $\mathbf{pk}$, $\mathcal{A}$ randomly picks two strings ω_0 and ω_1 of the same length as the length of the reference string used by (NIPM,NIVM) for inputs of length n. Then $\mathcal{A}$ receives as a challenge an encryption $\tilde{\xi}$ of ω_b computed using public key pk and $b \in \{0,1\}$. $\mathcal{A}$'s task is to guess $b \in \{0,1\}$ with a non-negligible advantage over $1/2$ (we assume that b is randomly chosen).

For all the sessions, $\mathcal{A}$ interacts with the an s-concurrent prover P^* mounting a concurrent attack, and simulates the verifier by computing the two messages as explained below. When $\mathcal{A}$ reaches session j^*, $\mathcal{A}$ outputs her guess for bit b.

1. Session $j \neq j^*$.

   At V-round-1, $\mathcal{A}$ sends an encryption ξ of a randomly chosen string σ_v computed with r_v as randomness and sends the first round of the witness-indistinguishable argument of knowledge a_1. Upon receiving message (σ_p, a_2, v) from P^*, $\mathcal{A}$ inverts the one-to-one length-preserving one-way function f on v obtaining $u = f^{-1}(v)$ by performing exhaustive search in $\{0,1\}^{k'}$. $\mathcal{A}$ then computes a_3 by running WI_3 on input instance (pk_i, v) and witness u and sends to P^* the triple (σ_v, a_3, r_v).

   Note that $\mathcal{A}$ plays round V-round-1 identically to the honest verifier while $\mathcal{A}$ plays round V-round-3 by using a different witness w.r.t. V for the non-interactive zero-knowledge argument of knowledge that however is concurrent witness indistinguishable.

2. Session j^*.

   At V-round-1, $\mathcal{A}$ computes the first message of the witness-indistinguishable argument of knowledge a_1 and sets ξ equal to the challenge encryption $\tilde{\xi}$. Then $\mathcal{A}$ sends (ξ, a_1) to V.

   At V-round-3, $\mathcal{A}$ cannot continue with this session since she does not know the decryption of ξ (remember that $\xi = \tilde{\xi}$) and thus can not play the third round. However, by assumption P^* can produce with non-negligible probability a string Π^* that is accepted by NIVM on input x and reference string $\rho_b^* = \omega_b \oplus \sigma_p$. Let τ be an upper bound on the length of such a non-interactive zero-knowledge argument. $\mathcal{A}$ checks, by exhaustive search, if there exists $\Pi_0 \in \{0,1\}^\tau$, such that NIVM accepts Π_0 on input x and ρ_0^* as reference string. Then $\mathcal{A}$ searches for a string $\Pi_1 \in \{0,1\}^\tau$ by considering ρ_1^* as reference string. If a proof Π_0 is found and no proof Π_1 is found then $\mathcal{A}$ outputs 0; in the opposite case $\mathcal{A}$ outputs 1; otherwise (that is, if both or neither proof exists) $\mathcal{A}$ randomly guesses the bit b.

   We note that the distribution of the first message of session j^* is still identical to the distribution of the honest verifier's message.

Let us now show that the probability that $\mathcal{A}$ correctly guesses b is non-negligibly larger that $1/2$. We have that

$$Pr[\mathcal{A} \text{ outputs } b] = Pr[\exists \Pi_b \wedge \not\exists \Pi_{1-b}] + \frac{1}{2}\left(Pr[\exists \Pi_b \wedge \exists \Pi_{1-b}] + Pr[\not\exists \Pi_b \wedge \not\exists \Pi_{1-b}]\right)$$

$$= \frac{1}{2} + \frac{1}{2}\left(Pr[\exists \Pi_b \wedge \not\exists \Pi_{1-b}] - Pr[\not\exists \Pi_b \wedge \exists \Pi_{1-b}]\right)$$

$$= \frac{1}{2} + \frac{1}{2}\left(Pr[\exists \Pi_b \wedge \not\exists \Pi_{1-b} \wedge x \notin L] - Pr[\not\exists \Pi_b \wedge \exists \Pi_{1-b} \wedge x \notin L]\right).$$

The last equality follows from the observation that, by the completeness of the NIZK, the events $\not\exists \Pi_{1-b}$ and $\not\exists \Pi_b$ can happen only if $x \notin L$. Now, we have

$$Pr[A \text{ outputs } b] = \frac{1}{2} + \frac{1}{2}\left(Pr[\exists \Pi_b \wedge x \notin L] - Pr[\exists \Pi_b \wedge \exists \Pi_{1-b} \wedge x \notin L] - \right.$$
$$\left. Pr[\not\exists \Pi_b \wedge \exists \Pi_{1-b} \wedge x \notin L]\right)$$
$$= \frac{1}{2} + \frac{p(n)}{2} - \frac{Pr[\exists \Pi_{1-b} \wedge x \notin L]}{2}.$$

Now, since the string ω_{1-b} is picked at random and P^* has no information about it, the string ρ^*_{1-b} is random and thus, by the soundness of (NiPM,NiVM), $Pr[\exists \Pi_{1-b} \wedge x \notin L]$ is negligible. Therefore, the probability that A correctly guesses b is non-negligibly larger than $1/2$.

We note that algorithm A takes time POLY$(n) \cdot (2^\tau + 2^{k'})$. Writing τ as $\tau = n^\gamma$, for some constant γ, we pick k and k' so that $n^\gamma < k^{\eta/2}$ and $k' < k^{\eta/2}$. We thus have that A breaks an η-secure verifiably binding cryptosystem in time bounded by POLY$(k^{\eta/2})(2^{k^{\eta/2}} + 2^{k^{\eta/2}}) = o(2^{k^\eta})$.

Therefore the existence of A contradicts the η-security of the cryptosystem.

In our proof we assumed that A knows the value j. If this is not the case that A can simply guess the values and the same analysis applies and the probability that A correctly guesses b decreases by a polynomial factor.

Resettable Zero Knowledge. Let V^* be an (s,t)-resetting verifier. We now present a probabilistic polynomial-time algorithm $S \equiv S^{V^*}$ that has black-box access to V^* and whose output is computationally indistinguishable from the view of the interactions between P and V^*.

We start with an informal discussion. The construction of S is very similar to the construction of the simulator for the constant-round (sequentially sound) resettable zero-knowledge argument for any NP language and in the BPK model, given in [3] (protocol 6.2). In particular, note that both the protocol of Figure 1 and protocol 6.2 in [3] can be abstractly described as follows. The prover and the verifier run a 3-round argument of knowledge, where the verifier, acting as a prover, proves knowledge to the prover, acting as verifier, of some trapdoor information. Knowledge of the trapdoor information allows for efficient simulation of the interaction between the prover and the verifier. In [3], the trapdoor information is the private key associated with the verifier's public key. In our protocol, the trapdoor information is either the private key associated with the verifier's public key (for the real verifier) or the inverse of an output of a one-to-one length-preserving one-way function sent from the prover to the verifier. Note that just to obtain round optimality we use a special witness-indistinguishable argument of knowledge where the statement is known only after that the second round is played while its size is known from the beginning. Due to this difference, our simulator only differs from the one of [3] in the fact that we need to prove that when the simulator runs the extractor of the argument of knowledge, with high probability it extracts the verifier's private key (rather than $f^{-1}(v)$). The rest of the construction of our simulator is conceptually identical to that of [3], but we still review a more precise description here for completeness.

First of all, without loss of generality, we make the following two simplifying assumptions. Recall that, since our protocol is a resettable zero-knowledge argument system, V^* is allowed to reset the prover. However, in [3] Canetti *et al.* proved that in such a setting a verifier that concurrently interacts with many incarnations of the prover does not get any advantage with respect to a sequential (resetting) verifier (that is, a verifier that runs a new session only after having terminated the previous one). Thus in this proof we will consider V^* as a sequential (resetting) verifier. A second assumption is that we can define S for a modification of our protocol in which the prover uses a truly random function rather than a pseudo-random one to compute her random bits. Proving that the two views are computationally indistinguishable is rather standard.

S runs the first stage of V^* so that the public file composed by $s(n)$ entries is obtained. In the second stage, the aim of the simulator is to obtain the private keys corresponding to the public keys of the public file. Let $V^*(F)$ be the state of V^* at the end of the first stage.

In the following, we say that a session is *solved* by S if S has the private key corresponding to the public key used by V^* in this session. The work of S in the second stage of the simulation is composed by at most $s(n)+1$ sequential phases. In each phase, either S has a chance of terminating the simulation or S learns one more private key. At the end of each phase S rewinds V^* to state $V^*(F)$. The simulation ends as soon as S manages to solve all sessions of a phase.

We describe now the work of S during a phase. Once a session is started, S receives the first message from V^*. Then there are two cases. If the session is solved by S then S can simulate the prover; otherwise, S tries to obtain the private key used in this session so that all future sessions involving this verifier will be solved by S.

Specifically, first consider the simpler case of a solved session. We distinguish two sub-cases. First, we consider the sub-case where the first message in the session (ξ, a_1) has not appeared before for the same incarnation of the prover, *i.e.*, (ξ, a_1) has not appeared before for the same prover oracle accessed by V^* with the same random tape, same witness and same theorem. Then S runs the simulator for (NiPM,NiVM) on input x and obtains a pair (σ^*, Π^*) and then forces σ equal to σ^* in the following way. Since S knows the verifier's secret-key (we are assuming in this sub-case that the session is solved), S can decrypt ξ and thus obtain the string σ_v computed by the verifier at the first round. Thus S sets $\sigma_p = \sigma_v \oplus \sigma^*$. Consequently, in round P-round-4, S will send "proof" Π^* (that is computationally indistinguishable from the proof computed by the real prover). We use here the binding property of the encryption scheme since S must decrypt ξ obtaining the same value σ_v that will be sent by V^* in round V-round-3.

Now we consider the sub-case where the first message in the session (ξ, a_1) has already appeared in such a phase for the same incarnation of the prover. Here S sends the same strings σ_p, a_2 and the same k'-bit string v that was sent in the previous session containing (ξ, a_1) as first message for the same incarnation of the prover. Even for the case of the third message of a session that has already appeared for the same incarnation of the prover, S replies with the same round P-round-4 played before.

We now consider the harder case of a session which is not solved by S. In this case S uses the argument of knowledge of V^* to obtain the private key used in this session. Specifically, in any unsolved session, the simulator uses the extractor E associated with the witness-indistinguishable argument of knowledge used by the verifier.

Recall that we denote by (ξ, a_1) the first message sent by the verifier in the current session, by pk_i the verifier's public key and by $v = f(u)$ the puzzle sent by the simulator when simulating the prover's first message. We now distinguish three possible cases.

Case 1: The message (ξ, a_1) has not yet appeared in a previous session for the same incarnation of the prover and the extractor E obtains sk_i as witness. Note that S obtains the verifier's private key by running E. This is the most benign of the three cases since the session is now solved.

Case 2: The message (ξ, a_1) has not yet appeared in a previous session for the same incarnation of the prover and the extractor E obtains $f^{-1}(v)$ as witness. Note however that the value v has been chosen by S itself. If this case happens with non-negligible probability then we can use V^* to invert the one-way function f. We stress that this case is the only conceptual difference between our proof and the proof of rZK of protocol 6.2 in [3].

Case 3: The message (ξ, a_1) has already appeared in a previous session for the same incarnation of the prover. Note that since we are assuming that the current session is not solved by S, this means that in at least one previous session, V^* sent (ξ_1, a_1) but then did not continue with such a session. This prevents S from simulating as in case 2 since the simulation would not be correct. (Specifically, as discussed in [3], in a real execution of the argument, the pseudo-random string used as random string for the prover's first message is determined by the previous uncompleted session (the input of R_s is the same in both cases and the seed s is taken from the same random string) and therefore cannot be reset by S to simulate this case by running an independent execution of E.) This problem is bypassed precisely as in [3]. That is, S tries to continue the simulation from the maximal sequence of executions which does not contain (ξ, a_1) as a first step of the verifier for such an incarnation of the prover, using a new random function.

The same analysis in [3] shows that this simulation strategy ends in expected polynomial time and returns a distribution indistinguishable from a real execution of the argument. $\square$

3-*Round WI Argument of Knowledge*. As already pointed out above, we can save one round (and thus obtain a 4-round argument system instead of 5-round one) by having the prover send the puzzle after the verifier has started the witness-indistinguishable argument of knowledge. In this argument of knowledge, the verifier acts as a prover and shows knowledge of either the secret key associated with his private key or of a solution of the puzzle. Consequently, the input statement of such an argument of knowledge is not known from the start and actually, when the first message is produced, only its length is known.

Next we briefly describe such an argument of knowledge by adapting to our needs the technique used by [16] to obtain a non-interactive zero-knowledge proof system for Hamiltonicity.

1. The prover commits to n randomly generated Hamiltonian cycles (each edge is hidden in a committed adjacency matrix of degree n);
2. the graph G is presented to the prover and the verifier and verifier sends an n-bit random challenge;
3. if the i-th bit of the challenge is 0 then the prover opens the i-th Hamiltonian cycle;
4. if the i-th bit of the challenge is 1 then the prover sends a permutation π_i and shows that each edge that is missing in the graph $\pi_i(G)$ corresponds to a commitment of 0 in the i-th committed Hamiltonian cycle.

Completeness, soundness and witness indistinguishability can be easily verified. The protocol is an argument of knowledge since an extractor that rewinds the prover and changes the challenge obtains a Hamiltonian cycle of G.

4   Sequentially Sound Sequential Zero Knowledge for $\mathcal{NP}$ in the BPK Model

In this section we give a 3-round sequentially sound sequential zero-knowledge argument in the BPK model for any language in $\mathcal{NP}$.

Assumptions. We start by listing the tools and the complexity-theoretic assumptions we need for the construction of this section.

1. We assume the existence of an η-secure signature scheme SS = (SigG, Sig, Ver). Here SigG denotes the key generator algorithm that receives the security parameter k (in unary) and returns a pair (pk, sk) of public keys; Sig is the signature algorithm that takes as input a message m and a private key sk and returns a signature s of m; and Ver is the signature verification algorithm that takes a message m, a signature s and a public key pk and verifies that s is a valid signature.

   The scheme SS is η-secure in the sense that no algorithm running in time $o(2^{k^\eta})$ that has access to a signature oracle but not to the private key can forge the signature of a message m for which it has not queried the oracle.

   It is well known that if sub-exponentially strong one-way functions exist then it is possible to construct secure signature schemes [18].

   We assume that signatures of k-bit messages produced by using keys with security parameter k have length k. This is not generally true as for each signature scheme we have a constant a such that signatures of k-bit messages have length k^a but this has the advantage of not overburdening the notation. It is understood that all our proofs continue to hold if this assumption is removed.

2. We assume the existence of a one-round perfectly binding computationally hiding γ-extractable commitment scheme. The scheme is γ-extractable in the sense that there exists an extractor algorithm E that on input a commitment, computes in time $O(2^{k^\gamma})$ the committed value.

   Such a commitment schemes are known to exist under the assumption of the existence of sub-exponentially strong one-to-one length-preserving one-way functions.

3. We also assume the existence of ZAPs for all $\mathcal{NP}$ (see [19]).

In sums, our construction is based on the existence of subexponentially strong one-to-one length preserving one-way functions and one-way trapdoor permutations.

We start by briefly describing the main idea of our protocol. The prover and the verifier play the following game: the prover picks a random message m_1, computes a commitment $\tilde{m}_1$ of m_1 and asks the verifier to sign $\tilde{m}_1$; the verifier signs the commitment and sends back to the prover such a signature and a message m_2. Finally the prover, constructs an extractable commitment com of a random message and proves to the verifier using a ZAP that either $x \in L$ or com is the extractable commitment of a signature of a commitment of m_2. Let us now informally argue about sequential soundness and sequential zero-knowledge of the argument system described. For the sequential soundness, we observe that, since m_2 is chosen at random by the verifier for each sequential execution of the protocol, it is unlikely that the prover knows the signature of a commitment of m_2. For the zero-knowledge property instead, the simulator, once m_2 is received, rewinds V^* and opens a new session with the verifier in which he sets $m_1 = m_2$, computes a commitment of $m_1 = m_2$ and sends it to the verifier that thus produces a signature of a commitment of m_2. Going back to the original session, the simulator has a witness for the ZAP and can thus complete the simulation.

Theorem 2. *If there exist subexponentially strong one-to-one length-preserving one-way functions and trapdoor permutations then there exists a 3-round sequentially sound sequential zero-knowledge argument for $\mathcal{NP}$ in the* BPK *model.*

Proof. Completeness and *Sequential soundness* can be easily proved. For the *Sequential Zero Knowledge*, we now describe a simulator S. We consider a malicious verifier V^* that in the first stage outputs the public file F and in the second stage interacts with P by considering $s(n)$ possible theorems and $s(n)$ possible entries of F. However V^* is now a sequential verifier and thus he cannot run twice the same incarnation of P, neither he can run two concurrent sessions with P. Thus the simulation proceeds session by session and we can focus only in the simulation of a generic session.

Let V_1^* be the state of V^* at the beginning of a given session. The simulator sends in the first round a message that is distributed identically w.r.t. the one of the prover. Then V^* replies by sending a message m_2, let V_2^* the state of V^* in such a step. The simulator rewinds V^* to state V_1^* and plays again the first round but this time he sets $m_1 = m_2$. The simulator repeats this first round with a different randomness as long as the verifier sends a valid second message that therefore contains a signature of a commitment of m_2. The simulator can use the signature of a commitment of m_2 as witness for the third round of the original proof, that can be given by rewinding V^* to state V_2^*. More precisely, S rewinds V^* to state V_2^* and computes $\tilde{a}$ as a commitment of a commitment of m_2 and $\tilde{b}$ as a commitment of the previously received signature. Then S has a witness for playing the ZAP.

The previously described rewind strategy allows the simulator to complete the simulation in expected polynomial-time and, moreover, the indistinguishability of the ZAP and the hiding of the commitment scheme guarantee that the distribution of the output is computationally indistinguishable from an interaction between a real prover and V^*.

We remark that it is possible to base our construction on primitives secure against polynomial-time adversaries by employing a 3-round witness indistinguishable argument where the statement is chosen by the prover before producing the third message.

5   Conclusions

In an asynchronous environment like the Internet resettable zero-knowledge protocols that are not concurrently sound in the BPK model cannot be considered secure and previous concurrently sound protocols required stronger assumptions than the BPK model.

In this work we have positively closed one of the main open problems regarding zero knowledge in the BPK model. We have shown that a constant-round concurrently sound resettable zero-knowledge argument system in the BPK model exists. In particular, we have shown a 4-round protocol which is optimal for the black-box model.

Acknowledgments

We would like to thank the anonymous referee for his/her very useful remarks on our submission.

References

1. Goldwasser, S., Micali, S., Rackoff, C.: The Knowledge Complexity of Interactive Proof-Systems. SIAM J. on Computing **18** (1989) 186–208
2. Dwork, C., Naor, M., Sahai, A.: Concurrent Zero-Knowledge. In: Proceedings of the 30th ACM Symposium on Theory of Computing (STOC '98), 409–418
3. Canetti, R., Goldreich, O., Goldwasser, S., Micali, S.: Resettable Zero-Knowledge. In: Proceedings of the 32nd ACM Symposium on Theory of Computing (STOC '00), 235–244
4. Canetti, R., Kilian, J., Petrank, E., Rosen, A.: Black-Box Concurrent Zero-Knowledge Requires $\omega(\log n)$ Rounds. In: Proceedings of the 33st ACM Symposium on Theory of Computing (STOC '01), 570–579

5. Barak, B.: How to Go Beyond the Black-Box Simulation Barrier. In: Proceeding of the 42nd Symposium on Foundations of Computer Science, (FOCS '01), 106–115
6. Blum, M., De Santis, A., Micali, S., Persiano, G.: Non-Interactive Zero-Knowledge. SIAM J. on Computing **20** (1991) 1084–1118
7. Micali, S., Reyzin, L.: Soundness in the Public-Key Model. In: Advances in Cryptology – Crypto '01. Volume 2139 of Lecture Notes in Computer Science. Springer-Verlag, 542–565
8. Barak, B., Goldreich, O., Goldwasser, S., Lindell, Y.: Resettably-Sound Zero-Znowledge and its Applications. In: Proceeding of the 42nd Symposium on Foundations of Computer Science, (FOCS '01), 116–125
9. Micali, S., Reyzin, L.: Min-round Resettable Zero-Knowledge in the Public-key Model. In: Advances in Cryptology – Eurocrypt '01. Volume 2045 of Lecture Notes in Computer Science. Springer-Verlag (2001) 373–393
10. Zhao, Y., Deng, X., Lee, C., Zhu, H.: Resettable Zero-Knowledge in the Weak Public-Key Model. In: Advances in Cryptology – Eurocrypt '03. Volume 2045 of Lecture Notes in Computer Science. Springer-Verlag (2003) 123–139
11. Goldreich, O.: Concurrent Zero-Knowledge with Timing, Revisited. In: Proceedings of the 34th ACM Symposium on Theory of Computing (STOC '02), ACM (2002) 332–340
12. Di Crescenzo, G., Ostrovsky, R.: On Concurrent Zero-Knowledge with Preprocessing. In: Advances in Cryptology – Crypto '99. Volume 1666 of Lecture Notes in Computer Science. Springer-Verlag (1999) 485–502
13. Damgard, I.: Efficient Concurrent Zero-Knowledge in the Auxiliary String Model. In: Advances in Cryptology – Eurocrypt '00. Volume 1807 of Lecture Notes in Computer Science. Springer-Verlag (2000) 418–430
14. Goldwasser, S., Micali, S.: Probabilistic encryption. J. of Comp. and Sys. Sci. **28** (1984) 270–299
15. El Gamal, T.: A public key cryptosystem and a signature scheme based on discrete logarithms. In: Advances in Cryptology – Crypto '84. Volume 196 of Lecture Notes in Computer Science. Springer Verlag (1984) 10–18
16. Feige, U., Lapidot, D., Shamir, A.: Multiple Non-Interactive Zero Knowledge Proofs Under General Assumptions. SIAM J. on Computing **29** (1999) 1–28
17. Pass, R.: Simulation in Quasi-Polynomial Time and Its Applications to Protocol Composition. In: Advances in Cryptology – Eurocrypt '03. Volume 2045 of Lecture Notes in Computer Science. Springer-Verlag (2003) 160–176
18. Rompel, J.: One-Way Functions are Necessary and Sufficient for Digital Signatures. In: Proceedings of the 22nd ACM Symposium on Theory of Computing (STOC '90). (1990) 12–19
19. Dwork, C., Naor, M.: Zaps and their applications. In: IEEE Symposium on Foundations of Computer Science. (2000) 283–293

Zero-Knowledge Proofs and String Commitments Withstanding Quantum Attacks

Ivan Damgård[1], Serge Fehr[2,*], and Louis Salvail[1]

[1] BRICS, FICS, Aarhus University, Denmark[**]
{ivan,salvail}@brics.dk
[2] CWI, Amsterdam, The Netherlands
fehr@cwi.nl

Abstract. The concept of zero-knowledge (ZK) has become of fundamental importance in cryptography. However, in a setting where entities are modeled by quantum computers, classical arguments for proving ZK fail to hold since, in the quantum setting, the concept of rewinding is not generally applicable. Moreover, known classical techniques that avoid rewinding have various shortcomings in the quantum setting.
We propose new techniques for building *quantum* zero-knowledge (QZK) protocols, which remain secure even under (active) quantum attacks. We obtain computational QZK proofs and perfect QZK arguments for any NP language in the common reference string model. This is based on a general method converting an important class of classical honest-verifier ZK (HVZK) proofs into QZK proofs. This leads to quite practical protocols if the underlying HVZK proof is efficient. These are the first proof protocols enjoying these properties, in particular the first to achieve perfect QZK.
As part of our construction, we propose a general framework for building unconditionally hiding (trapdoor) string commitment schemes, secure against quantum attacks, as well as concrete instantiations based on specific (believed to be) hard problems. This is of independent interest, as these are the first unconditionally hiding string commitment schemes withstanding quantum attacks.
Finally, we give a partial answer to the question whether QZK is possible in the plain model. We propose a new notion of QZK, *non-oblivious verifier* QZK, which is strictly stronger than honest-verifier QZK but weaker than full QZK, and we show that this notion can be achieved by means of efficient (quantum) protocols.

1 Introduction

Since its introduction by Goldwasser, Micali and Rackoff [14], the concept of *zero-knowledge* (ZK) proof has become a fundamental tool in cryptography. In-

---

[*] Research was carried out while at the Centre for Advanced Computing - Algorithms and Cryptography, Department of Computing, Macquarie University, Australia.

[**] BRICS stands for Basic Research in Computer Science (www.brics.dk) and FICS for Foundations in Cryptography and Security, both funded by the Danish Natural Sciences Research Council.

M. Franklin (Ed.): CRYPTO 2004, LNCS 3152, pp. 254–272, 2004.

formally, in a ZK proof of a statement, the verifier learns nothing beyond the validity of the statement. In particular, everything the verifier can do as a result of the interaction with the prover during the ZK proof, the verifier could also do "from scratch", i.e., without interacting with the prover. This is argued by the existence of an efficient *simulator* which produces a simulated transcript of the execution, indistinguishable from a real transcript. ZK protocols exist for any NP language if one-way functions exist [2, 3, 15], also more efficient solutions are known for specific languages like Quadratic-Residuosity [14] or Graph-Isomorphism [15].

From a theoretical point of view, it is natural to ask whether such classical protocols are still secure if cheating players are allowed to run (polynomial time bounded) quantum computers. But the question also has some practical relevance: although quantum computers may not be available to the general public in any foreseeable future, even a single large scale quantum computer could be used to attack the security of existing protocols.

To study this question, two issues are important. First, the computational assumption on which the protocol is based must remain true even if the adversary is quantum. This rules out many assumptions such as hardness of factoring or extracting discrete logs [23], but a few candidates still remain, for instance some problems related to lattices or error correcting codes. In general, it is widely believed that quantum one-way functions exist, i.e., functions that are easy to compute classically, but hard to invert, even on a quantum computer.

A second and more difficult question is whether the proof of security remains valid against a quantum adversary. A major problem in this context comes from the fact that in the classical definition of ZK, the simulator is allowed to *rewind* the verifier in order to generate a simulated transcript of the protocol execution. However, if prover and verifier are allowed to run quantum computers, rewinding is not generally applicable, as it was originally pointed out by Van de Graaf [27]. We discuss this in more detail later, but intuitively, the reason is that when a quantum computer must produce a classical output, such as a message to be sent, a (partial) measurement on its state must be done. This causes an irreversible collapse of the state, so that it is not generally possible to reconstruct the original state. Moreover, copying the verifier's state before the measurement is forbidden by the no-cloning theorem. Therefore, protocols that are proven ZK in the classical sense using rewinding of the verifier may not be secure with respect to a quantum verifier. This severe breakdown of the classical concept of ZK in a quantum world is the motivation of this work.

It is well known that rewinding can cause "problems" already in a classical setting. In particular, it has been realized that rewinding the verifier limits the composability of ZK protocols. As a result, techniques have been proposed that avoid rewinding the verifier, for instance the non-black-box ZK technique from [1], or – in the common reference string model – techniques providing concurrent ZK [13, 22, 9], non-interactive ZK [4] or universally-composable (UC) ZK [5, 6, 11] and related models [21]. One might hope that some of these ideas would translate easily to the quantum setting.

However, the non-black box technique from [1] is based on the simulator using the verifier's program and current state to predict its reaction to a given message. Doing so for a quantum verifier will collapse its state when a measurement is done to determine its next message, so it is not clear that this technique will generalize to a quantum setting. The known constructions of UCZK protocols and non-interactive ZK are all based on computational assumptions that are either false in a quantum setting or for which we have no good candidate for concrete instantiations: the most general sufficient assumption is the existence of one-way trapdoor permutations (i.e. as far as we know) but all known candidates are easy to invert on a quantum computer. Regardless of this type of problem, great care has to be taken with the security proof: despite the fact that the simulator in the UC model must not use rewinding, it is **not** true that a security proof in the UC model automatically implies security against quantum adversaries - we discuss this in more details later in the paper. Finally, the technique for concurrent ZK from [9] avoids rewinding the verifier but instead rewinds the prover to prove soundness, leading to similar problems.

Before describing our results, we note that quantum zero-knowledge proof systems were already studied from a complexity theoretic point of view by Watrous in [26]. The proof systems considered there all assume the prover to be computationally unbounded and the zero-knowledge condition is only enforced against honest verifiers. Clearly, these restrictions make those proof systems unsuitable for cryptographic applications. In this paper, we focus on efficient quantum zero-knowledge protocols in a cryptographic setting.

We propose three distinct techniques applicable to an important class of (classical) honest-verifier ZK (HVZK) proofs (in which the verifier is guaranteed to follow the protocol), namely so-called Σ-protocols (3-move public-coin protocols). We convert such protocols into *quantum zero-knowledge* (QZK) proofs, which are ZK (as well as sound) even with respect to (active) quantum attacks. In all cases, the new proof protocol proceeds in three moves like the underlying Σ-protocol, and its overhead in terms of communication is reasonable. To the best of our knowledge, these are the first (practical) zero-knowledge proofs withstanding active quantum attacks.

The first technique assumes the existence of an unconditionally hiding trapdoor string commitment scheme (secure against quantum attacks) and can be proven secure in the common-reference-string (CRS) model. It requires only classical computation and communication and achieves *perfect or statistical* QZK, assuming the underlying Σ-protocol was perfect or statistical HVZK, and is an interactive argument (computationally sound). The communication overhead of the new QZK protocol in comparison with the underlying Σ-protocol is essentially given by communicating and opening one string commitment. The technique directly implies perfect or statistical QZK arguments for NP.

This first approach requires addressing the problem of constructing unconditionally hiding and computationally binding trapdoor string commitment schemes withstanding quantum attacks. This is non-trivial since the classical definition of computational binding cannot be used for a quantum adversary as it was pointed out in [12] with respect to bit commitments and in [8] with respect to string commitments. In fact, it was not even clear how computational binding for a string commitment should be defined. In [8], a computational binding condition was introduced with their application in mind but no concrete instance was proposed.

We propose a new definition of computational binding that is strong enough for our (and other) applications. On the other hand, we propose a generic construction for schemes satisfying our definition based on special-sound Σ-protocols for hard-to-decide languages, and we give examples based on concrete intractability assumptions. Our construction yields the first unconditionally hiding string commitment schemes withstanding quantum attacks, under concrete as well as under general intractability assumptions. Moreover, since our definition implies the one from [8], our schemes can be used to provide secure quantum oblivious transfer.

The second technique assumes the existence of any quantum one-way function and is also secure in the CRS model. It requires classical communication and computation and produces computational QZK interactive proofs for any NP language. It can be efficiently instantiated under more specific complexity assumptions.

The last technique requires no computational assumption and is provably secure in the *plain model* (no CRS). However, it requires quantum computation and communication and does not achieve full QZK but what we call *non-oblivious verifier QZK*. This new notion is weaker than QZK but strictly stronger than honest-verifier QZK (as defined in [26]). Essentially, a non-oblivious verifier may arbitrarily deviate from the protocol but still generates all private and public classical random variables available to the honest verifier according the same distribution. The (quantum) communication complexity of the non-oblivious verifier QZK proof essentially equals the (classical) communication complexity of the underlying Σ-protocol.

The paper is organized as follows. In Sect. 2, we introduce some relevant notations. We also argue why rewinding causes a problem in a quantum setting and why UCZK does not imply QZK. In Sect. 3, we define and construct the unconditionally hiding (trapdoor) commitment schemes used in Sect. 4 for QZK proofs in the common-reference-string model. Finally, the non-oblivious verifier QZK proof in the plain model is presented in Sect. 5.

Due to space limitations, some descriptions and discussions appear in a shortened form in this proceedings version, they appear in full in the full version [10].

2   Preliminaries

2.1   Zero-Knowledge Interactive Proofs

The Classical Case: We assume the reader to be familiar with the classical notions of (HV)ZK interactive proofs (and arguments) and of (special-sound) Σ-protocols. We merely fix some notation and terminology here. For an introduction to these concepts we refer to the full version of this paper [10] or to the literature.

Let $R = \{(x, w)\}$ be a binary relation. Write $L_R = \{x \mid \exists w : (x, w) \in R\}$ for the language defined by R. For $x \in L_R$, any w such that $(x, w) \in R$ is called a *witness (for $x \in L$)*, and we write $W_R(x) = \{w \mid (x, w) \in R\}$ for the set of witnesses for $x \in L$. We assume that the size of the witnesses for $x \in L$ are polynomially bounded by the size of x, and that R is poly-time testable.

We refer to a Σ-protocol (P, V) for a language L by a triple (a, c, z), where we understand a, c and z as the processes of choosing/computing the first message a, the (random) challenge c and the corresponding answer z, respectively, as specified by the protocol (with some input $x \in L$), and we write $a \leftarrow a$, $c \leftarrow c$ and $z \leftarrow z_x(a, c)$, respectively, for the execution of these processes. Furthermore, we write verify_x for the verification predicate which is applied by V and whose output $\texttt{accept}$ or $\texttt{reject}$, respectively 0 or 1, determines whether V should accept the proof or not. We stress that when considering a computationally bounded (honest) prover P as we do here the answer z is typically not computed by P as a function of a, c and x (as the notation $z \leftarrow z_x(a, c)$ might suggest), but rather as a function of the randomness used to generate a, of the challenge c and of a witness $w \in W_R(x)$. Per default, we understand a Σ-protocol to be *unconditionally* sound. Clearly, for a fixed $x \notin L$, the soundness error ϵ of such a Σ-protocol is given by the maximum over all possible first messages a of the fraction of the possible challenges c for a that allow an answer z which is accepted by V.

It is known that statistical ZK Σ-protocols only exist for languages $L \in$ co-AM. Most of the well-known Σ-protocols are proof-system for languages that are trivial on a quantum computers. However, some languages like graph isomorphism (i.e. GI) have special sound Σ-protocols and are not known to be trivial on a quantum computer. This is also the case for some recently proposed lattice problems [19]. It is not known whether co-AM can be efficiently recognized by a quantum computer.

The Quantum Case: ZK quantum interactive proof systems are defined as the natural generalization of their classical counterpart and were introduced and first studied by Watrous [24, 26]. Quantum ZK (QZK) is defined as for the classical case except that the quantum simulator is required to produce a state that is exponentially close, in the trace-norm sense, to the verifier's view. Formal definitions for QZK proof systems can be found in the full version [10].

2.2   The Problem with Quantum Rewinding

Rewinding a party to a previous state is a common proof technique for showing the security of many different kinds of protocols in the computational model. In general, this technique cannot be applied when the party is modeled by a quantum computer. Originally observed by Van de Graaf [27], this implies that security proofs of many well-established classical protocols do not hold if one party is running a quantum computer even if the underlying assumption under which the security proof holds withstands quantum attacks.

Rewinding is in general not possible since taking a snapshot of a quantum memory is tantamount to quantum cloning. Unlike in the classical case, there is no way to copy a quantum memory regardless of what the memory contains. The only generic way to restore a quantum memory requires to re-generate it from scratch. Proceeding that way may not be possible efficiently.

One consequence of the *no quantum rewinding* paradigm is particularly relevant to us. Sequential repetitions of an HVZK Σ-protocol for a language L results in a ZK protocol for L with negligible soundness error. It follows that this straightforward construction is not guaranteed to be secure against quantum verifiers.

Another example is the use of rewinding for proving secure applications of computationally binding commitment schemes. Such a security proof is done by showing that an attacker that breaks the application can be used to compute two different openings of a commitment and thus to break the binding property of the commitment scheme. This reduction, however, requires typically to rewind of the attacker, and thus by the no quantum rewinding paradigm does not yield a valid security proof in a quantum setting.

More details can be found in the full version [10].

2.3   UCZK Does Not Imply QZK

In [5], Canetti proposes a new framework for defining and proving cryptographic protocols secure: the universal composability (UC) framework. This framework allows to define and prove secure cryptographic protocols as stand-alone protocols, while at the same time guaranteeing security in any application by means of a general composition theorem. The UC security definition essentially requires that the view of any adversary attacking the protocol can be simulated while in fact running an idealized version of the protocol, which essentially consists of a trusted party called *ideal functionality*. The simulation should be indistinguishable for any distinguisher, called *environment*, which may be *on-line*, and provides the inputs and receives the outputs. Furthermore, the UC definition explicitly *prohibits* rewinding of the environment and thus of the adversary (as it may communicate with the environment). This restriction is crucial for the proof of the composition theorem. We refer to [5] for more details.

Since the UC framework forbids rewinding the adversary, it seems that UCZK implies QZK, assuming the underlying computational assumption withstands quantum attacks. This intuition is false in general. The reason being that even though the UC framework does not allow the simulator to rewind the adversary, it is still allowed to use rewinding as a proof-technique in order to show that the simulator produces a "good" simulation. For instance, it is allowed to argue that if an environment can distinguish the simulation from a real protocol execution, then by rewinding the environment together with the adversary one can solve efficiently a problem assumed to be hard. We illustrate this on a concrete example in [10].

3   Unconditionally Hiding (Trapdoor) Commitments

In this section we study and construct classical (trapdoor) commitment schemes secure against quantum attacks. In contrast to quantum commitment schemes, such schemes do not require quantum computation (in order to compute, open or verify commitments), but they are guaranteed to remain secure even under quantum attacks. Our construction, which is based on hard-to-decide languages with special-sound Σ-protocols, yields the first unconditionally hiding string commitment schemes withstanding quantum attacks. In Sect. 4, we use these commitments to construct QZK proofs. A further application of our commitment schemes is given in [10], where it is shown how they give rise to quantumly secure oblivious transfer.

3.1 Defining Security in a Quantum Setting

Informally, by publishing a commitment $C = \mathsf{commit}_{pk}(s, \rho)$ for a random ρ, a *commitment scheme* allows a party to commit to a secret s, such that the commitment C reveals nothing about the secret s (*hiding property*) while on the other hand the committed party can *open* C to s by publishing (s, ρ) *but only to s* (*binding property*).

Formally, a commitment scheme (of the kind we consider) consists of two poly-time algorithms: A key-generation algorithm $\mathcal{G}$ which takes as input the security parameter ℓ and specifies an instance of the scheme by generating a *public-key pk*, and an algorithm commit which allows to compute $C = \mathsf{commit}_{pk}(s, \rho)$ given a public-key pk as well as s and ρ chosen from appropriate finite sets $\mathcal{S}$ and $\mathcal{R}$ (specified by pk). $\mathcal{S}$ is called the *domain* of the commitment scheme. Classically, the hiding property is formalized by the non-existence of a *distinguisher* which is able to distinguish $C = \mathsf{commit}_{pk}(s, \rho)$ from $C = \mathsf{commit}_{pk}(s', \rho')$ with non-negligible advantage, where $s, s' \in \mathcal{S}$ are chosen by the distinguisher and $\rho, \rho' \in \mathcal{R}$ are random. On the other hand, the binding property is formalized by the non-existence of a *forger* able to compute $s, s' \in \mathcal{S}$ and $\rho, \rho' \in \mathcal{R}$ such that $s \neq s'$ but $\mathsf{commit}_{pk}(s, \rho) = \mathsf{commit}_{pk}(s, \rho')$. If the distinguisher respectively the forger is restricted to be poly-time, then the scheme is said to be *computationally* hiding respectively binding, while without restriction on the distinguisher respectively the forger, it is said to be *unconditionally* hiding respectively binding.

In order to define security of such a commitment scheme $(\mathcal{G}, \mathsf{commit})$ in a quantum setting, the (computational or unconditional) hiding property can be adapted in a straightforward manner by allowing the distinguisher to be quantum. The same holds for the *unconditional* binding property, which is equivalent to requiring that every C uniquely defines s such that $C = \mathsf{commit}_{pk}(s, \rho)$ for some ρ. However, adapting the *computational* binding property in a similar manner simply by allowing the forger to be quantum results in a too weak definition. The reason being that in order to prove secure an application of a commitment scheme, which is done by showing that an attacker that breaks the application can be transformed in a black-box manner into a forger that violates the binding property, the attacker typically needs to be rewound, which cannot be justified in a quantum setting by the no-quantum-rewinding paradigm as discussed in Sect. 2.2. The following definition for the computational binding property of a commitment scheme with respect to quantum attacks is strong enough to prove secure applications (as in Sect. 4 and in [10]) based on the security of the underlying commitment scheme, but it is still weak enough in order to prove the binding property for concrete commitment schemes (see Sect. 3.2 and 3.3).

Let $(\mathcal{G}, \mathsf{commit})$ be a commitment scheme as introduced above, and let $\mathcal{S}$ denote its domain. Informally, we require that it is infeasible to produce a list of commitments and then open (a subset of) them in a certain specified way with a probability significantly greater than expected. We formalize this as follows. Let Q be a predicate of the following form. Q takes three inputs: (1) a non-empty set $A \subseteq \{1, \ldots, N\}$ where N is upper bounded by a polynomial in ℓ, (2) a tuple $\boldsymbol{s}_A = (s_i)_{i \in A}$ with $s_i \in \mathcal{S}$, and (3) an element $u \in \mathcal{U}$ where $\mathcal{U}$ is some finite set; and it outputs $Q(A, \boldsymbol{s}_A, u) \in \{0, 1\}$. We do *not* require Q to be efficiently computable. Consider a polynomially bounded quantum forger $\mathcal{F}$ in the following game: $\mathcal{F}$ takes as input pk, generated by $\mathcal{G}$, and announces commitments $C_1, \ldots, C_N$. Then, it is given a random $u \in \mathcal{U}$, and it outputs A, $\boldsymbol{s}_A = (s_i)_{i \in A}$ and $\boldsymbol{\rho}_A = (\rho_i)_{i \in A}$. $\mathcal{F}$ is said to win the game if $Q(A, \boldsymbol{s}_A, u) = 1$ and $C_i = \mathsf{commit}_{pk}(s_i, \rho_i)$ for every $i \in A$. We require that every forger has essentially the same success probability in winning the game as when using an *ideal* (meaning *unconditionally* binding) commitment scheme (where every C_i uniquely defines s_i). In the latter case, the success probability is obviously given by $p_{\text{IDEAL}} = \max_{\boldsymbol{s} \in \mathcal{S}^N} |sat_Q(\boldsymbol{s})| / |\mathcal{U}|$ with $sat_Q(\boldsymbol{s}) = \{u \in \mathcal{U} \mid \exists A : Q(A, \boldsymbol{s}_A, u) = 1\}$, where $\boldsymbol{s}_A$ stands for the restriction of $\boldsymbol{s}$ to its coordinates s_i with $i \in A$. In this definition, Q models a condition that must be satisfied by the opened value in order for the opening to be useful for the committer. For each application scenario, such a predicate can be defined.

Definition 1. *A commitment scheme $(\mathcal{G}, \mathsf{commit})$ is called computational Q-binding if for every predicate Q, every polynomially bounded quantum forger $\mathcal{F}$ wins the above game with probability $p_{\text{REAL}} = p_{\text{IDEAL}} + adv$, where adv, the advantage of $\mathcal{F}$, is (negative or) negligible (in ℓ).*

It is not hard to verify that in a classical setting (where $\mathcal{F}$ is allowed to be rewound), the classical computational binding property is equivalent to the above computational Q-binding property. Furthermore, it is rather obvious that the computational Q-binding property for a commitment scheme with domain $\mathcal{S}$ implies the computational Q-binding property for the natural extension of the scheme to the domain $\mathcal{S}^k$ (for any k) by committing componentwise. Note that this desirable preservation of the binding property does not hold for the binding property introduced in [8].

Finally, we define a *trapdoor* commitment scheme[1] as a commitment scheme in the above sense with the following additional property. Besides the public-key pk, the generator $\mathcal{G}$ also outputs a trapdoor τ which allows to break either the hiding or the binding property. Specifically, if the scheme is unconditionally binding, then τ allows to efficiently compute s from $C = \mathsf{commit}_{pk}(s, \rho)$, and if it is unconditionally hiding, then τ allows to efficiently compute commitments C and correctly open them to any s.

---

[1] Depending on its flavor, a trapdoor commitment scheme is also known as an extractable respectively as an equivocable or a chameleon commitment scheme.

3.2 A General Framework

In this section, we propose a general framework for constructing unconditionally hiding and computationally Q-binding (trapdoor) string commitment schemes. For that, consider a language $L = L_R$ and assume that

1. L admits a (statistical) HVZK *special-sound* Σ-protocol $\Pi = (\mathsf{a}, \mathsf{c}, \mathsf{z})$ [2],
2. there exists an efficient generator $\mathcal{G}_{\mathsf{yes}}$ generating $x \in L$ together with a witness $w \in W_R(x)$ (more precisely, $\mathcal{G}_{\mathsf{yes}}$ takes as input security parameter ℓ and outputs $x \in L$ of bit size ℓ and $w \in W_R(x)$), and
3. for all poly-size quantum circuits $\mathcal{D}$ and polynomials $p(\ell) > 0$, if ℓ is large enough then there exists $x_{\mathsf{no}} \notin L$ of bit size ℓ such that for x_{yes} generated by $\mathcal{G}_{\mathsf{yes}}$ (on input ℓ)

$$\big| \Pr\left(\mathcal{D}(x_{\mathsf{yes}}) = \mathsf{yes}\right) - \Pr\left(\mathcal{D}(x_{\mathsf{no}}) = \mathsf{yes}\right) \big| < 1/p(\ell).$$

Note that 3. only requires that for every *distinguisher* $\mathcal{D}$ it is hard to distinguish a randomly generated yes-instance $x \in L$ from *some* no-instance $x \notin L$, which in particular may depend on $\mathcal{D}$.

Given such L, the construction in Fig. 1 provides an unconditionally hiding trapdoor commitment scheme. We assume that c samples challenge c randomly from $\{0,1\}^t$ for some t.

$\mathcal{G}$ is given by $\mathcal{G}_{\mathsf{yes}}$, where the generated $x \in L$ is parsed as public key pk and $w \in W_R(x)$ as trapdoor τ. The domain $\mathcal{S}$ is defined to be $\mathcal{S} = \{0,1\}^t$.

commit_{pk}: To commit to $s \in \mathcal{S} = \{0,1\}^t$, use the HVZK simulator for Π to generate (a, c, z). Set $C = (a, s \oplus c)$ to be the commitment for s.

A commitment $C = (a, d)$ is opened to s by announcing the corresponding values c and z, and such an opening is accepted if and only if $s \oplus c = d$ and $\mathsf{verify}_x(a, c, z) = \mathsf{accept}$.

Fig. 1. Trapdoor commitment scheme $(\mathcal{G}, \mathsf{commit})$.

If Π is *special* HVZK, meaning that (a, c, z) can be simulated for a given c, then the commitment scheme can be slightly simplified: (a, c, z) is generated such that $c = s$ and C is simply set to be $C = a$.

Theorem 1. *Under assumption 3., $(\mathcal{G}, \mathsf{commit})$ in Fig. 1 is an unconditionally hiding and computationally Q-binding trapdoor commitment scheme.*

---
[2] As will become clear, the prover's efficiency in the Σ-protocol does not influence the efficiency of the resulting commitment scheme as far as the committer and the receiver are concerned. An efficient prover is only required if one wants to take advantage of the trapdoor.

As will become clear from the proof below, if the underlying Σ-protocol Π is *perfect* HVZK, then $(\mathcal{G}, \mathsf{commit})$ is *perfectly* binding in the sense that there exists no distinguisher with *non-zero* advantage, meaning that a commitment C for s is statistically independent of s.

Proof. It is clear that a correct opening is accepted. It is also rather obvious that the scheme is unconditionally hiding: The distribution of (a, c, z) generated by the HVZK simulator is statistically close to the distribution of (a, c, z) generated by the protocol. There, however, c is chosen independently of a. Therefore, a gives essentially no information on c and thus $C = (a, s \oplus c)$ gives essentially no information on s (as $s \oplus c$ acts as a one-time pad). The trapdoor property can be seen as follows. Knowing the trapdoor $\tau = w$, put $C = (a, d)$ where $a \leftarrow \mathsf{a}$ and d is randomly sampled from $\{0, 1\}^t$. Given arbitrary $s \in \{0, 1\}^t$, compute $c = d \oplus s$ and $z \leftarrow \mathsf{z}_x(a, c)$ using the witness w (and the randomness for the generation of a). It is obvious that (s, c, z) opens C correctly to s.

It remains to show the computational Q-binding property. We show that if there exists a forger $\mathcal{F}$ that can break the Q-binding property of the commitment scheme (without knowing the trapdoor) for some predicate Q according to Definition 1, then there exists a circuit $\mathcal{D}$ that contradicts assumption 3. $\mathcal{D}$ is illustrated in Figure 2 and is quantum if and only if $\mathcal{F}$ is.

$\mathcal{D}$: The input is x, either in L or not in L.

1. Invoke $\mathcal{F}$ with public-key $pk = x$ in order to get commitments $C_1, \ldots, C_N$,
2. Pick random $u \in \mathcal{U}$ and announce it to $\mathcal{F}$,
3. $\mathcal{F}$ announces $A \subseteq \{1, \ldots, N\}$ and, for $i \in A$, tries to open C_i to s_i such that $Q(A, s_A, u) = 1$ for $s_A = (s_i)_{i \in A}$,
4. Verify the openings and whether indeed $Q(A, s_A, u) = 1$, if successful then output **yes** and otherwise **no**.

Fig. 2. Distinguisher $\mathcal{D}$ for $x \in L$ versus $x \notin L$.

If x is generated by $\mathcal{G}_{\mathsf{yes}}$ then $pk = x$ is a valid public-key for the commitment scheme with the right distribution and thus $\Pr(\mathcal{D}(x) = \mathsf{yes}) = p_{\mathrm{REAL}} = p_{\mathrm{IDEAL}} + adv$ where adv is $\mathcal{F}$'s advantage. On the other hand, if $x \notin L$, then by the special soundness property of Π, given a there is only one c that allows an answer z such that $\mathsf{verify}_x(a, c, z) = \mathsf{accept}$. Hence, for any C_i there is only one $s_i \in \mathcal{S}$ to which C_i can be successfully opened. Therefore, $\Pr(\mathcal{D}(x) = \mathsf{yes}) \leq p_{\mathrm{IDEAL}}$. If adv is (positive and) non-negligible, then this contradicts 3. $\square$

We would like to point out once more that our definition of the (computational) binding property inherits the following feature. If a commitment scheme with domain $\mathcal{S}$ is computational Q-binding, then its natural extension to a commitment scheme with domain $\mathcal{S}^k$ by committing componentwise (with the same pk) is also computational Q-binding. In particular, any computational Q-binding *bit* commitment scheme gives rise to a computational Q-binding *string* commitment scheme.

3.3   Concrete Instantiations

We propose three concrete languages which are believed to be hard to decide as required in the above section and which admit HVZK special-sound Σ-protocols. The first language is based on a problem from coding theory: the Code-Equivalence (CE) problem. It requires to decide whether two generator matrices generate the same code up to a permutation of the coordinates, and it is known to be at least as hard (in the worst case) as the Graph-Isomorphism (GI) problem. Furthermore, it admits a similar Σ-protocol as GI. Finally, and in contrast to GI, there is a generator believed to produce hard yes-instances. More details are given in [10].

The next two languages are gap versions of the famous lattice problems Shortest-Vector and Closest-Vector, where the no-instances are promised to be "not too close" to the yes-instances. Σ-protocols for these problems were recently proposed in [19], where the generation of hard instances is also addressed. Again, more details are given in [10].

These languages give rise to concrete instantiations of the commitment scheme developed in the above section, based on concrete computational assumptions.

4   Quantum Zero-Knowledge Proofs

4.1   Common-Reference-String Model

The *common-reference-string* (CRS) model assumes that there is a string σ (honestly) generated according to some distribution and available to all parties from the start of the protocol. In the CRS model, an interactive proof (or argument) is (Q)ZK if there exists a simulator which can simulate the (possibly dishonest) verifier's view of the protocol execution together with a CRS σ having correct joint distribution as in a real execution.

4.2   Efficient QZK Arguments

We show how to convert any HVZK Σ-protocol into a quantum zero-knowledge (QZK) argument. The construction is based on a trapdoor commitment scheme and can be proven secure in the CRS model.

It is actually very simple. P and V simply execute the Σ-protocol, but instead of sending message a in the first move, P sends a *commitment* to a, which he then opens when he sends the answer z to the challenge c in the third move. The zero-knowledge property then follows essentially by observing that the simulator (who knows the trapdoor of the commitment scheme) can cheat in the opening of the commitment. So far, the strategy for the QZK proof is the same as in Damgård's concurrent ZK proof [9]; the proof of soundness however will be different since [9] requires to rewind the prover, which cannot be justified in our case by the no-quantum-rewinding paradigm. In order not to rely on the *special* HVZK property (as introduced and explained in Sect. 3.2), the protocol is slightly more involved than sketched here, though the idea remains.

Let a HVZK Σ-protocol $\Pi = (\mathsf{a}, \mathsf{c}, \mathsf{z})$ for a language $L = L_R$ be given. Let ϵ denote its soundness error. We assume without loss of generality that a and c sample first messages a and challenges c of fixed bit lengths r and t, respectively. Furthermore, let an unconditionally hiding and computationally Q-binding trapdoor commitment scheme $(\mathcal{G}, \mathsf{commit})$ be given (where the knowledge of the trapdoor allows to break the binding property of the scheme). We assume that its domain $\mathcal{S}$ contains $\{0,1\}^{r+t}$. Consider Protocol 1 illustrated in Fig. 3.

Protocol 1: V has input x, claimed to be in L; P has input x and $w \in W_R(x)$. The CRS is set to be pk where pk is generated by $\mathcal{G}$.

1. P computes $a \leftarrow \mathsf{a}$ and chooses $c_P \leftarrow \mathsf{c}$. Then it commits to the concatenation $a\|c_P$ of a and c_P by $C = \mathsf{commit}_{pk}(a\|c_P, \rho)$, and sends C to V.
2. V chooses $c_V \leftarrow \mathsf{c}$ and sends it to P.
3. P computes $z \leftarrow \mathsf{z}_x(a, c)$ for $c = c_P \oplus c_V$ and sends (a, c_P, ρ) and z to V.
4. V accepts iff $C = \mathsf{commit}(a\|c_P, \rho)$ and $\mathsf{verify}_x(a, c_P \oplus c_V, z) = \mathsf{accept}$.

Fig. 3. QZK proof protocol in the CRS model.

As mentioned above, Protocol 1 can be slightly simplified in case Π is *special* HVZK in that P commits to a (rather than to $a\|c_P$) and computes z with respect to the challenge $c = c_V$ provided by V.

Theorem 2. *Under the assumption that $(\mathcal{G}, \mathsf{commit})$ is an unconditionally hiding and computationally Q-binding trapdoor commitment scheme, Protocol 2 is a QZK (quantum) argument for L in the CRS model. Its soundness error is $\epsilon' = \epsilon + negl$ where negl is negligible (in the security parameter).*

Concerning the flavor of QZK, Protocol 2 is *computational* QZK if the underlying Σ-protocol Π is computational HVZK, and it is *statistical* QZK provided that Π is statistical or perfect HVZK. In case $(\mathcal{G}, \mathsf{commit})$ is perfectly (rather than unconditionally) hiding, the flavor of QZK of Protocol 2 is exactly given by the flavor of HVZK of Π.

Proof. As mentioned above, the zero-knowledge property is rather straight forward: The simulator generates a public-key for the commitment scheme together with a trapdoor and outputs the public-key as CRS. Then, on input $x \in L$, it generates a commitment C (which he can open to an arbitrary value using the trapdoor) and sends it to $\tilde{\mathsf{V}}$. On receiving c_V from $\tilde{\mathsf{V}}$, the simulator simulates an accepting conversation (a, c, z) for the original Σ-protocol using the HVZK property, it sets $c_P = c \oplus c_V$ and computes ρ such that $C = \mathsf{commit}(a\|c_P, \rho)$ using the trapdoor, and it sends (a, c_P, ρ) and z to $\tilde{\mathsf{V}}$.

For the soundness property, it has to be shown that given a (quantum) prover $\tilde{\mathsf{P}}$, which succeeds in making (honest) V accept the proof for an $x \notin L$ with a probability exceeding ϵ by a non-negligible amount, $\tilde{\mathsf{P}}$ can be used to break the Q-binding property of the commitment scheme for some predicate Q. Fix $x \notin L$. We define Q as follows. $N = 1$, and $\mathcal{U}$ is given by the set of all possible challenges c_V sampled by c. For $s \in S$ and $u = c_V \in \mathcal{U}$, where s is parsed as $s = a\|c_P$ with $a \in \{0,1\}^r$ and $c_P \in \{0,1\}^t$, we set $Q(\{1\}, s, u) = 1$ if and only if the challenge $c = c_P \oplus c_V$ for the first message a allows an answer z such that $\mathsf{verify}_x(a, c, z) = \mathsf{accept}$. Note that $A = \{1\}$ is the only legitimate choice for A. By construction of Q, making V accept the proof means that $\tilde{\mathsf{P}}$ opens C (correctly) to $a\|c_P$ such that $Q(\{1\}, a\|c_P, c_V) = 1$. Furthermore, $p_{\text{IDEAL}} = \epsilon$. It follows that if $\tilde{\mathsf{P}}$ succeeds in making V accept the proof with probability greater that ϵ by a non-negligible amount, then $\tilde{\mathsf{P}}$ is a forger $\mathcal{F}$ that breaks the Q-binding property of $(\mathcal{G}, \mathsf{commit})$. This completes the proof.   □

4.3   QZK Arguments for All of NP

Consider a (generic) ZK argument for an NP-complete language using (ordinary) unconditionally hiding commitments. For instance, consider the classical interactive proof for Circuit-Satisfiability due to Brassard, Chaum and Crépeau [3]: the prover "scrambles" the wires and the gates' truth tables of the circuit and commits upon it, and he answers the challenge $c = 0$ by opening all commitments and showing that the scrambling is done correctly and the challenge $c = 1$ by opening the (scrambled) wires and rows of the gates' truth tables that are activated by the satisfying input. Following the lines of the proof of Theorem 2 above, it is straightforward to prove that replacing the commitment scheme in this construction by an unconditionally hiding and computationally Q-binding commitment scheme results in a QZK argument in the CRS model for Circuit-Satisfiability, and thus for all languages in NP.

4.4   Computational QZK Proofs

We sketch how to construct rather efficient computational QZK proofs for languages that allow (computational) HVZK Σ-protocols based on specific intractability assumptions, as well as computational QZK proofs for all of NP based on any quantum one-way function.

Consider any of the languages $L = L_R$ with HVZK Σ-protocol on which the commitment construction from Sect. 3.2 is based, except that we allow the Σ-protocol to be *computational* HVZK. Assume in addition that there is also a generator $\mathcal{G}_{\text{no}}$ that produces no-instances that cannot be distinguished from the yes-instances produced by $\mathcal{G}_{\text{yes}}$.

Then, put a no-instance x_{no} in the reference string. The prover can now prove any statement S that can be proved by an HVZK Σ-protocol Π by using a standard witness-indistinguishable HVZK proof for proving that S is true or $x_{no} \in L$ [7]. Here, we allow the Σ-protocol Π to be *computational* HVZK, in particular Π might be the Σ-protocol for Circuit-Satisfiability sketched in Sect. 4.3 above but based on an unconditionally binding and computationally hiding commitment scheme (secure against quantum attacks), which can be constructed from any (quantum) one-way function (see below).

This is clearly unconditionally sound, and can be simulated, where the simulator uses a yes-instance x_{yes} in place of x_{no} and uses its witness $w \in W_R(x_{yes})$ to complete the protocol without rewinding. A distinguisher would have to contradict the HVZK property of one of the underlying Σ-protocols, or the indistinguishability of yes- and no-instances.

This can be instantiated efficiently if we are willing to assume about the coding or lattice problem or some other candidate problem that it also satisfies this stronger version of indistinguishability of yes- and no-instances. But it can also be instantiated in a version that can be be based on any one-way function: First, the (unconditionally binding and computationally hiding) commitment scheme of Naor [20] is also secure against quantum adversaries, and exists if any one-way function exists. So consider the language of pairs (pk, O) where pk is a public-key for the commitment scheme and O is a commitment of 0. This language has a computational HVZK Σ-protocol using generic ZK techniques, driven by Naor's commitments. Furthermore, the set of no-instances (pk, E) where E is a commitment to 1 is easy to generate and hard to distinguish from the yes-instances.

5   Relaxed Honest-Verifier Quantum Proofs

It is a natural question whether QZK proof systems exist without having to rely upon common reference strings. In this section, we answer this question partially. We define a quantum interactive proof system associated to any Σ-protocol. Our scheme is QZK against a relaxed version of honest verifiers that we call non-oblivious. Intuitively, a non-oblivious verifier is a verifier having access to the same classical variables than the honest verifier. We show that any HVZK Σ-protocol can be turned into a non-oblivious verifier QZK proof using quantum communication.

5.1   Quantum Circuits for Σ-Protocols

Assume $L = L_R$ has a classical HVZK Σ-protocol $\Pi = (a, c, z)$. We specify unitary transforms $Z_x(a)$, and $T_x(a)$, depending on $a \leftarrow a$, which implement quantum versions of the computations specified by z and verify. Throughout, we assume without loss of generality that c samples c uniformly from $\{0, 1\}^t$ for some t.

The answer $z \leftarrow z_x(a, c)$ to challenge c when a was announced during the first round can be computed quantumly through some unitary transform $Z_x(a)$ depending upon the initial announcement a. That is, provided quantum registers P and X, we have:

$$Z_x(a) : |c\rangle^P |y\rangle^X \mapsto |c\rangle^P |y \oplus z_x(a, c)\rangle^X.$$

Similarly, the testing process performed by V can also be executed by a quantum circuit $T_x(a)$ depending on the announcement of a. Transformation $T_x(a)$ stores the output of the verification process in an extra one-qubit register T:

$$T_x(a) : |z\rangle^X |c\rangle^V |t\rangle^T \mapsto |z\rangle^X |c\rangle^V |t \oplus \mathsf{verify}_x(a, c, z)\rangle^T.$$

If $z \leftarrow \mathsf{z}_x(a, c)$ and $\mathsf{verify}_x(a, c, z)$ can be classically computed in polynomial time (given the randomness of the computation of a and a witness $w \in W_R(x)$ for the former), circuits $Z_x(a)$ and $T_x(a)$ can be implemented by poly-size quantum circuits.

5.2   EPR-Pairs Based Proofs

The idea behind the protocol is as follows. P chooses $a \leftarrow \mathsf{a}$ and sends the answer to all possible challenges in quantum superposition to V. V then verifies quantumly that all answers in the superposition are correct. In a further step, P convinces V that the state contains the answer to more than one challenge. Since Π is assumed to be special sound, it follows that $x \in L$.

Concretely, P starts by choosing $a \leftarrow \mathsf{a}$ and by preparing t EPR pairs in state:

$$|\Omega_t\rangle^{P,V} = 2^{-t/2} \sum_{c \in \{0,1\}^t} |c\rangle^P |c\rangle^V = 2^{-t/2} \sum_{c \in \{0,1\}^t} |c\rangle^P_\times |c\rangle^V_\times. \tag{1}$$

The two equivalent ways of writing $|\Omega_t\rangle$ shows that it exhibits the same correlation between registers P and V in both the computational and the diagonal bases. This property will be used later in the protocol. Now, P adds an extra register X initially in state $|0\rangle^X$ before applying $Z_x(a)$ upon registers P and X. This results in state,

$$|\psi_a\rangle = 2^{-t/2} \sum_{c \in \{0,1\}^t} Z_x(a) |c\rangle^P |0\rangle^X \otimes |c\rangle^V = 2^{-t/2} \sum_{c \in \{0,1\}^t} |c\rangle^P |z\rangle^X \otimes |c\rangle^V, \tag{2}$$

where every z in the superposition is computed as $z \leftarrow \mathsf{z}_x(a, c)$. P then announces a and sends registers V and X to V allowing him to apply the verification quantum circuit $T_x(a)$ after adding an extra register T initially in state $|0\rangle^T$. That is,

$$|\psi_a^T\rangle = (\mathbb{I}_P \otimes T_x(a)) |\psi_a\rangle |0\rangle^T = 2^{-t/2} \sum_{c \in \{0,1\}^t} |c\rangle^P \otimes T_x(a) |z\rangle^X |c\rangle^V |0\rangle^T$$

$$= 2^{-t/2} \sum_{c \in \{0,1\}^t} |c\rangle^P \otimes |z\rangle^X |c\rangle^V |\mathsf{verify}_x(a, c, z)\rangle^T = |\psi_a\rangle \otimes |0\rangle^T.$$

V then measures register T in the computational basis and rejects if $|0\rangle^T$ is not observed. Provided P was honest, the test will always be successful by assumption on the original Σ-protocol Π, and the verification process does not affect the state $|\psi_a\rangle$. V then returns register X back to P, who can recover t shared EPR pairs by running $Z_x(a)^\dagger$, the inverse of $Z_x(a)$. Finally, P measures register P in the diagonal basis and announces the outcome to V. V does the same to register V and verifies that the same outcome is obtained. By the properties of EPR pairs (1), it follows that the measurements coincide provided P was honest. A compact description of the protocol is given by Protocol 2 in Fig. 4.

Protocol 2: V has input x, claimed to be in L; P has input x and $w \in W_R(x)$.

1. P computes a first message $a \leftarrow$ a and prepares the quantum state $|\psi_a\rangle^{P,X,V} = 2^{-t/2} \sum_c |c\rangle^P |z\rangle^X |c\rangle^V$ as in (2) where $z \leftarrow z_x(a,c)$, and he sends a and the registers X and V to V,
2. V runs the verification circuit $T_x(a)$ and rejects if a non-zero outcome is obtained. If the test was successful then V returns register X to P,
3. P runs $(Z_x(a)^\dagger \otimes \mathbb{I}_V)|\psi_a\rangle = |\Omega_t\rangle^{P,V} \otimes |0\rangle^X$, measures the P register in the diagonal basis and announces the outcome $c_P \in \{0,1\}^t$ to V.
4. V accepts iff register V measured in the diagonal basis produces outcome $c_V = c_P$.

Fig. 4. Non-oblivious verifier QZK proof.

5.3 Soundness

Consider $x \notin L$. We show that in Protocol 2, any prover $\tilde{P}$ has probability at most 2^{-t} to convince V, given that Π is special sound. Let a be announced by $\tilde{P}$ at step 1. By the special soundness property of Π, if $\tilde{P}$ passes the test at step 2. then the state shared between $\tilde{P}$ and V is of the following form: $|\tilde{\psi}_a\rangle = |\gamma_{a,x}\rangle^{P,X} \otimes |c\rangle^V |0\rangle^T$, where c is the unique challenge that can be answered given the announcement of a. Since after register X has been sent back to $\tilde{P}$, register V is in pure state, it follows that only one answer is possible when V is measured in the computational basis. That is, $|c\rangle$ is guaranteed to be observed. However, V's final test involves a measurement of that same register in the diagonal basis, and it is easy to see that the outcome of a measurement in the diagonal basis applied to $|c\rangle$ is uniformly distributed over $\{0,1\}^t$. This is a special case of the entropic uncertainty relations [18]. It follows:

Theorem 3. *If Π is a special-sound HVZK Σ-protocol for language $L = L_R$ where c samples in $\{0,1\}^t$, then Protocol 2 is a quantum interactive proof for L with soundness error 2^{-t}.*

It should be mentioned that Π being special sound is not a strict necessary condition for Protocol 2 to be sound. A more careful analysis can handle the case where Π is "not too far away" from special sound. For simplicity, in this paper we only address the case of special sound Σ-protocols.

5.4 Non-oblivious Verifier Quantum Zero-Knowledge

Classical Σ-protocols with large challenges are not known to be ZK against a dishonest verifier. This is due to the fact that rewinding allows the simulator to succeed only if it has a non-negligible probability to guess the challenge that the verifier will pick. This is true even with respect to verifiers that submit a uniformly distributed challenge $c \in \{0,1\}^t$ and are able to do the verification test as prescribed. To see this, let $\sigma : \{0,1\}^\ell \to \{0,1\}^\ell$ be a one-way permutation and let us assume for simplicity that $t = \ell$ and a samples a from $\{0,1\}^t$. If $\tilde{V}$ announces challenge $c = a \oplus \sigma(m)$ for random $m \in \{0,1\}^\ell$ and $a \leftarrow a$ announced by P as first message, then the simulator must generate (a, c, z, m) since it is part of $\tilde{V}$'s view. However, the simulator typically can compute a only after having picked c, which means that it has to compute m as $m = \sigma^{-1}(c \oplus a)$. Note that even though $c \oplus a$ is not necessarily uniformly distributed, it seems that the simulator has typically not enough control over the value $c \oplus a$ in order to compute m.

Notice that a verifier $\tilde{V}$ acting as described above rejects a false statement with the same probability and chooses the challenge c with the same distribution as an honest verifier, yet there is no known efficient simulator for $\tilde{V}$. In this section we show that Protocol 2 is quantum zero-knowledge provided that $\tilde{V}$ is non-oblivious of the value c_V needed for the verification at step 4. More generally, we define non-oblivious verifiers the following way:

Definition 2. *A verifier $\tilde{V}$ is said to be* non-oblivious *if it produces the same (public and private) variables as honest V according the same distribution.*

As illustrated above, in contrast to an honest verifier a non-oblivious verifier can produce his variables in an arbitrary manner, as long as they are correctly distributed.

In Protocol 2, a non-oblivious verifier $\tilde{V}$ has access to the string c_V so it can be made available to the simulator. Indeed, this allows to produce a simulation of the interaction between P and $\tilde{V}$. It is straightforward to verify that the simulator described in Fig. 5 generates the same view as when $\tilde{V}$ interacts with P:

Simulator: Input is $x \in L$.

1. Run the HVZK simulator for Π in order to get triplet (a, c, z), and send a together with the quantum state $|c\rangle|z\rangle$ to $\tilde{V}$,
2. If $\tilde{V}$ rejects P then halt, otherwise throw away the state sent by $\tilde{V}$,
3. Extract c_V using the non-obliviousness of $\tilde{V}$ and announce $c_P = c_V$.

Fig. 5. Simulator for Protocol 2.

Theorem 4. *Protocol 2 built from a special-sound (statistical/perfect) HVZK Σ-protocol Π is (statistical/perfect) QZK provided $\tilde{V}$ is non-oblivious.*

A weaker assumption about $\tilde{V}$'s behavior would be obtained if the only constraint was that $\tilde{V}$ detects false statements with the same probability as the honest verifier V. Let us say that such a verifier is *verification-enabled*. In general, a verification-enabled verifier $\tilde{V}$ is not necessarily non-oblivious since in order to verify $\tilde{P}$'s announcement, c_P does not necessarily have to be determined by $\tilde{V}$ without P's help. However, it can be shown that for Σ-protocols with challenges of polylogarithmic size, any verification-enabled $\tilde{V}$ in Protocol 2 is also non-oblivious.

Acknowledgements

The authors are grateful to Claude Crépeau for having introduced the problem to one of us and discussed its relevance. We would also like to thank Jesper Nielsen for enlightening discussions.

References

1. BARAK, B., *How to Go Beyond the Black-box Simulation Barrier*, in 42th Annual Symposium on Foundations of Computer Science (FOCS), 2001.
2. BRASSARD, G., and C. CRÉPEAU, *Zero-Knowledge Simulation for Boolean Circuits*, in Advances in Cryptology - CRYPTO 86, Lecture Notes in Computer Science, vol. 263, Springer-Verlag, 1987.
3. BRASSARD, G., D. CHAUM, and C. CRÉPEAU, *Minimum Disclosure Proofs of Knowledge*, JCSS, 37(2), 1988.
4. BLUM, M., P. FELDMAN and S. MICALI, *Non-Interactive Zero-Knowledge and Its Applications*, in 20th Annual Symposium on Theory Of Computing (STOC), 1988.
5. CANETTI, R., *Universally Composable Security: A New Paradigm for Cryptographic Protocols*, in 42th Annual Symposium on Foundations of Computer Science (FOCS), 2001.
6. CANETTI, R., and M. FISCHLIN, *Universally Composable Commitments*, in Advances in Cryptology - CRYPTO 01, Lecture Notes in Computer Science, vol. 2139, Springer-Verlag, 2001.
7. CRAMER, R., I. DAMGÅRD, and B. SCHOENMAKERS, *Proofs of Partial Knowledge and Simplified Design of Witness Hiding Protocols*, in Advances in Cryptology - CRYPTO 94, Lecture Notes in Computer Science, vol. 839, Springer-Verlag, 1994.
8. CRÉPEAU, C., P. DUMAIS D. MAYERS and L. SALVAIL, *Computational Collapse of Quantum State with Application to Oblivious Transfer*, in Advances in Cryptology – TCC 04, Lecture Notes in Computer Science, vol. 2951, Springer-Verlag, 2004.
9. DAMGÅRD, I., *Efficient Concurrent Zero-Knowledge in the Auxiliary String Model*, in Advances in Cryptology - EUROCRYPT 00, Lecture Notes in Computer Science, vol. 1807, Springer-Verlag, 2000.
10. DAMGÅRD, I.,S. FEHR, and L. SALVAIL, *Zero-Knowledge Proofs and String Commitments Withstanding Quantum Attacks*, full version of this paper, BRICS report nr. RS-04-9, available at www.brics.dk/RS/04/9, 2004.
11. DAMGÅRD, I., and J. NIELSEN, *Perfect Hiding and Perfect Binding Universally Composable Commitment Schemes with Constant Expansion Factor*, in Advances in Cryptology - CRYPTO 02, Lecture Notes in Computer Science, vol. 2442, Springer-Verlag, 2002.
12. DUMAIS, P., D. MAYERS, and L. SALVAIL, *Perfectly Concealing Quantum Bit Commitment From Any Quantum One-Way Permutation*, in Advances in Cryptology - EUROCRYPT 00, Lecture Notes in Computer Science, vol. 1807, Springer-Verlag, 2000.

13. DWORK, C., M. NAOR, and A. SAHAI, *Concurrent Zero-Knowledge*, in 30th Annual Symposium on Theory Of Computing (STOC), 1998.
14. GOLDWASSER, S., S. MICALI, and C. RACKOFF, *The Knowledge Complexity of Interactive Proof Systems*, in 17th Annual Symposium on Theory Of Computing (STOC), 1985.
15. GOLDREICH, O., S. MICALI, and A. WIGDERSON, *Proofs that Yield Nothing but their Validity, or All Languages in NP Have Zero-Knowledge Proof Systems*, J. ACM., 38(3), 1991.
16. FIAT, A., and A. SHAMIR, *How to Prove Yourself: Practical Solutions to the Identification and Signature Problem*, in Advances in Cryptology - CRYPTO 86, Lecture Notes in Computer Science, vol. 263, Springer-Verlag, 1987.
17. KITAEV, A., and J. WATROUS, *Parallelization, Amplification, and Exponential Time Simulation of Quantum Interactive Proof Systems*, in 32nd Annual Symposium on Theory of Computing (STOC), 2000.
18. MAASSEN, H., and J.B.M. UFFINK, *Generalized Entropic Uncertainty Relations*, Phys. Rev. Letters, vol. 60, 1988.
19. MICCIANCIO, D., and S. P. VADHAN, *Statistical Zero-Knowledge Proofs with Efficient Provers: Lattice Problems and More*, in Advances in Cryptology - CRYPTO 03, Lecture Notes in Computer Science, vol. 2729, Springer-Verlag, 2003.
20. NAOR, M., *Bit Commitment Using Pseudorandomness*, Journal of Cryptology, vol. 4, no. 2, 1991.
21. PFITZMANN, B., and M. WAIDNER,*Composition and Integrity Preservation of Secure Reactive Systems*, in 7th ACM Conference on Computer and Communications Security, 2000.
22. RICHARDSON, R. and J. KILIAN, *On the Concurrent Composition of Zero-Knowledge Proofs*, in Advances in Cryptology - EUROCRYPT 99, Lecture Notes in Computer Science, vol. 1592, Springer-Verlag, 1999.
23. SHOR, P., *Algorithms for Quantum Computation: Discrete Logarithms and Factoring*, in 35th Annual Symposium on Foundations of Computer Science (FOCS), 1994.
24. WATROUS, J,*PSPACE has Constant-Round Quantum Interactive Proof Systems*, in 40th Annual Symposium on Foundations of Computer Science (FOCS), 1999.
25. WATROUS, J.,*Succinct Quantum Proofs for Properties of Finite Groups*, Proceedings of the 41st Annual Symposium on Foundations of Computer Science, 2000.
26. WATROUS, J., *Limits on the Power of Quantum Statistical Zero-Knowledge*, in 43rd Annual Symposium on the Foundations of Computer Science (FOCS), 2002.
27. VAN DE GRAAF, J., *Towards a Formal Definition of Security for Quantum Protocols*, Ph.D. thesis, Computer Science and Operational Research Department, Université de Montréal, 1997.

The Knowledge-of-Exponent Assumptions and 3-Round Zero-Knowledge Protocols

Mihir Bellare and Adriana Palacio

Dept. of Computer Science & Engineering, University of California, San Diego
9500 Gilman Drive, La Jolla, CA 92093, USA
{mihir,apalacio}@cs.ucsd.edu
http://www-cse.ucsd.edu/users/{mihir,apalacio}

Abstract. Hada and Tanaka [11, 12] showed the existence of 3-round, negligible-error zero-knowledge arguments for NP based on a pair of non-standard assumptions, here called KEA1 and KEA2. In this paper we show that KEA2 is false. This renders vacuous the results of [11, 12]. We recover these results, however, under a suitably modified new assumption called KEA3. What we believe is most interesting is that we show that it is possible to "falsify" assumptions like KEA2 that, due to their nature and quantifier-structure, do not lend themselves easily to "efficient falsification" (Naor [15]).

1 Introduction

A classical question in the theory of zero knowledge (ZK) [10] is whether there exist 3-round, negligible-error ZK proofs or arguments for NP. The difficulty in answering this question stems from the fact that such protocols would have to be non-black-box simulation ZK [9], and there are few approaches or techniques to this end. A positive answer has, however, been provided, by Hada and Tanaka [11, 12]. Their result (a negligible-error, 3-round ZK argument for NP) requires a pair of non-standard assumptions that we will denote by KEA1 and KEA2.

THE ASSUMPTIONS, ROUGHLY. Let q be a prime such that $2q + 1$ is also prime, and let g be a generator of the order q subgroup of Z^*_{2q+1}. Suppose we are given input q, g, g^a and want to output a pair (C, Y) such that $Y = C^a$. One way to do this is to pick some $c \in \mathbb{Z}_q$, let $C = g^c$, and let $Y = (g^a)^c$. Intuitively, KEA1 can be viewed as saying that this is the "only" way to produce such a pair. The assumption captures this by saying that any adversary outputting such a pair must "know" an exponent c such that $g^c = C$. The formalization asks that there be an "extractor" that can return c. Roughly:

KEA1: For any adversary **A** that takes input q, g, g^a and returns (C, Y) with $Y = C^a$, there exists an "extractor" **Ā**, which given the same inputs as **A** returns c such that $g^c = C$.

Suppose we are given input q, g, g^a, g^b, g^{ab} and want to output a pair (C, Y) such that $Y = C^b$. One way to do this is to pick some $c \in \mathbb{Z}_q$, let $C = g^c$,

M. Franklin (Ed.): CRYPTO 2004, LNCS 3152, pp. 273–289, 2004.

and let $Y = (g^b)^c$. Another way is to pick some $c \in \mathbb{Z}_q$, let $C = (g^a)^c$, and let $Y = (g^{ab})^c$. Intuitively, KEA2 can be viewed as saying that these are the "only" ways to produce such a pair. The assumption captures this by saying that any adversary outputting such a pair must "know" an exponent c such that either $g^c = C$ or $(g^a)^c = C$. The formalization asks that there be an "extractor" that can return c. Roughly:

KEA2: For any adversary **A** that takes input q, g, g^a, g^b, g^{ab} and returns (C, Y) with $Y = C^b$, there exists an "extractor" $\bar{\mathbf{A}}$, which given the same inputs as **A** returns c such that either $g^c = C$ or $(g^a)^c = C$.

As per [11, 12], adversaries and extractors are poly-size families of (deterministic) circuits. See Assumption 2 for a formalization of KEA2, and Assumption 4 for a formalization of KEA1.

HISTORY AND NOMENCLATURE OF THE ASSUMPTIONS. KEA1 is due to Damgård [7], and is used by [11,12] to prove their protocol is ZK. To prove soundness of their protocol, Hada and Tanaka [11,12] introduce and use KEA2. (In addition, they make the Discrete Logarithm Assumption, DLA.) The preliminary version of their work [11] referred to the assumptions as SDHA1 and SDHA2 (**S**trong **D**iffie-**H**ellman **A**ssumptions 1 and 2), respectively. However, the full version [12] points out that the formalizations in the preliminary version are flawed, and provides corrected versions called non-uniform-DA1 and non-uniform-DA2. The latter are the assumptions considered in this paper, but we use the terminology of Naor [15] which we feel is more reflective of the content of the assumption: "KEA" stands for "**K**nowledge of **E**xponent **A**ssumption", the exponent being the value c above.

FALSIFYING KEA2. In this paper we show that KEA2 is false. What is interesting about this —besides the fact that it renders the results of [11, 12] vacuous— is that we are able to "falsify" an assumption whose nature, as pointed out by Naor [15], does not lend itself easily to "efficient falsification." Let us explain this issue before expanding more on the result itself.

The most standard format for an assumption is to ask that the probability that an adversary produces a certain output on certain inputs is negligible. For example, the Factoring assumption is of this type, asking that the probability that a polynomial-time adversary can output the prime factors of an integer (chosen by multiplying a pair of random primes) is negligible. To show such an assumption is false, we can present an "attack," in the form of an adversary whose success probability is not negligible. (For example, a polynomial-time factoring algorithm.) KEA1 and KEA2 are not of this standard format. They involve a more complex quantification: "For every adversary there exists an extractor such that ...". To show KEA2 is false, we must show there is an adversary *for which there exists no extractor*. As we will see later, it is relatively simple to identify an adversary for which there does not *appear* to exist an extractor, but how can we actually show that none of the infinite number of possible extractors succeeds?

AN ANALOGY. The difficulty of falsifying an assumption with the quantifier format of KEA2 may be better appreciated via an analogy. The definition of ZK has a similar quantifier format: "For every (cheating) verifier there exists a simulator such that ...". This makes it hard to show a protocol is not ZK, for, even though we may be able to identify a cheating verifier strategy that appears hard to simulate, it is not clear how we can actually show no simulator exists. (For example, it is hard to imagine how one could find a simulator for the cheating verifier, for Blum's ZK proof of Hamiltonian Cycle [5], that produces its challenges by hashing the permuted graphs sent by the prover in the first step. But there is to date no proof that such a simulator does not exist). However it has been possible to show protocols are not black-box simulation ZK [9], taking advantage of the fact that the quantification in this definition is different from that of ZK itself. It has also been possible to show conditional results, for example that the parallel version of the Fiat-Shamir [8] protocol is not ZK, unless there is no hash function that, when applied to collapse this protocol, results in a secure signature scheme [16]. Our result too is conditional.

FALSIFICATION RESULT. At an intuitive level, the weakness in KEA2 is easy to see, and indeed it is surprising this was not noted before. Namely, consider an adversary $\mathbf{A}$ that on input q, g, g^a, g^b, g^{ab} picks c_1, c_2 in some fashion, and outputs (C, Y) where $C = g^{c_1}(g^a)^{c_2}$ and $Y = (g^b)^{c_1}(g^{ab})^{c_2}$. Then $Y = C^b$ but this adversary does not appear to "know" c such that either $g^c = C$ or $(g^a)^c = C$. The difficulty, however, as indicated above, is to prove that there does not exist an extractor. We do this by first specifying a particular strategy for choosing c_1 and c_2 and then showing that if there exists an extractor for the resulting adversary, then this extractor can be used to solve the discrete logarithm problem (DLP). Thus, our result (cf. Theorem 1) is that if the DLP is hard then KEA2 is false. Note that if the DLP is easy, then KEA2 is true, for the extractor can simply compute a discrete logarithm of C and output it, and thus the assumption that it is hard is necessary to falsify KEA2.

REMARK. We emphasize that we have not found any weaknesses in KEA1, an assumption used not only in [7, 11, 12] but also elsewhere.

KEA3. Providing a 3-round, negligible-error ZK protocol for NP is a challenging problem that has attracted considerable research effort. The fact that KEA2 is false means that we "lose" one of the only positive results [11, 12] that we had on this subject. Accordingly, we would like to "recover" it. To this end, we propose a modification of KEA2 that addresses the weakness we found. The new assumption is, roughly, as follows:

KEA3: For any adversary $\mathbf{A}$ that takes input q, g, g^a, g^b, g^{ab} and returns (C, Y) with $Y = C^b$, there exists an "extractor" $\bar{\mathbf{A}}$, which given the same inputs as $\mathbf{A}$ returns c_1, c_2 such that $g^{c_1}(g^a)^{c_2} = C$.

Before proceeding to use this assumption, we note a relation that we consider interesting, namely, that KEA3 implies KEA1 (cf. Proposition 2)[1]. The relation means that KEA3 is a natural extension of KEA1. It also allows us to simplify result statements, assuming only KEA3 rather than both this assumption and KEA1.

RECOVERING THE ZK RESULT. Let HTP denote the 3-round protocol of Hada and Tanaka, which they claim to be sound (i.e., have negligible error) and ZK. The falsity of KEA2 invalidates their proof of soundness. However, this does not mean that HTP is not sound: perhaps it is and this could be proved under another assumption, such as KEA3. This turns out to be almost, but not quite, true. We identify a small bug in HTP based on which we can present a successful cheating prover strategy, showing that HTP is not sound. This is easily fixed, however, to yield a protocol we call pHTP (patched HTP). This protocol is close enough to HTP that the proof of ZK (based on KEA1) is unchanged. On the other hand, the proof of soundness of HTP provided in [12] extends with very minor modifications to prove soundness of pHTP based on KEA3 and DLA (cf. Theorem 2). In summary, assuming KEA3 and DLA, there exists a 3-round, negligible error ZK argument for NP.

STRENGTH OF THE ASSUMPTIONS. The knowledge-of-exponent assumptions are strong and non-standard ones, and have been criticized for assuming that one can perform what some people call "reverse engineering" of an adversary. These critiques are certainly valid. Our falsification of KEA2 does not provide information on this aspect of the assumptions, uncovering, rather, other kinds of problems. However, by showing that such assumptions can be falsified, we open the door to further analyses.

We also stress that in recovering the result of [12] on 3-round ZK we have not succeeded in weakening the assumptions on which it is based, for KEA3 certainly remains a strong assumption of the same non-standard nature as KEA1.

RELATED WORK. Since [11, 12] there has been more progress with regard to the design of non-black-box simulation ZK protocols [1]. However, this work does not provide a 3-round, negligible-error ZK protocol for NP. To date, there have been only two positive results. One is that of [11, 12], broken and recovered in this paper. The other, which builds a proof system rather than an argument, is reported in [14] and further documented in [13]. It also relies on non-standard assumptions, but different from the Knowledge of Exponent type ones. Roughly, they assume the existence of a hash function such that a certain discrete-log-based protocol, that uses this hash function and is related to the non-interactive OT of [3], is a proof of knowledge.

---

[1] KEA2 was not shown by [12] to imply KEA1. Our proof of Proposition 2 does extend to establish it, but the point is moot since KEA2 is false and hence of course implies everything anyway.

2   Preliminaries

If x is a binary string, then $|x|$ denotes its length, and if $n \geq 1$ is an integer, then $|n|$ denotes the length of its binary encoding, meaning the unique integer ℓ such that $2^{\ell-1} \leq n < 2^{\ell}$. The empty string is denoted ε. We let $\mathbb{N} = \{1, 2, 3, \ldots\}$ be the set of positive integers. If q is a prime number such that $2q+1$ is also prime, then we denote by G_q the subgroup of quadratic residues of $\mathbb{Z}^*_{2q+1}$. (Operations are modulo $2q+1$ but we will omit writing " mod $2q+1$" for simplicity.) Recall this is a cyclic subgroup of order q. If g is a generator of G_q then we let $\mathrm{DLog}_{q,g}\colon G_q \rightarrow \mathbb{Z}_q$ denote the associated discrete logarithm function, meaning $\mathrm{DLog}_{q,g}(g^a) = a$ for any $a \in \mathbb{Z}_q$. We let

$$GL = \{ (q,g) \,:\, q, 2q+1 \text{ are primes and } g \text{ is a generator of } G_q \} .$$

For any $n \in \mathbb{N}$ we let GL_n be the set of all $(q,g) \in GL$ such that the length of the binary representation of $2q+1$ is n bits, i.e.,

$$GL_n = \{ (q,g) \in GL \,:\, |2q+1| = n \} .$$

Assumptions and problems in [11,12] involve circuits. A family of circuits $\mathbf{C} = \{\mathbf{C}_n\}_{n \in \mathbb{N}}$ contains one circuit for each value of $n \in \mathbb{N}$. It is poly-size if there is a polynomial p such that the size of $\mathbf{C}_n$ is at most $p(n)$ for all $n \in \mathbb{N}$. Unless otherwise stated, *circuits are deterministic*. If they are randomized, we will say so explicitly. We now recall the DLA following [12].

Assumption 1. [DLA]   Let $\mathbf{I} = \{\mathbf{I}_n\}_{n \in \mathbb{N}}$ be a family of randomized circuits, and $\nu\colon \mathbb{N} \rightarrow [0,1]$ a function. We associate to any $n \in \mathbb{N}$ and any $(q,g) \in GL_n$ the following experiment:

Experiment $\mathbf{Exp}^{\mathrm{dl}}_{\mathbf{I}}(n,q,g)$

$a \xleftarrow{\$} \mathbb{Z}_q$; $A \leftarrow g^a$; $\bar{a} \xleftarrow{\$} \mathbf{I}_n(q,g,A)$; If $a = \bar{a}$ then return 1 else return 0

We let

$$\mathbf{Adv}^{\mathrm{dl}}_{\mathbf{I}}(n,q,g) = \Pr \left[\mathbf{Exp}^{\mathrm{dl}}_{\mathbf{I}}(n,q,g) = 1 \right]$$

denote the *advantage* of $\mathbf{I}$ on inputs n, q, g, the probability being over the random choice of a and the coins of $\mathbf{I}_n$, if any. We say that $\mathbf{I}$ has *success bound* ν if

$$\forall n \in \mathbb{N} \; \forall (q,g) \in GL_n \;:\; \mathbf{Adv}^{\mathrm{dl}}_{\mathbf{I}}(n,q,g) \leq \nu(n) .$$

We say that the *Discrete Logarithm Assumption (DLA) holds* if for every poly-size family of circuits $\mathbf{I}$ there exists a negligible function ν such that $\mathbf{I}$ has success bound ν. ∎

The above formulation of the DLA, which, as we have indicated, follows [12], has some non-standard features that are important for their results. Let us discuss these briefly.

First, we note that the definition of the success bound is not with respect to (q,g) being chosen according to some distribution as is standard, but rather makes the stronger requirement that the advantage of $\mathbf{I}$ is small for all (q,g).

Second, we stress that the assumption only requires poly-size families of *deterministic* circuits to have a negligible success bound. However, in their proofs, which aim to contradict the DLA, Hada and Tanaka [11, 12] build adversaries that are poly-size families of randomized circuits, and then argue that these can be converted to related poly-size families of deterministic circuits that do not have a negligible success bound. We will also need to build such randomized adversaries, but, rather than using ad hoc conversion arguments repeated across proofs, we note the following more general Proposition, which simply says that DLA, as per Assumption 1, implies that poly-size families of randomized circuits also have a negligible success bound. We will appeal to this in several later places in this paper.

Proposition 1. *Assume the DLA, and let* $\mathbf{J} = \{\mathbf{J}_n\}_{n \in \mathbb{N}}$ *be a poly-size family of randomized circuits. Then there exists a negligible function* ν *such that* $\mathbf{J}$ *has success bound* ν. ∎

As is typical in such claims, the proof proceeds by showing that for every n there exists a "good" choice of coins for $\mathbf{J}_n$, and by embedding these coins we get a deterministic circuit. For completeness, we provide the proof in the full version of this paper [4].

3   KEA2 Is False

We begin by recalling the assumption. Our presentation is slightly different from, but clearly equivalent to, that of [12]: we have merged the two separate conditions of their formalization into one. Recall that they refer to this assumption as "non-uniform-DA2," and it was referred to, under a different and incorrect formalization, as SDHA2 in [11].

Assumption 2. **[KEA2]** Let $\mathbf{A} = \{\mathbf{A}_n\}_{n \in \mathbb{N}}$ and $\bar{\mathbf{A}} = \{\bar{\mathbf{A}}_n\}_{n \in \mathbb{N}}$ be families of circuits, and $\nu \colon \mathbb{N} \to [0, 1]$ a function. We associate to any $n \in \mathbb{N}$, any $(q, g) \in GL_n$, and any $A \in G_q$ the following experiment:

Experiment $\mathbf{Exp}_{\mathbf{A}, \bar{\mathbf{A}}}^{\mathrm{kea2}}(n, q, g, A)$

$\quad b \xleftarrow{\$} \mathbb{Z}_q \, ; \, B \leftarrow g^b \, ; \, X \leftarrow A^b$

$\quad (C, Y) \leftarrow \mathbf{A}_n(q, g, A, B, X) \, ; \, c \leftarrow \bar{\mathbf{A}}_n(q, g, A, B, X)$

$\quad$ If $(Y = C^b \text{ AND } g^c \neq C \text{ AND } A^c \neq C)$ then return 1 else return 0

We let

$$\mathbf{Adv}_{\mathbf{A}, \bar{\mathbf{A}}}^{\mathrm{kea2}}(n, q, g, A) \;=\; \Pr\left[\mathbf{Exp}_{\mathbf{A}, \bar{\mathbf{A}}}^{\mathrm{kea2}}(n, q, g, A) = 1\right]$$

denote the *advantage* of $\mathbf{A}$ relative to $\bar{\mathbf{A}}$ on inputs n, q, g, A. We say that $\bar{\mathbf{A}}$ is a *kea2-extractor for* $\mathbf{A}$ *with error bound* ν if

$$\forall n \in \mathbb{N} \; \forall (q, g) \in GL_n \; \forall A \in G_q \; : \; \mathbf{Adv}_{\mathbf{A}, \bar{\mathbf{A}}}^{\mathrm{kea2}}(n, q, g, A) \leq \nu(n) \, .$$

We say that *KEA2 holds* if for every poly-size family of circuits $\mathbf{A}$ there exists a poly-size family of circuits $\bar{\mathbf{A}}$ and a negligible function ν such that $\bar{\mathbf{A}}$ is a kea2-extractor for $\mathbf{A}$ with error bound ν. ∎

We stress again that in the above formulations, following [12], both the adversary and the extractor are families of *deterministic* circuits. One can consider various variants of the assumptions, including an extension to families of randomized ciruits, and we discuss these variants following the theorem below.

Theorem 1. *If the DLA holds then KEA2 is false.* ∎

The basic idea behind the failure of the assumption, as sketched in Section 1, is simple. Consider an adversary given input q, g, A, B, X, where $A = g^a, B = g^b$ and $X = g^{ab}$. The assumption says that there are only two ways for the adversary to output a pair C, Y satisfying $Y = C^b$. One way is to pick some c, let $C = g^c$ and let $Y = B^c$. The other way is to pick some c, let $C = A^c$ and let $Y = X^c$. The assumption thus states that the adversary "knows" c such that either $C = g^c$ (i.e., $c = \mathrm{DLog}_{q,g}(C)$) or $C = A^c$ (i.e., $c = \mathrm{DLog}_{q,A}(C)$). This ignores the possibility of performing a linear combination of the two steps above. In other words, an adversary might pick c_1, c_2, let $C = g^{c_1} A^{c_2}$ and $Y = B^{c_1} X^{c_2}$. In this case, $Y = C^b$ but the adversary does not appear to necessarily know $\mathrm{DLog}_{q,g}(C) = c_1 + c_2 \mathrm{DLog}_{q,g}(A)$ or $\mathrm{DLog}_{q,A}(C) = c_1 \mathrm{DLog}_{q,A}(g) + c_2$.

However, going from this intuition to an actual proof that the assumption is false takes some work, for several reasons. The above may be intuition that there exists an adversary for which there would not exist an extractor, but we need to *prove* that there is no extractor. This cannot be done unconditionally, since certainly if the discrete logarithm problem (DLP) is easy, then in fact there is an extractor: it simply computes $\mathrm{DLog}_{q,g}(C)$ and returns it. Accordingly, our strategy will be to present an adversary **A** for which we can prove that if there exists an extractor **Ā** then there is a method to efficiently compute the discrete logarithm of A.

An issue in implementing this is that the natural adversary **A** arising from the above intuition is randomized, picking c_1, c_2 at random and forming C, Y as indicated, but our adversaries must be deterministic. We resolve this by designing an adversary that makes certain specific choices of c_1, c_2. We now proceed to the formal proof.

PROOF OF THEOREM 1. Assume to the contrary that KEA2 is true. We show that the DLP is easy.

The outline of the proof is as follows. We first construct an adversary **A** for the KEA2 problem. By assumption, there exists for it an extractor **Ā** with negligible error bound. Using **Ā**, we then present a poly-size family of randomized circuits $\mathbf{J} = \{\mathbf{J}_n\}_{n \in \mathbb{N}}$ and show that it does not have a negligible success bound. By Proposition 1, this contradicts the DLA.

The poly-size family of circuits $\mathbf{A} = \{\mathbf{A}_n\}_{n \in \mathbb{N}}$ is presented in Figure 1. Now, under KEA2, there exists a poly-size family of circuits $\bar{\mathbf{A}} = \{\bar{\mathbf{A}}_n\}_{n \in \mathbb{N}}$ and a negligible function ν such that $\bar{\mathbf{A}}$ is an extractor for **A** with error bound ν. Using $\bar{\mathbf{A}}$, we define the poly-size family of circuits $\mathbf{J} = \{\mathbf{J}_n\}_{n \in \mathbb{N}}$ shown in Figure 1.

Claim 1. For all $n \in \mathbb{N}$, all $(q, g) \in GL_n$ and all $A \in G_q$

$$\Pr\left[\bar{a} \xleftarrow{\$} \mathbf{J}_n(q, g, A) \; : \; g^{\bar{a}} \neq A\right] \leq \nu(n) . \quad ∎$$

$\mathbf{A}_n(q, g, A, B, X)$	$\mathbf{J}_n(q, g, A)$
$C \leftarrow gA$	$b \stackrel{\$}{\leftarrow} \mathbb{Z}_q \; ; \; B \leftarrow g^b \; ; \; X \leftarrow A^b$
$Y \leftarrow BX$	$c \leftarrow \bar{\mathbf{A}}_n(q, g, A, B, X)$
Return (C, Y)	$C \leftarrow gA$
	If $g^c = C$ then $\bar{a} \leftarrow (c - 1) \bmod q$ EndIf
	If $A^c = C$ then $\bar{a} \leftarrow (c - 1)^{-1} \bmod q$ EndIf
	Return $\bar{a}$

Fig. 1. Adversary $\mathbf{A} = \{\mathbf{A}_n\}_{n \in \mathbb{N}}$ for the KEA2 problem and adversary $\mathbf{J} = \{\mathbf{J}_n\}_{n \in \mathbb{N}}$ for the DLP, for the proof of Theorem 1.

Note the claim shows much more than we need. Namely, $\mathbf{J}$ does not merely have a success bound that is not negligible. In fact, it succeeds with probability almost one.

Proof (Claim 1). We let $\Pr[\cdot]$ denote the probability in the experiment of executing $\mathbf{J}_n(q, g, A)$. We first write some inequalities leading to the claim and then justify them:

$$\Pr\left[g^{\bar{a}} \neq A\right] \leq \Pr\left[g^c \neq C \wedge A^c \neq C\right] \tag{1}$$

$$\leq \mathbf{Adv}^{\text{kea2}}_{\mathbf{A}, \bar{\mathbf{A}}}(n, q, g, A) \tag{2}$$

$$\leq \nu(n). \tag{3}$$

We justify Equation (1) by showing that if $g^c = C$ or $A^c = C$ then $g^{\bar{a}} = A$. First assume $g^c = C$. Since $C = gA$, we have $g^c = gA$, whence $A = g^{c-1}$. Since we set $\bar{a} = (c - 1) \bmod q$, we have $A = g^{\bar{a}}$. Next assume $A^c = C$. Since $C = gA$, we have $A^c = gA$, whence $A^{c-1} = g$. Now observe that $c \neq 1$, because otherwise $A^c = A \neq gA$. (Since g is a generator, it is not equal to 1). Since $c \neq 1$ and q is prime, $c - 1$ has an inverse modulo q which we have denoted by $\bar{a}$. Raising both sides of the equation "$A^{c-1} = g$" to the power $\bar{a}$ we get $A = g^{\bar{a}}$.

$\mathbf{Exp}^{\text{kea2}}_{\mathbf{A}, \bar{\mathbf{A}}}(n, q, g, A)$ returns 1 exactly when $Y = C^b$ and $g^c \neq C$ and $A^c \neq C$. By construction of $\mathbf{A}$, we have $C = gA$ and $Y = BX$, and thus $Y = C^b$, so $\mathbf{Exp}^{\text{kea2}}_{\mathbf{A}, \bar{\mathbf{A}}}(n, q, g, A)$ returns 1 exactly when $g^c \neq C$ and $A^c \neq C$. This justifies Equation (2).

Equation (3) is justified by the assumption that $\bar{\mathbf{A}}$ is an extractor for $\mathbf{A}$ with error bound ν. ∎

Claim 1 implies that $\mathbf{J}$ does not have a negligible success bound, which, by Proposition 1, shows that the DLP is not hard, contradicting the assumption made in this Theorem. This completes the proof of Theorem 1.

EXTENSIONS AND VARIANTS. There are many ways in which the formalization of Assumption 2 can be varied to capture the same basic intuition. However, Theorem 1 extends to these variants as well. Let us discuss this briefly.

As mentioned above, we might want to allow the adversary to be randomized. (In that case, it is important that the extractor get the coins of the adversary as an additional input, since otherwise the assumption is clearly false.) Theorem 1 remains true for the resulting assumption, in particular because it is stronger than the original assumption. (Note however that the proof of the theorem would be easier for this stronger assumption.)

Another variant is that adversaries and extractors are uniform, namely standard algorithms, not circuits. (In this case we should certainly allow both to be randomized, and should again give the extractor the coins of the adversary.) Again, it is easy to see that Theorem 1 extends to show that the assumption remains false.

4   The KEA3 Assumption

The obvious fix to KEA2 is to take into account the possibility of linear combinations by saying this is the only thing the adversary can do. This leads to the following.

Assumption 3. [KEA3] Let $\mathbf{A} = \{\mathbf{A}_n\}_{n \in \mathbb{N}}$ and $\bar{\mathbf{A}} = \{\bar{\mathbf{A}}_n\}_{n \in \mathbb{N}}$ be families of circuits, and $\nu \colon \mathbb{N} \to [0, 1]$ a function. We associate to any $n \in \mathbb{N}$, any $(q, g) \in GL_n$, and any $A \in G_q$ the following experiment:

Experiment $\mathbf{Exp}^{\mathrm{kea3}}_{\mathbf{A}, \bar{\mathbf{A}}}(n, q, g, A)$

$\quad b \xleftarrow{\$} \mathbb{Z}_q \,;\, B \leftarrow g^b \,;\, X \leftarrow A^b$
$\quad (C, Y) \leftarrow \mathbf{A}_n(q, g, A, B, X) \,;\, (c_1, c_2) \leftarrow \bar{\mathbf{A}}_n(q, g, A, B, X)$
$\quad$ If $(Y = C^b \text{ AND } g^{c_1} A^{c_2} \neq C)$ then return 1 else return 0

We let

$$\mathbf{Adv}^{\mathrm{kea3}}_{\mathbf{A}, \bar{\mathbf{A}}}(n, q, g, A) \;=\; \Pr \left[\mathbf{Exp}^{\mathrm{kea3}}_{\mathbf{A}, \bar{\mathbf{A}}}(n, q, g, A) = 1 \right]$$

denote the *advantage* of $\mathbf{A}$ relative to $\bar{\mathbf{A}}$ on inputs n, q, g, A. We say that $\bar{\mathbf{A}}$ is a *kea3-extractor for* $\mathbf{A}$ *with error bound* ν if

$$\forall n \in \mathbb{N} \;\; \forall (q, g) \in GL_n \;\; \forall A \in G_q \;:\; \mathbf{Adv}^{\mathrm{kea3}}_{\mathbf{A}, \bar{\mathbf{A}}}(n, q, g, A) \leq \nu(n) \,.$$

We say that *KEA3 holds* if for every poly-size family of circuits $\mathbf{A}$ there exists a poly-size family of circuits $\bar{\mathbf{A}}$ and a negligible function ν such that $\bar{\mathbf{A}}$ is a kea3-extractor for $\mathbf{A}$ with error bound ν. ∎

We have formulated this assumption in the style of the formalization of KEA2 of [12] given in Assumption 2. Naturally, variants such as discussed above are possible. Namely, we could strengthen the assumption to allow the adversary to be a family of randomized circuits, of course then giving the extractor the adversary's coins as an additional input. We do not do this because we do not need it for what follows. We could also formulate a uniform-complexity version of the assumption. We do not do this because it does not suffice to prove the results that follow. However, these extensions or variations might be useful in other contexts.

In Appendix A we recall the formalization of KEA1 and prove the following:

Proposition 2. *KEA3 implies KEA1.* ∎

This indicates that KEA3 is a natural extension of KEA1.

5   Three-Round Zero Knowledge

The falsity of KEA2 renders vacuous the result of [11,12] saying that there exists a negligible-error, 3-round ZK argument for NP. In this section we look at recovering this result.

We first consider the protocol of [11,12], here called HTP. What has been lost is the proof of soundness (i.e., of negligible error). The simplest thing one could hope for is to re-prove soundness of HTP under KEA3 without modifying the protocol. However, we identify a bug in HTP that renders it unsound. This bug has nothing to do with the assumptions on which the proof of soundness was or can be based.

The bug is, however, small and easily fixed. We consider a modified protocol which we call pHTP. We are able to show it is sound (i.e., has negligible error) under KEA3. Since we have modified the protocol we need to re-establish ZK under KEA1 as well, but this is easily done.

ARGUMENTS. We begin by recalling some definitions. An argument for an NP language L [6] is a two-party protocol in which a polynomial-time prover tries to "convince" a polynomial-time verifier that their common input x belongs to L. (A party is said to be polynomial time if its running time is polynomial in the length of the common input.) In addition to x, the prover has an auxiliary input a. The protocol is a message exchange at the end of which the verifier outputs a bit indicating its decision to accept or reject. The probability (over the coin tosses of both parties) that the verifier accepts is denoted $\mathbf{Acc}_V^{P,a}(x)$. The formal definition follows.

Definition 1. A two-party protocol (P, V), where P and V are both polynomial time, is an *argument for L with error probability* $\delta : \mathbb{N} \to [0,1]$, if the following conditions are satisfied:

COMPLETENESS: For all $x \in L$ there exists $w \in \{0,1\}^*$ such that $\mathbf{Acc}_V^{P,w}(x) = 1$.

SOUNDNESS: For all probabilistic polynomial-time algorithms $\widehat{P}$, all sufficiently long $x \notin L$, and all $a \in \{0,1\}^*$, $\mathbf{Acc}_V^{\widehat{P},a}(x) \leq \delta(|x|)$.

We say (P, V) is a *negligible-error argument for L* if there exists a negligible function $\delta : \mathbb{N} \to [0,1]$ such that (P, V) is an argument for L with error probability δ. ∎

CANONICAL PROTOCOLS. The 3-round protocol proposed by [11,12], which we call HTP, is based on a 3-round argument $(\bar{P}, \bar{V})$ for an NP-complete language L with the following properties:

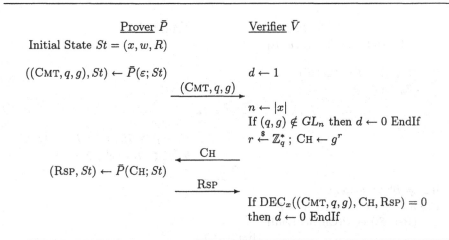

$$
\begin{array}{ll}
\text{Prover } \bar{P} & \text{Verifier } \bar{V} \\
\text{Initial State } St = (x, w, R) & \\
((\text{CMT}, q, g), St) \leftarrow \bar{P}(\varepsilon; St) & d \leftarrow 1 \\
\qquad\qquad \xrightarrow{(\text{CMT}, q, g)} & \\
& n \leftarrow |x| \\
& \text{If } (q, g) \notin GL_n \text{ then } d \leftarrow 0 \text{ EndIf} \\
& r \xleftarrow{\$} \mathbb{Z}_q^* \,;\, \text{CH} \leftarrow g^r \\
\qquad\qquad \xleftarrow{\text{CH}} & \\
(\text{RSP}, St) \leftarrow \bar{P}(\text{CH}; St) & \\
\qquad\qquad \xrightarrow{\text{RSP}} & \\
& \text{If } \text{DEC}_x((\text{CMT}, q, g), \text{CH}, \text{RSP}) = 0 \\
& \text{then } d \leftarrow 0 \text{ EndIf}
\end{array}
$$

Fig. 2. A 3-round argument. The common input is x. Prover $\bar{P}$ has auxiliary input w and random tape R, and maintains state St. Verifier $\bar{V}$ returns boolean decision d.

(1) The protocol is of the form depicted in Figure 2. The prover is identified with a function $\bar{P}$ that given an incoming message M_{in} (this is ε when the prover is initiating the protocol) and its current state St, returns an outgoing message M_{out} and an updated state. The initial state of the prover is (x, w, R), where x is the common input, w is an auxiliary input and R is a random tape. The prover's first message is called its *commitment*. This is a tuple consisting of a string CMT, a prime number q and an element g, where $(q, g) \in GL_{|x|}$. The verifier selects a *challenge* CH uniformly at random from G_q, and, upon receiving a *response* RSP from the prover, applies a deterministic *decision predicate* $\text{DEC}_x((\text{CMT}, q, g), \text{CH}, \text{RSP})$ to compute a boolean decision.

(2) For any $x \notin L$ and any commitment (CMT, q, g), where $(q, g) \in GL_{|x|}$, there is at most one challenge $\text{CH} \in G_q$ for which there exists a response $\text{RSP} \in \{0, 1\}^*$ such that $\text{DEC}_x((\text{CMT}, q, g), \text{CH}, \text{RSP}) = 1$. This property is called *strong soundness*.

(3) The protocol is honest-verifier zero knowledge (HVZK), meaning there exists a probabilistic polynomial-time *simulator* S such that the following two ensembles are computationally indistinguishable:

$$\{S(x)\}_{x \in L} \quad \text{and} \quad \{\mathbf{View}_{\bar{V}}^{\bar{P}, W(x)}(x)\}_{x \in L} \,,$$

where W is any function that given an input in L returns a witness to its membership in L, and $\mathbf{View}_{\bar{V}}^{\bar{P}, W(x)}(x)$, is a random variable taking value $\bar{V}$'s internal coin tosses and the sequence of messages it receives during an interaction between prover $\bar{P}$ (with auxiliary input $W(x)$) and verifier $\bar{V}$ on common input x.

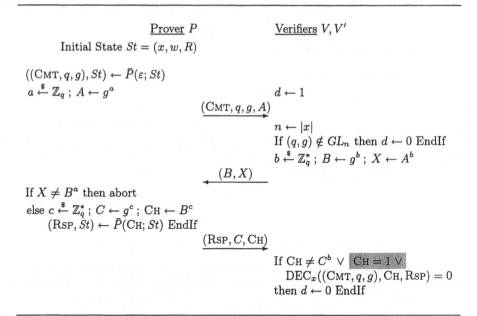

Fig. 3. HTP and pHTP. Verifier V of protocol HTP $= (P, V)$ does not include the highlighted portion. Verifier V' of protocol pHTP $= (P, V')$ does.

If $(\bar{P}, \bar{V})$ is a 3-round argument for an NP-complete language, meeting the three conditions above, then we refer to $(\bar{P}, \bar{V})$ as a *canonical argument*. In what follows, we assume that we have such canonical arguments. They can be constructed in various ways. For example, a canonical argument can be constructed by modifying the parallel composition of Blum's zero-knowledge protocol for the Hamiltonian circuit problem [5], as described in [11, 12].

THE HADA-TANAKA PROTOCOL. Let $(\bar{P}, \bar{V})$ be a canonical argument for an NP-complete language L, and let DEC be the verifier's decision predicate. The Hada-Tanaka protocol HTP $= (P, V)$ is described in Figure 3. Note V's decision predicate does not include the highlighted portion of its code.

We now observe that the HTP protocol is unsound. More precisely, there exist canonical arguments such that the HTP protocol based on them does not have negligible error. This is true for any canonical argument $(\bar{P}, \bar{V})$ satisfying the extra condition that for infinitely many $x \notin L$ there exists a commitment (CMT_x, q_x, g_x) for which there is a response RSP_x to challenge 1 that will make the verifier accept. There are many such canonical arguments. For instance, a canonical argument satisfying this condition results from using an appropriate encoding of group elements in Hada and Tanaka's modification of the parallel composition of Blum's zero-knowledge protocol for the Hamiltonian circuit problem.

Proposition 3. *Let HTP be the Hada-Tanaka protocol based on a canonical argument satisfying the condition stated above. Then there exists a polynomial-time prover for HTP that can make the verifier accept with probability one for infinitely many common inputs not in L.* ∎

Proof (Proposition 3). Let $(\bar{P}, \bar{V})$ be the canonical argument and let V be the verifier of the corresponding protocol HTP. Consider a cheating prover $\widehat{P}$ that on initial state $(x, ((\text{CMT}_x, q_x, g_x), \text{RSP}_x), \varepsilon)$ selects an exponent $a \in \mathbb{Z}_{q_x}$ uniformly at random, and sends $(\text{CMT}_x, q_x, g_x, g_x^a)$ as its commitment to verifier V. Upon receiving a challenge (B, X), it checks if $X = B^a$. If not, it aborts. Otherwise, it sends $(\text{RSP}_x, 1, 1)$ as its response to V. By the assumption about protocol $(\bar{P}, \bar{V})$, for infinitely many $x \notin L$ there exists an auxiliary input $y = ((\text{CMT}_x, q_x, g_x), \text{RSP}_x) \in \{0,1\}^*$ such that $\text{Acc}_V^{\widehat{P}, y}(x) = 1$. ∎

PROTOCOL pHTP. The above attack can be avoided by modifying the verifier to include the highlighted portion of the code in Figure 3. We call the resulting verifier V'. The following guarantees that the protocol pHTP $= (P, V')$ is sound under KEA3, if the DLP is hard.

Theorem 2. *If KEA3 holds, the DLA holds, and $(\bar{P}, \bar{V})$ is a canonical 3-round argument for an NP-complete language L, then pHTP $= (P, V')$ as defined in Figure 3 is a negligible-error argument for L.*

PROOF OF THEOREM 2. The proof is almost identical to that of [12]. For completeness, however, we provide it.

Completeness follows directly from the completeness of protocol $(\bar{P}, \bar{V})$. To prove soundness, we proceed by contradiction. Assume that pHTP is not sound, i.e., there is no negligible function δ such that the soundness condition in Definition 1 holds with respect to δ. We show that the DLP is easy under KEA3.

By the assumption that pHTP is not sound and a result of [2], there exists a probabilistic polynomial-time algorithm $\widehat{P}$ such that the function

$$\text{Err}_{\widehat{P}}(n) = \max\{ \text{Acc}_{V'}^{\widehat{P}, a}(x) : x \in \{0,1\}^n \wedge x \notin L \wedge a \in \{0,1\}^* \}^2$$

is not negligible. Hence there exists a probabilistic polynomial-time algorithm $\widehat{P}$, a polynomial p, and an infinite set $S = \{ (x, a) : x \in \{0,1\}^* \setminus L \wedge a \in \{0,1\}^* \}$ such that for every $(x, a) \in S$

$$\text{Acc}_{V'}^{\widehat{P}, a}(x) > 1/p(|x|), \tag{4}$$

and $\{ x \in \{0,1\}^* : \exists a \in \{0,1\}^* \text{ such that } (x, a) \in S \}$ is infinite.

Since $\widehat{P}$ takes an auxiliary input a, we may assume, without loss of generality, that $\widehat{P}$ is deterministic. We also assume that, if (CMT, q', g', A') is $\widehat{P}$'s commitment on input ε when the initial state is (x, a, ε), for some $x, a \in \{0,1\}^*$ with $|x| = n$, then $(q', g') \in GL_n$. (There exists a prover $\widehat{P}'$ for which $\text{Acc}_{V'}^{\widehat{P}', a}(x) =$

---

[2] We note that this set is finite since $\widehat{P}$ is a polynomial-time algorithm and $\text{Acc}_{V'}^{\widehat{P}, a}(x)$ depends only on the first $t_{\widehat{P}}(|x|)$ bits of a, where $t_{\widehat{P}}(\cdot)$ is the running time of $\widehat{P}$.

$\mathbf{A}_n(q, g, A, B, X)$ $//\, n \in K$
  $St \leftarrow (x, a, \varepsilon)$; $((\text{CMT}, q', g', A'), St) \leftarrow \widehat{P}(\varepsilon; St)$
  If $q' \neq q \vee g' \neq g \vee A' \neq A$ then return $(1, 1)$
  else $((\text{RSP}, C, \text{CH}), St) \leftarrow \widehat{P}((B, X); St)$; return (C, CH) EndIf

$\mathbf{A}_n(q, g, A, B, X)$ $//\, n \notin K$
  Return $(1, 1)$

$\mathbf{J}_n(q, g, A)$ $//\, n \in K$
  $St \leftarrow (x, a, \varepsilon)$; $((\text{CMT}, q', g', A'), St) \leftarrow \widehat{P}(\varepsilon; St)$
  If $q' \neq q \vee g' \neq g$ then return $\perp$ EndIf
  $b \stackrel{\$}{\leftarrow} \mathbb{Z}_q$; $B \leftarrow A \cdot g^b$; $X \leftarrow B^{a'}$
  $((\text{RSP}, C, \text{CH}), St_1) \leftarrow \widehat{P}((B, X); St)$; $(c_1, c_2) \leftarrow \bar{\mathbf{A}}_n(q, g, A', B, X)$
  If $\text{DEC}_x((\text{CMT}, q, g), \text{CH}, \text{RSP}) = 0 \vee \text{CH} \neq B^{c_1} X^{c_2}$ then return $\perp$ EndIf
  $b' \stackrel{\$}{\leftarrow} \mathbb{Z}_q$; $B' \leftarrow g^{b'}$; $X' \leftarrow B'^{a'}$
  If $B = B'$ then $\bar{a} \leftarrow b' - b \bmod q$; return $\bar{a}$ EndIf
  $((\text{RSP}', C', \text{CH}'), St_1') \leftarrow \widehat{P}((B', X'); St)$; $(c_1', c_2') \leftarrow \bar{\mathbf{A}}_n(q, g, A', B', X')$
  If $\text{DEC}_x((\text{CMT}, q, g), \text{CH}', \text{RSP}') = 0 \vee \text{CH}' \neq B'^{c_1'} X'^{c_2'}$ then return $\perp$ EndIf
  If $c_1 + a' c_2 \not\equiv 0 \pmod q$ then
    $\bar{a} \leftarrow (b' c_1' + b' a' c_2' - b c_1 - b a' c_2) \cdot (c_1 + a' c_2)^{-1} \bmod q$; return $\bar{a}$
  else return $\perp$ EndIf

$\mathbf{J}_n(q, g, A)$ $//\, n \notin K$
  Return $\perp$

Fig. 4. Adversary $\mathbf{A} = \{\mathbf{A}_n\}_{n \in \mathbb{N}}$ for the KEA3 problem and adversary $\mathbf{J} = \{\mathbf{J}_n\}_{n \in \mathbb{N}}$ for the DLP, for the proof of Theorem 2.

$\mathbf{Acc}_{V'}^{\widehat{P}, a}(x)$ for every $x, a \in \{0, 1\}^*$ and this assumption holds.) We will use $\widehat{P}$ to construct an adversary $\mathbf{A}$ for the KEA3 problem. By assumption, there exists for it an extractor $\bar{\mathbf{A}}$ with negligible error bound. Using $\bar{\mathbf{A}}$ and $\widehat{P}$, we then present a poly-size family of randomized circuits $\mathbf{J} = \{\mathbf{J}_n\}_{n \in \mathbb{N}}$ and show that it does not have a negligible success bound. By Proposition 1, this shows that the DLP is not hard.

Let $K = \{ n \in \mathbb{N} : \exists (x, a) \in S \text{ such that } |x| = n \}$. We observe that K is an infinite set. For each $n \in K$, fix $(x, a) \in S$ such that $|x| = n$. The poly-size family of circuits $\mathbf{A} = \{\mathbf{A}_n\}_{n \in \mathbb{N}}$ is presented in Figure 4. Now, under KEA3, there exists a poly-size family of circuits $\bar{\mathbf{A}} = \{\bar{\mathbf{A}}_n\}_{n \in \mathbb{N}}$ and a negligible function ν such that $\bar{\mathbf{A}}$ is an extractor for $\mathbf{A}$ with error bound ν. For each $n \in K$, let $a' = \text{DLog}_{q', g'}(A')$, where (CMT, q', g', A') is $\widehat{P}$'s commitment on input ε when the initial state is (x, a, ε). Using $\bar{\mathbf{A}}$, we define the poly-size family of circuits $\mathbf{J} = \{\mathbf{J}_n\}_{n \in \mathbb{N}}$ shown in Figure 4. The proof of the following is in [4].

Claim 2. For infinitely many $n \in \mathbb{N}$ there exists $(q, g) \in GL_n$ such that for every $A \in G_q$

$$\Pr\left[\bar{a} \xleftarrow{\$} \mathbf{J}_n(q, g, A) \ : \ g^{\bar{a}} = A \right] > \frac{1}{p(n)^2} - \frac{8}{2^n p(n)} - 2\nu(n) \ . \quad \blacksquare$$

Claim 2 implies that $\mathbf{J}$ does not have a negligible success bound, which, by Proposition 1, shows that the DLP is not hard, contradicting the assumption made in this Theorem.

ZERO KNOWLEDGE OF pHTP. Having modified HTP, we need to revisit the zero knowledge. Hada and Tanaka proved that if the canonical argument is HVZK (property (3) above) then HTP is zero knowledge under KEA1. However, we observe that pHTP modifies only the verifier, not the prover. Furthermore, only the decision predicate of the verifier is modified, not the messages it sends. This means that the view (i.e., the internal coin tosses and the sequence of messages received during an interaction with a prover P) of verifier V' of pHTP is identical to that of verifier V of HTP. Thus, zero knowledge of pHTP follows from zero knowledge of HTP, and in particular is true under the same assumptions, namely KEA1.

SUMMARY. In summary, pHTP is a 3-round protocol that we have shown is a negligible-error argument for NP assuming DLA and KEA3, and is ZK assuming KEA1. Given Proposition 2, this means we have shown that assuming DLA and KEA3 there exists a 3-round negligible-error ZK argument for NP.

Acknowledgments

Mihir Bellare is supported in part by NSF grants CCR-0098123, ANR-0129617 and CCR-0208842, and by an IBM Faculty Partnership Development Award. Adriana Palacio is supported in part by a NSF Graduate Research Fellowship.

Proposition 2 is due to Shai Halevi and we thank him for permission to include it. We thank the Crypto 2004 referees for their comments on the paper.

References

1. B. BARAK. How to go beyond the black-box simulation barrier. *Proceedings of the 42nd Symposium on Foundations of Computer Science*, IEEE, 2001.
2. M. BELLARE. A note on negligible functions. *Journal of Cryptology*, Vol. 15, No. 4, pp. 271–284, June 2002.
3. M. BELLARE AND S. MICALI. Non-interactive oblivious transfer and applications. *Advances in Cryptology – CRYPTO '89*, Lecture Notes in Computer Science Vol. 435, G. Brassard ed., Springer-Verlag, 1989.
4. M. BELLARE AND A. PALACIO. The Knowledge-of-Exponent assumptions and 3-round zero-knowledge protocols. Full version of the current paper, available via http://www-cse.ucsd.edu/users/mihir.
5. M. BLUM. How to prove a theorem so no one else can claim it. *Proceedings of the International Congress of Mathematicians*, pp. 1444–1451, 1986.

6. G. BRASSARD, D. CHAUM AND C. CRÉPEAU. Minimum disclosure proofs of knowledge. *J. Computer and System Sciences*, Vol. 37, No. 2, pp. 156–189, October 1988.

7. I. DAMGÅRD. Towards practical public-key cryptosystems provably-secure against chosen-ciphertext attacks. *Advances in Cryptology – CRYPTO '91*, Lecture Notes in Computer Science Vol. 576, J. Feigenbaum ed., Springer-Verlag, 1991.

8. A. FIAT AND A. SHAMIR. How to prove yourself: Practical solutions to identification and signature problems. *Advances in Cryptology – CRYPTO '86*, Lecture Notes in Computer Science Vol. 263, A. Odlyzko ed., Springer-Verlag, 1986.

9. O. GOLDREICH AND H. KRAWCZYK. On the Composition of Zero Knowledge Proof Systems. *SIAM J. on Computing*, Vol. 25, No. 1, pp. 169–192, 1996.

10. S. GOLDWASSER, S. MICALI AND C. RACKOFF. The knowledge complexity of interactive proof systems. *SIAM Journal of Computing*, Vol. 18, No. 1, pp. 186–208, February 1989.

11. S. HADA AND T. TANAKA. On the existence of 3-round zero-knowledge protocols. *Advances in Cryptology – CRYPTO '98*, Lecture Notes in Computer Science Vol. 1462, H. Krawczyk ed., Springer-Verlag, 1998. [Preliminary version of [12].]

12. S. HADA AND T. TANAKA. On the existence of 3-round zero-knowledge protocols. Cryptology ePrint Archive: Report 1999/009, March 1999. http://eprint.iacr.org/1999/009/. [Final version of [11].]

13. M. LEPINSKI. On the existence of 3-round zero-knowledge proofs. SM Thesis, MIT, June 2002.
http://theory.lcs.mit.edu/~cis/theses/lepinski-masters.ps.

14. M. LEPINSKI AND S. MICALI. On the existence of 3-round zero-knowledge proof systems. MIT LCS Technical Memo. 616, April 2001. http://www.lcs.mit.edu/publications/pubs/pdf/MIT-LCS-TM-616.pdf.

15. M. NAOR. On cryptographic assumptions and challenges. Invited paper and talk, *Advances in Cryptology – CRYPTO '03*, Lecture Notes in Computer Science Vol. 2729, D. Boneh ed., Springer-Verlag, 2003.

16. K. SAKURAI AND T. ITOH. On the discrepancy between serial and parallel of zero-knowledge protocols. *Advances in Cryptology – CRYPTO '92*, Lecture Notes in Computer Science Vol. 740, E. Brickell ed., Springer-Verlag, 1992.

A   KEA3 Implies KEA1

We recall KEA1, following [12], but applying the same simplifications as we did for KEA2 so as to merge their two conditions into one:

Assumption 4. [KEA1] Let $\mathbf{A} = \{\mathbf{A}_n\}_{n \in \mathbb{N}}$ and $\bar{\mathbf{A}} = \{\bar{\mathbf{A}}_n\}_{n \in \mathbb{N}}$ be families of circuits, and $\nu \colon \mathbb{N} \to [0,1]$ a function. We associate to any $n \in \mathbb{N}$, any $(q, g) \in GL_n$, and any $A \in G_q$ the following experiment:

Experiment $\mathbf{Exp}_{\mathbf{A}, \bar{\mathbf{A}}}^{\mathrm{kea1}}(n, q, g)$

$\quad b \xleftarrow{\$} \mathbb{Z}_q \,;\, B \leftarrow g^b$

$\quad (C, Y) \leftarrow \mathbf{A}_n(q, g, B) \,;\, c \leftarrow \bar{\mathbf{A}}_n(q, g, B)$

$\quad$ If $(Y = C^b \text{ AND } g^c \neq C)$ then return 1 else return 0

We let

$$\mathbf{Adv}_{\mathbf{A}, \bar{\mathbf{A}}}^{\mathrm{kea1}}(n, q, g) \;=\; \Pr\left[\, \mathbf{Exp}_{\mathbf{A}, \bar{\mathbf{A}}}^{\mathrm{kea1}}(n, q, g) = 1 \right]$$

denote the *advantage* of $\mathbf{A}$ relative to $\bar{\mathbf{A}}$ on inputs n, q, g. We say that $\bar{\mathbf{A}}$ is a *kea1-extractor for $\mathbf{A}$ with error bound ν* if

$$\forall n \in \mathbb{N} \ \forall (q, g) \in GL_n \ : \ \mathbf{Adv}^{\text{kea1}}_{\mathbf{A}, \bar{\mathbf{A}}}(n, q, g) \leq \nu(n) \ .$$

We say that *KEA1 holds* if for every poly-size family of circuits $\mathbf{A}$ there exists a poly-size family of circuits $\bar{\mathbf{A}}$ and a negligible function ν such that $\bar{\mathbf{A}}$ is a kea1-extractor for $\mathbf{A}$ with error bound ν. ∎

Proof (Proposition 2). Let $\mathbf{A}$ be an adversary (poly-size family of circuits) for KEA1. We need to show there exists a negligible function ν and a poly-size family of circuits $\bar{\mathbf{A}}$ such that $\bar{\mathbf{A}}$ is a kea1-extractor for $\mathbf{A}$ with error-bound ν.

We begin by constructing from $\mathbf{A}$ the following adversary $\mathbf{A}'$ for KEA3:

Adversary $\mathbf{A}'_n(q, g, A, B, X)$
  $(C, Y) \leftarrow \mathbf{A}_n(q, g, B)$
  Return (C, Y)

We have assumed KEA3. Thus there exists a negligible function ν and an extractor $\bar{\mathbf{A}}'$ such that $\bar{\mathbf{A}}'$ is a kea3-extractor for $\mathbf{A}'$ with error bound ν. Now we define an extractor $\bar{\mathbf{A}}$ for $\mathbf{A}$ as follows:

Extractor $\bar{\mathbf{A}}_n(q, g, B)$
  $a \xleftarrow{\$} \mathbb{Z}_q$; $A \leftarrow g^a$; $X \leftarrow B^a$
  $(c_1, c_2) \leftarrow \bar{\mathbf{A}}'(q, g, A, B, X)$
  $c \leftarrow c_1 + ac_2 \bmod q$
  Return c

We claim that $\bar{\mathbf{A}}$ is a kea1-extractor for $\mathbf{A}$ with error bound ν. To see this, assume $\bar{\mathbf{A}}'_n(q, g, A, B, X)$ is successful, meaning $g^{c_1} A^{c_2} = C$. Then $g^c = g^{c_1 + ac_2} = g^{c_1} A^{c_2} = C$ so $\bar{\mathbf{A}}_n(q, g, B)$ is successful as well. ∎

Near-Collisions of SHA-0

Eli Biham and Rafi Chen

Computer Science Department
Technion – Israel Institute of Technology
Haifa 32000, Israel
{biham,rafi_hen}@cs.technion.ac.il
http://www.cs.technion.ac.il/~biham/

Abstract. In this paper we find two near-collisions of the full compression function of SHA-0, in which up to 142 of the 160 bits of the output are equal. We also find many full collisions of 65-round reduced SHA-0, which is a large improvement to the best previous result of 35 rounds. We use the very surprising fact that the messages have many neutral bits, some of which do not affect the differences for about 15–20 rounds. We also show that 82-round SHA-0 is much weaker than the (80-round) SHA-0, although it has more rounds. This fact demonstrates that the strength of SHA-0 is not monotonous in the number of rounds.

1 Introduction

SHA-0 is a cryptographic hash function, which was issued as a Federal Information Processing Standard (FIPS-180) by NIST in 1993 [8]. It is based on the principles of MD4 [12] and MD5 [13]. The algorithm takes a message of any length up to 2^{64} bits and computes a 160-bit hash value. A technical revision, called SHA-1, which specifies an additional rotate operation to the algorithm, was issued as FIPS-180-1 [9] in 1995. The purpose of the revision according to NIST is to improve the security provided by the hash function.

Finding collisions of hash functions is not an easy task. The known cases of successful finding of collisions (such as the attack on Snefru [14, 2], and the attack on MD4 [12, 4]) are rare, and use detailed weaknesses of the broken functions. It is widely believed that finding near-collisions (i.e., two messages that hash to almost the same value, with a difference of only a few bits) are as difficult, or almost as difficult, as finding a full collision. The Handbook of Applied Cryptography [7] defines near-collision resistance by

> *near-collision resistance.* It should be hard to find any two inputs x, x' such that $h(x)$ and $h(x')$ differ in only a small number of bits.

and states that it may serve as a certificational property. In some designs of hash functions, such as SHA-2/224 [10], SHA-2/384 [11], and Tiger [1], the designers that wish to allow several hash sizes for their design, base the version with the smaller size on the one with the larger size, and discard some of the output bits, thus showing the confidence of the designers in the difficulty of finding near-collisions. Near-collisions were also used in the cryptanalysis of MD4 [15, 4].

M. Franklin (Ed.): CRYPTO 2004, LNCS 3152, pp. 290–305, 2004.
© International Association for Cryptologic Research 2004

Table 1. Comparison of Chabaud and Joux' Results to Our Results

| | Chabaud and Joux | | Our Results | |
	Rounds	Complexity	Rounds	Complexity
Optimized for	80	2^{61}	82	2^{43}
Best collision found	35	2^{14}	65	2^{29} (*)
Conforming rounds found	≈ 56 [6]	–	76	2^{40} (**)
Near collisions (18-bit diff)	–	–	80	2^{40} (**)

(*) About half an hour on a PC
(**) Our actual search took less than a day on a PC which is
    equivalent to a complexity of 2^{35}

Near-collisions are the simplest example of forbidden relations between outputs of the hash function. Another proposed forbidden relation of the hash results is *division intractability* [5] where finding messages hashed to a divisor of other hashes should be difficult.

In [3] Chabaud and Joux proposed a theoretical attack on the full SHA-0 with complexity of 2^{61}. Using their technique they found a collision of SHA-0 reduced to 35 rounds.

In this paper we improve over the results of [3], and present attacks with lower complexities. We present collisions of 65-round reduced SHA-0, and near-collisions of the full compression function of SHA-0 in which up to 142 of the 160 bits of the hash value are equal. We use the very surprising observation that many bits of the message are neutral bits, i.e., they do not affect the differences of the intermediate data for 15–20 rounds. We observe that the strength of SHA-0 is not monotonous, i.e., collisions of 82 rounds are easier to find than of 80 rounds, and use it in our search for near-collisions. We also present several observations on variants of SHA-0.

A comparison of Chabaud and Joux' results with our results is given in Table 1.

Table 2 shows the complexity of finding collisions of reduced and extended SHA-0, as a function of the number of rounds. The table demonstrates that the strength of SHA-0 is not monotonous with the number of rounds. In the complexity calculations we assume that for the extended SHA-0, the additional rounds after the original 80 rounds are performed with the f_i function being XOR, like in rounds 60, ..., 79 that preceed them. We also assume that the first 22 rounds can be gained for free by using the neutral bits.

A comparison between finding near-collisions using a generic attack and our attack is given in Table 3. Note that the generic attack hashes a large number of random messages, all of them are then kept in memory. Due to the birthday paradox, it is expected to have a collision or near-collision with complexity (number of messages) about

$$1.17\sqrt{2^{160}/\binom{160}{k}},$$

Table 2. The Complexity of Finding Collisions of Reduced/Extended SHA-0

Number of Rounds	Complexity	Number of Rounds	Complexity
64	2^{29}	80	2^{56}
65	2^{29}	81	2^{43}
68	2^{43}	82	2^{43}
74	2^{50}	83	2^{65}
75	2^{52}	84	2^{64}
76	$- (*)$	85	2^{71}
77	2^{66}	86	2^{95}
78	2^{56}	87	$- (*)$
79	2^{56}	92	2^{74}

(*) There is no disturbance vector for which the differences of the five registers after 76 or 87 rounds are zero and which no do not have consequent disturbances in the first 17 rounds

Table 3. The Complexities of Finding Near-Collisions of the Compression Function of SHA-0 by a Generic Attack and by Our Attack (the number of different bits is the Hamming distance of the five registers before the feed-forward)

Number of Diff. Bits	0	1	2	3	4	5	18
Generic (time & memory)	2^{80}	2^{76}	2^{73}	2^{70}	2^{68}	2^{65}	2^{41}
Ours (time, negligible memory)	2^{56}	2^{43}	2^{43}	2^{42}	2^{42}	2^{42}	2^{40}

where k is the Hamming weight of the difference. As this attack is generic, it uses no special properties on SHA-0, and thus cannot be used to gain insight on its design.

This paper is organized as follows: Section 2 describes the SHA-0 algorithm, and a few notations. Section 3 describes the attack of Chabaud and Joux. Our improved attack is presented in Section 4. Two pairs of near-collisions of the compression function of SHA-0 and full collision of 65-round reduced SHA-0 are given in section 5. Section 6 describes small variations of SHA-0 that largely affect its security. Finally, Section 7 summarizes the paper.

2   Description of SHA-0

SHA-0 hashes messages of any length in blocks of 512 bits, and produces a message digest of 160 bits.

1. The message is padded with a single bit '1', followed by 0–511 bits '0', followed by a 64-bit representation of the message length, where the number of zeroes is selected to ensure the total length of the padded message is a multiple of 512 bits. The padded message is divided to 512-bit blocks $M_1, \ldots, M_n$.
2. A 5-word buffer h_0 is initialized to

$$h_0 = (67452301_x, EFCDAB89_x, 98BADCFE_x, 10325476_x, C3D2E1F0_x).$$

Table 4. Functions and Constants

Rounds	$f_i(B, C, D)$	K_i
$0 \le i \le 19$	$BC \vee \bar{B}D$	$5A827999_x$
$20 \le i \le 39$	$B \oplus C \oplus D$	$6ED9EBA1_x$
$40 \le i \le 59$	$BC \vee BD \vee CD$	$8F1BBCDC_x$
$60 \le i \le 79$	$B \oplus C \oplus D$	$CA62C1D6_x$

3. Each block M_j in turn is subjected to the compression function, along with the current value of the buffer h_{j-1}. The output is a new value for h_j:

$$h_j = \text{compress}(M_j, h_{j-1}).$$

4. h_n is the output of the hash function.

The compression function is:

1. Divide the 512-bit block M_j to 16 32-bit words $W_0, W_1, \ldots, W_{15}$.
2. Expand the 16 words to 80 words by the recurrence equation:

$$W_i = W_{i-3} \oplus W_{i-8} \oplus W_{i-14} \oplus W_{i-16}, \qquad i = 16, \ldots, 79. \qquad (1)$$

We denote expansion of a block to 80 words by this equation by $\exp(\cdot)$, and note that $W = \exp(M_j)$.
3. Divide h_{j-1} to the five registers A, B, C, D, and E by

$$(A_0, \ B_0, \ C_0, \ D_0, \ E_0) = h_{j-1}$$

4. Iterate the following round function 80 times ($i = 0, \ldots, 79$)

$$A_{i+1} = (W_i + \text{ROL5}(A_i) + f_i(B_i, C_i, D_i) + E_i + K_i) \bmod 2^{32}, \qquad (2)$$

$$B_{i+1} = A_i, \quad C_{i+1} = \text{ROL30}(B_i), \quad D_{i+1} = C_i, \quad E_{i+1} = D_i,$$

where the functions and constants used in each round are described in Table 4.
5. The output of the compression function is

$$h_j = (A_0 + A_{80}, \ B_0 + B_{80}, \ C_0 + C_{80}, \ D_0 + D_{80}, \ E_0 + E_{80}).$$

In the remainder of the paper we consider only 512-bit messages and only the first application of the compression function. We denote the j'th bit of W_i by W_i^j, and similarly we denote the j'th bits of A_i, B_i, C_i, D_i, and E_i by A_i^j, B_i^j, C_i^j, D_i^j, and E_i^j. We also use the notation f_i to denote the output of $f_i(B_i, C_i, D_i)$ in round i, and f_i^j denotes the j'th bit of f_i.

Table 5. Single Difference and Corrections

		Disturbance	Correction	Rounds			
Round		i	$i+1$	$i+2$	$i+3$	$i+4$	$i+5$
Input	W^1	$0 \rightarrow 1$	$1 \rightarrow 0$				
Words	W^6		$1 \rightarrow 0$				
	W^{31}				$0 \leftrightarrow 1$	$0 \leftrightarrow 1$	$0 \leftrightarrow 1$
Desired	A^6	$0 \rightarrow 1$					
Results	f^1		$0 \rightarrow 1$				
	f^{31}				$0 \leftrightarrow 1$	$0 \leftrightarrow 1$	
	E^{31}						$0 \leftrightarrow 1$
Registers		A^1	B^1	C^{31}	D^{31}	E^{31}	
Differences		$0 \rightarrow 1$	$0 \rightarrow 1$	$0 \rightarrow 1$	$0 \rightarrow 1$	$0 \rightarrow 1$	None

3   Description of Chabaud and Joux Attack

In the attack of Chabaud and Joux [3] messages are constructed with specific differences, such that the effect of the differences of the messages on the difference of the registers $A, \ldots, E$ can be canceled within a few rounds. The cancellation is performed by applying correcting patterns by additional differences in the messages.

The attack is initiated by a selection of a difference Δ, that is later used as the difference of the two colliding messages. The difference is selected with various disturbances and corrections, where the corrections are additional differences used to correct the differences caused by the disturbances. The disturbances are always selected in bit 1 of the message words. Due to the rotations by 5 and 30 bits in the round function, corrections are made in bits 1, 6, and 31 of the words. These disturbances and corrections are aimed to limit the evolution of differences to other bits. The result is that in an expected run, A_i and A'_i can only differ in bit 1 (i.e., $A_i \oplus A'_i \in \{0, 00000002_x\}$), and each time they differ, they cause differences in the other registers in the following rounds, which are then corrected by differences of the messages (or W's).

A disturbance starts by setting bit 1 in one of the input words of M' as the complement of the corresponding bit of M. We now show how applying a correction sequence on bits 6, 1, 31, 31, 31 on the following words may cancel the differences at the end of the sequence. Suppose the initial disturbance is in $W_i^1 \neq W_i'^1$. This input difference causes registers A and A' to differ at bit 1. On each consequent round the difference moves to the next register (B, C, D or E), while the corrections of bits 6, 1, 31, 31, 31 in the input words $W_{i+1}', \ldots W_{i+5}'$, respectively, keep registers A and A' equal in these rounds. After this sequence of a single disturbance and five corrections, the registers' contents are equal. By generating M' from M by applying this mask, and calculating the difference of A and A' at each round we can get the differences described in Table 5 with a non negligible probability. The table describes a disturbance with $W_i^1 = 0$ and $W_i'^1 = 1$, and the required corrections. A similar disturbance and corrections can be applied for a '1' to '0' difference. The notation $0 \rightarrow 1$ refer to a change where a bit is '0' in W and '1' in W'. The notation $0 \leftrightarrow 1$ means that there is a change either from '0' to '1' or from '1' to '0'.

Let $\mathcal{D}$ be a vector of 80 words, which correspond to the 80 rounds of the compression function. Each word in the vector is set to '1' if there is a disturbance in the corresponding round, and is set to '0' otherwise. We call this vector the *disturbance vector*. Since getting a collision for the full function requires five correcting rounds, full collisions require the last five words of the disturbance vector to be zero (but for near-collisions this property is not required). Let $SR^l(\mathcal{D})$ be the vector of 80 words received by prepending l zero words to the first $80 - l$ words of $\mathcal{D}$ (i.e., a non-cyclic shift operation of the words). Then, the corrections are made in bit 6 in the rounds which correspond to non-zero words in $SR^1(\mathcal{D})$, in bit 1 in $SR^2(\mathcal{D})$, and in bit 31 in $SR^3(\mathcal{D})$ and $SR^4(\mathcal{D})$ and $SR^5(\mathcal{D})$. Thus, the expansion of Δ to 80 round can be written in the form

$$
\begin{aligned}
\exp(\Delta) = &\left((\mathcal{D} \oplus SR^2(\mathcal{D})) \lll 1 \right) \oplus \\
&\left(SR^1(\mathcal{D}) \lll 6 \right) \oplus \\
&\left((SR^3(\mathcal{D}) \oplus SR^4(\mathcal{D}) \oplus SR^5(\mathcal{D})) \lll 31 \right),
\end{aligned}
$$

where $\lll$ denotes shift of each word of the vector separately. In addition, since $\exp(\Delta)$ is expanded by the linear feedback shift register of Equation (1), the disturbance vector $\mathcal{D}$ is also generatable by this linear feedback shift register. See [3] for additional details on the attack, and the additional required constraints.

We expect that the value of $A_{i+1} \oplus A'_{i+1}$ be $\mathcal{D}_i \lll 1$ if all the corrections succeed (i.e., only disturbances in the current round affect the difference after the round). Thus, the vector of the expected values of $(A_{i+1} \oplus A'_{i+1})_{i=0,...,79}$, which we denote by δ is

$$
\delta = \mathcal{D} \lll 1
$$

(note that the indices of δ are $1, \ldots, 80$, rather than $0, \ldots, 79$).

As the correction process is probabilistic, and assuming each disturbance has the same probability for correction, we are interested in the disturbance vector with the least Hamming weight for getting the least search complexity (but note that the correction probabilities vary, and depend on the f_i's used in the correction rounds).

4   Our Improved Attack

Our attack is based on the attack of Chabaud and Joux with enhancements that increase the probability of finding collisions and near-collisions.

The main idea is to start the collision search from some intermediate round, thus eliminating the probabilistic behavior of prior rounds. In order to start the collision search from round r, we build a pair of messages M and M' with a difference $M \oplus M' = \Delta$, and with the two additional properties described below. Before we describe these properties we wish to make the following definitions:

Definition 1. Given the difference Δ of two messages, the attack of Chabaud and Joux defines the expected differences δ of the values of register A in each round. We say that a pair of messages *conforms to* δ_r if $A_i \oplus A_i' = \delta_i$ for every $i \in \{1, \ldots, r\}$ (which means that the differences at the output of the first r rounds 0, ..., $r - 1$ are as expected).

Definition 2. Let M and M' be a pair of messages that conforms to δ_r for some $r \geq 16$. We say that the i'th bit of the messages ($i \in \{0, \ldots, 511\}$) is a *neutral bit* with respect to M and M' if the pair of messages received by complementing the i'th bits of M and M' also conform to δ_r. We say that the pair of the i'th and j'th bits is *neutral* with respect to M and M' if all the pairs of messages received by complementing *any* subset of these bits ($\{i\}$, $\{j\}$, or $\{i, j\}$) in both messages M and M' also conform to δ_r. We say that a set of bits $S \subseteq \{0, \ldots, 511\}$ is *neutral* with respect to M and M' if all pairs of messages received by complementing *any* subset of the bits in S in both messages M and M' also conform to δ_r. We say that a subset $S \subseteq \{0, \ldots, 511\}$ of the bits of the messages is a *2-neutral set* with respect to M and M' if every bit in S is neutral, and every pair of bits in S is neutral.

We denote the size of the maximal 2-neutral set (for given messages and r) by $k(r)$. We are now ready to describe the two additional properties:

1. The message pair conforms to δ_r. Having the required sequence of $A \oplus A'$ implies that all other differences (i.e., $B \oplus B'$, $C \oplus C'$, $D \oplus D'$, $E \oplus E'$) are also as required.

2. The message pair has a large-enough 2-neutral set of bits. We expect that a large fraction of the subsets of the bits in the 2-neutral set are also neutral.

Given a pair of messages with these properties, we can construct a set of $2^{k(r)}$ message pairs by complementing subsets of the bits of the 2-neutral set. Since a large fraction of these pairs conform to δ_r, while the probability of random pairs is much smaller, it is advisable to use these pairs for the attack.

How r and $k(r)$ are determined? Starting the search from round r we can calculate the probability

$$p(r) = \prod_{i=r}^{79} p_i$$

of successful corrections in all the rounds given messages that conform to δ_r (where p_i is the probability of successful corrections in round i, or 1 if no correction is performed). When the disturbance vector has zeroes at the last five rounds, $p(r)$ is the probability for getting a collision (otherwise, a near-collision is expected). The number of conforming pairs we need to test is expected to be about $1/p(r)$. Since every subset of $k(r)$ neutral bits can be used, we can try $2^{k(r)}$ pairs using with these bits. Thus, we should select r that satisfies $2^{k(r)} \geq 1/p(r)$. In fact, we select the largest r that satisfies this inequality.

4.1   Finding 2-Neutral Sets of Bits of a Given Pair

The following algorithm finds a 2-neutral set of bits. The input to the algorithm is a pair of messages M, M' with a difference Δ that conforms to δ_r. The algorithm generates 512 candidate pairs by complementing single bits in M, M' (leaving their difference unchanged). Let e_i, $i \in \{0, \ldots, 511\}$, denote a message whose value has a single bit '1', and 511 bits '0', where the bit '1' is in the i'th location. The candidate pairs can be written by

$$(M \oplus e^i, M' \oplus e^i), \qquad\qquad i \in \{0, \ldots, 511\}.$$

Each candidate pair is tested to conform to δ_r. If a candidate pair conforms to δ_r, then bit i is a neutral bit.

In order to find a 2-neutral set of bits we define a graph whose vertices correspond to the neutral bits. We then add an edge for each pair of bits whose simultaneous complementations does not affect conformance. This graph describes all the bits whose complementation does not affect conformance, and all the pairs of these bits whose simultaneous complementations does not affect conformance. We are now interested to find the maximal clique (or an almost maximal clique) in this graph, i.e., the maximal subset of vertices for which any vertex in the subset is connected to any other vertex in the subset by an edge. Although in general finding a maximal clique is an NP-complete problem, in our case finding a large enough clique is not difficult, as many vertices are connected to all other vertices by edges.

We are now ready to make some very important observations, on which the success of our attack is based:

Observation 1 When we perform a search with the set of $2^{k(r)}$ message pairs, about 1/8 of the pairs (i.e., about $2^{k(r)-3}$ pairs) conform to δ_r.

Let

$$p(r \rightarrow r') = \prod_{j=r}^{r'-1} p_i$$

be the probability that a pair that conforms to δ_r also conforms to $\delta_{r'}$, and notice that $p(r) = p(r \rightarrow 80)$.

Observation 2 Let r and r' be some rounds where $p(r \rightarrow r') \approx 2^{-k(r)}$. By trying the $2^{k(r)}$ generated message pairs, we get the expected number of pairs conforming to $\delta_{r'}$, but surprisingly a fraction of the pairs that conform to $\delta_{r'}$ also conform to $\delta_{r'+l}$, which we would expect to get with a larger set of about $2^{k(r)+\alpha}$, where $2 \leq l \leq 4$ and $3 \leq \alpha \leq 8$.

In the actual attack we improve the algorithm further by searching for pairs of non-neutral bits whose simultaneous complementation create pairs that also conform to δ_r (and similarly search for triplets of bits, or larger sets of bits). Using this method we receive a larger number of neutral "bits" that can be used for our analysis with higher rounds.

Table 6. A Pair of Messages with 40 Neutral Bits and Simultaneous Neutral Bits for $r = 22$ (the bits are numbered in the range $0, \ldots, 511$)

M=	$19EF75A8_x$	$D2F24D9A_x$	$8F179A7D_x$	$1A295690_x$
	$2E84C143_x$	$D74B9DDC_x$	$18C10577_x$	$8107056E_x$
	$5B1A47ED_x$	$6212C3F2_x$	$3B2D04F8_x$	$F5581AB0_x$
	$26D8CDBC_x$	$AB3A3248_x$	$F347E871_x$	$46278F39_x$
M'=	$19EF75A8_x$	$D2F24D9A_x$	$8F179A7D_x$	$1A295692_x$
	$2E84C103_x$	$D74B9DDE_x$	$98C10577_x$	$0107056E_x$
	$DB1A47EF_x$	$6212C3B2_x$	$3B2D04F8_x$	$75581AF0_x$
	$A6D8CDBE_x$	$AB3A324A_x$	$7347E831_x$	$C6278F3B_x$

Singles:	388	457	458	459	464	484	485	489
	490	491	494	495	496	499	501	506
	507							

Pairs:	301 264		461 424		493 456		497 460	
	500 463		502 428					

Triplets:	296 175 138	341 220 183	376 255 218	386 265 228
	391 270 233	462 426 425	466 429 393	488 483 478
	492 334 297			

Quadru-	229 137 108 71	331 210 116 79	364 338 337 300
plets:	455 435 434 397	505 437 431 400	

Quintu-	471 470 469 433 395	487 465 344 343 306
plets:	504 480 451 438 420	

An example of a pair of messages with its neutral set of bits, is given in Table 6. In this example $r = 22$ and the size of the neutral set is $k(22) = 40$. In particular, the quadruplet 229 137 108 71 consists of bits of rounds 7, 4, 3, and 2, so the changes at round 2 are successfully corrected by the changes in the other rounds so the difference is unaffected for 20 rounds, and even from round 7 there are 15 additional rounds whose difference is not affected.

Observation 3 In many cases pairs of bits that are simultaneously neutral, but each bit is not, are of the form W_i^j, $W_{i-l}^{(j-5l) \bmod 32}$ for small l's. Similarly triplets (and quartets, etc.) of non-neutral bits, whose simultaneous complementation is neutral are of the same form, i.e., W_i^j, and $W_{i-l}^{(j-5l) \bmod 32}$ for two different small l's. We call such sets of bits *simultaneous-neutral sets*, and in case of pairs of bits *simultaneous-neutral pairs*.

4.2   Finding a Pair with a Larger 2-Neutral Set

For the attack, we are interested in finding a message pair with a maximal 2-neutral set of bits. Assume that we are already given a pair conforming to δ_r. We are now modifying this pair slightly in order to get another pair that conforms to δ_r with a larger 2-neutral set of bits.

This algorithm takes the given message pair as a base, modifies it in a certain way that we describe later, and calls the algorithm that finds the 2-neutral set of the new pair. If the size of this set is larger than the set of the base pair, the base pair is replaced by the new pair, and the algorithm restarts with the new pair as the base.

By modifying the current message pair we create a new pair that hopefully conforms to δ_r. The modifications are made in bits that maximize the probability of success. In order to create a new conforming pair, we modify several neutral bits (and simultaneously-neutral sets of bits), and check whether the resultant pair conforms to δ_r.

In some cases we can improve further. In rounds where bit 1 differs, i.e., $W_i^1 \neq W_i'^1$, the carry from bit 1 to the next can create a difference in the next bit. The probability for this carry to make this difference is $1/2$. In such case $A_{i+1} \oplus A_{i+1}' \neq 00000002_x$, and thus the new pair does not conform to δ_r.

Observation 4 If the differences of the carry is changed, the change can be canceled by complementing W_i^0 and $W_i'^0$, or by complementing other bits in the message that affect A_{i+1}^0 indirectly.

Such bits are also W_{i-1}^{27} and $W_{i-1}'^{27}$ (which affect A_i^{27}, and then A_{i+1}^0 after the rotate operation), or W_{i-2}^{22} and $W_{i-2}'^{22}$, or $W_{i-l}^{(32-5l) \bmod 32}$ and $W_{i-l}'^{(32-5l) \bmod 32}$ for other small l's. Each such complementation has probability $1/2$ to cancel the difference in the carry.

This algorithm can be simplified as follows: The algorithm takes as an input a message and modifies a few subsequent bits in several subsequent words, with the shift of five bits as mentioned above. For example, the modified bits cover all $2^{24} - 1$ (non-empty) subsets of $\{W_0^0, \ldots, W_0^3\} \cup \{W_1^5, \ldots, W_1^8\} \cup \ldots \cup \{W_5^{25}, \ldots, W_5^{28}\}$. Then, the pattern of modification is shifted by all 31 possible rotations. Finally, we proceed and make the same analysis starting from W_1, then W_2, etc. The modification process ends when the algorithm starts with W_{10}. This simplification lacks consideration of some optimizations and details given earlier, whose incorporation is vital for an optimized implementation.

4.3   Increasing the Number of Conforming Rounds

In order to start the search at a higher round we need to construct a pair that conforms to $\delta_{r'}$, where $r' > r$. This pair is constructed using the last pair with the maximal number of neutral bit we have. The pair undergoes small modifications of the form described above. Once a message conforms to $\delta_{r'}$ is found, we use the algorithms described in Subsections 4.1 and 4.2 to find a 2-neutral set, and then to find a pair with the largest 2-neutral set.

4.4   Final Search

After computing the 2-neutral set, we start the final search by complementing sequentially every subset of the bits in the 2-neutral set (a total of $2^{k(r)} - 1$ trials). Since a large fraction of the resulting pairs of messages conform to δ_r, then the search effectively starts at round r. If in addition $2^{k(r)} > 1/p(r)$, then we expect

Table 7. Probability Summary

Round Function	$-\log P_i$
0, ..., 19 IF	25
20, ..., 39 XOR	16
40, ..., 59 MAJ	15
60, ..., 79 XOR	15

to find a collision or a near-collision, depending on the expected difference after r rounds. If $2^{k(r)} > 1/p(r \to r')$ for some r', then we expect to find a collision (or a near-collision) of r' rounds reduced (or extended) SHA-0.

5 Results

In our search we used Δ that is optimized for finding 82-round collisions (thus also near-collisions of 80 rounds). This Δ is not suitable for finding full collisions of 80 rounds, as it has two disturbances at the last five rounds. However, its corresponding 80-round probability is much higher than the probability of a Δ that allows a full collision. Although this Δ cannot provide full collisions, it can lead to collisions of 65-round reduced SHA-0 and of 82-round extended SHA-0. The overall probability of successful corrections in 82-round SHA-0 is $p(0 \to 82) = 2^{-71}$. A probability summary for each set of 20 consecutive rounds (i.e., the IF, XOR, MAJ, XOR rounds) is described in Table 7 (in rounds 80 and 81 the probability is 1 if $f_{80} = f_{81} = $ XOR). Using our technique with $r = 22$ the overall probability is reduced to 2^{-43}. Our algorithm finds a 2-neutral set with 40 neutral and simultaneous-neutral bits (see Table 6), thus we expect to find near-collisions of the compression function after 73 rounds in two computation days on a PC. Our actual findings (using an earlier set of neutral bits) are near-collisions of the compression function with a difference of only three bits (of $A \oplus A', \ldots, E \oplus E'$) after 76 rounds (that still conform to δ_{76}), which are also near-collisions of the full compression function (but do not conform to δ_{80}), and full collisions of 65-round reduced SHA-0. The near collisions were found after about a day of computation for each pair, which is equivalent to a search with a complexity of 2^{35}. Finding 65-round near-collisions take about half an hour. Two such pairs of messages (in 32-bit hex words) are:

1. $M_1 = $ 310EEB32 AC418FC2 415D5A54 6FFA5AA9
5EE5A5F5 7621F42D 8AE2F4CA F7ACF74B
B144B4E1 5164DF45 C61AD50C D5833699
6F0BB389 B6468AC5 4D4323F9 86088694

   $M_1' = $ 310EEB32 AC418FC2 415D5A54 6FFA5AAB
5EE5A5B5 7621F42F 0AE2F4CA 77ACF74B
3144B4E3 5164DF05 C61AD50C 558336D9
EF0BB38B B6468AC7 CD4323B9 06088696

Table 8. Difference of the Hash Results Before and After the Feed-forward (i.e., $A_{80} \oplus A'_{80}, \ldots, E_{80} \oplus E'_{80}$ and $(A_0 + A_{80}) \oplus (A'_0 + A'_{80}), \ldots, (E_0 + E_{80}) \oplus (E'_0 + E'_{80}))$, and Their Hamming Weights

	Difference (in hex)	Weight
M_1 and M'_1:		
Before:	00401FA0 00060184 00000400 80000020 80000000	17
After:	01C061A0 00020084 00000C00 800001E0 80000000	19
M_2 and M'_2:		
Before:	00C030A4 000E0304 00000403 80000060 80000000	20
After:	004070A4 00020104 00000C07 80000020 80000000	18

2. $M_2 =$ EF567055 F0722904 009D8999 5AFB3337
   37D5D6A8 9E843D80 69229FB9 06D589AA
   4AD89B67 CFCCCD2C A9BAE20D 6F18C150
   43F89DA4 2E54FE2E AE7B7A15 80A09D3D

   $M'_2 =$ EF567055 F0722904 009D8999 5AFB3335
   37D5D6E8 9E843D82 E9229FB9 86D589AA
   CAD89B65 CFCCCD6C A9BAE20D EF18C110
   C3F89DA6 2E54FE2C 2E7B7A55 00A09D3F

The differences of the results of hashing M_1 and M_2 with the full SHA-0 are described in Table 8 along with the number of differing bits. Tables 9 and 10 show detailed information of the evolution of differences in each round of the compression function, including the expanded messages, their differences, the differences $A_{i+1} \oplus A'_{i+1}$, the probability of conformance of each round (in log form), and the rounds where the values collide, or the number of differing bits of the five registers. Both messages collide after 65 rounds, and have only small differences afterwards. If we consider SHA-0 reduced to 76 rounds, our results show a near collision with difference of only three bits before the feed forward and three and four bits difference after the feed forward when using M_1 and M_2.

6   SHA-0 Variants

In this section we analyze some variants of SHA-0 that show strengths and weaknesses of the hash function.

6.1   Increasing the Number of Rounds

There are Δ's that lead to collision after 82 rounds, whose probability $p(0 \rightarrow 82)$ is considerably larger than the probability $p(0 \rightarrow 80)$ of the best Δ that leads to an 80-round collision. Therefore, increasing the number of rounds of SHA-0 from 80 to 82 would make it much easier to find collisions.

Table 9. A Near-Collision and its Differences in the Various Rounds (M_1 and M_1' are formed by the first 16 words of W and W')

Round (i)	W_i	W_i'	$\exp(\Delta)$	$A_{i+1} \oplus A_{i+1}'$	$-\log P_i$	Diff Bits
0	$310EEB32_x$	$310EEB32_x$	00000000_x	00000000_x	0	collision
1	$AC418FC2_x$	$AC418FC2_x$	00000000_x	00000000_x	0	collision
2	$415D5A54_x$	$415D5A54_x$	00000000_x	00000000_x	0	collision
3	$6FFA5AA9_x$	$6FFA5AAB_x$	00000002_x	00000002_x	1	1
4	$5EE5A5F5_x$	$5EE5A5B5_x$	00000040_x	00000000_x	0	1
5	$7621F42D_x$	$7621F42F_x$	00000002_x	00000000_x	2	1
6	$8AE2F4CA_x$	$0AE2F4CA_x$	80000000_x	00000000_x	1	1
7	$F7ACF74B_x$	$77ACF74B_x$	80000000_x	00000000_x	1	1
8	$B144B4E1_x$	$3144B4E3_x$	80000002_x	00000002_x	1	1
9	$5164DF45_x$	$5164DF05_x$	00000040_x	00000000_x	0	1
10	$C61AD50C_x$	$C61AD50C_x$	00000000_x	00000002_x	3	2
11	$D5833699_x$	$558336D9_x$	80000040_x	00000000_x	1	2
12	$6F0BB389_x$	$EF0BB38B_x$	80000002_x	00000000_x	3	2
13	$B6468AC5_x$	$B6468AC7_x$	00000002_x	00000002_x	2	2
14	$4D4323F9_x$	$CD4323B9_x$	80000040_x	00000000_x	1	2
15	86088694_x	06088696_x	80000002_x	00000000_x	2	1
16	$77518F42_x$	$F7518F42_x$	80000000_x	00000000_x	1	1
17	$DF9C29D7_x$	$5F9C29D5_x$	80000002_x	00000002_x	2	2
18	$5FAAAC39_x$	$DFAAAC7B_x$	80000042_x	00000002_x	1	2
19	$BB09175F_x$	$BB09171F_x$	00000040_x	00000002_x	3	3
20	$6490CB61_x$	$E490CB21_x$	80000040_x	00000002_x	2	4
21	$6861259A_x$	$686125D8_x$	00000042_x	00000000_x	1	4
22	$CDEC748D_x$	$4DEC748F_x$	80000002_x	00000002_x	1	3
23	$445065FB_x$	$C45065F9_x$	80000002_x	00000002_x	1	3
24	$686ECB35_x$	$686ECB75_x$	00000040_x	00000002_x	0	2
25	$9697B486_x$	$1697B486_x$	80000000_x	00000002_x	2	2
26	$B2EBAF47_x$	$32EBAF05_x$	80000042_x	00000002_x	1	3
27	$B0A26036_x$	$30A26074_x$	80000042_x	00000000_x	1	3
28	$D04FEF97_x$	$D04FEF95_x$	00000002_x	00000000_x	1	2
29	$EAC4868C_x$	$EAC4868C_x$	00000000_x	00000000_x	0	2
30	$475CB800_x$	$475CB800_x$	00000000_x	00000000_x	0	1
31	$CD8B252F_x$	$4D8B252F_x$	80000000_x	00000000_x	0	collision
32	$AA516EC2_x$	$AA516EC0_x$	00000002_x	00000002_x	1	1
33	$B55E320E_x$	$B55E324C_x$	00000042_x	00000002_x	1	2
34	$445AED30_x$	$445AED70_x$	00000040_x	00000002_x	2	3
35	$C99B3C31_x$	$499B3C73_x$	80000042_x	00000002_x	1	3
36	$CC6D6275_x$	$CC6D6277_x$	00000002_x	00000000_x	1	3
37	$82AF2BDD_x$	$02AF2BDD_x$	80000000_x	00000000_x	0	2
38	$2B453B89_x$	$2B453B89_x$	00000000_x	00000000_x	0	1
39	$D3219627_x$	53219627_x	80000000_x	00000000_x	0	collision
40	$F27B216D_x$	$F27B216D_x$	00000000_x	00000000_x	0	collision
41	$B82EDD37_x$	$B82EDD37_x$	00000000_x	00000000_x	0	collision
42	$F5DF3BC7_x$	$F5DF3BC7_x$	00000000_x	00000000_x	0	collision
43	$6186FBE6_x$	$6186FBE6_x$	00000000_x	00000000_x	0	collision
44	$E350E8D5_x$	$E350E8D5_x$	00000000_x	00000000_x	0	collision
45	$503FB3B9_x$	$503FB3B9_x$	00000000_x	00000000_x	0	collision
46	$A7CE16AD_x$	$A7CE16AF_x$	00000002_x	00000002_x	1	1
47	$48A469D3_x$	$48A46991_x$	00000042_x	00000002_x	1	2
48	$4C4F1126_x$	$4C4F1164_x$	00000042_x	00000000_x	1	2
49	$6325C5A5_x$	$E325C5A7_x$	80000002_x	00000000_x	2	2
50	$354CDD51_x$	$354CDD51_x$	00000000_x	00000000_x	0	2
51	$66FDFD2C_x$	$66FDFD2C_x$	00000000_x	00000000_x	1	1
52	$675D748C_x$	$E75D748C_x$	80000000_x	00000000_x	0	collision
53	$34FDD312_x$	$34FDD312_x$	00000000_x	00000000_x	0	collision
54	$180DF165_x$	$180DF167_x$	00000002_x	00000002_x	1	1
55	$44F6564F_x$	$44F6560D_x$	00000042_x	00000002_x	1	2
56	$7F16D89E_x$	$7F16D8DC_x$	00000042_x	00000002_x	1	2
57	$A2801211_x$	22801211_x	80000000_x	00000002_x	3	3
58	$6735580C_x$	$6735584E_x$	00000042_x	00000000_x	1	4
59	$28526DED_x$	$28526DAF_x$	00000042_x	00000000_x	2	3
60	$814398E5_x$	$814398E7_x$	00000002_x	00000000_x	1	2
61	$4B535174_x$	$4B535174_x$	00000000_x	00000000_x	0	2
62	$DBDE9B03_x$	$DBDE9B03_x$	00000000_x	00000000_x	0	1
63	$EE3462DC_x$	$6E3462DC_x$	80000000_x	00000000_x	0	collision
64	$4D46459D_x$	$4D46459D_x$	00000000_x	00000000_x	0	collision
65	$7C86B19B_x$	$7C86B199_x$	00000002_x	00000002_x	1	1
66	$DB10930D_x$	$DB10934F_x$	00000042_x	00000002_x	1	2
67	$3714064E_x$	$3714060C_x$	00000042_x	00000000_x	1	2
68	$8295AC97_x$	$0295AC95_x$	80000002_x	00000000_x	1	2
69	$E0484724_x$	$E0484724_x$	00000000_x	00000000_x	0	2
70	$8BD1B4B6_x$	$8BD1B4B4_x$	00000002_x	00000000_x	1	2
71	$8AD78A15_x$	$0AD78A55_x$	80000040_x	00000000_x	0	1
72	$B52D822B_x$	$B52D822B_x$	00000000_x	00000002_x	2	2
73	$7D857AD1_x$	$FD857A93_x$	80000042_x	00000000_x	1	3
74	$B7B1D9F1_x$	$37B1D9B3_x$	80000042_x	00000000_x	1	3
75	$E138B8FC_x$	$E138B8FC_x$	00000000_x	00000002_x	2	3
76	$A58DD5A0_x$	$A58DD5E2_x$	00000042_x	00000082_x	1	5
77	$F29EAD7D_x$	$F29EAD3F_x$	00000042_x	00001000_x	1	5
78	$FC71D2D4_x$	$FC71D2D6_x$	00000002_x	00060184_x	1	9
79	$BDE88CF2_x$	$BDE88CF2_x$	00000000_x	$00401FA0_x$	0	17

Table 10. A Near-Collision and its Differences in the Various Rounds (M_2 and M'_2 are formed by the first 16 words of W and W')

Round (i)	W_i	W'_i	$\exp(\Delta)$	$A_{i+1} \oplus A'_{i+1}$	$-\log P_i$	Diff Bits
0	$EF567055_x$	$EF567055_x$	00000000_x	00000000_x	0	collision
1	$F0722904_x$	$F0722904_x$	00000000_x	00000000_x	0	collision
2	$009D8999_x$	$009D8999_x$	00000000_x	00000000_x	0	collision
3	$5AFB3337_x$	$5AFB3335_x$	00000002_x	00000002_x	1	1
4	$37D5D6A8_x$	$37D5D6E8_x$	00000040_x	00000000_x	0	1
5	$9E843D80_x$	$9E843D82_x$	00000002_x	00000000_x	2	1
6	$69229FB9_x$	$E9229FB9_x$	80000000_x	00000000_x	1	1
7	$06D589AA_x$	$86D589AA_x$	80000000_x	00000000_x	1	1
8	$4AD89B67_x$	$CAD89B65_x$	80000002_x	00000002_x	1	1
9	$CFCCCD2C_x$	$CFCCCD6C_x$	00000040_x	00000000_x	0	1
10	$A9BAE20D_x$	$A9BAE20D_x$	00000000_x	00000002_x	3	2
11	$6F18C150_x$	$EF18C110_x$	80000040_x	00000000_x	1	2
12	$43F89DA4_x$	$C3F89DA6_x$	80000002_x	00000000_x	3	2
13	$2E54FE2E_x$	$2E54FE2C_x$	00000002_x	00000002_x	2	2
14	$AE7B7A15_x$	$2E7B7A55_x$	80000040_x	00000000_x	1	2
15	$80A09D3D_x$	$00A09D3F_x$	80000002_x	00000000_x	2	1
16	$8B479C85_x$	$0B479C85_x$	80000000_x	00000000_x	1	1
17	$CB3EAD0A_x$	$4B3EAD08_x$	80000002_x	00000002_x	2	2
18	$1E522001_x$	$9E522043_x$	80000042_x	00000000_x	1	2
19	20205362_x	20205322_x	00000040_x	00000002_x	3	3
20	$D63179BF_x$	$563179FF_x$	80000040_x	00000002_x	2	4
21	$A8576A05_x$	$A8576A47_x$	00000042_x	-00000000_x	1	4
22	$ADA12DA9_x$	$2DA12DAB_x$	80000002_x	00000000_x	1	3
23	$9F88A004_x$	$1F88A006_x$	80000002_x	00000002_x	1	3
24	$C0728FEA_x$	$C0728FAA_x$	00000040_x	00000000_x	0	2
25	$C64B8CDF_x$	$464B8CDF_x$	80000000_x	00000002_x	2	2
26	$6B98FFAC_x$	$EB98FFEE_x$	80000042_x	00000000_x	1	3
27	$A11EE3F6_x$	$211EE3B4_x$	80000042_x	00000000_x	1	3
28	$FDF912D1_x$	$FDF912D3_x$	00000002_x	00000000_x	1	2
29	$6D3BF6BA_x$	$6D3BF6BA_x$	00000000_x	00000000_x	0	2
30	$298328CF_x$	$298328CF_x$	00000000_x	00000000_x	0	1
31	$29EF82E2_x$	$A9EF82E2_x$	80000000_x	00000000_x	0	collision
32	$385CC5D4_x$	$385CC5D6_x$	00000002_x	00000002_x	1	1
33	$04D65A78_x$	$04D65A3A_x$	00000042_x	00000000_x	1	2
34	$8A1424F0_x$	$8A1424B0_x$	00000040_x	00000002_x	2	3
35	$11351F45_x$	$91351F07_x$	80000042_x	00000000_x	1	3
36	$82BF1CBF_x$	$82BF1CBD_x$	00000002_x	00000000_x	1	3
37	$D0F0184B_x$	$50F0184B_x$	80000000_x	00000000_x	0	2
38	$556595C9_x$	$556595C9_x$	00000000_x	00000000_x	0	1
39	$F293B286_x$	$7293B286_x$	80000000_x	00000000_x	0	collision
40	$4346ADD9_x$	$4346ADD9_x$	00000000_x	00000000_x	0	collision
41	$36E6A098_x$	$36E6A098_x$	00000000_x	00000000_x	0	collision
42	$EEE67B0B_x$	$EEE67B0B_x$	00000000_x	00000000_x	0	collision
43	$9E56A7D0_x$	$9E56A7D0_x$	00000000_x	00000000_x	0	collision
44	60238639_x	60238639_x	00000000_x	00000000_x	0	collision
45	$7AC21718_x$	$7AC21718_x$	00000000_x	00000000_x	0	collision
46	$DAECDF02_x$	$DAECDF00_x$	00000002_x	00000002_x	1	1
47	$BF89EC25_x$	$BF89EC67_x$	00000042_x	00000000_x	1	2
48	$8BCC5BE5_x$	$8BCC5BA7_x$	00000042_x	00000000_x	1	2
49	$F9E93AA7_x$	$79E93AA5_x$	80000002_x	00000000_x	2	2
50	$59C4AF61_x$	$59C4AF61_x$	00000000_x	00000000_x	0	2
51	$D45FFB3B_x$	$D45FFB3B_x$	00000000_x	00000000_x	1	1
52	$4E1035E8_x$	$CE1035E8_x$	80000000_x	00000000_x	0	collision
53	$016512B4_x$	$016512B4_x$	00000000_x	00000000_x	0	collision
54	$18901C29_x$	$18901C2B_x$	00000002_x	00000002_x	1	1
55	$35ECCBD3_x$	$35ECCB91_x$	00000042_x	00000000_x	1	2
56	$27099F83_x$	$27099FC1_x$	00000042_x	00000000_x	1	2
57	$49C921C6_x$	$C9C921C6_x$	80000000_x	00000002_x	3	3
58	$E2ED9980_x$	$E2ED99C2_x$	00000042_x	00000002_x	1	4
59	$17C2D470_x$	$17C2D432_x$	00000042_x	00000000_x	2	3
60	$BD164D15_x$	$BD164D17_x$	00000002_x	00000000_x	1	2
61	$26C37009_x$	$26C37009_x$	00000000_x	00000000_x	0	2
62	$5E724CBE_x$	$5E724CBE_x$	00000000_x	00000000_x	0	1
63	$CE9A5044_x$	$4E9A5044_x$	80000000_x	00000000_x	0	collision
64	$D3C21B0E_x$	$D3C21B0E_x$	00000000_x	00000000_x	0	collision
65	$3A0DACE4_x$	$3A0DACE6_x$	00000002_x	00000002_x	1	1
66	$3BA3534D_x$	$3BA3530F_x$	00000042_x	00000002_x	1	2
67	$113A26F1_x$	$113A26B3_x$	00000042_x	00000000_x	1	2
68	$D19BC830_x$	$519BC832_x$	80000002_x	00000000_x	1	2
69	$29E9FA23_x$	$29E9FA23_x$	00000000_x	00000000_x	0	2
70	$70D1E9E5_x$	$70D1E9E7_x$	00000002_x	00000002_x	1	2
71	63247261_x	$E3247221_x$	80000040_x	00000000_x	0	1
72	$3FCFE72E_x$	$3FCFE72E_x$	00000000_x	00000002_x	2	2
73	$14D7B0B7_x$	$94D7B0F5_x$	80000042_x	00000002_x	1	3
74	$077CF5B9_x$	$877CF5FB_x$	80000042_x	00000002_x	1	3
75	$1FF465A6_x$	$1FF465A6_x$	00000000_x	00000002_x	2	3
76	$2628792C_x$	$2628796E_x$	00000042_x	00000182_x	1	6
77	$C6CC2FD7_x$	$C6CC2F95_x$	00000042_x	$00000100C_x$	1	8
78	$E295DBF3_x$	$E295DBF1_x$	00000002_x	$000E0304_x$	1	13
79	$B19BF7ED_x$	$B19BF7ED_x$	00000000_x	$00C030A4_x$	0	20

6.2   Different Order of Functions

Modifying the order of the f_i functions can reduce the complexity of the attack. For example, if the order would be IF, XOR, MAJ, XOR, ..., IF, XOR, MAJ, XOR, where in each round the function changes, the restrictions caused by two consecutive IF round would be removed, and thus Δ's with much higher probabilities could be chosen.

6.3   SHA-1

Since in SHA-1 Equation (1) is replaced by

$$W_i = \mathrm{ROL1}(W_{i-3} \oplus W_{i-8} \oplus W_{i-14} \oplus W_{i-16}), \qquad i = 16, \ \ldots, \ 79. \qquad (3)$$

which makes the mixing of the message bits much more effective, and since the techniques used in this paper uses the properties inherited from equation (1), the presented attacks are not applicable to SHA-1.

7   Summary

In this paper we described how to find near-collisions of SHA-0 using the surprising existence of many neutral bits. The near-collisions were found within a day on our PC. Our technique also improves the complexity of finding full collisions of SHA-0, but we concentrated on near-collisions due to the very low complexity of finding them. The observation that the strength of SHA-0 is not monotonous with the number of rounds is used here to find near-collisions of 80 rounds by applying the much more efficient attack on SHA-0 extended to 82 rounds. We expect that finding full collisions will take a month of computation time, and intend to check it in the continuation of our research. Due to the additional rotate operation, the results of this paper are not applicable to SHA-1.

References

1. Ross Anderson, Eli Biham, *Tiger: a Fast New Hash Function*, proceedings of Fast Software Encryption, LNCS 1039, pp. 89–97, Springer Verlag, 1996.
2. Eli Biham, Adi Shamir, *Differential Cryptanalysis of Snefru, Khafre, REDOC-II, LOKI and Lucifer*, Advances in Cryptology, proceedings of CRYPTO '91, LNCS 576, pp. 156–171, 1992.
3. Florent Chabaud, Antoine Joux, *Differential Collisions in SHA-0*, Advanced in Cryptology, proceedings of CRYPTO '98, LNCS 1462, pp. 56–71, Springer Verlag, 1999.
4. Hans Dobbertin, *Cryptanalysis of MD4*, Journal of Cryptology, Vol. 11 pp. 253–271, 1998.
5. Rosario Genaro, Shai Halevi, Tal Rabin, *Secure Hash-and-Sign Signatures Without the Random Oracle*, Advanced in Cryptology, proceedings of EUROCRYPT'99, LNCS 1592, pp. 123–139, 1999.

6. Antoine Joux, private communications, 2004.
7. Alfred Menezes, Paul van Oorschot, Scott Vanstone, *Handbook of Applied Cryptography*, CRC Press, 1997.
8. National Institute of Standards and Technologies, *Secure Hash Standard*, Federal Information Processing Standards Publication, FIPS-180, May 1993.
9. National Institute of Standards and Technologies, *Secure Hash Standard*, Federal Information Processing Standards, Publication FIPS-180-1, April 1995.
10. National Institute of Standards and Technologies, *FIPS 180-2 Secure Hash Standard, Change Notice 1*, Federal Information Processing Standards Publication, FIPS-180-2, December, 2003.
11. National Institute of Standards and Technologies, *Secure Hash Standard*, Federal Information Processing Standards Publication, FIPS-180-2, August 2002.
12. Ron Rivest, *The MD4 Message-Digest Algorithm*, Network Working Group Request for Comments:1186, October 1990.
13. Ron Rivest, *The MD5 Message-Digest Algorithm*, Network Working Group Request for Comments:1321, April 1992.
14. Ralph Merkle, *A Fast Software One-Way Hash Function*, Journal of Cryptology, Vol. 3, No. 1, pp. 43–58, 1990.
15. Serge Vaudenay, *On the Need for Multipermutation: Cryptanalysis of MD4 and SAFER*, proceedings of Fast Software Encryption, Second International Workshop, LNCS 1008, pp. 286–297, Springer-Verlag, 1995.

Multicollisions in Iterated Hash Functions. Application to Cascaded Constructions

Antoine Joux

DCSSI Crypto Lab
51, Bd de Latour-Maubourg
75700 Paris 07 SP, France
antoine.joux@m4x.org

Abstract. In this paper, we study the existence of multicollisions in iterated hash functions. We show that finding multicollisions, i.e. r-tuples of messages that all hash to the same value, is not much harder than finding ordinary collisions, i.e. pairs of messages, even for extremely large values of r. More precisely, the ratio of the complexities of the attacks is approximately equal to the logarithm of r. Then, using large multicollisions as a tool, we solve a long standing open problem and prove that concatenating the results of several iterated hash functions in order to build a larger one does not yield a secure construction. We also discuss the potential impact of our attack on several published schemes. Quite surprisingly, for subtle reasons, the schemes we study happen to be immune to our attack.

1 Introduction

One-Way hash functions are widely used cryptographic primitives, they operate on messages of almost arbitrary length[1] and output a fixed size value. Cryptographic hash functions should satisfy many security properties, such as the impossibility from a given hash to recover an associated message. However, the main security requirement for a hash function is its collision resistance. Informally, given a good hash function, no attacker should be able to find a pair of different messages M and M' leading to identical hash values. It is a well-known fact that all hash functions suffer from a generic birthday paradox based attack. More precisely, if H is a hash function that outputs n–bit values, then among the hash values of $2^{n/2}$ different messages, there exists a collision with non negligible probability. For this reason, hash functions that output values smaller than 160 bits are considered as deprecated. Yet, in the past, 128–bit hash functions were proposed and for legacy reasons they are still encountered in applications.

In practice, building a cryptographic function with an input of variable size is not a simple task. For this reason, most hash functions are based on an iterated construction that makes use of a so-called compression function, whose inputs have fixed sizes. Examples of such a construction are Snefru [7], MD4 [12], MD5 [13] or SHA [9]. In this paper, we specifically study one-way hash-functions built by iterating a compression function.

---

[1] The length is often bounded by a very large number such as 2^{64}. However, this is irrelevant for the attacks presented here.

M. Franklin (Ed.): CRYPTO 2004, LNCS 3152, pp. 306–316, 2004.

Our main goal is to solve a long standing open problem: Is the concatenation of two independent hash values more secure than a single hash-value ? This question is of general interest and has appeared in many contexts. As far as we know, this construction first appeared as a generic transform in the PhD thesis of B. Preneel [10] and was called cascading. It was presented there as a mean to increase the security level at the cost of a decreased performance.

In fact, this idea of cascading hash functions is likely to be encountered in applications, for example, a construction called SHA-1x was used at some point in PGP and involves the computation of two SHA-1 values with a different set of initial constants. Similarly, the authors of RIPEMD [4] propose optional extensions of their hash functions to 256 and 320 bits values. In this case, the use of two hashing is extremely efficient since the original 128 and 160 bits algorithms already involve two parallel hashing whose results are normally added together.

Yet, according to [4], one should not expect to improve the security level with these constructions since unwanted dependencies between two slightly different instances of the same hash function may yield unforeseen attacks. In the same vein, a length doubling transform is suggested in the hash function chapter of [14], together with a warning that while no attacks are known several people have serious reservations about the construct.

As a consequence, the security of hash functions cascading is not very clear. Roughly, the cryptographic folklore states that the construction is good when two "independent" hash functions are cascaded. Clearly, this is true for random oracles and the generalization seems natural. For reference about this folklore knowledge, the interested reader may look up fact 9-27 in [6], that states that such a cascade is secure and that one could hope for a security of the order of the product of the security of the initial hash functions. However, we show in section 4 that this construction is in fact insecure, whenever an iterated hash function is involved in the cascading. Even cascading a 160-bit iterated hash function and a 160-bit random oracle does not really increase security above the initial 2^{80} level for collision resistance and above 2^{160} for preimage (or second preimage) resistance.

In order to solve this problem and prove that cascading two hash values is in fact insecure, we first address the simpler question of constructed multicollisions in an iterated hash function. This notion of multicollisions was first used by Merkle in [8] to study the security of a hash function based on DES. A related security property, namely r-collision freeness, has been suggested as a useful tool for building efficient cryptographic primitives. It was used for the micropayment scheme Micromint of Rivest and Shamir [11], for identification schemes by Girault and Stern in [5] and for signature schemes by Brickell and al. in [1]. The intuition behind this problem is that constructing r different messages with the same hash values should be much harder than constructing only two such messages. Once again, this is true when using random oracles. However, when iterated hash functions are involved, this intuition is false and multicollisions can be easily constructed.

The paper is organized as follows. In section 2 we recall some basic facts about iterated hash function and the possible security properties of hash functions. In section 3, we describe the basic attack for constructing multicollisions in iterated hash functions. In section 4, we use this attack as a tool and show that the security obtained when cascading several hash values is far from optimal. Unintuitively, this attack works even when two completely unrelated hash functions are cascaded and does not stem from any unforeseen correlation between similar hash functions. Finally, in section 5, we study the impact of our construction on several concrete schemes that rely on cascading or multicollision resistance. Very surprisingly, in all published examples, we encounter some obstruction which prevents the attack from working.

2   Basic Facts About Iterated Hash Functions

An iterated hash function H is built by iterating a basic compression function. The compression function f takes two inputs, a chaining variable and a message block, it outputs the next value of the chaining variable. Before processing, the message is first padded and split into elementary blocks. The padding itself is generally performed by appending a single '1' bit, followed by as many '0' bits as needed. To avoid some attacks, the binary encoding of the message length can also be added to complete the padding. This is called a Merkle-Damgard strengthening [8, 3]. Once the padded message is split into ℓ blocks, M_1, ..., M_ℓ, the chaining variable is set to some fixed initial value and the iteration is performed. To summarize, the hashing process works as follows:

- Pad the original message and split it into blocks $M_1, \ldots, M_\ell$.
- Set H_0 to the initial value IV.
- For i from 1 to ℓ, let $H_i = f(H_{i-1}, M_i)$.
- Output $H(M) = H_\ell$.

Given such an iterated hash function, defining its security is a tricky matter. Ideally, the hash function is often seen as a concrete substitute for random oracles in cryptographic construction. Of course, it is well known (see [2]) that this extreme level of security is in fact impossible to reach. Thus, the security level of hash function is usually characterized by considering "easier" security goals. The most frequently encountered goal is the impossibility for a bounded adversary to find a collision in the hash function. We recall that a collision is a pair of different messages M and M' such that $H(M) = H(M')$. Due to the birthday paradox, there is a generic attack that find collisions after about $2^{n/2}$ evaluations of the hash function, where n is the size in bits of the hash values. The attack works by randomly choosing messages and computing their hash values until a collision occurs. Typically, with iterated hash functions, the size of messages' blocks is often larger than the size of the hash values themselves, and this attack usually works on the compression function itself. Other important security goals for hash functions are preimage resistance and second-preimage resistance. An attack against preimage resistance is an attack that, given some target value y, finds a message M such that $H(M) = y$. An attack against second preimage resistance, given a message M, finds another message such that $H(M) = H(M')$.

The best generic attacks against these security goals cost about 2^n evaluation of the function H.

The notion of collision can easily be generalized to that of r-way collision (or, for short, r-collision). A r-collision is simply a r-tuple of messages $M^{(1)}$, ..., $M^{(r)}$, such that $H(M^{(1)}) = \cdots = H(M^{(r)})$. Assuming as above that the hash values behave almost randomly, finding an r-collision could be done by hashing about $2^{n \cdot (r-1)/r}$ messages. When r becomes large, this tends to 2^n. Due to this fact, relying on r-collision freeness in cryptographic construction seems a good way to gain more security without increasing the size of the hash functions. This is very tempting in some applications such as identification schemes [5] and signature schemes [1]. The next section demonstrates that, in fact, r-collisions in iterated hash functions are not much harder to construct than ordinary collisions, even for very large values of r.

3   Constructing Multicollisions

In this section, we show that constructing multicollisions in iterated hash function can be done quite efficiently. More precisely, constructing 2^t-collisions costs t times as much as building ordinary 2-collisions. Before describing the attack, let us remark that the padding process can be ignored as long as we consider collisions between messages of the same length. Indeed, in that case, the blocks of padding are identical. Moreover, if the intermediate hash chaining values collide at some point in the hash computation of two messages, the following values remain equal as soon as the ends of the messages are identical. Thus, on messages of the same length, collisions without the padding clearly lead to collisions with the padding.

For simplicity of exposure, we assume that the size of the message blocks is bigger than the size of the hash (and chaining) values. However, the attack can be easily generalized to the other case. We also assume that we can access a collision finding machine C, that given as input a chaining value h outputs two different blocks B and B' such that $f(h, B) = f(h, B')$. This collision finding machine may use the generic birthday attack or any specific attack based on a weakness of f. The most relevant property is that C should work properly for all chaining values[2]. To illustrate the basic idea, we first show how 4-collisions can be obtained with two calls to C. Starting from the initial value IV, we use a first call to C to obtain two different blocks, B_0 and B_0' that yield a collision, i.e. $f(IV, B_0) = f(IV, B_0')$. Let z denotes this common value and using a second call to C, find two other blocks B_1 and B_1' such that $f(z, B_1) = f(z, B_1')$. Putting these two steps together, we obtain the following 4-collision:

$$f(f(IV, B_0), B_1) = f(f(IV, B_0), B_1') = f(f(IV, B_0'), B_1) = f(f(IV, B_0'), B_1').$$

We now claim that this basic idea can be extended to much larger collisions by using more calls to the machine C. More precisely, using t calls, we can build 2^t-collisions in H. The attack works as follows:

---

[2] Or at least on a fixed proportion of them.

- Let h_0 be equal to the initial value IV of H.
- For i from 1 to t do:
    - Call C and find B_i and B_i' such that $f(h_{i-1}, B_i) = f(h_{i-1}, B_i')$.
    - Let $h_i = f(h_{i-1}, B_i)$.
- Pad and output the 2^t messages of the form $(b_1, \dots, b_t, \text{Padding})$ where b_i is one of the two blocks B_i or B_i'.

Clearly, the 2^t different messages built as above all reach the same final value. In fact, they have an even stronger property. Namely, all the intermediate hash values are equal, since all of the 2^t hashing processes go through h_0, h_1, ..., h_t. A schematic representation of these 2^t messages together with their common intermediate hash values is drawn in figure 1.

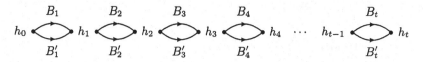

Fig. 1. Schematic representation of multicollision construction

Some generalizations. If f works on messages blocks which are smaller that the chaining values, the natural way to proceed is to group a few consecutive applications of f. For example, we can consider the function $f^{(2)}(h, B_1, B_2) = f(f(h, B_1), B_2)$ which composes two rounds of the compression function. As soon as the total size of the input blocks exceed the size of one chaining value, we can apply the original attack to the composed compression function.

Another generalization is to build 2^t-collisions from a 2-collision attack machine C that works only on a fixed proportion ϵ of the chaining values. Of course, this is not the case with the generic birthday attack, however, it may happen with some specific attacks. In that case, the basic attack described above only works with probability ϵ^t. Indeed, if any of the h_i does not belong to the set of chaining values that C can attack, we are in trouble. However, this bad behavior can be easily corrected by inserted a randomization step between two consecutive applications of C. Namely, after finding B_i and B_i' such that $f(h_{i-1}, B_i) = f(h_{i-1}, B_i')$, choose a random block R_i and let:

$$h_i = f(f(h_{i-1}, B_i), R_i).$$

If h_i fails to be in the scope of C, change R_i to get another candidate. Altogether, this randomization technique leads to a global complexity of the attack of the order of t/ϵ calls to C.

4   On the Security of Cascaded Hash Functions

A natural construction to build large hash values is to concatenate several smaller hashes. For example, given two hash functions F and G, it seems reasonable given a message M to form the large hash value $(F(M)\|G(M))$. In this construction, F and G can either be two completely different hash functions or two slightly different instances of the same hash function[3]. If F and G are good iterated hash functions with no attack better than the generic birthday paradox attack, we claim that the hash function $F\|G$ obtained by concatenating F and G is not really more secure that F or G by itself. Moreover, this result applies both to collision resistance, preimage resistance and second preimage resistance.

4.1   Collision Resistance

Assume that F outputs an n_f-bit hash value and G an n_g-bit value. Then, with respect to collision resistance, the security level of F is $2^{n_f/2}$ and the level of G is $2^{n_g/2}$. If $F\|G$ was a good hash function, the complexity of the best attack would be $2^{(n_f+n_g)/2}$. We claim that there exists a much better attack which find collisions on $F\|G$ with complexity of the order of $n_g 2^{n_f/2} + 2^{n_g/2}$ if $n_f \leq n_g$ (respectively $n_f 2^{n_g/2} + 2^{n_f/2}$ if $n_f \geq n_g$). Assuming than $n_f \leq n_g$, the attack works as follows.

First, using the multicollision algorithm of section 3 with t equal to $n_g/2$ rounded up, construct a 2^t-collision on F. This costs t calls to the basic birthday paradox attack on the compression function of f, i.e. about $t2^{n_f/2}$ operations. This yields 2^t different messages with the same hash value on the F side. Since $t \geq n_g/2$, we can perform direct application of the birthday paradox on this set of 2^t elements and, with reasonable probability, expect that a collision occurs among the n_g-bit hashes of these 2^t messages by G. To increase the probability of success, it suffices to increase the value of t and add a few more calls to the basic attack on F.

Note that when evaluating the complexity of the attack, one must take into account the contribution of applying G to 2^t different messages of size t. With a naive implementation, this would cost $t2^t$ calls to the compression function of G. However, using the tree structure of the messages, this can be reduced to 2^t evaluations, assuming that the compression functions of F and G operate on the same size of blocks. Otherwise, it is necessary to add some padding between blocks in order to resynchronize the two functions.

A very important fact about this attack is that it does not require of G to be an iterative hash function. Any hash function will do, and this attack on cascaded hash works even when G is replaced by a random oracle[4]. Since a random oracle is independent from any function, this shows that the folklore knowledge about cascading hash functions is false. Thus, at least in that case, cascading two good and independent hash functions does not significatively improve collision resistance.

---

[3] E.g., two instances of SHA-1 with different constants.

[4] The only difference in that case is the fact that the evaluations of G on the 2^t messages can no longer be simplified. As a consequence, assuming that the cost of calling the random oracle G is linear in the size of the message, the contribution of G to the complexity becomes $t2^t$

4.2   Preimage and Second-Preimage Resistance

Concerning preimage resistance, it is already known that cascading two hash functions is, at least in some cases, the cascade is no stronger than the weakest hash function. Indeed, assume that we are hashing messages from a relatively small set, say a set of 2^m messages. Clearly, the best generic attack to find a preimage in that case is to perform exhaustive search on the set of messages, which costs 2^m steps. Assume that the output of each the two hash functions being cascaded is larger than m bit and that on this set of messages, one of the two hash functions, say F, has a shortcut attack. Then, we can clearly use this attack to recover a candidate preimage. Once this is done, it suffices to check that this candidate is also a preimage for the other function. The new attack presented in this section deals with a different case, where the entropy of the message space is much larger. It shows that even then the cascaded hash is no more secure than F itself.

Assume again that F outputs an n_f-bit hash value and G an n_g-bit value. Then, with respect to preimage resistance, the security level of F is 2^{n_f} and the level of G is 2^{n_g}. Indeed, the best known generic algorithm to break preimage resistance is to try random messages until the expected hash value is reached. This amounts to exhaustive search on the set of possible hash values. If $F\|G$ was a good hash function, the complexity of this exhaustive search attack would be $2^{(n_f+n_g)}$. As with collision resistance, there exists a much better attack which find a preimage on $F\|G$ with complexity of the order of $n_g 2^{n_f/2} + 2^{n_f} + 2^{n_g}$ if $n_f \leq n_g$ (respectively $n_f 2^{n_g/2} + 2^{n_g} + 2^{n_f}$ if $n_f \geq n_g$). Assuming than $n_f \leq n_g$, the attack works as follows.

First, using the multicollision algorithm of section 3 with t equal to n_g, construct a 2^t-collision on F. This costs t calls to the basic birthday paradox attack on the compression function of f, i.e. about $t2^{n_f/2}$ operations. Then, search for an additional block that maps the last chaining value to the target value of F. Note that when looking for this additional block, we need to compute the output of the complete F function, including the padding of the message. However, this is a simple matter. After, this last step, we obtain 2^t different messages with the expected hash value on the F side. Since $t = n_g$, we expect that, with constant probability, at least one of these 2^t messages also match the expected t-bit value on the G side. Once again, the probability of success can be improved by adding a few more steps to the attack. Note that this attack on preimage resistance does not either require for G to be an iterative hash function. As before, it also works when G is replaced by a random oracle.

Clearly, the above attack finds a preimage for the target value which is essentially random. As a consequence, it can be applied directly without any change when a second preimage is requested.

4.3   Extensions and Open Problems

Given these attacks, it is natural to ask whether they generalizes to three or more concatenated hash values. In this section, we focus on the possibility of generalizing the collision search attack. We show that it does and that the generalization is almost straightforward. Indeed, assume that H is a third hash function on n_h bits, then using the above attack on $F\|G$ a couple of times, say $t \approx n_h/2$ times, it is possible as in section 3 to build a 2^t-collision on $F\|G$. Among these 2^t messages, we expect a collision of H. All in all, this yields a simultaneous collision on F, G and H. When $n_f = n_g = n_h = n$, the expression of the complexity simplifies and is of the order of $n^2 \cdot 2^{n/2}$. More generally, a simultaneous collision on q different n-bit iterative hash functions can be found with complexity $n^{q-1} \cdot 2^{n/2}$. Thus, the security of such a construction stays within a polynomial factor of the security of a single good iterative hash function of the same size. Similarly, variants of the attack on preimage resistance can be adapted to the case of q different hash functions. However, since they are more complicated, we do not present them here. One possible variant is described in appendix.

Another generalization of the above attack is also worth noting. In [14], B. Schneier described a different way of building a long hash from a hash function F. In this method, $F(M)$ is concatenated with $G(F(M)\|M)$ (or $G(M\|F(M))$). At first view, this is more complicated than the $F\|G$ construction. However, the very same attack can be applied. Indeed, when a 2^t-collision is found on $F(M)$, this fixes $F(M)$ in the first half of the big hash and also the copy of $F(M)$ in the call to G, thus a collision on the G part is expected exactly as before. The preimage attack also works as before. We leave open the problem of finding a related construction making q calls to n-bit hash function and with security higher than $n^{q-1} \cdot 2^{n/2}$, with respect to collision resistance.

One can also study a related question, how does the security of the concatenated hash $F\|G$ behaves, when F and G have non-generic attacks better than the birthday paradox collision search ? In that case, can $F\|G$ be significantly more secure than the best of F and G ?

For the sake of simplicity, assume once again that $n_f = n_g = n$. Then, if F has a collision finding algorithm C as in section 3 with complexity $2^{n/2}/n$ or better and G has no shortcut attack better than the birthday paradox, the security of $F\|G$ is essentially the same as the security of G itself. On the other hand, if G also admits a shortcut attack (as in section 3), it is unclear whether the two shortcut attacks may be used together to improve the composed attack against $F\|G$. Yet, some other type of attacks against G can be integrated into a better composed attack on $F\|G$. To give an example, let g denote the compression function of G. Assume that there exists a shortcut attack which given a large set $g_1, \ldots, g_N$ of chaining values finds a message block B and two indices i and j such that $g(g_i, B) = g(g_j, B)$ in time N. Clearly, such a merging attack could be used to turn an N-collision on F into a full collision on $F\|G$. Thus, it is safer to assume that $F\|G$ is essentially as secure as the best of F and G, no more.

5   Potential Applications

While the ideas of cascaded construction and multicollisions are frequently encountered in the cryptographic folklore, they are somewhat avoided in published papers. As a consequence, we were not able to find a single research paper that can be cryptanalyzed using the attacks presented here. In this section, we describe some published construction which were likely candidates and explain why the attacks failed.

Cascaded hash functions. Among the frequently encountered hash function, RIPEMD is the most suited to the cascaded construction. Indeed, the basic algorithm already consists of two separate hash computations which are put together at the end of the round function. Thus, using the result of the two chains to form a longer hash would be both natural and efficient. In fact, the authors of RIPEMD propose in [4] optional extensions to 256 and 320 bits by using this idea. These extensions are not fully specified, but a sketch is given. In this sketch, the authors of RIPEMD recommend to add to the basic cascade some interaction between the two parallel compression functions. More precisely, they propose to swap one register from the first chain and its counterpart in the second chain after each round of the compression function (5 rounds in RIPEMD-160 and 4 rounds in RIPEMD-128). This interaction was introduced as an additional security measure and, with respect to our attack, this countermeasure is very efficient and completely voids it.

Use of multicollisions. Among the published constructions that make use of multicollisions, one can cite the micropayment scheme Micromint [11], the identification scheme of Girault and Stern [5] and the signature scheme of Brickell and *al.* [1]. In these three applications, multicollisions are indeed used, however, in the proposed instances of the schemes, no iterated hash functions with a small internal memory is used. Instead, one encounters either a block cipher based compression function with a small block size but without iteration or a truncation of an iterated hash function with a relatively large output size. In both cases, our attack is unapplicable. In the first case, the required iteration is not available, in the second, the attack needs collisions on the full internal states of the hash function, rather than on the truncated states.

6   Conclusion

In this paper, we have shown that multicollisions in iterated hash functions are not really harder to find than ordinary collision. This yields the first effective attack against a natural construction that extend the size of hash values by concatenating several independent results. While considered suspect by some, especially when used with related hash functions, this construction had never been attacked before. The cryptanalysis we presented here yields attacks against collision resistance, preimage resistance and second preimage resistance. As a consequence, it leaves open the problem of constructing secure hash functions with variable-output length, which is a important primitive to instantiate some cryptographic paradigm such as the full domain hash.

Another important theoretical result is the fact that iterated hash functions cannot be used as entropy-smoothing functions on arbitrary sets of inputs. Devising good cryptographic entropy-smoothing functions would be a nice topic for future research.

References

1. E. Brickell, D. Pointcheval, S. Vaudenay, and M. Yung. Design validation for siscrete logarithm based signature schemes. In *PKC'2000*, volume 1751 of *Lecture Notes in Computer Science*, pages 276–292. Springer–Verlag, 2000.
2. R. Canetti, O. Goldreich, and S. Halevi. The random oracle methodology, revisited. In *Proc. 30th Annual ACM Symposium on Theory of Computing (STOC)*, pages 209–218, 1998.
3. I. Damgård. A design principle for hash functions. In *Advances in Cryptology – Crypto'89*, volume 435 of *Lecture Notes in Computer Science*, pages 416–427. Springer–Verlag, 1989.
4. H. Dobbertin, A. Bosselaers, and B. Preneel. RIPEMD-160, a strengthened version of RIPEMD. In *Fast Software Encryption*, volume 1039 of *Lecture Notes in Computer Science*, pages 71–82. Springer–Verlag, 1996.
5. M. Girault and J. Stern. On the length of cryptographic hash-values used in identification schemes. In *Advances in Cryptology – Crypto'94*, volume 839 of *Lecture Notes in Computer Science*, pages 202–215. Springer–Verlag, 1994.
6. A. Menezes, P. van Oorschot, and S. Vanstone. *Handbook of Applied Cryptography*. CRC Press, 1997. Available on line : http://www.cacr.math.uwaterloo.ca/hac.
7. R. Merkle. A fast software one-way hash function. *Journal of Cryptology*, 3(1):43–58, 1990.
8. R. C. Merkle. One way hash functions and DES. In *Advances in Cryptology – Crypto'89*, volume 435 of *Lecture Notes in Computer Science*, pages 428–446. Springer–Verlag, 1989.
9. Secure hash standard. Federal Information Processing Standard Publication 180-1, 1995.
10. B. Preneel. *Analysis and design of cryptographic hash functions*. PhD thesis, Katholieke Universiteit Leuven, January 1993.
11. R. Rivest and A. Shamir. PayWord and MicroMint – two simple micropayment schemes. *CryptoBytes*, 2(1):7–11, Spring 1996.
12. R. L. Rivest. The MD4 message digest algorithm. In *Advances in Cryptology – Crypto'90*, volume 537 of *Lecture Notes in Computer Science*, pages 303–311. Springer–Verlag, 1991.
13. R. L. Rivest. The MD5 message-digest algorithm. Network Working Group Request for Comments: 1321, April 1992.
14. B. Schneier. *Applied Cryptography*. John Wiley & Sons, second edition edition, 1996.

A Preimage Resistance with Many Hash Functions Cascaded

While the attack against collision resistance described in section 4.1 is easily generalized to q hash functions and yields an attack with complexity $n^{q-1} \cdot 2^{n/2}$, assuming that each function outputs n bits, this is not the case for the attack on preimage resistance. Indeed, the attack we described in section 4.2 is not straightforward to generalize. The goal of this section is to present a variant of the attack that can be easily generalized. The drawback is that this variant is slightly more complicated than the initial attack. In this variant, each hash function is attacked in two steps. Each of these steps is constructed in a way that ensures compatibility with the previous hash functions. The first step within each hash function is to find two different sequences of blocks that, through the iterated hash process, sends the initial value back to itself. The cost of this amounts to twice exhaustive search on the set of possible chaining values. The second step is to find a terminating sequence of blocks that sends this chaining value to the target value for the current hash function. This costs about one exhaustive search on the same set.

When processing the first message, we look for single block sequences. Moreover, the terminating block should be correctly padded. A slight technical problem is that padding requires a priori knowledge of the final message size. However, we show at the end of this section that this size can be fixed in advance before launching the attack. Let T denotes this size, then we consider as inputs to the first hash functions the 2^{T-1} messages formed of $T-1$ blocks, each chosen among the two basic blocks that send the initial value h_0 to itself and one final block which sends h_0 to the target value h_F. A representation of these messages is given in figure reffig:preim

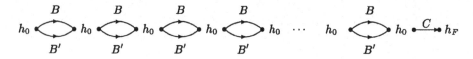

Fig. 2. First step of the preimage attack

With the second hash function, the looping sequences are constructed by concatenating n blocks chosen among the two (B and B') that makes the first hash function loop (a few more blocks can be added to increase the probability of finding good sequences). Clearly, when applying one of these sequences both the first and the second hash functions are going back to their initial values. The final sequence is constructed by concatenating many copies of the looping blocks B or B' and a single instance of the final block, in order to send the second hash function to its expected destination. Clearly, such a sequence also sends the first hash function to its target value. The advantage of this attack compared to that of section 4.2 is that additional hash functions can be processed by iterating the previous procedure. A notable exception is the computation of the last hash function which requires no looping part and can thus be simplified. The total runtime is clearly bounded by a polynomial ($O(n^q)$) times the cost of exhaustive search. The length T of the message can be easily predetermined and is of the order of n^{q-1}.

Adaptively Secure Feldman VSS and Applications to Universally-Composable Threshold Cryptography

Masayuki Abe[1] and Serge Fehr[2,*]

[1] NTT Laboratories, Japan
abe@isl.ntt.co.jp
[2] CWI, Amsterdam, The Netherlands
fehr@cwi.nl

Abstract. We propose the first distributed discrete-log key generation (DLKG) protocol from scratch which is adaptively-secure in the non-erasure model, and at the same time completely avoids the use of interactive zero-knowledge proofs. As a consequence, the protocol can be proven secure in a universally-composable (UC) like framework which prohibits rewinding. We prove the security in what we call the single-inconsistent-player UC model, which guarantees arbitrary composition as long as all protocols are executed by the same players. As an application, we propose a fully UC threshold Schnorr signature scheme.

Our results are based on a new adaptively-secure Feldman VSS scheme. Although adaptive security was already addressed by Feldman in the original paper, the scheme requires secure communication, secure erasure, and either a linear number of rounds or digital signatures to resolve disputes. Our scheme overcomes all of these shortcomings, but on the other hand requires some restriction on the corruption behavior of the adversary, which however disappears in some applications including our new DLKG protocol.

We also propose several new adaptively-secure protocols, which may find other applications, like a sender non-committing encryption scheme, a distributed trapdoor-key generation protocol for Pedersen's commitment scheme, or distributed-verifier proofs for proving relations among commitments or even any NP relations in general.

1 Introduction

A distributed key generation protocol is an essential component in threshold cryptography. It allows a set of n players to jointly generate a key pair, (pk, sk), that follows the distribution defined by the target cryptosystem, without the need for a trusted party. While the public-key pk is output in clear, the corresponding secret-key sk remains hidden and is maintained in a shared manner

---

* Research was carried out while at the Centre for Advanced Computing - Algorithms and Cryptography, Department of Computing, Macquarie University, Australia.

M. Franklin (Ed.): CRYPTO 2004, LNCS 3152, pp. 317–334, 2004.

among the players via a secret sharing scheme. This should allow the players to later use sk without explicitly having to reconstruct it. The distributed key-generation for discrete-log based schemes, DLKG in short, amounts to the joint generation of a random group element y as public-key and a sharing of its discrete-log (DL) $x = \log_g(y)$ as secret-key with regard to some given base g. A DLKG protocol must remain secure in the presence of a malicious adversary who may corrupt up to a minority of the players and make them behave in an arbitrary way. Informally, it is required that, for any adversary, y must be uniformly distributed, and the adversary must learn nothing about x beyond $y = g^x$.

DLKG was first addressed by Pedersen in [14]. Gennaro et al. pointed out that Pedersen's scheme is not secure against a rushing adversary (and even against a non-rushing adversary) and proposed a new (statically) secure scheme [12]. Then Frankel et al. and Canetti et al. introduced in [11] respectively [7] adaptively secure schemes in the erasure model, and Jarecki and Lysyanskaya improved the schemes to work in the non-erasure model and to remain secure under concurrent composition [13].

These DLKG protocols which are secure against an adaptive adversary rely heavily on the use of interactive zero-knowledge proofs. This poses the question whether this is an inherent phenomenon for adaptively secure DLKG. We answer this question in the negative. Concretely, we propose an adaptively-secure distributed key-generation protocol from scratch which completely avoids the use of interactive zero-knowledge proofs. As a consequence, the protocol can be and is proven secure in a relaxed version of Canetti's universally-composable (UC) framework [4], which prohibits rewinding. We show the usefulness of our distributed key-generation protocol by showing how it gives rise to a (fully) UC threshold Schnorr signature scheme. To the best of our knowledge, this is the first threshold scheme proven secure in the UC framework.

The relaxed UC framework, which we call the single-inconsistent-player (SIP) UC framework, coincides with the original UC framework, except that the simulator is allowed to fail in case the adversary corrupts some designated player P_{j^*}, which is chosen at random from the set of all players and announced to (and only to) the simulator. This relaxation still allows for a powerful composition theorem in that protocols may be arbitrary composed, as long as all subsidiary protocols involve the same set of players.

We stress once more that this relaxation only applies to the proposed distributed key-generation protocol but *not* to its application for the threshold Schnorr signature scheme.

Our DLKG protocol (and thus the threshold Schnorr signature scheme) is based on a new adaptively-secure version of Feldman's famous (statically secure) VSS scheme. Although adaptive security was already addressed by Feldman in the original paper [10], and besides the well known standard Feldman VSS scheme he also proposed an adaptively-secure version, the proposed scheme has several shortcomings: (1) it requires the players to be able to reliably erase data, (2) it either proceeds over a linear number of rounds or otherwise needs to incorporate signatures as we will point, and (3) it requires secure communication channels (or expensive non-committing encryption schemes). We propose a new variant of Feldman's VSS scheme which overcomes all of these limitations. Even though the proposed scheme is not fully adaptively secure but requires some restriction on the corruption behavior of the adversary, this restriction is acceptable in that it disappears in the above applications to threshold cryptography.

Furthermore, as building blocks for the above schemes or as related constructions, we also propose a sender non-committing encryption scheme, a new adaptively secure distributed trapdoor-key generation protocol for Pedersen's commitment scheme, as well as adaptively secure distributed-verifier zero-knowledge proofs, which all may very well find other applications. Finally, in the full version of this paper [3], we propose several additional applications and/or related adaptively-secure constructions of independent interest: a simple modification of Feldman's adaptively-secure VSS scheme which overcomes (1) and (2) above, though not (3), but is *fully* adaptively-secure, an adaptively secure version of Pedersen's VSS scheme as a *committed* VSS, a threshold version of the DSS signature scheme in the UC model, a threshold version of the Cramer-Shoup cryptosystem in the SIP UC model, and a common-reference-string generator with applications to zero-knowledge proofs in the UC model.

The paper is organized as follows. Section 2 reviews the model we are considering. It includes a short introduction to the UC framework of Canetti and the new SIP UC framework. In Sect. 3 we recall Feldman's statically and adaptively secure VSS schemes, and we point out an obstacle in the dispute resolution phase of the adaptive scheme, before we construct our version in Sect. 4. Finally, Sect. 5 shows the applications to adaptively-secure DLKG, universally-composable threshold cryptography and distributed-verifier proofs.

Due to space limitations, many definitions and proofs could only be sketched in this proceedings version of the paper; the formal treatment can be found in the full version [3].

2 Preliminaries

2.1 Communication and Adversary

We consider a *synchronized authenticated-link model* where a message from P_s to P_r is delivered within a constant delay and accepted by P_r if and only if it is sent from P_s to P_r. Moreover, we assume a broadcast channel with which every player sends a message authentically and all players receives the same message.

We consider a central adversary $\mathcal{A}$ which may corrupt players at will. Corrupting a player P_i allows $\mathcal{A}$ to read P_i's internal state and to act on P_i's behalf from that point on. In the non-erasure model, $\mathcal{A}$ additionally gains P_i's complete history. $\mathcal{A}$ is called t-*limited* if it corrupts at most t players. Furthermore, $\mathcal{A}$ is called *static* if it corrupts the players *before* the protocol starts, and $\mathcal{A}$ is called *adaptive* if it corrupts the players *during* the execution of the protocol, depending on what it has seen so far.

2.2 Canetti's Universally Composable Framework

In order to formally specify and prove the security of our protocols, we will use the universally composable (UC) framework of Canetti [4]. We briefly sketch this framework here; for a more elaborate description we refer to the full version of the paper [3] or to the literature. The UC framework allows to define and prove secure cryptographic protocols as stand-alone protocols, while at the same time guaranteeing security in any application by means of a general composition theorem. In order to define a protocol π secure, it is compared with an *ideal functionality* $\mathcal{F}$. Such a functionality can be thought of as a trusted party with whom every player can communicate privately and which honestly executes a number of specified commands. The UC security definition essentially requires that for every (real-life) adversary $\mathcal{A}$ attacking the protocol π, there exists an ideal-life adversary $\mathcal{S}$, also called *simulator*, which gets to attack the ideal-life scenario where only the players and $\mathcal{F}$ are present, such that $\mathcal{S}$ achieves "the same" as $\mathcal{A}$ could have achieved by an attack on the real protocol. In the framework, this is formalized by considering an *environment* $\mathcal{Z}$ which provides inputs to and collects outputs from the honest players and communicates in the real-life execution with $\mathcal{A}$ and in the ideal-life execution with $\mathcal{S}$. It is required that it cannot tell the difference between the real-life and the ideal-life executions, meaning that its respective outputs in the two cases are computationally indistinguishable.

As mentioned above, the UC framework provides a very general composition theorem: For any protocol ρ that securely realizes functionality $\mathcal{G}$ in the so-called $\mathcal{F}$-hybrid model, meaning that it may use $\mathcal{F}$ as a subroutine, composed protocol ρ^π that replaces $\mathcal{F}$ with a secure protocol π also securely realizes $\mathcal{G}$ (in the real-life model).

2.3 Single-Inconsistent-Player UC Framework

The single-inconsistent-player (SIP) technique of [7] is often used to achieve both adaptive security and efficiency. A protocol in the SIP model is secure (i.e. securely simulatable in the classical model of computation) if the adversary does not corrupt a designated player which is chosen independently at random before the protocol starts. Using the terms of the UC framework, it means that the simulator $\mathcal{S}$ is given as input the identity of a randomly chosen player P_{j^*}, and $\mathcal{S}$ is required to work well as long as P_{j^*} is uncorrupt. In the case of t-limited adversary with $t < n/2$, this reduces $\mathcal{S}$'s success probability by a factor of $1/2$. This still guarantees security in that whatever $\mathcal{A}$ can do in the real-life model, $\mathcal{S}$ has a good chance in achieving the same in the ideal-life model. Indeed, in the classical sense, a simulator is considered successful if it works better than with negligible probability. However, with such a simulator $\mathcal{S}$, the composition theorem no longer works in its full generality. To minimize the effect of the SIP approach, we have to limit the set of players to be the same in all subsidiary protocols. This way, P_{j^*} can be sampled once and for all, and the condition that P_{j^*} remains uncorrupt applies to (and either holds or does not hold) simultaneously for all protocols. With this limitation, the composition theorem essentially works as before. See also the full version of the paper [3].

2.4   Some Functionalities

We briefly introduce some functionalities we use throughout the paper. For formal descriptions and more elaborate discussions, we refer to [3].

Secure-Message-Transmission Functionalities: The secure-message-transmission functionality, as defined in [4], is denoted by $\mathcal{F}_{\text{SMT}}$. On receiving (*send, sid, P_r, m*) from P_s, $\mathcal{F}_{\text{SMT}}$ sends (*sid, P_s, m*) to P_r and (*sid, P_s, P_r*) to the (ideal-life) adversary $\mathcal{S}$ and halts. If the length of m may vary, it is also given to $\mathcal{S}$.

Due to some subtle technicalities, $\mathcal{F}_{\text{SMT}}$ cannot be securely realized in a synchronized communication model against an active adversary. The reason is that in any (interactive) candidate realization π_{SMT}, the adversary $\mathcal{A}$ can corrupt the sender during the execution of the protocol and change the message m to be securely transmitted (or abort the protocol), while this cannot be achieved by $\mathcal{S}$. Indeed, once $\mathcal{F}_{\text{SMT}}$ is invoked it is always completed with the initial input (and the output is delivered to the receiver). To overcome this problem, we introduce *spooled* SMT. This is captured by $\mathcal{F}_{\text{SSMT}}$, which first spools the sent message m and only delivers the spooled message m (or a possibly different m in case of a corrupt P_s) when receiving another (the actual) send command from P_s. This allows $\mathcal{S}$ to change m after $\mathcal{F}_{\text{SSMT}}$ has been launched simply by corrupting P_s after the spool- but before the send-command.

$\mathcal{F}_{\text{SSMT}}$ can be realized over a public network using non-committing encryption as introduced by Canetti *et al.* in [6]. However, this is rather expensive as the best known schemes [9] still bear a ciphertext expansion $O(\kappa)$. Instead, our results are based on efficient though not fully adaptively secure realizations.

In our construction, we will also use an extended version of the $\mathcal{F}_{\text{SSMT}}$ functionality which allows P_s, in case of a dispute, to convince the other players of the message m sent to P_r. This is specified by $\mathcal{F}_{\text{SSMTwO}}$, spooled SMT *with opening*, which works as $\mathcal{F}_{\text{SSMT}}$ except that it additionally allows an open-command sent from P_s (and only from P_s), upon which it announces the transmitted message m to all players.

Committed-VSS Functionalities: An advantage of using Feldman and Pedersen VSS in protocol design is that besides producing a (correct) sharing, they also commit the dealer to the shared secret. Often, this commitment can be and is used in upper-level protocols. However, in the definition of UC-secure VSS given in [4], such a commitment is hidden in the protocol and not part of the functionality, and thus not available for external protocols. We introduce the notion of *committed* VSS to overcome this inconvenience.

Let $\text{com}_K : S_K \times R_K \to Y_K$ be a (efficiently computable) commitment function, indexed by a *commitment key* K. Typically, K is sampled by a poly-time generator (on input the security parameter). A commitment for a secret $s \in S_K$ is computed as $y = \text{com}_K(s; r)$, where we use the semicolon ';' to express that the second argument, r, is chosen randomly (from R_K) unless it is explicitly given.

A committed VSS (with respect to commitment scheme com_K) is specified by functionality $\mathcal{F}_{\text{VSS}}^{\text{com}_K}$, which sends (*shared, sid, P_d, $\text{com}_K(s; r)$*) to all players and $\mathcal{S}$ on receiving (*share*, *sid*, (s, r)) from P_d, the dealer, and later, on receiving (*open*, *sid*) from $t+1$ distinct players, it sends (*opened*, *sid*, s) to all players and $\mathcal{S}$.

Due to the same technical problem as above, if the dealer may be adaptively corrupted, we need to incorporate spooling into the committed VSS functionality: $\mathcal{F}_{\text{svss}}^{\text{com}K}$ first spools (s, r) (and gives $\text{com}_K(s; r)$ to $\mathcal{S}$) before awaiting and executing the actual share-command (for the original or a new s).

We would like to mention that for certain candidate protocols π_{vss} for committed VSS (with spooling), whose security relies on the commitment scheme com_K, the generation of the key K needs to be added to the VSS functionality in order to be able to prove π_{vss} secure in the UC framework. This is for instance the case for Pedersen's VSS scheme as discussed in [3].

2.5   The Discrete-Log Setting

Let κ be a security parameter and q be a prime of size κ. Let G_q denote a group of order q, and let g be its generator. We use multiplicative notation for the group operation of G_q. Some of our constructions require G_q to be the order-q multiplicative subgroup of $\mathbb{Z}_p^*$ with prime $p = 2q + 1$. Unless otherwise noted, all arithmetics are done in $\mathbb{Z}_q$ or G_q and should in each case be clear from the context.

Throughout, we assume that such (G_q, q, g) is given to all players, and that the Decision Diffie-Hellman problem for (G_q, q, g) is intractable, meaning that the respective uniform distributions over $\text{DH} = \{(g^\alpha, g^\beta, g^\gamma) \in G_q{}^3 \mid \alpha \cdot \beta = \gamma\}$ and $\text{RND} = G_q{}^3$ are computationally indistinguishable. This assumption implies the discrete-log assumption for (G_q, q, g): given a random $h = g^\omega$, it is computationally infeasible to compute ω.

3   The Original Feldman VSS

The Basic Scheme: Let $\alpha_1, \ldots, \alpha_n \in \mathbb{Z}_q$ be distinct and non-zero. In order to share a (random) secret $s \in \mathbb{Z}_q$, the dealer selects a Shamir sharing polynomial $f(X) = s + a_1 X + \cdots + a_t X^t \in \mathbb{Z}_q[X]$ and sends $s_j = f(\alpha_j)$ privately to P_j. Additionally, he broadcasts $C_0 = g^s$ as well as $C_k = g^{a_k}$ for $k = 1, \ldots, t$. Each player P_j now verifies whether

$$g^{s_j} = \prod_{k=0}^{t} C_k^{\alpha_j^k}. \tag{1}$$

If it does not hold for some j, then player P_j broadcasts an *accusation* against the dealer, who has to respond by broadcasting s_j such that (1) holds. If he fails, then the execution is rejected, while otherwise P_j uses the new s_j as his share. Correct reconstruction is achieved simply by filtering out shares that do not satisfy (1).

This scheme is proved secure against a *static* adversary: Assume that $\mathcal{A}$ corrupts $P_{j_1}, \ldots, P_{j_t}$. Given $C_0 = g^s$, the simulator $\mathcal{S}$ simply chooses random shares $s_{j_i} \in \mathbb{Z}_q$ $(i = 1 \dots t)$ for the corrupted players, and it computes $C_1, \dots, C_t$ with the right distribution from g^s and $g^{s_{j_1}}, \dots, g^{s_{j_t}}$'s by applying appropriate Lagrange interpolation coefficients "in the exponent". Informally, this shows that $\mathcal{A}$ learns nothing about s beyond g^s.

This simulation-based proof though fails completely if the adversary may corrupt players *adaptively*, i.e., during or even after the execution of the protocol. The problem is that given $C_0 = g^s$, $\mathcal{S}$ needs to come up with $C_1, \dots, C_t$ such that if $\mathcal{A}$ corrupts some player P_j at some later point, $\mathcal{S}$ can serve $\mathcal{A}$ with s_j such that (1) is satisfied. However, it is not known how to successfully provide such s_j for any dynamic choice of j without knowing s, unless $\mathcal{A}$ corrupts the dealer to start with.

Adaptive Security with Erasure: Feldman addressed adaptive security by providing a set-up phase where the dealer assigns a *private* X-coordinate $\alpha_j \in \{1, \dots, n\}$ to every P_j. Additionally, he needs to convince the players of the uniqueness of their α_j. This is done in the following way. Let E be a semantically-secure public-key encryption function, with public-key chosen by the dealer.

1. The dealer computes an encryption $A_j = E(j; r_j)$ (with random r_j) for every $j \in \{1, \dots, n\}$, and he chooses $\alpha_1, \dots, \alpha_n$ as a random permutation of $1, \dots, n$. Then, he broadcasts $A_1, \dots, A_n$ ordered in such a way that A_j appears in α_j-th position, and he privately sends (α_j, r_j) to P_j.
2. Each P_j locates A_j in position α_j and verifies whether $A_j = E(j; r_j)$ and, if it holds, erases r_j. The dealer erases $r_1, \dots, r_n$, too.

After the erasure is completed, the dealer performs the basic Feldman VSS with X-coordinates $\alpha_1, \dots, \alpha_n$. We stress that it is important that the erasures of the r_j's must be done before entering to the sharing phase. On reconstruction, each player broadcasts (α_j, s_j).

Since each A_j can be opened only to j, player P_j is convinced of the uniqueness of α_j. Simulation against an adaptive adversary is argued separately for each phase. If a player gets corrupted in the set-up phase, the simulator $\mathcal{S}$ just honestly gives the internal state of the corrupt player to the adversary. Nothing needs to be simulated. Then, the sharing phase is simulated similar as for the static adversary, except that, since $\mathcal{S}$ does not know which players will be corrupted, it predetermines shares for a *random* subset of size t of the X-coordinates $\{1, \dots, n\}$, and whenever a player P_j gets corrupted one of these prepared X-coordinates is assigned to P_j as his α_j. Since r_j has already been erased, it is computationally infeasible to determine whether A_i in position α_j is an encryption of j or not.

An Obstacle in Dispute Resolution: We identify a problem in the dispute resolution of the above scheme[1]. Suppose that honest P_j accuses the dealer, and that

---

[1] No dispute resolution procedure is shown in [10]. It is said that a player simply rejects the dealer when he receives an incorrect share (and the dealer is disqualified if more than $t + 1$ players rejects). But this works only if $t < n/3$.

instead of publishing correct (α_j, s_j), the corrupt dealer responds by publishing (α_i, s_i) of another honest player P_i. Since r_j and r_i have been already erased, both P_j and P_i have no way to prove that the published α_i is different from the original assignment.

To efficiently settle such a dispute, digital signature scheme will be needed. That is, the dealer sends α_j with his signature in the set-up phase. This allows P_j to publish the signature when he accuses the dealer in the sharing phase. Without using digital signatures, $O(t)$ additional rounds are needed to settle the dispute: If P_i observes that his (α_i, s_i) is published to respond to the accusation from P_i, P_i also accuses the dealer and the dealer publishes the data for P_i this time. After repeating this accuse-then-publish process at most $t + 1$ times, the dealer either gets stuck or exposes $t + 1$ correct shares.

4   Adaptive Security Without Overheads and Erasures

The goal of this section is an adaptively secure Feldman VSS that provides (1) security without the need for reliably erasing data, (2) efficient dispute resolution without digital signatures, and (3) efficient realization over a public network, i.e. without secure channels (or expensive non-committing encryptions).

The first two goals are achieved by a simple modification of the original Feldman VSS. The idea is to replace the encryption function E with instantiations of a trapdoor commitment scheme with certain properties whose commitment keys are provided separately from each player so that the trapdoors are not known to the dealer. We show this modified Feldman VSS and the security proof in [3]. Since Pedersen's commitment scheme turns out to be good enough for this purpose, we have a scheme that meets (1) and (2) solely under the DL assumption. Furthermore, the modified scheme is more efficient in the number of communication rounds over the original adaptively-secure Feldman VSS.

Hence, what the secure-channels model is concerned, we are done. Unfortunately, we do not know how to efficiently implement the above scheme efficiently over a public network, even when limiting the power of the adversary as we do in Sect. 4.2 below. Therefore, we design a new scheme which allows to seamlessly install our efficient components for public communication presented later.

4.1   Construction in a Hybrid Model

Our approach is to let each player P_j select a random non-zero X-coordinate $\alpha_j \in \mathbb{Z}_q$ and send it privately to the dealer. When corrupted, a simulated player reveals a (fake) X-coordinate that has been prepared in advance to be consistent with the transcript, as in Feldman's approach. On the other hand, in case of a dispute, each player P_j should be able to convince the other players of his α_j. This is achieved by initially sending α_j to the dealer using secure message transmission *with opening*, as specified in Sect. 2.4 by functionality $\mathcal{F}_{\text{SSMTwO}}$. The scheme is detailed in Fig. 1 in the $(\mathcal{F}_{\text{SSMTwO}}, \mathcal{F}_{\text{SSMT}})$-hybrid model.

Consider Feldman's commitment scheme fcom_g with base g: a commitment for a secret $s \in \mathbb{Z}_q$ is computed as $\mathsf{fcom}_g(s; r) = \mathsf{fcom}_g(s) = g^s$ (without using r).

[Sharing Phase]

F-1. Each P_j selects $\alpha_j \leftarrow \mathbb{Z}_q^*$ and sends α_j to the dealer via $\mathcal{F}_{\text{SSMTwO}}$. The dealer replaces any α_j that happens to be 0 by $\alpha_j = 1$.

F-2. The dealer selects $f(X) = a_0 + a_1 X + \cdots + a_t X^t \leftarrow \mathbb{Z}_q[X]$ where $a_0 = s$ and computes $C_k = g^{a_k}$ for $k = 0, \ldots, t$ and broadcasts $(C_0, \ldots, C_t)$. For every P_i he sends $s_i = f(\alpha_i)$ by using $\mathcal{F}_{\text{SSMT}}$.

F-3. Each P_j verifies $g^{s_j} = \prod_{k=0}^t C_k^{\alpha_j^k}$ and broadcasts *verified* if it holds. Otherwise, P_j broadcasts *accuse dealer*. For every accusation, the following sub-protocol is executed in parallel.

   (a) P_j sends (*open, sid*) to $\mathcal{F}_{\text{SSMTwO}}$ and every player receives α_j. If $\alpha_j = 0$, then it is replaced by $\alpha_j = 1$.

   (b) The dealer broadcasts corresponding s_j.

   (c) If (α_j, s_j) satisfies the verification predicate, then P_j accepts (α_j, s_j) as his share, otherwise the players output a default sharing of $s = 0$. (Note that the players have agreement on the published values α_j and s_j.)

[Reconstruction Phase]

Each P_j broadcasts (α_j, s_j), identifies $\mathcal{Q} \subseteq \{1, \ldots, n\}$, $|\mathcal{Q}| \geq t + 1$, so that (α_i, s_i) satisfies the verification predicate for all $i \in \mathcal{Q}$, reconstructs secret s by Lagrange interpolation with regard to $\mathcal{Q}$, and then outputs s.

Fig. 1. Adaptively secure Feldman-VSS π_{XFVSS} in $(\mathcal{F}_{\text{SSMTwO}}, \mathcal{F}_{\text{SSMT}})$-hybrid model.

Proposition 1. *Protocol π_{XFVSS} shown in Fig. 1 securely realizes $\mathcal{F}_{\text{SVSS}}^{\text{fcom}_g}$ in the $(\mathcal{F}_{\text{SSMTwO}}, \mathcal{F}_{\text{SSMT}})$-hybrid model against t-limited adaptive adversary for $t < n/2$.*

The proof is given in [3]. Essentially, it uses the same idea as Feldman's version: the simulator prepares (random) shares $\tilde{s}_1, \ldots, \tilde{s}_t$ for t X-coordinates $\tilde{\alpha}_1, \ldots, \tilde{\alpha}_t$ and assigns to every newly corrupt player P_j one of these X-coordinates as α_j and the corresponding share as s_j.

4.2 Efficient Composition to the Real-Life Protocol

This section provides protocols that realize $\mathcal{F}_{\text{SSMT}}$ and $\mathcal{F}_{\text{SSMTwO}}$ over the public network with broadcast, i.e., without secure channels. Then, by applying the composition theorem, one can have adaptively secure Feldman VSS as a real-life protocol. As we shall see, these realizations are efficient but have some limitation on the adversary, which though can be successfully overcome in our applications.

Our constructions require an efficient bidirectional mapping between $\mathbb{Z}_q$ and G_q while the DDH problem should be hard to solve. This is the case when G_q is the order-q multiplicative subgroup of $\mathbb{Z}_p^*$ with prime $p = 2q+1$. Indeed, encoding $\mathbb{Z}_q \rightarrow G_q$ can be done by $m \mapsto M = m^2 \bmod p$, where $m \in \mathbb{Z}_q$ is identified with its representant in $\{1, \ldots, q\}$. This encoder is denoted by $M = Encode(m)$ and the corresponding decoder by $m = Decode(M)$.

Receiver Non-committing Message Transmission: By π_{RNC}, we denote a protocol that realizes $\mathcal{F}_{\text{SMT}}$ (or $\mathcal{F}_{\text{SSMT}}$) with receiver non-committing feature. That is, remains secure even if the receiver is adaptively corrupted (in the non-erasure model), while the sender may only be statically corrupted. Note that with such a restriction on the sender, $\mathcal{F}_{\text{SMT}}$ can be realized (without spooling). We review the construction by [13] (adapted to accept messages $m \in \mathbb{Z}_q$), which is originally designed in a classical model but can fit to the UC model. A proof is given in the full version of the paper [3].

A-0. (Initial step) Sender P_s chooses $h \leftarrow G_q$ and sends it to receiver P_r.
A-1. P_r selects $z_1, z_2 \leftarrow \mathbb{Z}_q$, computes $y = g^{z_1} h^{z_2}$, and sends y to P_s.
A-2. P_s computes $u = g^r$, $v = h^r$ and $c = Encode(m) y^r$, where $r \leftarrow \mathbb{Z}_q$, and sends (u, v, c) to P_r
A-3. P_r computes $m = Decode(cu^{-z_1}v^{-z_2})$.

Fig. 2. Protocol π_{RNC} for receiver non-committing transmission.

Lemma 1. *Under the DDH assumption, protocol π_{RNC} securely realizes $\mathcal{F}_{\text{SMT}}$ (or $\mathcal{F}_{\text{SSMT}}$) against an adaptive adversary if the sender is only statically corrupt and S is aware of ϖ with $Encode(m) = g^\varpi$.*

The assumption that the ideal-life adversary S is aware of the DL of $Encode(m)$ seems quite restrictive for π_{RNC} to be a general stand-alone tool. It is however acceptable for our purpose as m will be chosen by S in an upper-level protocol (playing the role of the to-be-corrupted sender) such that it knows the DL of $Encode(m)$. We stress that this assumption does not mean at all that S is given any kind of power to solve the DL problem.

Sender Non-committing Message Transmission with Opening: A protocol π_{SNC} that realizes $\mathcal{F}_{\text{SSMT}}$ with sender non-committing feature follows easily from π_{RNC}. The receiver P_r simply uses π_{RNC} to securely send a randomly chosen $k \in G_q$ to the sender P_s (precisely, P_r sends the message $Decode(k) \in \mathbb{Z}_q$), and then P_s sends $e = kEncode(m)$ to P_r, who computes m as $m = Decode(ek^{-1})$. We also consider the following variant of π_{SNC}, which we denote by π_{SNCwo}. All communication is done over the broadcast channel, and in an additional phase, the *opening phase*, the sender P_s publishes z_1 and z_2, privately sampled for the secure transmission of m, and every player verifies whether $g^{z_1} h^{z_2} = y$ and computes $k = cu^{-z_1}v^{-z_2}$ and $m = Decode(ek^{-1})$.

Lemma 2. *Under the DDH assumption, protocol π_{SNC} securely realizes $\mathcal{F}_{\text{SSMT}}$ and π_{SNCwo} securely realizes $\mathcal{F}_{\text{SSMTwo}}$ against an adaptive adversary if the receiver is only statically corrupt and S is aware of ϖ with $Encode(m) = g^\varpi$.*

The proof of Lemma 2 is similar to that of Lemma 1, although slightly more involved. For completeness, it is included in [3].

Composition with the Efficient Realizations: We now show that when the functionalities $\mathcal{F}_{\text{SSMTwo}}$ and $\mathcal{F}_{\text{SSMT}}$ in the hybrid-protocol π_{XFVSS} are implemented by

π_{SNCwO} and π_{RNC}, respectively, then the composed protocol securely realizes $\mathcal{F}_{\text{VSS}}^{\text{fcom}_g}$ (or $\mathcal{F}_{\text{SVSS}}^{\text{fcom}_g}$) in some weakened sense as stated below.

Theorem 1. *Implementing the functionality $\mathcal{F}_{\text{SSMTwO}}$ in step F-1 of the hybrid-protocol π_{XFVSS} from Fig. 1. by π_{SNCwO} and $\mathcal{F}_{\text{SSMT}}$ in step F-2 by π_{RNC} results in a secure realization of $\mathcal{F}_{\text{VSS}}^{\text{fcom}_g}$ (or $\mathcal{F}_{\text{SVSS}}^{\text{fcom}_g}$) in the real-life model, assumed that (1) the adversary corrupts the dealer only statically, and (2) the adversary corrupts players only before the reconstruction phase.*

Proof. The claim follows essentially from Proposition 1, Lemma 1 and 2, and the composition theorem. It remains to show that the assumptions for Lemma 1 and 2 are satisfied. By assumption (1) it is guaranteed that the receiver in π_{SNC} and the sender in π_{RNC} (which in both cases is the dealer) is only statically corrupt. Furthermore, by (2) and the way $\mathcal{S}$ works in the proof of Proposition 1, the messages, which are supposedly send through $\mathcal{F}_{\text{SSMTwO}}$ and $\mathcal{F}_{\text{SSMT}}$ and for which $\mathcal{S}$ has to convince $\mathcal{A}$ as being the messages sent through $\mathcal{F}_{\text{SSMTwO}}$ respectively $\mathcal{F}_{\text{SSMT}}$ are the values $\tilde{\alpha}_1, \ldots, \tilde{\alpha}_t$ and $\tilde{s}_1, \ldots, \tilde{s}_t$, all chosen randomly from $\mathbb{Z}_q$ (respectively $\mathbb{Z}_q^*$) by $\mathcal{S}$. Hence, $\mathcal{S}$ could sample them just as well by choosing $\varpi \leftarrow \mathbb{Z}_q$ and computing $Decode(g^\varpi)$ such that the conditions for Lemma 1 and 2 are indeed satisfied. Finally, as the dealer may only be statically corrupted, we do not need to care about spooling. Thus $\mathcal{F}_{\text{VSS}}^{\text{fcom}_g}$ and $\mathcal{F}_{\text{SVSS}}^{\text{fcom}_g}$ are equivalent here. $\square$

5   Applications to Threshold Cryptography

In this section, we propose several applications of the adaptively-secure Feldman VSS scheme from the previous section. Our main applications are a DLKG protocol and a UC threshold Schnorr signature scheme, though we also propose some related applications which might be of independent interest like a trapdoor-key generation protocol for Pedersen's commitment scheme and distributed-verifier UC proofs of knowledge. Interestingly, even though our Feldman VSS scheme has restricted adaptive security, the applications remain fully adaptively secure in the (SIP) UC model and do not underly restrictions as posed in Theorem 1.

To simplify terminology, from now on when referring to protocol π_{XFVSS}, we mean π_{XFVSS} from Fig. 1 with $\mathcal{F}_{\text{SSMTwO}}$ and $\mathcal{F}_{\text{SSMT}}$ replaced by π_{SNCwO} and π_{RNC} as specified in Theorem 1. Furthermore, it will at some point be convenient to use a different basis, say h, rather than the public parameter g in the core part of π_{XFVSS}, such that for instance h^s will be published as C. This will be denoted by $\pi_{\text{XFVSS}}[h]$, and obviously securely realizes $\mathcal{F}_{\text{VSS}}^{\text{fcom}_h}$. We stress that this modification is not meant to affect the sub-protocols π_{SNCwO} and π_{RNC}.

5.1   How to Generate the First Trapdoor Commitment-Key

In many protocols, a trapdoor commitment-key is considered as given by some trusted party so that the trapdoor information is unknown to any player. If the trusted party is replaced with multi-party computation, as we usually do, the protocol should be designed not to use any common trapdoor commitment-key. In this section, we show a protocol that meets this requirement.

The players execute protocol π_{HGEN} in Fig. 3. We assume that it is triggered by a player P_i who sends *init* to all players. The protocol outputs a (trapdoor) commitment-key $h \in G_q$ for Pedersen's commitment scheme. Note that the corresponding trapdoor $\log_g h = \sum_{j \in \mathcal{Q}} \chi_j$ is not shared among the players (in the usual way).

H-1. Every P_j chooses $\chi_j \leftarrow \mathbb{Z}_q$ and sends (*share*, *sid*, χ_j) to $\mathcal{F}_{\text{VSS}}^{\text{fcom}_g}$. Let $\mathcal{Q}$ be the set of players whose (*shared*, *sid*, P_j, Y_j) is published by $\mathcal{F}_{\text{VSS}}^{\text{fcom}_g}$. Remember that $Y_j = \text{fcom}_g(\chi_j) = g^{\chi_j}$.

H-2. Every player outputs $h = \prod_{j \in \mathcal{Q}} Y_j$.

Fig. 3. Commitment-key generation protocol π_{HGEN} in $\mathcal{F}_{\text{VSS}}^{\text{fcom}_g}$-hybrid model.

Unfortunately, one cannot expect h to be random as a rushing party can affect its distribution. However, the protocol inherits the following two properties which are sufficient for our purpose. (1) A simulator that simulates π_{HGEN} can compute the DL of h, and (2) given $Y \in G_q$, a simulator can embed Y into h so that given $\log_g h$, the simulator can compute $\log_g Y$. The latter in particular implies that the adversary is not able to compute the trapdoor $\log_g h$.

Our idea for formally capturing such a notion is that the ideal functionality challenges the adversary $\mathcal{S}$ by sending a random $h' \in G_q$ and allows $\mathcal{S}$ to randomize it so that h' is transformed into h such that $\mathcal{S}$ knows the trapdoor for h if and only if it knows it for h'. This clearly captures (1) and (2) above.

Definition 1 (Commitment-Key Generation Functionality: $\mathcal{F}_{\text{HGEN}}$).

1. *On receiving* (*generate*, *sid*) *from* P_i, *choose* $h' \leftarrow G_q$ *and send* (h', P_i) *to* $\mathcal{S}$.
2. *On receiving* $\gamma \in \mathbb{Z}_q$ *from* $\mathcal{S}$, *compute* $h = h' g^{\gamma}$ *and send* (*com-key*, *sid*, h) *to all players and* $\mathcal{S}$.

Proposition 2. *Protocol* π_{HGEN} *in Fig. 3 securely realizes* $\mathcal{F}_{\text{HGEN}}$ *against t-limited adaptive adversary for $t < n/2$ in the* $\mathcal{F}_{\text{VSS}}^{\text{fcom}_g}$*-hybrid SIP UC model.*

The proof is given in the full version of the paper. Essentially, on receiving (h', P_i) from $\mathcal{F}_{\text{HGEN}}$, $\mathcal{S}$ simulates the SIP P_{j*}'s call to $\mathcal{F}_{\text{VSS}}^{\text{fcom}_g}$ with input h', and it sends $\gamma = \sum_{j \neq j*} \chi_j$ to $\mathcal{F}_{\text{HGEN}}$.

We claim that $\mathcal{F}_{\text{VSS}}^{\text{fcom}_g}$ in π_{HGEN} can be securely realized by the protocol π_{XFVSS} from Theorem 1. This may look contradictory since π_{XFVSS} is secure only against static corruption of the dealer as stated in Theorem 1, while in π_{HGEN} every player acts as a dealer and may be adaptively corrupted. However, looking at the proof, except for the run launched by the SIP P_{j*}, $\mathcal{S}$ simulates all runs of $\mathcal{F}_{\text{VSS}}^{\text{fcom}_g}$ honestly with true inputs. Hence, for these simulations, the situation is exactly as in the case where the dealer is statically corrupted and the secret is known to the simulator at the beginning. Furthermore, the reconstruction phase of $\mathcal{F}_{\text{VSS}}^{\text{fcom}_g}$ is never invoked in π_{HGEN}. Thus, the following holds.

Theorem 2. *Implementing* $\mathcal{F}_{\mathrm{VSS}}^{\mathrm{fcom}_g}$ *in* π_{HGEN} *of Fig. 3 by* π_{XFVSS} *results in a secure realization of* $\mathcal{F}_{\mathrm{HGEN}}$ *against t-limited adaptive adversary for* $t < n/2$ *in the (real-life) SIP UC model.*

5.2   DL-Key Generation

This section constructs an adaptively secure protocol for DLKG, whose functionality is defined below. Clearly, from such a key-generation protocol (respectively functionality), one expects that it outputs a public-key y and in some hidden way produces the corresponding secret-key x (typically by having it shared among the players), such that x can be used to do some cryptographic task like signing or decrypting if enough of the players agree [16]. However, as we want to view our protocol as a generic building block for threshold schemes, we simply require that the secret-key x can be opened rather than be used for some specific task. In Sect. 5.3 we then show a concrete example threshold scheme based on our DLKG protocol.

Definition 2 (Threshold DL Key Generation Functionality: $\mathcal{F}_{\mathrm{DLKG}}$).

1. *On receiving* (generate, sid) *from* P_i, *select* $x \leftarrow \mathbb{Z}_q$, *compute* $y = g^x$, *and send* (key, sid, y) *to all players and* $\mathcal{S}$.
2. *On receiving* (open, sid) *from* $t+1$ *players, send* (private, sid, x) *to all players and* $\mathcal{S}$.

Our realization of $\mathcal{F}_{\mathrm{DLKG}}$ is illustrated in Fig. 5 below, and makes use of (ordinary) Pedersen's VSS scheme given in Fig. 4.

[Sharing Phase]

P-1.   The dealer selects $f(X) = a_0 + a_1 X + \cdots + a_t X^t \leftarrow \mathbb{Z}_q[X]$ and $f'(X) = b_0 + b_1 X + \cdots + b_t X^t \leftarrow \mathbb{Z}_q[X]$ where $a_0 = s$. Let $r = b_0$. The dealer then computes and broadcasts $C = C_0 = g^s h^r$ and $C_k = g^{a_k} h^{b_k}$ for $k = 1, \ldots, t$, and he sends $s_i = f(i)$ and $r_i = f'(i)$ to P_i using $\mathcal{F}_{\mathrm{SSMT}}$.

P-2.   Each P_i verifies whether $g^{s_i} h^{r_i} = E_i$ where $E_i = \prod_{k=0}^{t} C_k^{i^k}$. P_i broadcasts *verified* if it holds and else initiates the accusation sub-protocol which is the same as that of Feldman VSS with obvious modification.

[Reconstruction Phase]

Every player P_i publicly opens E_i to s_i. The secret s is reconstructed using Lagrange interpolation from the correctly opened s_i's.

Fig. 4. Pedersen's VSS scheme: $\mathsf{PedVSS}_{g,h}(s) \rightarrow (s_1, \ldots, s_n, r, C)$.

We do not prove Pedersen's VSS secure in the UC framework, and in fact it is not (as a committed VSS against an adaptive adversary). The only security requirement we need is covered by the following well-known fact.

Lemma 3. *Except with negligible probability, after the sharing phase of Pedersen's VSS, both the s_i's and r_i's of the uncorrupt players are correct sharings of s and r such that $g^s h^r = C$ and such that s is reconstructed in the reconstruction phase (and s and r coincide with the dealer's choice in case he remains honest), or otherwise $\log_g h$ can be efficiently extracted from the adversary.*

We write $\mathsf{PedVSS}_{g,h}^j(s) \to (s_1, \ldots, s_n, r, C)$ to denote an execution of the sharing phase of Pedersen's VSS with secret s and player P_j acting as dealer, and with values $s_1, \ldots, s_n, r, C$ generated as described in Fig. 4.

[Key-Generation Phase]

K-1. A player, P_i, sends $(\textbf{\textit{generate}}, sid)$ to $\mathcal{F}_{\text{HGEN}}$, and commitment-key h is obtained.

K-2. Each player P_j chooses a random $x_j \in \mathbb{Z}_q$ and executes the sharing phase of Pedersen's VSS with secret x_j and commitment-key h: $\mathsf{PedVSS}_{g,h}^j(x_j) \to (x_{j1}, \ldots, x_{jn}, r_j, C_j)$. If a player P_j refuses then a default Pedersen sharing of $x_j = 0$ is taken instead.

K-3. Each P_j sends $(\textbf{\textit{share}}, sid_j, x_j)$ to $\mathcal{F}_{\text{VSS}}^{\text{fcom}_g}$ and $(\textbf{\textit{share}}, sid'_j, r_j)$ to $\mathcal{F}_{\text{VSS}}^{\text{fcom}_h}$.

K-4. If P_i receives $(\textbf{\textit{shared}}, sid_j, P_j, C'_j)$ and $(\textbf{\textit{shared}}, sid'_j, P_j, C''_j)$, he verifies that $C_j = C'_j C''_j$ holds. (Note that $C'_j = g^{x_j}$ and $C''_j = h^{r_j}$.) If either of such messages has not been received or the relation does not hold, then x_j is reconstructed from its Pedersen sharing, and every P_i sets $C'_j = g^{x_j}$. Output of this phase is the public-key $y = \prod_{j=1}^n C'_j$, while each P_j stores x_j as his (additive) secret-key share, to which he is committed by C_j.

[Opening Phase]

Every player P_j publicly opens C_j by broadcasting x_j and r_j. If a player P_j fails to do so, x_j is reconstructed from its Pedersen sharing. Secret-key x is then computed as $x = \sum_{j=1}^n x_j$.

Fig. 5. Threshold DLKG protocol π_{DLKG} in $(\mathcal{F}_{\text{HGEN}}, \mathcal{F}_{\text{VSS}}^{\text{fcom}_g} \mathcal{F}_{\text{VSS}}^{\text{fcom}_h}, \mathcal{F}_{\text{SMT}})$-hybrid model.

Note that in π_{DLKG} the *additive* shares x_j are used to reconstruct the secret-key x, rather than the threshold-shares implicitly given by $\xi_j = \sum_i x_{ij}$. The reason is that even though using the threshold shares can be proven secure in the hybrid-model, it resists a security proof when the ideal functionality $\mathcal{F}_{\text{SSMT}}$ in Pedersen's VSS is replaced by π_{RNC} as we do (due to the DL condition from Lemma 1). In [3] we show how to modify the scheme in order to be able to use the threshold-shares as secret-key shares. Also note that using the terminology introduced in [2], based on the results in [1], step K-3 can be seen as a *distributed-verifier zero-knowledge proof* of knowledge of x_j and r_j such that $g^{x_j} = C'_j$ and $h^{r_j} = C''_j$ (see also Sect. 5.4).

Theorem 3. *Implementing in the DLKG protocol π_{DLKG} from Fig. 5 the functionalities $\mathcal{F}_{\text{HGEN}}$, $\mathcal{F}_{\text{SSMT}}$, $\mathcal{F}_{\text{VSS}}^{\text{fcom}_g}$ and $\mathcal{F}_{\text{VSS}}^{\text{fcom}_h}$ by π_{HGEN}, π_{RNC}, $\pi_{\text{XFVSS}}[g]$ and $\pi_{\text{XFVSS}}[h]$, respectively, results in a secure realization of $\mathcal{F}_{\text{DLKG}}$ against adaptive t-limited adversary for $t < n/2$ in the SIP UC model.*

Using the UC with joint state framework [8], one can prove using similar arguments that the commitment-key h can be generated once and for all invocations of π_{DLKG}. Furthermore, concerning efficiency, the communication complexity of the key-generation phase is comparable to that of the schemes by [13]: it requires $O(n^2\kappa)$ bits to be sent over the bilateral public channels and another $O(n^2\kappa)$ bits to be broadcast.

The full proof of Theorem 3 is given in the full version of the paper. We simply sketch its idea here. First, the simulator $\mathcal{S}$ simulates the generation of h such that it knows the DL of h, while step K-2 is executed as prescribed. Then, it reconstructs the x_i's of the corrupt players, and it computes C'_{j*} and C''_{j*} for the SIP P_{j*} such that $C'_{j*} \cdot \prod_{j \neq j*} g^{x_j} = y$ and $C_{j*} = C'_{j*} C''_{j*}$, where y is the value provided by $\mathcal{F}_{\text{DLKG}}$. Then it simulates the two Feldman VSSes with P_{j*} as dealer, while the other executions are followed as prescribed (with inputs x_j respectively r_j). As a result, the output of the key-generation phase is y. In the opening phase, having received $x = \log_g(y)$ from $\mathcal{F}_{\text{DLKG}}$, $\mathcal{S}$ simply adapts P_{j*}'s initial x_{j*} such that $\sum_j x_j = x$, and it uses the DL of h to open C_{j*} to (the new) x_{j*}. The only difference in the adversary's and thus the environment's view between the simulation and a real execution lies in the encrypted Pedersen shares of (the initial) x_{j*} given to the uncorrupt players. By the property of π_{RNC}, this cannot be distinguished by the environment.

From now on, when referring to protocol π_{DLKG}, we mean π_{DLKG} from Fig. 5 with the functionalities replaces by real-life protocols as specified in Theorem 3.

5.3 Universally-Composable Threshold Schnorr-Signatures

As an example application of our DL-key generation protocol, we propose a threshold variant of Schnorr's signature scheme [15], provable secure in the UC framework. The scheme is illustrated in Fig. 6. Recall, a Schnorr signature for message m under public-key $y = g^x$ consists of (c, s) such that $r = g^s/y^c$ satisfies $H(m, r) = c$, where H is a cryptographic hash-function. Such a signature is computed by the signer (in the single-signer variant), who knows the secret-key x, by choosing $k \leftarrow \mathbb{Z}_q$ and computing $r = g^k$, $c = H(m, r)$ and $s = k + cx$. Schnorr's signature scheme can be proven secure, in the sense of existential unforgability against chosen message attacks, in the *random oracle* model.

Consider the ideal threshold signature functionality $\mathcal{F}_{\text{TSIG}}$ by adapting the (single-signer) signature functionality $\mathcal{F}_{\text{SIG}}$ from [5] in the obvious way.

Theorem 4. *Protocol π_{TSIG} securely realizes $\mathcal{F}_{\text{TSIG}}$ against adaptive t-limited adversary for $t < n/2$ in the UC model, under the DDH assumption and under the assumption that the standard Schnorr signature scheme is secure.*

We stress that interestingly π_{TSIG} securely realizes $\mathcal{F}_{\text{TSIG}}$ in the *standard* rather than the SIP UC model.

Proof. (Sketch) The simulator $\mathcal{S}$ simply executes *honestly* π_{TSIG}. Note that the public-key y is not dictated by $\mathcal{F}_{\text{TSIG}}$, but rather $\mathcal{F}_{\text{TSIG}}$ asks $\mathcal{S}$ to provide it. In order to prove that this is a good simulation, we argue as follows. The only

[Key-Generation Phase]

The players execute the key-generation phase of π_{DLKG}, resulting in a public-key y, private (additive) secret-key shares $x_1, \ldots, x_n$ with corresponding commitments $C_1, \ldots, C_n$, and commitment-key h.

[Signing Phase]

In order to sign a message m, the following steps are executed.

S-1. The players once more invoke the key-generation phase of π_{DLKG}, but skipping the generation of h and taking h from the generation of y. Denote the output by r, the corresponding additive secret-key shares by $k_1, \ldots, k_n$, and the corresponding commitments by $K_1, \ldots, K_n$.

S-2. Each player P_j computes $c = H(m, r)$ and publicly opens $K_j C_j^c$ to $s_j = k_j + c x_j$. If a player P_j fails to do so, s_j is reconstructed from its Pedersen sharing (which is implicitly given by the Pedersen sharings of x_j and k_j). Signature (c, s) is completed by $s = \sum_j s_j$.

Fig. 6. Threshold Schnorr-signature scheme π_{TSIG}.

way $\mathcal{Z}$ may see a difference is when $\mathcal{A}$ breaks the signature scheme, i.e., when a player provides at some point a valid signature on a message that has not been signed. However, if there exist $\mathcal{Z}$ and $\mathcal{A}$ that can enforce such an event with non-negligible probability, then there exists a forger F that breaks the existential unforgability against chosen message attacks of the standard (single-signer) Schnorr signature scheme. F works as follows. F runs $\mathcal{Z}$ and $\mathcal{A}$, and it simulates the action of $\mathcal{S}$, i.e. the execution of π_{TSIG}, as follows. It uses the SIP simulator for the key-generation phase of π_{DLKG} to force the output of the key-generation to be the given public-key y. Furthermore, to sign a message m, it asks the signing oracle for a signatures (c, s) on m, it forces as above the outcome of S-1 to be $r = g^s / y^c$, and it uses a straightforward modification of the SIP simulator for the opening phase of π_{DLKG} to simulate the signing phase: the simulated P_{j*} opens $K_{j*} C_{j*}^c$ to $s - \sum_{j \neq j*} s_j$ in step S-2 (rather than to $k_{j*} + c x_{j*}$), forcing the output of the signing phase to be the given signature (c, s). Additionally, whenever a message-signature pair (m, σ) is asked to be verified, F first checks whether m was never signed before and if σ is a valid signature on m. Once such a pair (m, σ) is found, F outputs that pair and halts. Similar to the proof of Theorem 3, one can show that if $\mathcal{A}$ does not corrupt the SIP then $\mathcal{Z}$ cannot distinguish between the real execution of π_{TSIG} (executed by the simulator $\mathcal{S}$) and the SIP simulation (executed by the forger F). Hence, by assumption on $\mathcal{Z}$ and $\mathcal{A}$, F outputs a signature on a message not signed by the signing oracle with non-negligible probability. □

5.4 Adaptively Secure Distributed-Verifier Proofs

In designing threshold cryptography, it is quite common to prove some relation (or knowledge) about committed witnesses in zero-knowledge manner. In the UC framework, however, zero-knowledge proofs are extremely expensive components: they are realized by combining a generic non-interactive zero-knowledge proof with a common-reference string generator, or UC-secure commitment scheme (which anyway needs common reference string) with generic zero-knowledge proof system for an NP-complete language such as Hamiltonian. They are generic and powerful, but cannot be directly used for practical subjects such as showing equality of discrete-logs or knowledge of a representation.

Combining our results with techniques developed in [1, 2], one can construct adaptively secure efficient distributed-verifier zero-knowledge proofs in universally composable way for many practical subjects. We illustrate a concrete example. Suppose that a prover needs to show that a triple $(g^\alpha, g^\beta, g^\gamma)$ is in DH, i.e. satisfies $\alpha \cdot \beta = \gamma$. This can be done as follows. A prover shares α twice: once using the sharing phase of $\pi_{\text{XFVSS}}[g]$ and once using that of $\pi_{\text{XFVSS}}[g^\beta]$ with base g^β. Furthermore, in the second execution, the same sharing polynomial and X-coordinates as in the first execution are used. Hence the second execution is completed only by broadcasting a new commitment of the sharing polynomial, which is verified by the players by using the same share and X-coordinate received in the first execution. This guarantees that indeed the same secret, α, has been shared. Note that $(g^\beta)^\alpha$, supposed to be g^γ, is published in the second execution. Finally, the prover shares β (or γ) using the sharing phase of $\pi_{\text{XFVSS}}[g]$ with base g. If all sharing phases are accepted, the proof is accepted. Given $(g^\alpha, g^\beta, g^\gamma)$, S can simulate the prover by simulating the dealer in each execution of π_{XFVSS}. In the case of corrupt prover who completes the proof, S can extract α and β from the set of uncorrupt players. Hence the simulator can extract a witness (α, β) needed to invoke ideal zero-knowledge functionality.

The techniques of [1, 2] also apply to other commitment schemes that Feldman's, and allow to prove other relations as well like equality and additive and inverse relations among committed values. From these building blocks, one can even construct an adaptive distributed verifier proof for any NP relation by following the construction in [2].

Acknowledgements

We thank Jesper Buus Nielsen for suggesting to use spooling to overcome the problem addressed in Sect. 2.4. We also thank Stas Jarecki and the anonymous referees for their careful reading and invaluable comments.

References

1. M. Abe. Robust distributed multiplication without interaction. In M. Wiener, editor, *Advances in Cryptology — CRYPTO '99*, volume 1666 of *Lecture Notes in Computer Science*, pages 130–147. Springer-Verlag, 1999.
2. M. Abe, R. Cramer, and S. Fehr. Non-interactive distributed-verifier proofs and proving relations among commitments. In Y. Zheng, editor, *Advances in Cryptology – ASIACRYPT '02*, volume 2501 of *Lecture Notes in Computer Science*, pages 206–223. Springer-Verlag, 2002.

3. M. Abe and S. Fehr. Adaptively secure Feldman VSS and applications to universally-composable threshold cryptography. In Cryptology ePrint Archive, Report 2004/119, 2004. http://eprint.iacr.org.
4. R. Canetti. Universally composable security: A new paradigm for cryptographic protocols. In *Proceedings of the 42nd IEEE Annual Symposium on Foundations of Computer Science*, pages 136–145, 2001.
5. R. Canetti. On universally composable notions of security for signature, certification and authentication. In Cryptology ePrint Archive, Report 2003/239, 2003. http://eprint.iacr.org.
6. R. Canetti, U. Feige, O. Goldreich, and M. Naor. Adaptively secure multi-party computation. In *Proceedings of the 28th annual ACM Symposium on the Theory of Computing*, pages 639–648, 1996.
7. R. Canetti, R. Gennaro, S. Jarecki, H. Krawczyk, and T. Rabin. Adaptive security for threshold cryptosystems. In M. Wiener, editor, *Advances in Cryptology — CRYPTO '99*, volume 1666 of *Lecture Notes in Computer Science*, pages 98–115. Springer-Verlag, 1999.
8. R. Canetti and T. Rabin. Universal composition with joint state. In Cryptology ePrint Archive, Report 2003/047, 2002. http://eprint.iacr.org.
9. I. Damgård and J. Nielsen. Improved non-committing encryption schemes based on a general complexity assumption. In M. Bellare, editor, *Advances in Cryptology — CRYPTO 2000*, volume 1880 of *Lecture Notes in Computer Science*, pages 432–450. Springer-Verlag, 2000.
10. P. Feldman. A practical scheme for non-interactive verifiable secret sharing. In *Proceedings of the 28th IEEE Annual Symposium on Foundations of Computer Science*, pages 427–437, 1987.
11. Y. Frankel, P. D. MacKenzie, and M. Yung. Adaptively-secure distributed public-key systems. In J. Nesetril, editor, *European Symposium on Algorithms (ESA '99)*, volume 1643 of *Lecture Notes in Computer Science*, pages 4–27. Springer-Verlag, 1998.
12. R. Gennaro, S. Jarecki, H. Krawczyk, and T. Rabin. Secure distributed key generation for discrete-log based cryptosystems. In J. Stern, editor, *Advances in Cryptology — EUROCRYPT '99*, volume 1592 of *Lecture Notes in Computer Science*, pages 295–310. Springer-Verlag, 1999.
13. S. Jarecki and A. Lysyanskaya. Adaptively secure threshold cryptography: Introducing concurrency, removing erasures (extended abstract). In B. Preneel, editor, *Advances in Cryptology – EUROCRYPT 2000*, volume 1807 of *Lecture Notes in Computer Science*, pages 221–242. Springer-Verlag, 2000.
14. T. P. Pedersen. A threshold cryptosystem without a trusted party. In D. W. Davies, editor, *Advances in Cryptology — EUROCRYPT '91*, volume 547 of *Lecture Notes in Computer Science*, pages 522–526. Springer-Verlag, 1991.
15. C. P. Schnorr. Efficient signature generation for smart cards. *Journal of Cryptology*, 4(3):239–252, 1991.
16. D. Wikström. A universally composable mix-net. In *Proceedings of the First Theory of Cryptography Conference (TCC'04)*, volume 2951 of *Lecture Notes in Computer Science*, pages 315–335. Springer-Verlag, 2004.

Round-Optimal Secure Two-Party Computation

Jonathan Katz[1,*] and Rafail Ostrovsky[2,**]

[1] Dept. of Computer Science, University of Maryland
jkatz@cs.umd.edu
[2] Dept. of Computer Science, U.C.L.A.
rafail@cs.ucla.edu

Abstract. We consider the central cryptographic task of *secure two-party computation*: two parties wish to compute some function of their private inputs (each receiving possibly different outputs) where security should hold with respect to arbitrarily-malicious behavior of either of the participants. Despite extensive research in this area, the exact round-complexity of this fundamental problem (i.e., the number of rounds required to compute an *arbitrary* poly-time functionality) was not previously known.

Here, we establish the **exact** round complexity of secure two-party computation with respect to black-box proofs of security. We first show a lower bound establishing (unconditionally) that four rounds are *not* sufficient to securely compute the coin-tossing functionality for any super-logarithmic number of coins; this rules out 4-round protocols for other natural functionalities as well. Next, we construct protocols for securely computing *any* (randomized) functionality using only five rounds. Our protocols may be based either on certified trapdoor permutations or homomorphic encryption schemes satisfying certain additional properties. The former assumption is implied by, e.g., the RSA assumption for large public exponents, while the latter is implied by, e.g., the DDH assumption. Finally, we show how our protocols may be modified – without increasing their round complexity and without requiring erasures – to tolerate an *adaptive* malicious adversary.

1 Introduction

Round complexity measures the number of messages that parties need to exchange in order to perform some joint task. Round complexity is a central measure of efficiency for any interactive protocol, and much research has focused on improving bounds on the round complexity of various cryptographic tasks. As representative examples (this list is not exhaustive), we mention work on upper- and lower-bounds for zero-knowledge proofs and arguments [6, 7, 19, 27, 28, 40], concurrent zero-knowledge [13, 15, 17, 35, 41, 42], and secure two-party and multi-party computation [4, 5, 10, 11, 14, 21–23, 31–34, 37, 43]. The study of secure two-party computation is fundamental in this regard: not only does it encompasses

* Part of this work was supported by NSF Trusted Computing Grant #0310751.
** Part of this work was supported by a gift from the Teradata Corporation.

M. Franklin (Ed.): CRYPTO 2004, LNCS 3152, pp. 335–354, 2004.
© International Association for Cryptologic Research 2004

functionalities whose round-complexity is of independent interest (such as coin tossing or the zero-knowledge functionality), but it also serves as an important special case in the study of secure computation.

Yao [43] presented a constant-round protocol for secure two-party computation when the adversarial party is assumed to be *honest-but-curious* (or *passive*). Goldreich, Micali, and Wigderson [25, 29] extended Yao's result, and showed a protocol for secure multi-party computation (and two-party computation in particular) tolerating *malicious* (or *active*) adversaries. Unfortunately, their protocol does not run in a constant number of rounds. Recently, Lindell [37] gave the first constant-round protocol for secure two-party computation in the presence of malicious adversaries; he achieves this result by constructing the first constant-round coin-tossing protocol (for polynomially-many coins) and then applying the techniques of [29]. The number of rounds in the resulting protocol for secure two-party computation is not specified by Lindell, but is on the order of 20–30.

The above works all focus on the case of a non-adaptive adversary. A general methodology for constructing protocols secure against an *adaptive* adversary is known [12], and typically requires additional rounds of interaction.

Lower bounds on the round-complexity of secure two-party computation with respect to black-box[1] proofs of security have also been given. (We comment further on black-box bounds in Section 1.2.) Goldreich and Krawczyk [28] showed that, assuming $NP \not\subseteq BPP$, zero-knowledge (ZK) proofs or arguments for NP require 4 rounds. Since ZK proofs (of knowledge) are a particular example of a two-party functionality, this establishes a lower bound of 4 rounds for secure two-party computation. Under the same complexity assumption, Lindell [38] has shown that for some polynomial p, secure coin-tossing of $p(k)$ coins requires at least 4 rounds.

1.1   Our Results

Here, we *exactly characterize* the (black-box) round complexity of secure two-party computation by improving the known bounds. In particular:

Lower bound: We show that 5 rounds are necessary for securely tossing any super-logarithmic (in the security parameter) number of coins, with respect to black-box proofs of security. Thus implies a 5-round black-box lower bound for a number of other (deterministic) functionalities as well. Beyond the implications for the round complexity of secure computation, we believe the result is of independent interest due to the many applications of coin-tossing to other cryptographic tasks.

The result of Goldreich and Krawczyk [28] mentioned above implies a black-box lower bound of five rounds for the "symmetric" ZK functionality (where the parties simultaneously prove statements to each other) – and hence the same lower bound on the black-box round complexity of secure two-party computation

---

[1] Throughout this paper, "black-box" refers to black-box use *of an adversary's code/circuit* (and not black-box use *of a cryptographic primitive*, as in [30]). A definition of black-box proofs of security is given in Appendix A.

of general functionalities – assuming $NP \not\subseteq BPP$. In contrast, our lower bound holds *unconditionally*.

Matching upper bound: As our main result, we construct 5-round protocols for securely computing any (randomized) poly-time functionality in the presence of a *malicious* adversary. Our protocols may be based on various cryptographic assumptions, including certified, enhanced trapdoor permutations (see Definition 1 and Remark 1), or homomorphic encryption schemes satisfying certain additional properties. The former may be based on, for example, the RSA assumption for large public exponents, while the latter may be based on, for example, the decisional Diffie-Hellman (DDH) assumption in certain groups. Due to space limitations, we focus on the (more difficult) case of certified trapdoor permutations, and refer the reader to the full version for protocols based on alternate assumptions.

In Section 4.1, we sketch how our protocols can be extended – without increasing the round complexity and without requiring erasures – to tolerate an *adaptive* adversary. The necessary cryptographic assumptions are described in more detail there.

1.2   A Note on Black-Box Lower Bounds

Until the recent work of Barak [1, 2], a black-box impossibility result was generally viewed as strong evidence for the "true" impossibility of a given task. Barak showed, however, that non-black-box use of an adversary's code could, in fact, be used to circumvent certain black-box impossibility results [1]. Nevertheless, we believe *there is still an important place in cryptography for black-box impossibility results* for at least the following reasons:

1. A black-box impossibility result is useful insofar as it rules out a certain *class of techniques* for solving a given problem.
2. With respect to our current understanding, protocols constructed using non-black-box techniques, currently seem inherently less efficient than those constructed using black-box techniques.

It remains an interesting open question to beat the lower bound given in this paper using non-black-box techniques, or to prove that this is impossible.

1.3   Discussion

Yao's results [43] give a 4-round protocol secure against *honest-but-curious* adversaries, assuming the existence of enhanced [26, Sec. C.1] trapdoor permutations (an optimal 3-round protocol secure against honest-but-curious adversaries can be constructed based on the existence of homomorphic encryption schemes). Our lower bound shows that additional rounds are necessary to achieve security against the stronger class of *malicious* adversaries. Our upper bound, however, shows that (at least in the case of trapdoor permutations) a *single* (i.e., fifth) additional round suffices.

Our technique for achieving security against *adaptive* adversaries applies only to adversaries who corrupt at most one of the players. An interesting open question is to construct a *constant-round* protocol tolerating an adaptive adversary who can potentially corrupt *both* players.

2 Definitions and Cryptographic Preliminaries

We omit the (completely standard) definitions of security for two-party computation used in this work, which follow [9, 10, 25, 39]. However, we provide in Appendix A our definition of *black-box simulation* which is used to prove the lower bound of Section 3.

We assume the reader is familiar with the cryptographic tools we use and refer the reader elsewhere for definitions of non-interactive (perfectly binding) commitment schemes [24], 3-round witness-indistinguishable (WI) proofs of knowledge [20, 24], witness-extended emulation for proofs/arguments of knowledge [37], and the Feige-Shamir 4-round ZK argument of knowledge [18, 19]. We note that all the above may be constructed based on the existence of certified, enhanced trapdoor permutations.

To establish notation, we provide here our working definitions of *trapdoor permutations*, *hard-core bits*, and Yao's *garbled circuit technique*. We also discuss *equivocal commitment*, and show a new construction of this primitive.

Trapdoor permutations. For the purposes of the present abstract, we use the following simplified definition of trapdoor permutations (but see Remark 1):

Definition 1. *Let $\mathcal{F}$ be a triple of* PPT *algorithms* (Gen, Eval, Invert) *such that if* $\mathsf{Gen}(1^k)$ *outputs a pair* (α, td), *then* $\mathsf{Eval}(\alpha, \cdot)$ *is a permutation over* $\{0,1\}^k$ *and* $\mathsf{Invert}(\mathsf{td}, \cdot)$ *is its inverse.* $\mathcal{F}$ *is a* **trapdoor permutation family** *if the following is negligible in k for all poly-size circuit families $\{A_i\}$:*

$$\Pr[(\alpha, \mathsf{td}) \leftarrow \mathsf{Gen}(1^k); y \leftarrow \{0,1\}^k; x \leftarrow A_k(\alpha, y) : \mathsf{Eval}(\alpha, x) = y].$$

We additionally assume that $\mathcal{F}$ satisfies (a weak variant of) **"certifiability"**: *namely, given some α it is possible to decide in polynomial time whether* $\mathsf{Eval}(\alpha, \cdot)$ *is a permutation over* $\{0,1\}^k$.

For notational convenience, we let (α, td) be implicit and will simply let $\mathsf{f}(\cdot)$ denote $\mathsf{Eval}(\alpha, \cdot)$, and $\mathsf{f}^{-1}(\cdot)$ denote $\mathsf{Invert}(\mathsf{td}, \cdot)$ (where α, td are understood from the context). Of course, f^{-1} can only be efficiently evaluated if td is known.

Remark 1. The above definition is somewhat less general than others that have been considered (e.g., that of [24, Def. 2.4.5]); in particular, the present definition assumes a domain of $\{0,1\}^k$ and therefore no "domain sampling" algorithm is necessary. Furthermore, the protocol of Section 4 does *not* immediately generalize for trapdoor permutations requiring such domain sampling. Nevertheless, by introducing additional machinery it is possible to modify our protocol so that it may be based on any family of enhanced trapdoor permutations (cf. [26, Sec. C.1]) satisfying the certifiability condition noted above. For simplicity, however, we use the above definition in proving our results and defer the more complicated protocol (and proof) to the full version.

Hard-core bits. We assume the reader is familiar with the notion of hard-core bits for any trapdoor permutation family (see [24]), and thus we merely describe the notation we use. Let $H = \{h_k : \{0,1\}^k \to \{0,1\}\}$ be a hard-core bit for some trapdoor permutation family $\mathcal{F}$ (we will let k be implicit, and set $h = h_k$); thus (informally), $h(z)$ is "hard" to predict given $f(z)$. We extend this notation to a vector of k hard-core bits in the following way:

$$h(z) \stackrel{\text{def}}{=} h(z)|h(f(z))|\cdots|h(f^{k-1}(z)).$$

Now (informally), $h(z)$ "looks pseudorandom" given $f^k(z)$.

Yao's "garbled circuit". Our secure computation protocol uses as a building block the "garbled circuit" technique of Yao [43] which enables constant-round secure computation for *honest-but-curious adversaries*. We abstract Yao's technique, and only consider those aspects of it which are necessary for our proof of security. In what follows, F is a description of a two-input/single-output circuit whose inputs and output have the same length k (yet the technique may be generalized for inputs and output of arbitrary polynomial lengths). Yao's results give PPT algorithms $\mathsf{Yao}_1, \mathsf{Yao}_2$ for which:

- Yao_1 is a randomized algorithm which takes as input a security parameter 1^k, a circuit F, and a string $y \in \{0,1\}^k$. It outputs a "garbled circuit" circuit and *input-wire labels* $Z_{1,0}, Z_{1,1}, \ldots, Z_{k,0}, Z_{k,1} \in \{0,1\}^k$. The "garbled circuit" may be viewed as representing the function $F(\cdot, y)$.
- Yao_2 is a deterministic algorithm which takes as input 1^k, a "garbled circuit" circuit, and k values $Z_1, \ldots, Z_k \in \{0,1\}^k$. It outputs either an invalid symbol $\perp$, or a value $v \in \{0,1\}^k$.

(When k is clear from the context, we omit it.)

We briefly describe how the above algorithms may be used for secure computation in the honest-but-curious setting. Let player 1 (resp., 2) hold input x (resp., y), and assume that player 1 is to obtain the output $F(x, y)$. First, player 2 computes (circuit, $\{Z_{i,b}\}$) $\leftarrow \mathsf{Yao}_1(F, y)$ and sends circuit to player 1. Then, the two players engage in k instances of oblivious transfer: in the ith instance, player 1 enters with "input" x_i, player 2 enters with "input" $(Z_{i,0}, Z_{i,1})$, and player 1 obtains the "output" $Z_i \stackrel{\text{def}}{=} Z_{i,x_i}$. Player 1 then computes $v = \mathsf{Yao}_2(\text{circuit}, Z_1, \ldots, Z_k)$ and outputs v.

A 3-round protocol for oblivious transfer (OT) based on trapdoor permutations may be constructed as follows (we remark that using number-theoretic assumptions, 2-round OT is possible): Let player 1 have input b and player 2 have input strings $Z_0, Z_1 \in \{0,1\}^k$ (the goal is for player 1 to obtain Z_b). Player 2 begins by generating trapdoor permutation (f, f^{-1}) and sending f to player 1. Next, player 1 chooses random $z'_0, z'_1 \in \{0,1\}^k$, sets $z_b = f^k(z'_b)$ and $z_{\bar{b}} = z'_{\bar{b}}$, and sends z_0, z_1 to player 2. Finally, player 2 computes $W_0 = Z_0 \oplus h(f^{-k}(z_0))$, computes

W_1 analogously, and sends W_0, W_1 to player 1. Player 1 can then easily recover Z_b. (A proof of security for essentially the above protocol appears in [25].) Note that in the honest-but-curious setting it is secure to run polynomially-many executions of the above in parallel.

Putting everything together, we obtain the following 3-round protocol for secure computation of any single-output functionality in the honest-but-curious setting:

Round 1 Player 2 runs Yao$_1$ to generate (circuit, $\{Z_{i,b}\}$). He then sends circuit and the f's for oblivious transfer.

Round 2 Player 1 sends k pairs (z_0, z_1).

Round 3 Player 2 sends k pairs (W_0, W_1).

Output computation Player 1 can now recover the appropriate $\{Z_i\}$ and thus compute the output value v using Yao$_2$, as discussed above.

Finally, any protocol for secure computation of single-output functionalities can be used for secure computation of two-output functionalities using only one additional round [25, Prop. 7.2.11]. Furthermore, any protocol for secure computation of deterministic functionalities may be used for secure computation of randomized ones (with the same round complexity) [25, Prop. 7.4.4].

With the above in mind, we describe the properties required of Yao$_1$, Yao$_2$. We first require *correctness*: for any F, y, any output (circuit, $\{Z_i\}$) of Yao$_1(F, y)$, and any x we have $F(x, y) = $ Yao$_2$(circuit, $Z_{1,x_1}, \ldots, Z_{k,x_k}$). The algorithms also satisfy the following notion of *security*: there exists a simulator Yao-Sim which takes x, v as input, and which outputs circuit and a set of k input-wire labels $\{Z_i\}$; furthermore, the following distributions are computationally indistinguishable (by poly-size circuit families):

$$1. \ \left\{ (\text{circuit}, \{Z_{i,b}\}) \leftarrow \text{Yao}_1(F, y) : (\text{circuit}, \{Z_{i,x_i}\}) \right\}_{x,y}$$

$$2. \ \left\{ v = F(x, y) : \text{Yao-Sim}(x, v) \right\}_{x,y}.$$

Algorithms (Yao$_1$, Yao$_2$) satisfying the above definitions may be constructed assuming the existence of one-way functions.

Equivocal commitment. Although various notions of equivocal commitment have appeared previously, we present here a definition and construction specific to our application. Informally, an equivocal commitment scheme is an interactive protocol between a sender and a receiver which is computationally hiding and computationally binding in a real execution of the protocol. However, in a *simulated* execution of the protocol (where the simulator interacts with the receiver), the simulator is not bound to any particular value but can instead open the commitment to any desired value. Furthermore, for any (non-uniform) PPT receiver R and any string x, the view of R when the real sender commits/decommits to x is computationally indistinguishable from the view of R when the simulator "commits" in an equivocal way and later opens this commitment as x. We defer a formal definition, especially since one follows easily from the construction we now provide.

We construct an equivocal commitment scheme for a single bit in the following way: let Com be a non-interactive (perfectly binding) commitment scheme. To commit to a bit x, the sender chooses coins ω_1, ω_2 and computes $C = \mathsf{Equiv}(x; \omega_1, \omega_2) \overset{\text{def}}{=} \mathsf{Com}(x; \omega_1) \| \mathsf{Com}(x; \omega_2)$. It sends C to the receiver and performs a zero-knowledge proof/argument that C was constructed correctly (i.e., that there exist x, ω_1, ω_1 such that $C = \mathsf{Equiv}(x; \omega_1, \omega_2)$). The receiver rejects in case the proof/argument fails. To decommit, the sender chooses a bit b at random and reveals x, ω_b. Note that a simulator can "equivocate" the commitment by setting $C = \mathsf{Com}(x; \omega_1) \| \mathsf{Com}(\bar{x}; \omega_2)$ (where x is chosen at random in $\{0, 1\}$), simulating the zero-knowledge step, and then revealing ω_1 or ω_2 depending on x and the bit to be revealed. By committing bit-by-bit, the above extends easily to yield an equivocal commitment scheme for polynomial-length strings.

3   The Round Complexity of Coin Tossing

We show that any protocol for securely flipping a super-logarithmic number of coins (which is proven secure via black-box simulation) requires at least 5 rounds. (The reader is referred to Appendix A for a definition of black-box simulation.) More formally:

Theorem 1. *Let $p(k) = \omega(\log k)$, where k is the security parameter. Then there does not exist a 4-round protocol for tossing $p(k)$ coins which can be proven secure via black-box simulation.*

The above theorem refers to the case where *both* parties are supposed to receive the resulting coin as output.

Before starting our proof, we note that the above theorem is "tight" in the following two regards: first, for any $p(k) = O(\log k)$, 3-round protocols (proven secure using black-box simulation) for tossing $p(k)$ coins are known [8, 25, 29], assuming the existence of a non-interactive commitment scheme. Furthermore, our results of Section 4 imply a 5-round protocol (based on the existence of trapdoor permutations) for tossing any polynomial number of coins. In fact, we can also construct a 5-round protocol for tossing any polynomial number of coins based on the existence of a non-interactive commitment scheme; details will appear in the final version.

Proof (sketch). We assume (toward a contradiction) some 4-round protocol Π for tossing $p = p(k)$ coins. Without loss of generality, we may assume that player 1 sends the final message of Π (since in the ideal model, only player 1 has the ability to abort the trusted party); hence, player 2 must send the first message of Π. Consider a real-model adversary $\widetilde{A}_1$, corrupting player 1, who acts as follows: Let $\mathsf{Good} \subset \{0, 1\}^{p(k)}$ be some set of "small" but noticeable size, whose exact size we will fix later. $\widetilde{A}_1$ runs protocol Π honestly until it receives the third message, and then computes the value c of the tossed coin. If $c \in \mathsf{Good}$, then $\widetilde{A}_1$ completes execution of the protocol honestly and outputs some function of its view; otherwise, $\widetilde{A}_1$ aborts with output $\bot$.

Black-box security of Π implies the existence of a black-box ideal-model adversary $\widetilde{B}_1$ satisfying the following property (informally): conditioned upon receiving a coin $c \in$ Good from the trusted party, with all but negligible probability $\widetilde{B}_1$ "forces" an execution with $\widetilde{A}_1$ in which $\widetilde{A}_1$ does not abort and hence $\widetilde{A}_1$'s view is consistent with some coin $c' \in$ Good (for our proof, it does not matter whether $c' = c$ or not).

We next define a real-model adversary $\widetilde{A}_2$, corrupting player 2, acting as follows: $\widetilde{A}_2$ incorporates the code of $\widetilde{B}_1$ and – simulating the trusted party for $\widetilde{B}_1$ – feeds $\widetilde{B}_1$ a coin c randomly chosen from Good. By the above, $\widetilde{B}_1$ can with overwhelming probability "force" an execution with $\widetilde{A}_1$ in which $\widetilde{A}_1$ sees a view consistent with some $c' \in$ Good. We show that we can use $\widetilde{B}_1$ to "force" an execution with (the honest) A_1 in which A_1 outputs some $c' \in$ Good with sufficiently high probability. Of course, $\widetilde{A}_2$ (and hence $\widetilde{B}_1$) interacts with the honest A_1, and not with adversarial $\widetilde{A}_1$; thus, in particular, $\widetilde{A}_2$ (and hence $\widetilde{B}_1$) cannot rewind A_1. However, since $\widetilde{A}_1$ acts "essentially" like the honest A_1 (with the only difference being due to aborts), we can show that $\widetilde{A}_2$ "forces" A_1 to output a coin $c' \in$ Good with at least some inverse polynomial probability $1/q(k)$, where $q(k)$ relates to the number of queries $\widetilde{B}_1$ makes to its oracle for $\widetilde{A}_1$.

Choosing Good such that $|\text{Good}|/2^k \leq 1/2q(k)$, we derive a contradiction: in any ideal-model execution, an honest player 1 outputs a coin in Good with probability at most $1/2q(k)$; in the real world, however, $\widetilde{A}_2$ forces an honest A_1 to output a coin in Good with probability at least $1/q(k)$. This implies a simple, poly-time distinguisher with non-negligible advantage at least $1/2q(k)$.

Remark 2. Theorem 1 immediately extends to rule out 4-round, black-box protocols for other functionalities (when both parties are supposed to receive output), and in particular some natural, deterministic ones. For example, the theorem implies that 4 rounds are not sufficient for computing the "xor" functionality (i.e., $F(x, y) = x \oplus y$) on inputs of super-logarithmic length, since any such protocol could be used to toss a super-logarithmic number of coins (in the same number of rounds). This can be generalized in the obvious way.

4 A 5-Round Protocol for Secure Computation

Here, we prove the existence of a 5-round protocol for secure computation of general functionalities based on the existence of (certified) trapdoor permutations (see Definition 1 and Remark 1). To simplify matters, we describe a 4-round protocol for secure computation of *deterministic* functionalities in which *only the first party receives output*; this suffices for our main result since any such protocol can be used for secure computation of randomized functionalities in which both parties receive (possibly different) outputs, at the cost of one more (i.e., fifth) additional round [25, Propositions 7.2.11 and 7.4.4].

Before describing our protocol, we provide some intuition about the "high-level" structure of our protocol and highlight some techniques developed in the course of its construction. We stress that our protocol does *not* merely involve "collapsing" rounds by running things in parallel – new techniques are needed to obtain a round-optimal protocol. At the core of our protocol is Yao's 3-round protocol tolerating *honest-but-curious* adversaries (it will be helpful in what follows to refer to the description of Yao's "basic" protocol in Section 2). The standard way of adding robustness against *malicious* adversaries (see [25]) is to "compile" this protocol by having the parties (1) commit to their inputs; (2) run (modified) coin-tossing protocols, so each party ends up with a random tape and the other party receives a commitment to this tape; and (3) run the basic Yao protocol with ZK proofs/arguments of correct behavior (given the committed values of the input and random tape) at each round. We may immediately note this approach will not suffice to obtain a 4-round protocol, since a ZK proof/argument for the first round of Yao's protocol alone will already require 4 rounds. Instead, we briefly (and informally) summarize some of the techniques we use to achieve a 4-round protocol. In the following (but not in the more formal description that follows), we number the rounds from 0–3, where round 0 corresponds to an "initialization" round, and rounds 1–3 correspond to rounds 1–3 of Yao's basic protocol.

- We first observe that in Yao's protocol *a malicious player 2 gains nothing by using a non-random tape* and thus coin-tossing for this party is not needed.
- It is essential, however, that player 1 is unable to choose his coins in round two. However, full-blown coin-tossing is unnecessary, and *we instead use a 3-round sub-protocol which "forces" player 1 to use an appropriate set of coins.* (This sub-protocol is run in rounds 0–2.) This component and its analysis are based loosely on earlier work of Barak and Lindell [3].
- When compiling Yao's protocol, *player 1 may send his round-two message before the proof of correctness for round one (being given by player 2) is complete* (here, we use the fact that the trapdoor permutation family being used is "certifiable"). We thus construct our protocol so the proof of correctness for round one completes in round three. To obtain a proof of security, we require player 2 to delay revealing circuit until round three. Yet, a proof of security also requires player 2 to be committed to a circuit at the end of the round one. We resolve this dilemma by having player 2 commit to circuit in round one using an *equivocal commitment scheme.*
- Finally, use a *specific* WI proof of knowledge (from [36]; see also [18]) with the property that *the statement to be proved (and, by implication, a witness) need not be known until the last round of the protocol,* yet soundness, completeness, and witness-indistinguishability still hold. (The proof *of knowledge* aspect must be dealt with more carefully; see Appendix B.) Furthermore, this proof system has the property that the first message from the prover is computed *independently* of the statement being proved (as well as its witness); we use this fact when constructing an adaptively-secure protocol in Section 4.1.

We also construct a novel 4-round ZK argument of knowledge with similar properties (see Appendix B), by modifying the Feige-Shamir ZK argument of knowledge [19]. Our new protocol may be of independent interest.

Let $\mathcal{F} = \{F_k\}_{k \in \mathbb{N}}$ be a polynomial-size (deterministic) circuit family representing the functionality of interest, where F_k takes two k-bit inputs and returns a k-bit output to player 1. (Clearly, the protocol extends for arbitrary input/output lengths. We have also mentioned earlier how the protocol may be extended for randomized, two-output functionalities.) When k is understood, we write F instead of F_k. Let $x = x_1 \cdots x_k \in \{0, 1\}^k$ represent the input of player 1, let $y = y_1 \cdots y_k \in \{0, 1\}^k$ represent the input of player 2, and let $v = F(x, y)$. In the following, i always ranges from 1 to k, and b ranges from 0 to 1.

First round. The protocol begins with first player proceeding as follows:

1. Player 1 chooses $2k$ values $\{r_{i,b}\} \stackrel{\text{def}}{=} \{r_{1,0}, r_{1,1}, \ldots, r_{k,0}, r_{k,1}\}$ at random from $\{0, 1\}^k$. It then chooses $2k$ random coins $\{\omega_{i,b}\}$ and computes $\mathsf{Com}_{i,b} = \mathsf{Com}(r_{i,b}; \omega_{i,b})$, where Com is any perfectly-binding commitment scheme.
2. Player 1 also prepares the first message (which we call PoK_1) of a 3-round witness indistinguishable proof of knowledge (for a statement which will be fully determined in the third round; see the earlier remarks). For later reference, define $\mathsf{statement}_1$ as the following:

$$\exists \{(r_i, \omega_i)\}_{1 \leq i \leq k} \text{ s.t. } \forall i : (\mathsf{Com}_{i,0} = \mathsf{Com}(r_i; \omega_i) \vee \mathsf{Com}_{i,1} = \mathsf{Com}(r_i; \omega_i)).$$

   (Informally, $\mathsf{statement}_1$ represents the fact that player 1 "knows" either the decommitment of $\mathsf{Com}_{i,0}$ or the decommitment of $\mathsf{Com}_{i,1}$ for each i.)
3. Player 1 also prepares the first message (acting as the verifier) of the modified Feige-Shamir ZK argument of knowledge (see Appendix B). We denote this message by FS'_1.
4. The message sent by player 1 contains $\{\mathsf{Com}_{i,b}\}$, PoK_1, and FS'_1.

Second round. Player 2 proceeds as follows:

1. Player 2 generates $2k$ trapdoor permutations (denoted $\{(\mathsf{f}_{i,b}, \mathsf{f}_{i,b}^{-1})\}$) using $2k$ invocations of $\mathsf{Gen}(1^k)$, chooses $2k$ values $\{r'_{i,b}\}$ at random from $\{0, 1\}^k$, and prepares the second message (denoted PoK_2) for the WI proof of knowledge initiated by player 1 in the previous round.
2. Next, player 2 generates a "garbled circuit" (cf. Section 2) for the functionality F, based on its own input y. This involves choosing random coins Ω and computing (circuit, $\{Z_{i,b}\}$) = $\mathsf{Yao}_1(F, y; \Omega)$. Player 2 also computes commitments to the $\{Z_{i,b}\}$: that is, it chooses coins $\{\omega'_{i,b}\}$ and computes $\overline{\mathsf{Com}}_{i,b} = \mathsf{Com}(Z_{i,b}; \omega'_{i,b})$.
3. Player 2 next chooses random coins ζ and generates an equivocal commitment $\mathsf{Equiv} = \mathsf{Equiv}(\text{circuit}; \zeta)$.
4. Next, player 2 prepares the second message (denoted FS'_2) for the modified Feige-Shamir ZK argument of knowledge (for a statement which will be fully determined in the fourth round; cf. Appendix B). For future reference, let $\mathsf{statement}_2$ be the following: there exist $\left(y, \Omega, \text{circuit}, \{Z_{i,b}\}, \{\omega'_{i,b}\}, \zeta\right)$ s.t.:
   (a) (circuit, $\{Z_{i,b}\}$) = $\mathsf{Yao}_1(F, y; \Omega)$;
   (b) $\forall i, b$: $\overline{\mathsf{Com}}_{i,b} = \mathsf{Com}(Z_{i,b}; \omega'_{i,b})$; and

(c) Equiv = Equiv(circuit; ζ).

(Informally, statement$_2$ states that player 2 performed the preceding two steps correctly.)

5. The message includes $\{f_{i,b}\}$, $\{r'_{i,b}\}$, $\{\overline{\text{Com}}_{i,b}\}$, Equiv, PoK$_2$, and FS$'_2$.

Third round. Player 1 proceeds as follows:

1. If any of the $\{f_{i,b}\}$ are not valid[2], player 1 aborts. Otherwise, player 1 will use k parallel invocations of oblivious transfer to obtain the input-wire labels corresponding to its input x. Formally, for each i player 1 prepares values $(z_{i,0}, z_{i,1})$ in the following way:
   - If $x_i = 0$, choose random $z'_{i,0} \in \{0,1\}^k$ and set $z_{i,0} = f^k_{i,0}(z'_{i,0})$. Also, set $z_{i,1} = r_{i,1} \oplus r'_{i,1}$ (recall, $r_{i,1}$ was committed to by player 1 in the first round, and $r'_{i,1}$ was obtained from player 2 in the second round).
   - If $x_1 = 1$, choose random $z'_{i,1} \in \{0,1\}^k$, set $z_{i,1} = f^k_{i,1}(z'_{i,1})$, and set $z_{i,0} = r_{i,0} \oplus r'_{i,0}$.
2. Define statement$_3$ as follows:
   $\exists \{(r_i, \omega_i)\}_{1 \leq i \leq k}$ s.t. $\forall i$:
   - $\left(\text{Com}_{i,0} = \text{Com}(r_i; \omega_i) \wedge z_{i,0} = r_i \oplus r'_{i,0}\right)$ **or**
   - $\left(\text{Com}_{i,1} = \text{Com}(r_i; \omega_i) \wedge z_{i,1} = r_i \oplus r'_{i,1}\right)$.
   Informally, this says that player 1 correctly constructed the $\{z_{i,b}\}$ values.
3. Player 1 then prepares the final message (denoted PoK$_3$) for the proof of knowledge begun in round 1. The statement[3] to be proved is: statement$_1$ $\wedge$ statement$_3$. Player 1 also prepares the third message for the modified Feige-Shamir ZK protocol (denoted FS$'_3$).
4. The message includes $\{z_{i,b}\}$, PoK$_3$, and FS$'_3$.

Fourth round. The second player proceeds as follows:

1. If either PoK$_3$ or FS$'_3$ would cause rejection, player 2 aborts. Otherwise, player 2 completes the oblivious transfer in the standard way. Namely, for each $z_{i,b}$ sent in the previous round, player 2 computes $z'_{i,b} \stackrel{\text{def}}{=} f^{-k}_{i,b}(z_{i,b})$ and xor's the k resulting hard-core bits with the corresponding input-wire labels thusly: $W_{i,b} = h(z'_{i,b}) \oplus Z_{i,b}$.
2. Define statement$_4$ as follows:
   $\exists \{(Z_{i,b}, \omega'_{i,b}, z'_{i,b})\}_{1 \leq i \leq k; b \in \{0,1\}}$ s.t. $\forall i, b$:

$$\left(\overline{\text{Com}}_{i,b} = \text{Com}(Z_{i,b}; \omega'_{i,b})\right) \bigwedge \left(f^k_{i,b}(z'_{i,b}) = z_{i,b}\right) \bigwedge \left(W_{i,b} = h(z'_{i,b}) \oplus Z_{i,b}\right).$$

Informally, this says that player 2 performed the oblivious transfer correctly.
3. Player 2 prepares the final messages (denoted FS$'_4$) for the modified Feige-Shamir protocol. The statement to be proved is: statement$_2$ $\wedge$ statement$_4$.

---

[2] Recall (cf. Definition 1) that the trapdoor permutation family is certifiable.

[3] An honest player 1 actually knows multiple witnesses for statement$_1$. For concreteness, we have the player choose one of these at random to complete the proof.

4. Finally, player 2 decommits Equiv as circuit (recall from Section 2 how de-commitment is done for equivocal commitments).
5. The message includes the $\{W_{i,b}\}$, circuit (and the corresponding decommit-ment), and FS_4'.

Output computation. The first player concludes the protocol as follows: If FS_4' or the decommitment of circuit would cause rejection, player 1 aborts. Otherwise, by completing the oblivious transfer (in the standard way) player 1 obtains $Z_i \stackrel{\text{def}}{=} Z_{i,x_i}$ (recall, x is the input of player 1) and computes $v = \mathsf{Yao}_2(\text{circuit}, Z_1, \ldots, Z_k)$. If $v \neq \bot$, it outputs v. Otherwise, it aborts.

Sufficient assumptions. As noted in Section 2, every component of the above protocol may be based on the existence of a trapdoor permutation family (the certifiability property is only needed for the verification performed by player 1 at the beginning of the third round). Furthermore, as noted in Remark 1, although the description of the protocol (and its proof of security) use the definition of a trapdoor permutation family given by Definition 1, it is possible to adapt the protocol so that its security may be based on any family of (certified) enhanced trapdoor permutations, as per the definitions of [24, 26].

Theorem 2. *Assuming the existence of a trapdoor permutation family, the above protocol Π securely computes functionality F.*

Proof. We separately prove two lemmas dealing with possible malicious behavior of each of the parties; the theorem follows. We first consider the case when player 2 is malicious:

Lemma 1. *Let (A_1, A_2) be a pair of (non-uniform) PPT machines in which A_1 is honest. There exist a pair of (non-uniform) expected polynomial-time machines (B_1, B_2) such that*

$$\left\{ \mathrm{REAL}_{\Pi, \overline{A}(z)}(x,y) \right\}_{x,y,z} \stackrel{\mathrm{c}}{\equiv} \left\{ \mathrm{IDEAL}_{F, \overline{B}(z)}(x,y) \right\}_{x,y,z}. \tag{1}$$

Proof (sketch). Clearly, we may take B_1 to be honest. We assume that A_2 is deterministic, and construct B_2 using black-box access to A_2 as follows:

1. B_2 runs a copy of A_2 internally, passing to it any auxiliary information z. To emulate the first round of the protocol, B_2 acts exactly as an honest player 1, generates a first-round message, and passes this message to A_2. In return, B_2 receives a second-round message which includes, in particular, $\{r_{i,b}'\}$. If an honest player 1 would abort after receiving this second-round message, B_2 aborts (without sending any input to the trusted party) and outputs whatever A_2 outputs.
2. Otherwise B_2 generates a third-round message exactly as an honest player 1 would, with the following exception: for *all* i, b, it sets $z_{i,b} = r_{i,b} \oplus r_{i,b}'$. Note in particular that B_2 can easily compute PoK_3, since both statement$_1$ and statement$_3$ are true. It passes the third-round message to A_2, and receives in return a fourth-round message.

3. If an honest player 1 would abort after receiving the fourth-round message, B_2 aborts (without sending any input to the trusted party) and outputs whatever A_2 outputs. Otherwise, B_2 attempts to extract[4] from A_2 an input value y (cf. step 4 of the second round in the description of the protocol). If extraction fails, B_2 aborts and outputs fail.

4. Otherwise, B_2 sends y to the trusted party. It then stops and outputs whatever A_2 outputs.

We may note the following differences between the ideal world and the real world: (1) in the second round, B_2 sets $z_{i,b} = r_{i,b} \oplus r'_{i,b}$ for *all* i, b, whereas an honest player 1 does this only for i, b such that $x_i \neq b$; also (2) B_2 passes the input value y to the trusted party (and hence player 1 will receive the value $F(x, y)$ from this party), whereas in the real world player 1 will compute an output value based on the circuit and other values it receives from A_2 in the fourth round. Nevertheless, we claim that Equation (1) holds based on (1) the hiding property of the commitment scheme used in the first round and (2) the argument of knowledge (and hence soundness) property of the modified Feige-Shamir protocol (cf. Appendix B), as well as the correctness of the Yao "garbled circuit" construction. A complete proof appears in the full version.

Lemma 2. *Let (A_1, A_2) be a pair of (non-uniform)* PPT *machines in which A_2 is honest. There exist a pair of (non-uniform) expected polynomial-time machines (B_1, B_2) such that*

$$\left\{ \mathrm{REAL}_{\Pi, \overline{A}(z)}(x, y) \right\}_{x,y,z} \stackrel{c}{\equiv} \left\{ \mathrm{IDEAL}_{F, \overline{B}(z)}(x, y) \right\}_{x,y,z}. \tag{2}$$

Proof (sketch). Clearly, we may take B_2 to be honest. We assume that A_1 is deterministic, and construct B_1 using black-box access to A_1 as follows:

1. B_1 runs a copy of A_1 internally, passing to it any auxiliary information z and receiving a first-round message from A_1. Next, B_1 emulates the second round of the protocol as follows: it generates $\{f_{i,b}\}$, $\{r'_{i,b}\}$, and PoK_2 exactly as an honest player 2. All the commitments $\overline{\mathsf{Com}}_{i,b}$, however, are random commitments to 0^k. Furthermore, commitment Equiv is set up in an "equivocal" way (cf. Section 2) so that B_1 will later be able to open this commitment to any value of its choice. B_1 prepares FS'_2 using the ZK simulator for the modified Feige-Shamir protocol (cf. Appendix B). B_1 passes the second-round message thus constructed to A_1, and receives in return a third-round message. If an honest player 2 would abort after receiving this message, B_1 aborts (without sending any input to the trusted party) and outputs whatever A_1 outputs.

---

[4] Technically, B_2 runs a *witness-extended emulator* [37] for the modified Feige-Shamir proof system, which results in a transcript t and a witness w. This is what we mean when we informally say that B_2 "attempts to extract".

2. Otherwise, B_1 attempts to extract (cf. footnote 4) values $\{(r_i, \omega_i)\}_{1 \leq i \leq k}$ corresponding to (half) the commitments $\{\mathsf{Com}_{i,b}\}$ sent by A_1 in the first round. If extraction fails, B_1 outputs fail. Otherwise, let $b_i \in \{0,1\}$ be such that $\mathsf{Com}_{i,b_i} = \mathsf{Com}(r_i; \omega_i)$. B_1 then defines a string $x = x_1 \cdots x_k$ as follows:

$$\text{If } z_{i,b_i} = r_i \oplus r'_{i,b_i}, \text{ then } x_i = \bar{b}_i; \text{ otherwise, } x_i = b_i.$$

(x_i is B_1's "guess" as to which input-wire label A_1 is "interested in".) B_1 sends the string x thus defined to the trusted party, and receives a value v in return. It then runs $\mathsf{Yao\text{-}Sim}(x, v)$ to generate a garbled circuit circuit along with input-wire labels $\{Z_i\}$ (cf. Section 2). B_1 then prepares the "answers" $\{W_{i,b}\}$ to the oblivious transfer as follows, for each i: it correctly sets $W_{i,x_i} = h(\mathsf{f}_{i,x_i}^{-k}(z_{i,x_i})) \oplus Z_i$, but chooses $W_{i,\bar{x}_i}$ at random.
3. B_1 emulates the fourth round of the protocol as follows: it sends the $\{W_{i,b}\}$ as computed above, sends circuit as computed above (note that the corresponding decommitment can be given since Equiv was constructed in an "equivocal" way), and uses the simulator for the modified Feige-Shamir protocol to compute FS'_4 (cf. Appendix B). B_1 passes the final message thus constructed to A_1, and outputs whatever A_1 outputs.

We note (informally) the following differences between the ideal world and the real world: (1) $\{\overline{\mathsf{Com}_{i,b}}\}$ are commitments to 0^k rather than to "real" input-wire labels; (2) Equiv is set up so that B_1 can "equivocate" and later open this as any value it chooses; (3) the modified Feige-Shamir ZK argument is simulated rather than real; (4) the answers $\{W_{i,\bar{x}_i}\}$ are "garbage" (where x is B_1's guess as to the "input" of A_1); and (5) the garbled circuit is constructed using $\mathsf{Yao\text{-}Sim}$ rather than Yao_1. Nevertheless, we claim that Equation (2) holds. Due to lack of space, a complete proof appears in the full version.

4.1 Handling Adaptive Adversaries

We briefly sketch how the protocol above can be modified – without increasing the round complexity – to provide security against an *adaptive* adversary who can monitor communication between the parties and decide whom to corrupt at any point during the protocol based on this information. (We consider only an adversary who can corrupt at most *one* of the parties.) In brief, we modify the protocol by using a (public-key) adaptively-secure encryption scheme [12] to encrypt the communication between the two parties. Two issues arise:

1. The encryption scheme of [12] requires a key-generation phase which would necessitate additional rounds. We avoid this extra phase using the assumption of *simulatable public-key cryptosystems* [16] (see below). The existence of such cryptosystems is implied in particular by the DDH assumption [16]; see there for constructions based on alternate assumptions.
2. Regardless of the encryption scheme used, one additional round seems necessary just to exchange public keys. To avoid this, we do *not* encrypt the first message from player 1 to player 2. Nevertheless, the modified protocol

is adaptively-secure: the proof uses the fact that the first round (as well as the internal state after the first round) is identical in both the real execution and the simulation for a malicious player 2 (cf. the proof of Lemma 1).

The modified protocol. Before describing our construction of an adaptively-secure encryption scheme, we outline how it will be used to achieve adaptive security for our protocol. Let Π denote the protocol given in the previous section. Our adaptively-secure protocol Π' proceeds as follows: in the first round of Π', player 1 sends a message just as in the first round of Π, but also sends sufficiently-many public keys (for an adaptively-secure encryption scheme) to enable player 2 to encrypt the messages of rounds two and four. (The adaptively-secure encryption scheme we use only allows encryption of a single bit; therefore, the number of public keys sent by player 1 is equal to the bit-length of messages two and four in Π.) In the second round of Π', player 2 constructs a message as in Π, encrypts this message using the corresponding public keys sent in round one, and additionally sends sufficiently-many public keys (for an adaptively-secure encryption scheme) to enable player 1 to encrypt the messages of rounds three and five. Π' proceeds by having the players construct a message just as in the corresponding round of Π and then having them encrypt these messages using the appropriate public keys sent by the other player.

We defer a proof of security for this construction to the final version.

An adaptively-secure encryption scheme. Informally, a public-key cryptosystem for single-bit plaintexts is *simulatable* if (1) it is possible to obliviously generate a public key without learning the corresponding secret key, and also (2) given a public key, it is possible to obliviously sample a random (valid) ciphertext without learning the corresponding message. We assume further that if a ciphertext is obliviously sampled in this way, then the probability that the corresponding plaintext will be a 0 or 1 is equal (or statistically close). See [16, Def. 2] for a formal definition.

Given such a cryptosystem, our construction of an adaptively-secure encryption scheme for a single-bit is as follows: the receiver generates k pairs $(pk_{i,0}, pk_{i,1})$ of public keys by generating one key of each pair (selected at random) using the key-generation algorithm, and the other key using the oblivious sampling algorithm. This also results in a set of k secret keys (one for each pair of public keys). To encrypt a bit m, the sender proceeds as follows: for each index i, choose a random bit b_i, set $C_{b_i} \leftarrow \mathcal{E}_{pk_{i,b_i}}(m)$ and choose $C_{\bar{b}_i}$ using the oblivious sampling algorithm. Then send the k ciphertext pairs (C_0, C_1).

To decrypt, the receiver decrypts one ciphertext out of each pair using the secret key it knows, and sets the decrypted message equal to the majority of the recovered bits. Note that correctness holds with all but negligible probability since (on average) $3/4$ of the $2k$ ciphertexts constructed by the sender decrypt to the desired message m (namely, the k ciphertexts encrypted using the legitimate encryption algorithm, along with (on average) $1/2$ of the remaining k ciphertexts chosen via the oblivious sampling algorithm).

We defer a proof that this scheme is adaptively secure to the full version.

Acknowledgments

We thank Bill Aiello for a helpful discussion, and the anonymous referees for their useful remarks. Thanks also to Yuval Ishai for his useful suggestions towards improving the presentation of our results.

References

1. B. Barak. How to Go Beyond the Black-Box Simulation Barrier. *42nd IEEE Symposium on Foundations of Computer Science (FOCS)*, IEEE, pp. 106–115, 2001.
2. B. Barak. Constant-Round Coin-Tossing with a Man-in-the-Middle or Realizing the Shared Random String Model. *43rd IEEE Symposium on Foundations of Computer Science (FOCS)*, IEEE, pp. 345–355, 2002.
3. B. Barak and Y. Lindell. Strict Polynomial Time in Simulation and Extraction. *34th ACM Symposium on Theory of Computing (STOC)*, ACM, pp. 484–493, 2002.
4. J. Bar-Ilan and D. Beaver. Non-Cryptographic Fault-Tolerant Computing in Constant Number of Rounds of Interaction. *Principles of Distributed Computing*, ACM, pp. 201–209, 1989.
5. D. Beaver, S. Micali, and P. Rogaway. The Round Complexity of Secure Protocols. *22nd ACM Symposium on Theory of Computing (STOC)*, ACM, pp. 503–513, 1990.
6. M. Bellare, S. Micali, R. Ostrovsky. Perfect Zero-Knowledge in Constant Rounds STOC 1990: 482-493.
7. M. Bellare, S. Micali, R. Ostrovsky. The (True) Complexity of Statistical Zero Knowledge STOC 1990: 494-502
8. M. Blum. Coin Flipping by Phone. *IEEE COMPCOM*, pp. 133–137, 1982.
9. R. Canetti. Security and Composition of Multi-Party Cryptographic Protocols. *J. Cryptology* 13(1): 143–202, 2000.
10. R. Canetti, Y. Lindell, R. Ostrovsky, A. Sahai. Universally composable two-party and multi-party secure computation. STOC 2002: 494-503
11. R. Canetti, E. Kushilevitz, R. Ostrovsky, A. Rosen: Randomness versus Fault-Tolerance. J. Cryptology 13(1): 107-142 (2000)
12. R. Canetti, U. Feige, O. Goldreich, and M. Naor. Adaptively-Secure Multiparty Computation. *28th ACM Symposium on Theory of Computing (STOC)*, ACM, pp. 639–648, 1996.
13. R. Canetti, J. Kilian, E. Petrank, and A. Rosen. Concurrent Zero-Knowledge Requires $\widetilde{\Omega}(\log n)$ Rounds. *33rd ACM Symp. on Theory of Comp. (STOC)*, ACM, pp. 570–579, 2001.
14. R. Cramer and I. Damgård. Secure Distributed Linear Algebra in a Constant Number of Rounds. *Adv. in Cryptology – Crypto '01*, LNCS 2139, Springer-Verlag, pp. 119–136, 2001.
15. G. Di Crescenzo and R. Ostrovsky. On Concurrent Zero-Knowledge with Preprocessing. In CRYPTO 1999: pp. 485-502.
16. I. Damgård and J.B. Nielsen. Improved Non-Committing Encryption Schemes. *Adv. in Cryptology – Crypto 2000*, LNCS vol. 1880, Springer-Verlag, pp. 432–450, 2000.
17. A. De Santis, G. DiCrescenzo, R. Ostrovsky, G. Persiano, A. Sahai: Robust Non-interactive Zero Knowledge. CRYPTO 2001: 566-598

18. U. Feige. Alternative Models for Zero Knowledge Interactive Proofs. PhD thesis, Weizmann Institute of Science, 1990.

19. U. Feige and A. Shamir. Zero Knowledge Proofs of Knowledge in Two Rounds. *Adv. in Cryptology – Crypto 1989*, LNCS vol. 435, Springer-Verlag, pp. 526–544, 1989.

20. U. Feige and A. Shamir. Witness Indistinguishability and Witness Hiding Protocols. *22nd ACM Symposium on Theory of Computing (STOC)*, ACM, pp. 416–426, 1990.

21. M. Fitzi, J. Garay, U. Maurer, R. Ostrovsky: Minimal Complete Primitives for Secure Multi-party Computation. CRYPTO 2001: 80-100

22. R. Gennaro, Y. Ishai, E. Kushilevitz, and T. Rabin. The Round Complexity of Verifiable Secret Sharing and Secure Multicast. *33rd ACM Symposium on Theory of Computing (STOC)*, ACM, pp. 580–589, 2001.

23. R. Gennaro, Y. Ishai, E. Kushilevitz, and T. Rabin. On 2-Round Secure Multiparty Computation. *Adv. in Cryptology – Crypto 2002*, LNCS vol. 2442, Springer-Verlag, pp. 178–193, 2002.

24. O. Goldreich. *Foundations of Cryptography, vol. 1: Basic Tools*. Cambridge University Press, Cambridge, UK, 2001.

25. O. Goldreich. Draft of a Chapter on Cryptographic Protocols, June 2003. Available at http://www.wisdom.weizmann.ac.il/~oded/foc-vol2.html.

26. O. Goldreich. Draft of an Appendix Regarding Corrections and Additions, June 2003. Available at http://www.wisdom.weizmann.ac.il/~oded/foc-vol2.html.

27. O. Goldreich and A. Kahan. How to Construct Constant-Round Zero-Knowledge Proof Systems for NP. *J. Cryptology* 9(3): 167–190, 1996.

28. O. Goldreich and H. Krawczyk. On the Composition of Zero-Knowledge Proof Systems. *SIAM J. Computing* 25(1): 169–192, 1996.

29. O. Goldreich, S. Micali, and A. Wigderson. How to Play any Mental Game: a Completeness Theorem for Protocols with Honest Majority. *19th ACM Symposium on Theory of Computing (STOC)*, ACM, pp. 218–229, 1987.

30. R. Impagliazzo and S. Rudich. Limits on the Provable Consequences of One-Way Permutations. *21st ACM Symposium on Theory of Computing (STOC)*, ACM, pp. 44–61, 1989.

31. J.. Kilian, E. Kushilevitz, S. Micali, R. Ostrovsky: Reducibility and Completeness in Private Computations. SIAM J. Comput. 29(4): 1189-1208 (2000)

32. Y. Ishai and E. Kushilevitz. Randomizing Polynomials: A New Representation with Applications to Round-Efficient Secure Computation. *41st IEEE Symposium on Foundations of Computer Science (FOCS)*, IEEE, pp. 294–304, 2000.

33. J. Katz, R. Ostrovsky, and A. Smith. Round Efficiency of Multi-Party Computation with a Dishonest Majority. *Adv. in Cryptology – Eurocrypt 2003*, LNCS vol. 2656, Springer-Verlag, pp. 578–595, 2003.

34. E. Kushilevitz, R. Ostrovsky, A. Rosén: Amortizing Randomness in Private Multiparty Computations. SIAM J. Discrete Math. 16(4): 533-544 (2003)

35. J. Kilian and E. Petrank. Concurrent and Resettable Zero-Knowledge in Polylogarithmic Rounds. *31st ACM Symposium on Theory of Computing (STOC)*, ACM, pp. 560–569, 2001.

36. D. Lapidot and A. Shamir. Publicly-Verifiable Non-Interactive Zero-Knowledge Proofs. *Adv. in Cryptology – Crypto 1990*, LNCS vol. 537, Springer-Verlag, pp. 353–365, 1991.

37. Y. Lindell. Parallel Coin-Tossing and Constant-Round Secure Two-Party Computation. *Adv. in Cryptology – Crypto 2001*, LNCS vol. 2139, Springer-Verlag, pp. 171–189, 2001.

38. Y. Lindell. Personal communication, 2001.
39. S. Micali and P. Rogaway. Secure Computation. *Adv. in Cryptology – Crypto 1991*, LNCS vol. 576, Springer-Verlag, pp. 392–404, 1991.
40. M. Naor, R. Ostrovsky, R. Venkatesan, M.Yung. Perfect Zero-Knowledge Arguments for NP Can Be Based on General Complexity Assumptions. CRYPTO 1992: 196-214
41. M. Prabhakaran, A. Rosen, and A. Sahai. Concurrent Zero Knowledge with Logarithmic Round-Complexity. *43rd IEEE Symposium on Foundations of Computer Science (FOCS)*, IEEE, pp. 366–375, 2002.
42. R. Richardson and J. Kilian. On the Concurrent Composition of Zero-Knowledge Proofs. *Adv. in Cryptology – Eurocrypt 1999*, LNCS vol. 1592, Springer-Verlag, pp. 415–431, 1999.
43. A.C. Yao. How to Generate and Exchange Secrets. *27th IEEE Symposium on Foundations of Computer Science (FOCS)*, IEEE, pp. 162–167, 1986.

A   Black-Box Simulation

Typical definitions of security for two-party computation only require that for every pair of admissible real-world adversaries $\overline{A}$ there exists a pair of ideal-world adversaries $\overline{B}$ satisfying some relevant criterion (namely, indistinguishability of the resulting output distributions). Most work in this area, however, (and especially prior to the work of Barak [1]) proves the existence of such a $\overline{B}$ via what is known as a *black-box simulation*; this means that the ideal-model adversary B_i corresponding to the dishonest real-model adversary A_i is constructed using only oracle access to A_i.

More formally, a *black-box simulation for party 1* (with a completely analogous definition for black-box simulation for party 2) implies the existence of a simulator S_1 for which the following holds: For any real-model adversary A_1, let $B_1(x, z; r_A, r_S)$ (where x, z are the inputs of B_1 and $r = (r_A, r_S)$ are the random coins of B_1) be defined by $S_1^{A_1(x, z, \cdot; r_A)}(x; r_S)$, where $A_1(x, z, \cdot; r_A)$ denotes the next-message function of A_1 on the given inputs and random coins (we stress that S_1 is not explicitly given the auxiliary input z nor the random coins r_A). Then $\overline{A} = (A_1, A_2)$ and $\overline{B} = (B_1, B_2)$ (where A_2, B_2 are just the honest algorithms) satisfy the relevant criterion. Furthermore, S_1 runs in expected polynomial-time, where each oracle call to $A_1(x, z, \cdot; r_A)$ is counted as a single step. Finally (although this is not essential to our results), it is typical to assume that S_1 is a *uniform* algorithm. Note that if A_1 runs in strict polynomial time, the above implies that the entire algorithm B_1 runs in expected polynomial time; furthermore, if A_1 is uniform then so is B_1 (on the other hand, if A_1 is a non-uniform machine, then B_1 will be too). We say that a protocol is *proven secure via black-box simulation* if the simulations for both parties are black-box.

We stress a crucial point about the above: when we say S_1 runs in expected polynomial-time, we mean that there is a *fixed* polynomial $q(\cdot)$ such that the expected running time of S_1 on input x, when interacting with *any* A_1 (and counting queries to A_1 as a single step), is $q(|x|)$. On the other hand, the expected running time of B_1 (including the steps of A_1, and no longer counting each query to A_1 as a single step) cannot be bounded *a priori* by any fixed polynomial, as the running time of B_1 will of course depend on the running time of A_1. Of course, as noted above, if A_1 runs in strict polynomial time $q'(\cdot)$ then B_1 runs in expected time (at most) $q'(\cdot)q(\cdot)$, which is polynomial. (Note that this definition of black-box simulation avoids the technical problem of, e.g., [28] regarding the need for B_1 to feed A_1 coins r_A whose length depends on A_1 and is not bounded *a priori* by any polynomial.)

B   Proof Systems Used in This Work

We provide here a laconic sketch of the proof systems claimed in Section 4; further details and proofs will appear in the full version. We first describe the WI proof of knowledge of [36] (as described in [18]).

We will be working with the NP-complete language HC of graph Hamiltonicity, and thus assume statements to be proved take the form of graphs, while witnesses correspond to Hamilton cycles. If thm is a graph, we abuse notation and also let thm denote the statement "thm $\in HC$". We show how the proof system can be used to prove the following statement: thm $\wedge$ thm', where thm will be included as part of the first message, while thm' is only included in the last round (indeed, it will not be fixed until the third round begins). The proof system runs k parallel executions of the following 3-round protocol:

1. The prover commits to two adjacency matrices for two randomly-chosen cycle graphs C, C'. The commitment is done bit-by-bit using a perfectly-binding commitment scheme.
2. The verifier responds with a single bit b, chosen at random.
3. If $b = 0$, the prover opens all commitments. If $b = 1$, the prover sends two permutations mapping the cycle in thm (resp., thm') to C (resp., C'). For each non-edge in thm (resp., thm'), the prover opens the commitment at the corresponding position in C (resp., C').
4. The verifier checks that all commitments were opened correctly. If $b = 0$, the verifier additionally checks whether both decommitted graphs are indeed cycle graphs. If $b = 1$, the verifier checks whether each non-edge in thm (resp., thm') corresponds to a non-edge in C (resp., C').

Note that the prover does not need to know either thm or thm' (or the corresponding witnesses) until the beginning of the third round. However, we assume thm is fixed as part of the first-round message because this will enable us to claim stronger properties about the above proof system.

Very informally, we claim that the proof system above satisfies the following:

- It is complete and sound. In particular, the probability that an all-powerful prover can cause a verifier to accept when either thm or thm' are not true is at most 2^{-k}. We stress that this holds even if the prover can adaptively choose thm' after viewing the second-round message of the verifier.
- It is witness indistinguishable.

- It is a proof of knowledge for thm. (More formally, we can achieve a notion similar to that of witness-extended emulation [37] for thm.) We do not know whether such a claim holds for thm′.

Note also that the first round of the above proof system (as well as the internal state of the prover immediately following this round) is independent of thm or the associated witness. We rely on this fact in Section 4.1.

Next, we informally describe our modification of the Feige-Shamir ZK argument of knowledge [19] which will allow the prover to prove thm ∧ thm′, where thm is sent as part of the second round yet thm′ is only sent as part of the last round (indeed, it need not be known until the beginning of that round). *We use the notation used in the description of the Feige-Shamir protocol in [18, Prot. 8.2.62].* Our modified protocol proceeds as follows:

1. The first round is as in the original protocol, and includes values x_1, x_2.
2. The prover chooses a random $R \in \{0,1\}^{2k}$ and computes Equiv = Equiv(R, ζ) (cf. Section 2). Let ok denote the statement that Equiv was formed correctly.
3. Let $\widetilde{\text{thm}}$ denote the statement: (thm ∧ ok) ∨ $(f(w') = x_1)$ ∨ $(f(w') = x_2)$ (this statement is reduced to a single graph $\widetilde{\text{thm}}$). The prover sends Equiv and also the first message of the WI proof system described above.
4. The verifier's third message is as in the original protocol, except that the verifier additionally chooses and sends a random $R' \in \{0,1\}^{2k}$.
5. The prover decommits (as in Sec. 2) to R. Let prg be the statement that $r = R \oplus R'$ is pseudorandom (i.e., $\exists s$ s.t. $G(s) = r$, for G a PRG). Let $\widetilde{\text{thm}}'$ be the statement thm′ ∨ prg (reduced to a single graph $\widetilde{\text{thm}}'$). The prover completes the WI proof system, as above, for the statement $\widetilde{\text{thm}} \wedge \widetilde{\text{thm}}'$.
6. The verifier checks the decommitment of R, and verifies the proof as before.

We claim the following about the above proof system:

- It is complete and sound (for a poly-time prover) for thm and thm′. (As argued earlier, rounds 2–4 constitute a proof of knowledge for $\widetilde{\text{thm}}$. As in [18] – relying on the one-wayness of f – this implies that if a poly-time prover can cause a verifier to accept with "high" probability, then a witness for thm ∧ ok can be extracted with essentially the same probability. If ok is true, then with all but negligible probability prg will *not* be true. Soundness of the proof of knowledge sub-protocol then implies that $\widetilde{\text{thm}}'$ *is* true. But this means that thm′ is true.)
- It is zero-knowledge. (In addition to simulating for $\widetilde{\text{thm}}$ as in [18], the simulator also uses the equivocal commitment property to decommit to an R such that prg is true.)
- It is an argument of knowledge for thm (we have already argued as much above).

Security, Liberty, and Electronic Communications

Susan Landau

Sun Microsystems
susan.landau@sun.com

Dedicated to the memory of Dorrie Weiss.

1 Introduction

We live in perilous times. We live in times where a dirty bomb going off in lower Manhattan is not unimaginable. We live in times where the CIA interrogations of al Qaeda leaders were so harsh that the FBI would not let its agent participate [36]. We live in times when security and liberty are both endangered.

We also live in times of unimaginable technical creativity. It is faster to use Instant Messaging to query a colleague halfway across the world than it is to walk down the hallway and ask the question, when Google can search four billion web pages faster than the time it takes to pull the right volume of the Oxford English Dictionary off the library shelf. We live surrounded by a plethora of communicating and computing devices — telephones, PDAs, cell phones, laptops, PCs, computers — and this is only the beginning of the communications revolution.

September 11th presaged a radical change in terrorist intent, a radical change that few had anticipated. The U.S. government responded to September 11th in a number of ways, including the passage of the U.S.A. Patriot Act, which qualitatively extended the government's electronic-surveillance capabilities. The Patriot Act engendered strong debate (though not in Congress, where the law passed handily). The most controversial issue regarding the changes in electronic-surveillance law was that the requirement that foreign intelligence be a "primary" reason for a Foreign Intelligence Surveillance Act (FISA) wiretap was modified to foreign intelligence need only be a "significant" reason for a FISA tap.

Absent from the debates on the Patriot Act was an acknowledgement of the radical changes that had occurred in communications technologies since the passage of the first Federal wiretap statute in 1968. Communications technology has changed in numerous ways over the past forty years — there is now wide availability of mobile communications, a vast increase in connectivity, and packet-switched systems are being employed for telephony — but there has been no commensurate review of electronic-surveillance laws. We are in a peculiar state: we communicate using mobiles phones and laptops, but the laws governing electronic surveillance were developed at a time of fixed-location circuit-based switching systems. Instead of a full-scale reevaluation of surveillance laws, over the last two decades we have pursued a path of minor tweaks to the electronic-surveillance laws. The result is an electronic-surveillance regime that may be well

M. Franklin (Ed.): CRYPTO 2004, LNCS 3152, pp. 355–372, 2004.

out of sync with the times. This has serious implications for security, liberty, technology, and innovation. In this paper, we examine electronic-surveillance laws in light of current threats and new technologies. We begin by examining the climate in which wiretap laws came to be enacted.

2   The Political Climate at the Time of the Wiretap Act

The sixties were a time of turmoil in the United States, a time of political protest, and civil unrest. In 1963, President John Kennedy was assassinated in a motorcade in Dallas, Texas. In 1965 Malcolm X was killed as he delivered a speech in an auditorium in Harlem. In April 1968, Martin Luther King was killed, and two months later, Robert Kennedy, who was running for President, was shot moments after he learned he had won the California primary. There had been civil rights marches in Washington in the early 1960s, and anti-Vietnam protests in the latter half of the decade. In the summer of 1964, downtown Newark burned; in 1965, the Watts section of Los Angeles; in 1967, downtown Detroit.

It was against this backdrop that the President's Commission on Law Enforcement and Administration of Justice presented its report. Organized crime had been a problem in the United States since Prohibition, but, because FBI Director J. Edgar Hoover ignored it, so did the Federal government. Several events in the late 1950s and early 1960s changed that.

The first was the discovery, on November 15, 1957, by a New York state trooper, of a meeting of organized crime bosses. The trooper was doing routine morning rounds when he discovered far too many black limousines for the tiny upstate town of Apalachin. The trooper set up a roadblock; the crime bosses fled, and "the next day, the nation awoke to headlines like 'Royal Clambake for Underworld Cooled by Police,' and 'Police Ponder NY Mob Meeting; All Claim They Were Visiting Sick Friend' [13, pp. 168-9]. Meanwhile, while counsel to the Senate Select Committee on Improper Activities in the Labor or Management Field, Robert Kennedy had uncovered ties between the unions and organized crime. When he became attorney general, Kennedy made organized crime a priority [29]. And finally, an organized crime turncoat, Joseph Valachi, broke the code of silence by testifying to a Senate investigating committee in 1963.

This confluence of events made pursuing organized crime a law-enforcement priority in the late 1960s. The complications of investigating organized crime — the reluctance of victims to testify, so-called victimless crimes (e.g., prostitution), and the corruption of local law enforcement made electronic surveillance a particularly valuable tool. The Commission concluded, "A majority of the members of the Commission believe that legislation should be enacted granting carefully circumscribed authority for electronic surveillance to law enforcement officers..." [33, p. 203].

But, as noted in [13, p. 170],:

> Not all experts agreed with the commission's conclusions. Attorney General Clark prohibited all use of wiretaps by federal law-enforcement of-

ficers. He told Congress: 'I know of no Federal conviction based upon any wiretapping or electronic surveillance, and there have been a lot of big ones. ... I also think that we make cases effectively without wiretapping or electronic surveillance. I think it may well be that with the commitment of the same manpower to other techniques, even more convictions could be secured, because in terms of manpower, wiretapping, and electronic surveillance is very expensive." [8, p. 320] Clark pointed out that in 1967, without using wiretaps, federal strike forces had obtained indictments against organized-crime figures in nine states, and that "each strike force has obtained more indictments in its target city than all federal indictments in the nation against organized crime in as recent a year as 1960" [8, pp. 79-80]

President Johnson publicly supported Clark's opposition to wiretapping, and the President proposed limiting wiretapping to national-security cases [9, p. 222]. But political turmoil and the Crime Commission's report led Congress in a different direction, and in 1968 it passed the Omnibus Crime Control and Safe Streets Act of 1968 (18 USC §2510–2521), Title III of which legalized law-enforcement wiretaps in criminal investigations. Because of the very invasive nature of the search, wiretaps were limited to a list of twenty-six crimes specified in the act, including murder, kidnapping, extortion, gambling, counterfeiting, and sale of marijuana. The Judiciary Committee's report explained that "each offense was chosen because it was intrinsically serious or because it is characteristic of the operations of organized crime," [44, p. 97].

President Johnson was ambivalent about wiretaps. He had used them — on Martin Luther King during the Democratic convention in 1964 and on Vice President Humphrey in 1968 — but the President described the Title III provisions for wiretapping as undesirable [9, p. 1842]. Nonetheless Johnson signed the bill. Because of the invasive nature of electronic surveillance, Congress decided that there should be stringent oversight, and that review of a federal wiretap warrant application must be done by a federal district court judge.

The judge must determine that (i) there is probable cause to believe that an individual is committing, has committed, or is about to commit an indictable offense; (ii) there is probable cause to believe that communications about the offense will be obtained through the interception; (iii) normal investigative procedures have been tried and either have failed, appear unlikely to succeed, or are too dangerous; and (iv) there is probable cause to believe that the facilities subject to surveillance are being used or will be used in the commission of the crime (§2518 (3)(a-d)).

Title III covers procedures for obtaining wiretaps for law-enforcement investigation. In 1972, in a court case involving "domestic national-security issues," the Supreme Court ordered an end to warrantless wiretapping, even for national-security purposes. Because of Watergate, and the discovery of numerous so-called national-security wiretaps that were actually wiretaps for political purposes [42], it took until 1978 before Congress was actually able to frame and pass legislation authorizing procedures for obtaining wiretaps for national-security investigations: the Foreign Intelligence Surveillance Act. The judge, a member of the Foreign Intelligence Surveillance Court, a court of eleven judges appointed from seven of the United States judicial circuits (§1803 (a)), must determine (i) that there is probable cause that the target is a foreign or target of a foreign power, (ii) that there is probable cause that the targeted communications device is being used by the foreign power or its agent, that (iii) that a primary purpose of the surveillance is to obtain foreign intelligence information, and that (iv) such information cannot reasonably by obtained by other investigative techniques. [1]

Title III and FISA form the basis for U.S. wiretap law. There are also state statutes (approximately half of all criminal wiretaps in the United States are done under state wiretap warrants). The rules governing state wiretaps must be at least as restrictive as those governing Title III.

There have been several updates and modifications to the federal wiretap statutes, which will be discussed after examining the changes in communications technology over the last four decades.

3   Current Threats

In the U.S. we are currently seeing a strident debate on surveillance technologies, most especially datamining. This paper is not the place for a full discussion of the methods and means used in terrorist investigations. In the context of reexamining electronic-surveillance laws, however, it is useful to make some observations about terrorism and terrorist investigations.

By any measure, terrorism is a very difficult offense to investigate or prevent. In many cases, the first crime committed is the only crime. There is no trail. The investigative reporter, Seymour Hersh, described CIA efforts in southern Lebanon during the 1980s,

> ... when the C.I.A. started to go after the Islamic Jihad, a radical Lebanese group linked to a series of kidnappings in the Reagan years, 'its people systematically went through documents all over Beirut, even destroying student records.'

One of the hallmarks of modern terrorist groups is the shifting and diffuse organizational structure [39, p. 271]. On the one hand, this means that eliminating the leadership does not necessarily eliminate the problem. On the other, diffuse and ever-changing structures create weaknesses within the organization. One that can be exploited is the terrorists' need for communication.

In this situation, traffic analysis often proves more useful than wiretapping. Wiretaps can be confused by encryption, even encryption of a very simple sort. Seymour Hersh reported that,

---

[1] The law provides that "[N]o United States person may be considered a foreign power or an agent of a foreign power solely upon the basis of activities protected by the first amendment to the Constitution of the United States" (§1805(a)(3)(A)).

The terrorists coped with the American ability to intercept conversations worldwide by constantly changing codes — often doing little more than changing the meanings of commonly used phrases.

The problem is being unable to decode the language is not new. It can even occur without deliberate intent by the criminal or terrorist group. The National Research Council report, *Cryptography's Role in Securing the Information Society* described an FBI wiretap of police officers who were allegedly guarding a drug shipment. The FBI agents overhead a conversation in which the officers discussed murdering an individual who had filed a police brutality complaint. The bureau was unable to decode a participant's "street slang," and was thus unable to prevent the murder [10, p. 88].

The inability to understand surveilled conversations does not mean that the surveillance is useless. In particular, traffic analysis has become an extremely valuable aspect of surveillance, and one cannot confuse traffic-analysis efforts in quite the same way as one confuses content analysis. One example of the value of traffic analysis is that Osama Bin Laden stopped using a cell phone in late 2001 because of the tracking capabilities of U.S. intelligence.

Even "anonymous" cellphones can be used for tracking. In a case in 2002, investigators tracked al Qaeda members through terrorists use of prepaid Swisscom phonecards. These had been purchased in bulk — anonymously. But when investigators discovered through a wiretap on an intercepted call that "lasted less than a minute and involved not a single word of conversation" that they were on to an al Qaeda group, the agents tracked the users of the bulk purchase [45]. The result was the arrest of a number of operatives and the break-up of al Qaeda cells. You can run, but you can't hide. Anonymity is not all that it is cracked up to be.

One important aspect of terrorist investigations is to "follow the money." Many terrorist groups hide behind legitimate charitable groups, but these are groups with money trails [39, p. 274]. (We should note, however, that "following the money" is not a straightforward issue in terms of civil liberties. The Patriot Act section dealing with money laundering and terrorist financing is controversial admidst claims that its provisions have been applied to charitable groups with no ties to terrorist activities.) Money trails can be complicated to follow, and the terrorists do a good job of hiding trails by passing money through many intermediaries, but the fact is that there is a trail. Once there is trail, it can be investigated.

The current terrorist threat is very different from earlier terrorist movements. A different from earlier terrorism threats, such as the Russian nihilists of the nineteenth century or the Palestinian terrorists of the 1970s, is the huge reservoir of potential recruits. Globalization complicates the problem. (Indeed, one legitimately argue that globalization is a large part of the problem — but that is a topic for a different paper.) In the late 1990s, Senators Hart and Rudman chaired a national security commission study to examine emerging threats. In a prescient observation, the Hart-Rudman report in early 2001 warned of the likelihood of catastropic domestic attacks caused by international terrorism. The report observed, "All borders will become more porous." [41, p. 2] This has already happened in Europe. While the borders have become porous, apparently cooperation between different nations' law enforcement has not yet followed suit.

Terrorism is not a passing phenomenon. It will be with us for a long time. It is important that we respond to the threat in a way that simultaneously protects our security and our liberty.

4   Changing Communications Technology

The first hundred of years the telephone saw change: from local systems entirely mediated by operators to global networks entirely run by electronic switching systems. There was innovation: mobile phone, first deployed n 1946 [6, 215], faxes, and modems. There was development of infrastructure: optical fibers and communication satellites, as well the digitization of the backbone network.

Yet slightly more than a generation ago, the telephone remained a fixed device: a black machine with a rotary dial that transmitted voice (also data; from the beginning, the telephone was also a data-transmission network data, e.g., telegraph). In the sixties innovation was the introduction of the "Princess" phone (in colors!: white, beige, pink, blue, or turquoise) and Touchtone service (buttons instead of rotary dials), while industry got Centrex, an automatic switching exchange for large offices, and "data-phones" (modems) [6, p. 266]. What occurred in the first century was growth: ten million phone users in 1900, one hundred million in 1960, five hundred million in 1990[2].

The innovation of the first hundred years of the telephone pales in contrast to the growth and changes of the last decade and a half. There were 1.4 billion users in 2000, 400 million of those cell phone users. There probably has been as much innovation in telephony in the last quarter century as there had been in the previous one hundred years.

Recent telecommunications growth has been spurred by three technical developments: mobile technology, greater bandwidth, and the Internet. AT&T has had car phones since 1946 [6, p. 215], but such service was rare and expensive until the early 1990s. Mobile technology took off with the 1983 development of "cell" technology. In under a decade, cell phones have become ubiquitous, as has the wireless Internet. Once the Web appeared, the race to install broadband was on. In 1999, less than 10% of U.S. households had broadband; by early 2004, the percentage was 45% [32]. The shift to Internet communications is the most fundamental of the changes. The Internet enabled email, (which is the killer app of the Internet) [34], Instant Messaging, and the nascent technology: VoIP (voice over IP).

This is only the beginning of the communications revolution. We are moving from a circuit-based system based on transmitting voice to a high-speed, packet-switched network transmitting data. The pervasiveness of our communication systems will shift all that we do. These social and technological changes should be taken into account the discussion of electronic-surveillance laws.

---

[2] These numbers are international.

5   The 2004 Questions

5.1   What Is the Current Legal Framework?

Title III and FISA set the framework for U.S. electronic-surveillance laws. Since their passage (in 1968 and 1978 respectively), there have been three major Federal laws that affected wiretapping: the Electronic Communications Privacy Act (ECPA), the Communications Assistance for Law Enforcement Act (CALEA), and the U.S.A. Patriot Act.

ECPA updated Title III and FISA to apply to "electronic communications," defined as communications carried by wire or radio and not involving the human voice. ECPA was less strict about the type of crimes for which there could be interception: any federal felony may be investigated using interception of electronic communications. ECPA also modified the rules for electronic communications. In contrast to Title III and FISA, which required naming the device and person to be tapped, ECPA allowed for "roving wiretaps" — wiretaps with unspecified locations — if there was demonstration of probable cause that the subject was attempting to evade surveillance by switching telephones. In recognition of the greater ease in obtaining signalling information, ECPA provided for traffic analysis. Under ECPA, a subpoena is needed for all pen registers, which record all numbers dialed from a phone, and all trap-and-trace devices, which record all numbers dialed to a phone. Furthermore, under ECPA, law enforcement only needs a search warrant, rather than the more stringent wiretap warrant, to access stored communications (voice mail or email that has been read and then stored).

The Communications Assistance for Law Enforcement Act (CALEA) in 1994 was very controversial. In 1992 the FBI pressed for a "Digital Telephony" bill, which required that all telephone-switching equipment be designed to accommodate wiretapping. Civil-liberties groups and the telecommunications industry opposed the bill, and there were no sponsors of it.

The FBI returned to Congress in 1994 with a modified version, the "Communications Assistance for Law Enforcement Act," which included a $500 million authorization (but not appropriation) to the telecommunications companies for modifications to old equipment (this caused the telecommunications companies to drop their opposition). The bill required that any equipment deployed after January 1, 1995 would have to meet law-enforcement interception standard; the Department of Justice would determine which would be the standards-setting organization. This bill passed in the waning days of 1994 after certain civil-liberties groups dropped their opposition.

From the start, implementation of CALEA went badly. The Department of Justice put the FBI, an agency not known for expertise in telecommunications, in charge of setting the implementation standards. In October 1995 the FBI announced its requirements, which would have entailed capacity to simultaneously monitor thirty thousand lines [19] [20] [13, p. 197], a striking number at a time when the total number of annual Title III and FISA surveillances, including pen registers and trap-and-trace devices, was a quarter of that. (In 1995 the average Title III wiretap ran for 29 days [1, p. 13]. There is no public information about the length of FISA taps.) There were strong objections to the methodology the FBI used to arrive at this figure and the bureau decided to reexamine the capacity issue. Their new methodology required capacity to run sixty-thousand surveillances simultaneously[3] [20][13, p. 198]. Recognizing that the delay in developing compliance standards made it impossible for the telecommunication companies to meet the law's deadline (October 1, 1998, four years after the passage of CALEA), the FCC granted an extension til June 2000 [22].

There was also a fight about location information for cellular calls. During hearings on CALEA, FBI Director Freeh had promised that the bill would not expand wiretapping powers[24, p. 29], and the legislative report stated that "call-identifying information shall not include any information that may disclose the physical location of the subscriber" (CALEA §103 a2B). Nonetheless the FBI proposed that the cellular telecommunications group adopt a standard that would enable law enforcement to quickly establish the location of a wireless user [30]. In a 2000 decision, the U.S. Court of Appeals upheld the location standard implemented as a result of CALEA (*United States Telecommunications Association et al. v. FCC and U.S.*, 99-1442, U.S. Court of Appeals).

In CALEA, Congress defined "information services," distinguishing it from "telecommunications services." Information services were defined as "(A) mean[ing] the offering of a capability generating, acquiring, storing, transforming, processing, retrieving, utilizing, or making available information via telecommunications; and (B) includes– (i) a service that permits a customer to retrieve stored information from, or file information for storage in, information storage facilities; (ii) electronic publishing; and (iii) electronic messaging services; but (C) does not include any capability for a telecommunications carrier's internal management, control, or operation of its telecommunications network" (CALEA §102 (6)). The bill explicitly states that the interception requirements do *not* apply to information services (CALEA §103 (b)(2)(A)).

Over time, the list of crimes for which Title III is applicable grew substantially. It now lists 98 offenses, including computer fraud and abuse (18 U.S.C. §2516). Even though the vast majority of wiretapping investigations concentrate on drug trafficking and organized crime[2, Table 3], the law is not so tightly focused as had been at its inception.

5.2   How Exposed Is Personal Information?

Changes in technology as well as social norms means that individuals leave tracks wherever they go in modern society. A generation ago, individuals scrawled their names on a card inside the book they borrowed from a library; now book borrowing records library are entered into a central database. A generation ago, individuals received a hotel key; now the "key" is a plastic card that includes a

---

[3] In both cases, the proposed monitoring capacity appears as a percentage of phone lines. Thus, if number of phone lines increases, required monitoring capacity would do so proportionally.

strip that may or may not have the lodger's name and credit-card information on it. A generation ago, an individual gave a name for a plane ticket, and then may have sold the ticket to a friend; now the name on government-issued IDs must match name on the the ticket. As Jeffrey Rosen has observed, we are the "naked crowd" [37].

One significant change over the last several decades is the major loss of anonymity that has resulted from credit cards becoming the payment method of choice. The financial dossiers created enable tracking and identification of individuals in a way that plunking three hundred dollars down for a used car does not. Because credit cards have essentially become required for travel (at least for car rental and hotel reservations), credit-card records provide excellent after-the-fact records of where individuals have been, when (and, in some cases, with whom). Evidence of this is in the tracking of the September 11th hijackers. By September 14, 2001, law enforcement had put together a impressive dossier on the hijackers: where and how they had purchased their tickets, where they were living before the attacks, and where they had gone to and flight school (not all of them had) [23]. It was in the ubiquitous trail that individuals leave as part of modern life.

We leave video tracks not just at the airport and the ATM, but at totally unexpected stops. Timothy McVeigh had no intention of leaving a trail when he rented a truck in Junction City Kansas but, as noted in [13, p. 267], he had.

> Investigators ...used photos from several days before the explosion to prove that Timothy McVeigh was the "Robert D. Kling" who, on the afternoon of April 17, 1995, in Junction City, Kansas, rented the Ryder truck used in the bombing. Days and weeks after the bombing investigators meticulously reconstructed McVeigh's movements on April 17. Surveillance photos taken at a McDonalds about a mile from the Ryder agency showed McVeigh at the restaurant at 3:49 and 3:57 PM on that day. Shortly afterward, "Kling" rented the truck. When prosecutors claimed that the McDonalds's photo was of McVeigh, his lawyer did not dispute the point. The photo was taken several days before there was any hint it would be useful in a criminal case —and *then the evidence was available when needed*[5].

Imminent changes in technology will create even more detailed trails. Sensors, low-cost wireless devices, will monitor the environment and report back: "The elderly patient has a blood pressure of 110/70," "The room is at 75 degrees." RFID (Radio Frequency ID) devices will report about items an individual carries on his person: clothes, currency, a book. The sensor and RFID communications will often occur without the individual's knowledge[4].

It is not clear how an expiring milk carton informing the supermarket that it is time for a new dairy order will benefit tracking of terrorists and criminals. But one wouldn't necessarily have anticipated that an intercepted phone call in which no words were spoken and that was paid for via an anonymously-purchased

---

[4] The Internet will be the communications medium.

prepaid card would have led to a major breakthrough in a terrorist investigation either. The fact that data storage is dropping in price encourages the storage of transactional information, information that will be accessible to investigators.

It is not currently the case that an individual's data is arbitrarily subject to law enforcement perusal. The question of under what circumstances government can do data mining is currently a subject of much debate and some studies (e.g., [40]). In thinking about federal wiretap statutes, it is important to put the issue in context, and in particular to be cognizant that there is much more data easily accessible on individuals than there was at the time of the passage of the Wiretap Act. Under appropriate circumstances, that data is available to law-enforcement and national-security officials.

5.3   What Is the Effect of Communications Surveillance on Liberty?

We have briefly examined the changes in communications technology and in the accessibility of individual's private data at the dawn of the twenty-first century. We need to begin at the beginning, the time of the founding of the United States. As Whitfield Diffie has remarked,

> [P]rior to the electronic era conversing in complete privacy required nei- ther special equipment nor advanced planning. Walking a short distance away from other people and looking around to be sure that no one was hiding nearby was sufficient. Before tape recorders, parabolic mi- crophones, and laser interferometers, it was not possible to intercept a conversation held out of sight and earshot of other people. No matter how much George III might have wanted to learn the contents of Han- cock's private conversations with Adams, he had no hope of doing so unless he could induce one or the other to defect to the Crown[13, p. 2].

In the United States, the founders reacted to the broad searches by British solders under general writs of assistance by restricting government power through the Fourth Amendment of the U.S. Constitution,

> The right of the people to be secure in their persons, houses, papers and effects against unreasonable searches and seizures shall not be violated, and no Warrants shall issue but upon probable cause, supported by Oath or affirmation, and particularly describing the place to be searched, and the persons or things to be seized.

"No warrants shall issue but upon probable cause ... and particularly de- scribing the place to be searched, and the persons or things to be seized." This would be significant when it came time to apply the Fourth Amendment to com- munications surveillance. Justice Louis Brandeis wrote in his famous dissent in the Olmstead case,

> The evil incident to invasion of the privacy of the telephone is far greater than that involved in tampering with the mails. Whenever a telephone line is tapped, the privacy of the persons at both ends of the line is

invaded, and all conversations between them upon any subject, and although proper, confidential, and privileged, may be overheard. Moreover, the tapping of one man's telephone line involves the tapping of the telephone of every other person whom he may call, or who may call him. As a means of espionage, writs of assistance and general warrants are but puny instruments of tyranny and oppression when compared with wire tapping [4, pp. 475-6].

Experiences with government surveillance, extensively described elsewhere (see e.g., [13, pp. 137-150, 172-179, 271-2], [42], demonstrated serious dangers to political discourse and public expression. During the period from the 1940s to the 1970s, for example, Supreme Court justices, White House staffers, members of the National Security Council, Congressional staffers, civil-rights leaders, including Martin Luther King and Ralph Abernathy Jr, anti-Vietnam War protesters, and journalists were wiretapped. These breaches made Congress wary of providing law-enforcement and national-security investigators with such a potentially invasive tool. This is why the requirements for a wiretap warrant are significantly more stringent than those for a "normal" search warrant [5].

Wiretaps intrude on a conversation between two people and thus require the high level of wiretap search warrant before tapping can commence. But there is no similar level of protection for transactional information on what number is being called and what number is calling. The legal rationale is that such transactional information is already being shared with a third party (in this case, the telephone switch) and the communicating parties do not have any expectation of privacy on the data. Thus a subpoena, which can be obtained from a magistrate, suffices for pen registers and trap-and-trace devices[6].

---

[5] It is also why public reporting of Title III wiretaps is required; each year, the Administrative Office of the U.S. Courts produces a report listing each Title III wiretap of the previous year (ongoing taps are not reported until they have ceased to be used), including the D.A., the judge issuing the wiretap search warrant, the length of order, the "most" serious crime for which the wiretap was ordered (there may be more than one for a single wiretap), the number of incriminating and non-incriminating calls picked up on the wiretap, the cost of the surveillance, etc. (Except for annually reporting to Congress the number of surveillances, there are no public disclosure requirements for FISA wiretaps.)

[6] This paper concentrates on the technology side of the electronic-surveillance issues, not the policy. Nonetheless, we would be remiss if we did not point out that traffic analysis, though usually less intrusive than content surveillance, may nonetheless cause severe privacy breaches. One such example occurred in the 1980s FBI investigation of CISPES, the Committee in Solidarity with the People of El Salvador, an American group which supported the opposition to the El Salvadorian government. On the basis of an informer's information, the FBI started an investigation of CISPES, eventually culminating in files on more than twenty-three hundred individuals. Much of the information was obtained through phone records. The investigation was not justified; the group was not a terrorist organization, and in 1988, FBI Director William Sessions told Congress that, "[T]here was no reason ... to expand the investigation so widely" [38, p. 122].

In this paper we are focusing our discussion on technology implications of wiretapping rather than policy issues. Nonetheless, as we consider the role of surveillance in current communications technology, we must never lose sight of Brandeis's words, "As a means of espionage, writs of assistance and general warrants are but puny instruments of tyranny and oppression when compared with wire tapping [4, pp. 476]."

6   Telephony and the Internet: Two Different Architectures

The Public Switched Telephone Network (PSTN) was built to maximize the quality of voice transmissions and everything in the network was designed to that end. The Internet was designed for reliability, a very different quality. The PSTN uses circuit switching to transmit information from sender to receiver, the Internet, packet switching. The PSTN and the Internet have fundamentally different architectures. This simple fact means that many of the surveillance tasks do not directly translate from one domain to the other.

6.1   Electronic Surveillance on the Internet

Consider, for example, the effect of packet-based technology on the transmittal of transactional information. In telephony, signaling information appears at the beginning of the call and is separated from call contents. In packet-switched systems such as the Internet, because data is broken into "packets," each one of which has the addressing information, contents do not have the same physical separation from the "signalling" information (probably more properly called transactional information in this case).

Furthermore, electronic communications typically present more personally-identifiable information present in the so-called transactional information. At a minimum, this may simply include place of business, e.g., susan.landau@sun.com. But it may include much more, e.g., if the transactional information is the result of a google search, the URL will reveal the search terms [7].

---

[7] That "pen registers" and "trap-and-trace devices" garner additional information when used in packet-switching network systems than they do in traditional circuit-switched telephony systems did not escape the notice of technologists and civil-liberties groups. When the news of Carnivore, the FBI's Internet monitoring system became public in the summer of 2000, one of the criticisms of the system was that the transactional information that Carnivore was sweeping up was more than the government was entitled to under the limited subpoena power used for pen registers and trap-and-trace devices. Carnivore was quite controversial. In the summer of 2001, it looked as if there might Congressional action limiting Carnivore's use. Instead September 11th happened. The Patriot Act gave law enforcement explicit power to use subpoenas for pen registers and trap-and-trace devices on electronic communications (§216).

An even more crucial different between the PSTN and the Internet is that in the Internet, *the intelligence is at the endpoints.* The underlying network system is simple, while the endpoints can deploy complex systems. This fundamental architectural idea is what makes the Internet so versatile. Applications can be designed far beyond what the original designers of the Internet had in mind. And indeed, innovation has flourished because the endpoints competed and created new services. No one needs to depend on the infrastructure company to do the innovation for them.

The design flexibility comes at a price that we do not often think of as a price: the Internet is hard to control. This does not mean political or border controls (though those are also often difficult to implement on the Internet) but design control. This is not a bug; it is an extremely attractive feature. In a sharp, and deliberate, distinction from the telephony network, the Internet was designed to be loosely controlled. The layered approach to network design provides that effect and is what has enabled much of Internet innovation.

For those that choose to invest the effort, Internet communications can be fully protected. The Internet design of intelligence at the endpoints complicates wiretapping, which is useless if end systems adequately protect their communications (although a wiretapped encrypted conversation will still provide traffic information). In recent years, protecting the privacy of communications has become an important security goal. Indeed, the U.S. government has moved in the direction of simplifying the deployment of communications security in commercial equipment, partially as a result of the government's move to purchasing COTS (commercial off the shelf) equipment rather than the purchase of custom-designed systems. Instead of restricting the use of cryptography, the U.S. government has recently encouraged a number of security efforts, including the development of the 128-bit Advanced Encryption Standard and the deployment of Elliptic Curve Cryptosystems. Attempts to build wiretapping capabilities into Internet protocols would seem to go against these efforts.

At the same time, as an IETF Network Working Group studying the issue of architecting wiretap requirements into Internet protocols observed, "the use of existing network features, if deployed intelligently, provide extensive opportunities for wireteapping" [35].

6.2   The Risks Wiretapping Poses to Internet Security

Under CALEA, telecommunications systems deployed after January 1, 1995 must be built wiretap accessible. Suppose one were to call for that same requirement on the Internet. Does such an obligation make sense? Can it be architected in? What does it do to security requirements?

Wiretapping is an architected security breach. Saying that Internet communication protocols necessarily must have wiretapping requirements built in is to say that security loopholes must be built into communication protocols. It means that privacy of the communication must be deliberately violated and in a way that does not alert the sender or recipient.

Of course, U.S. law-enforcement and national-security agents are not the only ones interested in wiretapping the Internet; foreign governments are as well. Any technology that is designed to simplify Internet wiretapping by U.S. intelligence may well be exploited by foreign-intelligence services. During the discussions on CALEA, there were concerns about the security problems created by "building in" wiretapping capablities for digital telephony [15]. Such fears pale when measured against designing such capabilities for the Internet. Internet wiretapping technology, found and reverse engineered by foreign-intelligence services, could enable massive surveillance of U.S. "persons" (citizens and corporations). Used in combination with inexpensive automated search technology, this could lead to serious security breaches.

There is risk to the U.S. economy (the potential loss of corporate information). There is risk to U.S. national security (through the provision of cost-effective massive intelligence gathering). There is risk to the freedom of U.S. citizens. These are the risks [7] that the European governments responded to when, in 1999, they decided to liberalize their cryptographic export-control policy. As did the United States when it liberalized its cryptographic export-control policies shortly afterwards [14].

If we were to build access for U.S. law enforcement or national security into Internet communications, such protocol design would have be done very carefully. Can it be? It is highly doubtful. As the IETF Network Working Group observed, any protocol designed with wiretapping capabilities built in is inherently less secure than it would be without the wiretapping capability. Building wiretapping requirements into network protocols makes the protocols more complex. As is well known, complex protocols are prone to security flaws. The secure Internet is a challenge. Despite best efforts, security breaches slip into many protocols. No one wants to see deliberately-architected security breaches. In 2000 the IETF Network Working Group decided not to consider requirements for wiretapping as part of the IETF standards process [35].

7   What Is the Right Tradeoff for Communications Surveillance?

What are the costs to communications technology of continuing to enable wire-taps? A recent FBI petition to the FCC gives an illustration. The bureau argued that "CALEA's purpose is to help lawful electronic surveillance keep pace with changes in telecommunications technology as telecommunications services migrate to new technologies" [21, pp. 3-4] and stated that thus "CALEA is applicable not only to entities and services that employ circuit-mode technology, but also to entities and services that employ packet-mode technology" [21, p. 6]. The Bureau urged the FCC to declare that any service providing voice communications, including Voice over IP (VoIP), should be viewed as a "telecommunications carrier."

The breadth of this claim is startling. Were the FCC to grant the petition (unknown at the time of this writing), this would put the FBI squarely in the middle of designing IETF protocols. What would the technological cost of granting this petition be? One can scarcely imagine. At a minimum, granting the petition would "drive up costs, impair and delay innovation, threaten privacy, and force development of the latest Internet innovations offshore" according to a response filed by a coalition of industry and civil-liberties groups [26]. As we have observed earlier, it would also threaten security.

Does the value of wiretapping justify trying to preserve the tool? This, of course, depends on whom you ask. As the FBI was pressing the Digital Telephony bill in the early 1990s, the bureau argued that wiretapping was a critical tool in the fight against organized crime. The FBI presented claims that court-ordered wiretaps resulted in over seven thousand convictions, three hundred million dollars in fines levied, and over three-quarters of a billion dollars in recoveries, restitutions, and court-ordered forfeitures over a six-year period [18]. But White House staffers [3], the Treasury Department [28], and the Vice-President's office [31] all disputed the FBI numbers.

There is no question that wiretapping can be effective in some cases. Its most important value may be as a deterrent: knowing that law enforcement is listening in, criminals and terrorists stay off the line. Or they speak in code: "The big guy is coming. He will be here soon." [45] Making the use of electronic communications difficult for criminals and terrorists denies them one of the greatest technological advances of the last century.

As we have seen, greater surveillance value may come from traffic analysis, which has already shown remarkable benefits in the fight against terrorism. Given the U.S. government's shift on cryptographic export controls, one might reasonably argue that intelligence agencies have come to the same conclusion.

The debate about electronic surveillance must not occur in isolation. U.S. wiretapping laws were passed when the opportunity to easily obtain massive, automatically-created, data trails did not exist. Video cameras in McDonalds, at ATM machines, E-Z pass automatically recording the trip through the toll booths, sensors and RFID tags are all aspects of this changing technology. One has just to look at disappearance of pay phones[8] to realize how much the way we communicate, both in frequency and in mode, has substantially changed from only a generation ago.

If Congress were not to preserve law-enforcement's capability to wiretap, what investigative tools might be offered in trade? A clear one is easy access to communications transactional information. One of the non-controversial aspects of the Patriot Act is that it simplified the procedure for obtaining pen register and trap-and-trace orders, no longer requiring an application in each jurisdiction, but letting a single application suffice. Traffic analysis has become significantly easier to obtain and it may be appropriate to trade further capabilities in this direction. For example, the decreasing costs of storage have made record saving much less onerous. Might it be appropriate to require service providers to keep records of communications (which numbers, when, for how long) for a specified

---

[8] The new wing at Bradley Airport in Hartford, Connecticut, which has twelve gates, has exactly two pay phones.

period in exchange for deciding that communications systems will not be required to be built wiretap accessible?

The threat of terrorism will confront our society for a long time. But we should not necessarily be extending a 1960s wiretap law into the twenty-first century. Instead we should be examining first principles to determine what surveillance laws are appropriate for current challenges. Wiretapping became a law-enforcement tool in the late 1920s; its use was codified in the 1960s and 1970s. If attempting to preserve the tool in order to enable investigators to hold onto this capability would freeze communications in an antiquated technology, that may be the wrong route for our society to take. It may be that few security benefits accrue from the requirement that electronic communications be designed "wiretap accessible" while efforts to do so significantly impede innovation. It is time to fully examine electronic surveillance: it value, needs, and costs. Such a discussion is a necessity in our complicated times. It is crucial as we attempt to solve the current threats to security and liberty.

References

1. Administrative Office of the U.S. Courts, Washington D.C., *1995 Wiretap Report*.
2. Administrative Office of the U.S. Courts, Washington D.C., *2003 Wiretap Report*.
3. Anderson, Betsy and Todd Buchholz, "Memo for Jim Jukes," 22 May 1992 in [16].
4. Brandeis, Louis, Dissenting opinion in *Olmstead v. United States*, 277 U.S. 438, 1928.
5. Brooke, James, "Prosecutors in Bomb Trial Focus on Time Span and Truck Rental," *New York Times*, May 10, 1997, p. A1 and A10.
6. Brooks, John, *Telephone: the First Hundred Years*, Harper and Row, 1975.
7. Duncan Campbell, "Interception 2000: Development of Surveillance Technology and Risk of Abuse of Economic Information," Report to the Director General for Research of the European Parliament, Luxembourg, April 1999.
8. Clark, Ramsey. (1967), in [44, pp. 285-321].
9. Congressional Quarterly Weekly 1968b Congressional Quarterly Weekly, (1968b), Vol. 26, Washington, D.C., July 19.
10. Dam, Kenneth and Herbert Lin (eds.), Committee to Study National Cryptography Policy, Computer Science and Telecommunications Board, National Resource Council, *Cryptography's Role in Securing the Information Society*, National Academy Press, 1996.
11. Dempsey, James X., "Communications Privacy in the Digital Age: Revitalizing the Federal Wiretap Laws to Enhance Privacy," *Albany Law Journal of Science and Technology*, Vol. 8, No. 1, 1997.
12. Dempsey, James X. and David Cole, *Terrorism and the Constitution: Sacrificing Civil Liberties in the Name of National Security*, First Amendment Foundation, 1999.
13. Diffie, Whitfield and Susan Landau, *Privacy on the Line: The Politics of Wiretapping and Encryption*, MIT Press, 1998.
14. Department of Commerce, Bureau of Export Administration: 15 CFR Parts 734, 740, 742, 770, 772, and 774, Docket No. RIN: 0694-AC11, Revisions to Encryption Items. Effective January 14, 2000.

15. Electronic Frontier Foundation, *Analysis of the FBI Proposal Regarding Digital Telephony*, 17, September 1992.

16. Electronic Privacy Information Center, David Banisar (ed.), *1994 Cryptography and Privacy Sourcebook: Primary Documents on U.S. Encryption Policy, the Clipper Chip, the Digital Telephony Proposal, and Export Controls*, Diane Publishing, Upland, PA., 1994.

17. Electronic Privacy Information Center, *1996 EPIC Cryptography and Privacy Sourcebook: Documents on Wiretapping, Cryptography, the Clipper Chip, Key Escrow and Export Controls*, Diane Publishing Co., Upland, PA, 1996.

18. Federal Bureau of Investigation, "Benefits and Costs of Legislation to Ensure the Government's Continued Capability to Investigate Crime with the Implementation of New Telecommunications Technologies," in [16]

19. Letter to Telecommunications Industry Liaison Unit, Federal Bureau of Investigation, November 13, 1995, in [17, pp. B14-B20].

20. Federal Bureau of Investigation, "Implications of Section 104 of the Communications Act for Law Enforcement," in *Federal Register*, Vol. 62, Number 9, January 14, 1997, pp. 192-1911.

21. Federal Bureau of Investigation, *In the Matter of the United States Department of Justice, the Federal Bureau of Investigation, and the Drug Enforcement Agency: Joint Petition for Rulemaking to Resolve Various Outstanding Issues Concerning the Implementation of the Communications Assistance for Law Enforcement Act: Joint Petition for Expedited Rulemaking before the Federal Communications Commission*, 10 March 2004.

22. Federal Communications Commission, *Memorandum Opinion and Order*, 10 September 1998.

23. Firestone, David and Dana Canedy, "After the Attacks: The Suspects; FBI Documents Detail the Movements of 19 Men Believed to be Hijackers," *New York Times*, 15 September 2001, p. A1.

24. Freeh, Louis in United States Senate, Committee on the Judiciary, Subcommittee on Technology and the Law (Senate), and United States House of Representatives, Committee on the Judiciary, Subcommittee on Civil and Constitutional Rights, *Digital Telephony and Law Enforcement Access to Advanced Telecommunications Technologies and Services*, Joint Hearings on HR 4922 and S. 2375, March 18 and August 11, 1994, One Hundred Third Congress, Second Session.

25. Hersh, Seymour, "Annals of National Security: What Went Wrong," *The New Yorker*, 8 October 2001.

26. Joint Statement of Industry and Public Interest, *In the Matter of Joint Petition for Rulemaking to Resolve Various Outstanding Issues Concerning the Implementation of the Communications Assistance for Law Enforcement Act*, 27 April 2004.

27. Leone, Richard C. and Greg Anrig Jr. (eds.), *The War on our Freedoms: Civil Liberties in an Age of Terrorism*, Century Foundation, 2002.

28. Levy, Ron, "Memo for Doug Steiner," 26 May 1992, in [16].

29. Lewis, Anthony, "Robert Kennedy Vows in Georgia to Act on Rights, *New York Times*, May 7, 1961, p. 1.

30. Markoff, John, "Cellular Industry Rejects U.S. Plan for Surveillance," *New York Times*, September 20, 1996, p. A1.

31. McIntosh, David and James Gattuso, "Memo for Jim Jukes," 22 May 1992, in [16].

32. Neilsen/Net Ratings, "Broadband Growth Trend."

33. The President's Commission on Law Enforcement and the Administration of Justice, *The Challenge of Crime in a Free Society*, United States Government Printing Office, 1967.

34. Odlyzko, Andrew, "Content is not King,"
    http://www.dtc.umn.edu//simodlyzko/doc/networks.html
35. *NWG, RFC2804 — IETF Policy on Wiretapping*, May 2000.
36. Risen, James, David Johnston, and Neil A. Lewis, "Harsh CIA Methods Cited in Top Qaeda Interrogations," New York Times, 13 May 2004, p. A1.
37. Rosen, Jeffrey, *The Naked Crowd*, Random House, 2004.
38. Sessions, William, Testimony in [43].
39. Stern, Jessica, *Terror in the Name of God*, Harper Collins Publishers, 2003.
40. Technology and Privacy Advisory Committee, Department of Defense, *Safeguarding Privacy in the Fight Against Terrorism*, March 2004.
41. United States Commission on National Security/Twenty-First Century, *Road Map for National Security: Imperative for Change: Phase III Report of the U.S. Commission on National Security/Twenty-First Century*, 31 January 2001.
42. United States, Senate Select Committee to Study Governmental Operations with respect to Intelligence Activities, *Intelligence Activities and the Rights of Americans, Final Report, Book II*, Report 94-755, Ninety-fourth Congress, Second Session, April 23, 1976.
43. United States Senate, Select Committee on Intelligence, *Senate Select Committe on Intelligence: Inquiry into the FBI Investigation of the Committee in Solidarity wih the People of El Salvador*, Hearings on February 23, April 13, September 14 and 29, 1988, One Hundredth Congress, Second Session.
44. United States House of Representatives, Committee on the Judiciary, Subcommittee No. 5, *Anti-Crime Program*, Hearings on HR 5037, 5038, 5384, 5385 and 5386, March 15, 16, 22, 23, April 5, 7, 10, 12, 19, 20, 26 and 27, 1967, Ninetieth Congress, First Session, 1967.
45. Van Natta Jr., Don and Desmond Butler, "How Tiny Swiss Cellphone Chips Helped Track Global Terror Web," *New York Times*, 4 March 2004, p. A1.

An Improved Correlation Attack Against Irregular Clocked and Filtered Keystream Generators

Håvard Molland and Tor Helleseth

The Selmer Center*
Institute for Informatics,
University of Bergen,
Norway

Abstract. In this paper we propose a new key recovery attack on irregular clocked keystream generators where the stream is filtered by a nonlinear Boolean function. We show that the attack is much more efficient than expected from previous analytic methods, and we believe it improves all previous attacks on the cipher model.

Keywords: Correlation attack, Stream cipher, Boolean functions, Irregular clocked shift registers.

1 Introduction

In this paper we present a new key recovery correlation attack on ciphers based on an irregular clocked linear feedback shift register (*LFSR*) filtered by a Boolean function. The cipher model we attack is composed of two components, the clock control generator and the data generator and is shown in Fig. 1.

- *The data generator sub system* consists of $\mathrm{LFSR_u}$ of length l_u and the nonlinear multivariate function f. The internal state of $\mathrm{LFSR_u}$ is filtered by a Boolean function f. The output from f is the high linear complexity bit stream $\mathbf{v}$.
- *The clock control sub system* consists of $\mathrm{LFSR_s}$ of length l_s where the output from $\mathrm{LFSR_s}$ is sent through the clock function $D()$. The output from $D()$ is the clock control sequence of integers, $\mathbf{c}$, which is used to clock $\mathrm{LFSR_u}$.

The effect of the irregular clocking is that $\mathbf{v}$ is irregularly decimated and the positions of the bits in the stream are altered. The result from this decimation is the keystream $\mathbf{z}$. The secret key in this cipher is the $(l_u + l_s)$ initialization bits for $\mathrm{LFSR_u}$ and $\mathrm{LFSR_s}$ $(\mathbf{I_u}, \mathbf{I_s})$.

To attack this encryption scheme we need to know the positions the keystream bits $\mathbf{z}$ had in the stream $\mathbf{v}$ before $\mathbf{v}$ was irregularly decimated. The previous effective algorithms are not specially designed to attack irregular clocked and

---

* This work was supported by the Norwegian Research Council.

M. Franklin (Ed.): CRYPTO 2004, LNCS 3152, pp. 373–389, 2004.
© International Association for Cryptologic Research 2004

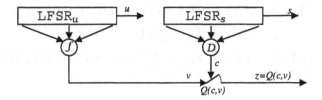

Fig. 1. The general cipher model we attack in this article

filtered generators. But there exist effective attacks on the data generator sub system[6, 1, 10, 3, 4]. To deal with the irregular clocking, one of two techniques are often used:

1. **Do the attack on the data generator 2^{l_s} times [7].** The attack is done one time for each guess for the 2^{l_s} possible initialization states for $LFSR_s$. If the attack on the sub system has complexity $O(K)$ the full attack will have complexity $O(K \cdot 2^{l_s})$.

2. **Ignore the clock control generator [3, 14, 4].** If the attack on the data generator subsystem needs M keystream bits, we can use the fact[14] that we know the original $\mathbf{v}$ position of every $2^{l_s} - 1$ bit in the keystream $\mathbf{z}$. Thus we can only use every $2^{l_s} - 1$ keystream bit in the attack, which means that we need $(2^{l_s} - 1) \cdot M$ keystream bits to succeed.

None of these techniques are optimal. The first one leads to large runtime complexity, the second leads to the need for a large number of keystream bits.

Our attack is not designed to attack the data generator subsystem only, but is especially aimed at irregular clocked and filtered keystream generators as one system. First we guess the initialization state $\mathbf{I}_s$ for $LFSR_s$. From this we can reconstruct the positions the bits in $\mathbf{z}$ had in $\mathbf{v}$. Using the iteration algorithm from[11] this reconstruction is done using just a couple of operations per guess, exploiting the cyclic redundancies in $LFSR_s$. This method is fully explained in Section 4.3. This method gives the guess $\mathbf{v}^* = (.., *, z_i, ..., z_j, ...*, ..., z_k, ..., *, ...)$, where z_i, z_j, z_k are some keystream bits and the stars are the deleted bits. Then we test $\mathbf{v}^*$ to see if it is likely that the stream is generated by the data generator subsystem $LFSR_u$ and f. Hence, we only use a distinguisher test on the the $\mathbf{v}^*$ stream to decide if the guess for $\mathbf{I}_s$ is correct. This is easier than to actually decode the $\mathbf{v}^*$ stream to find $\mathbf{I}_u$, and then decide if we have found the correct $\mathbf{I}_s$. When $\mathbf{I}_s$ is determined, we can use one of the previous attacks on the data generator sub system to determine $\mathbf{I}_u$.

The distinguisher test is to evaluate a large number m of low weight parity check equations on the bit stream $\mathbf{v}^*$. All equations are derived from one multiple $h(x)$ of weight 4 of the generator polynomial $g_u(x)$. Surprisingly this test works much better than expected from previous evaluation methods. In previous correlation attacks, the Piling up lemma[9] is often used to calculate the correlation[1, 7, 6] which the algorithm must decode. Since our algorithm only uses a distinguisher on $\mathbf{v}^*$ we can use a correlation property of the function f

which gives much higher correlation between $\mathbf{v}^*$ and the keystream $\mathbf{z}$. Thus we need fewer parity check equations. This correlation property exists even if the function is correlation immune in the normal sense.

Our attack has complexity $O(2^{l_s} \cdot m)$, independently of the length of LFSR$_u$. A cipher based on the model we attack in this paper is LILI-128. To attack the LILI-128 cipher our algorithm needs about 2^{23} parity check equations. In LILI-128, $l_s = 39$, thus the runtime for our attack is $2^{39+23} \approx 2^{62}$ parity checks, with virtually no precomputation. We have implemented and tested the attack, and it works on computers having under 300 MB of RAM, and needs only around 68 Mbyte of keystream data. The precomputation has low runtime complexity and is negligible. When $\mathbf{I_s}$ is found, we can use one of the previous algorithms to attack the data generator sub system.

A comparable previous correlation attack by Johansson and Jönsson is presented in [7]. The runtime for the attack is 2^{71} parity checks and the precomputations is 2^{79} table lookups. The keystream length is approximately 2^{30}. This attack uses the first technique to handle the irregular clocking.

Recently new algebraic attacks have been proposed by Courtois and Meier[3, 4]. This attack uses the second technique to handle the irregular clocking in LILI-128. Although the attack has an impressive runtime complexity $2^{31} \cdot C$ (an optimistic estimation for some unknown constant C), the attack needs about 2^{60} keystream bits to succeed, which is unpractical.

There is also a time-memory trade-off attack against LILI-128 by Markku-Juhani Olavi Saarinen[14]. This attack needs approximately $2^{51.4}$ bits of computer memory and 2^{46} keystream bits. The runtime complexity is claimed to be 2^{48} DES operations, which is not easy to compare with our runtime complexity. But the high use of computer memory and keystream bits also makes this attack unpractical.

2   A Correlation Property of Nonlinear Functions

Let $V = F_2^n$ and let f be a balanced Boolean function from V to F_2. We start by analyzing the boolean function $f(\mathbf{x})$ for a correlation property that we will use in the attack. A similar property is analyzed in [18] where they look at the nonhomomorphicity of functions. In this paper we identify the probability

$$p = P(f(\mathbf{x}_1) + f(\mathbf{x}_2) + f(\mathbf{x}_3) + f(\mathbf{x}_4) = 0 \mid \mathbf{x}_1 + \mathbf{x}_2 + \mathbf{x}_3 + \mathbf{x}_4 = \mathbf{0}) \quad (1)$$

which is crucial for our attacks success rate.

2.1   The Correlation Property

Let $q = 2^n$ and let $\mathbf{a} \cdot \mathbf{b} = \sum_{i=1}^{n} a_i b_i$ denote the inner product of $\mathbf{a} = (a_1, a_2, \ldots, a_n)$ and $\mathbf{b} = (b_1, b_2, \ldots, b_n)$. Define the Walsh coefficients of f by

$$\hat{f}(\mathbf{a}) = \sum_{\mathbf{x} \in V} (-1)^{f(\mathbf{x}) + \mathbf{a} \cdot \mathbf{x}}.$$

Lemma 1. *Let f be a function from $V = F_2^n$ to F_2 and let $\mathbf{x}_i \in F_2^n$ for $i = 1, 2, 3, 4$. Let $q = 2^n$ and let N denote the number of solutions of*

$$\mathbf{x}_1 + \mathbf{x}_2 + \mathbf{x}_3 + \mathbf{x}_4 = \mathbf{0} \tag{2}$$
$$f(\mathbf{x}_1) + f(\mathbf{x}_2) + f(\mathbf{x}_3) + f(\mathbf{x}_4) = 0. \tag{3}$$

Then

$$N = \frac{q^3}{2} + \frac{1}{2q} \sum_{\mathbf{a} \in V} \hat{f}(\mathbf{a})^4. \tag{4}$$

Proof. Each term in the sum below gives a contribution $2q$ for each solution of the system of equations, and zero otherwise. Therefore, we have

$$2qN = \sum_{\mathbf{x}_1, \mathbf{x}_2, \mathbf{x}_3, \mathbf{x}_4 \in V} \left(\sum_{\mathbf{a} \in V} (-1)^{\mathbf{a} \cdot (\mathbf{x}_1 + \mathbf{x}_2 + \mathbf{x}_3 + \mathbf{x}_4)} \right) \left(\sum_{y=0}^{1} (-1)^{y(f(\mathbf{x}_1) + f(\mathbf{x}_2) + f(\mathbf{x}_3) + f(\mathbf{x}_4))} \right)$$

$$= \sum_{\mathbf{a} \in V} \sum_{y=0}^{1} \sum_{\mathbf{x}_1, \mathbf{x}_2, \mathbf{x}_3, \mathbf{x}_4 \in V} (-1)^{yf(\mathbf{x}_1) + \cdots + yf(\mathbf{x}_4) + \mathbf{a} \cdot \mathbf{x}_1 + \cdots + \mathbf{a} \cdot \mathbf{x}_4}$$

$$= \sum_{\mathbf{a} \in V} \sum_{y=0}^{1} \left(\sum_{\mathbf{x} \in V} (-1)^{yf(\mathbf{x}) + \mathbf{a} \cdot \mathbf{x}} \right)^4$$

$$= q^4 + \sum_{\mathbf{a} \in V} \hat{f}(\mathbf{a})^4,$$

where the first term comes from the case $y = 0$ and $\mathbf{a} = \mathbf{0}$, and the last term from the case $y = 1$.

Corollary 1. *If $f(\mathbf{x})$ is a balanced function then the number of solutions N of the system of equations above is,*

$$N \geq \frac{q^3}{2} + \frac{q^3}{2(q-1)}.$$

Proof. Since $f(\mathbf{x})$ is balanced we obtain $\hat{f}(\mathbf{0}) = \sum_{\mathbf{x} \in V} (-1)^{f(\mathbf{x})} = 0$. It follows from Parseval's identity that the average value of $\hat{f}(\mathbf{a})^2$ is $\frac{q^2}{q-1}$. Hence, it follows from the Cauchy-Schwartz inequality that $\sum_{\mathbf{a} \in V} \hat{f}(\mathbf{a})^4 \geq (q-1) \frac{q^4}{(q-1)^2}$, which substituted in the lemma above gives the result.

Corollary 2. *The expected number of solutions N of the system of equations above is,*

$$E(N) = \frac{q^3}{2} + \frac{3q^2 - 2q}{2}.$$

Proof. An average estimate of N can be found as follows. When there exist two equal vectors $x_{i_1} = x_{i_2}$ in Equation (2), the two other vectors x_{i_3}, x_{i_4} will also be equal. When this occurs it follows that the Equation (3) will sum to zero.

This gives the unbalance that causes the high correlation. Equation (2) implies $x_4 = x_1 + x_2 + x_3$ Then there are $q(q-1)(q-2)$ triples in x_1, x_2, x_3 where all the x_i's are distinct and there are therefore $3q^2 - 2q$ triples with one or two pairs $x_{i_1} = x_{i_2}$. Using this fact and substituting Equation (2) into Equation (3), we can write

$$
\begin{aligned}
2N &= \sum_{x_1,x_2,x_3 \in V} \sum_{y=0}^{1} (-1)^{y(f(x_1)+f(x_2)+f(x_3)+f(x_1+x_2+x_3))} \\
&= q^3 + \sum_{x_1,x_2,x_3 \in V} (-1)^{f(x_1)+f(x_2)+f(x_3)+f(x_1+x_2+x_3)} \\
&= q^3 + (3q^2 - 2q) + \sum_{x_1,x_2,x_3 \text{ distinct} \in V} (-1)^{f(x_1)+f(x_2)+f(x_3)+f(x_1+x_2+x_3)}.
\end{aligned}
$$

Since for an arbitrary function f we can expect that $f(x_1)$, $f(x_2)$, $f(x_3)$, and $f(x_1 + x_2 + x_3)$ take on all binary quadruples approximately equally often when $x_1 \neq x_2 \neq x_3 \neq x_1$, we expect in the average the last term to be 0. This implies the result.

Corollary 3. *Let f be an arbitrary balanced function, and let p denote the probability*

$$
p = Prob(f(x_1) + f(x_2) + f(x_3) + f(x_4) = 0 \mid x_1 + x_2 + x_3 + x_4 = 0),
$$

then p is expected to be $E(p) = \frac{1}{2} + \frac{3q-2}{2q^2}$ and its minimum is $p_{min} \geq \frac{1}{2} + \frac{1}{2(q-1)}$.

Proof. Since Equation (2) has q^3 solutions, it follows from Corollary 1 that the expected probability is equal to $E(p) = \frac{E(N)}{q^3} = \frac{1}{2} + \frac{3q-2}{2q^2}$. Further from Corollary 2 we obtain that the minimum is $p_{min} \geq (\frac{q^3}{2} + \frac{q^3}{2(q-1)})/q^3 = \frac{1}{2} + \frac{1}{2(q-1)}$.

Corollary 4. *Given a specific balanced function f, the probability*

$$
p = Prob(f(x_1) + f(x_2) + f(x_3) + f(x_4) = 0 \mid x_1 + x_2 + x_3 + x_4 = 0),
$$

is $p = \frac{1}{2} + \frac{\sum_{a \in V} \hat{f}(a)^4}{2q^4}$

Proof. Using the N from Lemma 1 we get $p = \frac{N}{q^3} = \frac{1}{2} + \frac{\sum_{a \in V} \hat{f}(a)^4}{2q^4}$

It is straightforward to extend Lemma 1 to compute the number of common solutions of the two equations

$$
\begin{aligned}
x_1 + x_2 + \cdots + x_w &= 0 \\
f(x_1) + f(x_2) + \cdots + f(x_w) &= 0.
\end{aligned}
$$

and show that the corresponding probability

$$
Prob(f(x_1) + f(x_2) + \cdots + f(x_{w-1}) = 0 \mid x_1 + x_2 + \cdots + x_{w-1} = 0),
$$

equals $p = \frac{1}{2} + \frac{\sum_{a \in V} \hat{f}(a)^w}{2q^w}$, which reduces to the result of Corollary 4 when $w = 4$.

In the case $w = 3$, we can calculate the expected value of a balanced Boolean function, with a given $f(0)$, to be $E(p) = \frac{1}{2} + \frac{3q-2}{2q^2}(-1)^{f(0)}$. This implies that the bias is the same for the case $w = 3$ as for $w = 4$. Similar arguments for equations with $w \geq 5$ show that these equations give too low correlation, which would lead to a high runtime complexity for our attack. It turns out that for $w = 3$ the attack needs much more keystream bits to succeed, see the Sections 4.1 and 5.2. Since the correlation bias is exactly the same for $w = 3$ and $w = 4$ it is optimal to use $w = 4$.

2.2   Analysis of Some Functions

In Table 1 we have analyzed some functions using Corollary 4. This correlation is surprisingly high. Let $p_{app} = 0.53125$ be the best linear approximation to the LILI-128 function. Due to the design of the previous attacks[6, 7, 10] the channel noise has been independent of the stream $\mathbf{u}$ generated by LFSR_u. Thus the Piling up lemma [9], $p_{pil} = \frac{1}{2} + 2^{w-1}(\frac{1}{2} - p_{app})^w$, is used to evaluate the crossover correlation $1 - p_{pil}$ which the algorithms must be able to decode. Using the Piling up lemma for weight $w = 4$ equations, the correlation p_{pil} for LILI-128 will be $p_{pil} = 0.50000763$. From Table 1 we have the correlation $p = 0.501862$. The reason for the higher correlation, is that our attack only uses a distinguisher on the data generator sub system, and not a complete decoder. Hence, in our key recovery attack on the clock control system, we can use Corollary 4 from Section 2.1 to calculate the correlation. To test the corollary we generated 2000 random and balanced Boolean tables for $n = 10$, and calculated the average correlation. The result was that the average p was 0.501466 which is close to the theoretical expected $E(p) = 0.5001464$.

Table 1. The probability $P(f(\mathbf{x}_1) + f(\mathbf{x}_2) + f(\mathbf{x}_3) + f(\mathbf{x}_4) = 0 \mid \mathbf{x}_1 + \mathbf{x}_2 + \mathbf{x}_3 + \mathbf{x}_4 = 0)$ calculated for some given functions. $E(p)$ is the expected correlation for given $q = 2^n$ and p is the actual correlation for the given function

Function	Number of inputs bits n	Best linear approximation.	$E(p)$	p
Geffe function	2	0.75	0.671875	0.625
LILI-128	10	0.53125	0.501464	0.501862
LILI-II	12	0.51367	0.500366	0.500190

3   A General Model

Here we define a general model for irregular clocked and filtered stream ciphers, and some well known properties for the model.

3.1 General Model

Let $g_u(x)$ and $g_s(x)$ be the feedback polynomials for the shift registers LFSR$_u$ of length l_u and LFSR$_s$ of length l_s. We let $\mathbf{I}_s = (s_0, s_1, ..., s_{l_s-1})$ and $\mathbf{I}_u = (u_0, u_1, ..., u_{l_u-1})$ be the initialization states for LFSR$_s$ and LFSR$_u$. The initialization states $(\mathbf{I}_s, \mathbf{I}_u)$ define *the secret key* for the given cipher system.

From $g_s(x)$ we can calculate a clock control sequence $\mathbf{c}$ in the following way. Let $c_t = D(L_s^t(\mathbf{I}_s)) \in \{a_1, a_2, ..., a_A\}$, $a_j \geq 0$, be a function where the input $L_s^t(\mathbf{I}_s)$ is the inner state of LFSR$_s$ after t feedback shifts and A is the number of values that c_t can take. Let p_j be the probability $p_j = \text{Prob}(c_t = a_j)$.

LFSR$_u$ produces the stream $\mathbf{u} = (u_0, u_1, ...)$ which is filtered by f. The output from f is $v_k = f(u_{k+i_0}, u_{k+i_1}, ..., u_{k+i_{n-1}})$, or the equivalent $v_k = f(L_u^k(\mathbf{I}_u))$. The clock c_t decides how many times LFSR$_u$ is clocked before the output bit v_k is taken as keystream bit z_t. Thus the keystream z_t is produced by $z_t = v_{k(t)}$, where $k(t)$ is the total sum of the clock at time t, that is $k(t) \leftarrow k(t-1) + c_t$. This gives the following definition for the clocking of LFSR$_u$.

Definition 1. *Given bit stream $\mathbf{v}$ and clock control sequence $\mathbf{c}$, let $\mathbf{z} = Q(\mathbf{c}, \mathbf{v})$ be the function that generates $\mathbf{z}$ of length M by*

$$Q(\mathbf{c}, \mathbf{v}) : z_t \leftarrow v_{k(t)}, 0 \leq t < M$$

where $k(t) = \sum_{j=0}^t c_j - 1$.

If $a_j \geq 1$, $1 \leq j \leq A$, the function $Q(\mathbf{c}, \mathbf{v})$ can be considered as a deletion channel with input $\mathbf{v}$ and output $\mathbf{z}$. The deletion rate is

$$P_d = 1 - \frac{1}{\sum_{j=1}^A p_j a_j}. \tag{5}$$

The $D()$ function described above can in this model be among others the shrinking generator, the step-1/step-2 generator and the stop and go generator. Next we define the (not complete) reverse of Definition 1.

Definition 2. *Given the clock control sequence $\mathbf{c}$ and keystream $\mathbf{z}$, let the function $\mathbf{v}^* = Q^*(\mathbf{c}, \mathbf{z})$ be the (not complete) reverse of Q, defined as*

$$Q^*(\mathbf{c}, \mathbf{z}) : v_{k(t)}^* \leftarrow z_t, 0 \leq t < M,$$

where $k(t) = \sum_{j=0}^t c_j - 1$, and $v_k = *$ for the entries k in $\mathbf{v}^*$ where v_k^* is deleted. When this occurs we say that v_k^* is not defined.

The length of $\mathbf{v}^*$ will be $N^* = \sum_{j=0}^{M-1} c_j$. Given a stream $\mathbf{z}$ of length M, the expected length N of the stream $\mathbf{v}$ is

$$\text{E}(N) = \frac{M}{(1 - P_d)} = M \sum_{j=1}^A p_j a_j. \tag{6}$$

Note that the only difference between this definition and Definition 1, is that $\mathbf{v}$ and $\mathbf{z}$ have switched sides. Thus $Q^*(\mathbf{c}, \mathbf{z})$ is a reverse of $Q(\mathbf{c}, \mathbf{v})$. But since some bits are deleted, the reverse is not complete and we get the stream $\mathbf{v}^*$.

The probability for a bit v_k^* being defined is $\text{Prob}(v_k^*) = 1 - P_d$. This happens when $k = k(t)$ holds for some t, $0 \le t < M$. It follows that the sum $v_k^* + v_{k+j_1}^* + \dots + v_{k+j_{w-1}}^*$ will be defined if and only if all of the bits in the sum are defined. Thus the sum will be defined for given k in $\mathbf{v}^*$ with probability

$$P_{\text{def}} = (1 - P_d)^w. \tag{7}$$

4   The Attack

4.1   Equations of Weight 4

To succeed with our attack we need to find exactly one weight 4 equation

$$\lambda_u : u_k + u_{k+j_1} + u_{k+j_2} + u_{k+j_3} = 0 \tag{8}$$

that holds over all $\mathbf{u}$ generated by LFSR_u for $k \ge 0$. This corresponds to finding a multiple $h(x) = a(x)g_u(x)$ of weight 4. There exist several algorithms for finding such a multiple, see among others [13, 2, 5, 17, 12].

In this paper we use the fast search algorithm in [12, 11], which is a modified version of the David Wagner's Generalized Birthday Algorithm[17]. If the stream $\mathbf{u}$ has length N, this algorithm has runtime complexity $O(N \log N)$ and memory complexity $O(N)$, where N is of order $2^{l_u/3}$. The algorithm is effective in practice, and we have succeeded in finding multiples of the generator polynomial of high degree, see Section 6.3 for an example. We refer to Appendix C in [11] for the details for this search algorithm.

Next, we let the input vector $\mathbf{x}_k$ to the Boolean function $f(\mathbf{x})$ be

$$\mathbf{x}_k = (u_{k+i_0}, u_{k+i_1}, \dots, u_{k+i_{n-1}}), \tag{9}$$

where $(i_0, i_1, \dots, i_{n-1})$ defines the tapping positions from the internal state $L_u^k(\mathbf{I}_u)$ of LFSR_u after k feedback shifts. Substituting the vector (9) into the Equation (8) we have that $\mathbf{x}_k + \mathbf{x}_{k+j_1} + \mathbf{x}_{k+j_2} + \mathbf{x}_{k+j_3} = \mathbf{0}$ always holds for $k \ge 0$. Since $v_k = f(\mathbf{x}_k)$ we have from Corollary 4 that the equation

$$\lambda_v : v_k + v_{k+j_1} + v_{k+j_2} + v_{k+j_3} \approx 0, \tag{10}$$

will hold for $k \ge 0$ with probability $p = \frac{1}{2} + \frac{\sum_{\mathbf{a} \in V} \hat{f}(\mathbf{a})^4}{2q^4}$.

Remark 1. In [8] the multiple of $g_u(x)$ of weight $w = 3$ is exploited to define an iterative decoding attack on regularly clocked LFSRs filtered by Boolean functions. The constrained system

$$\sum_{a=0}^{w-1} \mathbf{x}_{k+j_a} = \mathbf{0} \tag{11}$$

$$z_{k+j_a} = f(\mathbf{x}_{k+j_a}), \, 0 \le a < w$$

is analyzed. This system is similar to the one we use in this paper, but it is used differently. Since there are limited solutions to this system, the *a posteriori* probabilities for each of the *input bits* $(u_{k+j_a+i_0}, u_{k+j_a+i_1}, ..., u_{k+j_a+i_{n-1}})$ in x_{k+j_a} can be calculated. Then these probabilities are put into a Gallager like probabilistic decoding algorithm(SOJA) which outputs I_u. However the correlation property in Corollary 4 is neither identified or exploited in [8].

4.2 Naive Algorithm

Let $\hat{I}_s$ be a guess for the initialization state I_s. Given the keystream z of length M, we generate $\hat{c}_t = D(L_s^t(\hat{I}_s))$, $0 \le t \le M$ and $\hat{v}^* = Q^*(\hat{c}, z)$ of length $N \approx \sum_{t=0}^{M-1} \hat{c}_t^i$. Then we test if $\hat{v}^*$ is likely to have been generated by LFSR$_u$ using the following method.

Find m entries in $\hat{v}^*$ where the equation is defined. From this we get a set of m equations. We test the m equations, and let the metric for the guess be the number of equations that hold. When we have the correct guess for I_s we expect pm of the equations to hold, where p is calculated using Corollary 4. Thus, this is a maximum likelihood decoding algorithm.

The runtime complexity for the attack will be of order $2^{l_s} \cdot (m + N)$, since we have to generate the bit stream $\hat{v}^*$ of length N for each of the 2^{l_s} guesses. In a real attack, N will be a large number and the naive algorithm will have very high runtime complexity.

4.3 Some Observations

If we use the technique in the previous section the attack has the runtime $2^{l_s} \cdot (m + N)$. In [11, Sec. 3.3] two important observations were made that reduce the complexity down to $2^{l_s} \cdot m$. Since $N \gg m$, these observations will speed up the attack considerably. We start with an initial guess $I_s^0 = (1, 0, ..., 0)$ and let the i'th guess be the internal state of LFSR$_s$ after i feedback shifts, that is $I_s^i = L_s^i(I_s^0)$.

Let $c^i = (c_0^i, c_1^i, ..., c_{M-1}^i)$ be the i'th guess for the clock control sequence defined by $c_t^i = D(L_s^{i+t}(1, 0, ..., 0))$, $0 \le t < M$. Let $v^i = Q^*(c^i, z)$ be the corresponding guess for v^* of length $N_i = \sum_{t=0}^{M-1} c_t^i$. We can now give a iterative method for generating v^{i+1} from v^i.

Lemma 2. *We can transform v^i into $v^{i+1} = Q^*(c^{i+1}, z)$ using the following method: Delete the first c_0^i entries $(*, ..., *, z_0)$ in v^i, append the $c_{M-1}^{i+1} = c_M^i$ entries $(*, ..., *, z_M)$ at the end, and replace z_t with z_{t-1} for $1 \le t \le M$.*

Proof. See Appendix B.1 in [11]. $\qquad\square$

Lemma 2 shows that we can generate each v^i using just a few operations instead of N operations, when implemented properly (See Appendix A.1 for the implementation details). This gives a fast method for generating all possible guesses for v^* given a keystream z. But using this lemma we still have to search for m

entries in $\mathbf{v}^*$ where the equations are defined. Since on average we must search through $1/P_{\text{def}}$ entries in $\mathbf{v}^*$ per equation, we want to avoid this search. In the next theorem we show how this can be done. The theorem proves that we can reuse the equation set for $\mathbf{v}^i$ in $\mathbf{v}^{i+1}$.

Theorem 1. *If the sum*

$$v_k + v_{k+k_1} + \dots + v_{k+k_w-1} = z_t + z_{t+j_1} + \dots + z_{t+j_{w-1}} = \gamma_{\mathbf{z},t}$$

is defined over $\mathbf{v}^i$, *then the sum*

$$v_{k-c_0^i} + \dots + v_{k+k_w-1-c_0^i} = z_{t-1} + z_{t+j_1-1} \dots + z_{t+j_{w-1}-1} = \gamma_{\mathbf{z},t-1}$$

is defined over $\mathbf{v}^{i+1}$.

Proof. See Appendix B.2 in [11]. $\qquad$

The main result from this theorem is that the equation set defined over $\mathbf{v}^i$ will be defined over $\mathbf{v}^{i+1}$ when we shift the equations c_0^i entries to the left over $\mathbf{v}^{i+1}$. This means that we can just shift the equations one entry to the left over $\mathbf{z}$, and we will have a sum that is defined for the guess $\hat{\mathbf{I}}_s = D(L_s^{i+1}(1,0,\dots,0))$. Thus, the theorem shows that we can avoid a lot of computations if we let the i'th guess for the inner state of LFSR_s be $L_s^i(1,0,\dots,0)$.

Remark 2. To use the lemma and theorem above we do not put the *actual bit values* z_t and restore them to the position $k(t)$ in $\mathbf{v}^*$ given by $Q^*(\mathbf{c},\mathbf{z})$. Instead we store the *index* z_t (the pointer to the position t in $\mathbf{z}$) in $v_{k(t)}$. This means that $v_{k(t)}^*$ holds the position t, which the keystream bit z_t have in $\mathbf{z}$. But when we evaluate an equation we use the indices to put in the actual bit values.

4.4   An Efficient Algorithm

Assume we have found an equation $\lambda_{\mathbf{v}} : v_k + v_{k+j_1} + v_{k+j_2} + v_{k+j_3} \approx 0$. The equation holds over $\mathbf{v}$ with probability p calculated using Corollary 4. Let the first guess for the initialization state for $\mathbf{s}$ be $\mathbf{I}_s^0 = (1,0,0,\dots,0)$, generate $\mathbf{c}^0$ by $c_t^0 = D(L_s^t(1,0,\dots 0))$, $t < M$, and $\mathbf{v}^0 = Q^*(\mathbf{c}^0,\mathbf{z})$. Next we try to find m entries $(k_1,k_2,\dots,k_m)$ in $\mathbf{v}^0$ where the equation $\lambda_{\mathbf{v}}$ is defined. From this we get the equation set

$$
\begin{aligned}
v_{k_1}^0 + v_{k_1+j_1}^0 + v_{k_1+j_2}^0 + v_{k_1+j_3}^0 &\approx 0 \\
v_{k_2}^0 + v_{k_2+j_1}^0 + v_{k_2+j_2}^0 + v_{k_2+j_3}^0 &\approx 0 \\
&\vdots \qquad\qquad \vdots \\
v_{k_m}^0 + v_{k_m+j_1}^0 + v_{k_m+j_2}^0 + v_{k_m+j_3}^0 &\approx 0.
\end{aligned}
\tag{12}
$$

Since every $v_{k_x+j_y}$ in this equation set is defined in $\mathbf{v}^0$ and $z_t = v_{k(t)}$, we can replace $v_{k_x+j_y}$ with the corresponding bit z_{t_x} from the keystream $\mathbf{z}$. Thus, $\mathbf{v}^0$ is a sequence of pointers to $\mathbf{z}$ and we can write the equations over $\mathbf{z}$ as the equation set Ω :

$$
\begin{aligned}
z_{t_{1,1}} + z_{t_{1,2}} + z_{t_{1,3}} + z_{t_{1,4}} &\approx 0 \\
z_{t_{2,1}} + z_{t_{2,2}} + z_{t_{2,3}} + z_{t_{2,w}} &\approx 0 \\
&\vdots \qquad\quad \vdots \\
z_{t_{m,1}} + z_{t_{m,2}} + z_{t_{m,3}} + z_{t_{m,w}} &\approx 0.
\end{aligned}
\tag{13}
$$

We are now finished with the precomputation. Let $metric_{\mathrm{best}}$ be the number of equations in Ω that hold. We iterate as follows:

Input The keystream $\mathbf{z}$ of length M, the equation λ, the equation set Ω, the index sequence $\mathbf{v}^0$, the states $L^0(1,0,...0)$ and $L^M(1,0...,0)$, and let $i \leftarrow 0$.

1. Calculate $c_{M-1}^{i+1} = c_M^i = D(L_s^{M+i}(1,0,...,0))$.
2. Use Lemma 2 to generate $\mathbf{v}^{i+1} = Q^*(\mathbf{c}^{i+1}, \mathbf{z})$ and lower all indexes in the equation set Ω by one. Theorem 1 guarantees that the equations are defined over $\mathbf{v}^{i+1}$.
3. If the first equation in Ω gets a negative index, then remove the equation from Ω. Find a new index at the end of $\mathbf{v}^{i+1}$ where λ is defined, and add the new equation over $\mathbf{z}$ to Ω.
4. Calculate $metric$ as the number of equations in Ω that hold.
5. If $metric_{\mathrm{best}} > metric$, set $metric_{\mathrm{best}} \leftarrow metric$ and $\mathbf{I}_s^i = L_s^i(10,0,...0)$.
6. Set $i \leftarrow i+1$ and go to step 1.
7. Output $\mathbf{I}_s^i$ as the initialization state for LFSR$_s$.

Remark 3. The algorithm is presented this way to make it readable and to show the basic idea. To reach the complexity $O(2^{l_s} \cdot m)$ a few technical details on the implementation of the algorithm are needed. These details are given in Appendix A.

5   Theoretical Properties

5.1   Success Formula

We can let an (unusual) encoder be defined by removing the Boolean function from the cipher. Then we can use coding theory to evaluate the attack. Let the initialization state $\mathbf{I}_s$ for LFSR$_s$ define the information bits in such a system.

Let $\mathbf{y} = (y_0, y_1, ..., y_{M-1})$ be the (not filtered) irregular clocked stream from LFSR$_u$, that is $\mathbf{y} = Q(\mathbf{c}, \mathbf{u})$ and $c_t = D(L_s^t(\mathbf{I}_s))$. Then the bitstream $\mathbf{y}$ defines the codeword that is sent over a noisy channel. Let the keystream $\mathbf{z} = Q(\mathbf{c}, \mathbf{v})$ (the filtered version of $\mathbf{y}$) be the received codeword.

Assume we have the wrong guess for $\mathbf{I}_s$, then approximately $m/2$ of the equations in the set (13) will hold. Now assume we have have guessed the correct $\mathbf{I}_s$. According to the observation in Section 2.1 the equations in the set (13) will hold with probability $p = \frac{1}{2} + \sum_{\mathbf{a} \in V} \hat{f}(\mathbf{a})^4 / 2q^4$, independently of the initialization bits $\mathbf{I}_u$.

Let p define the channel 'noise'. The uncertainty is defined by $H(p) = -p \log p - (1-p) \log(1-p)$, and the channel capacity is given by $C(p) = 1 - H(p)$. We can approximate $C(p)$ with $C(p) \approx 2(p - \frac{1}{2})^2 / \ln 2$. Following Shannon's noisy coding theorem we can set up this bound for success.

Proposition 1. *The attack will succeed with probability $> \frac{1}{2}$ if the number of parity check equations m is*

$$m > m_0 = \frac{l_{\mathrm{s}}}{C(p)} \approx \frac{0.347 l_{\mathrm{s}}}{(p - \frac{1}{2})^2}$$

where $p \approx \frac{1}{2} + \sum_{y \in V} \hat{f}(y)^4 / 2q^4$ and $q = 2^n$, where n is the number of input bits in $f(\mathbf{x})$.

When m is close to $2 \cdot m_0$ we expect the probability for success to be close to 1, see [15]. The simulations of our algorithm show that if we set $m = 2.1 \cdot m_0$ the success rate is approximately 99%.

5.2   Keystream Length

If the generator polynomial $g_{\mathrm{u}}(x)$ has weight $w > 4$, we must find a multiple $h(x)$ of $g_{\mathrm{u}}(x)$ of weight 4 and a degree l_{h}. We need at least the $\mathbf{v}$ stream to be of length l_{h}. In addition, to find m entries in $\mathbf{v}$ where the equation is defined $\mathbf{v}$ must at least have length

$$N > l_{\mathrm{h}} + m / P_{\mathrm{def}}. \tag{14}$$

From the expectation (6) of N we get $E(M) = N(1 - P_{\mathrm{d}}) = (1 - P_{\mathrm{d}}) l_{\mathrm{h}} + m / (1 - P_{\mathrm{d}})^3$, which proves the following proposition:

Proposition 2. *Let an equation over $\mathbf{v}$ be defined by $h(x)$ of weight 4 and degree l_{h}. To obtain an equation set Ω of m equations over $\mathbf{z}$, the length of the $\mathbf{z}$ stream must be*

$$M > (1 - P_{\mathrm{d}}) l_{\mathrm{h}} + m / (1 - P_{\mathrm{d}})^3. \tag{15}$$

The keystream length M depends on the number of equations m, the deletion rate P_{d} and the degree l_{h} of $h(x)$. The degree l_{h} is then again highly dependent on the search algorithm we use to find $h(x)$. When we use the search algorithm in [11, 17] the degree l_{h} of $g_{\mathrm{h}}(x)$ will be of order $l_{\mathrm{h}} = 2^{(2+l_{\mathrm{u}})/3}$, which is close to the theoretical expected degree $2^{l_{\mathrm{u}}/(w-1)}$ [5] for $w = 4$.

5.3   Runtime Complexity

The runtime complexity for our attack is

$$O(2^{l_{\mathrm{s}}} \cdot m) = O\left(\frac{2^{l_{\mathrm{s}}} \cdot l_{\mathrm{s}}}{(p - \frac{1}{2})^2}\right) \tag{16}$$

parity check tests, where p is calculated using Corollary 4. Note that the runtime is independent of the length l_{u} of LFSR$_{\mathrm{u}}$.

5.4   Memory Complexity

If we implement the attack directly as described in Sections 4.3 and 4.4 the algorithm will need around $32N + 4 * 32m$ bits of computer memory. The reason for the $32N$ term is that $\mathbf{v}^i = z_0, *, *, z_1, z_2, ..., *, z_{M-1}$ of length N is a sequence of pointers of 32 bits. In appendix A.2 we show how we can store $\mathbf{v}^i$ using N memory bits without affecting the runtime complexity. The total amount of memory *bytes* needed is then

$$\frac{N}{8} + 16m \tag{17}$$

6   Simulations of the Attack

The LILI-128 cipher[16] is based on the general model we attack in this paper. To be able to compare our attack with previous attacks, we have tested the attack on this cipher.

6.1   The LILI-128 Cipher

In the LILI cipher the clock control generator is defined by

$$g_s(x) = x^{39} + x^{35} + x^{33} + x^{31} + x^{17} + x^{15} + x^{14} + x^2 + 1,$$

and $c_t = D(s_{t+12}, s_{t+20}) = 1 + s_{t+12} + 2s_{t+20}$. The data generator sub system is

$$g_u(x) = x^{89} + x^{83} + x^{80} + x^{55} + x^{53} + x^{42} + x^{39} + x + 1,$$

and $v_k = f(u_k, u_{k+1}, u_{k+3}, u_{k+7}, u_{k+12}, u_{k+20}, u_{k+30}, u_{k+44}, u_{k+65}, u_{k+80})$, defined by a Boolean table of size 1024. Further on we get $P_d = 0.6$, and $P_{\text{def}} = 0.0256$ for $w = 4$, and $p = 0.501862$. The number of keybits in the secret key $(\mathbf{I}_s, \mathbf{I}_u)$ is $39 + 89 = 128$.

6.2   Simulations

We have done the simulations on some versions of the LILI-128 cipher with LFSRs of different lengths to empirically verify the success formula in Section 5.1. See Table 2 for the simulations. Note that we use the full size LFSR$_u$ from the LILI cipher in the three attacks in the bottom of the table. For $l_s = 11$ and $p = 0.501862$ we get $m_0 = 1.1 \cdot 10^6$.

We have implemented the attack in C code using the Intel icc compiler on a Pentium IV processor. Using the full 32-bit capability and all the implementation tricks explained in Appendix A our implementation uses only approximately 7 cycles per parity check test. Hence the algorithm works fast in practice and will take $7 \cdot 2^{l_s} m$ processor cycles.

Each attack is run 100 times, and the table shows that the estimated success rate holds and that the algorithm is efficient.

Table 2. We have tested the attack on the LILI-128 Boolean function with $p = 0.501862$. Note that the runtime for finding $\mathbf{I_s}$ is independent of the length l_u of LFSR$_u$, and the length M of the keystream. The attack on a full LFSR$_u$ of length 89 and reduced LFSR$_s$ of length 11 took 12 seconds

l_s	l_u	Keystream length M	Successes out of 100	m	Runtime	$2^{l_s} \cdot m$
11	60	$2^{24,1}$	59	m_0	6 sec.	2^{31}
11	60	$2^{25,1}$	100	$2.2 \cdot m_0$	13 sec.	2^{32}
11	40	$2^{24,0}$	51	m_0	6 sec.	2^{31}
11	40	$2^{25,0}$	100	$2.2 \cdot m_0$	13 sec.	2^{32}
10	89	2^{29}	99	$2.1 \cdot m_0$	6 sec	2^{32}
11	89	2^{29}	99	$2.1 \cdot m_0$	12 sec	2^{33}
12	89	2^{29}	99	$2.1 \cdot m_0$	24 sec	2^{34}

6.3   A Complete Attack on LILI-128

Preprocessing. For the LILI cipher, we have found a multiple $h(x) = a(x)g_u(x)$ which corresponds to the recursion $u_t + u_{t+139501803} + u_{t+210123252} + u_{t+1243366916} = 0$ and we have that

$$\text{Prob}(v_t + v_{t+139501803} + v_{t+210123252} + v_{t+1243366916} = 0) = 0.501862. \quad (18)$$

This precomputation took only 5 hours and 40 Gbyte hard disk space. We see that $l_h = 1243366916$.

Finding $\mathbf{I_s}$. We have $p = 0.501862$, and $m_0 = 39/C(0.501862) \approx 3.9 \cdot 10^6 \approx 2^{21.9}$. To be almost sure to succeed we use $m = 2.1m_0$ equations. Hence, the runtime for attacking LILI-128 is

$$2^{39} \cdot 2^{23} = 2^{62}$$

parity checks. Using our implementation this corresponds to $2^{62} \cdot 7$ processor cycles. Using Proposition 2 with $P_d = 0.6$ we need a keystream of length $M \approx 2^{29}$. The attack needs about 290 Mbyte of RAM. It can easily be parallelized and distributed among processors with virtually no overhead, since there is no need for communcation between the processor, and no need for shared memory. If we have 1024 Pentium IV 2.53 GHz processors, each having access to about 290 MB of memory, the attack would take about 4.5 months using 68 Mbyte of keystream data.

Finding $\mathbf{I_u}$ when $\mathbf{I_s}$ is known. Our attack only finds the initialization bits $\mathbf{I_s}$ for LFSR$_s$. It is possible to combine the Quick Metric from [12] with the previous attack against LILI in [7] to find $\mathbf{I_u}$ when $\mathbf{I_s}$ is given. Since this is not the scope of this paper we will not go into details, and we refer to [7, 12] for the exact description. The preprosessing stage will have complexity of order $2^{44.7}$ memory lookups, and runtime complexity of order $2^{42.5}$ parity checks. The complexity for the method above is much lower than the complexity for finding I_s and will therefore have little effect on the overall runtime for a full attack.

7 Conclusion

We have proposed a new key recovery correlation attack on irregular clocked keystream generators where the stream is filtered by a nonlinear Boolean function. Our attack uses a correlation property of Boolean functions, that gives higher correlation than previous methods. Thus we need fewer equations to succeed. The property holds even if the function is correlation immune. Using this property together with the iteration techniques from [11] we get a low runtime and low memory complexity algorithm for attacking the model. The algorithm outputs the initialization bits I_s for LFSR$_s$. Knowing I_s there exist previous algorithms which can determine I_u efficiently.

Acknowledgment

We would like to thank Matthew Parker, John Erik Mathiassen and the anonymous referees for many helpful comments.

References

1. V. Chepyzhov, T. Johansson, and B. Smeets. A simple algorithm for fast correlation attacks on stream ciphers. In *Fast Software Encryption, FSE 2000*, volume 1978 of *Lecture Notes in Computer Science*, pages 181–195. Springer-Verlag, 2001.
2. Philippe Chose, Antoine Joux, and Michel Mitton. Fast correlation attacks: An algorithmic point of view. In *Advances in Cryptology - EUROCRYPT 2002*, volume 2332 of *Lecture Notes in Computer Science*, pages 209–221. Springer-Verlag, 2002.
3. Nicolas Courtois. Fast algebraic attacks on stream ciphers with linear feedback. In *Advances in Cryptology-CRYPTO' 2003*, volume 2729 of *Lecture Notes in Computer Science*, pages 176–194, 2003.
4. Nicolas Courtois and Willi Meier. Algebraic attacks on stream ciphers with linear feedback. In *Advances in Cryptology - EUROCRYPT 2003*, volume 2656 of *Lecture Notes in Computer Science*, pages 345–359, 2003.
5. J.D Golić. Computation of low-weight parity-check polynomials. *Electronic Letters*, october 1996. 32(21):1981-1982.
6. T. Johansson and F. Jönsson. Theoretical analysis of a correlation attack based on convolutional codes. In *Proceedings of 2000 IEEE International Symposium on Information Theory*, IEEE Transaction on Information Theory, page 212, 2000.
7. Fredrik Jönsson and Thomas Johansson. A fast correlation attack on LILI-128. In *Inf. Process. Lett. 81(3)*, pages 127–132, 2002.
8. Sabine Leveiller, Gilles Zémor, Philippe Guillot, and Joseph Boutros. A new cryptanalytic attack for pn-generators filtered by a boolean function. In *Selected Areas in Cryptography: 9th Annual International Workshop, SAC 2002*, volume 2595 of *Lecture Notes in Computer Science*, pages 232–249. Springer-Verlag, 2003.

9. M. Matsui. Linear cryptanalysis method for DES cipher. In *Advances in Cryptology-EUROCRYPT'93*, volume 765 of *Lecture Notes in Computer Science*, pages 386–397. Springer-Verlag, 1994.
10. W. Meier and O. Staffelbach. Fast correlation attacks on stream ciphers. In *Advances in Cryptology-EUROCRYPT'88*, volume 330 of *Lecture Notes in Computer Science*, pages 301–314. Springer-Verlag, 1998.
11. Håvard Molland. Improved linear consistency attack on irregular clocked keystream generators. In *Fast Software Encryption, FSE 2004*, To appear in LNCS. Springer-Verlag, 2004. Available at http://www.ii.uib.no/~molland/crypto
12. Håvard Molland, John Erik Mathiassen, and Tor Helleseth. Improved fast correlation attack using low rate codes. In *Cryptography and Coding, IMA 2003*, volume 2898 of *Lecture Notes in Computer Science*, pages 67–81, 2003.
13. W.T. Penzhorn and G.J Kuhn. Computation of low-weight parity checks for correlation attacks on stream ciphers. In *Cryptography and Coding, IMA 1995*, volume 1025 of *Lecture Notes in Computer Science*, pages 74–83. Springer-Verlag, 1995.
14. Markku-Juhani Olavi Saarinen. A time-memory tradeoff attack against LILI-128. In *Fast Software Encryption, FSE 2002*, volume 2365 of *Lecture Notes in Computer Science*, pages 231–236, 2002.
15. T. Siegenthaler. Decrypting a class of stream ciphers using ciphertext only. *IEEE Trans. on Comp.*, C-34:81–85, 1985.
16. L. Simpson, E. Dawson, J. Golić, and W. Millan. LILI keystream generator. In *SAC'2000*, volume 2012 of *Lecture Notes in Computer Science*, pages 231–236. Springer-Verlag, 2002. Available at http://www.isrc.qut.edu.au/lili.
17. D. Wagner. A generalized birthday problem. In *Advances in cryptology-CRYPTO' 2002*, volume 2442 of *Lecture Notes in Computer Science*, pages 288–303, 2002.
18. Xian-Mo Zhang and Yuliang Zheng. The nonhomomorphicity of boolean functions. In *Selected Areas in Cryptography, SAC 98*, volume 1556 of *Lecture Notes in Computer Science*, pages 280–295. Springer-Verlag, 1998.

Appendix

A   Implementation Details

To reach the runtime complexity $O(2^{l_s} \cdot m)$ and memory complexity down to $N + 128m$ bits, the implementation of the algorithm has some tricks. Since not all of these tricks are obvious we give more detailed descriptions of them below.

A.1   Runtime Details

Sliding window. In Lemma 2 we get $\mathbf{v}^{i+1}$ by among other things deleting the c_0^i first bits of $\mathbf{v}^i$. This is done using the sliding window technique, which means that we move the viewing to the right instead of shifting the whole sequence to the left. This way the shifting can be done in just a couple of operations. To avoid heavy use of memory, we slide the window over an array of fixed length N, so that the entries that become free at the beginning of the array are reused. Thus, the left and right indexes of the sliding window after i iterations will be

$$(left, right) = (i \bmod N, i + N_i \bmod N),$$

where $N > N_i$, for all i, $0 \le i < 2^{l_s}$.

The same sliding window technique is also used on the equation set when equations are deleted and added to the equation set.

Updating the indices. In Lemma 2 every pointer z_{t+1} in $\mathbf{v}^*$ is replaced with z_t for every $0 \le t \le M$, which would take M operations. If we skip the replacements we note that after i iterations the entry z_t in $\mathbf{v}^*$ will become z_{t+i}. It is also important to note that when we write $\mathbf{v} = (..., z_0..., z_t, ..., z_M, ...)$, the entries $z_0, ..., z_t, ..., z_M$ are pointers from $\mathbf{v}^*$ to $\mathbf{z}$. They are not the actual key bits. Thus, in the implementation we do not replace z_t with z_{t-1}. But when we after i iterations in the search for equations find an equation $v_k^i + v_{k+j_1}^i + ... + v_{k+j_{w-1}}^i = 0$ that is defined, we replace the corresponding equation $z_{t_1} + z_{t_2} + ... + z_{t_w}$ with $z_{t_1-i} + z_{t_2-i} + ... + z_{t_w-i}$, to compensate.

Reducing the memory access time. When we test an equation we must use pointers to pointers to the keystream. Then each equation test will have high memory access time. We can reduce this significantly by testing the equations on 32 states simultaneously. This is possible since the next state $\mathbf{I}_s^{i+1}$ is tested by shifting all the equations one entry to the left over $\mathbf{z}$. We can now take the bits $z_{t_a}, z_{t_a+1}, ..., z_{t_a+31}$ for each of the term $1 < a \le 4$ in the equations and put them into 32 bit registers. Now we can test the states and add one to the metrics of the states that satisfy the equation. This speeds up the runtime by a factor of approximately 20.

A.2   Memory Details

Reducing the use of memory. Instead of storing all the pointers, we set 1 in $\mathbf{v}^i$ where the bits are defined and 0 otherwise. When we search in $\mathbf{v}^i$ to find entries where the equation $\lambda_{\mathbf{v}}$ is defined, we keep track of where in $\mathbf{z}$ the four terms in $\lambda_{\mathbf{v}}$ points to by counting the number of 1's we pass during the search. This is done for each of the 4 terms in the equation $\lambda_{\mathbf{v}}$. This way we always know where in $\mathbf{z}$ the given equation of $\mathbf{v}^i$ points to. Using this trick the number of memory bits needed during an attack is reduced from $32N + 128m$ bits to

$$N + 128m$$

Implementing this trick will not affect the runtime of the attack.

Rewriting Variables: The Complexity of Fast Algebraic Attacks on Stream Ciphers

Philip Hawkes[1] and Gregory G. Rose[1]

Qualcomm Australia, Level 3, 230 Victoria Rd, Gladesville, NSW 2111, Australia
{phawkes,ggr}@qualcomm.com

Abstract. Recently proposed algebraic attacks [2, 6] and fast algebraic attacks [1, 5] have provided the best analyses against some deployed LFSR-based ciphers. The process complexity is exponential in the degree of the equations. Fast algebraic attacks were introduced [5] as a way of reducing run-time complexity by reducing the degree of the system of equations. Previous reports on fast algebraic attacks [1, 5] have underestimated the complexity of substituting the keystream into the system of equations, which in some cases dominates the attack. We also show how the Fast Fourier Transform (FFT) [4] can be applied to decrease the complexity of the substitution step. Finally, it is shown that all functions of degree d satisfy a common, function-independent linear combination that may be used in the pre-computation step of the fast algebraic attack. An explicit factorization of the corresponding characteristic polynomial yields the fastest known method for performing the pre-computation step.

1 Introduction

Many popular stream ciphers are based on linear feedback shift registers (LF-SRs) [11]. Such ciphers include *E0* [3], LILI-128 [12] and Toyocrypt(see [10]). They consist of a memory register called the state that is updated (changed) every time a keystream output is produced, and an additional device, called the nonlinear combiner. The nonlinear combiner computes a keystream output as a function of the current LFSR state[1]. The sequence of states produced by an LFSR depends on the initial state of LFSR, which is always presumed to be secret. Since recovering this initial state allows prediction of unknown keystream, we follow the convention of [5] and call it **K** as if it was actually the key. Most practical stream ciphers initialize this state from the real key and a nonce. The advantages of LFSRs are many. LFSRs can be constructed very efficiently in hardware and some recent designs are also very efficient in software. LFSRs can be chosen such that the produced sequence has a high period and good statistical properties.

---

[1] Some LFSR-based stream ciphers have a non-linear filter that maintains some bits of memory, but research has shown that such ciphers can be analyzed in the same way as ciphers without memory. Some designs use multiple LFSRs, but again these are usually equivalent to a single LFSR. Some modern stream ciphers use units larger than bits, but this discussion applies equally to such ciphers, so we will talk only in terms of bits.

M. Franklin (Ed.): CRYPTO 2004, LNCS 3152, pp. 390–406, 2004.

While there are many approaches to the cryptanalysis of LFSR-based stream ciphers, this paper is concerned primarily with the recently proposed algebraic attacks [2,6] and fast algebraic attacks [1,5]. Such attacks have provided the best analyses against some theoretical and deployed ciphers.

An algebraic attack consists of three steps. The first step is to find a system of algebraic equations that relate the bits of the initial state $\mathbf{K}$ and bits of the keystream $\mathbf{Z} = \{z_t\}$. Some methods [2,6] have been proposed for finding "localized" equations (where the keystream bits are in a small range $z_t, \ldots, z_{t+\theta}$). This first step is a pre-computation: the attacker must compute these equations before attacking a key-stream. Furthermore, the computation need only be performed once, and the attacker can use the same equations for attacking multiple key-streams. The second and third steps are performed after the attacker has observed some keystream. In the second step, the observed keystream bits are substituted into the algebraic equations (from the first step) to obtain a system of algebraic equations in the bits of $\mathbf{K}$. The third step is to solve these algebraic equations to determine $\mathbf{K}$. This will be possible if the equations are of low degree in the bits of $\mathbf{K}$, and a sufficient number of equations can be obtained from the observed keystream.

The process complexity of the third step is exponential in the degree of the equations. Fast algebraic attacks were introduced by Courtois at Crypto 2003 [5] as a way of reducing run-time complexity by reducing the degree of the system of equations. This method requires an additional pre-computation step; this step determines a linear combination of equations in the initial system that cancels out terms of high degree (provided the algebraic equations are of a special form). This yields a second system of equations relating $\mathbf{K}$ and the keystream $\mathbf{Z}$ that contains only terms of low degree. In the second step, the appropriate keystream values are now substituted into this second system to obtain a new system of algebraic equations in the bits of $\mathbf{K}$. Solving the new system (in the third step) is easier than solving the old system because the new system contains only terms of low degree.

Courtois [5] proposes using a method based on the Berlekamp-Massey algorithm [8] for determining the linear combination obtained in the additional pre-computation step. The normal Berlekamp-Massey algorithm has a complexity of D^2, while an asymptotically-fast implementation has a complexity of $C \cdot D(\log D)$ for some large constant C. It is unclear which method would be best for the size of D considered in these attacks. Armknecht [1] provides a method for improving the complexity when the cipher consists of multiple LFSRs.

Contributions of this paper. The first contribution is to note that previous reports on fast algebraic attacks (such as [1,5]) appear to have underestimated the complexity of substituting the keystream into the second system of equations[2]. The complexity was originally underestimated as only $O(DE)$ [5], where D is the size of the linear combination and E is the size of the second system

---

[2] We are aware (via private communication) of other proposed algebraic attacks in which the substitution complexities were initially ignored. In one case, the complexity of simple substitution was almost the square of the complexity of solving the system.

of equations. Table 1 lists the values of $O(DE)$ for previously published attacks from [1, 5]. However, simple substitution would require a complexity of $DE^2/2$ (see Section 2.3), and no other method was suggested for reducing the complexity. It is true that E bitwise operations of the substitution can be performed in parallel, reducing the time complexity to $DE^2/2$, but in cases where E is large, the process complexity should still be considered $DE^2/2$ in the absence of specialized hardware. In many cases $DE^2/2$ actually exceeds the complexity of solving the system of equations, as shown in Table 1[3]. The second contribution of this paper is to show how the Fast Fourier Transform (FFT) [4] can be applied to decrease the complexity of the substitution step to $2ED\log_2 D$. The resulting complexities of the FFT approach are also listed in Table 1.

Table 1. Comparison of substitution complexities for published fast algebraic attacks.

Cipher	D	E	Data req	Substitution Claimed (wrong)	Simple	FFT	Solving System	Total Process Complexity
E0	2^{23}	2^{18}	2^{24}	2^{41}	2^{59}	2^{47}	2^{49}	2^{49}
LILI-128	2^{21}	2^{12}	2^{60}	2^{33}	2^{44}	2^{39}	2^{39}	2^{40}
Toyocrypt	2^{18}	2^7	2^{18}	2^{23}	2^{31}	2^{30}	2^{20}	2^{30}

The final contribution of this paper is to provide an efficient method for determining the linear combination obtained in the additional pre-computation step of the fast algebraic attack. First, we make the observation that all functions of degree d satisfy a common function-independent linear combination of length $D = \sum_{i=0}^{d} \binom{n}{d}$ that is defined exclusively by the LFSR. Then we provide a direct method for computing this linear combination (based on the work of Key [7]). This method requires $c \cdot D(n(\log n)^2 + (\log_2 D)^3)$ operations for small constant c. This is a significant improvement on the complexities of previous methods.

Table 2. The complexities of the pre-computation step for published attacks, where C represents a large constant and c represents a small constant.

Cipher	D	Courtois [5]: based on Berlekamp-Massey $C \cdot D(\log D)$	D^2	Armknecht [1] Parallel Method	Direction Computation $c \cdot D(n(\log n)^2 + (\log_2 D)^3)$
E0	2^{23}	$C \cdot 2^{28}$	2^{46}	2^{43}	2^{37}
LILI-128	2^{21}	$C \cdot 2^{26}$	2^{42}	-	2^{35}
Toyocrypt	2^{18}	$C \cdot 2^{23}$	2^{36}	-	2^{32}

This paper is organized as follows: Section 2 describes fast algebraic attacks. In Section 3 we discuss the complexity of substitution step for fast algebraic attacks. Section 4 reviews the Fast Fourier Transform and Section 5 describes

---

[3] The attack on LILI-128 requires only every $(2^{39} - 1)$-st bit from a keystream of length 2^{60}. The process complexity of selecting these bits is ignored in the literature, and could be an area for useful discussion.

how the FFT can speed up the substitution step. Section 6 contains some observations on the pre-computation step. Section 7 concludes the paper.

2 Fast Algebraic Attacks

The length of the LFSR is n-bits; that is the internal state of the LFSR is $\mathbf{K}_t \in GF(2)^n$. A state $\mathbf{K}_{t+1}$ is derived from the previous state $\mathbf{K}_t$ by applying an (invertible) linear mapping $L : GF(2)^n \to GF(2)^n$, with $\mathbf{K}_{t+1} = L(\mathbf{K}_t)$. The function L can be represented by an $n \times n$ matrix over $GF(2)$, which is called the *state update matrix*. Notice that we can write $\mathbf{K}_t = L^t(\mathbf{K})$. Each keystream bit is generated by first updating the LFSR state (by applying L) and then applying a Boolean function to the bits of the LFSR state. For the purposes of this paper, everything about the cipher is presumed to be known to the attacker, except the initial state of the LFSR and any subsequent state derived from it.

Linearization: Recall that the first two steps of the attack result in a system of nonlinear algebraic equations in a small number of unknown variables (these variables being the bits of the initial state). The most successful algebraic attacks (to date), have been based on linearization. The basis of this technique is to "linearize" a system of nonlinear algebraic equations by assigning a new unknown variable to each monomial term that appears in the system. The same monomial term appearing in distinct equations is assigned the same new unknown variable. The system of equations then changes from a system of non-linear equations (with few unknown variables) into a system of linear equations (with a large number of unknown variables). If the number of linear equations exceeds the number of new unknown variables, then an attacker can solve the system to obtain the new unknown variables of the linear system (which will in turn reveal the unknown variables of the non-linear system). The advantage of linearization is that the attacker can use the large body of knowledge about the solution of linear systems.

2.1 The Monomial State

This section introduces some notation that is useful for describing linearization. For a given value of the state $\mathbf{K}_t$ and for a given degree d, we shall let $\mathbf{M}_d(t)$ (the *monomial state*) denote the $GF(2)$ column vector with each component being a corresponding monomial of degree d or less. The number of such monomials is $D = \sum_{i=0}^{d} \binom{n}{i} \approx \binom{n}{d}$, so $\mathbf{M}_d(t)$ contains D components. The initial monomial state $\mathbf{M}_d$ corresponds to the initial state $\mathbf{K}$.

Example 1. If $n = 4$ (that is, $\mathbf{K}_t = (k_3, k_2, k_1, k_0)$) and $d = 2$, then there are $D = 11$ monomials of degree 2:

$$\mathbf{M}_d(t) = (m_0, m_1, m_2, m_3, m_4, m_5, m_6, m_7, m_8, m_9, m10)^T$$
$$= (1, k_0, k_1, k_2, k_3, k_0 k_1, k_0 k_2, k_0 k_3, k_1 k_2, k_1 k_3, k_2 k_3)^T,$$

where "T" denotes the transpose of the matrix to make a column vector. For $\mathbf{K}_t = (0, 1, 1, 1)$, the values of the monomial components of $\mathbf{M}_d(t)$ are:

$$\mathbf{M}_d(t) = (1, k_0 = 1, k_1 = 1, k_2 = 1, \ldots, k_1 k_2 = 1, k_1 k_3 = 0, k_2 k_3 = 0)^T$$
$$= (1, 1, 1, 1, 0, 1, 1, 0, 1, 0, 0)^T.$$

The ordering of the monomial components is arbitrary; for consistency we will enumerate using lower subscripts first, as shown.   □

Expressing Functions of the LFSR State. We can express any Boolean function of the LFSR state as a product of the matrix $\mathbf{M}_d(t)$ with a row vector.

Example 2. Consider a Boolean function of the state $f(\mathbf{K}_t) = k_2 + k_1 k_3$ (using the LFSR state from Example 1). This function can be expressed:

$$f(\mathbf{K}_t) = k_2 + k_1 k_3$$
$$= 0 \times 1 + 0 \times k_0 + 0 \times k_1 + 1 \times k_2 + 0 \times k_3 + 0 \times k_0 k_1$$
$$+ 0 \times k_0 k_2 + 0 \times k_0 k_3 + 0 \times k_1 k_2 + 1 \times k_1 k_3 + 0 \times k_2 k_3$$
$$= (0, 0, 0, 1, 0, 0, 0, 0, 0, 1, 0) \cdot \mathbf{M}_d(t) = \mathbf{f} \cdot \mathbf{M}_d(t), \tag{1}$$

where the addition and multiplication operations are performed in $GF(2)$. We have now expressed the Boolean function $f(\mathbf{K}_t)$ as the product of the matrix $\mathbf{M}_d(t)$ with a row vector

$$\mathbf{f} = (f_0, \ldots, f_{D-1}) = (0, 0, 0, 1, 0, 0, 0, 0, 0, 1, 0)$$

that selects the values of the specific monomials required to evaluate $f(\mathbf{K}_t)$.   □

The row vector $\mathbf{f}$ depends only on the function f, and is independent of the LFSR feedback polynomial, the value of the initial state, and the index t.

The Monomial State Rewriting Matrix. The mapping from one LFSR state to the next LFSR state can be expressed as a matrix product $\mathbf{K}_{t+1} = L \cdot \mathbf{K}_t$. It is also possible to determine the mapping from one monomial state to the next monomial state as a matrix product $\mathbf{M}_d(t + 1) = \mathbf{R}_d \cdot \mathbf{M}_d(t)$.

Example 3. Consider a 4-bit LFSR as in Example 1 with monomial state $\mathbf{M}_d(t) = (1, k_0, k_1, k_2, k_3, k_0 k_1, k_0 k_2, k_0 k_3, k_1 k_2, k_1 k_3, k_2 k_3)^T$. If the LFSR has is of the form $s_{t+4} = s_{t+1} + s_t$, then the next state $\mathbf{K}_{t+1}$ has a corresponding next monomial state $\mathbf{M}_d(t + 1) = (m'_0, m'_1, m'_2, m'_3, m'_4, m'_5, m'_6, m'_7, m'_8, m'_9, m'_{10})^T$ which is related to the original monomial state as follows (only some relationships have been shown in order to save space):

$$
\begin{aligned}
m'_0 &= 1 & &= 1 & &= m_0 & &= (1, 0, 0, 0, 0, 0, 0, 0, 0, 0, 0) \cdot \mathbf{M}_d(t), \\
m'_1 &= k'_0 & &= k_1 & &= m_2 & &= (0, 0, 1, 0, 0, 0, 0, 0, 0, 0, 0) \cdot \mathbf{M}_d(t), \\
m'_4 &= k'_3 & &= k_0 + k_1 & &= m_1 + m_2 & &= (0, 1, 1, 0, 0, 0, 0, 0, 0, 0, 0) \cdot \mathbf{M}_d(t), \\
m'_{10} &= k'_2 k'_3 & &= k_3(k_0 + k_1) & &= m_7 + m_9 & &= (0, 0, 0, 0, 0, 0, 0, 1, 0, 1, 0) \cdot \mathbf{M}_d(t).
\end{aligned}
$$

Each component of the next monomial state is a linear function of the original monomial state. These linear functions for $m'_0, \ldots, m'_{D-1}$ can be combined into a matrix $\mathbf{R}_d$ (the "rewriting matrix") such that $\mathbf{M}_d(t+1) = \mathbf{R}_d \cdot \mathbf{M}_d(t)$.

$$
\mathbf{R}_d =
\begin{bmatrix}
1 & 0 & 0 & 0 & 0 & 0 & 0 & 0 & 0 & 0 & 0 \\
0 & 0 & 1 & 0 & 0 & 0 & 0 & 0 & 0 & 0 & 0 \\
0 & 0 & 0 & 1 & 0 & 0 & 0 & 0 & 0 & 0 & 0 \\
0 & 0 & 0 & 0 & 1 & 0 & 0 & 0 & 0 & 0 & 0 \\
0 & 1 & 1 & 0 & 0 & 0 & 0 & 0 & 0 & 0 & 0 \\
0 & 0 & 0 & 0 & 0 & 0 & 0 & 0 & 1 & 0 & 0 \\
0 & 0 & 0 & 0 & 0 & 0 & 0 & 0 & 0 & 1 & 0 \\
0 & 0 & 1 & 0 & 0 & 1 & 0 & 0 & 0 & 0 & 0 \\
0 & 0 & 0 & 0 & 0 & 0 & 0 & 0 & 0 & 0 & 1 \\
0 & 0 & 0 & 0 & 0 & 0 & 1 & 0 & 1 & 0 & 0 \\
0 & 0 & 0 & 0 & 0 & 0 & 0 & 1 & 0 & 1 & 0
\end{bmatrix}
$$

$\square$

Notice that the matrix $\mathbf{R}_d$ depends only on the LFSR and the degree d. This example generalizes: for every LFSR and degree d there is a "monomial state rewriting matrix" $\mathbf{R}_d$ such that $\mathbf{M}_d(t+1) = \mathbf{R}_d \cdot \mathbf{M}_d(t)$. Moreover, for every t, the monomial state after t clocks of the LFSR can be expressed as a $GF(2)$ matrix operation

$$
\mathbf{M}_d(t) = \mathbf{R}_d^t \cdot \mathbf{M}_d, \tag{2}
$$

where $\mathbf{M}_d$ is the initial monomial state. Combining equations (1) and (2), we get another expression for $f(\mathbf{K}_t)$:

$$
f(\mathbf{K}_t) = f(\mathbf{R}_d^t \cdot \mathbf{M}_d) = (\mathbf{f}(t) \cdot \mathbf{R}_d^t) \cdot \mathbf{M}_d = \mathbf{f}(t) \cdot \mathbf{M}_d, \tag{3}
$$

where the vector $\mathbf{f}(t) \stackrel{\text{def}}{=} \mathbf{f} \cdot \mathbf{R}_d^t$ depends solely on the function f, the monomial state update matrix $\mathbf{R}_d$ and the number of clocks t (all of which are known to the attacker). For example, the vectors $\mathbf{f}(t)$, $t \in \{0,1,2\}$ corresponding to the function f in Example 2 are:

$$
\begin{aligned}
\mathbf{f}(0) &= \mathbf{f} \cdot \mathbf{R}_d^0 = \mathbf{f} \cdot \mathbf{I} &&= (0,0,0,1,0,0,0,0,0,1,0), \\
\mathbf{f}(1) &= \mathbf{f} \cdot \mathbf{R}_d^1 = \mathbf{f}(0) \cdot \mathbf{R}_d &&= (0,0,0,0,1,0,1,0,1,0,0), \\
\mathbf{f}(2) &= \mathbf{f} \cdot \mathbf{R}_d^2 = \mathbf{f}(1) \cdot \mathbf{R}_d &&= (0,1,1,0,0,0,0,0,0,1,1).
\end{aligned}
$$

2.2   Algebraic Attacks

We always assume that the monomial state is unknown; it is the goal of algebraic attacks to determine the initial monomial state $\mathbf{M}_d$ (and thereby determine the initial LFSR state).

Step 1. The first step in an algebraic attack is to find a Boolean function h such that the equation

$$
h(\mathbf{K}_t, z_t, \ldots, z_{t+\theta}) = 0, \tag{4}
$$

is true for all clocks (or indices) t. The degree of h with respect to the bits of $\mathbf{K}_t$ we shall denote by d. Various methods have been proposed for finding such equations (see [2, 6]). These equations typically have small values for θ. For simplicity we shall hereafter combine the keystream bit values $z_t, \ldots, z_{t+\theta}$ into a keystream vector $\mathbf{z}_t$.

For the linearization approach, it is convenient to obtain an expression for $h(\mathbf{K}_t, \mathbf{z}_t)$ in terms of keystream bits and bits of the initial monomial state $\mathbf{M}_d$.

1. Express $h(\mathbf{K}_t, \mathbf{z}_t)$ as the inner product of $\mathbf{M}_d(t)$ and a keystream-dependent vector $\mathbf{h}(\mathbf{z}_t)$:

$$h(\mathbf{K}_t, \mathbf{z}_t) = \mathbf{h}(\mathbf{z}_t) \cdot \mathbf{M}_d(t). \tag{5}$$

2. Now, Equation (2) can be substituted into Equation (5);

$$h(\mathbf{K}_t, \mathbf{z}_t) = \mathbf{h}(\mathbf{z}_t) \cdot (\mathbf{R}_d^t \cdot \mathbf{M}_d) = (\mathbf{h}(\mathbf{z}_t) \cdot \mathbf{R}_d^t) \cdot \mathbf{M}_d = \mathbf{h}(t) \cdot \mathbf{M}_d,$$

where $\mathbf{h}(t) \overset{\text{def}}{=} \mathbf{h}(\mathbf{z}_t) \cdot \mathbf{R}_d^t$.

3. Equation (4) is thereby transformed to the form:

$$\mathbf{h}(t) \cdot \mathbf{M}_d = 0. \tag{6}$$

The components of $\mathbf{h}(t) = (h_0(t), \ldots, h_{D-1}(t))$ depend on: (a) the function h; (b) the monomial state rewriting matrix $\mathbf{R}_d$ associated with the monomials of degree of degree d or less; (c) the number of clocks t; and (d) the small keystream vector $\mathbf{z}_t$. An attacker has access to all of this information, so the attacker is able to compute all of the components of $\mathbf{h}(t)$. This means that the only unknowns in Equation (6) are the components of the initial monomial state $\mathbf{M}_d$.

Step 2. The second step of an algebraic attack consists of substituting the observed keystream vector $\mathbf{z}_t$ into the components of the vectors $\mathbf{h}(\mathbf{z}_t)$ and then computing the vector $\mathbf{h}(t) = \mathbf{h}(\mathbf{z}_t) \cdot \mathbf{R}_d^t$. The vectors $\mathbf{h}(t)$ are evaluated for many indices t. Each of the evaluated vectors $\mathbf{h}(t)$ provides the attacker with a linear equation in the D unknown bits of the initial monomial state $\mathbf{M}_d$. Since there are D unknowns, around D linear equations will be required to obtain a solvable system. An initial choice of D equations may contain linearly dependent equations, so more than D equations may be required in order to get a completely solvable system. It is thought that not many more than D equations will be required in practice (see remark at the end of section 5.1 of [5]), so we will assume D equations are sufficient.

Step 3. The third step recovers $\mathbf{M}_d$ by solving the resulting system of linear equations. The system can be solved by Gaussian elimination or more efficient methods [13]. The complexity of solving such a system of equations is estimated to be $O(D^\omega)$, where ω (known as the Gaussian coefficient) is estimated to be $\omega = 2.7$. In general, D will be about $\binom{n}{d} = O(n^d)$.

Complexities. The complexities of an algebraic attack are as follows:

- The complexity of finding the equation $h(\mathbf{K}_t, \mathbf{z}_t)$ depends on many factors and is beyond the scope of this paper.
- The amount of keystream required for the second step (the data complexity) is $D = O(n^d)$.
- The complexity of the second step (substituting the keystream into the equations) is $O(D^2) = O(n^{2d})$, assuming that the functions $h(\mathbf{K}_t, \mathbf{z}_t)$ are relatively simple functions of the keystream; and
- the complexity of solving the system in the third step is $O(n^{2.7}d)$.

Note 1. The complexity is exponential in the degree d. Hence, a low degree d is required for an efficient attack. Therefore, an attacker using an algebraic attack will always try to find a system of low degree equations.

2.3 Fast Algebraic Attacks

Courtois [5] proposed "fast algebraic attacks", as a method for decreasing the degree of a given system of equations. For fast algebraic attacks, we presume that the function h can be written in the form

$$h(\mathbf{K}_t, z_t, \ldots, z_{t+\theta}) = u(\mathbf{K}_t) + v(\mathbf{K}_t, z_t) \;=\; 0, \tag{7}$$

where u is of degree d in the bits of $\mathbf{K}_t$, v is of degree $e < d$ in the bits of $\mathbf{K}_t$ and *only v depends on the keystream*. Since the functions u and v are of two distinct degrees (in the bits of $\mathbf{K}_t$), it is simplest to consider them as depending on distinct monomial states $\mathbf{M}_d$ and $\mathbf{M}_e$, with corresponding monomial state rewriting matrices $\mathbf{R}_d$ and $\mathbf{R}_e$. There are $D = \sum_{i=0}^{d} \binom{n}{i} \approx \binom{n}{d}$ monomials of degree d or less, and $E = \sum_{i=0}^{e} \binom{n}{i} \approx \binom{n}{e}$ monomials of degree e or less.

A fast algebraic attack gains an advantage over the normal algebraic attacks by including an additional pre-computation step in which the attacker determines linear combinations of equation (7) that will cancel out the high-degree monomials of degree $e+1, e+2, \ldots, d$ that occur in $u(\mathbf{K}_t)$, but not in $v(\mathbf{K}_t, \mathbf{z}_t)$. As in equation (3), $u(\mathbf{K}_t)$ and $v(\mathbf{K}_t, \mathbf{z}_t)$ are written as vector inner-products:

- $u(\mathbf{K}_t) = \mathbf{u} \cdot \mathbf{M}_d(t) = \mathbf{u}(t) \cdot \mathbf{M}_d$, where $\mathbf{u}(t) = (u \cdot \mathbf{R}_d^t)$ is a vector with D components (all of which are *independent* of the keystream); and
- $v(\mathbf{K}_t, \mathbf{z}_t) = \mathbf{v}(\mathbf{z}_t) \cdot \mathbf{M}_d(t) = \mathbf{v}(t) \cdot \mathbf{M}_e$, where $\mathbf{v}(t) = \mathbf{v}(\mathbf{z}_t) \cdot \mathbf{R}_e^t)$ is a vector with E components (some of which are *dependent* on the keystream).

Equation (7) is then transformed to:

$$\mathbf{u}(t) \cdot \mathbf{M}_d + \mathbf{v}(t) \cdot \mathbf{M}_e = 0. \tag{8}$$

In the fast algebraic attack pre-computation step, the attacker finds $(D + 1)$ coefficients $b_0, \ldots, b_D \in \{0, 1\}$, such that

$$\sum_{i=0}^{D} b_i \cdot \mathbf{u}(t + i) = 0, \; \forall t. \tag{9}$$

Equations (8) and (9) can be combined:

$$\sum_{i=0}^{D} b_i \cdot (\mathbf{u}(t+i) \cdot \mathbf{M}_d + \mathbf{v}(t+i) \cdot \mathbf{M}_e) = \left(\sum_{i=0}^{D} b_i \cdot \mathbf{v}(t+i)\right) \cdot \mathbf{M}_e .$$

Thus, we obtain a linear expression in $\mathbf{M}_e$:

$$\mathbf{v}'(t) \cdot \mathbf{M}_e = 0, \text{ where} \tag{10}$$
$$\mathbf{v}'(t) = \sum_{i=0}^{D} b_i \cdot \mathbf{v}(t+i). \tag{11}$$

The second step of a fast algebraic attack is to evaluate many vectors $\mathbf{v}'(t+i)$, by substituting observed keystream vectors $\mathbf{z}_{t+i}$ into the vectors $\mathbf{v}(t+i)$ in equation (11). Each of the evaluated vectors $\mathbf{v}'(t)$ provides the attacker with a linear equation in the E unknown bits of the initial monomial state $\mathbf{M}_e$. Equation (10) involves fewer unknowns than the initial equation (7); this means that the fast algebraic attack requires fewer equations in order to solve for the unknowns. *Reducing the number of unknowns and equations significantly improves the third step of the attack as solving the system of E equations (10) takes significantly less time than solving the system of D equations of (6).* The complexity of the third step is now $O(E^\omega)$.

Courtois [5] and Armknecht [1] have proposed efficient methods for finding the coefficients of equation (9). The details are not relevant to this paper, but the complexities are provided in Table 2 for the purposes of comparison with the method proposed in Section 6 of this paper.

Data Complexity. Evaluating the vector $\mathbf{v}'(t)$ (for each equation (10)) requires substituting the bits from the D keystream vectors $\mathbf{z}_{t+i}$, $0 \le i \le D$. Obtaining E equations (10) can be achieved using the set of keystream vectors $\{\mathbf{z}_{t+i}, 0 \le i \le D, 1 \le t \le E\} = \{\mathbf{z}_t, 1 \le t \le D + E\}$. These keystream vectors can be obtained from the keystream bits $z_{t+i}, 1 \le t \le D + E + \theta$. Hence, the attack can be performed using as few as $(D + E + \theta) = O(D)$ keystream bits.

3   Substitution Complexity of Fast Algebraic attacks

Normal algebraic attacks and fast algebraic attacks differ in the complexity of substituting the keystream into the equations in Step 2. The vector $\mathbf{h}(t)$ is a function of a small number of keystream bits z_{t+i}, $0 \le i \le \theta$, but the vector $\mathbf{v}'(t)$ is a function of a large number of keystream bits z_{t+i}, $0 \le i \le D + \theta$. As discussed in the introduction, a misunderstanding resulted the attacks [1, 5] failing to account for this difference.

The naïve approach to substituting the keystream is to compute the vectors $\mathbf{v}(t)$ first and then substitute these vectors into the equations (11) individually[4]. Computing a single component of the vector $\mathbf{v}'(t) = \sum_{i=0}^{D} b_i \mathbf{v}(t+i)$ for a single value of t will require complexity $D/2$, since (on average) half of the coefficients

---

[4] We ignore the cost of computing $\mathbf{v}(t)$ as this cost is independent of the cost of determining $\mathbf{v}'(t)$ from the values of $\mathbf{v}(t)$.

b_i are expected to be zero. There are E components in each vector $\mathbf{v}'(t)$, so the complexity of substituting the keystream to obtain a single vector $\mathbf{v}'(t)$ using equation (11) is $E \times (D/2) = ED/2$. That is, obtaining a single equation (11) has complexity is $ED/2$. Since E equations are required in order to solve the system, the total cost of simple substitution will be $E \times (ED/2) = E^2 D/2$. Table 3 lists the complexity of simple substitution for the fast algebraic attacks in the literature. *Note that simple substitution is significantly more complex than solving the linear system of equations in these cases.*

Table 3. Comparing the claimed complexities of substitution, the complexities of simple substitution and the complexity of solving the linear system.

Cipher	n	d, e	D	E	Claimed Subs $O(ED)$	Simple Subs $E^2D/2$	Solving Linear System (E^ω)	Dominant Term
E0	2^7	4,3	2^{23}	2^{18}	2^{41}	2^{59}	2^{49}	2^{59}
LILI-128	89	4,2	2^{21}	2^{12}	2^{33}	2^{44}	2^{39}	2^{44}
Toyocrypt	2^7	3,1	2^{18}	2^7	2^{23}	2^{31}	2^{20}	2^{31}

4   The Discrete Fourier Transform

Real Spectral Analysis: First, we'll consider a quick tangential topic. A common tool in analyzing a real-valued function $a(x)$ (such as a sound wave) evaluated on a real domain $x \in [0, P]$ is to represents the function $a(x)$ as a sum of simple periodic functions (cosine and sine curves) where the function $a(x)$ is specified by the amplitudes of these periodic functions:

$$a(x) = A_0 + \sum_{\phi=1}^{\infty} A_\phi \cdot \cos\left(\frac{2\pi\phi}{P} \cdot x\right) + \sum_{\phi=1}^{\infty} A_\phi^* \cdot \sin\left(\frac{2\pi\phi}{P} \cdot x\right).$$

with *amplitudes* A_ϕ and A_ϕ^* assigned for each *frequency* ϕ. The sequences $\{A_\phi\}$ and $\{A_\phi^*\}$ are the *Fourier series* for $a(x)$, and evaluating the amplitudes is called a *spectral analysis*.

Discrete Spectral Analysis: Suppose $a(t)$ is a function defined at discrete values $t = \{1, 2, 3, \ldots, P\}$, and the values of $a(t)$ lie in a field $\mathcal{F}$. Such discrete functions are equivalent to sequences written $a = \{a(t)\}$. Discrete spectral analysis of a, like real spectral analysis, represents a using simple periodic sequences with period P. These periodic sequences are of the form $\Lambda_\phi = \{\Lambda_\phi(t) = \lambda^{\phi \cdot t}\}$, where $0 \le \phi \le (P-1)$, and λ is an element of multiplicative order P in some field $\mathcal{G}$; these functions are analogous to the sine and cosine curves.

In some cases, $\mathcal{F}$ has elements of multiplicative order P, and λ can be an element of $\mathcal{F}$; that is, $\mathcal{G} = \mathcal{F}$. In other cases, λ must be chosen in an a larger field $\mathcal{G}$ that is an extension field of $\mathcal{F}$. In either case, the field $\mathcal{G}$ is a vector space over $\mathcal{F}$; that is, elements of $\mathcal{G}$ are of the form $x = \sum_{i=1}^{p} x_i \nu_i$ for some basis $\{\nu_1, \ldots, \nu_p\}$, $p \ge 1$. Elements $x \in \mathcal{F}$ are mapped to elements $\bar{x} = x\bar{I} \in \mathcal{G}$ where

$\bar{I}$ is the identity element of $\mathcal{G}$. Thus, the sequence a of elements of $\mathcal{F}$ is mapped to the sequence $\bar{a}$ with elements of $\mathcal{G}$. A discrete spectral analysis determines a sequence of P "amplitudes" $A = \{A_\phi \in \mathcal{G}, 0 \le \phi \le (P-1)\}$, such that the sequence $\bar{a}$, can be expressed as:

$$\bar{a}(t) = \sum_{\phi=0}^{P-1} A_\phi \cdot \Lambda_\phi(t). \tag{12}$$

In this way, each sequence value $\bar{a}(t)$ is represented as a linear combination of sum of P periodic sequences Λ_ϕ, $0 \le \phi \le (P-1)$. It is well known that the sequence of amplitudes A can be computed directly from the sequence $\bar{a}$ as:

$$A_\phi = \frac{1}{P} \cdot \sum_{t=1}^{P} \bar{a}(t) \cdot \Lambda_\phi(-t). \tag{13}$$

The calculation of A from $\bar{a}$ as in (13) is called the *Discrete Fourier Transform* (DFT), while the calculation of $\bar{a}$ from A is the Inverse DFT. The most efficient method for performing the DFT, known as the Fast Fourier Transform (FFT) [4], requires a total of $P \log_2 P$ operations in the field $\mathcal{G}$. There is also an Inverse FFT that uses the same amount of computation to invert the DFT.

Convolutions and the DFT. The convolution of two discrete sequences a and b of period P is another sequences y of period P with $y(t) = \sum_{i=1}^{P} a(t) \cdot b(i - t(\mathrm{mod}\ P))$, $\forall t$. These are sequences of elements from the field $\mathcal{F}$. It is common to write $y(t) = (a*b)(t)$. Computing the convolution according to first principles would take P^2 multiplication and addition operations in the field $\mathcal{F}$. However, the Convolution Property provides us with an alternative method.

Convolution Property $y(t) = (a*b)(t)$, $\forall t$ if and only if $Y_\phi = A_\phi \cdot B_\phi$, $\forall \phi$.

The convolution can be computed by applying the FFT to a and b to form A and B, forming $\{Y_\phi = A_\phi \cdot B_\phi\}$; and finally applying the inverse FFT to Y in order to form y. The total complexity is $3(P \log_2 P) + P = P(3 \log_2 P + 1)$ operations in the field $\mathcal{G}$. In the cases where $\mathcal{G} = \mathcal{F}$, the FFT method is faster by a factor of $P/(3 \log_2 P + 1)$. In other cases, computations in $\mathcal{G}$ cost more than computations in $\mathcal{F}$ and the advantage is less. This "trick" has been applied in many areas such as the fast multiplication of larger numbers and polynomials (the product of two polynomials is the convolution of the two corresponding sequences of coefficients). We shall use this trick in the next two sections.

5   Applying the FFT to the Substitution Step

The calculation in equation (11) is performed component-wise, so we will begin by focussing on the sequence of values for only one of the monomial components $v_\mu(t)$ of the vectors $\mathbf{v}(t) = (v_0(t), \dots, v_{E-1}(t))$, $0 \le \mu \le (E-1)$, and the corresponding components $v'_\mu(t)$ of the vectors $\mathbf{v}'(t)$. Assume that the attacker has observed a sufficient amount of keystream, evaluated the values of $v_\mu(t)$ in equation (8) for $1 \le t \le (D+E)$, and determined the values $b_0, b_1, \dots, b_D$. The attacker now needs fast way to determine the values $v'_\mu(t) = \sum_{i=0}^{D} b_i \cdot v_\mu(t+i)$, $1 \leq t \leq E$, (see equation (11)) from the values $v_\mu(t)$.

The inefficiency of using simple substitution is indicated by two things:

- Equations (11) often re-use the same values of $v_\mu(t)$ when computing $v'_\mu(t)$, $1 \leq t \leq E$.
- Equations (11) all use the same linear combination;

This problem appears similar to computing $(\beta * v_\mu)(t) = v'_\mu(t)$ for an appropriate sequence β. Indeed, if β is defined as $\beta(t) = 0, 1 \leq t \leq P-D-1$, and $\beta(t) = b_{P-t}$, $P-D \leq t \leq P$, then $(\beta * v_\mu)(t) = v'_\mu(t)$ for $1 \leq t \leq P-D$, Thus, the FFT may be combined with the Convolution Property for computing $v'_\mu(t)$. The sequences v_μ and β are defined on the field $\mathcal{F} = GF(2)$, so $\mathcal{G}$ will be a field of the form $GF(2^p)$. We choose p to be the smallest value such that $2^p > (D+E)$, and define $P = 2^p - 1$. This choice seems best because it uses the smallest number of bits to represent elements of $\mathcal{G}$.

Basic DFT-Based Substitution Algorithm

1. Map the sequences β and v_μ in $\mathcal{F}$ to sequences $\overline{\beta}$ and $\overline{v}_\mu$ in $\mathcal{G}$.
2. Apply the DFT to obtain the sequence of amplitudes B from $\overline{\beta}$.
3. Apply the DFT to obtain the sequence of amplitudes V from $\overline{v}_\mu$.
4. Compute $Q_\phi = B_\phi \cdot V_\phi$, $0 \leq \phi \leq (P-1)$.
5. Apply the inverse DFT to obtain $\overline{q}$ from Q. Note $\overline{q}(t) = (\overline{\beta} * \overline{v}_\mu)(t)$.
6. Extract $v'_\mu(t)$ from $\overline{q}(t)$, $1 \leq t \leq E$; $v'_\mu(t) = 1$ if $\overline{q}(t) = I$, else $v'_\mu(t) = 0$.

Complexity. This may seem like a strange way to compute $v'_\mu(t)$, but the algorithm is very efficient when the FFT is used to compute the DFTs:

- The values B_ϕ are computed via the FFT using $P(\log_2 P)$ field operations (operations in the field $\mathcal{G}$). For given values $b_0, b_1, \ldots, b_D$, the same sequence $\overline{\beta}$ is used for each monomial component and for each attacked keystream. The attacker should pre-compute and store B to save time.
- The values V_ϕ are computed via the FFT using $P(\log_2 P)$ field operations.
- The values $Q_\phi = V_\phi \cdot B_\phi$ are computing using $(D+E)$ field multiplications.
- The sequence $\overline{q}$ can be obtained from $\{Q_\phi\}$ by applying the standard Inverse FFT; this requires in time $P(\log_2 P)$ field operations.

The pre-computation of B requires $P(\log_2 P)$ field operations. The run-time total complexity for computing the value of $v'_\mu(t)$ from the values $v_\mu(t)$ is approximately, $2P \log_2 P$ field operations. These field operations are more complex than $GF(2)$ (logical) operations. To a good approximation, each field operation is equal in complexity to $\log_2 P$ logical operations (much of this can be parallelized). Thus, for our calculations, the run-time complexity of the above algorithm is equivalent to around $2P(\log_2 P)^2$ logical operations.

Improvement 1. The above algorithm computes all $(D+E)$ values of $\overline{q}$, but only E of these values are ever used. An efficient alternative is to divide the linear combination into δ segments of length $D' = D/\delta$ and perform the FFTs on these segments using a smaller field $GF(2^{p'})$ where $2^{p'} - 1 = P' \geq E + D'$. If we define appropriate sequences $\beta[j]$, $1 \leq j \leq \delta$, then we may write:

$$v'_\mu(t) = \sum_{j=0}^{\delta-1}\sum_{i=0}^{D'} b_{D'j+i} \cdot v_\mu(t + D'j + i) = \sum_{j=0}^{\delta-1}(\beta[j] * v_\mu[j])(t).$$

with sub-sequences $\{v_\mu[j](t) = v_\mu(t + D'j)\}$. Now, $q(t) = \sum_{j=0}^{\delta-1}(\beta[j] * v_\mu[j])(t)$ if and only if $Q_\phi = \sum_{j=0}^{\delta-1} B[j]_\phi \cdot V[j]_\phi$. Thus, computing $q(t) = v'_\mu(t)$ requires: pre-computing the FFTs of $\beta[j]$ computing the FFTS of $v_\mu[j]$; computing Q_ϕ; and applying the inverse FFT to Q to obtain q and thus $v'_\mu(t)$. The FFTs and Inverse FFT dominate the complexity, requiring $(\delta + 1)P'(\log P')^2$ logical operations at run-time, where $P' \approx \frac{D}{\delta} + E$. The basic algorithm above uses $\delta = 1$. The optimal choice for δ (providing the lowest complexity) depends on D and E.

Improvement 2. The DFT-based substitution algorithm computes the values of $v'_\mu(t)$, $1 \leq t \leq E$, for only one component of the vectors $\mathbf{v}'(t)$, $1 \leq t \leq E$. There are a total of E monomial components $v'_\mu(t)$, $0 \leq \mu \leq E$ for each such vector; thus if each component is computed separately, then the total complexity of computing all components of every vector $\mathbf{v}'(t)$ would be approximately $E(\delta + 1)P'(\log P')$ field operations, or $E(\delta + 1)P'(\log P')^2$ logical operations. Fortunately, p' monomial components can be packed into each computation. For a set $\{\mu_1, \ldots, \mu_p\}$ of monomial indices we define a sequence $\overline{v} \in \mathcal{G}$: $\overline{v}(t) = (v_{\mu'_p}(t), v_{\mu_{p'-1}}(t), \ldots, v_{\mu_1}(t))$. Then

$$\overline{v}'(t) = (\overline{\beta} * \overline{v})(t) = (v'_{\mu'_p}(t), v'_{\mu_{p'-1}}(t), \ldots, v'_{\mu_1}(t)),$$

provides p' values $v'_{\mu_i}(t)$ for the price of one, dividing the total complexity by p'.

Improved DFT-Based Substitution Algorithm
Inputs: $v_{\mu_i}(t)$, $1 \leq t \leq (D + E)$, $1 \leq i \leq p'$; $B[j]$, $0 \leq j \leq \delta$.
Outputs: $v_{\mu_i}(t)$, $1 \leq t \leq E$, $1 \leq i \leq p'$.

1. Form $\overline{v}(t) = (v_{\mu'_p}(t), v_{\mu_{p'-1}}(t), \ldots, v_{\mu_1}(t))$, $1 \leq t \leq (D + E)$.
2. Form sub-sequences $\overline{v}[j] = \{\overline{v}(D'j + 1), \ldots, \overline{v}(D'j + P')\}$, $0 \leq j \leq \delta$.
3. Apply DFT to obtain $V[j]$ from $\overline{v}[j]$, $0 \leq j \leq \delta$.
4. Compute $Q_\phi = \sum_{j=0}^{\delta-1} B[j]_\phi \cdot V[j]_\phi$, $0 \leq \phi \leq (P - 1)$.
5. Apply inverse DFT to obtain $\overline{q}$ from Q.
6. Set $\overline{v}'(t) = \overline{q}(t)$, $1 \leq t \leq E$.
7. Output $\overline{v}'(t) = (v'_{\mu'_p}(t), v'_{\mu_{p-1}}(t), \ldots, v'_{\mu_1}(t))$, $1 \leq t \leq E$.

The run-time complexity, after applying these two improvements, is

$$(\delta + 1)EP'(\log P')^2 \times \frac{E}{\log P'} = (\delta + 1)EP'(\log P').$$

Table 4 shows the complexities of the FFT method for substitution for the current fast algebraic attacks in literature. In the case of $E0$, the improvement has been significant, and the substitution step no-longer dominates the run-time complexity. The improvement in the substitution complexity is less noticeable required for LILI-128, and insignificant for Toyocrypt. The substitution step still comprises a significant portion of the complexity for these attacks.

In all cases, the first improvement did not affect the complexity significantly; the largest improvement was by a factor of 4. Interestingly, the optimal value of δ was $\delta \approx D/E$, for which $D' = E$ and $P' \approx 2E$. The corresponding complexity is around $D/E \cdot E \cdot 2E \cdot (\log_2 E + 1) \approx 2DE(\log_2 E)$

Table 4. Comparing the complexities of substitution using the FFT method against the complexities of simple substitution.

Cipher	D	E	Substitution			Solving System	Total
			Simple $E^2D/2$	FFT Basic ($\delta = 1$)	$(\delta+1)EP'(\log P')$ Optimal Choice		
E0	2^{23}	2^{18}	$2^{59.2}$	2^{48}	$2^{46.8}$ ($\delta = 2^4$)	2^{49}	2^{49}
LILI-128	2^{21}	2^{12}	$2^{44.2}$	$2^{39.4}$	$2^{37.8}$ ($\delta = 2^8$)	2^{39}	2^{40}
Toyocrypt	2^{18}	2^7	$2^{31.4}$	$2^{31.2}$	$2^{29.2}$ ($\delta = 2^{10}$)	2^{20}	2^{29}

6   Improving the Pre-computation Step

A square matrix satisfies its characteristic polynomial. That is, if $p^{(d)}(x) = \sum_{i=0}^{D} p_i x^i$ is the characteristic polynomial of $\mathbf{R}_d$, then

$$p^{(d)}(\mathbf{R}_d) = \sum_{i=0}^{D} p_i \cdot \mathbf{R}_d^i = \mathbf{0}, \qquad (14)$$

where $\mathbf{0}$ represents the all-zero matrix. Suppose the coefficients $b_0, \ldots, b_D$ of Equation (9) are assigned the values of the coefficients $p_0, \ldots, p_D$ of the characteristic polynomial of $\mathbf{R}_d$. Then, for any function $u(\mathbf{K}_t)$ of degree d,

$$\sum_{i=0}^{D} b_i \cdot u(t+i) = \sum_{i=0}^{D} p_i \cdot (u \cdot \mathbf{R}_d^{t+i}) = u \cdot \mathbf{R}_d^t \cdot \left(\sum_{i=0}^{D} p_i \cdot \mathbf{R}_d^i \right) = u \cdot \mathbf{R}_d^t \cdot \mathbf{0} = 0.$$

The characteristic polynomial of $\mathbf{R}_d$ depends on the LFSR and the degree d, and is otherwise independent of the function $u(\mathbf{K}_t)$. Thus, the coefficients $p_0, \ldots, p_D$ (of the characteristic polynomial of $\mathbf{R}_d$) can be substituted for the coefficients $b_0, \ldots, b_D$ in equations (9) and (11) *for all functions* $u(\mathbf{K}_t)$ *of degree* d.

Most functions u of degree d have a minimal polynomial of degree $D = \sum_{i=0}^{d} \binom{n}{i}$ (see [9, Fact 6.55]); the minimal polynomial for these functions will be $p^{(d)}(x)$. However, there are functions the minimal polynomial is a smaller factor of $p^{(d)}(x)$. For example, in the attack on E0 [1], $p^{(d)}(x)$ has length $D = 11,017,633$; while the minimal polynomial of the Boolean function used in the attack is of slightly smaller length $D' = 8,822,188$. Using $p^{(d)}(x)$ in the attack on E0 (instead of using the minimal polynomial) would increase the complexity by a small amount. An advantage of the methods proposed in [1, 5] is that those method will find the minimal polynomial for a specified Boolean function, even if the minimal polynomial is smaller than $p^{(d)}(x)$.

It is not difficult to show that the polynomial $p^{(e)}(x)$ divides $p^{(d)}(x)$ This suggests that the linear combination $\sum_{i=0}^{D} p_i \cdot \mathbf{v}(t+i)$ may also be zero, thus resulting in a trivial equation $0 \cdot \mathbf{M}_e = 0$ that provides no information about the initial monomial state. The probability of this cancellation occurring is small; the vectors $\mathbf{v}(t+i) = \mathbf{v}(\mathbf{z}_{t+i}) \cdot \mathbf{R}_d^{t+i}$, in the sum $\sum_{i=0}^{D-E} p_i' \cdot \mathbf{v}(t+i) = \mathbf{v}'(t)$ depend on the keystream. Nonetheless, this suggests that a better approach would be to cancel only those components corresponding to monomials of degree greater than e using the polynomial

$$ p^{(d/e)}(x) \stackrel{\text{def}}{=} p^{(d)}(x)/p^{(e)}(x) = \sum_{i=0}^{D-E} p_i' \cdot x^i. $$

The linear combination $\sum_{i=0}^{D-E} p' \cdot \mathbf{u}(t+i)$ cancels components corresponding to monomials of degree in the range $[e+1, d]$. but will not cancel components corresponding to monomials of degree e or less. Hence, $\mathbf{u}'(t) = \sum_{i=0}^{D-E} p' \cdot \mathbf{u}(t+i)$ can be considered as an E-dimensional vector $\mathbf{u}'(t) = \mathbf{u}' \cdot \mathbf{R}_e^t$ for some vector $\mathbf{u}'$. Equation 10 would then become

$$ (\mathbf{u}'(t) + \mathbf{v}'(t)) \cdot \mathbf{R}_e = 0. \tag{15} $$

The probability that $\mathbf{u}'(t) = \mathbf{v}'(t)$ will depend on the probability that $\mathbf{u}' = \mathbf{v}(\mathbf{z})$ for a random $\mathbf{z}$. Unless $\mathbf{v}(\mathbf{z})$ is constant, this probability $\mathbf{u}' + \mathbf{v}(\mathbf{z}) = 0$ will be less than $1/2$. After substitution of many such vectors, the probability that $\mathbf{u}'(t) + \mathbf{v}'(t) = 0$ will be very small and Equation (15) is highly unlikely to be trivial.

6.1   Direct Computation of the Linear Combination

Suppose an LFSR state $\mathbf{K}_t$ of length n is updated according to state update matrix L, and the characteristic polynomial of L is primitive[5]. The following theorem, while not explicitly stated by Key [7], is a fairly obvious consequence of Key's ideas, so no proof is given. This result provides a direct method for computing $p^{(d)}(x)$.

Theorem 1. *(Largely due to Key [7])* If $\gamma \in GF(2^n)$ is a root of the characteristic polynomial of the LFSR state update matrix, then the characteristic polynomial of $\mathbf{R}_d$ is $p^{(d)}(x) = \prod_{\psi:w(\psi)\le d} (x - \gamma^{\psi})$, where $\psi \in GF(2^n)$ and $w(\psi)$ denotes the Hamming weight of ψ (that is, the number of 1's in the radix-2 representation of the integer ψ).   □

Factoring $p^{(d)}(x)$ into $GF(2)$ polynomials $y_{\psi}(x)$. Computing the entire product $\prod_{\psi:w(\psi)\le d} (x - \gamma^{\psi})$ while in $GF(2^n)$ would be costly. Fortunately, the

---

[5] This approach can be extended to cases where the characteristic polynomial is not primitive; for example, when the keystream is a function of more than one LFSR. See Key [7] for more details.

factors in $GF(2^n)$ can be easily grouped into $GF(2)$ polynomials of degree n or less. We define an equivalence relation "$\doteq$" where $\psi' \doteq \psi$ if and only if $\psi' \equiv 2^j \psi \pmod{2^n - 1}$, for some value j. Since the set $\{\psi' \doteq \psi\}$ is closed under multiplication by 2, the polynomial $y_\psi(x) = \prod_{\psi' \doteq \psi} (x - \gamma^{\psi'})$, has coefficients that are either 0 or the identity element I. That is, the product $y_\psi(x)$ can be represented as a $GF(2)$ polynomial. Thus, $p^{(d)}(x)$ can compute in two phases:

1. Compute the $GF(2)$ polynomials $y_\psi(x)$ for all ψ of weight d or less.
2. Multiply the $GF(2)$ polynomials $y_\psi(x)$ to form $p^{(d)}(x)$.

Computing $y_\psi(x)$. The FFT over $GF(2^n)$ may be used to compute the polynomials $y_\psi(x)$. First, apply the FFT to the sequences corresponding to $(x - \gamma^{\psi'})$ for $\psi' \doteq \psi$ to obtain sequences $\Gamma^{(\psi')}$, $\psi' \doteq \psi$. Second, form sequence Ω with $\Omega_\phi = \prod_{\psi' \doteq \psi} \Gamma_\phi^{(\psi')}$. Finally, apply the inverse FFT to Ω to obtain $y_\psi(x)$. The first step is the most costly; it requires $n(\log n)^2$ logical operations for each factor. There are D factors, so the total combined cost is $Dn(\log n)^2$ logical operations.

Multiplying $y_\psi(x)$ to form $p^{(d)}(x)$. The second phase has polynomials with coefficients in $GF(2)$ and uses FFT's in extension fields of $GF(2)$. Multiplying two $GF(2)$ polynomials in to get a product of degree less than $J = 2^j$ can be performed (via the FFT) using $J(1 + 3\log_2 J) = 2^j(1 + 3j)$ operations in the extension field $GF(2^j)$; this is equal to $2^j(1 + 3j)j$ logical operations[6]. Use the FFT to first multiply pairs of polynomials of degree n to get polynomials of degree $2n$. Then multiply pairs of polynomials of degree $2n$ to get polynomials of degree $4n$ and so forth until $p^{(d)}(x)$ is formed. The total complexity is

$$\sum_{j=\log_2 n}^{\log_2 D} \frac{D}{2^j} \cdot 2^j (1 + 3j)j = D \sum_{j=\log_2 n}^{\log_2 D} (3j^2 + j) \approx D(\log_2 D)^3.$$

The combined complexity for the two phases is $D[n(\log n)^2 + (\log_2 D)^3]$. In Table 2, the complexity of this method is compared against the previous methods.

7 Conclusion

We have shown that some published "fast algebraic attacks" on stream ciphers underestimate the process complexity of one of the steps, and we provide correct complexity estimates for these cases. We then show an improved method, using Fast Fourier Transforms, for substituting keystream bits into the system of equations needing to be solved. We also made some observations about the linear combination used in the pre-computation step of the fast algebraic attack. In particular, we found the fastest known method for performing the pre-computation. The fast algebraic attack remains an extremely powerful technique for analyzing LFSR-based stream ciphers.

---

[6] The attacker can "pack" multiple $GF(2)$ polynomials into a single $GF(2^j)$ sequence and thereby compute the convolution of multiple pairs $GF(2)$ polynomials using the same amount of computation. This reduces complexity by a relatively small factor.

Acknowledgements

Acknowledgement is due to Nicolas Courtois for inspirational discussions and fine French cuisine during which the results in Section 6 came to light. The reviewers also provided many useful insights; in particular, the two improvements to the basic DFT-based substitution algorithm.

References

1. F. Armknecht: *Improving Fast Algebraic Attacks,* to be presented at FSE2004 Fast Software Encryption Workshop, Delhi, India, February 5-7, 2004.
2. F. Armknecht and M. Krause: *Algebraic Attacks on Combiners with Memory,* proceedings of Crypto2003, Lecture Notes in Computer Science, vol. 2729, pp. 162-176, Springer 2003.
3. Bluetooth CIG, *Specification of the Bluetooth system, Version 1.1, February 22, 2001.* Available from www.bluetooth.com.
4. J. W.Cooley and J. W. Tukey: *An algorithm for the machine calculation of complex Fourier series,* Mathematics of Computation, 19, 90, pp. 297-301, 1965.
5. N. Courtois: *Fast Algebraic Attacks on Stream Ciphers with Linear Feedback,* Crypto2003, Lecture Notes in Computer Science, vol. 2729, pp. 177-194, Springer 2003.
6. N. Courtois and W. Meier: *Algebraic Attacks on Stream Ciphers with Linear Feedback,* EUROCRYPT 2003, Warsaw, Poland, Lecture Notes in Computer Science, vol. 2656, pp. 345-359, Springer, 2003.
7. E. Key: *An Analysis of the Structure and Complexity of Nonlinear Binary Sequence Generators,* IEEE Transactions on Information Theory, vol. IT-22, No. 6, November 1976.
8. J. Massey: *Shift Register synthesis and BCH decoding,* IEEE Transactions on Information Theory, IT-15 (1969), pp. 122-127.
9. A. Menezes, P.van Oorschot and A. Vanstone *Handbook of Applied Cryptography,* CRC Press series on discrete mathematics and its applications, CRC Press LLC, 1997.
10. M. Mihaljevic and H. Imai: *Cryptanalysis of Toyocrypt-HS1 stream cipher,* IEICE Transactions on Fundamentals, vol E85-A, pp. 66-73, January 2002. Available at www.csl.sony.co.jp/ATL/papers/IEICEjan02.pdf.
11. R. Rueppel: *Stream Ciphers;* Contemporary Cryptology: The Science of Information Integrity. G. Simmons ed., IEEE Press, New York, 1991.
12. L. Simpson, E. Dawson, J. Golic and W. Millan: *LILI Keystream Generator;* Selected Areas in Cryptography SAC'2000, Lecture Notes in Computer Science, vol. 1807, Springer, pp. 392-407.
13. V. Strassen: *Gaussian Elimination is Not Optimal;* Numerische Mathematik, vol. 13, pp. 354-356, 1969.

Faster Correlation Attack
on Bluetooth Keystream Generator E0

Yi Lu* and Serge Vaudenay

EPFL
http://lasecwww.epfl.ch

Abstract. We study both distinguishing and key-recovery attacks against E0, the keystream generator used in Bluetooth by means of correlation. First, a powerful computation method of correlations is formulated by a recursive expression, which makes it easier to calculate correlations of the finite state machine output sequences up to 26 bits for E0 and allows us to verify the two known correlations to be the largest for the first time. Second, we apply the concept of convolution to the analysis of the distinguisher based on all correlations, and propose an efficient distinguisher due to the linear dependency of the largest correlations. Last, we propose a novel maximum likelihood decoding algorithm based on fast Walsh transform to recover the closest codeword for any linear code of dimension L and length n. It requires time $O(n + L \cdot 2^L)$ and memory $\min(n, 2^L)$. This can speed up many attacks such as fast correlation attacks. We apply it to E0, and our best key-recovery attack works in 2^{39} time given 2^{39} consecutive bits after $O(2^{37})$ precomputation. This is the best known attack against E0 so far.

1 Background

Correlation properties play an important role in the security of nonlinear LFSR-based combination generators in stream ciphers. As name implies, the word *correlation* in stream ciphers is frequently referred to as the intrinsic relation between the keystream and a subset of the LFSR subsequences. The earliest studies dated back to [21, 25, 27] in the 80's and the concept of correlation immunity was proposed as a security criterion. In the 90's Meier-Staffelbach [22] analyzed correlation properties of combiners with one memory bit, followed by Golić [12] focusing on correlation properties of a general combiner with m-bit memory. Recently, a series of fast correlation attacks sprang up, to name but a few [5–7, 16, 24]. Thereupon we dedicate this paper to the generalized correlation attacks against E0, a combiner with 4-bit memory used in the short-range wireless technology Bluetooth. Prior to our work, existed various attacks [1, 8, 10, 11, 13–15, 17, 26] against E0. The best key-recovery attacks are algebraic attacks [1,

---

* Supported in part by the National Competence Center in Research on Mobile Information and Communication Systems (NCCR-MICS), a center of the Swiss National Science Foundation under the grant number 5005-67322.

M. Franklin (Ed.): CRYPTO 2004, LNCS 3152, pp. 407–425, 2004.

8], whose basic approach is to use the polynomial canceling all memory bits and involving only key bits, instead of considering the multiple polynomial to cancel the key bits in the distinguishing attack; besides, [9, 13, 14] discussed correlations of E0. In [14], Hermelin-Nyberg for the first time presented a rough computation method to compute the correlation (called bias for our purpose), but neither did they formalize the computation systematically, nor did they attempt to find a larger correlation. In [9, 13], two larger correlations for a short sequence of up to 6 bits were exposed. However, due to the limit of the computation method, no one was certain about the existence of a larger correlation for a longer sequence, which is critical to the security of E0.

Our first contribution in the paper is that based on Hermelin-Nyberg [14] we formulate a powerful computation method of correlations by a recursive expression, which makes it easier to calculate correlations of the Finite State Machine (FSM) output sequences up to 26 bits for E0 (and allows us to prove the two known correlations to be the only largest for the first time). Second, we apply the concept of convolution to the analysis of the distinguisher based on all correlations, which allows us to build an efficient distinguisher that halves the data complexity of the basic uni-bias-based distinguisher due to the linear dependency of the two largest biases. Our best distinguishing attack takes 2^{43} time given 2^{43}-bit keystream with $O(2^{45})$ precomputation[1]. Finally, by means of Fast Walsh Transform (FWT), we propose a novel Maximum Likelihood Decoding (MLD) algorithm to recover the closest codeword for any linear code. Our proposed algorithm can be easily applied to speed up a class of fast correlation attacks. Furthermore the algorithm is optimal when the length n of the code and the dimension L satisfy the relation $n \geq 2^L$, which is the case when we apply it to recover R_1 for E0. Our best key-recovery attack works in 2^{39} time given 2^{39} consecutive bits after $O(2^{37})$ precomputation. Compared with the minimum time complexity $O(2^{49})$ in algebraic attacks [1, 8], this is the best known attack against E0.

This paper is structured as follows: in Section 2, a description of E0 is given. In Section 3, we analyze the bias inside E0 systematically. Then based on one largest bias, we build a primary distinguisher for E0 in Section 4; an efficient way is shown in Section 5 that makes full use of all the largest biases to advance the distinguisher. In Section 6 we investigate the MLD algorithm for a linear code; the result is then applied to a key-recovery attack against E0 in Section 7. Finally we conclude in Section 8.

2   Description of the Bluetooth Keystream Generator E0

As specified in [3], the keystream generator E0 used in Bluetooth belongs to a combination generator with four memory bits[2], denoted by $\sigma_t = (c_{t-1}, c_t)$ at time t, where $c_t = (c_t^1, c_t^0)$. The whole system (Fig.1) uses four Linear Feedback

---

[1] Throughout this paper, $O(\cdot)$ is used to provide a rough estimate on complexities, eg. $O(2^{45})$ here means $c \cdot 2^{45}$ operations, where c is a small constant.

[2] The description of E0 (sometimes called one-level E0) here only involves the keystream generation after the initialization.

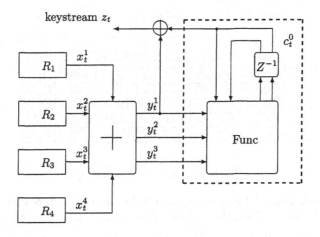

keystream z_t

Fig. 1. Outline of E0

Shift Registers (LFSRs) denoted by $R_1, \ldots, R_4$ with lengths $L_1 = 25$, $L_2 = 31$, $L_3 = 33$, $L_4 = 39$ and primitive feedback polynomials

$$
\begin{aligned}
p_1(x) &= x^{25} + x^{20} + x^{12} + x^8 + 1, \\
p_2(x) &= x^{31} + x^{24} + x^{16} + x^{12} + 1, \\
p_3(x) &= x^{33} + x^{28} + x^{24} + x^4 + 1, \\
p_4(x) &= x^{39} + x^{36} + x^{28} + x^4 + 1,
\end{aligned}
$$

respectively. At clock cycle t, the four LFSRs' output bits x_t^i, $i = 1, \ldots, 4$, will be added as integers. The sum $y_t \in \{0, \ldots, 4\}$ is represented in the binary system. Let y_t^i denote its i-th least significant bit ($i = 1, 2, 3$). A 16-state machine (the dashed box in Fig.1) emits one bit c_t^0 out of its state $\sigma_t = (c_{t-1}, c_t)$ and takes the input y_t to update σ_t by σ_{t+1}. Finally, the keystream z_t is obtained by xoring y_t^1 with c_t^0. That is,

$$
x_t^1 \oplus x_t^2 \oplus x_t^3 \oplus x_t^4 \oplus c_t^0 = z_t. \tag{1}
$$

The detailed mechanism of the FSM is beyond the scope of the paper except the fact that the embedded delay cell (the box labeled Z^{-1} in Fig.1) makes c_t^0 depend only on the initial state σ_0 of the FSM as well as the past vectors $y_{t-1}, y_{t-2}, \ldots, y_0$. For completeness, we briefly outline it: given y_t together with the state σ_t, the FSM moves into the state σ_{t+1}. Table 1 shows the state transition of the FSM, where the four-bit state is represented in the quaternary system (e.g. the FSM changes from $\sigma_t = 13$ into $\sigma_{t+1} = 32$ by the input $y_t = 2$).

More formally, by introducing two temporary bits $s_{t+1} = (s_{t+1}^1, s_{t+1}^0)$ each clock, the following indirect iterative expressions between s_{t+1} and c_{t+1} suffice to update c_t:

$$
s_{t+1} = \left\lfloor \frac{y_t + 2 \cdot c_t^1 + c_t^0}{2} \right\rfloor, \tag{2}
$$

$$
c_{t+1}^1 = s_{t+1}^1 \oplus c_t^1 \oplus c_{t-1}^0, \tag{3}
$$

Table 1. State transition of σ_{t+1} given y_t and σ_t

		σ_t																
		00	01	02	03	10	11	12	13	20	21	22	23	30	31	32	33
	0	00	11	23	32	03	12	20	31	01	10	22	33	02	13	21	30
	1	00	10	23	31	03	13	20	32	01	11	22	30	02	12	21	33
y_t	2	01	10	20	31	02	13	23	32	00	11	21	30	03	12	22	33
	3	01	13	20	30	02	10	23	33	00	12	21	31	03	11	22	32
	4	02	13	21	30	01	10	22	33	03	12	20	31	00	11	23	32

$$c_{t+1}^0 = s_{t+1}^0 \oplus c_t^0 \oplus c_{t-1}^1 \oplus c_{t-1}^0. \tag{4}$$

One can check Table 1 by those equations. We denote λ_t hereafter the content of LFSRs at time t. Then the state of E0 at time t is fully represented by the pair (λ_t, σ_t).

3   Biases Inside E0

Property 1. Assuming $y_t = 2$ holds for $t = t_0, t_0 + 1, \ldots, t_0 + 3$, then

$$c_{t_0}^0 \oplus c_{t_0+1}^0 \oplus c_{t_0+2}^0 \oplus c_{t_0+3}^0 \oplus c_{t_0+4}^0 = 1.$$

Proof. It's easy to verify that the state transition given $y_t = 2$ (the third bottom row in Table 1) is indeed a linear transformation over $GF(2)^4$, that actually satisfies the recurrence relation: $\sigma_{t+1} = A \times \sigma_t \oplus 01$, where states σ_{t+1} are represented by column vectors of $(c_t^1, c_t^0, c_{t+1}^1, c_{t+1}^0)$, and A is the following 4×4 square matrix over $GF(2)$:

$$A = \begin{pmatrix} 0 & 0 & 1 & 0 \\ 0 & 0 & 0 & 1 \\ 0 & 1 & 0 & 0 \\ 1 & 1 & 1 & 1 \end{pmatrix}.$$

Note that $x^4 + x^3 + x^2 + x + 1$ is the minimal polynomial of A, from which we deduce $\sigma_{t_0} \oplus \sigma_{t_0+1} \oplus \sigma_{t_0+2} \oplus \sigma_{t_0+3} \oplus \sigma_{t_0+4} = 33$.   □

Remark 2. Since $\Pr(y_t = 2) = \frac{6}{16}$, this seemingly suggests that

$$\Pr(c_t^0 \oplus c_{t+1}^0 \oplus c_{t+2}^0 \oplus c_{t+3}^0 \oplus c_{t+4}^0 = 1) \approx \frac{1}{2} + \left(\frac{6}{16}\right)^4 = \frac{1}{2} + \frac{81}{4096}.$$

As mentioned in [9, 13] (without relating to the above special case), this bit exhibits a much higher bias as shown later in Corollary 7. We will now introduce essential material in order to find a systematic algorithm to compute biases.

Proposition 3. *If (λ_0, σ_0) is random and uniformly distributed, then for any t*

  - *(λ_t, σ_t) is random and uniformly distributed,*
  - *$(c_t, c_{t-1}, \ldots, c_{t-24})$ is independent of y_t.*

Proof (sketch). The former half of the theorem is justified by the fact that (λ_t, σ_t) is a permutation of (λ_0, σ_0) for any t. About the latter half of theorem, first, we know that $(\lambda_{t-24}, \sigma_{t-24})$ is random and uniformly distributed by previous conclusion. Thus, $y_{t-24}, \ldots, y_{t-1}$ are i.i.d. random variables all independent of both σ_{t-24} and y_t. By Eq.(2,3,4), we complete the proof. $\square$

Interestingly, we deduce that if (λ_0, σ_0) is uniformly distributed, then any sequence of 39-bit consecutive E0 keystream is uniformly distributed; in particular, no better key-recovery attack against E0 exists other than tradeoffs given a sequence of 39-bit consecutive keystream.

The following definition is derived from *normalized correlation* [22, p.71].

Definition 4. *The bias of a random Boolean variable X is defined as*

$$\Delta(X) = \Pr(X = 0) - \Pr(X = 1) = E[(-1)^X].$$

The normalized correlation between two random Boolean variables X and Y is just the bias of $X \oplus Y$. Assuming that y_t is the sum of four balanced independent random bits and that c_t is uniformly distributed, then we know that $\Delta(a \cdot s_{t+1} \oplus w \cdot c_t)$ is a constant for any $a, w \in GF(2)^2$, denoted by $\Omega(a, w)$. Table 2 shows $\Omega(a, w)$ computed by Eq.(2), where dashed entries are zeros. The following important lemma (see Appendix A for proof) inspired by [14], gives an easy way of computing the bias for iterative structures.

Table 2. Bias of all linear combination of s_{t+1} and c_t: $\Omega(a, w)$

$\Omega(a,w)$		w		
		0	1	2	3
	0	-	-	-	-
a	1	-	-	$-1/4$	-
	2	-	$1/4$	$5/8$	-
	3	$-5/8$	-	-	$1/4$

Lemma 5. *Given $f : \mathcal{E} \times GF(2)^k \to GF(2)$ and $g : GF(2)^m \to GF(2)^k$, let X and Y be two independent random variables in $\mathcal{E}$ and $GF(2)^m$ respectively. Assuming that $g(Y)$ is uniformly distributed in $GF(2)^k$, for any $v \in GF(2)^m$, we have*

$$\Delta(f(X, g(Y)) \oplus v \cdot Y) = \sum_{w \in GF(2)^k} \Delta(f(X, g(Y)) \oplus w \cdot g(Y)) \cdot \Delta(w \cdot g(Y) \oplus v \cdot Y).$$

Corollary 6. *We set $h : (x^1, x^0) \mapsto (x^0, x^1 \oplus x^0)$ to be a permutation defined over $GF(2)^2$, and $\delta(a_1, \ldots, a_d) = \Delta(a_1 \cdot c_1 \oplus \cdots \oplus a_d \cdot c_d)$, where $a_1, \ldots, a_d \in GF(2)^2$. Assuming (λ_0, σ_0) is uniformly distributed, for any $d \leq 26$, we have*

$$\delta(a_1, \ldots, a_d) = \sum_{w \in GF(2)^2} \Omega(a_d, w) \cdot \delta(a_1, \ldots, a_{d-3}, a_{d-2} \oplus h(a_d), a_{d-1} \oplus a_d \oplus w).$$

Proof. By Eq.(2,3,4) we have

$$\delta(a_1,\ldots,a_d) = \Delta(a_d \cdot s_d \oplus a_1 \cdot c_1 \oplus \cdots \oplus (a_{d-2} \oplus h(a_d)) \cdot c_{d-2} \oplus (a_{d-1} \oplus a_d) \cdot c_{d-1}).$$

Then we apply Lemma 5 with $X = y_{d-1}$, $Y = (c_1,\ldots,c_{d-1})$, $g(Y) = c_{d-1}$, $f(X,g(Y)) = a_d \cdot s_d$ and $v = (a_1,\ldots,a_{d-3}, a_{d-2} \oplus h(a_d), a_{d-1} \oplus a_d)$, and we obtain

$$\delta(a_1,\ldots,a_d) = \sum_w \Omega(a_d, w)\delta(a_1,\ldots,a_{d-3}, a_{d-2} \oplus h(a_d), a_{d-1} \oplus a_d \oplus w).$$

Note that the assumption of Lemma 5 holds by Proposition 3. □

Now we use Corollary 6 iteratively to deduce some important biases of $\{c_t^0\}$ with Table 2 and the initial values $\delta(0,0) = 1$, and $\delta(a,b) = 0$ for $(a,b) \neq (0,0)$. A full list of nonzero triplets is given below for illustration:

$$\delta(0,0,0) = 1, \qquad\qquad \delta(1,3,2) = \tfrac{1}{4}, \qquad\qquad \delta(2,3,3) = -\tfrac{5}{8},$$
$$\delta(1,0,2) = \tfrac{5}{8}, \qquad\qquad \delta(2,0,3) = \tfrac{1}{4}, \qquad\qquad \delta(3,3,1) = -\tfrac{1}{4}.$$

Corollary 7. *Assuming (λ_0, σ_0) is random and uniformly distributed, we have*

$$\Pr(c_t^0 \oplus c_{t+1}^0 \oplus c_{t+2}^0 \oplus c_{t+3}^0 \oplus c_{t+4}^0 = 1) = \frac{1}{2} + \frac{25}{512},$$
$$\Pr(c_t^0 = c_{t+5}^0) = \frac{1}{2} + \frac{25}{512}.$$

Note that both biases were mentioned in [9, 13] (without formal proof). Now by Corollary 6, we can easily prove it as shown next.

Proof. We show the equivalent $\delta(1,1,1,1,1) = -\frac{25}{256}$ of the first bias as follows:

$$\delta(1,1,1,1,1) = \sum_w \Omega(1,w) \cdot \delta(1,1,2,w)$$
$$= -\frac{1}{4}\delta(1,1,2,2)$$
$$= -\frac{1}{4}\sum_w \Omega(2,w) \cdot \delta(1,0,w)$$
$$= -\frac{1}{4} \times \left(\frac{1}{4}\delta(1,0,1) + \frac{5}{8}\delta(1,0,2)\right)$$
$$= -\frac{25}{256}.$$

The second bias is similarly proved from $\delta(1,0,0,0,0,1) = \frac{25}{256}$. □

Also, we computed all ℓ-tuple biases for $\ell \leq 26$ and found that $\delta(1,1,1,1,1)$, $\delta(1,0,0,0,0,1)$ are the only largest ones. All biases for $\ell = 6$ are listed in Table 14, Appendix C. Throughout the paper we let

$$\gamma = \delta(1,0,0,0,0,1) = -\delta(1,1,1,1,1) = \frac{25}{256}.$$

4   A Primary Distinguisher for E0

4.1   The Connection Polynomial of the Equivalent Single LFSR

Let θ_i be the order of the connection polynomial $p_i(x)$ of R_i, for $i = 1, 2, 3, 4$. Since all $p_i(x)$ are primitive polynomials, $\theta_i = 2^{L_i} - 1$; furthermore, by Lemma 6.57 of [18, p.218], the equivalent LFSR to generate the same sequence of the sum of the four original LFSRs outputs over GF(2) has the connection polynomial $\prod_{i=1}^{4} p_i(x)$ with order $\theta = \text{lcm}(\theta_1, \theta_2, \theta_3, \theta_4) \approx 2^{125}$ (by Lemma 6.50, [18, p.214]) and degree $L = \sum_{i=1}^{4} L_i = 128$.

4.2   Finding the Multiple Polynomial with Low Weight

Let d_0 be the degree of a general polynomial $p(x)$. We use the standard approximation to estimate the minimal weight w_d of multiples of $p(x)$ with degree at most d by the following constraint: w_d is the smallest w such that

$$\frac{1}{2^{d_0}} \times \binom{d}{w-1} \geq 1. \tag{5}$$

Listed in Table 3 is the estimated[3] w_d corresponding to d with $p(x) = \prod_{i=1}^{4} p_i(x)$ ($d_0 = 128$) by solving Inequality (5).

To find multiples with low weight, efficient algorithms like [4] exist provided the degree is low, say, less than 2000, which does not apply to E0. So we can use the conventional birthday paradox to find $Q(x)$ with the minimal d (i.e. $w = w_d$), which takes precomputation time $PT \approx O(d^{\lceil \frac{w-1}{2} \rceil})$; or we apply the generalized birthday problem [29] to find $Q(x)$ of same weight but higher degree with much less precomputation as tradeoff. Table 4 compares the two algorithms. In Appendix B, we also provide some non-optimal multiples as examples, including $Q_4(x)$ with $w = 4$ and $d \approx 2^{65}$.

Table 3. The estimated minimal weight w_d of multiples of $p_1(x)p_2(x)p_3(x)p_4(x)$ with degree d by (5)

d	128	247	458	855	1749	2387	2^{18}	2^{23}	2^{27}	2^{33}	2^{44}	2^{65}	θ
w_d	$= 49$	≈ 31	≈ 24	≈ 20	≈ 17	≈ 16	≈ 9	≈ 7	≈ 6	≈ 5	≈ 4	≈ 3	$= 2$

4.3   Building a Uni-Bias-Based Distinguisher for E0

Let $Q(x) = \sum_{i=1}^{w} x^{q_i}$ be the normalized multiple of $\prod_{i=1}^{4} p_i(x)$ with degree d and weight w, where $0 = q_1 < q_2 < \ldots < q_w = d$. As $\oplus_{i=1}^{w} y_{t_0+q_i}^1 = 0$ holds for all t_0, by Eq.(1), we deduce

$$\oplus_{i=1}^{w}(z_{t_0+q_i+5} \oplus z_{t_0+q_i}) = \oplus_{i=1}^{w}(c_{t_0+q_i+5}^0 \oplus c_{t_0+q_i}^0). \tag{6}$$

---

[3] Two special cases occur for $d = d_0$ and $d = \theta$ because we know the exact value of w_d.

Table 4. Complexity of finding multiple of $p_1(x)p_2(x)p_3(x)p_4(x)$ with degree d, weight w

	birthday problem						tradeoff	
	with minimal d							
d	2^{18}	2^{23}	2^{27}	2^{33}	2^{44}	2^{65}	2^{32}	2^{43}
w	9	7	6	5	4	3	9	5
$\log_2 PT$	72	69	68	66	66	65	35	45

By the Piling-up Lemma [20] and Corollary 7, we know the right-hand side of Eq.(6) is equal to zero with probability $\frac{1}{2} + \frac{1}{2} \cdot \gamma^w$. With standard linear cryptanalysis techniques, we can distinguish the keystream $\{z_t\}$ of E0 from a truly random sequence with $\gamma^{-2 \cdot w}$ samples, simply by checking the left-hand side of Eq.(6) equals zero most of the time. Based on $Q(x)$ with d and w, we minimize the data complexity n by choosing $n = \gamma^{-2 \cdot w} + d$. Table 5 shows the minimum $n = 2^{34}$ is achieved with $d = 2^{33}$, $w = 5$. Table 6 summarizes the best performance of our primary distinguisher for E0 based on either the use of $Q_4(x)$ with weight 4 in Appendix B, or a search of $Q(x)$.

Table 5. Data complexity of the primary distinguisher for E0

d	L	247	458	855	1749	2387	2^{18}	2^{23}	2^{27}	2^{33}	2^{44}	2^{65}	2^{32}	2^{43}
w	49	31	24	20	17	16	9	7	6	5	4	3	9	5
$\log_2 n$	329	209	162	135	115	108	61	47	41	34	44	65	61	43

Table 6. Summary of the best primary distinguisher for E0

Type		d	w	Precomputation	Data	Time
use $Q(x) = Q_4(x)$		2^{65}	4	-		2^{65}
find $Q(x)$	minimal d	2^{33}	5	2^{66}		2^{34}
with	tradeoff	2^{43}	5	2^{45}		2^{43}

5   The Advanced Multi-bias-Based Distinguisher for E0

5.1   Preliminaries

Definition 8. *Given* $f, g : GF(2)^\ell \to \mathbf{R}$, *for* $a \in GF(2)^\ell$, *we define*

1. $(f \otimes g)(a) = \displaystyle\sum_{b \in GF(2)^\ell} f(b) \cdot g(a \oplus b); \; f^{\otimes w}(a) = \underbrace{(f \otimes \cdots \otimes f)}_{w \text{ times}}(a)$

2. $\hat{f}(a) = \displaystyle\sum_{b \in GF(2)^\ell} (-1)^{a \cdot b} f(b)$

3. $\|f\| = \sqrt{\displaystyle\sum_{a \in GF(2)^\ell} f^2(a)}$

4. $\Delta(f) = 2^{\frac{\ell}{2}} \|f - \frac{1}{2^\ell} \cdot \mathbf{1}\|$, *where* $\mathbf{1}$ *denotes a constant function equal to 1*

Note that the first two definitions correspond to convolution and Walsh transform respectively. We recall these basic facts: for any $f, g : GF(2)^\ell \to \mathbf{R}$, we have

- $\widehat{f \otimes g}(a) = \hat{f}(a) \cdot \hat{g}(a)$, for $a \in GF(2)^\ell$;

- $2^\ell \|f\|^2 = \|\hat{f}\|^2$;

- if f is a distribution, i.e. $\sum_a f(a) = 1$ and $f(a) \geq 0$ for all $a \in GF(2)^\ell$, then the distribution of the XOR of w i.i.d. random vectors with distribution f is $f^{\otimes w}$, moreover, $\Delta^2(f) = \sum_{a \neq 0} \hat{f}^2(a)$;
- If the random Boolean variable A follows the distribution f, then $\Delta(f) = \Delta(A)$, where $\Delta(A)$ is defined in Definition 4.

5.2   An Efficient Way to Deploy Multi-biases in E0 Simultaneously

Given a linear mapping $h : GF(2)^\ell \to GF(2)^r$ of rank r, we define r-bit vectors $A_t = h(c_{rt}^0, \ldots, c_{rt+\ell-1}^0)$ and $B_t = \oplus_{i=1}^w A_{t+q_i}$. Note that B_t can be derived from the keystream $\{z_t\}$ directly. Except for accidentally bad choices of h, we make a heuristic assumption that all A_t's are independent. Let $\mathcal{D}$ be the probability distribution of the ℓ-bit vector $(c_{rt}^0, \ldots, c_{rt+\ell-1}^0)$, and let $\mathcal{D}_A$ be the probability distribution of the r-bit vector A_t. The Walsh transforms of $\mathcal{D}_A$ and $\mathcal{D}$ are linked by

$$\hat{\mathcal{D}}_A(b) = \hat{\mathcal{D}}\left(h^t(b)\right), \text{ for all } \mathrm{b} \in GF(2)^r.$$

Now we discuss how to design h in order to reduce data complexity. From Baignères [2, Theorem 3, p.10], we know that we can distinguish a distribution f of r-bit random vectors from a uniform distribution with $1/\Delta^2(f)$ samples. Here, the distribution of B_t is $f = \mathcal{D}_A^{\otimes w}$. So the modified distinguisher needs data complexity

$$n = \frac{r}{\Delta^2(\mathcal{D}_A^{\otimes w})} + d \text{ (bits)}.$$

Let k be the number of the largest Walsh coefficients $\hat{\mathcal{D}}_A(b)$ over all nonzero b with absolute value[4] η. Since $\Delta^2(\mathcal{D}_A^{\otimes w}) \approx k\eta^{2w}$, we obtain

$$n \approx \frac{r}{k}\eta^{-2w} + d.$$

In order to lower n, it's necessary to have $r < k$. This implies the k largest coefficients are linearly dependent, which happens to be true in E0: recall that the 6-bit vectors of the three largest biases satisfy the linear relation,

$$(1,1,1,1,1,0) \oplus (0,1,1,1,1,1) = (1,0,0,0,0,1).$$

As a simple solution we may just pick $\ell = 6$, $r = 2$, $\alpha_1 = (1,1,1,1,1,0)$ and $\alpha_2 = (0,1,1,1,1,1)$ (where α_i denotes the i-th row of h), then we obtain

---

[4] Note that from Subsection 4.3 we have $\eta \leq \gamma$ for $\ell \leq 26$ regardless of r and h.

$k=3$. And n is reduced to a factor of $\frac{2}{3}$ for negligible d. Indeed, recall that we proved by computation that the largest Walsh coefficient for $\ell \leq 26$ are either $(0,\ldots,0,1,1,1,1,1,0,\ldots,0)$ or $(0,\ldots,0,1,0,0,0,0,1,0,\ldots,0)$. Thus $k \leq (\ell-4)+(\ell-5) = 2\ell - 9$. This leads to a more general solution, if we pick $\ell = r + 4$, and the i-th row of h as

$$\alpha_i = (\underbrace{0,\ldots,0}_{i-1 \text{ zeros}},1,1,1,1,1,\underbrace{0,\ldots,0}_{\ell-i-4 \text{ zeros}}) \text{ for } i = 1,\ldots,r,$$

then we obtain $k = 2r - 1$. And so the improved factor $\frac{r}{2r-1}$ of data complexity tends to $\frac{1}{2}$ for negligible d when r goes to infinity; however, because of the underlying assumption for E0, ℓ is restricted to no larger than 26, i.e. $r \leq 22$. To conclude, we show that the modified distinguisher (Algorithm 1) needs data complexity

$$n \approx \frac{r}{2r - 1} \cdot \gamma^{-2w} + d, \text{ for } 1 \leq r \leq 22. \tag{7}$$

Observe that Section 4 actually deals with the special case of $r = 1$. Table 7 shows the best improvement achieved with $r = 22$. We see that the minimum n drops from previous 2^{34} to 2^{33}.

Algorithm 1 The advanced distinguisher for E0
___

Parameters:

  $r \in [1, 22], \ell = r + 4$

  $h : GF(2)^\ell \to GF(2)^r$

  $\mathcal{D}_A$: the probability distribution of the r-bit vector A_t

  $Q(x) = \sum_{i=1}^{w} x^{q_i}$: the multiple polynomial of $p_1(x)p_2(x)p_3(x)p_4(x)$ with degree d

  n: the sample size by Eq.(7)

Input:

  keystream $z_0 z_1 \cdots z_{n-1}$ of either a truly random source $\mathcal{S}_0$ or the output $\mathcal{S}_1$ generated by E0

  initialize counters $u_0, u_1, \ldots, u_{2^r-1}$

  for $t = 0, 1, \ldots, \lfloor \frac{n-d-4}{r} \rfloor - 1$ **do**

    compute $b = \oplus_{i=1}^{w} h(z_{rt+q_i}, \cdots, z_{rt+q_i+\ell-1})$

    increment u_b

  end for

  if $\sum_b u_b \cdot \log\left(2^r \cdot \mathcal{D}_A^{\otimes w}(b)\right) > 0$ **then**

    accept $\mathcal{S}_1$ as the source

  else

    accept $\mathcal{S}_0$ as the source

  end if
___

6   A Maximum Likelihood Decoding Algorithm

We restate the MLD problem for a general linear code (see [19] for details) of length n and dimension L with generator matrix G (let G_t denote the t-th

Table 7. Data complexity of the advanced distinguisher for E0

d	L	247	458	855	1749	2387	2^{18}	2^{23}	2^{27}	2^{33}	2^{44}	2^{65}	2^{32}	2^{43}
w	49	31	24	20	17	16	9	7	6	5	4	3	9	5
$\log_2 n$	328	208	161	134	114	107	60	46	40	33	44	65	60	43

column vector of G): find the closest codeword $(x_1, \ldots, x_n)$ to the received vector $(s_1, \ldots, s_n)$, and decode the message $\mathbf{r} = (r_1, \ldots, r_L)$ such that $x_t = \mathbf{r}G_t$, i.e. find such $\mathbf{r}$ that minimizes $N(\mathbf{r}) = \sum_{t=1}^{n}(s_t \oplus x_t)$.

6.1   The Time-Domain Analysis

Obviously, the trivial approach (yet common in most correlation attacks) to find $\mathbf{r}$ is an exhaustive search in time-domain: for every message $\tilde{\mathbf{r}}$, we compute the distance $N(\tilde{\mathbf{r}})$ and keep the smallest. The final record leads to $\mathbf{r}$. The time complexity is $O(n \cdot 2^L)$ with memory n-bits.

6.2   The Frequency-Domain Analysis

We introduce an integer-valued function,

$$\mathcal{W}(x) = \sum_{1 \leq t \leq n : G_t = x^\top} (-1)^{s_t}, \qquad (8)$$

for all $x \in GF(2)^L$, where $\top$ denotes the matrix transpose. We compute the Walsh transform $\hat{\mathcal{W}}$ of $\mathcal{W}$ as follows:

$$\hat{\mathcal{W}}(\mathbf{r}) = \sum_{x \in GF(2)^L} (-1)^{\mathbf{r} \cdot x} \mathcal{W}(x) = \sum_{t=1}^{n}(-1)^{s_t \oplus \mathbf{r}G_t} = \sum_{t=1}^{n}(-1)^{s_t \oplus x_t} = n - 2N(\mathbf{r}).$$

We thereby reach the theorem below.

Theorem 9.

$$N(\mathbf{r}) = \frac{1}{2}\left(n - \hat{\mathcal{W}}(\mathbf{r})\right),$$

for all $\mathbf{r} \in GF(2)^L$, *where* $\mathcal{W}$ *is defined by Eq.(8).*

This generalizes the result [19, p. 414] of a special case when $n = 2^L$ and $G_t^\top$ corresponds to the binary representation of t. So we just compute the table of $\mathcal{W}$, perform FWT [30], and find the maximal $\hat{\mathcal{W}}(\mathbf{r})$. The time and memory complexities of FWT are $O(L \cdot 2^L)$, $O(2^L)$ respectively. Since the precomputation of $\mathcal{W}$ takes time $O(n)$ with memory $O(n)$, we conclude that our improved MLD algorithm runs in $O(n + L \cdot 2^L)$ with memory $O(2^L)$ (additionally, using linear transformation allows to compute FWT over $GF(2)^k$ with memory $O(2^k)$ where $k = \lceil \log_2 n \rceil$). Note that when $n \geq 2^L$, the time complexity corresponds to $O(n)$, which is optimal in the sense that it stands on the same order of magnitude as the data complexity does. Table 8 compares the original exhaustive search algorithm with the improved frequency transformation algorithm. Note that the technique of FWT was used in another context [7] to speed up other kinds of fast correlation attacks. In the next section we will see how it helps to speed up the attack [10] by a factor of 2^{24}. We estimate similar correlation attacks like [6] can be speeded up by a factor of 10; undoubtedly, some other attacks can be significantly improved by our algorithm as well.

Table 8. Maximum likelihood decoding algorithms

	time	memory
Exhaustive Search	$n \cdot 2^L$	n
Frequency Transformation	$n + L \cdot 2^L$	$\min(n, 2^L)$

6.3   A More Generalized MLD Algorithm

We further generalize the preceding problem by finding the L-bit vector $\mathbf{r}$ such that given a sequence of ℓ-bit vectors $S_1, \ldots, S_k$ and $f : GF(2)^\ell \to \mathbf{R}$ together with matrices $G_1, \ldots, G_k$ of size L by ℓ, the sequence of ℓ-bit vectors $X_1, \ldots, X_k$ defined by $X_t = \mathbf{r}G_t$ minimizes $N(\mathbf{r}) = \sum_{t=1}^k f(S_t \oplus X_t)$. Note that previous subsections are merely a special case of $\ell = 1$, $k = n$ and $f(a) = a$ for $a \in GF(2)$.

Define a real function $\mathcal{W}$ by:

$$\mathcal{W}(x) = \frac{1}{2^\ell} \sum_{1 \le t \le k, a \in GF(2)^\ell : aG_t^\top = x} (-1)^{a \cdot S_t} \hat{f}(a),$$

for all $x \in GF(2)^L$. We compute the Walsh transform $\hat{\mathcal{W}}$ of $\mathcal{W}$ as follows:

$$\hat{\mathcal{W}}(\mathbf{r}) = \sum_{x \in GF(2)^L} (-1)^{\mathbf{r} \cdot x} \mathcal{W}(x)$$

$$= \frac{1}{2^\ell} \sum_{t=1}^k \sum_{a \in GF(2)^\ell} (-1)^{a \cdot (\mathbf{r}G_t \oplus S_t)} \hat{f}(a)$$

$$= \sum_{t=1}^k f(\mathbf{r}G_t \oplus S_t)$$

$$= N(\mathbf{r}).$$

Algorithm 2 directly follows above computation. The total running time of our algorithm is $O(k\ell L2^\ell + L2^L)$ with memory $O(2^L)$. To speed up the computation of $\mathcal{W}$, we could precompute the inner products of all pairs of ℓ-bit vectors in time $O(2^{2\ell})$ with memory $O(2^{2\ell})$. Thus, the total running time of the algorithm is $O(2^{2\ell} + kL2^\ell + L2^L)$ with memory $O(2^{2\ell} + 2^L)$.

Algorithm 2 The generalized MLD algorithm

Parameter:
f, ℓ
Input:
$G = (G_1, \ldots, G_k)$: the generator matrix
vector stream $S_1, S_2, \cdots, S_k$
Processing:
apply FWT to compute the table of $\hat{f}$
initialize the table of $\mathcal{W}$ to 0
for all ℓ-bit a **do**
  for $t = 1, \ldots, k$ **do**
    increment $\mathcal{W}(aG_t^\top)$ by $\frac{1}{2^\ell}(-1)^{a \cdot S_t} \hat{f}(a)$
  end for
end for
apply FWT to find $\mathbf{r}$ that achieves the minimal $\hat{\mathcal{W}}(\mathbf{r})$
output $\mathbf{r}$

In the special case that is applicable to E0 (as is done in the next section): $G_{t+1} = AG_t$ for $t = 1, \ldots, k$, we precompute another table to map any L-bit vector x to $xA^\top$. It takes time 2^L with memory 2^L. The total time of the algorithm is thus $O\left(2^{2\ell} + (L+k)\,2^\ell + L2^L\right)$, with memory $O(2^{2\ell} + 2^L)$.

7   The Key-Recovery Attack Against E0

We approach similarly as in [10] to transform our distinguisher of Subsection 4.3 into a key-recovery attack. Our main contribution, however, is to decrease the time complexity by applying the preceding algorithm.

Let $Q(x) = \sum_{i=1}^{w} x^{q_i}$ be the multiple polynomial of $p_2(x)p_3(x)p_4(x)$ with degree d and weight w. Using techniques in Subsection 4.2 to find $Q(x)$ with (precomputation) complexity PC, we list the corresponding triplets (w, d, PC) for small w in Table 9.

Table 9. Complexity of finding multiple of $p_2(x)p_3(x)p_4(x)$ with degree d, weight w

	birthday problem				tradeoff
	with min. d				
weight w	5	4	3	2	5
degree d	2^{27}	2^{36}	2^{52}	2^{100}	$2^{34.3}$
Precomputation PC	2^{54}	2^{54}	2^{52}	-	$2^{36.3}$

Let $\tilde{\mathbf{x}}^1$ be a guess for $\mathbf{x}^1$, the initial state of R_1 which generates the keystream $\{z_t\}$ together with the other three fixed LFSRs. Denote $\tilde{x}_t^1$ the output bit of R_1 with the initial state $\tilde{\mathbf{x}}^1$ at time t. We define

$$b_t(\tilde{\mathbf{x}}^1) = \oplus_{i=1}^{w}(z_{t+q_i} \oplus z_{t+q_i+5}) \oplus \oplus_{i=1}^{w}(\tilde{x}_{t+q_i}^1 \oplus \tilde{x}_{t+q_i+5}^1). \qquad (9)$$

It can be shown that the second addend in Eq.(9) is also an m-sequence generated by the same LFSR. For brevity, we set

$$r_t = \oplus_{i=1}^{w}(\tilde{x}_{t+q_i}^1 \oplus \tilde{x}_{t+q_i+5}^1),$$
$$s_t = \oplus_{i=1}^{w}(z_{t+q_i} \oplus z_{t+q_i+5}),$$

for $t = 1, \ldots, n$ (it corresponds to the data complexity $n+d$). We rewrite Eq.(9) as

$$b_t(\tilde{\mathbf{x}}^1) = s_t \oplus r_t.$$

Given n-bit sequence of $b_t(\tilde{\mathbf{x}}^1)$'s, we count the occurrences[5] $N(\tilde{\mathbf{x}}^1)$ of ones, i.e. $N(\tilde{\mathbf{x}}^1) = \sum_{t=0}^{n-1} b_t(\tilde{\mathbf{x}}^1)$. Using the analysis of [28], we estimate $N(\mathbf{x}^1)$ is the smallest of all $N(\tilde{\mathbf{x}}^1)$ with

$$n \approx \frac{4L_1 \log 2}{\gamma^{2w}}. \tag{10}$$

Note that this estimated figure is actually comparable to the conventional estimation [6, 16] on critical data complexity n_0 in correlation attacks, where

$$n_0 = \frac{L_1}{1 - h(\frac{1}{2} + \frac{1}{2}\gamma^w)} \approx \frac{2L_1 \log 2}{\gamma^{2w}}, \tag{11}$$

and h is the binary entropy function[6]. According to [6] simulations showed the probability of success is close to 1 (resp. $\frac{1}{2}$) for $n = 2n_0$ (resp. $n = n_0$) which is consistent with our analysis. Table 10 shows our estimated minimal n for $N(\mathbf{x}^1)$ to achieve a top rank corresponding to w. Now define $G_t = (a_0, \ldots, a_{L_1-1})^\top$, where $a_0 + a_1 x + \cdots + a_{L_1-1}x^{L_1-1} = x^t \mod p_1(x)$. Clearly our current problem to recover R_1 right fits into the MLD problem in Subsection 6.2. So we use the preceding MLD algorithm to recover $\mathbf{r}$ first, then apply linear transform to solve $\mathbf{x}^1$. Finally we conduct the same analysis as in Section 5 to decrease data complexity down to $O(\frac{r}{2^r-1} \times \frac{4L_1 \log 2}{\gamma^{2w}})$; and we apply the technique introduced in Subsection 6.3 to obtain the reduced time complexity $O(n + \theta_1 \cdot 2^r + L_1 \cdot 2^{L_1})$. So, choosing $r = 12$, we can halve the time and data complexities. The attack complexities to recover R_1 for E0 are listed in Table 11. Once we recover R_1, we target R_2 next based on multiple of $p_3(x)p_4(x)$. Last, we use the technique of guess and determine in [11] to solve R_3 and R_4 with knowledge of the shortest two LFSRs. The detailed complexities of each step are shown in Table 12. A comparison of our attacks with the similar attack[7] [10] and the best two algebraic attacks [1, 8] is shown in Table 13.

---

[5] w is fixed in the attack, so we omit it in the notation $N(\tilde{\mathbf{x}}^1)$.

[6] $h(p) = -p \log_2 p - (1-p) \log_2(1-p)$ for $0 < p < 1$.

[7] The estimate of data complexity in [10] uses a different heuristic formula than ours. However we believe that their estimate and ours in Attack B are essentially the same.

Table 10. The estimated minimal n for $N(\mathbf{x^1})$ to top the rank from Eq.(10)

weight w	5	4	3	2	1
n	2^{40}	2^{33}	2^{27}	2^{20}	2^{14}

Table 11. Summary of primary partial key-recovery attacks against R_1 for E0

	w	d	n	data	precomputation	time	memory
Attack A	5	$2^{34.3}$	2^{39}	2^{39}	$2^{36.3}$	2^{39}	2^{25}
Attack B	4	2^{36}	2^{33}	2^{36}	2^{54}	2^{36}	2^{25}

Table 12. Detailed complexities of our key-recovery attack against E0

	w	d	n	data	precomputation	time	memory
R_1	5	$2^{34.3}$	2^{39}	2^{39}	$2^{36.3}$	2^{39}	2^{25}
R_2	3	2^{36}	2^{27}	2^{36}	2^{37}	2^{36}	2^{27}
R_3 and R_4	-	-	-	76	-	2^{33}	-
total	-	-	-	2^{39}	2^{37}	2^{39}	2^{27}

Table 13. Complexities comparison of our attacks with the similar attack and algebraic attacks

		Precomputation	Time	Data	Memory
Algebraic	[1]	-	$2^{67.58}$	$2^{23.07}$	$2^{46.14}$
Attacks	[8]	2^{28}	2^{49}	$2^{23.4}$	2^{37}
Attack [10]		2^{54}	2^{63}	2^{34}	2^{34}
Our	A	2^{37}	2^{39}	2^{39}	2^{27}
Attacks	B	2^{54}	2^{37}	2^{36}	2^{27}

8   Conclusions

This paper formulates a systematic computation method of correlations by a recursive expression, which makes it easier to calculate correlations of the FSM output sequences up to 26 bits for E0 (and allows us to prove for the first time that the two known biases are the only largest). Then we successfully apply the concept of convolution to the analysis of the distinguisher based on all correlations, which allows us to build an efficient distinguisher that halves the data complexity of the basic uni-bias-based distinguisher due to the linear dependency of the two largest biases. Finally, by means of FWT, we propose a novel MLD algorithm to recover the closest codeword for any linear code. Our proposed algorithm can be easily adapted to speed up a class of fast correlation attacks. Furthermore the algorithm is optimal when the length n of the code and the dimension L satisfy the relation $n \geq 2^L$, which is the case when we apply it to recover R_1 for E0. This results in the best known key-recovery attack against E0. Considering a maximal keystream length of 2745 bits for practical E0 in

Bluetooth, our results still remain the academic interest. Meanwhile, our attack successfully illustrates the attack methodology of Baignères et al.[8]

Acknowledgment

Words fail the first author when she would like to express her heartfelt thanks to her forever faithful bosom friend Aoife Hegarty, who, in the face of a rough time in her own life, has been generously offering continuous one-way help through all those years ...

References

1. F. Armknecht, M. Krause, *Algebraic Attacks on Combiners with Memory*, Advances on Cryptography - CRYPTO 2003, Lecture Notes in Computer Science, vol.2729, D. Boneh ed., Springer-Verlag, pp. 162-175, 2003
2. T. Baignères, *A Generalization of Linear Cryptanalysis*, Diploma Thesis, EPFL, 2003
3. Bluetooth^{TM}, *Bluetooth Specification*, version 1.2, pp. 903-948, November, 2003, available at http://www.bluetooth.org
4. A. Canteaut, F. Chabaud, *A New Algorithm for Finding Minimum-weight Words in a Linear Code: Application to Primitive Narrow-sense BCH Codes of Length 511*, INRIA, technical report, No. 2685, 1995
5. A. Canteaut, M. Trabbia, *Improved Fast Correlation Attacks Using Parity-check Equations of Weight 4 and 5*, Advances in Cryptology - EUROCRYPT 2000, Lecture Notes in Computer Science, vol.1807, B. Preneel ed., Springer-Verlag, pp. 573-588, 2000
6. V. Chepyzhov, T. Johansson, B. Smeets, *A Simple Algorithm for Fast Correlation Attacks on Stream Ciphers*, Fast Software Encryption 2000, Lecture Notes in Computer Science, vol.1978, B. Schneier ed., Springer-Verlag, pp. 181-195, 2000
7. P. Chose, A. Joux, M. Mitton, *Fast Correlation Attacks: An Algorithmic Point of View*, Advances in Cryptology - EUROCRYPT 2002, Lecture Notes in Computer Science, vol.2332, L. R. Knudsen ed., Springer-Verlag, pp. 209-221, 2002
8. N. T. Courtois, *Fast Algebraic Attacks on Stream Ciphers with Linear Feedback*, Advances on Cryptography - CRYPTO 2003, Lecture Notes in Computer Science, vol.2729, D. Boneh ed., Springer-Verlag, pp. 176-194, 2003
9. P. Ekdahl, T. Johansson, *Some Results on Correlations in the Bluetooth Stream Cipher*, Proceedings of the 10th Joint Conference on Communications and Coding, Austria, 2000
10. P. Ekdahl, *On LFSR Based Stream Ciphers (Analysis and Design)*, Ph.D. Thesis, Lund Univ., Nov. 2003
11. S. Fluhrer, S. Lucks, *Analysis of the E0 Encryption System*, Selected Areas in Cryptography- SAC 2001, Lecture Notes in Computer Science, vol. 2259, S. Vaudenay and A. Youssef eds., Springer-Verlag, pp. 38-38, 2001
12. J. D. Golić, *Correlation Properties of a General Binary Combiner with Memory*, Journal of Cryptology, vol. 9, pp. 111-126, Nov. 1996

---

[8] Personal communication. Prior work available in [2].

13. J. D. Golić, V. Bagini, G. Morgari, *Linear Cryptanalysis of Bluetooth Stream Cipher*, Advances in Cryptology - EUROCRYPT 2002, Lecture Notes in Computer Science, vol. 2332, L. R. Knudsen ed., Springer-Verlag, pp. 238-255, 2002

14. M. Hermelin, K. Nyberg, *Correlation Properties of the Bluetooth Combiner*, Information Security and Cryptology- ICISC'99, Lecture Notes in Computer Science, vol. 1787, JooSeok. Song ed., Springer-Verlag, pp. 17-29, 2000

15. M. Jakobsson, S. Wetzel, *Security Weakness in Bluetooth*, Topics in Cryptology - CT-RSA 2001, Lecture Notes in Computer Science, vol. 2020, D. Naccache ed., Springer-Verlag, pp. 176-191, 2001

16. T. Johansson, F. Jonsson, *Improved Fast Correlation Attacks on Stream Ciphers via Convolutional Codes*, Advances on Cryptography - CRYPTO'99, Lecture Notes in Computer Science, vol.1666, M. Wiener ed., Springer-Verlag, pp. 181-197, 1999

17. M. Krause, *BDD-Based Cryptanalysis of Keystream Generators*, Advances in Cryptology - EUROCRYPT 2002, Lecture Notes in Computer Science, vol. 2332, L. R. Knudsen ed., Springer-Verlag, pp. 222-237, 2002

18. R. Lidl, H. Niederreiter, *Introduction to Finite Fields and Their Applications*, Cambridge, 1986

19. F. J. MacWilliams, N. J. A. Sloane, *The Theory of Error-correcting Codes*, North-Holland, 1996

20. M. Matsui, *Linear Cryptanalysis Method for DES Cipher*, Advances in Cryptology - EUROCRYPT'93, Lecture Notes in Computer Science, vol.765, Springer-Verlag, pp. 386-397, 1993

21. W. Meier, O. Staffelbach, *Fast Correlation Attacks on Certain Stream Ciphers*, Journal of Cryptology, vol. 1, pp. 159-176, Nov. 1989

22. W. Meier, O. Staffelbach, *Correlation Properties of Combiners with Memory in Stream Ciphers*, Journal of Cryptology, vol. 5, pp. 67-86, Nov. 1992

23. A. J. Menezes, P. C. van. Oorschot, S. A. Vanstone, *Handbook of Applied Cryptography*, CRC, 1996

24. W. Penzhorn, *Correlation Attacks on Stream Ciphers: Computing Low Weight Parity Checks based on Error Correcting Codes*, Fast Software Encryption 96, Lecture Notes in Computer Science, vol.1039, D. Gollmann ed., Springer-Verlag, pp. 159-172, 1996

25. R. A. Rueppel, *Correlation Immunity and the Summation Generator*, Advances in Cryptology - CRYPTO'85, Lecture Notes in Computer Science, Springer-Verlag, pp. 260-272, 1986

26. M. Saarinen, *Re: Bluetooth and E0*, Posted at `sci.crypt.research`, 02/09/00

27. T. Siegenthaler, *Correlation-Immunity of Nonlinear Combining Functions for Cryptographic Applications*, IEEE Transactions on Information Theory, vol. 30, pp. 776-780, 1984

28. S. Vaudenay, *An Experiment on DES - Statistical Cryptanalysis*, Proceedings of the 3rd ACM Conferences on Computer Security, pp. 139-147, 1996

29. D. Wagner, *A Generalized Birthday Problem*, Advances in Cryptology - CRYPTO 2002, Lecture Notes in Computer Science, vol.2442, Springer-Verlag, pp. 288-304, 2002

30. R. K. Yarlagadda, J. E. Hershey, *Hadamard Matrix Analysis and Synthesis with Applications to Communications and Signal/Image Processing*, Kluwer Academic, pp. 17-22, 1997

Appendix

A. Proof of Lemma 5

Let $Z \in GF(2)^k$ be a random variable independent of X with uniform distribution. We have

$$\sum_w \Delta(f(X,Z) \oplus w \cdot Z) \cdot \Delta(w \cdot g(Y) \oplus v \cdot Y)$$

$$= \sum_w E[(-1)^{f(X,Z) \oplus w \cdot Z}] \cdot E[(-1)^{w \cdot g(Y) \oplus v \cdot Y}]$$

$$= \sum_{w,x,y,z} \Pr(x,z) \cdot \Pr(y) \cdot (-1)^{f(x,z) \oplus v \cdot y \oplus w \cdot (z \oplus g(y))}$$

$$= 2^k \cdot \sum_{x,y} \Pr(X=x, Z=g(y)) \cdot \Pr(Y=y) \cdot (-1)^{f(x,g(y)) \oplus v \cdot y}$$

$$= \sum_{x,y} \Pr(x,y) \cdot (-1)^{f(x,g(y)) \oplus v \cdot y}$$

$$= E[(-1)^{f(X,g(Y)) \oplus v \cdot Y}],$$

which is $\Delta(f(X, g(Y)) \oplus v \cdot Y)$.

B. Examples of Multiple Polynomials $Q(x)$

Example of $Q(x)$ with Low Degree. Here is a multiple polynomial of degree less than 855 with weight 31:

$$Q_1(x) = x^{668} + x^{579} + x^{553} + x^{313} + x^{262} + x^{121} + x^{117} + x^{109} + x^{106} + x^{101} +$$
$$x^{100} + x^{97} + x^{94} + x^{87} + x^{82} + x^{76} + x^{72} + x^{71} + x^{57} + x^{47} +$$
$$x^{40} + x^{37} + x^{34} + x^{32} + x^{23} + x^{21} + x^{17} + x^{16} + x^3 + x^2 + 1.$$

Observe that $Q_1(x)$ is not optimal as $w_{855} = 20$ from Table 3.

Examples of $Q(x)$ with Weight Four. Recall that $\theta_i = 2^{L_i} - 1$ is the order of $p_i(x)$ for $i = 1, 2, 3, 4$. By definition, $p_i(x)|x^{\theta_i} + 1$. On the other hand, $p_i(x)p_j(x)|\mathrm{lcm}(x^{\theta_i} + 1, x^{\theta_j} + 1) = x^{\mathrm{lcm}(\theta_i, \theta_j)} + 1$ for $i \neq j$, hence we deduce the following three multiple polynomials of $p(x)$ with weight 4 with ease:

$$Q_2(x) = (x^{\mathrm{lcm}(\theta_1, \theta_2)} + 1)(x^{\mathrm{lcm}(\theta_3, \theta_4)} + 1),$$
$$Q_3(x) = (x^{\mathrm{lcm}(\theta_1, \theta_3)} + 1)(x^{\mathrm{lcm}(\theta_2, \theta_4)} + 1),$$
$$Q_4(x) = (x^{\mathrm{lcm}(\theta_1, \theta_4)} + 1)(x^{\mathrm{lcm}(\theta_2, \theta_3)} + 1),$$

where

$$\mathrm{lcm}(\theta_1, \theta_2) = 2^{56} - 2^{31} - 2^{25} + 1, \quad \mathrm{lcm}(\theta_1, \theta_3) = 2^{58} - 2^{33} - 2^{25} + 1,$$
$$\mathrm{lcm}(\theta_1, \theta_4) = 2^{64} - 2^{39} - 2^{25} + 1, \quad \mathrm{lcm}(\theta_2, \theta_3) = 2^{64} - 2^{33} - 2^{31} + 1,$$
$$\mathrm{lcm}(\theta_2, \theta_4) = 2^{70} - 2^{39} - 2^{31} + 1, \quad \mathrm{lcm}(\theta_3, \theta_4) = (2^{39} - 1)\sum_{k=0}^{10} 2^{3k}.$$

The degrees of $Q_2(x), Q_3(x), Q_4(x)$ are approximately $2^{69}, 2^{70}, 2^{65}$ respectively. Note that we may also expect optimal multiples with degree in the same order of magnitude and weight 3.

C. Table of $\log_2 |\hat{\mathcal{D}}(a)|$ for $\ell = 6$

Table 14. $\log_2 |\hat{\mathcal{D}}(c_t^0, c_{t+1}^0, \ldots, c_{t+5}^0)|$ where dashed entries denote $-\infty$

		$c_{t+3}^0 c_{t+4}^0 c_{t+5}^0$							
		000	001	010	011	100	101	110	111
$c_t^0 c_{t+1}^0 c_{t+2}^0$	000	0	-	-	-	-	-	-	-
	001	-	-	-	-4	-	-	-	-
	010	-	-	-	-	-	-	-4	-
	011	-	-	-	-	-	-6	-	-3.356
	100	-	-3.356	-	-	-	-	-	-8
	101	-	-	-	-	-4	-	-	-
	110	-	-8	-	-5.356	-	-5.356	-	-5.356
	111	-	-	-6	-	-	-	-3.356	-

A New Paradigm of Hybrid Encryption Scheme

Kaoru Kurosawa[1] and Yvo Desmedt[2,3]

[1] Ibaraki University, Japan
kurosawa@cis.ibaraki.ac.jp
[2] Dept. of Computer Science, University College London, UK
[3] Florida State University, USA

Abstract. In this paper, we show that a key encapsulation mechanism (KEM) does not have to be IND-CCA secure in the construction of hybrid encryption schemes, as was previously believed. That is, we present a more efficient hybrid encryption scheme than Shoup [12] by using a KEM which is not necessarily IND-CCA secure. Nevertheless, our scheme is secure in the sense of IND-CCA under the DDH assumption in the standard model. This result is further generalized to universal$_2$ projective hash families.

Keywords: hybrid encryption, KEM, standard model

1 Introduction

1.1 Background

Cramer and Shoup showed the first provably secure practical public-key encryption scheme in the standard model [3, 6]. It is secure against adaptive chosen ciphertext attack (IND-CCA) under the Decisional Diffie-Hellman (DDH) assumption. They further generalized their scheme to projective hash families [4]. (In the random oracle model [1], many practical schemes have been proven to be IND-CCA, for example, OAEP+ [13], SAEP [2] , RSA-OAEP [8], etc. [7]. However, while the random oracle model is a useful tool, it does not rule out all possible attacks.)

On the other hand, a hybrid encryption scheme uses public-key encryption techniques to derive a shared key that is then used to encrypt the actual messages using symmetric-key techniques.

For hybrid encryption schemes, Shoup formalized the notion of a key encapsulation mechanism (KEM), and an appropriate notion of security against adaptive chosen ciphertext attack [12,6]. A KEM works just like a public key encryption scheme, except that the encryption algorithm takes no input other than the recipient's public key. The encryption algorithm can only be used to generate and encrypt a key for a symmetric-key encryption scheme. (One can always use a public-key encryption scheme for this purpose. However, one can construct a KEM in other ways as well.) A secure KEM, combined with an appropriately secure symmetric-key encryption scheme, yields a hybrid encryption scheme which is secure in the sense of IND-CCA [12].

M. Franklin (Ed.): CRYPTO 2004, LNCS 3152, pp. 426–442, 2004.

Shoup presented a secure KEM under the DDH assumption [12]. As a result, his hybrid encryption scheme is secure in the sense of IND-CCA under the DDH assumption in the standard model [12].

1.2 Our Contribution

In order to prove the security of hybrid encryption schemes, one has believed that it is essential for KEM to be secure in the sense of IND-CCA, as stated in [6, Remark 7.2, page 207].

In this paper, however, we disprove this belief. That is, it is shown that KEM does not have to be CCA secure, as was previously believed. On a more concrete level, we present a more efficient hybrid encryption scheme than Shoup [12] by using a KEM which is not necessarily secure in the sense of IND-CCA. Nevertheless, we prove that the proposed scheme is secure in the sense of IND-CCA under the DDH assumption in the standard model.

In a typical implementation, the underlying Abelian group may be a subgroup of Z_p^*, where p is a large prime. In this case, the size of our ciphertexts is $|p|$ bits shorter than that of Shoup [12]. The number of exponentiations per encryption and that of per decryption are also smaller. (Further, our scheme is more efficient than the basic Cramer-Shoup scheme [3, 6].)

This shows that one can start with a weak KEM and repair it with a hybrid construction. Eventually, more efficient hybrid encryption schemes could be obtained.

Our KEM is essentially a universal$_2$ projective hash family [4]. We present a generalization of our scheme to universal$_2$ projective hash families also.

The only (conceptual) cost one pays is that one needs to assume a simple condition on the symmetric encryption scheme. Namely, any fixed ciphertext is rejected with overwhelming probability, where the probability is taken over keys K. This property is already satisfied by the symmetric encryption scheme SKE which is used in the hybrid construction of Shoup [12]. Hence the SKE can be used in our hybrid construction too.

Our result gives new light to Cramer-Shoup encryption schemes [3, 4, 6] and opens a door to design more efficient hybrid encryption schemes.

2 Preliminaries

We denote by λ a security parameter. PPT denotes probabilistic polynomial time.

2.1 Notation and Definitions

$|S|$ denotes the cardinality of S if S is a set. $|m|$ denotes the bit length of m if m is a string or a number. If $A(\cdot, \cdot, \cdots)$ is a probabilistic algorithm, then $x \xleftarrow{R} A(x_1, x_2, \cdots)$ denotes the experiment of running A on input $x_1, x_2, \cdots$ and letting x be the outcome. If S is a set, $x \xleftarrow{R} S$ denotes the experiment of choosing $x \in S$ at random.

2.2   Public-Key Encryption Scheme (PKE)

A public-key encryption scheme is a three tuple of algorithms $\mathsf{PKE} = (\mathcal{K}_p, \mathcal{E}_p, \mathcal{D}_p)$. The key generation algorithm $\mathcal{K}_p$ generates a pair $(pk, sk) \stackrel{R}{\leftarrow} \mathcal{K}_p(1^\lambda)$, where pk is a public key and sk is a secret key. The encryption algorithm $\mathcal{E}_p$ takes a public key pk and a plaintext m, and returns a ciphertext $c \stackrel{R}{\leftarrow} \mathcal{E}_p(pk, m)$. The decryption algorithm $\mathcal{D}_p$ takes a secret key sk and a ciphertext c, and returns m or *reject*.

The chosen plaintext attack (IND-CPA) game is defined as follows. We imagine a PPT adversary A that runs in two stages. In the "find" stage, A takes a public key pk and queries a pair of equal length messages m_0 and m_1 to an encryption oracle. The encryption oracle chooses $b \stackrel{R}{\leftarrow} \{0,1\}$ and computes a challenge ciphertext c^* of m_b randomly. In the "guess" stage, given c^*, A outputs a bit $\tilde{b}$ and halts.

The adaptive chosen ciphertext attack (IND-CCA) game is defined similarly. The difference is that the adversary A is given access to a decryption oracle, where A cannot query the challenge ciphertext c^* itself in the guess stage.

Definition 1. *We say that* PKE *is secure in the sense of IND-CCA if* $|\Pr(\tilde{b} = b) - 1/2|$ *is negligible in the IND-CCA game for any PPT adversary A.*

In particular, we define the IND-CCA advantage of A as follows:

$$\mathsf{Adv}_{\mathsf{PKE}}^{cca}(A) = |\Pr(\tilde{b} = b) - 1/2|. \tag{1}$$

For any t and q_d, define $\mathsf{Adv}_{\mathsf{PKE}}^{cca}(t, q_d) = \max_A \mathsf{Adv}_{\mathsf{PKE}}^{cca}(A)$, where the maximum is taken over all A which runs in time t and makes at most q_d queries to the decryption oracle.

2.3   Diffie-Hellman Assumptions

Let G be an Abelian group of order Q, where Q is a large prime. Let g_1 be a generator of G. Let

$$DH = \{(g_1, g_2, g_1^r, g_2^r) \mid r \in Z_Q, g_2 = g_1^w, w \in Z_Q\}$$
$$Random = \{(g_1, g_2, g_1^{r_1}, g_2^{r_2}) \mid r_1 \in Z_Q, r_2 \in Z_Q, g_2 = g_1^w, w \in Z_Q\}$$

The decisional Diffie-Hellman (DDH) assumption claims that DH and $Random$ are indistinguishable.

For a distinguisher D, consider the following two experiments. In experiment 0, let $(g_1, g_2, u_1, u_2) \stackrel{R}{\leftarrow} DH$. In experiment 1, let $(g_1, g_2, u_1, u_2) \stackrel{R}{\leftarrow} Random$. Define

$$\mathsf{Adv}_G^{ddh}(D) \stackrel{\triangle}{=} |p_0 - p_1|,$$

where

$$p_0 \stackrel{\triangle}{=} \Pr(D = 1 \text{ in experiment 0}), \quad p_1 \stackrel{\triangle}{=} \Pr(D = 1 \text{ in experiment 1})$$

For any t, define $\mathsf{Adv}_G^{ddh}(t) \stackrel{\triangle}{=} \max_A \mathsf{Adv}_G^{ddh}(D)$, where the maximum is taken over all D which runs in time t.

2.4   Target Collision Resistant Hash Function

The notion of target collision resistant TCR family of hash functions was shown by Cramer and Shoup [6]. It is a special case of universal one-way hash function UOWH family introduced by Naor and Yung [10], where a UOWH family can be built from arbitrary one-way functions [10, 11].

In a TCR family, given a randomly chosen tuple of group elements x ($\in G^n$ for some n) and a randomly chosen hash function H, it is infeasible for an adversary A to find $y \neq x$ such that $H(x) = H(y)$. (In a UOWH family, x is chosen by the adversary.) In practice, one can use a dedicated cryptographic hash function, like SHA-1. Define

$$\mathrm{Adv}^{hash}_{\mathsf{TCR}}(A) \triangleq \Pr(A \text{ succeeds}).$$

For any t, define $\mathrm{Adv}^{hash}_{\mathsf{TCR}}(t) \triangleq \max_A \mathrm{Adv}^{hash}_{\mathsf{TCR}}(A)$, where the maximum is taken over all A which runs in time t.

3   Previous Results on KEM

It is known that by combining a KEM and a one-time symmetric encryption scheme which are both secure in the sense of IND-CCA, we can obtain a hybrid encryption scheme which is secure in the sense of IND-CCA.

3.1   KEM [12][6, Sec.7.1]

A key encapsulation mechanism KEM consists of the following algorithms.

- A key generation algorithm KEM.Gen that on input 1^λ outputs a public/secret key pair (pk, sk).
- An encryption algorithm KEM.Enc that on input 1^λ and a public key pk, outputs a pair (K, ψ), where K is a key and ψ is a ciphertext.
  A key K is a bit string of length KEM.Len(λ), where KEM.Len(λ) is another parameter of KEM.
- A decryption algorithm KEM.Dec that on input 1^λ, a secret key sk, a string (in particular a ciphertext) ψ, outputs either a key K or the special symbol reject.

KEM.Gen and KEM.Enc are PPT algorithms and KEM.Dec is a deterministic polynomial time algorithm.

In the chosen ciphertext attack (IND-CCA) game, we imagine a PPT adversary A that runs in two stages. In the find stage, A takes a public key pk and queries an encryption oracle. The encryption oracle computes:

$$(K^*, \psi^*) \xleftarrow{R} \mathsf{KEM.Enc}(1^\lambda); K^+ \xleftarrow{R} \{0,1\}^k; \tau \xleftarrow{R} \{0,1\};$$

$$\text{if } \tau = 0 \text{ then } K^\dagger \leftarrow K^* \text{ else } K^\dagger \leftarrow K^+$$

where $k = \mathsf{KEM.Len}(\lambda)$, and responds with the pair $(K^\dagger, \psi^*)$. In the guess stage, given $(K^\dagger, \psi^*)$, the adversary A outputs a bit $\tilde{\tau}$ and halts.

The adversary A is also given access to a decryption oracle. For each decryption oracle query, the adversary A submits a ciphertext ψ, and the decryption oracle responds with $\mathsf{KEM.Dec}(1^\lambda, sk, \psi)$, where A cannot query the challenge ciphertext ψ^* itself in the guess stage.

Definition 2. *We say that* KEM *is secure in the sense of IND-CCA if* $|\Pr(\tilde{\tau} = \tau) - 1/2|$ *is negligible in the above game for any PPT adversary A.*

3.2   One-Time Symmetric-Key Encryption [6, Sec.7.2]

A one-time symmetric-key encryption scheme SKE consists of two algorithms:

- A deterministic polynomial time encryption algorithm SKE.Enc that takes as input 1^λ, a key K and a message m, and outputs a ciphertext χ.
- A deterministic polynomial time decryption algorithm SKE.Dec that takes as input 1^λ, a key K and a ciphertext χ, and outputs a message m or the special symbol reject.

The key K is a bit string of length $\mathsf{SKE.Len}(\lambda)$, where $\mathsf{SKE.Len}(\lambda)$ is a parameter of the encryption scheme.

In the passive attack game, we imagine a PPT adversary A that runs in two stages. In the "find" stage, A takes 1^λ, and queries a pair of equal length messages m_0 and m_1 to an encryption oracle. The encryption oracle generates a random key K of length $\mathsf{SKE.Len}(\lambda)$, along with random $\sigma \stackrel{R}{\leftarrow} \{0,1\}$, and encrypts m_σ using the key K. In the "guess" stage, given the resulting ciphertext χ^*, A outputs a bit $\tilde{\sigma}$ and halts.

In the chosen ciphertext attack (IND-CCA) model, the adversary A is also given access to a decryption oracle in the guess stage. In each decryption oracle query, A submits a ciphertext $\chi \neq \chi^*$, and obtains the decryption of χ under the key K.

Definition 3. *We say that* SKE *is secure in the sense of IND-CCA if* $|\Pr(\tilde{\sigma} = \sigma) - 1/2|$ *is negligible in the IND-CCA game for any PPT adversary A.*

In particular, we define the IND-CCA advantage of A as follows.

$$\mathsf{Adv}^{cca}_{\mathsf{SKE}}(A) = |\Pr(\tilde{\sigma} = \sigma) - 1/2|. \tag{2}$$

For any t and q_d, define $\mathsf{Adv}^{cca}_{\mathsf{SKE}}(t, q_d) = \max_A \mathsf{Adv}^{cca}_{\mathsf{SKE}}(A)$, where the maximum is taken over all A which runs in time t and makes at most q_d queries to the decryption oracle.

3.3   Construction of SKE

Shoup showed a construction of a one-time symmetric-key encryption scheme as follows [12, page 281]. Let PRBG be a pseudo-random bit generator which stretches l-bit strings to strings of arbitrary (polynomial) length. We assume that $1/2^l$ is a negligible quantity. In a practical implementation, it is perfectly reasonable to stretch the key K_0 by using it as the key to a dedicated block cipher, and then evaluate the block cipher at successive points (so called "counter mode") to obtain a sequence of pseudo-random bits [6, Sec.7.2.2].

Let AXUH be a hash function which is suitable for message authentication, i.e., an almost XOR-universal hash function [9]. We assume that AXUH is keyed by an l'-bit string and hashes arbitrary bit string to l-bit strings. Many efficient constructions for AXUH exist that do not require any intractability assumptions.

To encrypt a message m by using a key $K = (K_0, K_1, K_2)$, we apply PRBG to K_0 to obtain an $|m|$-bit string f. Then we compute

$$e = f \oplus m, \tag{3}$$
$$a = \text{AXUH}(K_1, e) \oplus K_2. \tag{4}$$

The ciphertext is $\chi = (e, a)$, where a is called a tag. (We can generate K by applying PRBG to a shorter key.)

To decrypt $\chi = (e, a)$ using a key $K = (K_0, K_1, K_2)$, we first test if eq.(4) holds. If it does not hold, then we *reject*. Otherwise, we output $m = e \oplus f$.

3.4   A Hybrid Construction

Let KEM be a key encapsulation mechanism and let SKE be a one-time symmetric key encryption scheme such that $\text{KEM.Len}(\lambda) = \text{SKE.Len}(\lambda)$ for all λ. Let HPKE be the hybrid public-key encryption scheme obtained from KEM and SKE.

Proposition 1. *[6, Theorem 7.2] If KEM and SKE are secure in the sense of IND-CCA, then so is HPKE.*

4   Proposed Hybrid Encryption Scheme

In this section, we show a more efficient hybrid encryption scheme than before [12,6] by using a KEM which is not necessarily secure in the sense of IND-CCA. Nevertheless, we prove that the proposed scheme is secure in the sense of IND-CCA under the DDH assumption in the standard model.

4.1   Overview

A KEM works just like a public key encryption scheme, except that the encryption algorithm takes no input other than the recipient's public key. Instead, the encryption algorithm generates a pair (K, ψ), where K is a key of SKE and ψ is an encryption of K. The decryption algorithm applied to ψ yields K. In our hybrid encryption scheme, $\psi = (u_1, u_2) = (g_1^r, g_2^r)$.

The notion of IND-CCA is adapted to KEM as follows. The adversary does not give two messages to the encryption oracle. Rather, the encryption oracle runs the KEM encryption algorithm to obtain a pair (K, ψ). The encryption oracle then gives the adversary either (K, ψ) or (K^+, ψ), where K^+ is an independent random bit string; the choice of K versus K^+ depends on the value of the random bit b chosen by the encryption oracle.

Up to now, in order to prove the security of the hybrid encryption scheme, it has been believed to be essential for KEM to be secure in the sense of IND-CCA, as stated in [6, Remark 7.2, page 207].

However, we know of no way to prove that our KEM is secure in the sense of IND-CCA. Nevertheless, we prove that the proposed hybrid encryption scheme is secure in the sense of IND-CCA. This shows that one can start with a weak KEM and repair it with a hybrid construction. Eventually, more efficient hybrid encryption schemes could be obtained.

A generalization of our scheme to universal$_2$ projective hash families [4] will be given in Sec.8.

4.2 ϵ-Rejection Secure

We require that a one-time symmetric-key encryption scheme SKE satisfies the following property: any bit string χ is rejected by the decryption algorithm with overwhelming probability. Formally, we say that SKE is ϵ-rejection secure if for any bit string χ,

$$\Pr(\text{SKE.Dec}(1^\lambda, K, \chi) = reject) \geq 1 - \epsilon,$$

where the probability is taken over K.

This property is already satisfied by the one-time symmetric-key encryption scheme shown in Sec.3.3. Indeed, for any fixed $\chi = (e, a)$, eq.(4) holds with probability $1/2^l$ because K_2 is random. Therefore, this encryption scheme is ϵ-rejection secure for $\epsilon = 1/2^l$.

4.3 Proposed Scheme

The proposed hybrid encryption scheme is based on the basic Cramer-Shoup scheme [3,6]. However, it does not use v as the validity check as in [3,6], but rather it is used to derive the encapsulated key K. This saves the value h which was previously used to encapsulate the key, and one exponentiation encryption/decryption. It also makes the public key and the secret key one element shorter.

Let G be an Abelian group of order Q, where Q is a large prime. Let SKE be a one-time symmetric-key encryption scheme.

Let $H : G \to \{0,1\}^k$ be a hash function, where $k = \text{SKE.Len}(\lambda)$. We assume that $H(v)$ is uniformly distributed over $\{0,1\}^k$ if v is uniformly distributed over G. This is a very weak requirement on H, and we can use SHA-1, for example.

Key Generation. Generate two distinct generators g_1, g_2 of G at random. Choose $(x_1, x_2, y_1, y_2) \in Z_Q^4$ at random. Compute

$$c = g_1^{x_1} g_2^{x_2}, \ d = g_1^{y_1} g_2^{y_2}.$$

Finally, a random κ indexing a target collision resistant hash function TCR (see Sec.2.4) is chosen. The public-key is $pk = (g_1, g_2, c, d, \kappa)$ and the secret key is $sk = (x_1, x_2, y_1, y_2)$.

Encryption. To encrypt a message m, choose $r \in Z_Q$ at random and compute

$$u_1 = g_1^r, \ u_2 = g_2^r, \alpha = \mathsf{TCR}(\kappa; u_1, u_2),$$

$$v = c^r d^{r\alpha}, \ K = H(v), \ \chi = \mathsf{SKE.Enc}(1^\lambda, K, m).$$

The ciphertext is (u_1, u_2, χ). (In the ciphertext, the KEM part is $\psi = (u_1, u_2)$.)

Decryption. For a ciphertext $C = (u_1, u_2, \chi)$, compute

$$\alpha = \mathsf{TCR}(\kappa; u_1, u_2), \ v = u_1^{x_1 + y_1\alpha} u_2^{x_2 + y_2\alpha}, \ K = H(v).$$

Then decrypt χ under K using SKE.Dec, and output the resulting decryption z. (z may be reject.)

4.4 Security

Theorem 1. The proposed hybrid encryption scheme Hybrid is secure in the sense of IND-CCA under the DDH assumption if SKE is secure in the sense of IND-CCA and it is ϵ-rejection secure for negligible ϵ. In particular,

$$\mathsf{Adv}_{\mathsf{Hybrid}}^{cca}(t, q_d) \leq \mathsf{Adv}_{\mathsf{G}}^{ddh}(t_1) + \mathsf{Adv}_{\mathsf{TCR}}^{hash}(t_2) + \mathsf{Adv}_{\mathsf{SKE}}^{cca}(t_3, q_d) + q_d(\epsilon + \frac{1}{Q}) + \frac{4}{Q}.$$

where t_1, t_2, t_3 are essentially the same as t.

A proof will be given in the next section.

4.5 Efficiency Comparison

In the hybrid encryption scheme of Shoup [12] and in the Cramer-Shoup scheme [3],

- $v \in G$ is included in the ciphertext C to check the validity of C.
- $h \in G$ is included in a public-key to generate a key K of SKE.

In our scheme, on the other hand,

- v is not included in the ciphertext, but it is used to derive a key K of SKE.
- h is not necessary at all.

In a typical implementation, the underlying Abelian group G may be a subgroup of Z_p^*, where p is a large prime. Table 1 shows an efficiency comparison among the proposed hybrid encryption scheme, the hybrid encryption scheme of Shoup [12] and the basic Cramer-Shoup scheme [3]. (In the table, a denotes the tag of SKE as shown in Sec.3.3.)

We can see that

We can see that
- the size of our ciphertext is $|p|$ bits shorter than that of Shoup [12].
- the size of our public-key is $|p|$ bits shorter than that of Shoup [12].
- The number of exponentiations per encryption and that of per decryption of our scheme are also smaller.

Further, our scheme is more efficient than the Cramer-Shoup scheme [3] for $|m| < 2|p| - |a|$. Moreover, in Cramer-Shoup [3] m must belong to G (so $|m| \leq |q|$), while in ours and Shoup's [12] $m \in \{0,1\}^*$ (polynomial length).

Table 1. Efficiency Comparison

	ciphertext	public-key	exp/enc	exp/dec
Cramer-Shoup [3]	$4 \cdot |p|$	$5|p| + |\kappa|$	5	3
Shoup [12]	$3 \cdot |p| + |m| + |a|$	$5|p| + |\kappa|$	5	3
Proposed	$2 \cdot |p| + |m| + |a|$	$4|p| + |\kappa|$	4	2

5   Proof of Theorem 1

5.1   Outline

The following lemma is simple but useful.

Lemma 1. *[6, Lemma 6.2] Let S_1, S_2 and F be events defined on some probability space. Suppose that the event $S_1 \vee \neg F$ occurs if and only if $S_2 \vee \neg F$ occurs. Then*

$$|\Pr(S_1) - \Pr(S_2)| \leq \Pr(F).$$

Let A be an adversary who breaks the proposed scheme in the sense of IND-CCA. The attack game is as described in Sec.2.2. Suppose that the public key is (g_1, g_2, c, d, κ) and the secret key is (x_1, x_2, y_1, y_2). The target ciphertext is denoted by $C^* = (u_1^*, u_2^*, \chi^*)$. We also denote by r^*, α^*, v^*, K^* the values corresponding with r, α, v, K related to C^*.

Suppose that A queries at most q_1 times to the decryption oracle in the find stage, and at most q_2 times to the decryption oracle in the guess stage, where $q_d = q_1 + q_2$. We say that a ciphertext $C = (u_1, u_2, \chi)$ is valid if $u_1 = g_1^r$ and $u_2 = g_2^r$ for some r. Otherwise, we say that C is invalid.

Let $\log(\cdot)$ denote $\log_{g_1}(\cdot)$ and let $w = \log g_2$. Then

$$\log c = x_1 + w x_2 \tag{5}$$
$$\log d = y_1 + w y_2 \tag{6}$$

Let $\mathbf{G_0}$ be the original attack game, let $\tilde{b} \in \{0,1\}$ denote the output of A, and let T_0 be the event that $b = \tilde{b}$ in $\mathbf{G_0}$. Therefore,

$$\mathsf{Adv}_{\mathsf{Hybrid}}^{cca}(A) = |\Pr[T_0] - 1/2|.$$

We shall define a sequence $\mathbf{G_1}, \cdots, \mathbf{G_\ell}$ of modified attack games. For any $1 \leq i \leq \ell$, we let T_i be the event that $b = \tilde{b}$ in game $\mathbf{G_i}$.

In game $\mathbf{G_1}$, we modify the encryption oracle as follows: $v^* = c^{r^*} d^{r^* \alpha^*}$ is replaced by

$$v^* = (u_1^*)^{x_1 + y_1 \alpha^*} (u_2^*)^{x_2 + y_2 \alpha^*}.$$

This change is purely conceptual, and $\Pr[T_1] = \Pr[T_0]$.

In game $\mathbf{G_2}$, we modify the encryption oracle again, so that (u_1^*, u_2^*) is replaced by a random pair $(g_1^{r_1^*}, g_2^{r_2^*})$ with $r_1^* \neq r_2^*$. Under the DDH assumption, A will hardly notice, and $|\Pr[T_2] - \Pr[T_1]|$ is negligible. More precisely, we have

Lemma 2. *There exists a PPT algorithm A_1, whose running time is essentially the same as that of A, such that*

$$|\Pr[T_2] - \Pr[T_1]| \leq \mathsf{Adv}_G^{ddh}(A_1) + 3/Q.$$

The proof is the same as that of [6, Lemma 6.3].

In game $\mathbf{G_3}$, we modify the decryption oracle, so that it applies the following special rejection rule: In the guess stage, if the adversary submits a ciphertext $(u_1, u_2) \neq (u_1^*, u_2^*)$ but $\alpha = \alpha^*$, then the decryption oracle immediately outputs reject and halts. Let R_3 be the event that the decryption oracle in game $\mathbf{G_3}$ rejects a ciphertext using the special rejection rule. It is clear that games $\mathbf{G_2}$ and $\mathbf{G_3}$ proceed identically until the event R_3 occurs. In particular, the event $T_2 \wedge \neg R_3$ and $T_3 \wedge \neg R_3$ are identical. So by Lemma 1, we have

$$|\Pr[T_3] - \Pr[T_2]| \leq \Pr[R_3].$$

Lemma 3. *There exists a PPT algorithm A_2, whose running time is essentially the same as that of A, such that*

$$\Pr[R_3] \leq \mathsf{Adv}_{TCR}^{hash}(A_2) + 1/Q.$$

The proof is the same as that of [6, Lemma 6.5].

In game $\mathbf{G_4}$, we modify the decryption oracle, so that it rejects all invalid ciphertexts C in the find stage. Let R_4 be the event that a ciphertext is rejected in $\mathbf{G_4}$ that would not have been rejected under the rules of game $\mathbf{G_3}$. It is clear that games $\mathbf{G_3}$ and $\mathbf{G_4}$ proceed identically until the event R_4 occurs. In particular, the event $T_3 \wedge \neg R_4$ and $T_4 \wedge \neg R_4$ are identical. So by Lemma 1, we have

$$|\Pr[T_4] - \Pr[T_3]| \leq \Pr[R_4].$$

Lemma 4. $\Pr[R_4] \leq q_1 \cdot \epsilon.$ *(For the proof, see Section 5.2.)*

In game $\mathbf{G_5}$, we modify the encryption oracle as follows. $(u_1^*, u_2^*) = (g_1^{r_1^*}, g_2^{r_2^*})$ is randomly chosen in such a way that an event R_5 does not occur, where R_5 is the event that $(u_1^*, u_2^*) = (u_1, u_2)$ for some invalid ciphertext (u_1, u_2, χ) which A queries in the find stage. It is clear that the event $T_4 \wedge \neg R_5$ and $T_5 \wedge \neg R_5$ are identical. So by Lemma 1, we have

$$|\Pr[T_5] - \Pr[T_4]| \leq \Pr[R_5].$$

Lemma 5. $\Pr[R_5] \leq q_1/Q$.   *(For the proof, see Section 5.3.)*

In game $\mathbf{G_6}$, we modify the decryption oracle, so that it rejects all invalid ciphertexts C in the guess stage. Let R_6 be the event that a ciphertext is rejected in $\mathbf{G_6}$ that would not have been rejected under the rules of game $\mathbf{G_5}$. It is clear that games $\mathbf{G_5}$ and $\mathbf{G_6}$ proceed identically until the event R_6 occurs. In particular, the event $T_5 \wedge \neg R_6$ and $T_6 \wedge \neg R_6$ are identical. So by Lemma 1, we have

$$|\Pr[T_6] - \Pr[T_5]| \leq \Pr[R_6].$$

Lemma 6. $\Pr[R_6] \leq q_2 \cdot \epsilon$.   *(For the proof, see Section 5.4.)*

In game $\mathbf{G_7}$, we modify the encryption oracle and the decryption oracle, so that K^* is replaced by a random key K^+.

Lemma 7. $\Pr[T_6] = \Pr[T_7]$.   *(For the proof, see Section 5.5.)*

Lemma 8. *There exists a PPT algorithm A_3, whose running time is essentially the same as that of A, such that*

$$\mathsf{Adv}_{\mathsf{SKE}}^{cca}(A_3) = |\Pr[T_7] - 1/2|.$$

For the proof, see Section 5.6.

From the above results, we immediately obtain that

$$\mathsf{Adv}_{\mathsf{Hybrid}}^{cca}(A_3) \leq \mathsf{Adv}_{\mathsf{G}}^{ddh}(A_1) + \mathsf{Adv}_{\mathsf{TCR}}^{hash}(A_2) + \mathsf{Adv}_{\mathsf{SKE}}^{cca}(A_3) + q_d(\epsilon + \frac{1}{Q}) + \frac{4}{Q}.$$

5.2   Proof of Lemma 4

From the A's view, (x_1, x_2, y_1, y_2) is a random point satisfying eq.(5) and eq.(6). Suppose that A queries an invalid ciphertext (u_1, u_2, χ) to the decryption oracle, where $\log_{g_1}(u_1) = r_1$ and $\log_{g_2}(u_2) = r_2$ with $r_1 \neq r_2$. Let $v = u_1^{x_1+y_1\alpha} u_2^{x_2+y_2\alpha}$, where $\alpha = \mathsf{TCR}(\kappa; u_1, u_2)$. Then

$$\log v = r_1(x_1 + \alpha y_1) + r_2 w(x_2 + \alpha y_2). \tag{7}$$

It is clear that eq.(5),(6) and (7) are linearly independent. This means that v can take any value. In other words, v is uniformly distributed over G. Hence $K = H(v)$ is uniformly distributed over $\{0,1\}^k$. Now since SKE is ϵ-rejection secure, the decryption oracle accepts (u_1, u_2, χ) with probability at most ϵ. Consequently, we obtain this lemma.

5.3   Proof of Lemma 5

For any fixed (u_1, u_2),

$$\Pr[(u_1^*, u_2^*) = (u_1, u_2)] = \frac{1}{Q(Q-1)} \leq 1/Q$$

because $(r_1^*, r_2^*) \in Z_Q^2$ is randomly chosen in such a way that $r_1^* \neq r_2^*$.

5.4   Proof of Lemma 6

As the worst case, we assume that A knows v^*. Then from the A's view, (x_1, x_2, y_1, y_2) is a random point satisfying eq.(5), (6) and

$$\log v^* = r_1^*(x_1 + \alpha^* y_1) + r_2^* w(x_2 + \alpha^* y_2). \tag{8}$$

In the guess stage, suppose that A queries an invalid ciphertext (u_1, u_2, χ) to the decryption oracle, where $\log u_1 = r_1$ and $\log u_2 = r_2$ with $r_1 \neq r_2$. Let $v = u_1^{x_1 + y_1 \alpha} u_2^{x_2 + y_2 \alpha}$, where $\alpha = \mathsf{TCR}(\kappa; u_1, u_2)$. Then

$$\log v = r_1(x_1 + \alpha y_1) + r_2 w(x_2 + \alpha y_2). \tag{9}$$

Now

$$\begin{vmatrix} 1 & 0 & w & 0 \\ 0 & 1 & 0 & w \\ r_1^* & \alpha^* r_1^* & w r_2^* & \alpha^* w r_2^* \\ r_1 & \alpha r_1 & w r_2 & \alpha w r_2 \end{vmatrix} = w^2 (r_2^* - r_1^*)(r_2 - r_1)(\alpha - \alpha^*) \neq 0$$

Therefore, eq.(5), (6), (8) and (9) are linearly independent. This means that v is uniformly distributed over G. Hence $K = H(v)$ is uniformly distributed over $\{0, 1\}^k$. Now since SKE is ϵ-rejection secure, the decryption oracle accepts (u_1, u_2, χ) with probability at most ϵ.

Consequently, we obtain this lemma.

5.5   Proof of Lemma 7

In game $\mathbf{G_6}$, from the A's view, (x_1, x_2, y_1, y_2) is a random point satisfying eq.(5) and eq.(6). Further, it is clear that eq.(5),(6) and (8) are linearly independent. This means that v^* can take any value. In other words, v^* is uniformly distributed over G. Hence $K^* = H(v^*)$ is uniformly distributed over $\{0, 1\}^k$. Consequently, we obtain this lemma.

5.6   Proof of Lemma 8

We describe Algorithm A_3. Algorithm A_3 provides an environment for A as follows. First, A_3 runs the key generation algorithm of Hybrid to generate a public-key $pk = (g_1, g_2, c, d, \kappa)$ and the secret-key $sk = (x_1, x_2, y_1, y_2)$. In particular, A_3 chooses $w \in Z_Q$ randomly and computes $g_2 = g_1^w$. It then gives pk to A.

In the find stage, whenever A submits a ciphertext C to the decryption oracle, A_3 applies the decryption rule of game $\mathbf{G_7}$, using the secret-key sk and w.

When A submits (m_0, m_1) to the encryption oracle, A_3 submits (m_0, m_1) to her encryption oracle.

The encryption oracle of A_3 chooses a random key $K^+ \in \{0, 1\}^k$ along with a random bit σ, and encrypts m_σ using the key K^+. It then returns the resulting ciphertext χ^* to A_3.

A_3 generates (u_1^*, u_2^*) according to the encryption rule of game $\mathbf{G_7}$. It then returns the target ciphertext $C^* = (u_1^*, u_2^*, \chi^*)$ to A.

In the guess stage, suppose that A submits a ciphertext $C = (u_1, u_2, \chi)$ to the decryption oracle. If $(u_1, u_2) \neq (u_1^*, u_2^*)$, then A_3 applies the decryption rule of game $\mathbf{G_7}$, using the secret-key sk and w. Otherwise, A_3 queries χ to her decryption oracle, where the decryption oracle decrypts χ by using K^+. A_3 then returns the answer to A.

When A outputs $\tilde{\sigma}$, A_3 outputs $\tilde{\sigma}$ and halts. That completes the description of A_3.

It is clear that A_3 perfectly simulates the environment of A. Therefore,

$$\Pr[T_7] = \Pr(\sigma = \tilde{\sigma}).$$

On the other hand,

$$\mathsf{Adv}_{\mathsf{SKE}}^{cca}(A_3) = |\Pr(\sigma = \tilde{\sigma}) - 1/2|.$$

Consequently, we obtain this lemma.

6 Discussion

We have argued that a KEM does not have to be CCA-secure in the construction of hybrid encryption schemes, as was previously believed.

In the IND-CCA definition of hybrid encryption schemes, the decryption oracle returns the message m for a queried ciphertext $C = (\psi, \chi)$, where ψ is the KEM part and χ is the symmetric encryption ciphertext. On the other hand, in the IND-CCA definition of KEM, the decryption oracle returns the symmetric key K for a queried ψ. Hence, the IND-CCA definition of KEM is too demanding because the decryption oracle reveals much more information than the decryption oracle of the hybrid encryption scheme does.

Then one may consider to define a weaker condition on KEM such that when coupled with CCA-secure symmetric encryption (with the extra condition of Section 3.4), it would yield a CCA-secure hybrid encryption scheme. However, it seems to be impossible because the security of KEM and that of the symmetric encryption scheme are intertwined (as in our scheme).

7 Hash Proof System

Cramer and Shoup introduced a notion of Hash Proof System (HPS) [4, 5] in order to generalize their encryption scheme based on the DDH assumption [3]. By using HPS, they showed new CCA-secure encryption schemes under Quadratic Residuosity assumption and Paillier's Decision Composite Residuosity assumption, respectively.

In this section, we give the definition of a slight variant of HPS, where ϵ-universal$_2$ is replaced by *strongly universal$_2$*.

7.1   Subset Membership Problem [4, 5]

A subset membership problem Mem specifies a collection $\{\text{Instance}_n\}_{n\in\mathbb{N}}$ such that for every n, Instance_n is a probability distribution over problem instances Λ. Each Λ specifies the following:

- Define, non-empty sets, X, L and W such that $L \subset X$.
- A binary relation $R \subset X \times W$ such that $x \in L$ iff $(x, w) \in R$ for some witness $w \in W$.

We require that the following PPT algorithms exist.

1. *Instance sampling*: samples an instance Λ according to Instance_n on 1^n.
2. *Subset sampling*: outputs a random $x \in L$ together with a witness $w \in W$ for x on input 1^n and $\Lambda[X, L, W, R]$.
3. *Element sampling*: outputs a random $x \in X$.

We say that Mem is hard if (Λ, x_0) and (Λ, x_1) are indistinguishable for a random $x_0 \in L$ and a random $x_1 \in X \setminus L$.

7.2   Projective Hash Family

Let X and Π be finite, non-empty sets. Let $F = \{f_i : X \to \Pi\}_{i\in I}$ be a set of functions indexed by I. We call (F, I, X, Π) a universal hash family [4, 5].

Let $L \subset X$. Let S be a finite, non-empty set, and let $\alpha : I \to S$ be a function. Set $\text{Project} = (F, I, X, L, \Pi, S, \alpha)$.

Definition 4. *[4, 5]* $\text{Project} = (F, I, X, L, \Pi, S, \alpha)$ *is called a projective hash family if for all $i \in I$, the action of f_i on L is determined[3] by $\alpha(i)$.*

In other words, the value $f_i(x)$ is determined by $\alpha(i)$ if $x \in L$. We next define the notion of strongly universal$_2$ projective hash, a variant of Cramer-Shoup's ϵ-universal$_2$ projective hash.

Definition 5. Let $\text{Project} = (F, I, X, L, \Pi, S, \alpha)$ be a projective hash family. Consider the probability space defined by choosing $i \in I$ at random. We say that Project is *strongly* universal$_2$ if

- for all $s \in S, x \in X \setminus L$, and $\pi \in \Pi$,

$$\Pr[f_i(x) = \pi \mid \alpha(i) = s] = 1/|\Pi|,$$

- and for all $s \in S, x, x^* \in X \setminus L$ with $x \neq x^*$, and $\pi, \pi^* \in \Pi$,

$$\Pr[f_i(x) = \pi \mid f_i(x^*) = \pi^* \wedge \alpha(i) = s] = 1/|\Pi|.$$

Project is strongly universal$_2$ means that for any $x \notin L$, the value of $f_i(x)$ is uniformly distributed over Π conditioned on a fixed value of $\alpha(i)$, and it is also uniformly distributed over Π conditioned on fixed values of $\alpha(i)$ and $f_i(x^*)$ for $x^* \notin L$ with $x^* \neq x$.

---

[3] For a further clarification, see Section 7.3.

7.3   Hash Proof System [4, 5]

Let Mem be a subset membership problem. A hash proof system (HPS) P for Mem associates with each instance $\Lambda[X, L, W, R]$ of Mem a projective hash family Project $= (F, I, X, L, \Pi, S, \alpha)$.

P provides several algorithms to carry out basic operations: $i \overset{R}{\leftarrow} I$ and computing $\alpha(i) \in S$ given $i \in I$. The private evaluation algorithm for P computes $f_i(x) \in \Pi$ given $i \in I$ and $x \in X$. The public evaluation algorithm for P computes $f_i(x) \in \Pi$ given $\alpha(i) \in S, x \in X$ and $w \in W$, where w is a witness for x.

8   Proposed Hybrid Construction Based on HPS

In this section, we generalize our hybrid encryption scheme of Sec.4.3 by using the variant of HPS shown above. Then efficient hybrid encryption schemes are obtained which are secure in the sense of IND-CCA under Quadratic Residuosity assumption and Paillier's Decision Composite Residuosity assumption, respectively, in the standard model.

8.1   Hybrid Construction

Let Mem be a subset membership problem and P be a hash proof system for Mem. Let SKE be a one-time symmetric-key encryption scheme.

Key Generation. Generate an instance $\Lambda[X, L, W, R]$ using the instance sampling algorithm of Mem. Suppose that P associates with $\Lambda[X, L, W, R]$ a projective hash family Project $= (F, I, X, L, \Pi, S, \alpha)$. Choose $i \in I$ at random and compute $s = \alpha(i)$.

The public key is s and the secret key is i. Let $H : \Pi \rightarrow \{0, 1\}^k$ be a hash function, where $k = \mathsf{SKE.Len}(\lambda)$. We assume that $H(v)$ is uniformly distributed over $\{0, 1\}^k$ if v is uniformly distributed over Π. This is a very weak requirement on H, and we can use SHA-1, for example.

Encryption. To encrypt a message m, generate $x \in L$ at random together with a witness $w \in W$ for x using the subset sampling algorithm of Mem. Compute $\pi = f_i(x)$ using the public evaluation algorithm for P on inputs s, x and w. Compute $K = H(\pi)$ and $\chi = \mathsf{SKE.Enc}(1^\lambda, K, m)$. The ciphertext is (x, χ).

Decryption. To decrypt a ciphertext (x, χ), compute $\pi = f_i(x)$ using the private evaluation algorithm for P on inputs i and x. Then decrypt χ under K using SKE.Dec, and outputs the resulting decryption z. (z may be reject.)

8.2   Security

Theorem 2. In the above construction, suppose that Mem is hard, and the associated projective hash family Project $= (F, I, X, L, \Pi, S, \alpha)$ is strongly universal$_2$ for each instance $\Lambda[X, L, W, R]$ of Mem. Moreover, suppose that the one-time symmetric-key encryption scheme SKE is secure in the sense of IND-CCA and it is ϵ-rejection secure for negligible ϵ. Then the proposed hybrid encryption scheme is secure in the sense of IND-CCA.

A proof is a generalization of that of Theorem 1. Roughly speaking, in the proof, if the challenge ciphertext is based upon application of the projective universal hash function f_i to an element $x^* \in L$, then the attack works as in the real case.

If $x^* \notin L$, then the following happens: At the beginning of the CCA attack, $\pi^* = f_i(x^*)$ (which is used as the symmetric key by $K^* = H(\pi^*)$) is totally uniform and secret from the point of view of the adversary. This is due to the strongly universal$_2$ property of the projective hash family Project. This information theoretic property of the symmetric key K^* remains as the attack progresses due to the fact that invalid queries are not decrypted due to the ϵ-rejection property of the SKE, where a ciphertext $C = (x, \chi)$ is invalid if $x \notin L$.

8.3 Examples

From [4, 5]. Let G be an Abelian group of order Q, where Q is a large prime. Let $X = G^2, W = Z_Q, L = \{(g_0^r, g_1^r) \mid r \in Z_Q\}$, where g_0, g_1 are two distinct generators of G. Then it is clear that the related membership problem Mem is hard if and only if the DDH assumption holds.

Let $\Gamma : G^2 \to Z_Q^n$ be an injective function for some n. Let $\Pi = S = G$ and $I = Z_Q^{2(n+1)}$. Define

$$\alpha(i_0, i_1, \cdots, i_n, j_0, j_1, \cdots, j_n) = (s_0, s_1, \cdots, s_n),$$

where $s_u = g_0^{i_u} g_1^{j_u}$ for $0 \le u \le n$. For $(x_0, x_1) \in X$, let $\Gamma(x_0, x_1) = (a_0, \cdots, a_n)$ and define

$$f_{(i_0, i_1, \cdots, i_n, j_0, j_1, \cdots, j_n)}(x_0, x_1) = x_0^{i_0 + a_1 i_1 \cdots + a_n i_n} x_1^{j_0 + a_1 j_1 \cdots + a_n j_n}.$$

(1) Project $= (F, I, X, L, \Pi, S, \alpha)$ is a projective hash family because if $(x_0, x_1) = (g_0^r, g_1^r)$, then

$$\pi = f_{(i_0, i_1, \cdots, i_n, j_0, j_1 \cdots, j_n)}(x_0, x_1) = (s_0 s_1^{a_1} \cdots s_n^{a_n})^r. \tag{10}$$

(2) Consider the probability space defined by choosing $(i_0, i_1, \cdots, i_n, j_0, j_1, \cdots, j_n) \in Z_Q^{2(n+1)}$ at random. For the example of [4, 5] we now have:

- For any $(x_0, x_1) \in X \setminus L$, $f_{(i_0, \cdots, i_n, j_0, \cdots, j_n)}(x_0, x_1)$ is uniformly distributed over G conditioned on fixed values of $(s_0, s_1, \cdots, s_n)$.
- For any $(x_0, x_1), (x_0^*, x_1^*) \in X \setminus L$ with $(x_0, x_1) \neq (x_0^*, x_1^*)$, we easily see that: $f_{(i_0, \cdots, i_n, j_0, \cdots, j_n)}(x_0, x_1)$ is uniformly distributed over G conditioned on fixed values of $(s_0, s_1, \cdots, s_n)$ and $\pi^* = f_{(i_0, \cdots, i_n, j_0, \cdots, j_n)}(x_0^*, x_1^*)$.

Hence Project is strongly universal$_2$.

Now from Sec.8.1, a concrete hybrid encryption scheme is obtained such that the ciphertext is $(g_0^r, g_1^r, \mathsf{SKE.Enc}(1^\lambda, K, m))$, where $K = H(\pi)$ and π is given by eq.(10). From Theorem 2, it is secure in the sense of IND-CCA if SKE satisfies the condition of the theorem. (This scheme is a TCR-free variant of Sec.4.3.)

Similarly, we can obtain efficient hybrid encryption schemes which are secure in the sense of IND-CCA under Quadratic Residuosity assumption and Paillier's Decision Composite Residuosity assumption, respectively.

Acknowledgment

We would like to thank Ronald Cramer and the anonymous reviewers for their helpful comments. The second author was supported by JPSP fellowship for research in Japan.

References

1. M. Bellare and P. Rogaway: Random Oracles are Practical: A Paradigm for Designing Efficient Protocols. ACM Conference on Computer and Communications Security 1993: 62-73
2. D.Boneh, Simplified OAEP for the RSA and Rabin Functions. CRYPTO 2001, pp.275-291 (2001)
3. R. Cramer and V. Shoup, "A practical public key cryptosystem provably secure against adaptive chosen ciphertext attack," *Advances in Cryptology - CRYPTO'98*, Lecture Notes in Computer Science Vol. 1462, H. Krawczyk ed., Springer-Verlag, 1998.
4. R. Cramer and V. Shoup, "Universal hash proofs and a paradigm for chosen ciphertext secure public key encryption, against adaptive chosen ciphertext attack," *Advances in Cryptology - Eurocrypt'02*, Lecture Notes in Computer Science Vol. 2332, Springer-Verlag, 2002.
5. Full length version of [4]. http://shoup.net/papers/uhp.pdf
6. R. Cramer and V. Shoup, Design and analysis of practical public-key encryption schemes secure against adaptive chosen ciphertext attack, SIAM Journal on Computing, Volume 33, Number 1, pp. 167-226 (2003)
7. E. Fujisaki and T. Okamoto, Secure Integration of Asymmetric and Symmetric Encryption Schemes, Advances in Cryptology - CRYPTO '99, Lecture Notes in Computer Science Vol. 1666, M. Wiener ed., Springer-Verlag, 1999.
8. E.Fujisaki, T.Okamoto, D.Pointcheval, J.Stern, RSA-OAEP Is Secure under the RSA Assumption. CRYPTO 2001, pp.260-274 (2001)
9. H.Krawczyk, LFSR-based Hashing and Authentication, CRYPTO 1994, pp.129-139 (1994)
10. M.Naor and M.Yung, Universal One-Way Hash Functions and their Cryptographic Applications, STOC 1989, pp.33-43 (1989)
11. J.Rompel: One-Way Functions are Necessary and Sufficient for Secure Signatures, STOC 1990, pp.387-394 (1990)
12. V. Shoup, Using Hash Functions as a Hedge against Chosen Ciphertext Attack, EUROCRYPT 2000, pp.275-288 (2000)
13. V. Shoup, OAEP Reconsidered, CRYPTO 2001, pp.239-259 (2001)

Secure Identity Based Encryption
Without Random Oracles

Dan Boneh[1,*] and Xavier Boyen[2]

[1] Computer Science Department, Stanford University, Stanford CA 94305-9045
dabo@cs.stanford.edu
[2] Voltage Security, Palo Alto, California
xb@boyen.org

Abstract. We present a fully secure Identity Based Encryption scheme whose proof of security does not rely on the random oracle heuristic. Security is based on the Decision Bilinear Diffie-Hellman assumption. This solves an open problem posed by Boneh and Franklin in 2001.

1 Introduction

Identity Based Encryption (IBE) provides a public key encryption mechanism where a public key is an arbitrary string such as an email address or a telephone number. The corresponding private key can only be generated by a Private Key Generator (PKG) who has knowledge of a master secret. In an IBE system, users authenticate themselves to the PKG and obtain private keys corresponding to their identities. Although Identity based encryption was proposed two decades ago [Sha84], and a few early precursors suggested over the years [Tan87,MY96], it is only recently that the first working implementations were proposed. Boneh and Franklin [BF01,BF03] defined a security model for Identity Based Encryption and gave a construction based on the bilinear Diffie-Hellman problem. Cocks [Coc01] describes another construction using quadratic residues modulo a composite. The security of these systems requires cryptographic hash functions that are modeled as random oracles, i.e., these systems are proven secure in the random oracle model [BR93]. The same holds for several other identity based systems featuring signatures [CC03], key exchange [SOK00], hierarchical identities [GS02], and signcryption [Boy03].

It is natural to ask whether secure IBE systems can exist in the standard model, i.e., without resorting to the random oracle heuristic. This question is especially relevant in light of several uninstantiable random oracle cryptosystems [CGH98,BBP04], which are secure in the random oracle model, but are trivially insecure under any instantiation of the oracle. Towards this goal, several recent results [CHK03,BB04,HK04] construct IBE systems secure without random oracles in weaker versions of the Boneh-Franklin model. However, until now, building a fully secure IBE remained open.

---

* Supported by NSF and the Packard Foundation.

M. Franklin (Ed.): CRYPTO 2004, LNCS 3152, pp. 443–459, 2004.
© International Association for Cryptologic Research 2004

In this paper we construct an IBE system that is secure in the Boneh-Franklin model without using random oracles. Security is based on the decisional version of the bilinear Diffie-Hellman assumption. Our system demonstrates that fully secure IBE systems can exist without random oracles. The main shortcoming of the proposed system is that it is inefficient; consequently, we mostly view our construction as an existence proof.

2   Preliminaries

Before presenting our results we briefly review a definition of security for an IBE system. We also review the definition for groups with a bilinear map. First, we introduce some notation.

2.1   Notation

For a finite set S we use $x \xleftarrow{\text{R}} S$ to define a random variable x that picks an element of S uniformly at random. For a randomized algorithm $\mathcal{A}$ we use $x \xleftarrow{\text{R}} \mathcal{A}(y)$ to define a random variable x that is the output of algorithm $\mathcal{A}$ on input y. We let $\Pr[b(x) : x \leftarrow \mathcal{A}(y)]$ denote the probability that the predicate $b(x)$ is true where x is the random variable defined by $x \leftarrow \mathcal{A}(y)$. For a vector $z \in \Sigma^n$ we use $z|_i$ to denote the i'th component of z.

2.2   Secure IBE Systems

Recall that an Identity Based Encryption system (IBE) consists of four algorithms [Sha84,BF01]: *Setup, KeyGen, Encrypt, Decrypt*. The *Setup* algorithm generates system parameters, denoted by *params*, and a master key *master-key*. The *KeyGen* algorithm uses the master key to generate the private key corresponding to a given identity. The encryption algorithm encrypts messages for a given identity (using the system parameters) and the decryption algorithm decrypts ciphertexts using the private key.

Boneh and Franklin [BF01] define chosen ciphertext security for IBE systems under a chosen identity attack. In their model the adversary is allowed to adaptively chose the public key it wishes to attack (the public key on which it will be challenged). More precisely, security for an IBE system is defined using the following two probabilistic experiments CCA-Exp$_\mathcal{A}(0)$ and CCA-Exp$_\mathcal{A}(1)$.

Experiment CCA-Exp$_\mathcal{A}(b)$: for an algorithm $\mathcal{A}$ and a bit $b \in \{0, 1\}$ define the following game between a challenger and $\mathcal{A}$:

Setup: A challenger runs the *Setup* algorithm. It gives $\mathcal{A}$ the resulting system parameters *params*. It keeps the corresponding *master-key* to itself.

Phase 1: Algorithm $\mathcal{A}$ issues queries $q_1, \ldots, q_m$ where each query q_i is one of:
   - Private key query for an identity ID_i. The challenger responds by running algorithm *KeyGen* to generate the private key d_i corresponding to the public key ID_i. It sends d_i to $\mathcal{A}$.

- Decryption query for a ciphertext C_i and an identity ID_i. The challenger responds by running algorithm *KeyGen* to generate the private key d_i corresponding to ID_i. It then runs algorithm *Decrypt* to decrypt the ciphertext C_i using the private key d_i. It gives $\mathcal{A}$ the resulting plaintext.

These queries may be asked adaptively, that is, each query q_i may depend on the replies to $q_1, \ldots, q_{i-1}$.

Challenge: Once $\mathcal{A}$ decides that Phase 1 is over it outputs an identity ID^* and two equal length plaintexts $M_0, M_1 \in \mathcal{M}$ that it wishes to be challenged on, under the constraint that it had not previously asked for the private key of ID^*. The challenger sets the challenge ciphertext to $C^* = Encrypt(params, \mathsf{ID}^*, M_b)$. It sends C^* as the challenge to $\mathcal{A}$.

Phase 2: Algorithm $\mathcal{A}$ issues more queries $q_{m+1}, \ldots, q_n$ where q_i is one of:

- Private key query for any identity ID_i where $\mathsf{ID}_i \neq \mathsf{ID}^*$. The challenger responds as in Phase 1.
- Decryption query C_i for identity ID^* where $C_i \neq C^*$. The challenger responds as in Phase 1.

These queries may be asked adaptively as in Phase 1.

Guess: Finally, $\mathcal{A}$ outputs a guess $b' \in \{0,1\}$.

We call b' the output of the game and define the random variable $\mathsf{CCA\text{-}Exp}_{\mathcal{A}}(b)$ as $\mathsf{CCA\text{-}Exp}_{\mathcal{A}}(b) = b'$. The probability is over the random bits used by the challenger and the adversary. We define adversary $\mathcal{A}$'s advantage in attacking the IBE system $\mathcal{E}$ as:

$$\mathrm{Adv}_{\mathcal{E},\mathcal{A}} = \left| \Pr\left[\mathsf{CCA\text{-}Exp}_{\mathcal{A}}(0) = 1\right] - \Pr\left[\mathsf{CCA\text{-}Exp}_{\mathcal{A}}(1) = 1\right] \right|.$$

Definition 1. *We say that an IBE system $\mathcal{E}$ is $(t, q_{ID}, q_C, \epsilon_{IBE})$-adaptive chosen ciphertext secure under a chosen identity attack if for any t-time* IND-ID-CCA *adversary $\mathcal{A}$ that makes at most q_{ID} chosen private key queries and at most q_C chosen decryption queries we have that $Adv_{\mathcal{E},\mathcal{A}} < \epsilon_{IBE}$. As shorthand, we say that $\mathcal{E}$ is $(t, q_{ID}, q_C, \epsilon_{IBE})$-*IND-ID-CCA *secure.*

Semantic Security. As usual, we define chosen plaintext security for an IBE system as in the game above, except that the adversary is not allowed to issue any decryption queries. The adversary may still issue adaptive private key queries. The resulting system is semantically secure under an adaptive chosen identity attack.

Definition 2. *We say that an IBE system $\mathcal{E}$ is $(t, q_{ID}, \epsilon_{IBE})$ chosen plaintext secure under a chosen identity attack if $\mathcal{E}$ is $(t, q_{ID}, 0, \epsilon_{IBE})$-chosen ciphertext secure under a chosen identity attack. As shorthand, we say that $\mathcal{E}$ is $(t, q_{ID}, \epsilon_{IBE})$-*IND-ID-CPA *secure.*

For $b \in \{0,1\}$ we use $\mathsf{CPA\text{-}Exp}_{\mathcal{A}}(b)$ to denote the experiment $\mathsf{CCA\text{-}Exp}_{\mathcal{A}}(b)$ where $\mathcal{A}$ cannot make any decryption queries.

2.3 Bilinear Groups

We briefly review the necessary facts about bilinear maps and bilinear map groups.

1. $\mathbb{G}$ and $\mathbb{G}_1$ are two (multiplicative) cyclic groups of prime order p;
2. g is a generator of $\mathbb{G}$.
3. e is a bilinear map $e : \mathbb{G} \times \mathbb{G} \to \mathbb{G}_1$.

Let $\mathbb{G}$ and $\mathbb{G}_1$ be two groups as above. A bilinear map is a map $e : \mathbb{G} \times \mathbb{G} \to \mathbb{G}_1$ with the following properties:

1. Bilinear: for all $u, v \in \mathbb{G}$ and $a, b \in \mathbb{Z}$, $e(u^a, v^b) = e(u, v)^{ab}$.
2. Non-degenerate: $e(g, g) \neq 1$.

We say that $\mathbb{G}$ is a bilinear group if the group action in $\mathbb{G}$ can be computed efficiently and there exists a group $\mathbb{G}_1$ and an efficiently computable bilinear map $e : \mathbb{G} \times \mathbb{G} \to \mathbb{G}_1$ as above. Note that $e(\cdot, \cdot)$ is symmetric since $e(g^a, g^b) = e(g, g)^{ab} = e(g^b, g^a)$.

3   Complexity Assumptions

Let $\mathbb{G}$ be a bilinear group of prime order p and g be a generator of $\mathbb{G}$. We review the standard Bilinear Diffie-Hellman (BDH) assumption as well as the definition for binary biased Pseudo Random Functions (PRF's) and collision resistant functions.

3.1 Bilinear Diffie-Hellman Assumption

The BDH problem [Jou00,BF01] in $\mathbb{G}$ is as follows: given a tuple $g, g^a, g^b, g^c \in \mathbb{G}$ as input, output $e(g, g)^{abc} \in \mathbb{G}_1$. An algorithm $\mathcal{A}$ has advantage ϵ_{BDH} in solving BDH in $\mathbb{G}$ if

$$\Pr\left[\mathcal{A}(g, g^a, g^b, g^c) = e(g, g)^{abc}\right] \geq \epsilon_{\text{BDH}}$$

where the probability is over the random choice of a, b, c in $\mathbb{Z}_p$ and the random bits of $\mathcal{A}$.

Similarly, we say that an algorithm $\mathcal{B}$ that outputs $b \in \{0, 1\}$ has advantage ϵ_{BDH} in solving the *decision* BDH problem in $\mathbb{G}$ if

$$\left|\Pr\left[\mathcal{B}(g, g^a, g^b, g^c, e(g, g)^{abc}) = 0\right] - \Pr\left[\mathcal{B}(g, g^a, g^b, g^c, T) = 0\right]\right| \geq \epsilon_{\text{BDH}} \quad (1)$$

where the probability is over the random choice of a, b, c in $\mathbb{Z}_p$, the random choice of $T \in \mathbb{G}_1$ and the random bits of $\mathcal{B}$. We use the following notation:

- We denote the distribution over 5-tuples in the left term of (1) by $\mathcal{P}_{\text{BDH}}$.
- We denote the distribution over 5-tuples in the right term of (1) by $\mathcal{R}_{\text{BDH}}$.

Definition 3. *We say that the $(t, \epsilon_{\text{BDH}})$-(Decision) BDH assumption holds in $\mathbb{G}$ if no t-time algorithm has advantage at least ϵ_{BDH} in solving the (decision) BDH problem in $\mathbb{G}$.*

Occasionally we drop the t and ϵ_{BDH} and refer to the BDH and Decision BDH assumptions in $\mathbb{G}$.

3.2   Biased Binary Pseudo-random Functions

Next we review the definition of a Pseudo Random Function (PRF) with bias δ. Let F be a function $F : \{0,1\}^w \to \{0,1\}$. We say that F has bias $\delta \in [0,1]$ if the expectation of F over all inputs in $\{0,1\}^w$ is δ, i.e., $(1/2^w) \sum_{x \in \{0,1\}^w} F(x) = \delta$.

We let Ω_δ denote the set of all functions $F : \{0,1\}^w \to \{0,1\}$ with bias δ. We also let K_1 denote a set of keys. For an algorithm $\mathcal{A}$ we define the following value:

$$\text{Exp}_{\mathcal{A}}^{\Omega_\delta} = \Pr\left[\mathcal{A}^F(k_1) = 1 \; : \; F \xleftarrow{\text{R}} \Omega_\delta, \; k_1 \xleftarrow{\text{R}} K_1 \right]$$

Here $\mathcal{A}^F(k_1)$ denotes the output of algorithm $\mathcal{A}$ when it is given oracle access to the function F and input k_1. The input k_1 is a dummy input needed only so that $\mathcal{A}$ takes the same input as the $\mathcal{A}$ below.

The biased Pseudo Random Functions that we will be using are parameterized by two random values, say $k_0 \in K_0$ and $k_1 \in K_1$. The parameter k_0 is kept secret while k_1 is public. To capture this concept we consider a set of functions $\mathcal{F} = \{F_{k_0,k_1} : \{0,1\}^w \to \{0,1\}\}_{k_0 \in K_0, k_1 \in K_1}$. For such a family of functions $\mathcal{F}$ and an algorithm $\mathcal{A}$ we define the following value:

$$\text{Exp}_{\mathcal{A}}^{\mathcal{F}} = \Pr\left[\mathcal{A}^{F_{k_0,k_1}}(k_1) = 1 \; : \; k_0 \xleftarrow{\text{R}} K_0, \; k_1 \xleftarrow{\text{R}} K_1 \right]$$

Note that $\mathcal{A}$ is given k_1 but is not given k_0.

Definition 4. *Let $\mathcal{F} = \{F_{k_0,k_1} : \{0,1\}^w \to \{0,1\}\}_{k_0 \in K_0, k_1 \in K_1}$ be a set of functions. We say that $\mathcal{F}$ is a $(\delta, t, \epsilon_{PRF}, q)$-biased-PRF if for any t-time oracle algorithm $\mathcal{A}$ making at most q queries to its oracle we have:*

$$\left| \text{Exp}_{\mathcal{A}}^{\Omega_\delta} - \text{Exp}_{\mathcal{A}}^{\mathcal{F}} \right| < \epsilon_{PRF}$$

We say that the parameter k_0 is kept secret while k_1 is public.

3.3   Collision Resistance

We briefly review the definition of collision resistant hash functions.

Definition 5. *Let Σ be an alphabet of size s and let n be some positive integer. We say that a family of functions $\mathcal{H} = \{H_k : \{0,1\}^w \to \Sigma^n\}_{k \in K}$ is (t, ϵ_H)-collision resistant if for any t-time algorithm $\mathcal{A}$ we have*

$$\Pr\left[H_k(x) = H_k(y) \text{ and } x \neq y \; : \; k \xleftarrow{\text{R}} K; \; (x,y) \xleftarrow{\text{R}} \mathcal{A}(k) \right] < \epsilon_H$$

It is well known that collision resistant hash functions can be constructed from a finite cyclic group for which the discrete log problem is intractable. Since the Decision BDH assumption in $\mathbb{G}$ implies that discrete-log in $\mathbb{G}$ is intractable it follows that the existence of collision resistant hash functions is implied by the Decision BDH assumption. Consequently, rather than saying that our construction depends on both Decision BDH and collision-resistance we can say that our construction depends on Decision BDH alone for security. Nevertheless, in our security theorems we state collision resistance as an explicit assumption so that one can use any cryptographic hash function such as SHA-1, if so desired.

4   Secure IBE Construction

Before presenting our secure IBE system we first introduce a specific construction for a biased binary PRF from any collision resistant hash function. Later, in Section 5, we prove that it is indeed a PRF with overwhelming probability.

4.1   A Special Biased Binary PRF

Let Σ be an alphabet of size s, and let $\Sigma_\perp = \Sigma \cup \{\perp\}$. For $0 \leq m \leq n$, denote by $\Sigma^{(n,m)}$ the set of vectors in $\Sigma_\perp^n$ that have exactly m components in Σ. For any vector $K \in \Sigma^{(n,m)}$ with $n \geq m > 0$, and any function $H : \{0,1\}^w \to \Sigma^n$ with $w > 0$, we define the *bias map* $F_{K,H} : \{0,1\}^w \to \{0,1\}$ as

$$F_{K,H}(x) = \begin{cases} 0 & \text{if } \exists i \in \{1,\ldots,n\} : H(x)|_i = K|_i \\ 1 & \text{if } \forall i \in \{1,\ldots,n\} : H(x)|_i \neq K|_i \end{cases}$$

Observe that when H is a random function, the bias map $F_{K,H}$ has an expectation of $(1-1/s)^m$ over the inputs $x \in \{0,1\}^w$.

Definition 6. *Let n, m, w be positive integers with $m \leq n$. Let Σ be an alphabet of size s and set $\delta = (1-1/s)^m$. We say that a hash function family $\{H_k : \{0,1\}^w \to \Sigma^n\}_{k \in \mathcal{K}}$ is a $(t, \epsilon_{PRF}, q, m)$-admissible if the function family $\{F_{K,H_k}\}_{K \in \Sigma^{(n,m)}, k \in \mathcal{K}}$ is a $(\delta, t, \epsilon_{PRF}, q)$-biased PRF. Here k is public and K is secret.*

In Section 5 we show how an admissible hash function family can be constructed given a collision resistant hash function family. In the rest of this section, we show how to use admissible hash functions to construct a secure IBE in the standard model.

4.2   Secure IBE Using Admissible Hash Functions

We are now ready to present our secure IBE system. It is based on a recent HIBE construction without random oracles by Boneh and Boyen [BB04] (secure in a selective identity attack model), itself inspired from a random oracle HIBE construction due to Gentry and Silverberg [GS02].

The system makes use of a collision resistant hash function and security is based on the Decision BDH assumption. Let $\mathbb{G}$ be a bilinear group of prime order p and g be a generator of $\mathbb{G}$. Let $e : \mathbb{G} \times \mathbb{G} \to \mathbb{G}_1$ be the bilinear map. We assume that the messages to be encrypted are elements of $\mathbb{G}_1$.

Throughout the section we let $\Sigma = \{1, \ldots, s\}$ be an alphabet of size s, although later we restrict our attention to the binary case $s = 2$. We also let $\{H_k : \{0,1\}^w \to \Sigma^n\}_{k \in \mathcal{K}}$ be a family of hash functions. For now, we assume that public keys (ID) are elements in $\{0,1\}^w$. We later extend the construction to public keys over $\{0,1\}^*$ by first hashing ID using a collision resistant hash $\tilde{H} : \{0,1\}^* \to \{0,1\}^w$. The IBE system works as follows:

Setup: To generate system parameters the algorithm picks a random $\alpha \in \mathbb{Z}_p$ and sets $g_1 = g^\alpha$. Next, it picks a random $g_2 \in \mathbb{G}$ and a random $n \times s$ matrix $U = (u_{i,j})$ where each $u_{i,j}$ is random in $\mathbb{G}$. Finally, the algorithm picks a random $k \in \mathcal{K}$ as a hash function key. The system parameters are $params = (g, g_1, g_2, U, k)$. the master key is $master\text{-}key = g_2^\alpha$.

KeyGen(params, ID, master-key): To generate the private key for an identity $\mathsf{ID} \in \{0,1\}^w$ the algorithm lets $\bar{a} = H_k(\mathsf{ID}) = a_1 \dots a_n \in \Sigma^n$ and picks random $r_1, \dots, r_n \in \mathbb{Z}_p$. The private key d_{ID} is:

$$d_{\mathsf{ID}} = \left(g_2^\alpha \cdot \prod_{i=1}^{n} u_{i,a_i}^{r_i}, \ g^{r_1}, \ \dots, \ g^{r_n} \right) \in \mathbb{G}^{n+1}$$

Encrypt(params, ID, M): To encrypt a message $M \in \mathbb{G}_1$ under the public key $\mathsf{ID} \in \{0,1\}^w$, first set $\bar{a} = H_k(\mathsf{ID}) = a_1 \dots a_n \in \Sigma^n$, then pick a random $t \in \mathbb{Z}_p$ and output

$$C = \left(e(g_1, g_2)^t \cdot M, \ g^t, \ u_{1,a_1}^t, \ \dots, \ u_{n,a_n}^t \right)$$

Note that $e(g_1, g_2)$ can be precomputed so that encryption does not require any pairing computations.

Decrypt(params, d_{ID}, C): To decrypt a ciphertext $C = (A, B, C_1, \dots, C_n)$ using the private key $d_{\mathsf{ID}} = (d_0, d_1, \dots, d_n)$, output:

$$A \cdot \frac{\prod_{j=1}^{n} e(C_j, d_j)}{e(B, d_0)} = M$$

Let $\bar{a} = H_k(\mathsf{ID}) = a_1 \dots a_n \in \Sigma^n$. Then, indeed, for a valid ciphertext we have:

$$\frac{\prod_{j=1}^{n} e(C_j, d_j)}{e(B, d_0)} = \frac{\prod_{j=1}^{n} e(u_{j,a_j}, g)^{t r_j}}{e(g, g_2)^{t\alpha} \prod_{j=1}^{n} e(g, u_{j,a_j})^{t r_k}} = \frac{1}{e(g_1, g_2)^t}$$

This completes the description of the system.

4.3 Security

We now turn to proving security of the IBE above. The system makes use of an admissible hash function family and security is based on the Decision BDH assumption. We prove security in the standard model, i.e., without random oracles.

Theorem 1. *Let $|\Sigma| = s$. Suppose the (t, ϵ_{BDH})-Decision BDH assumption holds in $\mathbb{G}$. Furthermore, suppose $\{H_k : \{0,1\}^w \to \Sigma^n\}_{k \in \mathcal{K}}$ is a $(t, \epsilon_{PRF}, q + 1, m)$-admissible family of hash functions. Set $\delta = (1 - 1/s)^m$ and $\Delta = \delta(1 - \delta)^q$. Assume that $\Delta > \epsilon_{PRF}$. Then the IBE system above is (t, q, ϵ_{IBE})-chosen plaintext (IND-ID-CPA) secure for any $\epsilon_{IBE} \geq 2\epsilon_{BDH}/(\Delta - \epsilon_{PRF})$.*

We note that taking $m = \Theta(s \log q)$ leads to $\Delta = \Theta(1/q)$. Then, ignoring ϵ_{PRF}, we have that $\epsilon_{\mathrm{IBE}} = \Theta(q\epsilon_{\mathrm{BDH}})$. Hence, in groups where $(t, \epsilon_{\mathrm{BDH}})$-Decision BDH holds we obtain a $(t, q, \Theta(q\epsilon_{\mathrm{BDH}}))$ secure IBE system without random oracles.

To prove the theorem we need to show that for any t-time algorithm $\mathcal{A}$ that makes at most q private key queries we have

$$\left| \Pr\left[\mathsf{CPA\text{-}Exp}_{\mathcal{A}}(0) = 1\right] - \Pr\left[\mathsf{CPA\text{-}Exp}_{\mathcal{A}}(1) = 1\right] \right| < \epsilon_{\mathrm{IBE}}$$

To do so we first define two additional experiments.

Experiment 1: $\mathsf{BDH\text{-}Exp}_{\mathcal{A}}(b, (g, g_1, g_2, g_3, T))$. Let $\mathcal{A}$ be an algorithm, b be a bit in $\{0, 1\}$, and (g, g_1, g_2, g_3, T) a 5-tuple where $g, g_1, g_2, g_3 \in \mathbb{G}$ and $T \in \mathbb{G}_1$. Define the following game between a simulator and $\mathcal{A}$:

Setup: To start, the simulator generates system parameters by first picking a random vector $V = v_1 \ldots v_n \in \Sigma^{(n,m)}$. It then generates an $n \times s$ matrix $U = (u_{i,j})$ as follows. For each $i = 1, \ldots, n$ and $j = 1, \ldots, s$ it picks a random $\alpha_{i,j} \in \mathbb{Z}_p$ and sets

$$u_{i,j} = \begin{cases} g_2 \cdot g^{\alpha_{i,j}} & \text{if } v_i = j, \text{ and} \\ g^{\alpha_{i,j}} & \text{otherwise} \end{cases}$$

Next, the simulator picks a random $k \in \mathcal{K}$ as a hash function key. It gives $\mathcal{A}$ the system parameters $params = (g, g_1, g_2, U, k)$. Note that the corresponding (unknown) master key is $master\text{-}key = g_2^{\alpha}$ where $\alpha = \log_g g_1$.

Phase 1. $\mathcal{A}$ issues up to q private key queries. Consider a query for the private key ID $\in \{0, 1\}^w$. Let $\bar{a} = H_k(\mathrm{ID}) = a_1 \ldots a_n \in \Sigma^n$. If $a_i \neq v_i$ for all $i = 1, \ldots, n$ then the simulator terminates the experiment and outputs abort. Otherwise, there exists an i such that $a_i = v_i \in \Sigma$. The simulator derives the private key for ID by first picking random elements $r_1, \ldots, r_n \in \mathbb{Z}_p$ and then setting

$$d_0 = g_1^{-\alpha_{i,v_i}} \prod_{j=1}^{n} u_{j,a_j}^{r_j}, \quad d_1 = g^{r_1}, \quad \ldots, \quad d_i = g^{r_i}/g_1, \quad \ldots, \quad d_n = g^{r_n} \quad (2)$$

We note that $(d_0, d_1, \ldots, d_n) \in \mathbb{G}^{n+1}$ is a valid random private key for ID. To see this, let $\tilde{r}_i = r_i - \alpha$. Then we have that

$$g_1^{-\alpha_{i,v_i}} \prod_{j=1}^{n} u_{j,a_j}^{r_j} = g_2^{\alpha} \cdot (g_2 \, g^{\alpha_{i,v_i}})^{-\alpha} \cdot \prod_{j=1}^{n} u_{j,a_j}^{r_j} = g_2^{\alpha} \cdot u_{i,a_i}^{\tilde{r}_i} \cdot \prod_{j=1, j \neq i}^{n} u_{j,a_j}^{r_j}$$

It follows that the key $(d_0, d_1, \ldots, d_n)$ defined in (2) satisfies:

$$d_0 = g_2^{\alpha} \cdot (u_{i,a_i}^{\tilde{r}_i} \cdot \prod_{j=1, j \neq i}^{n} u_{j,a_j}^{r_j}), \quad d_1 = g^{r_1}, \quad \ldots, \quad d_i = g^{\tilde{r}_i}, \quad \ldots, \quad d_n = g^{r_n}$$

where $r_1, \ldots, \tilde{r}_i, \ldots, r_n$ are uniform in $\mathbb{Z}_p$. This matches the definition for a private key for ID and hence $(d_0, d_1, \ldots, d_n)$ is a valid private key for ID. The simulator gives this key to $\mathcal{A}$.

Challenge. $\mathcal{A}$ outputs an identity ID^* and two messages $M_0, M_1 \in \mathbb{G}_1$. Let $\bar{a} = H_k(\mathsf{ID}^*) = a_1 \ldots a_n \in \Sigma^n$. If there exists an i such $a_i = v_i$ then the simulator terminates the experiment and outputs abort. Otherwise, the simulator responds with the challenge ciphertext

$$C = (M_b \cdot T, \; g_3, \; g_3^{\alpha_{1,a_1}}, \; \ldots, \; g_3^{\alpha_{n,a_n}})$$

Suppose $g_3 = g^c$. Then, we note that since $u_{i,a_i} = g^{\alpha_{i,a_i}}$ for all i, we have that:

$$C = (M_b \cdot T, \; g^c, \; u_{1,a_1}^c, \; \ldots, \; u_{n,a_n}^c)$$

Hence, if $T = e(g,g)^{abc} = e(g_1, g_2)^c$ then the challenge C is a valid encryption of M_b under ID^*.

Phase 2. $\mathcal{A}$ issues more private key queries for identities $\mathsf{ID} \neq \mathsf{ID}^*$. The simulator responds as before.

Guess. Finally, $\mathcal{A}$ outputs a guess $b' \in \{0,1\}$. The simulator outputs b' as the result of the experiment.

We define $\mathsf{BDH\text{-}Exp}_{\mathcal{A}}(b, (g, g_1, g_2, g_3, T))$ to be the random variable denoting the simulator's output in the above experiment. It takes one of three values: 0, 1, or abort.

Experiment 2: $\mathsf{PRF\text{-}Exp}_{\mathcal{A}}(b, F, k)$. Let $\mathcal{A}$ be an algorithm, b be a bit in $\{0,1\}$, F be a function $F : \{0,1\}^w \rightarrow \{0,1\}$, and $k \in \mathcal{K}$. Define the following game between a simulator and $\mathcal{A}$:

Setup: To generate system parameters the simulator picks a random $\alpha \in \mathbb{Z}_p$ and sets $g_1 = g^\alpha$. Next, it picks random $g_2 \in \mathbb{G}$ and a random $n \times s$ matrix $U = (u_{i,j})$ where each $u_{i,j} \in \mathbb{G}$. It gives $\mathcal{A}$ the system parameters $params = (g, g_1, g_2, U, k)$ and keeps to itself the master key $master\text{-}key = g_2^\alpha$.

Phase 1: $\mathcal{A}$ issues up to q adaptive private key queries. Consider a query for the private key $\mathsf{ID} \in \{0,1\}^w$. If $F(\mathsf{ID}) = 1$ the simulator terminates the experiment and outputs abort. Otherwise, the simulator uses $master\text{-}key$ to generate the private key for ID and gives the result to $\mathcal{A}$.

Challenge. $\mathcal{A}$ outputs an identity ID^* and two messages $M_0, M_1 \in \mathbb{G}_1$. If $F(\mathsf{ID}) = 0$ the simulator terminates the experiment and outputs abort. Otherwise, the simulator creates the encryption of M_b and gives the resulting challenge ciphertext to $\mathcal{A}$.

Phase 2. $\mathcal{A}$ issues more private key queries for identities $\mathsf{ID} \neq \mathsf{ID}^*$. The simulator responds as before (aborting as necessary).

Guess. Finally, $\mathcal{A}$ outputs a guess $b' \in \{0,1\}$. The simulator outputs b' as the result of the experiment.

We define $\mathsf{PRF\text{-}Exp}_{\mathcal{A}}(b, F)$ to be the random variable denoting the simulator's output in the above experiment. It takes one of three values: 0, 1, or abort.

Next, we state four facts about these experiments, which we prove in the full version of the paper. The proof of Theorem 1 will follow immediately from these facts. We define the following notation:

1. Define the random variable $Z = (g, g_1, g_2, g_3, T) \overset{\text{R}}{\leftarrow} \mathcal{P}_{\text{BDH}}$.
2. For $b = 0, 1$ define the random variable $T_b = \text{BDH-Exp}_{\mathcal{A}}(b, Z)$.
3. For $b = 0, 1$ define the value $t_b = \Pr[T_b = 1 \mid T_b \neq \text{abort}]$.
4. We let $\{F_{K,H_k}\}$ denote the distribution sampled by the following algorithm: pick a random $k \in \mathcal{K}$ and a random $K \in \Sigma^{(n,m)}$, and output the (function, key) pair (F_{K,H_k}, k).
5. We set $\delta = (1 - 1/s)^m$ and $\Delta = \delta(1 - \delta)^q$.

Claim 1. Consider $(F, k) \overset{\text{R}}{\leftarrow} \{F_{K,H_k}\}$. Then for $b = 0, 1$ the random variable $T_b = \text{BDH-Exp}_{\mathcal{A}}(b, Z)$ is identical to the random variable $\text{PRF-Exp}_{\mathcal{A}}(b, F, k)$.

Claim 2. For $b = 0, 1$ we have that t_b is equal to $\Pr[\text{CPA-Exp}_{\mathcal{A}}(b) = 1]$.

Claim 3. Let $(F, k) \overset{\text{R}}{\leftarrow} \{F_{K,H_k}\}$. Then for $b = 0, 1$:

$$\Pr[\text{PRF-Exp}_{\mathcal{A}}(b, F, k) = \text{abort}] < 1 - \Delta + \epsilon_{\text{PRF}}$$

Claim 4. We have that $|t_0 - t_1| < 2\epsilon_{\text{BDH}}/(\Delta - \epsilon_{\text{PRF}})$.

The proofs of these claims are given in the full version of the paper. The main theorem follows easily.

Proof (Proof of Theorem 1). The theorem follows directly from Claims 2 and 4. The two claims together show that for any t-time algorithm $\mathcal{A}$ that makes at most q private key queries, we have

$$\left| \Pr[\text{CPA-Exp}_{\mathcal{A}}(0) = 1] - \Pr[\text{CPA-Exp}_{\mathcal{A}}(1) = 1] \right| = |t_0 - t_1| < 2\epsilon_{\text{BDH}}/(\Delta - \epsilon_{\text{PRF}})$$

as required.   □

5   Constructing Admissible Hash Functions

It remains to show how an admissible hash function family can be constructed given a collision resistant hash function family. We do this in two steps: we first present some idealized sufficient conditions for a hash function family to be admissible, then show how these conditions can be achieved in the case of a binary alphabet given a family of collision resistant hash functions. As previously mentioned, the Decision BDH assumption can be used to realize collision resistance, although we are free to use more practical hash functions.

For simplicity, we define the following shorthand notation. We let $\Sigma^{(n,m)}$ be the universe of the possible values of the secret index K. For a hash function H, we respectively define the H-*null-set* and the H-*kernel* of any $x \in \{0,1\}^w$ as:

$$Z_H(x) = \{K \in \Sigma^{(n,m)} : F_{K,H}(x) = 0\}, \quad Y_H(x) = \{K \in \Sigma^{(n,m)} : F_{K,H}(x) = 1\}$$

Clearly, for any x the sets $Z_H(x)$ and $Y_H(x)$ form a partition of $\Sigma^{(n,m)}$ such that $|Z_H(x)| = \binom{n}{m}(s^m - (s-1)^m)$ and $|Y_H(x)| = \binom{n}{m}(s-1)^m$. For binary alphabets, we have

$$(s = 2) \implies |\Sigma^{(n,m)}| = \binom{n}{m}2^m, \quad |Z_H(x)| = \binom{n}{m}(2^m - 1), \quad |Y_H(x)| = \binom{n}{m}$$

Before delving into the construction, we need to precise the following notions.

Adversarial Uncertainty. We formalize the information made available to the adversary using the notion of knowledge state. At any time during the interaction of an algorithm $\mathcal{A}^F$ with a bias map oracle $F_{K,H}$ where H is public and K is secret, the algorithm's available knowledge about the oracle is captured by a distribution of the secret K. Initially the distribution is uniform over $\Sigma^{(n,m)}$ since K is chosen uniformly in this set. Now, suppose that prior to the next interaction with the oracle the distribution is uniform over some set S, then the distribution after the next oracle query $F_{K,H}(x_i)$ is uniform over a subset $S' \subseteq S$ such that

$$S' = \begin{cases} S \cap Z_H(x) & \text{if } F_{K,H}(x_i) = 0 \\ S \cap Y_H(x) & \text{if } F_{K,H}(x_i) = 1 \end{cases}$$

It follows that after learning the responses $\{F_{K,H}(x_i) : i = 1, \ldots, j\}$ to any set of queries $\{x_i : i = 1, \ldots, j\}$, the algorithm's knowledge state regarding K is completely captured by the uniform distribution over the set S_j given by

$$S_j = \left(\Sigma^{(n,m)}\right) \cap \underbrace{\bigcap_{\substack{i \in \{1,\ldots,j\} \\ F_{K,H}(x_i)=0}} Z_H(x_i)}_{S_j^{(0)}} \cap \underbrace{\bigcap_{\substack{i \in \{1,\ldots,j\} \\ F_{K,H}(x_i)=1}} Y_H(x_i)}_{S_j^{(1)}}$$

Here, $S_j^{(0)}$ and $S_j^{(1)}$ are respectively defined as the sets of values of $K \in \Sigma^{(n,m)}$ that are compatible with the "negative" and the "positive" responses from the set of oracle responses $\{F_{K,H}(x_i) : i = 1, \ldots, j\}$. Notice that reordering the queries has no effect on the knowledge state.

Hamming Separation Property. For two vectors $x, y \in \Sigma^n$, we write $d(x, y)$ for the Hamming distance between x and y. We say that a hash function family $\{H_k : \{0,1\}^w \to \Sigma^n\}_{k \in \mathcal{K}}$ satisfies the v-*Hamming separation* property if $\forall k \in \mathcal{K}$ and $\forall x, y \in \{0,1\}^w$ such that $H_k(x) \neq H_k(y)$, it also holds that $d(H_k(x), H_k(y)) \geq v$. In other words, any distinct $H_k(x)$ and $H_k(y)$ must take differing values in at least v coordinates (and thus have at most $n - v$ coordinates in common).

In Section 5.2, we show how to achieve the Hamming separation property from collision resistance using coding theory.

5.1 Sufficient Conditions for Admissibility

The following theorem gives a set of sufficient conditions for a hash family to be admissible as defined in Definition 6. We focus on binary alphabets ($s = 2$).

Theorem 2. *Let n, m, v, w be positive integers such that $m \leq n$ and $v \leq n$. Let Σ be an alphabet of size $s = 2$, and let $\delta = (1 - 1/s)^m = 2^{-m}$. Assume that $\mathcal{H} = \{H_k : \{0,1\}^w \to \Sigma^n\}_{k \in \mathcal{K}}$ is some (t, ϵ_H)-collision resistant hash function family that satisfies the v-Hamming separation property. Pose $\theta = (1 - v/n)^m$. If $\theta \leq \kappa \delta$ for some arbitrary $\kappa \in (1, \infty)$ then the family $\mathcal{H}$ is $(t, \epsilon_{PRF}, q, m)$-admissible provided that $\epsilon_{PRF} \geq \epsilon_H + \frac{13}{2}\gamma^2/\kappa$ and $q \leq \gamma/\kappa\delta$ for some arbitrary $\gamma \in (0, \frac{1}{2})$.*

Proof. It suffices to show that, in the view of any algorithm $\mathcal{A}$ interacting with a bias map oracle F_{K,H_k} for random $k \in \mathcal{K}$ and $K \in \Sigma^{(n,m)}$ where K is secret, the first q outputs of the oracle are distributed identically to the first q outcomes of a binomial random process of expectation δ, with probability at least $1 - \epsilon_{\mathrm{PRF}}$.

We henceforth omit the subscripts K and H_k since there is no ambiguity, and write $F(x)$ for $F_{K,H_k}(x)$. We use the abbreviations $Y_i = Y_{H_k}(x_i)$, $Z_i = Z_{H_k}(x_i)$, $h_i = H_k(x_i)$, and $F_i = F(x_i)$.

We compute the distribution of the first q oracle answers under the stated assumptions, treating the algorithm $\mathcal{A}$ as an adversary that adaptively selects the q points $x_1, \ldots, x_q$ at which F is queried. For now, we assume that $\forall i \neq j$: $x_i \neq x_j \Rightarrow h_i \neq h_j$ (and by the v-Hamming separation property, $d(h_i, h_j) \geq v$). By the $(t, \epsilon_{\mathrm{H}})$-collision resistance assumption on $\mathcal{H}$, this is true with probability at least $1 - \epsilon_{\mathrm{H}}$. We correct for this assumption at the end.

Suppose that before step $j \in \{1, \ldots, q\}$ the adversary has learned the $j - 1$ values respectively taken by $F(x)$ at arbitrary query points $x = x_1, \ldots, x_{j-1}$. Our goal is to find lower and upper bounds on the conditional probability that $F(x_j) = 1$ given the history of past queries and answers, in the adversary's view, uniformly for all choices of the next query point $x_j \notin \{x_1, \ldots, x_{j-1}\}$.

Let $X_i = \{x_1, \ldots, x_i\} = X_i^{(0)} \cup X_i^{(1)}$ where $X_i^{(0)} = \{x \in X_i : F(x) = 0\}$ and $X_i^{(1)} = \{x \in X_i : F(x) = 1\}$, and write $P_j = \Pr[F(x_j) = 1 \mid X_{j-1}^{(0)}, X_{j-1}^{(1)}]$ for the probability we seek to bound. Observe that the two sets $X_{j-1}^{(0)}$ and $X_{j-1}^{(1)}$ together capture all relevant information about the query history just before the j-th query, since the order of the queries is irrelevant. We have

$$
P_j = \Pr[F(x_j) = 1 \mid X_{j-1}^{(0)}, X_{j-1}^{(1)}] = \frac{|Y_j \cap S_{j-1}|}{|S_{j-1}|} = \frac{|Y_j \cap S_{j-1}^{(1)} \cap S_{j-1}^{(0)}|}{|S_{j-1}^{(1)} \cap S_{j-1}^{(0)}|}
$$

$$
= \frac{\left|Y_j \cap \left(\bigcap_{x \in X_{j-1}^{(1)}} Y(x)\right) \cap \left(\bigcap_{x \in X_{j-1}^{(0)}} Z(x)\right)\right|}{\left|\left(\bigcap_{x \in X_{j-1}^{(1)}} Y(x)\right) \cap \left(\bigcap_{x \in X_{j-1}^{(0)}} Z(x)\right)\right|} = \frac{\left|Y_j \cap Y_{\cap,j-1}^{(1)} \setminus Y_{\cup,j-1}^{(0)}\right|}{\left|Y_{\cap,j-1}^{(1)} \setminus Y_{\cup,j-1}^{(0)}\right|}
$$

where we have posed $Y_{\cap,j-1}^{(1)} = \left(\bigcap_{x \in X_{j-1}^{(1)}} Y(x)\right)$ and $Y_{\cup,j-1}^{(0)} = \left(\bigcup_{x \in X_{j-1}^{(0)}} Y(x)\right)$.

We can use this general expression and the v-Hamming separation property to bound P_j for query histories that contain either zero or one positive answer. We later show that the other cases are together very unlikely. Namely, we seek:

1. a uniform bounding interval on P_j for all query histories with $|X_{j-1}^{(1)}| = 0$ (i.e., containing only negative answers);
2. a uniform upper bound on P_j for all query histories such that $|X_{j-1}^{(1)}| = 1$ (i.e., containing one positive answer).

We obtain non-trivial uniform bounds of three different kinds, given by

$$
\forall X_{j-1}^{(0)}, X_{j-1}^{(1)} \ \text{s.t.} \ |X_{j-1}^{(1)}| = 0 : \quad (1 - \gamma)\delta \leq P_j \leq (1 + 2\gamma)\delta
$$

$$
\forall X_{j-1}^{(0)}, X_{j-1}^{(1)} \ \text{s.t.} \ |X_{j-1}^{(1)}| = 1 : \quad P_j \leq 2\kappa\delta
$$

Detailed calculations for these bounds are given in the full version of the paper.

Subject to the above inequalities, we set out to bound the probability that the biased PRF oracle F deviates from a sequence of q outcomes from a genuine memoryless binomial process of expectation δ over a sequence of length q.

Consider R, a binomial process of expectation δ. We construct a modified process R' whose i-th outcome is defined as $R'_i = R_i \oplus M_i$. Here, M is a control process whose purpose is to randomly decide whether R'_i should assume the value of R_i or its opposite, with a probability that depends on the previous outcomes $R'_1, \ldots, R'_{i-1}$ and the current drawing R_i. By properly choosing M, we can make R' behave exactly as F, i.e., have the q-prefixes of R' achieve the same joint distribution as the q-prefix of F. In particular, this means that the event that the processes R and F behave similarly over a sequence of length q is at least as likely as the event that $M_i = 0$ for all $i = 1, \ldots, q$, since in this case R and R' have the same first q outcomes. It remains to bound such probability. Here is the gist of the argument.

The goal is to devise an R' that perfectly simulates any q-prefix of $F = F_{K,H}$ for (unknown) random K, and bound the influence of M needed to do so. Suppose that for some query history $X^{(0)}_{j-1}, X^{(1)}_{j-1}$, the conditional expectation $P_j = \Pr[F_j = 1 \mid X^{(0)}_{j-1}, X^{(1)}_{j-1}]$ of F_j as viewed by the adversary exceeds the expectation $\Pr[R_j = 1] = \delta$ of the binomial process R_j. One can make the simulated process R'_j assume the expected law of F_j conditionally on this specific history by letting the control process take $M_j \leftarrow 1$ with conditional probability $(P_j - \delta)/(1 - \delta)$ when $R_i = 0$, and with probability 0 when $R_i = 1$. More generally, we find that for the process R' to perfectly simulate F, it suffices that for $j = 1, \ldots, q$, the conditional law of M_j given $R'_1, \ldots, R'_{j-1}, R_j$ satisfies

$$\Pr[M_j = 1, R_j = 0 \mid X^{(0)}_{j-1}, X^{(1)}_{j-1}] = \max\{\, 0, \ P_j - \delta \,\}$$

$$\Pr[M_j = 1, R_j = 1 \mid X^{(0)}_{j-1}, X^{(1)}_{j-1}] = \max\{\, 0, \ \delta - P_j \,\}$$

Let us write E_j for the event $[\exists i \leq j, M_i \neq 0]$. We outline how to use the above results to upper bound the unconditional probability $\Pr[E_j]$ for $j \leq q$. First, from the law of M we get $\Pr[M_j = 1 \mid X^{(0)}_{j-1}, X^{(1)}_{j-1}] \leq |P_j - \delta| \leq 1$, which we can bound further using our previous bounds on P_j in the cases where $|X^{(1)}_{j-1}| = 0, 1$. Next, we need to bound the probabilities $\Pr[X^{(0)}_{j-1}, X^{(1)}_{j-1}]$ of the conditioning events. The difficultly here is that the random variables $X^{(0)}_{j-1}, X^{(1)}_{j-1}$ derive from the complicated process R'. Fortunately, conditionally on the event $\neg E_{j-1}$, the process R' identifies with the binomial process R so that these probabilities have nice expressions in function of j and $|X^{(1)}_{j-1}|$. Note that these probabilities vanish quickly as $|X^{(1)}_{j-1}|$ increases, which is why we bounded P_j for $|X^{(1)}_{j-1}| = 0, 1$ only.

Thus, we have just reduced the upper bound computation of $\Pr[E_j]$ to that of $\Pr[E_{j-1}]$. Carrying this idea through, after some calculations we obtain

$$\Pr[E_q] \ = \ \Pr[\exists i \leq j, M_i \neq 0] \ = \ \sum_{j=1}^{q} \Pr[M_j = 1, \neg E_{j-1}] \ \leq \ \frac{13}{2}\gamma^2/\kappa$$

The formal derivation of this result may be found in the full version of the paper.

To conclude, we correct for the probability ϵ_H of finding a hash collision in the allotted time t, which in the worst scenario could yield an infallible discriminator between F and R. It follows that the probability that the F and R oracles can be distinguished admits the upper bound $\epsilon_H + \frac{13}{2}\gamma^2/\kappa \leq \epsilon_{PRF}$, as required.   □

5.2 Admissibility from Collision Resistance

We now show how to construct an admissible hash function family $\mathcal{H} = \{H_k : \{0,1\}^w \to \Sigma^n\}_{k\in\mathcal{K}}$ in the sense of Theorem 2, given an "ordinary" family of (t,ϵ_H)-collision resistant hash functions $\bar{\mathcal{H}} = \{\bar{H}_k : \{0,1\}^w \to \{0,1\}^\beta\}_{k\in\mathcal{K}}$. We give an explicit construction for the specific case of a binary alphabet ($s = 2$).

Theorem 3. *Let $\bar{\mathcal{H}} = \{\bar{H}_k : \{0,1\}^w \to \{0,1\}^\beta\}_{k\in\mathcal{K}}$ be an efficiently computable (t,ϵ_H)-collision resistant hash function family. Then for any $r \in (0, \frac{1}{2})$ there exists an efficiently computable function family $\mathcal{H} = \{H_k : \{0,1\}^w \to \{0,1\}^n\}_{k\in\mathcal{K}}$ that satisfies both the (t,ϵ_H)-collision resistance property and the bitwise v-Hamming separation property, where $\beta \leq n \leq 2\beta^2/(1-2r)^2$ and $v/n > r$.*

Proof. Let t be the smallest positive integer such that $2^t \geq \lceil\beta/t\rceil/(1-2r)+1$, and define $\ell = \lceil\beta/t\rceil$.

Let $\mu' : \{0,1\}^t \to \mathbb{F}_{2^t}$ be any bijection. Define the injection $\mu : \{0,1\}^\beta \to \mathbb{F}_{2^t}^\ell$ that, on input $z \in \{0,1\}^\beta$, partitions z in ℓ fragments of t bits each (padding the last fragment as necessary), applies the map μ' to each fragment, and concatenates all the outputs.

Let $\rho : \mathbb{F}_{2^t}^\ell \to \mathbb{F}_{2^t}^{2^t-1}$ be a Reed-Solomon error correcting code with parameters $[2^t-1, \ell, 2^t-\ell]$, i.e., a linear code that takes input words of size ℓ over the alphabet $\mathbb{F}_{2^t}$ and produces codewords of length $2^t - 1$ with minimum pairwise Hamming distance $2^t - \ell$.

Let $\eta' : \mathbb{F}_{2^t} \to \{0,1\}^{2^t}$ be the injection that maps any field element $i \in \{0,\ldots,2^t-1\}$ to the 2^t-bit vector given by the i-th row of a $2^t \times 2^t$ Hadamard matrix. Recall that a binary $m \times m$ Hadamard matrix is such that any two distinct rows or columns agree on exactly $m/2$ coordinates; it is well known that a $2^t \times 2^t$ Hadamard matrix exists and is easy to construct for all $t \geq 1$. Define the function $\eta : \mathbb{F}_{2^t}^{2^t-1} \to \{0,1\}^{2^t(2^t-1)}$ that applies η' individually to each coordinate of its input word and concatenates the resulting Hadamard vectors.

The desired hash family is then given by $\mathcal{H} = \{H_k : \{0,1\}^w \to \Sigma^n\}_{k\in\mathcal{K}}$ where $H_k = \eta \circ \rho \circ \mu \circ \bar{H}_k$.

It remains to show that $\mathcal{H}$ has the desired properties.

First, since $\eta\circ\rho\circ\mu$ is an injection, the (t,ϵ_H)-collision resistance of $\bar{H}_k$ entails the same for H_k.

Next, by the stated properties of the Reed-Solomon code, ρ produces codewords of size 2^t-1 with minimum pairwise Hamming distance $2^t-\ell$ in $\mathbb{F}_{2^t}$. Since η turns any two distinct elements of $\mathbb{F}_{2^t}$ into 2^t-bit vectors that differ in 2^{t-1} positions, it follows that $\eta\circ\rho$ produces binary vectors of size $n = 2^t(2^t-1)$ with minimum pairwise Hamming distance $v = 2^{t-1}(2^t-\ell)$ in $\mathbb{F}_2$. The corresponding ratio v/n is bounded as follows. Since t is chosen such that $(2^t - 1)(1 - 2r) \geq \ell$, we have $2^t - \ell \geq 2r(2^t - 1) + 1$, hence $(2^t - \ell)/(2^t - 1) > 2r$. It follows that $v/n > r$, as claimed.

Last, we have that $\beta \leq n = 2^t(2^t - 1) \leq 2\lceil \beta/t \rceil^2/(1 - 2r)^2 \leq 2\beta^2/(1 - 2r)^2$, as required. □

5.3   Putting It All Together – Concrete Bounds

It is useful to assign a more concrete meaning to the values taken by the parameters intervening in Theorems 2 and 3. We assume to be given ϵ_H (the adversarial advantage against the collision resistant hash functions), β (the collision resistant hash output length in bits), and q (the allowed number of PRF queries), under the "birthday paradox" constraint that $q \ll \sqrt{2^\beta}$. Our task is to find a suitable set of parameters so that (1) the security ϵ_IBE of the IBE system of Section 4.2 is within a polynomial factor of ϵ_BDH, and (2) the complexity of the four IBE operations takes polynomial time in the security parameters. For $s = 2$, we require that $\epsilon_\mathrm{IBE}/\epsilon_\mathrm{BDH} \leq O(\mathrm{poly}(q))$ and $n \leq O(\mathrm{poly}(\beta, \log(q), \log(1/\epsilon_\mathrm{IBE})))$.

We describe two settings of the parameters; one favoring security, the other favoring performance.

Favoring security. We first show how to satisfy the requirements for the PRF construction with a binary alphabet ($s = 2$) when the intrinsic PRF error probability (defined as $\epsilon'_\mathrm{PRF} = \epsilon_\mathrm{PRF} - \epsilon_\mathrm{H}$ in the notation of Theorem 2) is pegged to $\epsilon'_\mathrm{PRF} = \epsilon_\mathrm{H}$. We arbitrarily choose $\kappa \leftarrow 2$ and successively derive: $\gamma \leftarrow \sqrt{\epsilon'_\mathrm{PRF}}/2$, $m \leftarrow \lfloor \log_2(2q/\gamma) \rfloor$, $\delta \leftarrow 2^{-m} \approx \gamma/2q$, $r \leftarrow 1 - \sqrt[m]{2}/2 \approx \frac{1}{2} - \frac{1}{3m}$, $t \leftarrow$ least s.t. $2^t \geq \lceil \beta/t \rceil/(1 - 2r) + 1$, $\ell \leftarrow \lceil \beta/t \rceil$, $n \leftarrow 2^t(2^t - 1)$, $v \leftarrow 2^{t-1}(2^l - \ell)$, and $\theta \leftarrow (1 - \frac{v}{n})^m < \kappa\delta$. Evidently, the total PRF loss $\epsilon_\mathrm{PRF} = \epsilon_\mathrm{H} + \epsilon'_\mathrm{PRF} = 2\epsilon_\mathrm{H}$ is negligible and the bandwidth coefficient $n = O(\log_2^2(q/\sqrt{\epsilon_\mathrm{PRF}})\beta^2)$ is polynomial in $\log q$ and β. The price to pay for such a low value of ϵ_PRF is a fairly large n.

Favoring performance. We can attain better bounds by adjusting the PRF loss to best match the intrinsic loss incurred by the IBE construction itself, in function of q, as follows. Assuming that the loss ϵ_H due to hash collisions is negligible, under the $(t, \epsilon_\mathrm{BDH})$-Decision BDH assumption Theorem 1 gives a $(t, q, \epsilon_\mathrm{IBE})$-secure IBE such that $\epsilon_\mathrm{IBE} = 2\,\epsilon_\mathrm{BDH}/(\delta(1 - \delta)^q - \epsilon_\mathrm{PRF}) \approx 2\,\epsilon_\mathrm{BDH}/(\sqrt{\epsilon_\mathrm{PRF}}/4q - \epsilon_\mathrm{PRF})$. We can minimize ϵ_IBE for a prescribed value of q by seeking $\epsilon_\mathrm{PRF} \leftarrow (1/8q)^2$. For $\kappa \leftarrow 2$ this gives us a total IBE security loss $\epsilon_\mathrm{IBE} \approx 64\,q^2\,\epsilon_\mathrm{BDH} = \Theta(q^2\epsilon_\mathrm{BDH})$ under the improved bandwidth requirement $n \leq (9 + 4\log_2 q)^2\beta^2 = \Theta((\log_2 q)\beta^2)$.

We note that the optimal value of κ varies and is tied to the coding construction. We defer to the full paper the question of optimizing for all parameters.

6   Extensions

We very briefly outline a few simple extensions of the IBE system of Section 4.2.

Hierarchical IBE. Introduced by Horowitz and Lynn [HL02], HIBE was first constructed by Gentry and Silverberg [GS02] in the random oracle model. The IBE system of Section 4.2 generalizes naturally to give a semantically secure HIBE under an adaptive chosen identity attack (IND-ID-CPA) without random oracles. For a hierarchy of depth ℓ, both the ciphertext and private key contain ℓ blocks where each block contains n components. Thus, a private key at depth ℓ is an element of $\mathbb{G}^{\ell n+1}$. As our IBE, the HIBE uses collision resistant hash functions and is provably secure without random oracles whenever the Decision BDH assumption holds. The construction is similar to the construction of a (selective identity secure) HIBE without random oracles based on Decision BDH recently proposed by Boneh and Boyen [BB04]. The details are deferred to the full version of the paper.

Chosen Ciphertext Security. A recent result of Canetti et al. [CHK04] gives an efficient way to build a chosen ciphertext IBE (IND-ID-CCA) from a chosen plaintext 2-HIBE (IND-ID-CPA). Thus, by the previous paragraph, we obtain a full chosen identity, chosen ciphertext IBE (IND-ID-CCA) that is provably secure without random oracles. More generally, by starting from an $(\ell + 1)$-HIBE, a fully secure ℓ-HIBE can be similarly constructed without random oracles.

Arbitrary Identities. We can extend our IBE system to handle identities $\mathsf{ID} \in \{0,1\}^*$ (as opposed to $\mathsf{ID} \in \{0,1\}^w$) by first hashing ID using a collision resistant hash function $\tilde{H} : \{0,1\}^* \to \{0,1\}^w$ prior to key generation and encryption. A standard argument shows that if the scheme of Section 4.2 is IND-ID-CPA secure then so is the scheme with the additional hash. This holds for the HIBE and the chosen ciphertext secure system and as well.

7   Conclusions

We presented an Identity Based cryptosystem and proved its security without using the random oracle heuristic under the decisional Bilinear Diffie-Hellman assumption. Our results prove that secure IBE systems exist in the standard model. This resolves an open problem posed by Boneh and Franklin in 2001. However, the present system is not very practical and mostly serves as an existence proof. It is still a wonderful problem to find a practical IBE system secure without random oracles based on Decision BDH or a comparable assumption.

References

[BB04]   Dan Boneh and Xavier Boyen. Efficient selective-ID identity based encryption without random oracles. In *Advances in Cryptology—EUROCRYPT '04*, volume 3027 of *LNCS*, pages 223–238. Springer-Verlag, 2004.

[BBP04] Mihir Bellare, Alexandra Boldyreva, and Adriana Palacio. An uninstantiable random-oracle-model scheme for a hybrid-encryption problem. In *Advances in Cryptology—EUROCRYPT '04*, volume 3027 of *LNCS*, pages 171–188. Springer-Verlag, 2004.

[BF01]   Dan Boneh and Matt Franklin. Identity-based encryption from the Weil pairing. In *Advances in Cryptology—CRYPTO '01*, volume 2139 of *LNCS*, pages 213–229. Springer-Verlag, 2001.

[BF03]   Dan Boneh and Matt Franklin. Identity-based encryption from the Weil pairing. *SIAM J. of Computing*, 32(3):586–615, 2003.

[Boy03]   Xavier Boyen. Multipurpose identity-based signcryption: A Swiss Army knife for identity-based cryptography. In *Advances in Cryptology—CRYPTO '03*, volume 2729 of *LNCS*, pages 383–99. Springer-Verlag, 2003.

[BR93]   Mihir Bellare and Phil Rogaway. Random oracle are practical: A paradigm for designing efficient protocols. In *Proceedings of ACM Conference on Computer and Communications Security—CCS '93*, pages 62–73, 1993.

[CC03]   Jae Choon Cha and Jung Hee Cheon. An identity-based signature from gap Diffie-Hellman groups. In *Practice and Theory in Public Key Cryptography—PKC '03*, volume 2567 of *LNCS*. Springer-Verlag, 2003.

[CGH98]   Ran Canetti, Oded Goldreich, and Shai Halevi. The random oracle model revisited. In *Proceedings of ACM Symposium on Theory of Computing—STOC '98*. ACM, 1998.

[CHK03]   Ran Canetti, Shai Halevi, and Jonathan Katz. A forward-secure public-key encryption scheme. In *Advances in Cryptology—EUROCRYPT '03*, volume 2656 of *LNCS*, pages 255–271. Springer-Verlag, 2003.

[CHK04]   Ran Canetti, Shai Halevi, and Jonathan Katz. Chosen-ciphertext security from identity-based encryption. In *Advances in Cryptology—EUROCRYPT '04*, volume 3027 of *LNCS*, pages 207–222. Springer-Verlag, 2004.

[Coc01]   Clifford Cocks. An identity based encryption scheme based on quadratic residues. In *Proceedings of the 8th IMA International Conference on Cryptography and Coding*, pages 26–28, 2001.

[GS02]   Craig Gentry and Alice Silverberg. Hierarchical ID-based cryptography. In *Advances in Cryptology—ASIACRYPT '02*, volume 2501 of *LNCS*, pages 548–566. Springer-Verlag, 2002.

[HK04]   Swee-Huay Heng and Kaoru Kurosawa. *k*-resilient identity-based encryption in the standard model. In *Topic in Cryptology—CT-RSA '04*, volume 2964 of *LNCS*, pages 67–80, 2004.

[HL02]   Jeremy Horwitz and Benjamin Lynn. Towards hierarchical identity-based encryption. In *Advances in Cryptology—EUROCRYPT '02*, pages 466–481, 2002.

[Jou00]   Antoine Joux. A one round protocol for tripartite Diffie-Hellman. In *Proceedings of ANTS IV*, volume 1838 of *LNCS*, pages 385–394. Springer-Verlag, 2000.

[MY96]   Ueli M. Maurer and Yacov Yacobi. A non-interactive public-key distribution system. *Designs, Codes and Cryptography*, 9(3):305–316, November 1996.

[Sha84]   Adi Shamir. Identity-based cryptosystems and signature schemes. In *Advances in Cryptology—CRYPTO '84*, volume 196 of *LNCS*, pages 47–53. Springer-Verlag, 1984.

[SOK00]   Ryuichi Sakai, Kiyoshi Ohgishi, and Masao Kasahara. Cryptosystems based on pairings. In *Proceedings of Symposium on Cryptography and Information Security—SCIS '00*, Japan, 2000.

[Tan87]   Hatsukazu Tanaka. A realization scheme for the identity-based cryptosystem. In *Advances in Cryptology—CRYPTO '87*, volume 293 of *LNCS*, pages 341–349. Springer-Verlag, 1987.

Non-interactive Timestamping
in the Bounded Storage Model

Tal Moran[1], Ronen Shaltiel[2,*], and Amnon Ta-Shma[1]

[1] Department of Computer Science, Tel-Aviv University, Tel-Aviv, Israel
[2] Department of Computer Science and Applied Mathematics,
Weizmann Institute of Science, Rehovot, Israel

Abstract. A timestamping scheme is *non-interactive* if a stamper can stamp a document without communicating with any other player. The only communication done is at validation time. Non-Interactive timestamping has many advantages, such as information theoretic privacy and enhanced robustness. Unfortunately, no such scheme exists against polynomial time adversaries that have *unbounded storage* at their disposal.

In this paper we show non-interactive timestamping is possible in the bounded storage model. In this model it is assumed that all parties participating in the protocol have small storage, and that in the beginning of the protocol a very long random string (which is too long to be stored by the players) is transmitted. To the best of our knowledge, this is the first example of a cryptographic task that is possible in the bounded storage model, but is *impossible* in the "standard cryptographic setting", even assuming cryptographic assumptions.

We give an explicit construction that is secure against all bounded storage adversaries, and a significantly more efficient construction secure against all bounded storage adversaries that run in polynomial time.

Keywords: timestamping, bounded storage model, expander graphs, extractors

1 Introduction

The date on which a document was created is often a significant issue. Patents, contracts, wills and countless other legal documents critically depend on the date they were signed, drafted, etc. A timestamp for a document provides convincing proof that it existed at a certain time. For physical documents, many methods are known and widely used for timestamping: publication, witnessed signing and placing copies in escrow are among the most common. Techniques for timestamping digital documents, which are increasingly being used to replace their physical counterparts, have also become necessary.

Loosely speaking, a timestamping scheme consists of two mechanisms: A *stamping mechanism* which allows a user to stamp a document at some specific time t, and a *verification mechanism* which allows a recipient to verify at a later time $t' > t$ that the document was indeed stamped at time t.

---

* Research supported by the Koshland Scholarship.

M. Franklin (Ed.): CRYPTO 2004, LNCS 3152, pp. 460–476, 2004.

Previous Work

Digital timestamping systems were first introduced in Haber and Stornetta [16], where three timestamping systems are described. In the *naïve timestamping protocol*, the stamper sends the document to all the verifiers during timestamp generation. In the *linking* scheme, the stamper sends a one-way hash of the document to a trusted timestamping server. The server holds a *current hash*, which it updates by hashing it with the value sent by the stamper. This links the document to the previous documents and to the succeeding ones. In the *distributed trust* scheme, the document is used to select a subset of verifiers, to which the stamper sends a hash of the document. Bayer, Haber and Stornetta [3] improve upon the linking scheme, reducing the communication and storage requirements of the system and increasing its robustness, by replacing the linear list with a tree. Further work [17, 7, 6, 8, 5, 4] is mainly focused on additional improvements in terms of storage, robustness and reducing the trust required in the timestamping server(s).

One common feature of all the above protocols is that they require the stamper to send messages to a central authority (or a distributed set of servers) at timestamp generation.

Non-interactive Timestamping

We call a timestamping scheme *non-interactive* if it does not require the stamper to send messages at timestamp generation. Non-interactive timestamping schemes, if they exist, have a number of obvious advantages over active schemes. However, the notion of non-interactive timestamping seems self-contradictory. How can we prevent an adversary from faking timestamps, if no action is taken at timestamp generation? More precisely, suppose that an adversary "learns" some document at time $t' > t$ and wants to convince a verifier that he stamped the document at time t. He can simulate the behavior of an "honest stamper" who signs the document at time t and generate a timestamp for the document. Note that the "honest stamper" does not need to send any messages before time t' and therefore the adversary will be able to convince a verifier that the document was stamped at time t.

A crucial point in the argument above is that in order to perform this simulation the adversary must store all the information available to the "honest stamper" at time t. We show that non-interactive timestamping is possible in a scenario in which parties have bounded storage.

The Bounded Storage Model

In contrast to the usual approach in modern Cryptography, Maurer's *bounded storage model* [19] bounds the *storage* (memory size) of dishonest players rather than their running time.

In a typical protocol in the bounded storage model a long random string r of length R is initially broadcast and the interaction between the polynomial-time participants is conducted based on storing small portions of r. The security of such protocols should be guaranteed even against dishonest parties which have a lot of storage (much more than the honest parties) as long as they cannot store the whole string. Most of the previous work on the bounded storage model concentrated on private key encryption [19, 10, 2, 1, 14, 15, 18, 25], Key Agreement [10] and Oblivious Transfer [9, 12, 13]. In contrast, the notion of non-interactive timestamping cannot be implemented in the "standard cryptographic setting". To the best of our knowledge this is the first example of a protocol in the bounded storage model which achieves a task that is impossible in the "standard cryptographic setting".

Non-interactive Timestamping in the Bounded Storage Model

We now explain our setting for non-interactive timestamping in the bounded storage model. We assume that there are ℓ rounds and at every round $1 \leq t \leq \ell$, a long random string r of length R is transmitted[1].

The Stamping Mechanism: To stamp a document doc at time t, the scheme specifies a function Stamp(doc, r) whose output is short. To stamp the document doc, the stamper stores Stamp(doc, r). Intuitively, an adversary (who does not know doc at time t) is not able to store the relevant information and therefore is unable to stamp doc.

The Verification Mechanism: The verifier stores a short "sketch" of r (denoted by Sketch(r)) for every time t. At a later time the stamper can send the timestamp Stamp(doc, r) and the verifier checks whether this timestamp is "consistent" with his sketch.

Efficiency of a Timestamping Scheme: We say that a timestamping scheme is (T, V) efficient if the stamper's algorithm runs online (that is, in one pass) using space T and polynomial time and the verifier's algorithm runs online using space V and polynomial time. We want T and V to be small as functions of R.

Our Notion of Security

Loosely speaking, we want to ensure that even an adversary with a lot of storage (say storage $M = \delta R$ for some constant $\delta < 1$) cannot forge a timestamp. Note, however, that a stamper with storage $M > T$ can easily stamp $k = M/T$ documents by running the stamping mechanism on some k documents and storing the generated timestamp (which is of length at most T). We will therefore say that a scheme is secure if no adversary with space M can successfully stamp significantly more than M/T documents.

---

[1] One can imagine that random bits are transmitted at high rate continuously by a trusted party, and that the string r consists of the bits transmitted between time t and time $t + 1$.

One can also consider a probabilistic notion of security: given a randomly chosen document, after the random string has passed, the adversary will not be able to stamp the document with more than negligible probability. We note that our notion of security implies this probabilistic notion as well.

Security of a Timestamping Scheme: Given a (T, V)-efficient timestamping scheme. Let M_{max} be the bound on the storage of the most powerful adversary. The scheme is α-optimal ($\alpha > 1$) if, for every $M \leq M_{max}$, no adversary with space M can successfully stamp more than $\alpha \frac{M}{T}$ documents (for a formal definition of "successful stamping" see definition 4).

Notice that the definition above requires α–optimality for every $M \leq M_{max}$. Requiring α–optimality for M_{max} only, would have allowed adversaries with $M \ll M_{max}$ to produce $\frac{M_{max}}{T}$ stamped documents, contradicting the definition's spirit. The definition in its current form assures us that any adversary, weak or strong, with at most M_{max} memory, can honestly stamp *the same number of documents* if given slightly more resources (storage αM instead of M).

Our Results

In this paper we give two explicit constructions of non-interactive timestamping schemes in the bounded storage model. The first is secure in an information-theoretic sense (in the spirit of previous constructions in the bounded storage model). It requires no unproven assumptions and is secure against any adversary with arbitrary computational power as long as its storage capability is bounded. We now state this result (precise definitions appear in Section 3).

Theorem 1. *For every $\eta > 0$ and large enough R there exists a timestamping scheme that is $(T = O(R^{1/2+\eta}), V = O(R^{1/2+\eta}))$-efficient and $O(1)$-optimal. More precisely, every adversary with space $M^* \leq M_{max}^* = \Omega(R)$ has probability at most $2^{-R^{\Omega(1)}}$ to successfully stamp more than $O(M^*/T)$ documents. The timestamping scheme allows stamping documents of length $R^{\Omega(1)}$ and allows $R^{\Omega(\eta)}$ rounds.*

Our second system is more efficient. To achieve this efficiency it relies on cryptographic assumptions and is therefore secure only against adversaries that, in addition to being storage bounded, are required to run in polynomial time.

Theorem 2. *Assume that there exist collision resistant hash functions. There exists a timestamping scheme that is $(T = 2^{(\log \log R)^{O(1)}}, V = 2^{(\log \log R)^{O(1)}})$-efficient and $O(\log R)$-optimal. More precisely, every adversary with space $M^* \leq M_{max}^* = \Omega(R)$ and running time polynomial in R has negligible probability to successfully stamp more than $O(\log R \cdot M^*/T)$ documents. The timestamping scheme allows stamping documents of length R and allows R rounds.*

We remark that our technique can potentially reduce T and V to $\log^{O(1)} R$. This improvement requires an explicit construction of certain "expander graphs" that is not known today. More details will appear in the full version of the paper.

Advantages of Our Non-interactive Timestamping Scheme
Non-interactive timestamp systems have some significant advantages over the interactive systems known to date. We summarize some of these below:

- The only communication made before the verification process is the transmission of the random string r. This allows the timestamp system to be used in situations where communication is infeasible or undesirable. E.g., communication may be asymmetric: one central agency can broadcast all other users, while the users can not send messages to the agency.
- Everyone can stamp and everyone can verify and no central control or acquaintance between stamper and verifier is needed. The decentralized nature of this scheme overcomes many of the "trust" problems with interactive timestamp systems. Even in distributed interactive systems, some measure of trust must be given to third parties. Our non-interactive timestamp system requires only that the random string be truly random and receivable by all parties.
- Privacy. The scheme hides the fact that timestamping occurred at all, e.g., an inventor can safeguard her inventions without revealing even the fact of their existence. This also ensures privacy in an information-theoretic sense.
- Our schemes solve some of the robustness problems that plague interactive timestamping systems. In particular, it is much more difficult to mount a denial-of-service attack: there is no central point that can shut down the system, and even temporarily shutting down communications will not prevent the creation of new timestamps. The lack of communication also makes it difficult for an attacker to tell whether such an attack has succeeded.

Overview of the "Information-Theoretic" Construction
The setup is the following: A string r of length R is transmitted and the stamper wants to convince a verifier that he "knew" a document prior to the transmission of this string.

Using the Document to Select Indices: We implement the function Stamp(doc, r) as follows: Each document doc specifies some D indices that the stamper will remember from the long string. For that we use a bipartite graph where the left-hand vertices are all possible documents, the right-hand vertices are indices $1 \leq i \leq R$ and every left vertex has D neighbors. The indices selected by a document doc are the neighbors of doc. We want to force a stamper who would like to stamp k documents to store many indices. Intuitively, this is equivalent to the requirement that every k documents on the left have many different neighbors. This naturally leads to using an expander graph. (A bipartite graph is a (K, c)-expander if every $k < K$ vertices on the left have at least kc neighbors on the right)[2].

---

[2] We stress that we need to use *unbalanced graphs* (graphs which have many more vertices on the left than on the right-hand side). Such graphs were constructed in [24, 23]. However, we need graphs with somewhat different parameters. We construct such graphs by combining the constructions of [24] and a slight modification of [23, 21] (which in turn relies on explicit constructions of "randomness extractors" from [21, 22]).

To stamp a document *doc*, the stamper stores the content of the long string at the indices specified by *doc*. We use graphs with expansion $c \approx D$, therefore to correctly stamp k documents simultaneously an honest stamper must store roughly kD bits.

Using Random Sets for Verification: The function Sketch(r) is implemented as follows. The verifier chooses a random subset of size $|\mathcal{H}| \approx R/D$ from the indices of r and stores the content of r at these indices. After the transmission of the random string r, a stamper may send a timestamp of a document *doc* (that consists of the content of r at the D indices defined by *doc*). By the birthday problem, with high probability (over the choice of the verifier's random set) some of these indices were also stored by the verifier. The verifier checks that the content sent by the stamper is consistent with what he stored.

For a fixed string r and document *doc*, we say that a timestamp is "incorrect" if it differs from the "correct" timestamp of *doc* in many indices. The verification process we described guarantees that, with high probability, the verifier will reject an "incorrect" timestamp.

A Sketch of the Security Proof: The basic intuition for the security proof is the following: Suppose that an adversary is able to successfully stamp some k documents. This means that he correctly stamped these k documents (as otherwise he is caught by the verifier). However, correctly stamping k documents requires storing kD indices, therefore if the storage of the adversary is $kD \leq M < (k+1)D$ he can successfully stamp at most k documents. This is the best we can hope for (by our notion of security) as he could have stamped k documents by simply running the "stamping mechanism" on any k documents.

However, the argument above is not sufficient. It does not rule out the possibility that the adversary can stamp many documents such that the identity of these documents *depend* on the random string r. Our security definition requires that for every adversary, with high probability (over the choice of r) there do not exist k documents which the adversary can successfully stamp. To prove the security of our scheme we use a "reconstruction argument" and show that any adversary which breaks the security guarantee can be used to compress the string r into a shorter string in a way that does not lose a lot of information. As the string r is random, we get a contradiction. The details are given in Section 4.

Overview of the "Computationally-Bounded" Construction

In the previous construction we chose $|\mathcal{H}| \approx R/D$ so that a random subset of size $|\mathcal{H}|$ in $[R]$ would intersect a subset of size D. We chose $|\mathcal{H}| = D \approx \sqrt{R}$, allowing both the honest stamper and the verifier to store only $\sqrt{R}$ bits. We now show how to increase the efficiency and reduce the storage of honest parties to only $2^{(\log \log R)^{O(1)}}$ bits.

We use the same index selection mechanism as before. However, this time we choose $D = 2^{(\log \log R)^{O(1)}}$ (this precise choice of parameters corresponds to certain expander graphs). The verifier stores a short "hash" of the string r. When stamping a document the stamper also supplies a short "proof" that the indices he sent are consistent with the hashed value held by the verifier. We implement such a hashing scheme using *Merkle trees* [20]. We show that if collision resistant hash functions exist then a polynomial time adversary with bounded storage cannot produce an incorrect timestamp of a document. More precisely, we show that after the transmission of the random string r, no polynomial time adversary can *generate* many documents and stamp them correctly.

Hashing Documents Before Stamping Them: A bottleneck of our scheme is that when using expanders of degree D we can only handle documents of length D^3. However, in a computational setting (as we have already assumed the existence of collision resistant hash functions) we can stamp longer documents by first hashing them to shorter strings and then stamping them.

2 Preliminaries

2.1 Notation

The following conventions will be used throughout the paper.

Random String: We refer to the random string as r, its length is denoted by R, and we think of it as composed of N blocks of length n denoted $r_1, \ldots, r_N$. For any subset $S \subseteq [N]$, the expression $r_{|S}$ will be taken to mean the string generated by concatenating the blocks r_i for all $i \in S$.

Hamming Distance: The *Hamming Distance* between two strings r_1 and r_2 is the number of blocks on which the two strings differ.

Online Space: For a family of functions F, we denote by $Space(F)$ the maximum space used by any function in F. We say a function f can be computed *online* with space s if there is an algorithm using space at most s which reads its input bits one by one and computes f in one pass.

2.2 Unbalanced Expander Graphs

A graph is expanding if every sufficiently small set has a lot of neighbors. Our timestamping scheme relies on *unbalanced expanders*.

Definition 1 (unbalanced expander graphs). *A bipartite graph $G = (V_1, V_2, E)$ is (K_{max}, c)-expanding if, for any set $S \subset V_1$ of cardinality at most K_{max}, the set of its neighbors $\Gamma(S) \subseteq V_2$ is of size at least $c|S|$.*

Note that we do not require that $|V_1| = |V_2|$. In fact, in our timestamping scheme we will use graphs in which $|V_1| \gg |V_2|$. In this paper we need unbalanced expanders with very specific requirements. Loosely speaking we want a

---

[3] This is because in unbalanced expander graphs, the degree must be logarithmic in the number of left-hand vertices. Thus, shooting for degree D we can at most get that the left-hand set (which is the set of documents) is of size 2^D.

$(K_{max}, \Omega(D))$-expanding graph with as small as possible degree D and right-hand side of size roughly $K_{max}D$. We use some existing constructions of unbalanced graphs [24] as well as a modification of [23] to prove the next theorem (the proof will appear in the full version of the paper).

Theorem 3. *There exists a fixed constant $\beta > 0$ such that for every $K_{max} \leq |V_1|$, there exists a bipartite graph $G = (V_1, V_2, E)$ with left degree D that is $(K_{max}, c = \beta D)$ expanding with $D = 2^{O(\log \log V_1 + (\log \log K_{max})3)}$, and $|V_2| = 4\beta K_{max}D$. Furthermore, this graph is explicit in the sense that given a vertex $v \in V_1$ and an integer $1 \leq i \leq D$ one can compute the i'th neighbor of v in time polynomial in $\log |V_1| + \log D$.*

3   One Round Timestamping: The Model

In this section we formally define our model for timestamping in the bounded storage model. The definitions are only for a single round. Definitions for multiple rounds are straightforward generalizations and will appear in the full version.

A long random string r of length R is transmitted. The verifier takes a short sketch Sketch(r) of the random string and remembers it. An honest stamper, who wants to stamp a document $doc \in \mathcal{DOC}$, calculates $y =$ Stamp(doc, r). When, at a later stage, the stamper wants to prove he knew the document doc at stamping time, he sends y to the verifier who computes Verify(Sketch$(r), doc, y)$ and decides whether to accept or reject. More formally,

Definition 2 (Non-Interactive timestamping scheme). *A non-interactive timestamping scheme consists of three functions:*

 - *A stamping function* Stamp(doc, r).
 - *A sketch function* Sketch(r) *(we allow* Sketch *to be a probabilistic function).*
 - *A verification function* Verify(Sketch$(r), doc, y)$.

We require that for every string r and document doc, the function Verify(Sketch$(r), doc,$ Stamp$(doc, r))$ *accepts.*

We define efficiency:

Definition 3 (Efficiency). *A non-interactive timestamping scheme is (T, V)-efficient if* Stamp *can be computed online in space $T = T(R)$ and time polynomial in R, and* Sketch *can be computed online in space $V = V(R)$ and time polynomial in R.*

An honest stamper with space M can easily stamp M/T documents by running the function Stamp in parallel. We require that no adversary with memory M^* can successfully stamp significantly more than M^*/T documents. We first define our model for adversaries:

Definition 4 (adversary). *An adversary consists of two functions:* Store$^*(r)$, *which produces a short string b, and* Stamp$^*(doc, b)$ *which, given a document* *doc and b, attempts to produce a timestamp for doc. The space M^* of an adversary is the maximal length of* Store*$^*(r)$* [4]. *An adversary γ-successfully stamps a document doc at (some fixed) r if*

$$\Pr[\text{Verify}(\text{Sketch}(r), doc, \text{Stamp}^*(doc, \text{Store}^*(r))) = \text{Accept}] \geq \gamma$$

Note that this probability is over the coin tosses of Sketch *and the internal random coins of the adversary. Note that when the adversary is not computationally bounded, we can assume w.l.o.g. that the adversary is deterministic (does not use random coins)*

We define security as:

Definition 5 (Security). *We say that a (T, V)-efficient timestamping scheme is α-optimal (for $\rho > 0$, $\alpha \geq 1$, $\gamma > 0$ and $M^*_{max} < R$) if for every $M^* \leq M^*_{max}$ and every adversary A with space M^*,*

$$\Pr_r \left[A \text{ can } \gamma\text{-successfully stamp } \alpha \frac{M^*}{T} \text{ documents at } r \right] \leq \rho$$

Definition 5 is very strong. It guarantees that whenever the sketch size is small, no matter how powerful the adversary is, the number of documents the adversary can successfully stamp is very small.

3.1 Security Against Feasibly Generated Documents

Until now, we have allowed the adversary to run in arbitrary time. When the adversary is time-bounded, we can imagine scenarios where Definition 5 does not hold, yet the system is secure because the adversary does not have the computational power to find the documents he can illegally stamp. It makes sense to require security only against "feasibly generated documents". We model feasibly generated documents by a probabilistic polynomial time machine Generate^*_c which, on input r and an integer k, outputs k documents (all different).

Definition 6 (Security against feasibly generated documents). *We say that a (T, V)-efficient timestamping scheme is α-optimal (for $\rho > 0$, $\alpha \geq 1$, $\gamma > 0$ and $M^*_{max} < R$) against feasibly generated documents, if for every $M^* \leq M^*_{max}$, every adversary A with space M^*, and every polynomial time machine Generate^*_c:*

$$\Pr \left[A \text{ } \gamma\text{-successfully stamps at } r \text{ the documents } \text{Generate}^*_c(r, \alpha \frac{M^*}{T}) \right] \leq \rho$$

where the probability is over the choice of r and the random coins of Generate^*_c *and A.*

---

[4] Note that the adversary is not required to run *online* in space M^*. The function Store*$^*(r)$ can be an arbitrary function of r.

4   A Scheme with Information-Theoretic Security

In this section we describe a timestamping scheme which is information theoretically secure against arbitrary adversaries with small storage.

4.1   The Stamping Scheme

Let R, N and n be integers such that $R = N \cdot n$. Given a string $r \in \{0,1\}^R$, we partition it into N blocks of n bits. We use r_i to denote the i'th block of r. Let $\mathcal{DOC}$ denote the set of all documents which can be stamped. Let G be a $(K_{max}, \beta D)$ bipartite expander (V_1, V_2, E) with left degree D, where the "left" set V_1 is $\mathcal{DOC}$ and the "right" set V_2 is $[N]$. We define the three procedures Sketch, Stamp and Verify:

- Stamp$(doc, r) = r_{|\Gamma(doc)}$.
- Sketch$(r) = \mathcal{H}, r_{|\mathcal{H}}$ where $\mathcal{H}$ has $|\mathcal{H}|$ elements selected at random from $[N]$.
- Verify$(\text{Sketch}(r), doc, y) = \begin{cases} Accept & \text{Sketch}(r)_{|\mathcal{H} \cap \Gamma(doc)} = y_{|\mathcal{H} \cap \Gamma(doc)} \\ Reject & otherwise \end{cases}$

Notice that Sketch(r) contains the restriction of r to the indices of $\mathcal{H}$, and therefore in particular contains the restriction of r to the indices of $\mathcal{H} \cap \Gamma(doc)$, and y contains the restriction of r to $\Gamma(doc)$ and therefore in particular contains the restriction of r to the indices of $\mathcal{H} \cap \Gamma(doc)$.

Theorem 4. *Let G be a $(K_{max}, \beta D)$-expanding graph, $\gamma > 0$ and $g = \frac{1}{5}$. Fix n large enough such that $n > \log N \geq \frac{1}{g \cdot \beta}$ and assume that $\log |\mathcal{DOC}| \leq g \beta Dn$. If $D|\mathcal{H}| \geq \frac{N}{g\beta} \ln(\frac{1}{\gamma})$ then the scheme is $(T = Dn, V = |\mathcal{H}|(n + \log N))$-efficient and α-optimal for $\alpha = \frac{1}{(1-4g)\cdot\beta}$, $\rho = 2^{-g\beta nD}$, γ and $M_{max}^* = (1 - 4g)\beta DnK_{max}$.*

Plugging in parameters, a corollary of this is:

Corollary 1. *For every $\eta > 0$ and large enough R we construct a timestamping scheme which is $(T = R^{1/2+\eta}, V = R^{1/2+\eta})$-efficient and $O(1)$-optimal with $\rho = 2^{-R^{\Omega(1)}}$, $\gamma = 2^{-R^{\Omega(n)}}$ and $M_{max}^* = \Omega(R)$. The timestamping scheme allows stamping documents of length $R^{\Omega(1)}$.*

We prove the corollary in the full version of the paper. We remark that a probabilistic argument shows that there exist bipartite graphs of degree D which have expansion $(1 - o(1))D$ and using such non-explicit graphs in our construction (and setting $g = o(1)$) gives $\alpha = (1 + o(1))$ optimality (whereas the theorem below only achieves $\alpha = O(1)$). In the remainder of the section we prove Theorem 4.

4.2   Efficiency

The verifier first chooses a random set $\mathcal{H}$ and stores it, and then stores $R_\mathcal{H}$. This can indeed be done online with space $V = |\mathcal{H}|(n + \log N)$. We now explain how the stamper can run online in space $T = Dn$. Observe that it can calculate the indices it will need to store *before* the random string goes by (since it knows *doc* before it sees the random string). As the indices take $D \log N < Dn$ space, it can work in place, replacing each index with the contents of the block as it goes by. We now turn to proving security.

4.3   Security

In Definition 4 we defined "successful stamping". Without loss of generality, we assume the adversary is deterministic. Let $\mathcal{R}_{successful}(k) = \mathcal{R}^{\gamma}_{successful}(k)$ denote the set of random strings r on which the adversary γ-successfully stamps at least k documents. We would like to prove that $\mathcal{R}_{successful}$ has small probability. We first define a similar notion of "correct stamping":

Definition 7. *An adversary* correctly stamps *a document doc at r if* $\mathrm{Stamp}^*(doc, \mathrm{Store}^*(r)) = r_{|\Gamma(doc)}$. *An adversary* correctly stamps *a document doc at r with at most err errors, if the Hamming distance between* $\mathrm{Stamp}^*(doc, \mathrm{Store}^*(r))$ *and $r_{|\Gamma(doc)}$ is at most err.*

We let $\mathcal{R}_{correct}(k) = \mathcal{R}^{err=g\beta D}_{correct}(k)$ denote the set of random strings r for which there are at least k documents that the adversary correctly stamps with at most err errors.

The security proof has two parts.

Lemma 1. *Assume $\log |\mathcal{DOC}| \le g\beta Dn$, $err = g\beta D$ and $n \ge \frac{1}{g \cdot \beta}$. For every $k \le K_{max}$ and any adversary with space $M^* \le (1 - 4g)\beta k Dn$ we have $\Pr_r[r \in \mathcal{R}_{correct}(k)] \le 2^{-g\beta nDk}$.*

We then relate $\mathcal{R}_{correct}$ and $\mathcal{R}_{successful}$:

Lemma 2. *Assume $D|\mathcal{H}| \ge \frac{N}{g\beta} \ln(\frac{1}{\gamma})$. For every k, $\mathcal{R}_{successful}(k) \subseteq \mathcal{R}_{correct}(k)$.*

Together,

Proof. (of Theorem 4) We need show that no adversary with space M^* can γ-successfully stamp more than $k = \alpha \frac{M^*}{Dn}$ documents. Notice that for $M^* \le M^*_{max}$, $k = \frac{\alpha M^*}{Dn} \le \frac{\alpha M^*_{max}}{Dn} = K_{max}$ and $M^* = \frac{kDn}{\alpha} = kDn(1 - 4g)\beta$. Hence, $\Pr_r[r \in \mathcal{R}_{successful}(k)] \le \Pr_r[r \in \mathcal{R}_{correct}(k)] \le 2^{-g\beta nDk} \le 2^{-g\beta nD}$ where the first inequality follows by Lemma 2 and the second inequality follows by Lemma 1. The third inequality is because $k \ge 1$.

4.4   The Proof of Lemma 1

We first define a compression function $\mathrm{Com}(r)$ for $r \in \mathcal{R}_{correct}(k)$. Let $r \in \mathcal{R}_{correct}(k)$. Suppose $doc_1, \dots, doc_k$ are the documents that the adversary correctly stamps at r with at most err errors. Denote $\Gamma = \cup_{1 \le i \le k} \Gamma(doc_i)$, that is the set of all indices which are selected by one of the k documents. Denote

$\mathcal{BAD} = \cup_{1 \le i \le k}\mathcal{BAD}(doc_i)$, that is the set of all indices which are bad for at least one of the k documents. We call an index $j \in \Gamma \setminus \mathcal{BAD}$ useful. We choose $\mathrm{Com}(r)$ to be:

$$\mathrm{Com}(r) = (doc_1, \ldots, doc_k; \mathrm{Store}^*(r); r_{|\bar{\Gamma}}; \mathcal{BAD}; r_{|\mathcal{BAD}})$$

We define a "decompression" function $\mathrm{Dec}(a)$ that gets as input $\mathrm{Com}(r)$ and tries to recover r. Let r be a string from $\mathcal{R}_{\mathrm{correct}}$, i.e., a string on which the stamper correctly stamps k-documents with at most err errors. From $doc_1, \ldots, doc_k$, that appear in $\mathrm{Com}(r)$, we recover the set Γ, and from $\mathrm{Com}(r)$ we learn which indices are in the subset $\mathcal{BAD} \subset \Gamma$. Now, for every $1 \le j \le N$ we recover r_j as follows:

- If $j \notin \Gamma$ then we use the information in $r_{|\bar{\Gamma}}$ to find r_j.
- If $j \in \mathcal{BAD}$ then we use the information in $r_{|\mathcal{BAD}}$ to find r_j.
- If $j \in \Gamma \setminus \mathcal{BAD}$ then we find an i such that
  $j \in \Gamma(doc_i)$. We run $\mathrm{Stamp}^*(doc_i, \mathrm{Store}^*(r))$ and take r_j from its output.

The only case where we do not take the value of r_j directly from $\mathrm{Com}(r)$ is for $j \in \Gamma \setminus \mathcal{BAD}$. However, all such indices j are useful, and therefore we correctly decode them. Therefore, for every $r \in \mathcal{R}_{\mathrm{correct}}$ we have $\mathrm{Dec}(\mathrm{Com}(r)) = r$.

We now analyze the output length of the compression function Com. The documents $doc_1, \ldots, doc_k$ take $k \log |\mathcal{DOC}|$ bits space. $|\mathrm{Store}^*(r)| \le M^*$, by definition. As G is expanding and $k \le K_{max}$, $|\Gamma| \ge \beta k D$ and therefore $r_{|\bar{\Gamma}} \le R - \beta k D n$. We represent $\mathcal{BAD} \subseteq \Gamma$ by a binary vector of length $|\Gamma| \le k D$ which has a "one" for indices in $\Gamma \cap \mathcal{BAD}$ and a "zero" for indices in $\Gamma \setminus \mathcal{BAD}$. Each of the k documents is correctly stamped at r with at most err errors, and therefore for every such document doc_i we have $|\mathcal{BAD}_{doc_i}| \le err$ and $|\mathcal{BAD}| \le k \cdot err$. The representation of $r_{|\mathcal{BAD}}$ is therefore bounded by $k \cdot err \cdot n$. We conclude that the total length of the output of Com is at most $k \log |\mathcal{DOC}| + M^* + R - \beta k D n + k D + k \cdot err \cdot n$. We denote this quantity $R - \Delta$.

As every $r \in \mathcal{R}_{\mathrm{correct}}$ has a small description (of length $R - \Delta$) we have $|\mathcal{R}_{\mathrm{correct}}| \le 2^{R-\Delta}$ and therefore $Pr[r \in \mathcal{R}_{\mathrm{correct}}] \le 2^{-\Delta}$. We have $kD \le g\beta k D n$ (for large enough n). We also have $err = g\beta D$ and by our assumption $\log |\mathcal{DOC}| \le g\beta n D$. Altogether, $R - \Delta \le R - \beta D k n [1 - 3g] + M^*$. We get that $\Delta \ge (1 - 3g)\beta D k n - M^*$. As $M^* \le (1 - 4g)\beta k D n$ we get $\Delta \ge g\beta k D n$ as desired.

4.5   The Proof of Lemma 2

Claim. Fix an adversary, a string r and a document doc. If the adversary γ-successfully stamps doc at r then it correctly stamps doc at r with at most $\frac{N}{|\mathcal{H}|} \ln(\frac{1}{\gamma})$ errors.

Proof. We prove the contrapositive. Suppose for some $doc \in \mathcal{DOC}$ and r, the timestamp provided by the adversary for doc has $err^* > err$ incorrect indices. Denote by $\mathcal{BAD}_{doc} \subset [N]$ the set of incorrect indices. The verifier catches the adversary iff $\mathcal{BAD}_{doc} \cap \mathcal{H} \neq \emptyset$, i.e. if one of the incorrect indices is in $\mathcal{H}$ (the set of indices stored by Sketch). For each index in $\mathcal{H}$, the probability that it hits $\mathcal{BAD}_{doc}$ is $\frac{err^*}{N}$, and the probability that none of them hits $\mathcal{BAD}_{doc}$ is $(1 - \frac{err^*}{N})^{|\mathcal{H}|} \leq e^{-\frac{err^* |\mathcal{H}|}{N}}$ (assuming the set $\mathcal{H}$ is chosen with repetition). Hence, the adversary γ-successfully stamps doc with $\gamma \leq e^{-\frac{err |\mathcal{H}|}{N}}$. Turning that around, if the adversary γ-successfully stamps doc, then $err \leq \frac{N}{|\mathcal{H}|} \ln(\frac{1}{\gamma})$.

In particular, for every r and doc for which the stamper is γ successful, $err \leq \frac{N}{|\mathcal{H}|} \ln(\frac{1}{\gamma}) \leq g\beta D$. Hence, the stamper correctly stamps doc at r with at most $err = g\beta D$ errors. It follows that $\mathcal{R}_{\text{successful}}(k) \subseteq \mathcal{R}_{\text{correct}}(k)$ as desired.

5   An Efficient Scheme Secure Against Polynomial Time Adversaries

The scheme suggested in Section 4 requires the honest parties (stamper and verifier) to store many bits, namely $TV >> R$ where T is the stamp size, V the sketch size and R the random string length. In other words, if the stamp size is very small then the sketch size V is almost all of the random string. Our second scheme has small sketch *and* stamp size. This is achieved by using the previous stamping scheme with a small T and using a different verification method that allows the verifier to use much less storage. This verification method is valid only against computationally bounded adversaries and takes advantage of the bounded computational capabilities of the cheating party. In this section we briefly describe the scheme and give a sketch of the proof. Due to space constraints, the exact details will appear in the full version. We assume the reader has some familiarity with collision resistant hash functions[5] [11] (CRHFs) and Merkle trees [20].

5.1   The Stamping Scheme

Let $H = \{h: \{0,1\}^{2n} \mapsto \{0,1\}^n\}$ be a family of CRHFs[6] and $R = \log H + Nn$. We partition a string $r \in \{0,1\}^R$ into $N+1$ blocks, denoted $r_0, r_1, \ldots, r_N$ where r_0 is of length $\log H$ and for $i > 0$, r_i is of length n. The string r_0 (which didn't appear in the previous scheme) serves as a "key" to the "hash function". We use the same "index selection" mechanism as in Section 4; G is a bipartite graph with left degree D, where the left set is the set $\mathcal{DOC}$ and the right set is the set $[N]$. We now describe the stamp, sketch and verify procedures:

---

[5] Also called "collision intractable" or "collision free" hash functions

[6] Informally, this means that no computationally bounded adversary can find $x_1 \neq x_2$ such that $h(x_1) = h(x_2)$ when given a random function h in the family. In this paper we require hash functions which are hard even for adversaries which run in time slightly super-polynomial in n. This is because the adversary runs in time polynomial in R, whereas n can be very small compared to R.

Sketch$_c$ The verifier stores r_0 and the root of a Merkle tree whose leaves are $r_1, \ldots, r_N$, using the hash function specified by r_0 [7]. Note that Sketch$_c$ is *deterministic* (unlike the case of the previous section where Sketch is probabilistic).

Stamp$_c$ Given a document $doc \in \mathcal{DOC}$ the stamper uses the function Stamp of the previous section, and for every $j \in \Gamma(doc)$ stores r_j along with the Merkle-path from r_j to the root of the tree[8].

Verify$_c$ Given a document doc, a "root" a and a stamp y composed of D Merkle-paths, the function Verify$_c(a, doc, y)$ accepts iff all paths are valid (that is, the label of the tree root computed from the Merkle-paths is consistent with that stored by the verifier).

We note that both Sketch$_c$ and Stamp$_c$ can be computed online in small space, using the standard method for computing Merkle-trees online. For our choice of parameters, this gives the required efficiency. Using the expander construction of Theorem 3 for G, we obtain a scheme with efficiency $2^{(\log \log R)^{O(1)}}$ (and thus prove Theorem 2). It is possible to get an even more efficient scheme with $T = V = (\log R)^{O(1)}$. However, this result requires a better graph than the one constructed in Theorem 3. It is folklore that such graphs exist by a probabilistic argument. However, at this point no such explicit construction is known. In the remainder of the section we sketch the proof of security of the scheme (The complete proof of Theorem 2 appears in the full version of the paper).

5.2 Security

We follow the outline of the correctness proof of the information-theoretic version of Section 4, except that now we work with security for generated documents. We show that if the adversary *successfully stamps* many documents then he *correctly stamps* many documents which is impossible by the "reconstruction argument" of the previous section.

Fix some adversary with memory M^* and running time polynomial in R. We use *coins* to denote the concatenation of the random coins used by Stamp$_c^*$ and Generate$_c^*$. We define $\mathcal{R}_{\text{correct}}^{\text{comp}}(k)$ to be the set of pairs $(r, coins)$ such that, for every $doc \in$ Generate$_c^*(r, k)$, the Merkle paths output by Stamp$_c^*(\text{Store}_c^*(r), doc)$ are correct (i.e. that they are actual paths in the Merkle tree whose leaves are the blocks of r). In particular, this implies that the leaves of the paths are a "correct" timestamp for the k documents output by Generate$_c^*(r, k)$ in the sense of Section 4.

---

[7] Informally, a Merkle tree of $r_1, \ldots, r_N$ using the hash function h is a labeled binary tree, where the leaves are labeled by $r_1, \ldots, r_N$ and the label of each internal node is given by applying h to the concatenation of its children's labels.

[8] A Merkle-path from r_j consists of r_j along with the labels of the siblings of all nodes on the path from r_j to the root of the Merkle tree. Such as sequence contains sufficient information to compute the labels of all nodes on the path to the root node (by repeatedly applying the hash function).

We now want to define the computational analogue of $\mathcal{R}_{\text{successful}}$ and relate it to $\mathcal{R}_{\text{correct}}^{\text{comp}}$. We define $\mathcal{R}_{\text{successful}}^{\text{comp}}(k)$ to be the set of all pairs $(r, coins)$ for which the adversary successfully stamps the k documents output by $\text{Generate}_c^*(r, k)$ (i.e. for all $doc \in \text{Generate}_c^*(r, k)$, the the Merkle paths output by $Stamp_c^*(Store_c^*(r), doc)$ are accepted by the verifier). This definition of success corresponds to the notion of security in Definition 6.

We prove that $Pr_{r, coins}[(r, coins) \in \mathcal{R}_{\text{successful}}^{\text{comp}}(k) \setminus \mathcal{R}_{\text{correct}}^{\text{comp}}(k)] \leq \text{neg}$ (where neg is a negligible function of n). This is because we can imagine a machine which, when given a random hash function r_0, uniformly selects the pair $(r, coins)$ and runs the adversary. The claim follows, as for every pair $(r, coins) \in \mathcal{R}_{\text{successful}}^{\text{comp}}(k) \setminus \mathcal{R}_{\text{correct}}^{\text{comp}}(k)$, this machine can find a collision for r_0. Thus we have a computational analogue of Lemma 2.

We then show (using Lemma 1) that every random string for which the adversary can correctly stamp many documents can be compressed, which gives a bound on the probability that this occurs.

6   Discussion and Open Problems

Dealing with Errors: Most protocols in the Bounded Storage Model, and ours among them, assume the broadcast random string is received identically and without errors by all parties. However, in many natural implementations of such protocols, this assumption may not be realistic (e.g. when the random string has a natural source).

Our information-theoretic scheme can be made to work even with errors (provided the error rate is low enough) by allowing the verifier to accept a timestamp even if the the blocks in the intersection differ by a small amount. The proof of Lemma 1 already allows the adversary to make some errors when stamping, and still be considered successful. Increasing the error rate by a small amount will not invalidate the lemma (although the parameters suffer slightly).

The computational scheme, on the other hand, currently requires the random string to be received perfectly by all parties. It is an interesting open question whether this requirement can be removed.

Removing the Need for Constant Monitoring: Our timestamping schemes require the verifier to run the Sketch function in every round for which it may, someday, want to verify documents. The verifier must therefore constantly monitor the random string, which is too much to ask from a casual user of the system.

An implementation of our timestamp systems can overcome this difficulty by using "verification centers": dedicated third parties who act as verifiers. In some sense, such third parties appear in all previous timestamp protocols. This raises the issue of how much trust the user must place in the verification center.

In the computational version of our protocol, the verification center is also easily auditable by casual users: the verifier is deterministic and has no secret information. Any user can act as a verifier for a single round, and compare its state to that of the verification center: any inconsistency will be instantly visible.

Online Versus Locally-Computable: The strategies for the honest players are efficient in the sense that they work online using small space and polynomial time. A stronger notion of efficiency called "locally-computable" was suggested in [25]. It requires the honest players to store a small substring of the string r. More precisely, the players need to choose a subset $S \subseteq [R]$ before the random string is transmitted and only store $r|_S$. We point out that the "information-theoretic" scheme (Section 4) has this additional property, whereas the "computationally-bounded" scheme (Section 5) does not[9]. Natural open problems are whether the "information-theoretic" scheme can be improved to yield better parameters, and whether the "computationally-bounded" scheme can be improved to run with strategies that are locally computable.

Acknowledgements

We thank Danny Harnik, Moni Naor and Muli Safra for helpful discussions, and the anonymous reviewers for their constructive comments.

References

1. Y. Aumann, Y.Z. Ding, and M. O. Rabin. Everlasting security in the bounded storage model. *IEEE Transactions on Information Theory*, 48, 2002.
2. Y. Aumann and M. O. Rabin. Information theoretically secure communication in the limited storage space model. In *Advances in Crypology — CRYPTO '99*, volume 1666, pages 65–79, 1999.
3. D. Bayer, S. Haber, and W. S. Stornetta. Improving the efficiency and reliability of digital time-stamping. In R. M. Capocelli et al., editor, *Sequences II: Methods in Communication, Security and Computer Science*, pages 329–334. Springer-Verlag, Berlin Germany , New York, 1992.
4. J. Benaloh and M. de Mare. Efficient broadcast time-stamping. Technical Report 1, Clarkson University Department of Mathematics and Computer Science, August 1991.
5. Josh Cohen Benaloh and Michael de Mare. One-way accumulators: A decentralized alternative to digital signatures. *Lecture Notes in Computer Science*, 765:274, 1994.
6. Ahto Buldas and Peeter Laud. New linking schemes for digital time-stamping. In *Information Security and Cryptology*, pages 3–13, 1998.
7. Ahto Buldas, Peeter Laud, Helger Lipmaa, and Jan Villemson. Time-stamping with Binary Linking Schemes. In Hugo Krawczyk, editor, *Advances on Cryptology — CRYPTO '98*, volume 1462 of *Lecture Notes in Computer Science*, pages 486–501, Santa Barbara, USA, August 1998. Springer-Verlag.
8. Ahto Buldas, Helger Lipmaa, and Berry Schoenmakers. Optimally efficient accountable time-stamping. In *Public Key Cryptography*, pages 293–305, 2000.

---

[9] In the "computationally-bounded" scheme , both stamper and verifier read blocks of the string r online, and need to "hash them" quickly before reading the next incoming blocks. Thus, to implement this scheme, one needs to use hash functions which can be computed very efficiently.

9. Christian Cachin, Claude Crepeau, and Julien Marcil. Oblivious transfer with a memory-bounded receiver. In *IEEE Symposium on Foundations of Computer Science*, pages 493–502, 1998.

10. Christian Cachin and Ueli Maurer. Unconditional security against memory-bounded adversaries. In Burton S. Kaliski Jr., editor, *Advances in Cryptology — CRYPTO '97*, volume 1294 of *Lecture Notes in Computer Science*, pages 292–306. Springer-Verlag, 1997.

11. I.B. Damgard. Collision free hash functions and public-key signature schemes. In *Advances in Cryptology — EUROCRYPT '87, Proceedings*, volume 304, pages 203–216. Springer-Verlag, 1987.

12. Yan Zong Ding. Oblivious transfer in the bounded storage model. *Lecture Notes in Computer Science*, 2139:155, 2001.

13. Yan Zong Ding, Danny Harnik, Alon Rosen, and Ronen Shaltiel. Constant-round oblivious transfer in the bounded storage model. In *Theory of Cryptography — TCC '04*, volume 2951, Cambridge, MA, USA, February 2004. Springer-Verlag. To appear.

14. Y.Z. Ding and M.O. Rabin. Hyper-encryption and everlasting security. In *Annual Symposium on Theoretical Aspects of Computer Science (STACS)*, pages 1–26, 2002.

15. Stefan Dziembowski and Ueli Maurer. Tight security proofs for the bounded-storage model. In *Proceedings of the 34th Annual ACM Symposium on Theory of Computing*, pages 341–350. ACM, May 2002.

16. Stuart Haber and W. Scott Stornetta. How to time-stamp a digital document. *Lecture Notes in Computer Science*, 537:437, 1991.

17. Stuart Haber and W. Scott Stornetta. Secure names for bit-strings. In *ACM Conference on Computer and Communications Security*, pages 28–35, 1997.

18. C. Lu. Hyper-encryption against space-bounded adversaries from on-line strong extractors. In *Advances in Cryptology — CRYPTO '02*, volume 2442, pages 257–271. Springer, 2002.

19. U. Maurer. Conditionally-perfect secrecy and a provably-secure randomized cipher. *Journal of Cryptology*, 5(1):53–66, 1992.

20. Ralph C. Merkle. A certified digital signature. In *Proceedings on Advances in cryptology*, pages 218–238. Springer-Verlag New York, Inc., 1989.

21. R. Raz, O. Reingold, and S. Vadhan. Error reduction for extractor. In *40th IEEE Symposium on Foundations of Computer Science*, pages 191–201, 1999.

22. A. Srinivasan and D. Zuckerman. Computing with very weak random sources. *SIAM Journal on Computing*, 28:1433–1459, 1999.

23. Amnon Ta-Shma. Storing information with extractors. *Information Processing Letters*, 83(5):267–274, September 2002.

24. Amnon Ta-Shma, Christopher Umans, and David Zuckerman. Loss-less condensers, unbalanced expanders, and extractors. In ACM, editor, *Proceedings of the 33rd Annual ACM Symposium on Theory of Computing: Hersonissos, Crete, Greece, July 6–8, 2001*, pages 143–152, New York, NY, USA, 2001. ACM Press.

25. S.P. Vadhan. On constructing locally computable extractors and cryptosystems in the bounded storage model. In *Advances in Cryptology — CRYPTO '03*. Springer, 2003.

IPAKE: Isomorphisms for Password-Based Authenticated Key Exchange

Dario Catalano[1], David Pointcheval[1], and Thomas Pornin[2]

[1] CNRS–LIENS, Ecole Normale Supérieure, Paris, France
{Dario.Catalano,David.Pointcheval}@ens.fr
[2] Cryptolog, Paris, France
Thomas.Pornin@cryptolog.com

Abstract. In this paper we revisit one of the most popular password-based key exchange protocols, namely the OKE (for Open Key Exchange) scheme, proposed by Luck in 1997. Our results can be highlighted as follows. First we define a new primitive that we call *trapdoor hard-to-invert isomorphisms*, and give some candidates. Then we present a generic password-based key exchange construction, that admits a security proof assuming that these objects exist. Finally, we instantiate our general scheme with some concrete examples, such as the Diffie-Hellman function and the RSA function, but more interestingly the modular square root function, which leads to the first scheme with security related to the integer factorization problem. Furthermore, the latter variant is very efficient for one party (the server). Our results hold in the random-oracle model.

1 Introduction

Shortly after the introduction of the revolutionary concept of asymmetric cryptography, proposed in the seminal paper by Diffie and Hellman [9], people realized that properly managing keys is not a trivial task. In particular private keys tend to be pretty large objects, that have to be safely stored in order to preserve whatever kind of security. Specific devices have thus been developed in order to help human beings in storing their secrets, but it is clear that even the most technologically advanced device may become useless if lost or stolen. In principle the best way to store a secret is to keep it in mind. In practice, however, human beings are very bad at remembering large secrets (even if they are passwords or pass-phrases) and very often they need to write passwords down on a piece of paper in order to be able to keep track of them. As a consequence, either one uses a short (and memorable) password, or writes/stores it somewhere. In the latter case, security eventually relies on the mode of storage (which is often the weakest part in the system: a human-controlled storage). In the former case, a short password is subject to exhaustive search.

Indeed, by using a short password, one cannot prevent a brute force on-line exhaustive search attack: the adversary just tries some passwords of its own choice in order to try to impersonate a party. If it guesses the correct password,

M. Franklin (Ed.): CRYPTO 2004, LNCS 3152, pp. 477–493, 2004.

it can get in, otherwise it has to try with another password. In many applications, however, the number of such active attacks can be limited in various ways. For example one may impose some delay between different trials, or even closing the account after some fixed number of consecutive failures. Of course the specific limitations depend very much on the context – other kind of attacks, such as Denial of Service ones, for example, should be made hard to mount either. In any case, the important point we want to make here is that the impact of on-line exhaustive search can be limited. However on-line attacks are not the only possible threats to the security of a password-based system. Imagine for example an adversary who has access to several transcripts of communication between a server and a client. Clearly the transcript of a "real" communication somehow depends on the actual password. This means that a valid transcript (or several ones) could be used to "test" the validity of some password: the adversary chooses a random password and simply checks if the produced transcript is the same as the received one. In this way it is possible to mount an (off-line) exhaustive search attack that can be much more effective than the on-line one, simply because, in this scenario, the adversary can try all the possible passwords just until it finds the correct one. Such an off-line exhaustive search is usually called "dictionary attack".

1.1   Related Work

A password-based key exchange is an interactive protocol between two parties A and B, who initially share a short password pw, that allows A and B to exchange a session key sk. One expects from this key to be semantically secure w.r.t. any party, but A and B who should know it at the end of the protocol. The study of password-based protocols resistant to dictionary attacks started with the seminal work of Bellovin and Merritt [3], where they proposed the so-called *Encrypted Key Exchange* protocol (EKE). The basic idea of their solution is the following: A generates a public key and sends it to B encrypted – using a symmetric encryption scheme – with the common password. B uses the password to decrypt the received ciphertext. Then it proceeds by encrypting some value k using the obtained public key. The resulting ciphertext is then re-encrypted (once again using the password) and finally sent to A. Now A can easily recover k, using both his own private key and the common password. A shared session key is then derived from k using standard techniques.

A classical way to break password-based schemes is the partition attack [4]. The basic idea is that if the cleartexts encrypted with the password have any redundancy, or lie in a strict subset, a dictionary attack can be successfully mounted: considering one flow (obtained by eavesdropping) one first chooses a password, decrypts the ciphertext and checks whether the redundancy is present or not (or whether the plaintext lies in the correct range.) This technique allows to quickly select probable passwords, and eventually extract the correct one.

The partition attack can be mounted on many implementations of EKE, essentially because a public key usually contains important "redundancy" (as a matter of fact a public key – or at least its encoding – is not in general a random-looking string). Note that in the described approach (for EKE), the same symmetric encryption (using the same password) is used to encrypt both the public key, and the ciphertext generated with this key. This may create additional problems basically because these two objects (i.e. the public key and the ciphertext) are very often defined on completely unrelated sets. A nice exception to this general rule are ElGamal keys [12]. This is thus the sole effective application of EKE.

As noticed by the original authors [3], and emphasized by Lucks [17], it is "*counter-intuitive (...) to use a secret key to encrypt a public key*". For this reason Lucks [17] proposed OKE, (which stands for Open Key Exchange). The underlying idea of this solution is to send the public key in clear and to encrypt the second flow only. Adopting this new approach, additional public-key encryption schemes can be considered (and in particular RSA [23] for instance). However, one has to be careful when using RSA. The problem is that the RSA function is guaranteed to be a permutation only if the user behaves honestly and chooses his public key correctly. In real life, however, a malicious user may decide to generate keys that do not lead to a permutation at all. In such a case a partition attack becomes possible: an RSA-ciphertext would lie in a strict subset if $\mathbb{Z}_n^\star$. For this reason Lucks proposed a variant of his scheme, known as *Protected OKE*, to properly deal with the case of RSA. Later on, however, MacKenzie *et al.* [19, 18] proved that the scheme was flawed by presenting a way to attack it. At the same time they showed how to repair the original solution by proposing a new protocol they called SNAPI (for Secure Network Authentication with Password Identification), for which they provided a full proof of security in the random-oracle model. This proof, however, is specific to RSA, in the random-oracle model, and very intricate.

Interestingly enough, in the standard model, the problem of secure password-based protocols was not treated rigorously until very recently. The first rigorous treatment of the problem was proposed by Halevi and Krawczyk [15] who, however, proposed a solution that requires other setup assumptions on top of that of the human password. Later on, Goldreich and Lindell [14] proposed a very elegant solution that achieves security without any additional setup assumption. The Goldreich and Lindell proposal is based on sole existence of trapdoor permutations and, even though very appealing from a theoretical point of view, is definitely not practical. The first practical solution was proposed by Katz, Ostrovsky and Yung [16]. Their solution is based on the Decisional Diffie-Hellman assumption and assumes that all parties have access to a set of public parameters (which is of course a stronger set-up assumption than assuming that only human passwords are shared, but still a weaker one with respect to the Halevi-Krawczyk ones for example). Even more recently Gennaro and Lindell [13] presented an abstraction of the Katz, Ostrovsky and Yung [16] protocol that allowed them to construct a general framework for authenticated password-based key exchange in the common reference string model.

We note here that even though from a mathematical point of view a proof in the standard model is always preferable to a proof in the random-oracle model, all the constructions in the standard model presented so far are *way* less efficient with respect to those known in the random-oracle model. It is true that a proof in the random-oracle model should be interpreted with care, more as a heuristic proof than a real one. On the other hand in many applications efficiency is a big issue and it may be preferable to have a very efficient protocol with a heuristic proof of security than a much less efficient one with a complete proof of security.

1.2   Our Contributions

In this paper, we revisit the generic OKE construction by clearly stating the requirements about the primitive to be used: we need a family of isomorphisms with some specific computational properties that we call *trapdoor hard-to-invert isomorphisms* (see next section for a formal definition for these objects). Very roughly a trapdoor hard-to-invert isomorphism, can be seen as an isomorphic function that is in general hard to invert, unless some additional information (the trapdoor) is provided. Note that such an object is different with respect to traditional trapdoor functions. A trapdoor one-way function is always easy to compute, whereas a trapdoor hard-to-invert function may be not only hard to invert, but – at least in some cases – also hard to compute [10]. As it will become apparent in the next sections, this requirement is not strong because basically all the classical public-key encryption schemes fit it (RSA [23], Rabin with Blum moduli [22], ElGamal [12], and even the recent Okamoto-Uchiyama's [20] and Paillier's schemes [21]). More precisely our results can be described as follows.

First, after having described our security model, we present a very general construction – denoted **IPAKE** for *Isomorphism for Password-based Authenticated Key Exchange* – and we prove it is secure. Our security result relies on the computational properties of the chosen trapdoor hard-to-invert isomorphism family, in the random-oracle model. As a second result we pass instantiating the general construction with specific encryption schemes. We indeed show that trapdoor hard-to-invert isomorphisms can be based on the Diffie-Hellman problem, on the RSA problem, and even on integer factoring.

For lack of space, we refer to the full version [8] for the two first applications, since they are not really new. Plugging ElGamal directly leads to one of the AuthA variants, proposed to IEEE P1363 [2], or to PAK [5]. The security has already been studied in several ideal models [5–7]. The case of RSA leads to a scheme similar to RSA-OKE, SNAPI [19, 18], or to the scheme proposed by Zhu *et al.* [26].

More interestingly using such methods we can construct a very efficient solution from the Rabin function. To our knowledge this is the first efficient password-based authenticated key exchange scheme based on factoring.

2   Preliminaries

Denote with $\mathbb{N}$ the set of natural numbers and with $\mathbb{R}^+$ the set of positive real numbers. We say that a function $\varepsilon : \mathbb{N} \to \mathbb{R}^+$ is *negligible* if and only if for every polynomial $P(n)$ there exists an $n_0 \in \mathbb{N}$ such that for all $n > n_0$, $\varepsilon(n) \leq 1/P(n)$.

If A is a set, then $a \leftarrow A$ indicates the process of selecting a at random and uniformly over A (which in particular assumes that A can be sampled efficiently).

2.1 Trapdoor Hard-to-Invert Isomorphisms

Let I be a set of indices. Informally a family of *trapdoor hard-to-invert isomorphisms* is a set $F = \{f_m : X_m \rightarrow Y_m\}_{m \in I}$ satisfying the following conditions:

1. one can easily generate an index m, which provides a description of the function f_m – a morphism –, its domain X_m and range Y_m (which are assumed to be isomorphic groups), and a trapdoor t_m;
2. for a given m, one can efficiently sample pairs $(x, f_m(x))$, with x uniformly distributed in X_m;
3. for a given m, one can efficiently decide Y_m;
4. given the trapdoor t_m, one can efficiently invert $f_m(x)$, and thus recover x;
5. without the trapdoor, inverting f_m is hard.

This is almost the same definition as for trapdoor one-way permutations with homomorphic properties. There is a crucial difference however: one can sample pairs, but may not necessarily be able to compute $f_m(x)$ for a given x (point 2 above). As a consequence, the function is hard-to-invert, but it may be hard to compute as well.

More formally we say that F defined as above is a family of *trapdoor hard-to-invert isomorphisms* if the following conditions hold:

1 – There exist a polynomial p and a probabilistic polynomial time Turing Machine Gen which on input 1^k (where k is a security parameter) outputs pairs (m, t_m) where m is uniformly distributed in I and $|t_m| < p(k)$. The index m defines X_m and Y_m, which are isomorphic groups, an isomorphism f_m from X_m onto Y_m and a set R_m of values uniformly samplable, which will be used to sample $(x, f_m(x))$ pairs. The information t_m is referred as the *trapdoor*.

2.1 – There exists a polynomial time Turing Machine Sample^x which on input $m \in I$ and $r \in R_m$ outputs $x \in X_m$. Furthermore, for any m, the machine Sample$^x(m, \cdot)$ implements a bijection from R_m onto X_m.

2.2 – There exists a polynomial time Turing Machine Sample^y, such that on input $m \in I$ and $r \in R_m$ it outputs $f_m(x)$ for $x = $ Sample$^x(m, r)$. Therefore, Sample$^y(m, r) = f_m(\text{Sample}^x(m, r))$.

3 – There exists a polynomial time Turing Machine Check^y which, on input $m \in I$ and any y, answers whether $y \in Y_m$ or not.

4 – There exists a (deterministic) polynomial time Turing Machine Inv such that Inv$(m, t_m, f_m(x)) = x$, for all $x \in X_m$ and for all $m \in I$.

5 – For every probabilistic polynomial time Turing Machine $\mathcal{A}$ we have that, for large enough k,

$$\Pr[m \overset{R}{\leftarrow} I \,;\, x \overset{R}{\leftarrow} X_m \,;\, y = f_m(x) : \mathcal{A}(m, y) = x] \leq \varepsilon(k),$$

where $\varepsilon(\cdot)$ is a negligible function.

The last property is our formal *hard-to-invert* notion, which is quite similar to the usual *one-way* notion: they just differ if Sample$^x(m, \cdot)$ is one-way.

2.2   Verifiable Sub-family of Trapdoor Hard-to-Invert Isomorphisms

In the above definition, it is clear that for any $m \in I$, the function f_m is an isomorphism from the group X_m onto Y_m. However, in practice, the family of functions $\{f_m\}_m$ maybe indexed by a potentially larger set S (i.e. $I \subseteq S$), for which there may exist some indices that do not lead to an isomorphism. Therefore, we require more properties to be satisfied.

- there exists a large subset $I \subseteq S$, such that $F = \{f_m : X_m \to Y_m\}_{m \in I}$ is a family of *trapdoor hard-to-invert isomorphisms*;
- there exists a set J, of indices which provide an isomorphism – such that $I \subseteq J \subseteq S$ –, which admits an efficient zero-knowledge proof of membership.

The last property turns out to be crucial for the application we have in mind. In our setting the client has to choose the specific function to use in the protocol. This means that a dishonest client (i.e. one that does not share a password with the server) could propose an index whose corresponding function is not an isomorphism. This would give him the ability to run a partition attack (as already explained for RSA). For this reason we require the client to produce a function f together with a proof that it is actually an isomorphism.

2.3   Zero-Knowledge Proofs of Membership

As noticed above, the only property we want to be able to verify is the isomorphic one, and thus the fact that the index m actually lies in J: we just want the adversary not to be able to prove a wrong statement, we do not care about malleability [11]. One second point is that the zero-knowledge property will be required in the security proof: a valid index m is given, one tries to use the adversary to solve a hard problem related to m. Thus, we need to be able to provide a proof of validity of m, without any witness. Note however that the simulation is performed for valid statements only, and thus *simulation soundness* [24] is not required. Moreover, since we just have to simulate one proof without the witness (other executions will be performed as in an actual execution) *concurrent zero-knowledge* is not needed either.

For efficiency reasons, we will focus on a specific class of zero-knowledge proofs: for a given statement m, the verifier sends a random seed seed and then the prover non-interactively provides a proof $p = \mathsf{Prove}^m(m, w, \mathsf{seed})$ using a witness w that $m \in J$, w.r.t. the random seed seed; the proof can be checked without the witness $\mathsf{Check}^m(m, \mathsf{seed}, p)$. In our protocol, honest players will sample $m \in I$, and thus together the trapdoor t_m. This trapdoor will generally be a good witness. More formally we require:

- Completeness – Prove^m and Check^m are two efficient (polynomial time) algorithms, and for any $m \in J$ and any challenge seed, a witness helps to build a proof $p = \mathsf{Prove}^m(m, w, \mathsf{seed})$ which is always accepted: $\mathsf{Check}^m(m, \mathsf{seed}, p)$ accepts;

- Soundness – for any $m \notin J$, the probability for any adversary (on its random tape and the random seed seed) to forge a valid proof (accepted by the Check^m algorithm) is negligible within time t: $\mathsf{Succ}^{\mathsf{forge}}(t)$ will denote the maximal success probability for any adversary within time t;
- ROM-simulatability – granted the programmability of the random oracle, for any $m \in I$ and any seed, there exists an efficient way to perfectly simulate an accepted proof.

2.4   Concrete Examples

The Diffie-Hellman Family. The most natural example of family of trapdoor hard-to-invert isomorphisms is the Diffie-Hellman one. The machine Gen, on input the security parameter k, does as follows. First it chooses a random prime q of size k, and a prime p such that q divides $p-1$. Next, it chooses a subgroup $\mathbb{G}$ of order q in $\mathbb{Z}_p^*$ and a corresponding generator g. Finally it chooses a random element a in $\mathbb{Z}_q$, it sets $h = g^a \bmod p$ and outputs the pair (m, t_m) where $t_m = a$ and m is an encoding of (g, p, q, h). This defines our set I.

Now f_m is instantiated as follows. Set $X_m = Y_m = \mathbb{G} \backslash \{1\}$, $R_m = \mathbb{Z}_q$ and $\mathsf{Sample}^x : \mathbb{Z}_q \to \mathbb{G}$ is defined[1] as $\mathsf{Sample}^x(x) = g^x \bmod p$. Moreover f_m is defined as (for any $X \in \mathbb{G} \backslash \{1\}$): $f_m(X) = X^a \bmod p$.

Clearly, to efficiently evaluate f_m on a random point X, one should know either the trapdoor information a or any x such that $\mathsf{Sample}^x(x) = X$ (assuming, of course, that the computational Diffie-Hellman problem is infeasible in $\mathbb{G}$): $\mathsf{Sample}^y(x) = h^x$. Similarly knowledge of the trapdoor is sufficient to invert f_m on a random point Y: $\mathsf{Inv}(a, Y) = Y^{1/a}$. However inverting the function without knowing the trapdoor seems to be infeasible. Nevertheless, $Y_m = \mathbb{G}$ is efficiently decidable: $\mathsf{Check}^y(y)$ simply checks whether $y^q = 1 \bmod p$ or not.

For our functions to be isomorphisms, one just needs a to be co-prime with q, where q is actually the order of g. For better efficiency, the group informations (g, p, q) can be fixed, and considered as common trusted parameters. Therefore, Gen just chooses a and sets $h = g^a \bmod p$: one just needs to check that $h \neq 1 \bmod p$ and $h^q = 1 \bmod p$, no witness is required, nor additional proof: Prove^m does not need any witness for outputting any proof, since Check^m simply checks the above equality/inequality.

The RSA Family. Another natural example is the RSA permutation. In this case the machine Gen on input the security parameter k does as follows. First it chooses two random primes p, q of size $k/2$ and sets $n = pq$. Next, it chooses a public exponent e such that $\gcd(e, \varphi(n)) = 1$. Finally it outputs the pair (m, t_m) where $t_m = (p, q)$ and m is an encoding of (n, e). This defines our set I.

The function f_m is instantiated as follows. Set $X_m = Y_m = R_m = \mathbb{Z}_n^*$, and $\mathsf{Sample}^x : \mathbb{Z}_n^* \to \mathbb{Z}_n^*$ is the identity function, i.e. $\mathsf{Sample}^x(x) = x$. The function

---

[1] Note that we allow a slight misuse of notation here. Actually the function Sample^x should be defined as $\mathsf{Sample}^x : I \times \mathbb{Z}_q \to \mathbb{G}$. However we prefer to adopt a simpler (and somehow incorrect) notation for visual comfort.

f_m is defined as (for any $x \in \mathbb{Z}_n^\star$): $f_m(x) = x^e \bmod n$. Hence, $\mathsf{Sample}^y(x) = x^e \bmod n$. The Inv algorithm is straightforward, granted the trapdoor. And the Check^y algorithm simply has to check whether the element is prime to n.

As already noticed, since Sample^x is easy to invert, the RSA family is not only a trapdoor hard-to-invert isomorphism family, but also a trapdoor one-way permutation family. However, to actually be an isomorphism, (n, e) does not really need to lie in I, which would be very costly to prove (while still possible). It just needs to satisfy $\gcd(e, \varphi(n)) = 1$, which defines our set J. An efficient proof of validity is provided in the full version [8], where both Prove^m and Check^m are formally defined.

The Squaring Family. As a final example, we suggest the squaring function which is defined as the RSA function with the variant that $e = 2$. A problem here arises from the fact that squaring is not a permutation over $\mathbb{Z}_n^\star$, simply because 2 is not co-prime with $\varphi(n)$. However, if one considers *Blum* moduli (i.e. composites of the form $n = pq$, where $p \equiv q \equiv 3 \bmod 4$) then it is easy to check that the squaring function becomes an automorphism onto the group of quadratic residues modulo n (in the following we refer to this group as to Q_n.) However this is not enough for our purposes. An additional difficulty comes from the fact that we need an efficient way to check if a given element belongs to Y_m (which would be Q_n here): the need of an efficient algorithm Check^y. The most natural extension of Q_n is the subset J_n of $\mathbb{Z}_n^\star$, which contains all the elements with Jacobi symbol equal to $+1$. Note that for a Blum modulus $n = pq$, this set is isomorphic to $\{-1, +1\} \times Q_n$ (this is because -1 has a Jacobi symbol equal to $+1$, but is not a square). By these positions we get the *signed squaring*[2] isomorphism:

$$f_n : \{-1, +1\} \times Q_n \to J_n$$
$$(b \ , \ x) \mapsto b \times x^2 \bmod n.$$

For this family, the machine Gen, on input the security parameter k, does as follows. First it chooses two random Blum primes p, q of size $k/2$ and sets $n = pq$. Then it outputs the pair (m, t_m) where $t_m = (p, q)$ and m is an encoding of n. This thus defines our set I. The function f_m is instantiated as follows. Set $X_m = R_m = \{-1, +1\} \times Q_n$, $Y_m = J_n$ and $\mathsf{Sample}^x : \{-1, +1\} \times Q_n \to \{-1, +1\} \times Q_n$ is the identity function, i.e. $\mathsf{Sample}^x(b, x) = (b, x)$. The function f_m is defined as (for any $(b, x) \in \{-1, +1\} \times Q_n$): $f_m(b, x) = b \times x^2 \bmod n$. Hence, $\mathsf{Sample}^y(b, x) = f_m(b, x)$. The Inv algorithm is straightforward, granted the trapdoor. And the Check^y algorithm simply computes the Jacobi symbol.

As above, since Sample^x is easy to invert, the squaring family is not only a trapdoor hard-to-invert isomorphism family, but also a trapdoor one-way permutation family. However, to actually be an isomorphism, n does not really need to be a Blum modulus, which would be very costly to prove. What we need is just that -1 has Jacobi symbol $+1$ and any square in $\mathbb{Z}_n^\star$ admits exactly 4 roots. A validity proof is provided, with the mathematical justification, in the section 6, which thus formally defines both Prove^m and Check^m.

---

[2] By *signed*, we mean that the output of the function has a sign (plus or minus).

3 The Formal Model

3.1 Security Model

Players. We denote by A and B two parties that can participate in the key exchange protocol P. Each of them may have several *instances* called oracles involved in distinct, possibly concurrent, executions of P. We denote A (resp. B) instances by A^i (resp. B^j), or by U when we consider any user instance. The two parties share a low-entropy secret pw which is drawn from a small dictionary Password, according to a distribution $\mathcal{D}$. In the following, we use the notation $\mathcal{D}(n)$ for the probability to be in the most probable set of n passwords:

$$\mathcal{D}(n) = \max_{P \subseteq \mathsf{Password}} \left\{ \Pr_{pw \overset{R}{\leftarrow} \mathcal{D}} [pw \in P \,|\, \mathsf{Card}(P) \leq n] \right\}.$$

If we denote by $\mathcal{U}_N$ the uniform distribution among N passwords, $\mathcal{U}_N(n) = n/N$.

Queries. We use the security model introduced by Bellare *et al.* [1], to which paper we refer for more details. In this model, the adversary $\mathcal{A}$ has the entire control of the network, which is formalized by allowing $\mathcal{A}$ to ask the following queries:

- Execute(A^i, B^j): This query models passive attacks, where the adversary gets access to honest executions of P between the instances A^i and B^j by eavesdropping.
- Reveal(U): This query models the misuse of the session key by any instance U (use of a weak encryption scheme, leakage after use, etc). The query is only available to $\mathcal{A}$ if the attacked instance actually "holds" a session key and it releases the latter to $\mathcal{A}$.
- Send(U, m): This query models $\mathcal{A}$ sending a message to instance U. The adversary $\mathcal{A}$ gets back the response U generates in processing the message m according to the protocol P. A query Send($A^i, \mathtt{Start}$) initializes the key exchange algorithm, and thus the adversary receives the flow A should send out to B.

In the active scenario, the Execute-query may seem rather useless: after all the Send-query already gives the adversary the ability to carry out honest executions of P among parties. However, even in the active scenario, Execute-queries are essential to properly deal with dictionary attacks. Actually the number q_s of Send-queries directly asked by the adversary *does not* take into account the number of Execute-queries. Therefore, q_s represents the number of flows the adversary may have built by itself, and thus the number of passwords it may have tried. Even better, $q_a + q_b$ is an upper-bound on the number of passwords it may have tried, where q_a (and q_b resp.) is the number of A (B resp.) instances involved in the attack. For the sake of simplicity, we restricted queries to A and B only. One can indeed easily extend the model, and the proof, to the more general case, keeping in mind that we are interested in the security of executions involving at least A or B, with the password pw shared by them. Additional queries would indeed use distinct passwords, which could be assumed public in the security analysis (known to our simulator).

3.2   Security Notions

Two main security notions have been defined for key exchange protocols. The first one is the semantic security of the key, which means that the exchanged key is unknown to anybody else than the players. The second one is unilateral or mutual authentication, which means that either one, or both, of the participants actually know the key.

AKE Security. The semantic security of the session key is modeled by an additional query $\mathsf{Test}(U)$. The Test-query can be asked at most once by the adversary $\mathcal{A}$ and is only available to $\mathcal{A}$ if the attacked instance U is *Fresh*. The freshness notion captures the intuitive fact that a session key is not "obviously" known to the adversary. An instance is said to be *Fresh* if the instance has accepted (i.e. the flag accept is set to true) and neither it nor its partner (i.e. the other instance with same session tag —or SID— which is defined as the view the player has of the protocol —the flows— before it accepts) have been asked for a Reveal-query. The Test-query is answered as follows: one flips a (private) coin b and forwards sk (the value $\mathsf{Reveal}(U)$ would output) if $b = 1$, or a random value if $b = 0$.

We denote the **AKE advantage** as the probability that $\mathcal{A}$ correctly guesses the value of b. More precisely we define $\mathsf{Adv}_P^{\mathrm{ake}}(\mathcal{A}) = 2\Pr[b = b'] - 1$, where the probability space is over the password, all the random coins of the adversary and all the oracles, and b is the output guess of $\mathcal{A}$ for the bit b involved in the Test-query. The protocol P is said to be (t, ε)-**AKE-secure** if $\mathcal{A}$'s advantage is smaller than ε for any adversary $\mathcal{A}$ running with time t.

Entity Authentication. Another goal of the adversary is to impersonate a party. We may consider unilateral authentication of either A (A-Auth) or B (B-Auth), thus we denote by $\mathsf{Succ}_P^{A-\mathrm{auth}}(\mathcal{A})$ (resp. $\mathsf{Succ}_P^{B-\mathrm{auth}}(\mathcal{A})$) the probability that $\mathcal{A}$ successfully impersonates an A instance (resp. a B instance) in an execution of P, which means that B (resp. A) terminates (i.e. the terminate flag is set to true) even though it does not actually share the key with any accepting partner A (resp. B).

A protocol P is said to be (t, ε)-**Auth-secure** if $\mathcal{A}$'s success for breaking either A-Auth or B-Auth is smaller than ε for any adversary $\mathcal{A}$ running with time t. This protocol then provides *mutual authentication*.

4   Algorithmic Assumptions

In this section we state some algorithmic assumptions we need in order to construct an **IPAKE** protocol. As already sketched in section 1.2, our basic building block is a family of trapdoor hard-to-invert bijections $\mathcal{F}$. More precisely each bijection $f \in \mathcal{F}$ needs to be a group isomorphism from a group $(X_f, \oplus_f)$ into a group $(Y_f, \otimes_f)$, where $\ominus_f$ (resp. $\oslash_f$) is the inverse operation of $\oplus_f$ (resp. $\otimes_f$)[3]. As additional assumption we require the existence of a generalized full-domain hash function $\mathcal{G}$, which on a new input (f, q), outputs a uniformly distributed element in Y_f. This is the reason why we need the decidability of Y_f: in practice, $\mathcal{G}$ will be implemented by iterating a hash function until the output is in Y_f.

The non-invertibility of the functions in the family $\mathcal{F}$ is measured by the "ability", for any adversary $\mathcal{A}$, in inverting a random function (in $\mathcal{F}$) on a random point, uniformly drawn from Y_f:

$$\mathsf{Succ}_{\mathcal{F}}^{\mathsf{NI}}(\mathcal{A}) = \Pr[f \overset{R}{\leftarrow} \mathcal{F}, x \overset{R}{\leftarrow} X_f : \mathcal{A}(f, f(x)) = x].$$

More precisely, we denote by $\mathsf{Succ}_{\mathcal{F}}^{\mathsf{NI}}(t)$ the maximal success probability for all the adversaries running within time t. A simpler task for the adversary may be to output a list of n elements which contains the solutions:

$$\mathsf{SuccInSet}_{\mathcal{F}}^{\mathsf{NI}}(\mathcal{A}) = \Pr[f \overset{R}{\leftarrow} \mathcal{F}, x \overset{R}{\leftarrow} X_f, S \leftarrow \mathcal{A}(f, f(x)) : x \in S].$$

As above, we denote by $\mathsf{SuccInSet}_{\mathcal{F}}^{\mathsf{NI}}(n, t)$ the maximal success probability for all the adversaries running within time t, which output sets of size n.

4.1   The RSA Family: $\mathcal{F} = \mathsf{RSA}$

As described in section 2.4 the function f is defined by n and e, $Y_f = X_f = \mathbb{Z}_n^*$. And, for any $x \in \mathbb{Z}_n^*$, $f(x) = x^e \bmod n$. For a correctly generated n and a valid e (i.e an e such that $\gcd(\varphi(n), e) = 1$) the non-invertibility of the function is equivalent to the, widely conjectured, one-wayness of RSA. This leads to the following

$$\mathsf{Succ}_{\mathsf{RSA}}^{\mathsf{ow}}(t + nT_{\mathsf{exp}}) = \mathsf{Succ}_{\mathsf{RSA}}^{\mathsf{NI}}(t + nT_{\mathsf{exp}}) \geq \mathsf{SuccInSet}_{\mathsf{RSA}}^{\mathsf{NI}}(n, t) = \mathsf{SuccInSet}_{\mathsf{RSA}}^{\mathsf{ow}}(n, t)$$

where T_{exp} is an upper-bound on the time required to perform an exponentiation.

4.2   The Diffie-Hellman Family: $\mathcal{F} = \mathsf{DH}$

Let $\mathbb{G} = \langle g \rangle$ be any cyclic group of (preferably) prime order q. As sketched in section 2.4, the function f is defined by a point $P = g^x$ in $\mathbb{G}\backslash\{1\}$ (and thus $x \neq 0 \bmod q$), and $X_f = Y_f = \mathbb{G}$. For any $Q = g^y \in \mathbb{G}$, $f(Q) = g^{xy}$.

A (t, ε)-$\mathsf{CDH}_{g,\mathbb{G}}$ attacker, in the finite cyclic group $\mathbb{G}$ of prime order q, generated by g, is a probabilistic machine Δ running in time t such that

$$\mathsf{Succ}_{g,\mathbb{G}}^{\mathsf{cdh}}(\Delta) = \Pr_{x,y}[\Delta(g^x, g^y) = g^{xy}] \geq \varepsilon$$

---

[3] For visual comfort in the following we adopt the symbols f, X_f, Y_f rather than (respectively) f_m, X_m, Y_m.

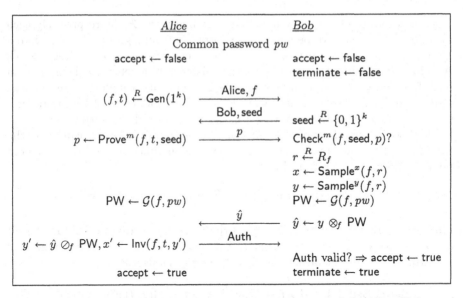

Fig. 1. An execution of the **IPAKE** protocol: Auth is computed by Alice (Bob resp.) as $\mathcal{H}_1(\text{Alice}\|\text{Bob}\|f\|\hat{y}\|pw\|x)$ ($\mathcal{H}_1(\text{Alice}\|\text{Bob}\|f\|\hat{y}\|pw\|x')$ resp.), and sk is computed by Alice (Bob resp.) as $\mathcal{H}_0(\text{Alice}\|\text{Bob}\|f\|\hat{y}\|pw\|x)$ ($\mathcal{H}_0(\text{Alice}\|\text{Bob}\|f\|\hat{y}\|pw\|x')$ resp.)

where the probability is taken over the random values x and y in $\mathbb{Z}_q$. As usual, we denote by $\mathsf{Succ}^{\mathsf{cdh}}_{g,\mathbb{G}}(t)$ the maximal success probability over every adversary running within time t. Then, when g and $\mathbb{G}$ are fixed, $\mathsf{Succ}^{\mathsf{NI}}_{\mathsf{DH}}(t) = \mathsf{Succ}^{\mathsf{cdh}}_{g,\mathbb{G}}(t)$. Using Shoup's result [25] about "self-correcting Diffie-Hellman", one can see that if $\mathsf{SuccInSet}^{\mathsf{NI}}_{\mathsf{DH}}(n,t) \geq \varepsilon$, then $\mathsf{Succ}^{\mathsf{NI}}_{\mathsf{DH}}(t') \geq 1/2$ for some $t' \leq 6/\varepsilon \times (T + nT_{\mathsf{exp}})$.

4.3   The Squaring Family: $\mathcal{F} = $ Rabin

As discussed in section 2.4 if one assumes that the modulus n is the product of two Blum primes, the signed squaring function f becomes an isomorphism from $\{-1, +1\} \times Q_n$ onto J_n. Furthermore, for a correctly generated n the non-invertibility of f is trivially equivalent to the one-wayness of factoring Blum composites. This leads us to the following inequality

$$\mathsf{Succ}^{\mathsf{ow}}_{\mathsf{Rabin}}(t + nT_{\mathsf{exp}}) = \mathsf{Succ}^{\mathsf{NI}}_{\mathsf{Rabin}}(t + nT_{\mathsf{exp}}) \geq \mathsf{SuccInSet}^{\mathsf{ow}}_{\mathsf{Rabin}}(n, t),$$

which provides a very tight bound because, in this case, T_{exp} represents the time required to perform a single modular multiplication (i.e. to square).

5   Security Proof for the **IPAKE** Protocol

5.1   Description and Notations

In this section we show that the **IPAKE** protocol distributes session keys that are semantically secure and provides unilateral authentication for the client A.

The specification of the protocol can be found on Figure 1. Some remarks, about notation, are in order

- We assume $\mathcal{F}$ to be a correct family, with a verifiable sub-family of trapdoor hard-to-invert isomorphisms f from X_f into Y_f. In the following, we identify m to f_m, and thus f. We denote by s the size of I. Furthermore, we denote by q a lower bound on the size of any Y_f.
- For this choice of parameters for the family $\mathcal{F}$, we can define the function $\mathcal{G}$ which is assumed to behave like a generalized full-domain random oracle. In particular we model $\mathcal{G}$ as follows: on input a couple (f, q) it outputs a random element, uniformly distributed in Y_f.

Since we only consider unilateral authentication (of A to B), we just introduce a terminate flag for B.

5.2 Security Proof

Theorem 1 (AKE/UA Security). *Let us consider the protocol* **IPAKE**, *over a family $\mathcal{F}$ of trapdoor hard-to-invert isomorphisms, with parameter (s, q), where* Password *is a dictionary equipped with the distribution $\mathcal{D}$. For any adversary $\mathcal{A}$ within a time bound t, with less than q_s active interactions with the parties (*Send*-queries) and q_p passive eavesdroppings (*Execute*-queries), and asking q_g and q_h hash queries to $\mathcal{G}$ and any $\mathcal{H}_i$ respectively:* $\mathsf{Adv}^{\mathsf{ake}}_{\mathsf{ipake}}(\mathcal{A}) \leq 4\varepsilon$ *and* $\mathsf{Adv}^{\mathsf{A-auth}}_{\mathsf{ipake}}(\mathcal{A}) \leq \varepsilon$, *with ε upper-bounded by*

$$3\mathcal{D}(q_a + q_b) + 6q_a \mathsf{SuccInSet}^{\mathsf{NI}}_{\mathcal{F}}(q_h^2, t + 2q_h^2 \tau_{law}) + q_b \mathsf{Succ}^{\mathsf{forge}}(t) + \frac{q_b}{2^{\ell_1}} + \frac{Q^2}{2q} + \frac{Q_P^2}{2s},$$

where q_a and q_b denote the number of A and B instances involved during the attack (each upper-bounded by $q_p + q_s$), $Q \leq q_g + q_h + 2q_p + q_s$ and Q_P denotes the number of involved instances ($Q_P \leq 2q_p + q_s$), and τ_{law} is the time needed for evaluating one law operation. Let us remind that ℓ_1 is the output length of $\mathcal{H}_1$ (the authenticator.)

For lack of space, we refer to the full version [8] for the full proof, here we justify the main terms in the security result.

Ideally, when one considers a password-based authenticated key exchange, one would like to prove that the two above success/advantage are upper-bounded by $\mathcal{D}(q_a + q_b)$, plus some negligible terms. For technical reasons in the proof (to get a *clear* proof) we have a small additional constant factor. This main term is indeed the basic attack one cannot avoid: the adversary guesses a password and makes an on-line trial. Other ways for it to break the protocol are:

- use a function f that is not a permutation, and in particular not a surjection. With the view of $\hat{y}$, the adversary tries all the passwords, and only a strict fraction leads to y in the image of f: this is a partition attack. But for that, it has to forge a proof of validity for f. Hence the term $q_b \times \mathsf{Succ}^{\mathsf{forge}}(t)$;

- use the authenticator Auth to check the correct password. But this requires the ability to compute $f^{-1}(\mathsf{PW})$. Hence the term $q_a \times \mathsf{SuccInSet}_{\mathcal{F}}^{\mathsf{NI}}(\cdot, \cdot)$.
- send a correct authenticator Auth, but being lucky. Hence the term $q_b/2^{\ell_1}$.

Additional negligible terms come from very unlikely collisions. All the remaining kinds of attacks need some information about the password.

6   A Concrete Example: The SQRT-IPAKE Protocol

An important contribution of this work (at least from a practical point of view) is the first efficient and provably secure password-based key exchange protocol based on factoring. The formal protocol appears in Figure 2. Here we describe the details of this specific implementation.

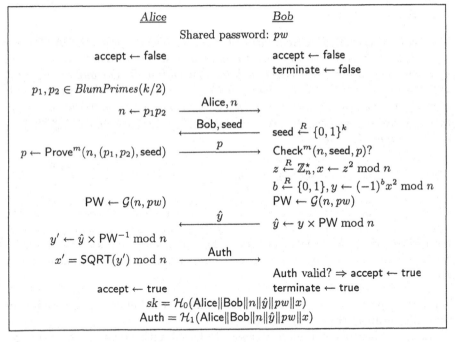

Fig. 2. SQRT – **IPAKE** protocol

6.1   Description of the SQRT-IPAKE Protocol

In order for the protocol to be correct we need to make sure that the adopted function is actually an isomorphism. As seen in section 2.4 this is the case if one assumes that the modulus n is the product of two Blum primes, and $f_n : \{-1, +1\} \times Q_n \to J_n$ is the signed squaring function.

We thus set $X_f = \{-1, +1\} \times Q_n$ and $Y_f = J_n$, and, of course, the internal law is the multiplication in the group $\mathbb{Z}_n^\star$. In order for the password PW to be generated correctly, we need a $\mathcal{G}(n, \cdot)$ hash function onto J_n. Constructing such a function is pretty easy: we start from a hash function onto $\{0, 1\}^k$, and we iterate it until we get an output in J_n. The details of this technique are deferred to the full version of this paper [8]. Here we stress that if $n \geq 646$ then very few iterations are sufficient. As already noticed, we require Alice to prove the following about the modulus n, so that the function is actually an isomorphism:

- The modulus n is in the correct range ($n \geq 646$);
- The Jacobi symbol of -1 is $+1$ in $\mathbb{Z}_n^\star$ (this is to make sure that f_n is actually a morphism);
- The signed squaring function is actually an isomorphism from $\{-1, +1\} \times Q_n$ onto J_n (this is to make sure that any square in $\mathbb{Z}_n^\star$ has exactly 4 roots).

Proving the first two statements is trivial. For the third one we need some new machinery.

6.2 Proof of Correct Modulus

With the following theorem (whose proof can be found in the full version of this paper [8]) we show that if n is a composite modulus (with at least two different prime factors) then the proposed function is an isomorphism.

Theorem 2. *Let n be a composite modulus containing at least two different prime factors and such that -1 has Jacobi symbol $+1$ in $\mathbb{Z}_n^\star$. Moreover let f_n be the morphism defined above. The following facts are true*

1. *If f_n is surjective then it is an isomorphism.*
2. *If f_n is not surjective, then at most half of the elements in J_n have a pre-image.*

The theorem above leads to the protocol Prove-Surjective (see Figure 3). The basic idea of this protocol is that we prove that our function is a bijection by proving it is surjective. Soundness follows from the second statement. However, in order to fall into the hypotheses of the theorem, we need to make sure n is actually a composite modulus of the required form (i.e. with at least two distinct prime factors). We achieve this with the Prove-Composite protocol (see Figure 3). The correctness (completeness, soundness and zero-knowledge properties) of these protocols is deferred to the full version of this paper [8].

Remark 3. We point out that our protocol is very efficient, for the verifier, in terms of modular multiplications. It is also possible for Alice to use the same modulus for different sessions.

Acknowledgments

We thank the anonymous referees for their fruitful comments.

Protocol Prove-Composite	Protocol Prove-Surjective
$\mathcal{H}_2(n,\cdot,\cdot)$ and $\mathcal{H}_4(n,\cdot,\cdot)$ are full-domain hash functions onto J_n $\mathcal{H}_3$ ($\mathcal{H}_5$ resp.) is a random oracle onto $\{0,1\}^k$ ($\{0,1\}^\ell$ resp.) Bob chooses a random seed **seed** and sends it to Alice	

For $i \leftarrow 1$ to ℓ, Alice

1. Sets $y_i = \mathcal{H}_2(n, \mathsf{seed}, i) \in J_n$	1. Sets $z_i = \mathcal{H}_4(n, \mathsf{seed}, i) \in J_n$
2. Computes $(\beta_i, \alpha_{i,0}, \alpha_{i,1}, \alpha_{i,2}, \alpha_{i,3})$ such that	2. Computes $(b_i, x_i) = f^{-1}(z_i)$ such that $(b_i, x_i) \in \{-1, +1\} \times Q_n$
$\quad - \ \alpha_{i,0} = -\alpha_{i,1} \bmod n$	3. Computes a value γ_i such that
$\quad - \ \alpha_{i,2} = -\alpha_{i,3} \bmod n$	$\gamma_i^2 = x_i \bmod n$ (this is to make
$\quad - \ \alpha_{i,j}^2 = y_i\beta_i \bmod n \ (j = 0, \dots, 3),$ where $\beta_i \in \{-1, +1\}$	sure that x_i is actually in Q_n);
3. Sets $h_{i,j} = \mathcal{H}_3(n, \alpha_{i,j}) \ (j = 0, \dots, 3)$	
One defines $c_1 \dots, c_\ell = \mathcal{H}_5(n, \mathsf{seed}, \{h_{i,j}\})$	

Alice answers with, for $i = 1, \dots, \ell$,

$(\beta_i, \alpha_{i,2c_i}, \alpha_{i,2c_i+1})$	(γ_i, b_i)

Bob checks that, for each $i = 1, \dots, \ell$,

1. the $h_{i,j}$, for $j = 0, \dots, 3$, are all distinct	$b_i\gamma_i^4 = \mathcal{H}_4(n, \mathsf{seed}, i) \bmod n$
2. $\alpha_{i,2c_i} = -\alpha_{i,2c_i+1} \bmod n$	
3. $h_{i,2c_i} = \mathcal{H}_3(n, \alpha_{i,2c_i})$ and $h_{i,2c_i+1} = \mathcal{H}_3(n, \alpha_{i,2c_i+1})$	
4. $\mathcal{H}_2(n, \mathsf{seed}, i) = \beta_i\alpha_{i,2c_i}^2 \bmod n$	

Fig. 3. Proof of Correct Modulus

References

1. M. Bellare, D. Pointcheval, and P. Rogaway. Authenticated Key Exchange Secure Against Dictionary Attacks. In *Eurocrypt '00*, LNCS 1807, pages 139–155. Springer-Verlag, Berlin, 2000.
2. M. Bellare and P. Rogaway. The AuthA Protocol for Password-Based Authenticated Key Exchange. Contributions to IEEE P1363. March 2000.
3. S. M. Bellovin and M. Merritt. Encrypted Key Exchange: Password-Based Protocols Secure against Dictionary Attacks. In *Proc. of the Symposium on Security and Privacy*, pages 72–84. IEEE, 1992.
4. C. Boyd, P. Montague, and K. Nguyen. Elliptic Curve Based Password Authenticated Key Exchange Protocols. In *ACISP '01*, LNCS 2119, pages 487–501. Springer-Verlag, Berlin, 2001.
5. V. Boyko, P. MacKenzie, and S. Patel. Provably Secure Password Authenticated Key Exchange Using Diffie-Hellman. In *Eurocrypt '00*, LNCS 1807, pages 156–171. Springer-Verlag, Berlin, 2000. Full version available at: http://cm.bell-labs.com/who/philmac/research/.
6. E. Bresson, O. Chevassut, and D. Pointcheval. Security Proofs for Efficient Password-Based Key Exchange. In *Proc. of the 10th CCS*, pages 241–250. ACM Press, New York, 2003.

7. E. Bresson, O. Chevassut, and D. Pointcheval. New Security Results on Encrypted Key Exchange. In *PKC '04*, LNCS, pages 145–159. Springer-Verlag, Berlin, 2004.
8. D. Catalano, D. Pointcheval, and T. Pornin. IPAKE: Isomorphisms for Password-based Authenticated Key Exchange. In *Crypto '04*, LNCS. Springer-Verlag, Berlin, 2004. Full version available from http://www.di.ens.fr/users/pointche/.
9. W. Diffie and M. E. Hellman. New Directions in Cryptography. *IEEE Transactions on Information Theory*, IT–22(6):644–654, November 1976.
10. Y. Dodis, J. Katz, S. Xu, and M. Yung. Strong Key-Insulated Signature Schemes. In *PKC '03*, LNCS, pages 130–144. Springer-Verlag, Berlin, 2003.
11. D. Dolev, C. Dwork, and M. Naor. Non-Malleable Cryptography. *SIAM Journal on Computing*, 30(2):391–437, 2000.
12. T. El Gamal. A Public Key Cryptosystem and a Signature Scheme Based on Discrete Logarithms. *IEEE Transactions on Information Theory*, IT–31(4):469–472, July 1985.
13. R. Gennaro and Y. Lindell. A Framework for Password-Based Authenticated Key Exchange. In *Eurocrypt '03*, LNCS 2656, pages 524–543. Springer-Verlag, Berlin, 2003.
14. O. Goldreich and Y. Lindell. Session-Key Generation Using Human Passwords Only. In *Crypto '01*, LNCS 2139, pages 408–432. Springer-Verlag, Berlin, 2001.
15. S. Halevi and H. Krawczyk. Public-Key Cryptography and Password Protocols. In *Proc. of the 5th CCS*. ACM Press, New York, 1998.
16. J. Katz, R. Ostrovsky, and M. Yung. Efficient Password-Authenticated Key Exchange Using Human-Memorizable Passwords. In *Eurocrypt '01*, LNCS 2045, pages 475–494. Springer-Verlag, Berlin, 2001.
17. S. Lucks. Open Key Exchange: How to Defeat Dictionary Attacks Without Encrypting Public Keys. In *Proc. of the Security Protocols Workshop*, LNCS 1361. Springer-Verlag, Berlin, 1997.
18. P. MacKenzie, S. Patel, and R. Swaminathan. Password-Authenticated Key Exchange Based on RSA. In *Asiacrypt '00*, LNCS 1976, pages 599–613. Springer-Verlag, Berlin, 2000.
19. P. MacKenzie and R. Swaminathan. Secure Network Authentication with Password Identification. Submission to IEEE P1363a. August 1999.
20. T. Okamoto and S. Uchiyama. A New Public Key Cryptosystem as Secure as Factoring. In *Eurocrypt '98*, LNCS 1403, pages 308–318. Springer-Verlag, Berlin, 1998.
21. P. Paillier. Public-Key Cryptosystems Based on Discrete Logarithms Residues. In *Eurocrypt '99*, LNCS 1592, pages 223–238. Springer-Verlag, Berlin, 1999.
22. M. O. Rabin. Digitalized Signatures. In R. Lipton and R. De Millo, editors, *Foundations of Secure Computation*, pages 155–166. Academic Press, New York, 1978.
23. R. Rivest, A. Shamir, and L. Adleman. A Method for Obtaining Digital Signatures and Public Key Cryptosystems. *Communications of the ACM*, 21(2):120–126, February 1978.
24. A. Sahai. Non-Malleable Non-Interactive Zero-Knowledge and Chosen-Ciphertext Security. In *Proc. of the 40th FOCS*. IEEE, New York, 1999.
25. V. Shoup. Lower Bounds for Discrete Logarithms and Related Problems. In *Eurocrypt '97*, LNCS 1233, pages 256–266. Springer-Verlag, Berlin, 1997.
26. F. Zhu, A. H. Chan, D. S. Wong, and R. Ye. Password Authenticated Key Exchange based on RSA for Imbalanced Wireless Network. In *Proc. of ISC '02*, LNCS 2433, pages 150–161. Springer-Verlag, Berlin, 2002.

Randomness Extraction and Key Derivation Using the CBC, Cascade and HMAC Modes*

Yevgeniy Dodis[1], Rosario Gennaro[2], Johan Håstad[3],
Hugo Krawczyk[4], and Tal Rabin[2]

[1] New York University
dodis@cs.nyu.edu
[2] IBM Research
{rosario,talr}@watson.ibm.com
[3] Royal Institute, Sweden
johanh@nada.kth.se
[4] Technion, Israel, and IBM Research
hugo@ee.technion.ac.il

Abstract. We study the suitability of common pseudorandomness modes associated with cryptographic hash functions and block ciphers (CBC-MAC, Cascade and HMAC) for the task of "randomness extraction", namely, the derivation of keying material from semi-secret and/or semi-random sources. Important applications for such extractors include the derivation of strong cryptographic keys from non-uniform sources of randomness (for example, to extract a seed for a pseudorandom generator from a weak source of physical or digital noise), and the derivation of pseudorandom keys from a Diffie-Hellman value.

Extractors are closely related in their applications to pseudorandom functions and thus it is attractive to (re)use the common pseudorandom modes as randomness extractors. Yet, the crucial difference between pseudorandom generation and randomness extraction is that the former uses random *secret* keys while the latter uses random *but known* keys. We show that under a variety of assumptions on the underlying primitives (block ciphers and compression functions), ranging from ideal randomness assumptions to realistic universal-hashing properties, these modes induce good extractors. Hence, these schemes represent a more practical alternative to combinatorial extractors (that are seldom used in practice), and a better-analyzed alternative to the common practice of using SHA-1 or MD5 (as a single un-keyed function) for randomness extraction. In particular, our results serve to validate the method of key extraction and key derivation from Diffie-Hellman values used in the IKE (IPsec's Key Exchange) protocol.

1 Introduction

1.1 Key Derivation and Randomness Extractors

Key derivation is a central functionality in cryptography concerned with the process of deriving secret and random cryptographic keys from some source of

---

* Extended abstract. Full version available at eprint.iacr.org/2004/

M. Franklin (Ed.): CRYPTO 2004, LNCS 3152, pp. 494–510, 2004.
© International Association for Cryptologic Research 2004

semi-secret randomness. In general, it is sufficient to derive a single random and secret key (say of length 128) which can then be used to key a pseudorandom function (or a pseudorandom generator) to obtain further pseudorandom keys as needed. Thus, a basic question, which motivates the work presented here, is how to derive such a random and secret key when all that is given is an imperfect source of randomness which contains some good amount of secret (computational) entropy, but this entropy is not presented in a direct form of uniformly (or pseudorandomly) distributed secret bits. This problem arises in a variety of scenarios such as when deriving keys from a non-uniform source of noise (as used, for example, by physical random generators) or from semi-random data (say, coming from user's input or the sampling of computer events, etc.). This is also the case when deriving keys from a Diffie-Hellman (DH) exchange. Let us elaborate on the latter case.

Let's assume that two parties run a DH protocol in order to agree on a shared secret key, namely, they exchange DH exponentials g^x and g^y and compute the DH value g^{xy}. In this case, g^x and g^y as seen by the attacker fully determine g^{xy}. Yet, it is assumed (by the Decisional Diffie-Hellman, DDH, assumption) that a computationally-bounded attacker cannot distinguish g^{xy} from a random element in the group generated by g. Thus, one can assume that g^{xy} contains $t = \log_2(order(g))$ bits of computational entropy relative to the view of the attacker (for a formal treatment of computational entropy in the DH context see [GKR04]). However, this entropy is spread over the whole value g^{xy} which may be significantly longer than t^1. Thus, we are in a situation similar to that of an imperfect source of randomness as discussed above. In particular, g^{xy} cannot be used directly as a cryptographic key, but rather as a source from which to *extract* a shorter string (say, of length 128) of full computational entropy which can then be used as a cryptographic key.

The tools used to derive a uniform key from these sources of imperfect randomness are often referred to as *randomness extractors*. The amount of theoretical results in this area is impressive; moreover, some of the constructions that have proven extraction guarantees are also efficient (see [Sha02] for a recent survey). One such example is the so called "pairwise independent universal hash functions" (also called "strongly universal") [CW79] which have quite efficient implementations and provable extraction properties. In particular, [HILL99] shows (see also [Lub96,Gol01]) that if an input distribution has sufficient min-entropy (meaning that no single value is assigned a too-large probability even though the distribution may be far from uniform) then hashing this input into a (sufficiently) shorter output using a function chosen at random from a family of strongly universal hash functions results in an output that is statistically-close to uniform. (This result is often referred to as the "Leftover Hash Lemma".)

---

[1] For example, consider that g is an element of prime order q in Z_p^* (i.e., p and q are primes and $q/p-1$), and that $|p| = 1024$ and $|q| = 512$. In this case the DDH assumption guarantees that the value g^{xy} hides (from the attacker) 512 bits of computational entropy, yet these bits are spread in some unknown way among the 1024 bits of g^{xy}.

Yet, in spite of these results and an extensive literature studying their application to real cryptographic problems (such as those mentioned earlier, and in particular for the DH case [GKR04]) one seldom encounters in practice the use of strong universal hashing or other proven extractors. Instead, the common practice is to use cryptographic hash functions (such as MD5 and SHA-1) for the purpose of randomness extraction. A main reason for this practice, justified by engineering considerations, is that cryptographic hash functions are readily available in software and hardware implementations, and are required by most cryptographic applications for purposes other than randomness extraction (e.g., as pseudorandom functions). Therefore, it is attractive and convenient to use them for key extraction as well. Also, the common perception that these hash functions behave as random functions (formalized via the notion of "random oracles") make them intuitively appealing for the extraction applications.

1.2   Randomness Extraction via Common Chaining Modes

In this paper we attempt to bridge between the world of provable extraction and the common practice of relying on idealized hash functions. The question that we ask is what is the best way to use cryptographic hash functions, or other widely available cryptographic tools such as block ciphers, for the task of randomness extraction. Specifically, we consider three common modes of operation: CBC chaining, cascade (or Merkle-Damgard) chaining, and HMAC, and analyze the appropriateness of these modes as extraction tools. Since the goal is to provide as general (and generic) as possible results, we do not investigate the extraction properties of specific functions (say SHA-1 or AES) but rather abstract the basic primitives (the compression functions in the case of the cascade and HMAC modes, and block ciphers in the case of CBC), as random functions or permutations[2].

Before going on with the description of our results, it is worth considering the following issue. Given that the common practice is to extract randomness using a hash function modeled as a random oracle, then how much do we gain by analyzing the above modes under the weaker, but still idealized, randomness assumption on the underlying basic primitives. There are several aspects to this question.

The first thing to note is that modeling the *compression function of SHA-1*, for example, as a random function, or as a family of random functions, is a strict relaxation to modeling *SHA-1* (as a single un-keyed function) as a random function. This is easily seen from the fact that even if one starts with a random function as the compression function the result of the cascade chaining (which is how SHA-1 is derived) *is not a random function*. (For example, in the cascade construction, the probability that two L-block inputs that differ only in their first block are mapped to the same k-bit output is $L/2^k$, while for

---

[2] In the case of HMAC we obtain results based on non-ideal assumptions on the underlying basic primitives (see Section 1.3).

a random function this probability is $1/2^k$). Another important point is that cryptographic design work focuses on building the basic blocks, i.e. a compression function or a block cipher. Thus, making an assumption on these primitives will represent the design goals which there can be an attempt to satisfy. Also analysis of the type presented here, rather than implying the security of any specific implementation of these functions, serves to validate the suitability of the corresponding chaining modes for some defined goal (in our case the goal is randomness extraction). Indeed, the common approach for analyzing such modes (e.g., [Dam89,BKR94,BCK96a,BCK96b]) is to make some assumption on the basic primitive (for example, assuming the underlying compression function to be a pseudorandom function, or a secure MAC, or a collision-resistant hash function) and then proving that these or other properties are preserved or implied by the chaining operation.

In addition, the "monolithic" randomness assumption on a single (unkeyed) function such as SHA-1 is inappropriate for the setting of randomness extraction as no single function (even if fully random) can extract a close-to-uniform distribution from arbitrary high-entropy input distributions. This is so, since once the function is fixed (even if to purely random values) then there are high-entropy input distributions that will be mapped to small subsets of outputs[3]. Therefore, the viable approach for randomness extraction is to consider a *family* (or collection) of functions indexed by a set of keys. When an application requires the hashing of an input for the purpose of extracting randomness, then a random element (i.e., a function) from this family is chosen and the function is applied to the given input. While there may be specific input distributions that interact badly with specific functions in the family, a good randomness-extraction family will make this "bad event" happen with very small probability. Universal hash families, mentioned before, are examples of this approach. An *important point* here is that, while the choice of a function from the family is done by selecting a random index, or key, this key does not need to be kept secret (this is important in applications that use extraction to generate secret keys; otherwise, if we required this index to be secret then we would have a "chicken and egg" problem).

In our setting, families of keyed functions come up naturally with block ciphers and compression functions (for the latter we consider, as in HMAC, the variable IV as the key to the function). These functions are defined on fixed length inputs (e.g., 512 bits in the case of compression function of SHA-1, or 128 in the case of AES). Then, to hash arbitrarily long inputs, we extend these families by the appropriate chaining mode: cascade chaining (or HMAC) for compression functions, and CBC-MAC in the case of block ciphers. What makes the analysis of these functions challenging (in the setting of randomness extraction) is that, as discussed before, the key to the function is random but known. For ex-

---

[3] For example, let F be a random function from ℓ to k bits and let S denote the subset of $\{0,1\}^\ell$ that is mapped by F to outputs with a low-order bit of zero. If we consider the uniform distribution on S as the input distribution, then this distribution has almost full entropy, yet the output of F on S is trivially distinguishable from uniform.

ample, the fact that the above functions are widely believed to be pseudorandom does not help much here since, once the key is revealed, the pseudorandom properties may be lost completely (see full paper). Yet, as we will see in Section 4.2, we do use the pseudorandom property in some of our analysis. Also worth noting is that using families that are pseudorandom for extraction is particularly convenient since these same functions can then be used by the same application (for example, a key-exchange protocol, a random generator, etc.) for further key derivation (using the extracted key to key the pseudorandom function).

The last question is how to generate the random known keys used by the extractor. Technically this is not hard, as the parties can generate the appropriate randomness, but the exact details depend on the application. For example, in the DH key exchange discussed earlier, the parties exchange in the clear randomly chosen values, which are then combined to generate a single key κ for the extractor family $\{H_\kappa\}$ (e.g. HMAC-SHA1). The shared key is set to $H_\kappa(g^{xy})$. We note that this is substantially the procedure in place in the IKE protocol [RFC2409,IKEv2] (see also [Kra03]), and this paper presents the first formal analysis of this design.

A similar DH key extraction step is required in non-interactive scenarios, such as ElGamal or Cramer-Shoup encryption. There the extractor key κ can be chosen either by the encryptor and appended to the ciphertext, or chosen by the decryptor and included in the public key (this choice is mandatory in case we want CCA-security, as we don't want to leave the choice of κ in the hands of the adversary). For a different example, consider a cryptographic hardware device, containing a physical random generator that samples some imperfect source of noise. In this case the application can choose a random hash function in the family and wire-in its key into the randomness generation circuit [BST03]. Notice that by using our results, it will be possible to perform the extraction step using circuitry (such as a block-cipher or a cryptographic hash function) which is very likely to already be part of the device.

1.3   Our Results

The Extraction Properties of CBC-MAC Mode. We show, in Section 3, that if f is a random permutation over $\{0,1\}^k$ and $\mathcal{X}$ is an input distribution with min-entropy of at least $2k$, then the statistical distance between $F(\mathcal{X})$ (where F represents the function f computed in CBC-MAC mode over L blocks) and the uniform distribution on $\{0,1\}^k$ is $L \cdot 2^{-k/2}$. As an example, in the application (discussed before) in which we use the CBC-MAC function F to hash a Diffie-Hellman value computed over a DDH group of order larger than 2^{2k}, we get that the output distribution $F(g^{xy})$ is computationally indistinguishable from a distribution whose distance from uniform is at most $L2^{-k/2}$, hence proving (under DDH) that the k-bit output from $F(g^{xy})$ is computationally indistinguishable from uniform (and thus suitable for use as a cryptographic key). Note that if one works over Z_p^* for 1024-bit p and $k = 128$, then all we need to assume is a min-entropy of 256 out of the 1024 bits of g^{xy}. In the full paper we show that for input distributions with particularly high entropy (in particular those that contain a k-bit block of almost-full entropy) the CBC-MAC mode guarantees an almost-uniform output for *any* family of permutations.

The Extraction Properties of Cascade Chaining. In Section 4 we study the cascade (or Merkle-Damgard) chaining used in common hash functions such as MD5 and SHA-1. We show these families to be good extractors when modeling the underlying compression function as a family of random functions. However, in this case we need a stronger assumption on the entropy of the input distribution. Specifically, if the output of the compression function is k-bit long (typically, $k = 128$ or 160) we assume a min-entropy of $2k$ over the whole input, and "enough" min-entropy over the distribution induced on the last block of input (typically of length 512 bits). For example, if the last block has k bits of min-entropy, and we assume L blocks, then the statistical distance between the output of the cascade construction and the uniform distribution (on $\{0,1\}^k$) is at most $L \cdot 2^{-k/2}$. We note that the above restriction on the last-block distribution is particularly problematic in the case of practical functions such as MD5 and SHA since the input-padding conventions of these functions may cause a full *fixed* block to be added as the last block of input. In this case, the output distribution is provably far from uniform. Fortunately, we show that our analysis is applicable also to the padded-input case. However, instead of proving a negligible statistical distance, what we show is that the output of the "padded cascade" is *computationally indistinguishable* from uniform, a result that suffices for the cryptographic applications of extraction. Finally, we prove that when every block of input has large-enough min-entropy (conditioned on the distribution of previous blocks), then the above extraction results hold under the sole (*and non-ideal*) assumption that the underlying compression function is a family of δ-AU functions (for sufficiently small δ).

The Extraction Properties of HMAC. HMAC is the most widely used pseudorandom mode based on functions such as MD5 or SHA, thus proving its extraction properties is extremely important. Our main result concerning the good extraction properties of HMAC is proven on the basis of a high min-entropy ($2k$ bits) in the input distribution $\mathcal{X}$, without relying on any particular entropy in the last block of input. Specifically, let F denote the keyed hash function underlying an instantiation of HMAC (e.g., F is SHA-1 with random IV) and let f be the corresponding outer compression function. Then we show that if F is collision resistant and f is modeled as a random function then the output of HMAC (on input drawn from the distribution $\mathcal{X}$) is indistinguishable from uniform for any attacker that is restricted in the number of queries to the function f. Moreover, if the compression function itself is a good extractor, then HMAC is a good extractor too. However, in this latter case if we are interested in an output of ℓ close-to-uniform bits then the key to the underlying compression function needs to be sufficiently larger than ℓ. As a concrete example, if $\ell = 160$ (e.g., we need to generate a pseudorandom key of length 160) then we can use HMAC with SHA2-512. Note that this result is particularly interesting in the sense that it uses no idealized assumptions, and yet the output of HMAC is provably close to uniform (even against completely unbounded attackers, including attackers that can break the collision resistance of F).

Remark *(Pseudorandom functions with known keys)*. It is tempting to use the pseudorandomness properties enjoyed by the modes studied here as a basis to claim good (computational) extraction properties. For example, in spite of the fact that the output of these functions may be statistically very-far from uniform, it is still true that no (efficient) *standard* statistical test will be able to tell apart this output from random (simply because such a test does not use the knowledge of the key even if this key is known.) Yet, for cryptographic applications using a family of functions as extractors, based solely on the assumption that the family is pseudorandom, may be totally insecure. We illustrate this point by showing, in the full paper, an example of a secure pseudorandom family whose output is trivially distinguishable from randomness once the key is known.

All proofs appear in the full version of the paper.

2   Universal Hashing and Randomness Extraction

Preliminaries. For a probability distribution $\mathcal{X}$ we use the notation $x \in_R \mathcal{X}$ to mean that x is chosen according to the distribution $\mathcal{X}$. For a set S, $x \in_R S$ is used to mean that x is chosen from S with uniform probability. Also, for a probability distribution $\mathcal{X}$ we use the notation $\Pr_{\mathcal{X}}(x)$ to denote the probability assigned by $\mathcal{X}$ to the value x. (We often omit the subscript when the probability distribution is clear from the context.) Throughout the paper, we will use $d(L)$ to denote the maximal numbers of divisors of any number smaller or equal to L. As a *very* crude upper bound, we will sometimes use the fact that $d(L) \leq 2\sqrt{L}$.

MIN-ENTROPY AND COLLISION PROBABILITY. For a probability distribution $\mathcal{X}$ over $\{0,1\}^\ell$, we define its *min-entropy* as the minimum integer m such that for all $x \in \{0,1\}^\ell, \Pr_{\mathcal{X}}(x) \leq 2^{-m}$. We denote the min-entropy of such $\mathcal{X}$ by $\mathbf{H}_\infty(\mathcal{X})$. The *collision probability* of $\mathcal{X}$ is $\mathsf{Col}(\mathcal{X}) = \Pr_{x,x' \in_R \mathcal{X}}(x = x') = \sum_x \Pr(\mathcal{X} = x)^2$, and the *Renyi (or collision) entropy* of $\mathcal{X}$ is $\mathbf{H}_2(\mathcal{X}) = -\log_2 \mathsf{Col}(\mathcal{X})$. It is easy to see that these two notions of entropy are related: $\mathbf{H}_\infty(\mathcal{X}) \leq \mathbf{H}_2(\mathcal{X}) \leq 2\mathbf{H}_\infty(\mathcal{X})$. In particular, we will frequently use the fact that $\mathsf{Col}(\mathcal{X}) = 2^{-\mathbf{H}_2(\mathcal{X})} \leq 2^{-\mathbf{H}_\infty(\mathcal{X})}$.

STATISTICAL DISTANCE. Let $\mathcal{X}_1, \mathcal{X}_2$ be two probability distributions over the set S. The *statistical distance* between the distributions $\mathcal{X}_1$ and $\mathcal{X}_2$ is defined as $\mathbf{SD}(\mathcal{X}_1, \mathcal{X}_2) = \frac{1}{2}\sum_{s \in S} |\Pr_{\mathcal{X}_1}(s) - \Pr_{\mathcal{X}_2}(s)|$. If two distributions have statistical distance of (at most) ϵ, then we refer to them as ϵ-close. We note that ϵ-close distributions cannot be distinguished with probability better than ϵ even by a computationally unbounded adversary. It is also well known that if $\mathcal{Y}$ has support on some set S and U is the uniform distribution over this set, then

$$\mathbf{SD}(\mathcal{Y}, U) \leq \frac{1}{2}\sqrt{|S| \cdot \mathsf{Col}(\mathcal{Y}) - 1} \tag{1}$$

Definition 1. *Let k and ℓ be integers, and $\{h_\kappa\}_{\kappa \in \mathcal{K}}$ be a family of hash functions with domain $\{0,1\}^\ell$, range $\{0,1\}^k$ and key space $\mathcal{K}$. We say that the family $\{h_\kappa\}_{\kappa \in \mathcal{K}}$ is δ-almost universal (δ-AU) if for every pair of different inputs x, y from $\{0,1\}^\ell$ it holds that $\Pr(h_\kappa(x) = h_\kappa(y)) \leq \delta$, where the probability is taken over $\kappa \in_R \mathcal{K}$. For a given probability distribution $\mathcal{X}$ on $\{0,1\}^\ell$, we say that $\{h_\kappa\}_{\kappa \in \mathcal{K}}$ is δ-AU w.r.t. $\mathcal{X}$ if $\Pr(h_\kappa(x) = h_\kappa(y)) \leq \delta$ where the probability is taken over $\kappa \in_R \mathcal{K}$ and $x, y \in_R \mathcal{X}$ conditioned to $x \neq y$.*

Clearly, a family $\{h_\kappa\}_{\kappa \in \mathcal{K}}$ is δ-AU if it is δ-AU w.r.t. all distributions on $\{0,1\}^\ell$. The notion of universal hashing originates with the seminal papers by Carter and Wegman [CW79,WC81]; the δ-AU variant used here was first formulated in [Sti94]. The main usefulness of this notion comes from the following lemma whose proof is immediately obtained by conditioning on whether the two independent samples from $\mathcal{X}$ collide or not (below $\mathbf{E}$ denotes the expected value).

Lemma 1. *If $\{h_\kappa\}_{\kappa \in \mathcal{K}}$ is δ-AU w.r.t. $\mathcal{X}$, then $\mathbf{E}_\kappa[\mathsf{Col}(h_\kappa(\mathcal{X}))] \leq \mathsf{Col}(\mathcal{X}) + \delta$.*

Now, using the above lemma and Eq. (1), the lemma below extends the well-known "Leftover Hash Lemma" (LHL) from [HILL99] in two ways. First, it relaxes the pairwise-independence condition assumed by that lemma on the family of hash functions, and allows for "imperfect" families in which the collision probability is only ϵ-close to perfect (i.e., $2^{-k} + \epsilon$ instead of 2^{-k}). Second, it allows for the collision probability to depend on the input distribution rather than being an absolute property of the family of hash functions. We use these straightforward extensions of the LHL in an essential way for achieving our results. We also note that the standard LHL can be obtained from Lemma 2 below by setting $\epsilon = 0$.

Lemma 2. *Let ℓ and k be integers, let $\mathcal{X}$ be a probability distribution over $\{0,1\}^\ell$, and let $\{h_\kappa\}_{\kappa \in \mathcal{K}}$ be a family of hash function with domain $\{0,1\}^\ell$ and range $\{0,1\}^k$. If $\{h_\kappa\}_{\kappa \in \mathcal{K}}$ is $(\frac{1}{2^k} + \epsilon)$-almost universal w.r.t. $\mathcal{X}$, U is uniform over $\{0,1\}^k$ and κ is uniform over $\mathcal{K}$, then*

$$\mathbf{SD}((\kappa, h_\kappa(\mathcal{X})), (\kappa, U)) \leq \frac{1}{2} \cdot \sqrt{2^k \cdot (\mathsf{Col}(\mathcal{X}) + \epsilon)} \leq \frac{1}{2} \cdot \sqrt{2^k \cdot (2^{-\mathbf{H}_\infty(\mathcal{X})} + \epsilon)} \quad (2)$$

Remark 1. It is important to note that for the above lemma to be useful one needs $\epsilon \ll 1/2^k$, or otherwise the derived bound on the statistical closeness approaches 1. Moreover, this fact is not a result of a sub-optimal analysis but rather there are examples of families with $\epsilon = 1/2^k$ (i.e., $(2/2^k)$-AU families) that generate outputs that are easily distinguishable from uniform. For example, if $\{h_\kappa\}$ is a family of pairwise independent hash functions with k-bit outputs, and we define a new family $\{h'_\kappa\}$ which is identical to $\{h_\kappa\}$ except that it replaces the last bit of output with 0, then the new family has collision probability of $2/2^k$ yet its output (which has a fixed bit of output) is trivially distinguishable from uniform. The fact that we need $\epsilon \ll 1/2^k$ (say, $\epsilon \approx 1/2^{2k}$) makes the analysis of CBC and the cascade construction presented in the next sections non-trivial. In particular, the existing analyses of these functions (such as [BKR94,BCK96b]) are too weak for our purposes as they yield upper bounds on the collision probability of these constructions that are larger than $2/2^k$.

3   The CBC-MAC Construction

Here we study the suitability of the CBC-MAC mode as a randomness extractor. Recall that for a given permutation f on $\{0,1\}^k$, the CBC-MAC computation of f on an input $x = (x_1, x_2, \ldots x_L)$, with L blocks in $\{0,1\}^k$, is defined as $\bar{x}_L$ where the latter value is set by the recursion: $\bar{x}_0 = 0$, $\bar{x}_i = f(x_i \oplus \bar{x}_{i-1})$ for $1 \leq i \leq L$. We denote the output of the above CBC-MAC process by $F(x)$.

Our main result in this section states the extraction properties of CBC-MAC for a *random* permutation f on $K = 2^k$ elements. To state it more compactly, we let $\epsilon(L, K) = (d(L))^2 L K^{-2} + L^6 K^{-3}$, and notice that $\epsilon(L, K) = O(L^2/K^2)$ when $L < K^{1/4}$ (here we use the fact that $d(L) \leq 2\sqrt{L}$).

Theorem 1. *Let F denote the CBC-MAC mode over a* random *permutation f on $\{0,1\}^k$, and let $\mathcal{X}$ be an input distribution to F defined over L-block strings. Then the statistical distance between $F(\mathcal{X})$ and the uniform distribution on $\{0,1\}^k$ is at most*

$$\sqrt{K \cdot 2^{-\mathbf{H}_\infty(\mathcal{X})} + O(K \cdot \epsilon(L, K))}$$

In particular, assuming $L < 2^{k/4}$ and $\mathbf{H}_\infty(\mathcal{X}) \geq 2k$, the above statistical distance is at most $O(L/2^{k/2})$.

The proof of the theorem follows from Lemma 2 in combination with the following lemma that shows that CBC-MAC mode with a random permutation is δ-AU for sufficiently small δ.

Lemma 3. *Let F denote the CBC-MAC mode over a permutation f on $\{0,1\}^k$. For any $x, y \in \{0,1\}^{Lk}$, if $x \neq y$ then $\Pr[F(x) = F(y)] \leq \frac{1}{K} + O(\epsilon(L, K))$ where the probability is over the choice of a* random *permutation f.*

4   The Cascade Construction

We first recall the Merkle-Damgard approach to the design of cryptographic hash functions and introduce some notation and terminology. For given integers k and b, $b > k > 0$, let $\{f_\kappa : \kappa \in \mathcal{K}\}$ be a family of functions such that $\mathcal{K} = \{0,1\}^k$, and for all $\kappa \in \mathcal{K}$ the function f_κ maps b bits into k bits. On the basis of this family we build another family $\{F_\kappa : \kappa \in \mathcal{K}\}$ that works on inputs of any length which is a multiple of b and produces a k-bit output. For each $\kappa \in \mathcal{K}$, the function F_κ is defined as follows. Let $x = (x_1, \ldots, x_L)$, for some $L \geq 1$ and $x_i \in \{0,1\}^b$ (for all i), denote the input to F_κ; we define L variables (each of length k)

$\bar{x}_1, \ldots, \bar{x}_L$ as $\bar{x}_0 = \kappa, \bar{x}_{i+1} = f_{\bar{x}_i}(x_{i+1})$, and set $F_\kappa(x) = \bar{x}_L$. For processing inputs of arbitrary length one needs a rule for padding inputs to a length that is a multiple of b. Specific functions that build on the above approach, such as MD5 and SHA-1, define their specific padding; we expand on this issue in Section 4.2. For the moment we assume inputs of length Lb for some L. Some more notation: Sometimes we use F to denote the family $\{F_\kappa\}_{\kappa \in \mathcal{K}}$, and we write $F(x)$ to denote the random variable $F_\kappa(x)$ for $\kappa \in_R \mathcal{K}$. Finally, we use K to denote 2^k.

The family $\{f_\kappa\}_{\kappa \in \mathcal{K}}$ is called the "compression function(s)" and the family $\{F_\kappa\}_{\kappa \in \mathcal{K}}$ is referred to as the "cascade construction" (over the compression function $\{f_\kappa\}_{\kappa \in \mathcal{K}}$). Typical values for the parameters of the compression function are $b = 512$ and $k \in \{128, 160\}$.

4.1   The Basic Cascade Construction

The main result of this section is the following. Assume $\mathcal{X}$ is an input distribution with $2k$ bits of (overall) min-entropy and "enough" bits of min-entropy in its last (b-bit) block ("enough" will be quantified below). For the cascade constructions we model the underlying family of compression functions as a family of random functions (with k-bit outputs). Then the output of F on the distribution $\mathcal{X}$ is statistically close to uniform. This result is formalized in the following theorem. As in Section 3, we let $\epsilon(L, K) = (d(L))^2 LK^{-2} + L^6 K^{-3}$, and notice that $\epsilon(L, K) = O(L^2/K^2)$ when $L < K^{1/4}$.

Theorem 2. *Let $F = \{F_\kappa\}$ be the cascade construction defined, as above, over a family of random functions $\{f_\kappa\}$. Let $\mathcal{X}$ be the input distribution to F defined over L-block strings, and $\mathcal{X}_L$ denote the probability distribution induced by $\mathcal{X}$ on the last block x_L for $x \in_R \mathcal{X}$. Then, if U is the uniform distribution over $\{0,1\}^k$, we have*

$$\mathbf{SD}(F(\mathcal{X}), U) \leq \sqrt{K \cdot 2^{-\mathbf{H}_\infty(\mathcal{X})} + L \cdot 2^{-\mathbf{H}_\infty(\mathcal{X}_L)} + O(K \cdot \epsilon(L, K))} \qquad (3)$$

In particular, if $\mathbf{H}_\infty(\mathcal{X}) \geq 2k$, $\mathbf{H}_\infty(\mathcal{X}_L) \geq k$, and $L \leq 2^{k/4}$, then $\mathbf{SD}(F(\mathcal{X}), U) \leq O(L/2^{k/2})$.

The proof of the theorem follows from Lemma 2 in combination with the following lemma that shows that the cascade construction with a random family of compression functions is δ-AU for sufficiently small δ.

Lemma 4. *Let $F = \{F_\kappa\}$ be the cascade construction defined over a family of random functions $\{f_\kappa\}$. Let $\mathcal{X}$ be an input distribution as assumed in Theorem 2, where $\mathbf{H}_\infty(\mathcal{X}_L) > \log L$. Then, the family F is $(\frac{1}{K} + \frac{L}{K 2^{\mathbf{H}_\infty(\mathcal{X}_L)}} + O(\epsilon(L, K)))$-AU w.r.t. $\mathcal{X}$.*

The proof of this lemma is based on the following two propositions: the first analyzes the collision probability of the cascade function F over random compression functions on inputs that differ (at least) in the last block (Proposition 1);

then we extend the analysis to the general case, i.e. for any two different x and y (Proposition 2). All proofs appear in the final paper.

Proposition 1. *Let $F = \{F_\kappa\}$ be the cascade construction defined over a family of random functions $\{f_\kappa\}$. Let x, y be two inputs to F that differ (at least) in the last block, namely, $x_L \neq y_L$, and let $\kappa \in \mathcal{K}$ be any value of the initial key. Then $\mathrm{Pr}_F(F_\kappa(x) = F_\kappa(y)) \leq \frac{1}{K} + O(\epsilon(L, K))$, where the probability is taken over the choice of random functions in F.*

Proposition 2. *Let F be defined as above, let x, y be two different inputs to F, and let κ be any value of the initial key. Then $\mathrm{Pr}_F(F(x) = F(y)) \leq \frac{L}{K} + O(L\epsilon(L, K))$, where the probability is taken over the choice of random functions in F.*

THE VALUE OF THE INITIAL KEY κ. We note that the above analysis holds for any value of the initial key κ, when the functions are truly random, which means that in principle κ can be fixed to a constant. However, in practice not all the functions of the function family $\{f_\kappa\}$ satisfy this requirement. Thus, choosing κ at random allows, for example, to extend our analysis to the situation where a negligible fraction of functions in F are not close to random functions.

NECESSITY OF MIN-ENTROPY IN THE LAST BLOCK. We argue that assuming non-trivial min-entropy in the last block is required, even if the family $\{f_\kappa\}$ of compression functions is completely random. Assume an input distribution in which the last block is fixed to a value B. The first $L-1$ blocks induce some distribution for the last key in the sequence. Examining the distribution on $f_\kappa(B)$, induced by (any) distribution on κ it is easy to see that this distribution is statistically far from the uniform distribution. In particular, we expected with high probability a constant fraction of elements will not appear in the distribution.

4.2   The Cascade Construction with Input Padding

The conditions imposed by Theorem 2 on the input distribution $\mathcal{X}$ conflict with a technical detail of the practical implementations of the cascade paradigm (such as MD5 and SHA-1): rather than applying the cascade process to the input $x \in_R \mathcal{X}$, these functions modify x by concatenating enough padding bits as to obtain a new input x' whose length is a full multiple of the block length. In some cases this padding results in a full *fixed* block appended to x. Therefore, even if $\mathcal{X}$ has the property that the last block of input has relatively high entropy (as required by Theorem 2) the actual input x' to the cascade does not have this property any more. This fact is sufficient to make our main result from Section 4.1 irrelevant to these real-world functions; luckily, however, we show here that this serious obstacle can be lifted.

In order to better understand this problem we first describe the actual way this padding is performed. We consider the concrete value $b = 512$ used as the block length in these functions. Let n denote the length of the input x, and let

$n' = n \bmod 512$. If $n' < 448$ then x is padded with the binary string $10^{447-n'}$ followed by a 64-bit representation of n. If $n' \geq 448$ then *a whole new block* is added with a padding similar to the one described above (with the binary representation of n occupying the last 64 bits of the added block). From this description, we see that if, for example, the original input x had a length which was an exact multiple of 512, then $x' = x \parallel B$ where B is a whole new block (and $\parallel$ represents the concatenation operation). Moreover, the value of B is the same (and fixed) for all inputs of the length of x. In particular, if we consider the case in which we hash Diffie-Hellman values of length of 1024 or 2048 bits (this is the common case when working over Z_p^* groups), then we get that the padded input x' will always have a fixed last block. In other words, regardless of the entropy existing in the original input x the actual input to the cascade process now has a zero-entropy last block.

For this case, in which the last block is fixed, we show here that a somewhat weaker (but still very useful) result holds. Specifically, combining Theorem 2 with the assumption that the family of (compression) functions $\{f_\kappa\}$ is pseudorandom, we can prove that the output from the cascade construction is *pseudorandom*, i.e., computationally indistinguishable from uniform (and thus sufficient for most cryptographic applications) This result holds even though the key to the cascade function F is revealed! We note that the assumption that the family $\{f_\kappa\}$ is pseudorandom is clearly implied by the modeling (from the previous subsection) of these functions as random. But we also note that assuming the compression function (family) of actual schemes such as MD5 or SHA-1 to be pseudorandom is a standard and widely-used cryptographic assumption (see [BCK96b] for some analytical results on these PRF constructions).

Lemma 5. *Let $\{f_\kappa\}_{\kappa \in \mathcal{K}}$ be a family of pseudorandom functions which is $\epsilon_p(T)$-indistinguishable from random for attackers restricted to time T and a single query. Let $F = \{F_\kappa\}_{\kappa \in \mathcal{K}}$ denote the cascade construction over the family $\{f_\kappa\}_{\kappa \in \mathcal{K}}$. Further, let $\mathcal{X}$ be a probability distribution on L-block strings from which the inputs to F are chosen, and B be a fixed b-bit block. If the output distribution $F_\kappa(\mathcal{X})$, with random but known key κ, is ϵ_d-statistically close to uniform, then the distribution $F_\kappa(\mathcal{X} \parallel B)$ (for random but known κ) is $(\epsilon_d + \epsilon_p(T))$-indistinguishable from uniform by attackers that run time T.*

The above lemma together with Theorem 2 show that if $\{f_\kappa\}$ is a family of random functions then the cascade construction with a fixed last block block is indistinguishable from random, provided that the original input distribution (before padding!) satisfies the conditions of Theorem 2.

It is also worth noting that Lemma 5 can be generalized to input distributions $\mathcal{X}^*$ that can be described as the concatenation of two probability distributions $\mathcal{X}$ and $\mathcal{Y}$, where $\mathcal{X}$ satisfies the conditions of Theorem 2, and $\mathcal{Y}$ is an arbitrary (polynomial-time samplable) distribution *independent* from $\mathcal{X}$.

A PRACTICAL CONSIDERATION. Note that the application of Lemma 5 on an input distribution $\mathcal{X}$ still requires the last block of $\mathcal{X}$ (before padding) to have relatively high min-entropy. To maximize this last-block min-entropy it is advisable that any input x whose length is not a multiple of $b = 512$ be "shifted to the right" (to a block boundary) by prepending a sufficient number of bits (say, all zeros) to the beginning of x. This way, the resultant string x' is of length a multiple of b and, more importantly, its last block contains the full entropy of the last b bits in x. Also, this shifting forces the appended padding described earlier to add a full block as assumed in Lemma 5[4].

4.3   Modeling the Compression Function as a δ-AU Family

In Section 4.1 we presented an analysis of the basic cascade construction under the modeling of the compression function as a family of random functions. Here we study the question of what can be guaranteed on the output distribution of the cascade under the simple assumption that *the family of compression functions is a good extractor* (or more generally that this family is δ-AU). Clearly this is a more realistic assumption on the underlying compression function. On the other hand, in order to prove a close-to-uniform output in this case we are going to require a stronger assumption on the input distribution. Specifically, we are going to assume that the distribution on every block of input has a high min-entropy (e.g., $2k$ bits of min-entropy out of the b bits in the block), conditioned on the distribution of the previous blocks. We prove below that under these conditions the output of the cascade function is statistically close to uniform.

We note that the above requirement on the input distribution, while stringent, is met in some important cases, such as applications that extract keys from a Diffie-Hellman value computed over a high-entropy group. In particular, this requirement is satisfied by the DH groups in use with the IKE protocol.

CONDITIONAL ENTROPY. Let $\mathcal{X}$ and $\mathcal{Y}$ be two probability distributions over $\{0,1\}^a$ and $\{0,1\}^b$ respectively. If $y \in \{0,1\}^b$ we denote with $\mathcal{X}|y$ the distribution $\mathcal{X}$ conditioned to the event that the string y is selected according to $\mathcal{Y}$. Then we can define the conditional min-entropy of $\mathcal{X}|y$ (and denote it as $\mathbf{H}_\infty(\mathcal{X}|y)$) as the minimum integer m such that for all $x \in \{0,1\}^a$, $\mathrm{Pr}_{\mathcal{X}|y}(x) \leq 2^{-m}$.

We define the conditional min-entropy of $\mathcal{X}$ with respect to $\mathcal{Y}$ as the expectation over $\mathcal{Y}$ of $\mathbf{H}_\infty(\mathcal{X}|y)$: $\mathbf{H}_\infty(\mathcal{X}|\mathcal{Y}) = \sum_{y \in \{0,1\}^b} \mathrm{Pr}_{\mathcal{Y}}(y) \cdot \mathbf{H}_\infty(\mathcal{X}|y)$.

Lemma 6. *Assume that the family of compression functions $\{f_\kappa\}_{\kappa \in \mathcal{K}}$ from b to k bits has the property that for any probability distribution $\mathcal{B}$ defined over $\{0,1\}^b$ with min-entropy of m, the output distribution $f_\kappa(\mathcal{B})$, for $\kappa \in_R \mathcal{K}$ and $\mathcal{B} \in_R \mathcal{B}$, is ϵ-close to uniform (for some given $\epsilon = \epsilon(b,k,m)$). Further, assume that $\mathcal{X}$ is an input distribution on L b-bit blocks with the property that for each $i = 1,\ldots,L$,*

---

[4] For example, assume the inputs from the distribution $\mathcal{X}$ to be of length 1800 bits. Given such an input x we prepend to it 248 '0's resulting in a 4-block string $x' = 0^{248} \parallel x$. Now when this input is processed by, say, SHA-1 an additional fifth block is added to x'. The important thing is that the last block of x' receives as much entropy from the last 512 bits of x as possible.

the distribution $\mathcal{X}_i$ induced by $\mathcal{X}$ on the i-th block has conditional min-entropy m with respect to the distribution induced by $\mathcal{X}$ on blocks $1, \ldots, i-1$, and that $\mathcal{X}_i$ Then, the cascade construction over the family $\{f_\kappa\}$ applied to the distribution $\mathcal{X}$ is $(L \cdot \epsilon)$-close to uniform.

In particular, if we assume that the family $\{f_\kappa\}$ is $(2^{-k} + 2^{-2k})$-AU and the min-entropy of each input block (as defined above) is at least $m = 2k$, then we get (using Lemma 2) a statistical distance between the cascade construction on L blocks and the uniform distribution on $\{0,1\}^k$ of at most $L2^{-k/2}$.

Combining Lemmas 6 and 5 we get that, under the above assumption on the input distribution, if the family of compression functions is both δ-AU and pseudorandom then the output of the padded cascade (see Section 4.2) is pseudorandom (i.e. indistinguishable from uniform).

5 HMAC Construction

We now turn to the study of HMAC [BCK96a] as a randomness extraction family. HMAC (and its underlying family NMAC) is defined using the cascade construction over a family of compression functions $\{f_\kappa\}_{\kappa \in \mathcal{K}}$ with domain $\{0,1\}^b$, range $\{0,1\}^k$ and $\mathcal{K} = \{0,1\}^k$ (as usual we denote $K = 2^k$). The family of functions NMAC uses two independent keys drawn from $\mathcal{K}$ and is defined over $\{0,1\}^*$ as $\mathrm{NMAC}_{\kappa_1,\kappa_2}(x) = f_{\kappa_2}(F_{\kappa_1}(x))$ where both x and the result from $F_{\kappa_1}(x)$ are padded as described in Section 4.2. On the basis of NMAC one defines the family HMAC as $\mathrm{HMAC}_{\kappa_1,\kappa_2}(x) = \mathrm{NMAC}_{\kappa'_1,\kappa'_2}(x)$ where $\kappa'_1 = f_{iv}(\kappa_1 \oplus pad_1)$ and $\kappa'_2 = f_{iv}(\kappa_2 \oplus pad_2)$, the value iv is fixed to the IV defined by the underlying hash function, and pad_1, pad_2 are two different fixed strings of length b. The analysis of HMAC is based on that of NMAC under the specialized assumption that the keys κ'_1 and κ'_2 are "essentially independent". We keep this assumption and develop our analysis on NMAC. (The reason of this form of derivation of the keys κ'_1, κ'_2 in HMAC is to allow for the use, without modification, of the underlying hash function; in particular, without having to replace the fixed IV with a variable value.)

We start by observing that if one considers the family $\mathrm{NMAC}_{\kappa_1,\kappa_2}$ as a δ-AU family then we get $\delta > 2/K$. This is so since for any two inputs x, y the probability that NMAC sends both values to the same output is the sum of the probability that $F_{\kappa_1}(x) = F_{\kappa_1}(y)$ (which is at least $1/K$) plus the probability that $F_{\kappa_1}(x) \neq F_{\kappa_1}(y)$ but f_{κ_2} maps these two different results to the same value (which is also at least $1/K$). Therefore we cannot apply the results of Section 2 directly to the analysis of NMAC.

However, we provide three analyses, which, under different assumptions, establish the security of NMAC as a randomness extractor.

DROPPING SOME OUTPUT BITS. Specifically, we assume the "outer" function f_{κ_2} outputs $k' = k - c$ bits (e.g., in case f_{κ_2} is a random function outputting k bits, one can simply drop the last c bits and view it as a random function outputting k' bits). In this case, a straightforward application of Lemma 1 and Lemma 2 shows that if the family $\{F_{\kappa_1}\}$ is δ_1-AU w.r.t. $\mathcal{X}$ and $\{f_{\kappa_2}\}$ is $(1/2^{k'} + \delta_2)$-AU, then NMAC extracts k' bits of statistical distance at most $\sqrt{2^{k'}(\mathsf{Col}(\mathcal{X}) + \delta_1 + \delta_2)}$ from uniform. Assuming now that both families consist of random functions, then $\delta_2 = 0$ and Proposition 2 implies that $\delta_1 \leq L/K + O(L\epsilon(L, K))$. This means that if $\mathbf{H}_\infty(\mathcal{X}) \geq k$, $k' < k - \log L - 2\log(1/\gamma)$ and $L < 2^{k/4}$, then NMAC extracts k' bits which are γ-close to uniform. In fact, the same is true even if the outer function family $\{f_{\kappa_2}\}$ is merely a good extractor (e.g., if it is pairwise independent). In any case, we get that dropping roughly $(\log L + 160)$ bits from the output of the NMAC construction makes it a good extractor. To make this result meaningful, however, we must consider compression functions whose key is non-trivially larger than 160 bits, such as the compression function for SHA2-512.

COMPUTATIONAL SECURITY. Our second approach to analyzing NMAC is similar to the analysis of the padded cascade from Lemma 5. We will present it in the full version.

MODELING f_{κ_2} AS A RANDOM ORACLE. As we remarked, even if f_{κ_2} is truly random, the value $f_{\kappa_2}(F_{\kappa_1}(\mathcal{X}))$ cannot be statistically close to uniform, even if $F_{\kappa_1}(\mathcal{X})$ was perfectly uniform. This was argued under an extremely strong distinguisher that can evaluate $f = f_{\kappa_2}$ at all of its 2^k inputs. This is different from the typical modeling of f as a random *oracle*. Namely, in the random oracle model it is assumed that the adversary can evaluate f at most a bounded number of points, $q \ll 2^k$. This assumption can be seen as restrictive, but in fact a realistic characterization of the adversary's capabilities. Thus, we show that when we model the outer function of NMAC, f, as a random oracle then the construction is a good randomness extractor. We start by showing the general result about the quality of using a random oracle f as an extractor, and then apply it to the NMAC construction.

5.1 Random Oracle as an Extractor

In this section, we show that by utilizing the power of the random oracle to the fullest, we can provide some provable guarantees on the quality of the random oracle as a randomness extractor. Our precise modeling of the random oracle $f : \{0,1\}^b \to \{0,1\}^k$ is the following. The adversary is allowed to adaptively query the random oracle f at upto q points, and possibly make the distribution $\mathcal{X}$ depend on these q queries. However, we assume that the remaining "unqueried" $(2^b - q)$ values of f are chosen randomly and *independently* of $\mathcal{X}$, and are never given to the adversary[5]. Finally, given a distribution $\mathcal{X}$ and a number q, we let $W_q(\mathcal{X})$ denote the probability mass of the q heaviest elements under $\mathcal{X}$.

---

[5] We stress that this is very different from our modeling of a random *function* from before, where the adversary first chooses the distribution $\mathcal{X}$, after which f is chosen at random (independently from $\mathcal{X}$) and given to the adversary *in its entirety*.

Lemma 7. *Assume f is a random oracle from b bits to k bits, and the adversary can evaluate f in at most q points. Then, for any distribution $\mathcal{X}$ on $\{0,1\}^b$, the maximal probability the adversary can distinguish $f(\mathcal{X})$ from the uniform distribution over $\{0,1\}^k$ is at most*

$$W_q(\mathcal{X}) \leq \min\left(q \cdot 2^{-\mathbf{H}_\infty(\mathcal{X})}, \sqrt{q \cdot \mathsf{Col}(\mathcal{X})} \right) \tag{4}$$

Remark 2. We already remarked that no single function can be a universally good extractor, which means that one has to use a function family instead, indexed by some key κ. On the other hand, in the idealized random oracle model, we manage to use a single random oracle f in Lemma 7. This is not a contradiction since we critically assumed that the adversary cannot read the entire description of the random oracle. In essence, the choice of the random oracle f can be viewed as a key κ, but the distinguisher cannot read the entire key (although it has a choice which parts of it to read) and therefore cannot adversarially choose a bad distribution $\mathcal{X}$. Put differently, in our analysis we could assume that a large part of the key (i.e., f) is chosen independently of $\mathcal{X}$, which is consistent with the conventional extractors such as those obtained by the LHL. However, unlike the conventional extractors, we (restrictively) assume that the adversary never learns the entire description of the key (i.e., the unqueried parts of f), which allowed us to get a much stronger bound that what we could get with the LHL. For example, LHL required $\mathsf{Col}(\mathcal{X}) \ll 2^{-k}$, while Lemma 7 only requires $\mathsf{Col}(\mathcal{X}) \ll 1/q$.

We will also use the following Corollary of Eq. (4) and Lemma 1.

Corollary 1. *If a family of functions $\{h_\kappa\}_{\kappa \in \mathcal{K}}$ is δ-almost universal w.r.t. $\mathcal{X}$, f is a random oracle, U is the uniform distribution of $\{0,1\}^k$, and the adversary can make at most q queries to the random oracle, then the maximal probability the adversary can distinguish the pair $(\kappa, f(h_k(\mathcal{X})))$ from (κ, U) is at most $\sqrt{q \cdot (\mathsf{Col}(\mathcal{X}) + \delta)}$.*

The above corollary implies that the composition of a collision-resistant hash function and a random oracle could be viewed as a relatively good extractor. This is because a (computational) collision-resistant function must be (information-theoretically) almost universal. More precisely, if a function family $\mathcal{H} = \{h_\kappa\}$ is collision-resistant with exact security δ against non-uniform adversaries running in linear time, it must also be δ-almost universal. For uniform adversaries running in time T, $\mathcal{H}$ must be δ-almost universal w.r.t. any $\mathcal{X}$ which is samplable in time $T/2$.

APPLICATION TO NMAC. We can now apply Corollary 1 to the case of NMAC assuming that the outer function $f = f_{\kappa_2}$ is a random oracle which can be evaluated in at most q places. By Proposition 2, the family $\{F_{\kappa_1}\}$ is δ-AU when the function family $\{f_{\kappa_1}\}$ is chosen at random, for $\delta < (L+1)/2^k$ (when $L < 2^{k/4}$). Thus, Corollary 1 implies that NMAC extracts k bits which cannot be distinguished from random with probability more than $\sqrt{q(\mathsf{Col}(\mathcal{X}) + (L+1)2^{-k})}$, which is negligible if $q \ll \min[2^{\mathbf{H}_2(\mathcal{X})}, 2^k/(L+1)]$.

Acknowledgments

We thank David Wagner for valuable discussions.

References

[BST03]   B. Barak, R. Shaltiel, and E. Tromer. True Random Number Generators Secure in a Changing Environment. In *CHESS '03*, pages 166–180, 2003.

[BCK96a]   M. Bellare, R. Canetti, and H. Krawczyk. Keying hash functions for message authentication. In *Crypto '96*, pages 1–15, 1996. LNCS No. 1109.

[BCK96b]   M. Bellare, R. Canetti, and H. Krawczyk. Pseudorandom Functions Revisited: The Cascade Construction and Its Concrete Security. In *Proc. 37th FOCS*, pages 514–523. IEEE, 1996.

[BKR94]   M. Bellare, J. Kilian, and P. Rogaway. The Security of Cipher Block Chaining. In *Crypto '94*, pages 341–358, 1994. LNCS No. 839.

[CW79]   L. Carter and M. N. Wegman. Universal Classes of Hash Functions. *JCSS*, 18(2):143–154, April 1979.

[Dam89]   I. Damgard. A Design Principle for Hash Functions. In *Crypto '89*, pages 416–427, 1989. LNCS No. 435.

[GKR04]   R. Gennaro, H. Krawczyk, and T. Rabin. Secure Hashed Diffie-Hellman over Non-DDH Groups. In *Eurocrypt '04*, 2004. LNCS No.

[Gol01]   Oded Goldreich. *Foundations of Cryptography*. Cambridge University Press, 2001. Preliminary version *http://philby.ucsd.edu/cryptolib.html/*.

[HILL99]   J. Håstad, R. Impagliazzo, L. Levin, and M. Luby. Construction of a Pseudo-random Generator from any One-way Function. *SIAM. J. Computing*, 28(4):1364–1396, 1999.

[IKEv2]   IKEv2. Internet Key Exchange (IKEv2) Protocol, draft-ietf-ipsec-ikev2-13.txt, (to be published as an RFC). Editor: C. Kaufman, March 2004.

[Kra03]   H. Krawczyk. SIGMA: The 'SIGn-and-MAc' Approach to Authenticated Diffie-Hellman and Its Use in the IKE Protocols. In *Crypto '03*, pages 400–425, 2003. LNCS No. 2729.

[Lub96]   Michael Luby. *Pseudorandomness and Cryptographic Applications*. Princeton Computer Science Note, Princeton University Press, January 1996.

[RFC2409]   RFC2409. The Internet Key Exchange (IKE). Editors: D. Harkins and D. Carrel, Nov 1998.

[Sha02]   R. Shaltiel. Recent developments in Extractors. Bulletin of the European Association for Theoretical Computer Science, Volume 77, June 2002, pages 67-95. Available at:
            http://www.wisdom.weizmann.ac.il/~ronens/papers/survey.ps

[Sti94]   D. R. Stinson. Universal Hashing and Authentication Codes. Designs, Codes and Cryptography 4 (1994), 369-380.

[WC81]   M.N. Wegman and L. Carter. New Hash Functions and Their Use in Authentication and Set Equality. *JCSS*, 22(3):265–279, July 1981.

Efficient Tree-Based Revocation
in Groups of Low-State Devices

Michael T. Goodrich[1,*], Jonathan Z. Sun[1,*], and Roberto Tamassia[2,**]

[1] Dept. of Computer Science, Univ. of California, Irvine, CA 92697-3425
{goodrich,zhengsun}@ics.uci.edu
[2] Dept. of Computer Science, Brown Univ., Providence, RI 02912
rt@cs.brown.edu

Abstract. We study the problem of broadcasting confidential information to a collection of n devices while providing the ability to revoke an arbitrary subset of those devices (and tolerating collusion among the revoked devices). In this paper, we restrict our attention to low-memory devices, that is, devices that can store at most $O(\log n)$ keys. We consider solutions for both zero-state and low-state cases, where such devices are organized in a tree structure T. We allow the group controller to encrypt broadcasts to any subtree of T, even if the tree is based on an multi-way organizational chart or a severely unbalanced multicast tree.

1 Introduction

In the group broadcast problem, we have a group S of n devices and a *group controller* (GC) that periodically broadcasts messages to all the devices over an insecure channel [8]. Such broadcast messages are encrypted so that only valid devices can decrypt them. For example, the messages could be important instructions from headquarters being sent to PDAs carried by employees in a large corporation. We would like to provide for *revocation*, that is, for an arbitrary subset $R \subset S$, we would like to prevent any device in R from decrypting the messages.

We are interested in schemes that work efficiently with low-memory devices, that is, devices that can store at most $O(\log n)$ secret keys. Such a scenario models the likely situation where the devices are small and the secure firmware dedicated to storing keys is smaller still. We refer to this as the *log-key* restriction. We consider two variants of this model.

- A *static* or *zero-state* version: the $O(\log n)$ keys on each device cannot be changed once the device is deployed. For example, memory for the devices could be written into secure firmware at deployment.

* This work was supported in part by NSF Grants CCR-0312760, CCR-0311720, CCR-0225642, and CCR-0098068.
** This work was supported in part by NSF Grants CCR-0311510, CCR-0098068, and IIS-0324846 and by a research gift from Sun Microsystems.

M. Franklin (Ed.): CRYPTO 2004, LNCS 3152, pp. 511–527, 2004.

 – The *dynamic* or *low-state* version: any of the $O(\log n)$ keys on each device can be updated in response to broadcast messages. For example, such devices might have a small tamper-resistant secure cache in which to store and update secret keys.

Organizing Devices Using Trees. The schemes we consider organize the set of n devices in a tree structure, associating each device with a different leaf in the tree. In fact, we consider three possible kinds of trees that the devices can conceptually be organized into.

1. A *balanced d-ary tree*. In this case, the devices are associated with the leaves of a balanced tree where each internal node has a constant d number of children; hence, each is at depth $O(\log n)$. This tree is usually chosen purely for the sake of efficiency, and, in fact, has been the only tree considered in previous related work we are familiar with. For example, it forms the basis of the Logical Key Hierarchy (LKH) scheme [26, 28], the One-way Function Tree (OFT) scheme [21], the Subset-Difference Revocation (SDR) scheme [16], and the Layered Subset Difference (LSD) scheme [10].
2. An *organizational chart*. In this case, the devices are associated with the leaves of a tree that represents an organizational chart, such as that of a corporation or university. For example, internal nodes could correspond to campuses, colleges, and departments. The height of this tree is assumed to be $O(\log n)$ but the number of children of an internal is not assumed to be bounded by a constant. Thus, the straightforward conversion of this tree into an equivalent bounded-degree tree may cause the height to become $\Omega(\log^2 n)$.
3. A *multicast tree*. In this case, the devices are associated with the nodes of a multicast tree rooted at the group controller. The logical structure of this tree could be determined in an ad hoc manner so that no bound is assumed on either the tree height or the degree of internal nodes. Thus, this tree may be quite imbalanced and could in fact have height that is exponentially greater than the number of keys each device can hold.

In using trees, particularly in the latter two cases, we feel it is important to provide the capability to the group controller of encrypting a message so that it may be decrypted only by the devices associated with nodes in a certain subtree. For instance, a sporting event might be broadcast to just a single region, or a directive from headquarters might be intended just for a single division. We call such a broadcast a *subtree broadcast*, which can also be modeled by multiple GCs, each assigned to a different subtree. We continue in this case to assume the network transmits a message to the entire group, even the revoked devices, but it should only be readable by the (unrevoked) devices in the specified subtree when the message is sent in a subtree broadcast. The motivation for organizing devices into trees and allowing for subtree broadcasts is derived from the way many organizations are naturally structured. For example, the ICS Company may have several departments divided into groups, and groups may in turn have divisions located in different cities.

After a secure broadcast system is set up, we need to have the ability to revoke devices to avoid revealing messages beyond the current members. (We also consider the complexities of adding new devices, but the need for revocation is better motivated, since additions will typically be done in large blocks.) Thus, we are interested in the following complexity measures for a set of n devices.

- *Broadcast cost*: the number of messages the group controller (GC) must send in order to reach a subtree containing r revoked devices.
- *Revocation cost*: the number of messages the GC must send in order to revoke a device. Note that this cost is zero in the zero-state case.
- *Insertion cost*: the number of messages the GC must send in order to add a device. Note that this cost parameter does not apply to the zero-state case.

Related Work. Broadcast/multicast encryption was first formally studied by Fiat and Naor [8], for the model where all the device keys are dynamic. Their algorithms satisfy the log-key restriction, however, only if no more than a constant number of revoked devices collude, which is probably not a realistic assumption. Several subsequent approaches have therefore strengthened the collusion resistance for broadcast encryption, and have done so using approaches where the group is represented by a fixed-degree tree with the group controller (GC) being the root and devices (users) being associated with leaves [3–7, 11, 13–15, 23, 24, 26, 28].

Of particular note is the *logical key hierarchy* (LKH) scheme proposed by Wallner *et al.* [26] and by Wong and Lam [28], which achieves $O(1)$ broadcast cost and $O(\log n)$ revocation cost under the log-key restriction (for the dynamic case). The main idea of the LKH scheme is to associate devices with the leaves of a complete binary tree, assign unique secret keys to each node in this tree, and store at each device x the keys stored in the path from x's leaf to the root. Some improvements of this scheme within the same asymptotic bounds are given by Canetti *et al.* [4, 5]. Using Boolean function minimization techniques, Chang *et al.* [6] deal with cumulative multi-user revocations and reduces the space complexity of the GC, i.e., the number of keys stored at the GC, from $O(n)$ to $O(\log n)$. Wong *et al.* [27] generalize the results from binary trees to key graphs. In addition, Sherman and McGrew [21] improve the constant factors of the LKH scheme using a technique they call one-way function trees (OFT), to reduce the size of revocation messages. Naor and Pinkas [17] and Kumar *et al.* [12] also study multi-user revocations withstanding coalitions of colluding users, and Pinkas [18] studies how to restore an off-line user who has missed a sequence of t group modifications with $O(\log t)$ message size. Also of note is work of Rodeh *et al.* [19], who describe how to use AVL trees to keep the LKH tree balanced. Thus, the broadcast encryption problem is well-studied for the case of fully-dynamic keys and devices organized in a complete or balanced k-ary tree (noticing that a k-ary tree can transform to binary with constant times of height increasing). We are not familiar with any previous work that deals with unbalanced trees whose structure must be maintained for the sake of subtree broadcasts, however.

There has also been some interesting recent work on broadcast encryption for zero-state devices (the static case). To begin, we note that several researchers have observed (e.g., see [10]) that the LKH approach can be used in the zero-state model under the log-key restriction to achieve $O(r \log(n/r))$ broadcast cost. (We will review the LKH approach in more detail in the next section.) Naor, Naor, and Lotspiech [16] introduce an alternative approach to LKH, which they call the *subset-difference revocation* (SDR) approach. They show that if devices are allowed to store $O(\log^2 n)$ static keys, then the group controller can send out secure broadcasts using $O(r)$ messages, i.e., the broadcast cost of their approach is $O(r)$. Halevy and Shamir [10] improve the performance of the SDR scheme, using an approach they call layered subset difference (LSD). They show how to reduce the number of keys per device to be $O(\log^{1+\epsilon} n)$ while keeping the broadcast cost $O(r)$. They also show how to further extend their approach to reduce the number of keys per device to be $O(\log n \log \log n)$ while increasing the broadcast cost to be $O(r \log \log n)$. These latter results are obtained using a super-logarithmic number of device keys; hence, they violate the log-key restriction.

Our Results. We provide several new techniques for broadcast encryption under the log-key restriction. We study both the static (zero-state) and dynamic (low-state) versions of this model, and present efficient broadcast encryption schemes for devices organized in tree structures. We study new solutions for balanced trees, organizational charts, and multicast trees. We show in Table 1 the best bounds on the broadcast, insertion and revocation cost for each of the possible combinations of state and tree structure we consider, under the log-key restriction.

Table 1. Best bounds for broadcast encryption among n devices under the log-key restriction, where each device can store only $O(\log n)$ keys.

		Balanced Tree	Org. Chart	Multicast Tree
static	broadcast cost	$O(r)$	$O(r)$	$O(r \log n)$
(zero-state)		(new)	(new)	(new)
dynamic	broadcast cost	$O(1)$	$O(1)$	$O(\log n)$
(low state)	revocation cost	$O(\log n)$	$O(\log n)$	$O(\log n)$
	insertion cost	$O(\log n)$	$O(\log n)$	$O(\log n)$
		LKH [19, 26, 28]	(new)	(new)

So, for example, we are able to match the log-key bound of the static LKH scheme while also achieving the $O(r)$ broadcast encryption complexity of the SDR scheme. Indeed, our scheme for this case, which we call the *stratified subset difference* (SSD) scheme, is the first scheme we are aware of for zero-state devices that simultaneously achieves both of these bounds. Moreover, we are able to match the best bounds for balanced trees, even for unbalanced high-degree organizational charts, which would not be possible using the natural conversion to a binary tree. Instead, we use biased trees [1] to do this conversion. But this approach is nevertheless limited, under the log-key restriction, to cases where the organizational chart has logarithmic height. Thus, for multicast trees, which can be very unbalanced (we even allow for height that is $O(n)$), we must take a different approach. In particular, in these cases, we extend the linking and cutting dynamic trees of Sleator and Tarjan [22] to the context of broadcast encryption, showing how to do subtree broadcasts in this novel context. This implies some surprisingly efficient performance bounds for broadcast encryption in multicast trees, for in severely unbalanced multicast trees the number of ancestors of the leaf associated with some device can be exponentially greater than the number of keys that device is allowed to store.

2   Preliminaries

The LKH Scheme for a Single Group. Let us briefly review the LKH scheme [26, 28], which is well known for key management in single groups. The LKH scheme organizes a group of n devices as a complete binary tree with the GC represented by the root and each user (that is, device) by a leaf, with a key stored at each node. Each device, as a leaf, knows the path from the root to itself and all the keys on this path. The GC, as the root, knows the whole tree and all the keys. (See Figure 1.)

To revoke a device x, the GC updates every key on the path from itself to x so that: (a) x cannot receive any updated key; and (b) any device other than x can receive an updated key if and only if it knows the old value of that key. The key updating is bottom-up, from the parent of x to the root. To distribute the new key at a node v, if v is the parent of x, then the GC encrypts the new key with the current key of the sibling of x; otherwise, GC encrypts the new key with the current keys of the two children of v, respectively. This procedure guarantees (a) and (b). The total number of messages is $O(\log n)$. Broadcasting to a subtree simply involves encrypting a message using the key for the root of that subtree; hence, the broadcast cost is $O(1)$.

In the static case, no updating is allowed. So, the GC must encrypt a broadcast using the root of every maximal subtree containing no revoked devices. Thus, in the static case, LKH has broadcast cost $O(r \log(n/r))$. (Recall that r is the number of revoked devices.) In both the static and dynamic case, however, the number of keys per device remains $O(\log n)$.

Subset Difference Revocation (SDR). The *subset difference revocation* (SDR) approach of Naor, Naor, and Lotspiech [16] is also based on associating all the devices with the leaves of a complete binary tree T. Define a subtree B as the union of all the paths from the root to leaves associated with revoked devices. Some internal nodes in B have one child and some two. Mark each internal node v in B with two children as a "cut vertex" and imagine that we cut out from T the edges from v to its two children. This would leave us with $O(r)$ rooted subtrees, each containing some number of valid devices and one revoked leaf (which may have previously been an internal node). Each such subtree is

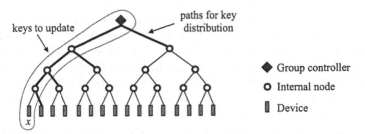

Fig. 1. The LKH scheme for key management in single groups.

therefore uniquely identified by its root, v, and its descendent node w that is revoked. The GC associates a secret key with each node v, and defines a label $L_v(w)$, for each node in the subtree, T_v, of T rooted at v. $L_v(v)$ is v's secret key, and for any internal node u in T_v, with left child x and right child y, we define $L_v(x) = f(L_v(u))$ and $L_v(y) = g(L_v(u))$, where f and g are collision-resistant one-way hash functions that maintain the size of input strings. (Here we use the abstract model of f and g; Naor, Naor, and Lotspiech use in [16] a pseudo-random generator G that triples the size of input, and take the left $1/3$ and right $1/3$ of the output to be the values of f and g.) Each leaf z in T_v stores the values of all the L_v labels of the nodes that are siblings of the path from z to v (that is, not on the path itself, but are siblings of a node on the path). The key used to encode a subtree rooted at v with a revoked node w inside is $L_v(w)$. Note that no descendent of w knows this value and no node outside of T_v can compute this value, which is what makes this a secure scheme. However, this scheme requires each device to hold $O(\log^2 n)$ keys, which violates the log-key restriction.

3   Improved Zero-State Broadcast Encryption

To improve the storage requirements for stateless broadcast encryption, so as to satisfy the log-key restriction, we take a data structuring approach. We begin with the basic approach of the subset difference (SDR) method. Without loss of generality, we assume that we are given a complete binary tree T with n leaves such that each leaf of T is associated with a different user. For any node v in T, let T_v denote the subtree rooted at v. In addition, for any node v and a descendent w of v, we let $T_{v,w}$ denote tree $T_u - T_w$, that is, all the nodes that are descendents of v but not w. Given a set of revoked users, we can use the same approach as SDR to partition T into at most $2r - 1$ subtrees $T_{v,w}$, such that union of all these trees represent the complete set of unrevoked users.

A Linear-Work Solution. As a warm-up for our efficient broadcast encryption scheme, we first describe a scheme that uses $O(\log n)$ keys per device and $O(r)$ messages per broadcast, but requires $O(n)$ work per device to decrypt messages (we will then show how to improve the device work bound keeping the other two asymptotic bounds unchanged).

The main idea is that the GC needs a way of encoding a message so that every leaf node in $T_{v,w}$ can decrypt this message, but not other user (or group of users) can decrypt it. We note as an additional space saving technique, we can name each node in T according to a level-numbering scheme (e.g., see [9]), so that the full structure of any tree $T_{v,w}$ can be completely inferred using just the names of v and w. Moreover, any leaf x in $T_{v,w}$ can determine its relative position in $T_{v,w}$ immediately from its own name, x, and the names of v and w.

Let us focus on a specific subtree T_v, for a node v in T. We define a set of *leftist* labels, $L_v(x)$, and *rightist* labels, $R_v(x)$, for each node of T_v. In particular, let us number the nodes in T_v two ways—first according to a left preorder numbering (which visits left children before right children) and second according to a right preorder numbering (which visits right children before left children) [9]. For a non-root node b in T_v, let a_l denote the predecessor of b in the left preorder numbering of the nodes in T_v. We define $L_v(b)$ to be $f(L_v(a_l))$, where f is a collision-resistant one-way hash function. Likewise, we let a_r denote the predecessor of b in the right preorder numbering of the nodes in T_v. We define $R_v(b)$ to be $g(R_v(a_r))$, where g is a (different) collision-resistant one-way hash function. We initialize these two hash chains by setting $L_v(v)$ and $R_v(v)$ to random seeds known only to the GC.

For each leaf node b in T_v, let c_l and c_r respectively denote the successors of b (if they exist) in the left and right preorder numberings of the nodes in T_v. The keys we store at b for T_v are $L_v(c_l)$ and $R_v(c_r)$. (Note that we specifically do not store $L_v(b)$ nor $R_v(b)$ at b.) For the complete key distribution, we store these two keys for each subtree T_v containing b (there are $\log n$ such subtrees). Given this key distribution, to encrypt a message for the nodes in $T_{v,w}$, a GC encrypts the message twice—once using $L_v(w)$ and once using $R_v(w)$.

Decryption. Let us next consider how a leaf node b in $T_{v,w}$ can decrypt a message sent to this subtree from the GC. Since w is not an ancestor of b, there are two possibilities: either w comes after b in the left preorder numbering of T_v or w comes after b in the right preorder numbering. Since b can determine the complete structure of T_v and b's relative position with w in this subtree from the names of v, b, and w, it can implicitly represent $T_{v,w}$ and know which of these two cases apply. So suppose the first case applies (as the second case is symmetric with the first). In this case, b starts with the label $L_v(c_l)$ it stores, where c_l is b's successor in the left preorder numbering of T_v. It then continues a left preorder traversal of T_v (which it can perform implicitly if memory is tight) until it reaches w. With each new node b encounters in this traversal, b makes another application of the one-way function f, computing the L_v labels of each visited node. Thus, when b visits w in this traversal, it will have computed $L_v(w)$ and can then decrypt the message. This computation takes at most $|T_{v,w}|$ hash function computations.

Security. Let us next consider the security of this scheme. First, observe that any node outside of T_v has no information that can be used to help decode a message for the nodes in some tree $T_{v,w}$, since $L_v(v)$ and $R_v(v)$ are chosen as random seeds and nodes outside of T_v receive no function of $L_v(v)$ or $R_v(v)$.

So the security risk that remains is that leaf descendents of w might be able to decrypt a message sent to the nodes in $T_{v,w}$. Let D_w denote the set of leaf descendents of w. For each node b in D_w, with successors c_l and c_r in the two preorder numberings, we store $L_v(c_l)$ and $R_v(c_r)$ at b. But none of these values for the nodes in D_w are useful for computing $L_v(w)$ or $R_v(w)$, without inverting a one-way function, since, in any preorder traversal, all the ancestors of a node are visited before the node is visited.

Thus, we have a key distribution strategy for the zero-state case that uses $O(\log n)$ keys per device and $O(r)$ messages per broadcast, albeit with work at each device that could be $O(n)$. In the remainder of this section, we describe how we can reduce this work bound while keeping the other asymptotic bounds unchanged.

The Stratified Subset Difference (SSD) Method. Given a constant k, we can decrease the work per device to be $O(n^{1/k})$, while increasing the space and message bounds by at most a factor of k, which should be a good trade-off in most applications. For example, when n is less than one trillion, $n^{1/8}$ is less than $\log n$. The method involves a stratified version of the scheme described above, giving rise to a scheme we call the *stratified subset difference (SSD)* method.

We begin by marking each node at a depth that is a multiple of $\lceil (\log n)/k \rceil$ as "red;" the other nodes are colored "blue." (See Figure 2.) Imagine further that we partition the tree T along the red nodes, subdividing T into maximal trees whose root and leaves are red and whose internal nodes are blue. Call each such tree a *blue tree* (even though its root and leaves are red). We then apply the method described above in each blue tree, as follows. For each leaf b in T, let $b_1, \cdots, b_k$ be the red ancestors of b, in top-down order. For $i = 1, \cdots k$, let T_i be the blue tree rooted at b_i and note that b_{i+1} is a leaf of T_i.

We store at node b labels $L_{b_i}(c_l)$ and $R_{b_i}(c_r)$ $(i = 1, \cdots k$ in $T)$, where c_l and c_r are the left and right preorder successors of b_{i+1} in T_i, respectively. Storing these labels increases the space per device by a factor of k.

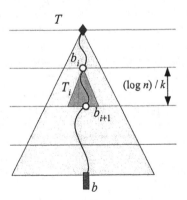

Fig. 2. Illustration of the stratified subset difference (SSD) scheme.

To encrypt a message, the GC first performs the subdivision of T into the subtrees $T_{v,w}$ as before. Then, the GC further partitions each tree $T_{v,w}$ at the red levels, and encodes the broadcast message, using the previously described scheme, for each blue subtree rooted at a node on the path from v to w. This increases the broadcast size by at most a factor of k, but now the work needed by each device is reduced to computing the L or R labels in a blue tree, which has size at most $n^{1/k}$. Thus, the work per device is reduced to $O(n^{1/k})$ in this SSD scheme.

Theorem 1. *Given a balanced tree T with n devices, for zero-state broadcast encryption, the stratified subset difference (SSD) scheme for T uses $O(\log n)$ keys per device and has $O(r)$ broadcast cost, where r is the number of revoked devices in the subtree receiving the broadcast. The work per device can be made to be $O(n^{1/k})$ for any fixed constant k.*

Moreover, as we have noted, the security of this scheme is as strong as that for SDR and LKH, i.e., it is resilient to collusions of any set of revoked devices.

4   A Biased Tree Scheme for an Organizational Chart

We recall that in the organizational chart structure for n devices, we have a hierarchical partition of the devices induced by a tree T of $k = O(\log n)$ height but with unbounded branches at each internal node. Namely, the leaves of T are associated with the devices and an internal node v of T represents the group (set) of devices associated with the leaves of the subtree rooted at v. Thus, sibling nodes of T are associated with disjoint groups and each device belongs to a unique sequence of $O(\log n)$ groups whose nodes are on the path from the device's leaf to the root of T. Without loss of generality, we assume that an internal node of T has either all internal children (subgroups) or all external children (devices), and its group is called an interior group or exterior group accordingly. We consider four types of update operations: *insertion* and *deletion* (revocation) of a device or of an empty group. After each modification, we want to maintain both forward and backward security.

Biased Trees. Biased trees, introduced by Bent *et al.* [1], are trees balanced by the weights of leaves (typically set as access frequencies). There are two versions of biased trees: locally biased and globally biased. We denote by $p(x)$, $l(x)$ and $r(x)$ the parent, left child and right child of a node x of a tree, and we use these denotations cumulatively. E.g., $lpp(x)$ is the left child of the grandparent of x. The following definitions are taken from [1].

A *biased search tree* is a full binary search tree such that each node x has a weight $w(x)$ and a rank $s(x)$. The weight of a leaf is initially assigned, and the weight of an internal node is the sum of the weights of its children. The rank $s(x)$ of a node x is a positive integer such that

1. $s(x) = \lfloor \log w(x) \rfloor$ if x is a leaf.
2. $s(x) \le s(p(x)) - 1$ if x is a leaf.
3. $s(x) \le s(p(x))$ and $s(x) \le s(pp(x)) - 1$.

A *locally biased* search tree has the following additional property:

Local bias. For any x with $s(x) \leq s(p(x)) - 2$,
1. if $x = lp(x)$, then either $rp(x)$ or $lrp(x)$ is a leaf with rank $s(x) - 1$; if $x = rp(x)$, then either $lp(x)$ or $rlp(x)$ is a leaf with rank $s(x) - 1$; and
2. if $x = lp(x)$, $p(x) = rpp(x)$ and $s(p(x)) = s(pp(x))$, then either $lpp(x)$ or $rlpp(x)$ is a leaf with rank $s(x) - 1$; if $x = rp(x)$, $p(x) = lpp(x)$ and $s(p(x)) = s(pp(x))$, then either $rpp(x)$ or $lrpp(x)$ is a leaf with rank $s(x) - 1$.

A *globally biased* search tree has the following additional property:

Global bias. For any x with $s(x) \leq s(p(x)) - 2$, both of the two neighboring leaves of x, i.e., the right-most leaf on the left and the left-most leaf on the right, have rank at least $s(x) - 1$.

Group Hierarchies and Biased Trees. Given an organizational chart T that represents a group hierarchy, we have to convert T to a binary tree before applying any encryption scheme for key management. Without loss of generality, we convert T to a binary tree B_T that preserves the original group hierarchy. Each internal node of T, representing a group G_i, becomes a special internal node in B_T that still represents G_i and accommodates a GC. Additional internal nodes are added between G_i and its children in T (i.e., subgroups or devices) for the purpose of binarization. As result, node G_i plus all its children in T and the paths between them in B_T form a binary subtree B_i in B_T with G_i being the root and each of its children in T being a leaf. Note that, without special care, B_T is likely to have super-logarithm height and balancing such a tree using standard techniques would destroy the group hierarchy.

Given a group hierarchy tree T, we assign a unit weight to each leaf and calculate the weights of other nodes in T accordingly, i.e., the weight of each internal node x is the number of devices in the subtree of T rooted at x. We replace each node x with a biased binary tree having the children of x as its leaves (using the weights of these nodes for the biasing). Thus, each subtree B_i representing a group G_i rooted at a node x in T can be initialized into a biased tree without affecting the structure of group hierarchy. Since $w(G_i)$ for each G_i is an invariant, i.e., the weights of the root and leaves in every B_i are invariant, the initialization is well defined and can be done in each B_i independently. That is, combining all the biased B_i's into B_T will not change the structure of the original hierarchy represented by T. (See Figure 3)

Key Assignment. After initializing the biased B_i's, we still assign a key to each node of B_T as in the LKH, and inform the keys to devices and GC's by the following security properties:

1. each device x knows all but only the keys on the path from G_0 to itself.
2. the GC of each G_i knows all but only the keys of G_i's descendants in B_T and those on the path from G_0 to G_i.

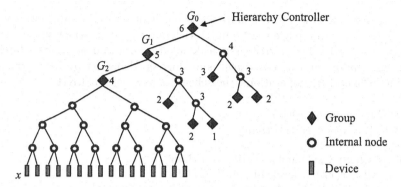

Fig. 3. Binary tree B_T consisting of biased trees B_0, B_1 and B_2. The ranks of the nodes in B_0 and B_1 are shown.

Broadcast and Multicast. Using the above security properties and appropriate signature or authentication mechanism [2, 4, 20, 25], the GC of each G_i can send a message securely with one key encryption to G_i or any subgroup or super-group of G_i, without any ambiguity.

Key Update and Tree Rebalance. As in the LKH scheme, keys should be updated after each *insertion* or *deletion* (revocation) of a device or group so that the security properties 1 and 2 are maintained. Moreover, we should also rebalance B_T to preserve the bias properties in each B_i. Assume that we can insert a leaf, delete a leaf, or update the weight of a leaf in B_i (by *insert(x)*, *delete(x)* and *reweight(x)*, respectively) while preserving both the security and bias properties. Then inserting or deleting a device $x \in G_k \subset G_{k-1} \subset \cdots \subset G_0$ can be done in three steps:

1. insert or delete a leaf in the exterior tree B_k;
2. update the weights $w(G_k), w(G_{k-1}), \cdots, w(G_1)$ in the interior trees B_{k-1}, $B_{k-2}, \cdots, B_0$ accordingly; and
3. update the keys on the path from x to G_0 bottom-up, as in the LKH scheme.

To insert or delete a group $G_{k+1} \subset G_k \subset \cdots \subset G_0$ is a similar process except starting with an insertion or deletion in an interior B_k. Therefore *insert*, *delete* and *reweight* in each B_i suffice all our hierarchy modifications in B_T. Such operations preserving the bias properties were already given and analyzed in [1], we now describe how to modify them to preserve the security properties, too.

Recall that the biased tree operations, including *insert*, *delete* and *reweight*, recursively call an operation *tilt* as the only subroutine to rebalance the biased tree structure [1]. Operation *tilt* performs a single rotation associated with rank modification. Since a node loses descendants during a rotation if it is rotated down and losing descendants is the only chance of key leaks in the LKH scheme. To maintain the security properties 1 and 2 after any rotation in B_i, it is necessary and sufficient to update the key at the node rotated down. Observing that updating a single key and distributing the result of a rotation are both easy in our scheme, we can replace the *tilt* in [1] with our *secure-tilt* which preserves the security properties 1 and 2.We give a detailed description of *secure-tilt-left* in Figure 4. Operation *secure-tilt-right* is analogous. Using *secure-tilt* as the sub-routine in biased tree operations, the scheme is as secure as LKH.

Algorithm *secure-tilt-left*(x)
   if $s(l(x)) = s(x) = s(r(x))$ then
     $s(x) \leftarrow s(x) + 1$
   else if $s(l(x)) < s(x)$ and $s(r(x)) = s(x)$ then
     let $x, l(x), r(x), lr(x), rr(x)$ be A, B, C, D, E.
     $p(C) \leftarrow p(A)$, $l(C) \leftarrow A$ and $r(A) \leftarrow D$. {left rotation at x}
     update $key(A)$
     distribute $key(A)$ and $key(C)$ to their descendants
     $x \leftarrow C$
   end if
   return x

Fig. 4. The algorithm for operation *secure-tilt-left*(x).

Efficiency of the Scheme. The *insert*, *delete* and *reweight* operations in biased trees are implemented as follows: *join* and *split* are the two basic biased tree operations. *join*(x, y) has global and local versions, which will merge two global or local biased trees with roots x and y and return the root of the resulting tree, and both versions work by recursively calling *secure-tilt*. *split*(T, x) will split T into two biased trees T_1 and T_2, each containing all the leaves of T with their binary search keys less than x and greater than x, respectively. *split* calls *local-join* as a subroutine and is applicable to both local and global biased trees.

Other operations are based on *join* and *split*: operation *insert*(x) splits T by x and then joins T_1, x and T_2 together; operation *delete*(x) splits T by x and then joins T_1 and T_2 back ignoring x; and operation *reweight*(x) splits T by x, updates the weight of x, and then joins T_1, x and T_2 back into T.

The correctness and efficiency of our hierarchy modifications in B_T follow those of biased tree operations. Notice that our *secure-tilt* takes constant message size as well as the constant-time *tilt* in [1], all time bounds in [1] also hold as bounds of message size in our scheme.

This gives us the following.

Theorem 2. *Given an organizational chart tree T with height k and n devices, under the log-key restriction, the dynamic biased binary tree scheme for T has has $O(1)$ broadcast cost and $O(k + \log n)$ revocation and insertion cost.*

Proof. We show how to access a device $x \in G_k \subset G_{k-1} \subset \cdots \subset G_0$ from G_0. The analysis of other operations is similar. Since the root of B_i is a leaf of B_{i-1}, and each biased tree B_i, $i = 0, 1, \cdots, k$, has the ideal access time, the time to access x from G_0 is

$$O\left(\left\lceil \log \frac{w(G_0)}{w(G_1)} \right\rceil + \cdots + \left\lceil \log \frac{w(G_{k-1})}{w(G_k)} \right\rceil + \left\lceil \log \frac{w(G_k)}{w(x)} \right\rceil\right) = O(\log n + k). \qquad \square$$

Thus, we satisfy the log-key restriction for any organizational chart with $k = O(\log n)$ height. We also note that applying our SSD approach to a static application of the techniques developed in this section results in a scheme using $O(\log n)$ keys per device and $O(r)$ messages per broadcast for an organization chart with height $O(\log n)$.

5   A Dynamic Tree Scheme for a Multicast Tree

Let us next consider the multicast tree structure, which, for the sake of broadcast encryption, is similar to the organizational chart, except that the height of a multicast tree can be much larger than logarithmic (we even allow for linear height). For a multicast tree T with n devices and m groups, we give a scheme with $O(\log m)$ broadcast cost and $O(\log n)$ update cost, irrespectively of the depth of T.

Dynamic Trees. Dynamic trees were first studied by Sleator and Tarjan [22] and used for various tree queries and network flow problems. The key idea is to partition a highly unbalanced tree into paths and associate a biased tree structure, which is in some sense balanced, to each path. Thus any node in the tree can be accessed and any update to the tree can be done in $O(\log n)$ time through the associated structure, regardless the depth of node or the height of tree. The dynamic tree used in our scheme is specified by taking the partition by weight (size) approach and not having cost on each edge. The following definition refers to this specification.

A *dynamic tree* T is a weighted binary search tree where the weight $w_T(x)$ is initially assigned if x is a leaf, or $w_T(x) = w_T(l(x)) + w_T(r(x))$ if x is an internal node. The edges of T are partitioned into solid and dashed edges so that each node links with its heavier child by a solid edge and with the lighter child by a dashed edge. Thus T is partitioned into solid paths P_j's linked by dashed edges. We denote by $h(P_j)$ the deepest node in P_j and $t(P_j)$ the upper-most one[1]. Then the edge between any $t(P_j)$ and its parent must be dashed, and vice versa. For $O(\log n)$ operations, each solid path P_j is further organized as a global biased tree, denoted by $B(P_j)$, so that the nodes from $h(P_j)$ to $t(P_j)$ become leaves of $B(P_j)$ from left to right, and the weight of a leaf x in $B(P_j)$ is assigned as $w_{B(P_j)}(x) = w_T(y)$ where y is the dashed child of x in T. Then T consists of these $B(P_j)$'s by linking the root of each $B(P_j)$ with the parent of $t(P_j)$, unless $t(P_j)$ is the root of T. (See Figure 5.) To show that such structure of T is well defined, let the root of $B(P_j)$ be x and the parent of $t(P_j)$ be $y \in P_{j-1}$, then we have that $w_{B(P_j)}(x) = w_T(t(P_j)) = w_{B(P_{j-1})}(y)$. Thus, x can replace $t(P_j)$ as a child of y.

---

[1] $h(P_j)$ must be a leaf of T by the "partition by weight" approach.

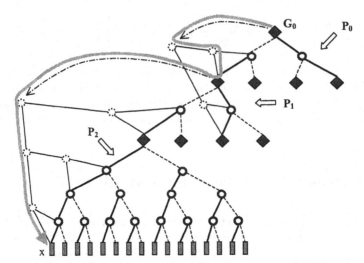

Fig. 5. Partition of tree B_T and the accessing path to x.

Group Hierarchies and Dynamic Trees. We convert a multicast tree T to a binary tree B_T that preserves the group hierarchy in T as same as in the biased tree scheme. Instead of using a biased tree, we simply use a complete binary tree for each B_i, then assign a unit weight $w_T(x) = 1$ to each device and partition B_T into a dynamic tree as above. A key is assigned to each node of each $B(P_j)$. Since the root of $B(P_j)$ becomes child of a leaf of $B(P_{j-1})$, each device becomes a descendant of a unique string of biased trees of paths $B(P_j), B(P_{j-1}), \cdots, B(P_0)$. The way a device is accessed is not through the real path in B_T but through the path in the string of $B(P_j)$'s. (See Figure 5.)

Broadcast and Multicast. Broadcast in a group G_i becomes a little more complicate because, although device x is a descendant of G_i in T, G_i may not be on the accessing path from G_0 to x. However, if $G_i \in P_j$, then the accessing path to any descendant of G_i must pass a node in the prefix of P_j from $h(P_j)$ to G_i. So, to broadcast in G_i, it is sufficient to encrypt the message by the keys in $B(P_j)$ that cover this prefix of P_i. In the full version, we show that, with the dynamic tree scheme, it takes $O(\log |P_j|)$ encryptions to broadcast a message in any group $G_i \in P_j$, either in worst case or in average.

Key Updates. We follow the dynamic tree operations in [22] to modify the hierarchy, and update the keys in the accessing path of the updated item as in the LKH scheme. Dynamic tree operations dynamically change the solid path partition to guarantee the $O(\log n)$ running time, and such change is carried out by the biased tree operations among $B(P_j)$'s. Therefore, operation *secure-tilt* preserves the security properties along any accessing path. The dynamic tree operations we use are as follows:

- *splice*(P_j): extend P_j by converting the edge from $t(P_j)$ to its parent solid, and the edge between *sibling*($t(P_i)$) and its parent dashed.

- *slice(P_j)*: Let (x, y) be the upper most edge in P_j such that y is not the heavier child of x, if there exist such edges in P_j. Then cut P_j by converting (x, y) into dashed and $(x, sibling(y))$ into solid.
- *expose(x)*: make the path from x to G_0 (the real path in B_T) into a single solid path by a series of *splices*.
- *conceal(P_j)*: convert every edge in P_j who does not link to a heavier child of parent into dashed by a series of *slices*.
- *link(x, y)*: combine two dynamic trees by making y the parent of x, where x is the root of the first tree and y is a node in the second.
- *cut(x)*: divide a dynamic tree into two by deleting the edge between x and $p(x)$.

Inserting or deleting a device or a group corresponds to a *link* or *cut* operation, respectively. Such dynamic tree operation take $O(\log n)$ time and can be reduced to a series of *join* and *split* operation on biased trees. The algorithmic template for a dynamic tree operation is the *expose-and-conceal* strategy, described as follows:

1. perform *expose(x)* on a node x;
2. if the above *expose* operation violates the "partition by weight" property, restore the property by executing *conceal(P_j)* on the appropriate path P_j.

Since all the dynamic tree operations reduce to a series of biased tree operations, operation *secure-tilt* is still the only subroutine that adjusts the partition of $T(P_j)$'s. Notice that the structure B_T is never adjusted, but the accessing path to each device x are adjusted through operations. From [22], we know that, with partition by weight and representing the solid paths as global biased trees, any dynamic tree operation takes $O(\log n)$ time. Since a hierarchy modification consists of a dynamic tree operation plus updating the keys in an access path, which is also of length $O(\log n)$, the efficiency of key updating for hierarchy modifications follows.

Theorem 3. *Given a multicast tree T with n devices, under the log-key restriction, structured in m groups, the dynamic tree scheme for T has $O(\log m)$ broadcast cost and $O(\log n)$ revocation and insertion cost.*

A zero-state version can also be developed, which uses the biased trees and broadcast scheme to send messages to the unrevoked leaves in a multicast tree T using $O(r \log n)$ broadcasts for devices storing $O(\log n)$ keys each, where r is the number of revoked devices.

References

1. S. W. Bent, D. D. Sleator, and R. E. Tarjan. Biased search trees. *SIAM J. Comput.*, 14(3):545–568, Aug. 1985.
2. D. Boneh, G. Durfee, and M. Franklin. Lower bounds for multicast message authentication. In *Proc. EUROCRYPT 2001, LNCS 2045*, pages 437–452, May 2001.

3. B. Briscoe. Marks: Zero side effect multicast key management using arbitrarily revealed key sequences. In *Proc. of First International Workshop on Networked Group Communication(NCGC'99)*, 1999.

4. R. Canetti, J. Garay, G. Itkis, D. Micciancio, M. Naor, and B. Pinkas. Multicast security: A taxonomy and some efficient constructions. In *Proc. INFOCOM '99*, volume 2, pages 708–716, New York, Mar. 1999.

5. R. Canetti, T. Malkin, and K. Nissim. Efficient communication — storage tradeoffs for multicast encryption. In *Advances in cryptology (EUROCRYPT'99), LNCS 1592*, pages 459–474, 1999.

6. I. Chang, R. Engel, D. Kandlur, D. Pendarakis, and D. Saha. Key management for secure Internet multicast using boolean function minimization techniques. In *Proc. IEEE INFOCOM*, volume 2, pages 689–698, 1999.

7. G. D. Crescenzo and O. Kornievskaia. Efficient kerberized multicast in a practical distributed setting. In *4th International Conference Information Security (ISC'01), LNCS 2200*, pages 27–45, Oct. 2001.

8. A. Fiat and M. Naor. Broadcast encryption. In *Advances in Cryptology - CRYPTO'93*, pages 480–491. LNCS 773, 1994.

9. M. T. Goodrich and R. Tamassia. *Algorithm Design: Foundations, Analysis and Internet Examples*. John Wiley & Sons, New York, NY, 2002.

10. D. Halevy and A. Shamir. The LSD broadcast encryption scheme. In *Advances in Cryptology (CRYPTO 2002)*, volume 2442 of *LNCS*, pages 47–60. Springer-Verlag, 2002.

11. E. Jung, A. X. Liu, and M. G. Gouda. Key bundles and parcels: Secure communication in many groups. In *Proceedings of the 5th International Workshop on Networked Group Communications (NGC-03), LNCS 2816*, pages 119–130, Munich, Germany, September 2003.

12. R. Kumar, R. Rajagopalan, and A. Sahai. Goding constructions for blackliting problems without computational assumptions. In *Advances in cryptology (CRYPTO'99), LNCS 1666*, pages 609–623, 1999.

13. D. A. McGrew and T. Sherman. Key establishment in large dynamic groups using one-way function trees. Technical Report 0755, TIS Labs at Network Associates Inc., Glenwood, MD, May 1998.

14. D. Micciancio and S. Panjwani. Optimal communication complexity of generic multicast key distribution. In *EUROCRYPT 2004*, pages 153–170, 2004.

15. M. J. Mihajevic. Key management schemes for stateless receivers based on time varying heterogeneous logical key hierarchy. In *ASIACRYPT 2003, LNCS 2894*, pages 137–154, 2003.

16. D. Naor, M. Naor, and J. Lotspiech. Revocation and tracing schemes for stateless receivers. In *CRYPTO'01*, volume 2139 of *LNCS*, pages 41–62. Springer-Verlag, 2001.

17. M. Naor and B. Pinkas. Efficient trace and revoke schemes. In *Proc. Financial Crypto 2000*, Feb. 2000.

18. B. Pinkas. Efficient state updates for key management. In *Proc. ACM Workshop on Security and Privacy in Digital Rights Management*, 2001.

19. O. Rodeh, K. P. Birman, and D. Dolev. Using AVL trees for fault tolerant group key management. *International Journal on Information Security*, pages 84–99, 2001.

20. B. Schneier. *Applied Cryptography, 2nd Ed.* John Wiley - Sons, 1996.

21. A. T. Sherman and D. A. McGrew. Key establishment in large dynamic groups using one-way function trees. *IEEE Trans. Software Engineering*, 29(5):444–458, 2003.

22. D. D. Sleator and R. E. Tarjan. A data structure for dynamic trees. *J. Computer and System Sciences*, 26:362–391, 1983.
23. J. Snoeyink, S. Suri, and G. Varghese. A lower bound for multicast key distribution. In *IEEE INFOCOM 2001*, volume 1, pages 422–431, 2001.
24. R. Tamassia and N. Triandopoulos. Computational bounds on hierarchical data processing with applications to information security. Technical report, Center for Geometric Computing, Brown University, 2004.
25. H. F. Tipton and M. Krause, editors. *Information Security Management Handbook, 4th Ed.* Auerbach, 1999.
26. D. M. Wallner, E. G. Harder, and R. C. Agee. Key management for multicast: issues and architecture. In *internet draft draft-waller-key-arch-01.txt*, Sep. 1998.
27. C. K. Wong, M. Gouda, and S. S. Lam. Secure group communications using key graphs. In *Proc. ACM SIGCOMM'98*, volume 28, pages 68–79, 1998.
28. C. K. Wong and S. S. Lam. Digital signatures for flows and multicasts. *IEEE/ACM Transactions on Networking*, 7:502–513, 1999.

Privacy-Preserving Datamining
on Vertically Partitioned Databases

Cynthia Dwork and Kobbi Nissim

Microsoft Research, SVC, 1065 La Avenida, Mountain View CA 94043
{dwork,kobbi}@microsoft.com

Abstract. In a recent paper Dinur and Nissim considered a statistical database in which a trusted database administrator monitors queries and introduces noise to the responses with the goal of maintaining data privacy [5]. Under a rigorous definition of breach of privacy, Dinur and Nissim proved that *unless the total number of queries is sub-linear in the size of the database*, a substantial amount of noise is required to avoid a breach, rendering the database almost useless.

As databases grow increasingly large, the possibility of being able to query only a sub-linear number of times becomes realistic. We further investigate this situation, generalizing the previous work in two important directions: multi-attribute databases (previous work dealt only with single-attribute databases) and vertically partitioned databases, in which different subsets of attributes are stored in different databases. In addition, we show how to use our techniques for datamining on published noisy statistics.

Keywords: Data Privacy, Statistical Databases, Data Mining, Vertically Partitioned Databases.

1 Introduction

In a recent paper Dinur and Nissim considered a statistical database in which a trusted database administrator monitors queries and introduces noise to the responses with the goal of maintaining data privacy [5]. Under a rigorous definition of breach of privacy, Dinur and Nissim proved that *unless the total number of queries is sub-linear in the size of the database*, a substantial amount of noise is required to avoid a breach, rendering the database almost useless[1]. However, when the number of queries is limited, it is possible to simultaneously preserve privacy and obtain some functionality by adding an amount of noise that is a function of the number of queries. Intuitively, the amount of noise is sufficiently large that nothing specific about an individual can be learned from a relatively small number of queries, but not so large that information about sufficiently strong statistical trends is obliterated.

---

[1] For unbounded adversaries, the amount of noise (per query) must be linear in the size of the database; for polynomially bounded adversaries, $\Omega(\sqrt{n})$ noise is required.

M. Franklin (Ed.): CRYPTO 2004, LNCS 3152, pp. 528–544, 2004.

As databases grow increasingly massive, the notion that the database will be queried only a sub-linear number of times becomes realistic. We further investigate this situation, significantly broadening the results in [5], as we describe below.

Methodology. We follow a cryptography-flavored methodology, where we consider a database access mechanism private only if it provably withstands any adversarial attack. For such a database access mechanism any computation over query answers clearly preserves privacy (otherwise it would serve as a privacy breaching adversary). We present a database access mechanism and prove its security under a strong privacy definition. Then we show that this mechanism provides utility by demonstrating a datamining algorithm.

Statistical Databases. A statistical database is a collection of samples that are somehow representative of an underlying population distribution. We model a database as a matrix, in which rows correspond to individual records and columns correspond to attributes. A query to the database is a set of indices (specifying rows), and a Boolean property. The response is a noisy version of the number of records in the specified set for which the property holds. (Dinur and Nissim consider one-column databases containing a single binary attribute.) The model captures the situation of a traditional, multiple-attribute, database, in which an adversary knows enough partial information about records to "name" some records or select among them. Such an adversary can target a selected record in order to try to learn the value of one of its unknown sensitive attributes. Thus, the mapping of individuals to their indices (record numbers) is not assumed to be secret. For example, we do not assume the records have been randomly permuted.

We assume each row is independently sampled from some underlying distribution. An analyst would usually assume the existence of a single underlying row distribution $\mathcal{D}$, and try to learn its properties.

Privacy. Our notion of privacy is a relative one. We assume the adversary knows the underlying distribution $\mathcal{D}$ on the data, and, furthermore, may have some a priori information about specific records, e.g., "p – the a priori probability that at least one of the attributes in record 400 has value 1 – is .38". We anlyze privacy with respect to any possible underlying (row) distributions $\{\mathcal{D}_i\}$, where the ith row is chosen according to D_i. This partially models a priori knowledge an attacker has about individual rows (i.e. $\mathcal{D}_i$ is $\mathcal{D}$ conditioned on the attacker's knowledge of the ith record). Continuing with our informal example, privacy is breached if the a posteriori probability (after the sequence of queries have been issued and responded to) that "at least one of the attributes in record 400 has value 1" differs from the a priori probability p "too much".

Multi-attribute Sub-linear Queries (SuLQ) Databases. The setting studied in [5], in which an adversary issues only a sublinear number of queries (SuLQ) to a single attribute database, can be generalized to multiple attributes in several natural ways. The simplest scenario is of a single k-attribute SuLQ database, queried by specifying a set of indices and a k-ary Boolean function. The response is a noisy version of the number of records in the specified set for which the function, applied to the attributes in the record, evaluates to 1. A more involved scenario is of multiple single-attribute SuLQ databases, one for each attribute, administered independently. In other words, our k-attribute database is *vertically partitioned* into k single-attribute databases. In this case, the challenge will be datamining: learning the statistics of Boolean functions of the attributes, using the single-attribute query and response mechanisms as primitives. A third possibility is a combination of the first two: a k-attribute database that is vertically partitioned into two (or more) databases with k_1 and k_2 (possibly overlapping) attributes, respectively, where $k_1 + k_2 \geq k$. Database i, $i = 1, 2$, can handle k_i-ary functional queries, and the goal is to learn relationships between the functional outputs, eg, "If $f_1(\alpha_{1,1}, \ldots, \alpha_{1,k_1})$ holds, does this increase the likelihood that $f_2(\alpha_{2,1} \ldots, \alpha_{2,k_2})$ holds?", where f_i is a function on the attribute values for records in the ith database.

1.1   Our Results

We obtain positive datamining results in the extensions to the model of [5] described above, while maintaining the strengthened privacy requirement:

1. Multi-attribute SuLQ databases: The statistics for every k-ary Boolean function can be learned[2]. Since the queries here are powerful (any function), it is not surprising that statistics for any function can be learned. The strength of the result is that statistics are learned while maintaining privacy.

2. Multiple single-attribute SuLQ databases: We show how to learn the statistics of any 2-ary Boolean function. For example, we can learn the fraction of records having neither attribute 1 nor attribute 2, or the conditional probability of having attribute 2 given that one has attribute 1. The key innovation is a procedure for testing the extent to which one attribute, say, α, implies another attribute, β, *in probability*, meaning that $\Pr[\beta|\alpha] = \Pr[\beta] + \Delta$, where Δ can be estimated by the procedure.

3. Vertically Partitioned k-attribute SuLQ Databases: The constructions here are a combination of the results for the first two cases: the k attributes are partitioned into (possibly overlapping) sets of size k_1 and k_2, respectively, where $k_1 + k_2 \geq k$; each of the two sets of attributes is managed by a multi-attribute SuLQ database. We can learn all 2-ary Boolean functions of the outputs of the results from the two databases.

We note that a single-attribute database can be simulated in all of the above settings; hence, in order to preserve privacy, the sub-linear upper bound on queries must be enforced. How this bound is enforced is beyond the scope of this work.

---

[2] Note that because of the noise, statistics cannot be learned exactly. An additive error on the order of $n^{1/2-\varepsilon}$ is incurred, where n is the number of records in the database. The same is true for single-attribute databases.

Datamining on Published Statistics. Our technique for testing implication in probability yields surprising results in the real-life model in which confidential information is gathered by a trusted party, such as the census bureau, who publishes aggregate statistics. Describing our results by example, suppose the bureau publishes the results of a large (but sublinear) number of queries. Specifically, for every, say, triple of attributes $(\alpha_1, \alpha_2, \alpha_3)$, and for each of the eight conjunctions of literals over three attributes $(\bar{\alpha}_1 \bar{\alpha}_2 \bar{\alpha}_3, \bar{\alpha}_1 \bar{\alpha}_2 \alpha_3, \ldots, \alpha_{k-2} \alpha_{k-1} \alpha_k)$, the bureau publishes the result of several queries on these conjunctions. We show how to construct approximate statistics for any binary function of *six* attributes. (In general, using data published for ℓ-tuples, it is possible to approximately learn statistics for any 2ℓ-ary function.) Since the published data are the results of SuLQ database queries, the total number of published statistics must be sublinear in n, the size of the database. Also, in order to keep the error down, several queries must be made for each conjunction of literals. These two facts constrain the values of ℓ and the total number k of attributes for which the result is meaningful.

1.2 Related Work

There is a rich literature on confidentiality in statistical databases. An excellent survey of work prior to the late 1980's was made by Adam and Wortmann [2]. Using their taxonomy, our work falls under the category of *output perturbation*. However, to our knowledge, the only work that has exploited the opportunities for privacy inherent in the fact that with massive of databases the actual number of queries will be sublinear is Sect. 4 of [5] (joint work with Dwork). That work only considered single-attribute SuLQ databases.

Fanconi and Merola give a more recent survey, with a focus on aggregated data released via web access [10]. Evfimievski, Gehrke, and Srikant, in the Introduction to [7], give a very nice discussion of work in randomization of data, in which data contributors (e.g., respondents to a survey) independently add noise to their own responses. A special issue (Vol.14, No. 4, 1998) of the *Journal of Official Statistics* is dedicated to disclosure control in statistical data. A discussion of some of the trends in the statistical research, accessible to the non-statistician, can be found in [8].

Many papers in the statistics literature deal with generating simulated data while maintaining certain quantities, such as marginals [9]. Other widely-studied techniques include cell suppression, adding simulated data, releasing only a subset of observations, releasing only a subset of attributes, releasing synthetic or partially synthetic data [13, 12], data-swapping, and post-randomization. See Duncan (2001) [6].

R. Agrawal and Srikant began to address privacy in datamining in 2000 [3]. That work attempted to formalize privacy in terms of confidence intervals (intuitively, a small interval of confidence corresponds to a privacy breach), and also showed how to reconstruct an original distribution from noisy samples (i.e., each sample is the sum of an underlying data distribution sample and a noise sample), where the noise is drawn from a certain simple known distribution.

This work was revisited by D. Agrawal and C. Aggarwal [1], who noted that it is possible to use the outcome of the distribution reconstruction procedure to significantly diminish the interval of confidence, and hence breach privacy. They formulated privacy (loss) in terms of k mutual information, taking into account (unlike [3]) that the adversary may know the underlying distribution on the data and "facts of life" (for example, that ages cannot be negative). Intuitively, if the mutual information between the sensitive data and its noisy version is high, then a privacy breach occurs. They also considered reconstruction from noisy samples, using the EM (expectation maximization) technique. Evfimievsky, Gehrke, and Srikant [7] criticized the usage of mutual information for measuring privacy, noting that low mutual information allows complete privacy breaches that happen with low but significant frequency. Concurrently with and independently of Dinur and Nissim [5] they presented a privacy definition that related the a priori and a posteriori knowledge of sensitive data. We note below how our definition of privacy breach relates to that of [7, 5].

A different and appealing definition has been proposed by Chawla, Dwork, McSherry, Smith, and Wee [4], formalizing the intuition that one's privacy is guaranteed to the extent that one is not brought to the attention of others. We do not yet understand the relationship between the definition in [4] and the one presented here.

There is also a very large literature in secure multi-party computation. In secure multi-party computation, *functionality* is paramount, and privacy is only preserved to the extent that the function outcome itself does not reveal information about the individual inputs. In privacy-preserving statistical databases, *privacy* is paramount. Functions of the data that cannot be learned while protecting privacy will simply not be learned.

2   Preliminaries

Notation. We denote by $\mathsf{neg}(n)$ (read: negligible) a function that is asymptotically smaller than any inverse polynomial. That is, for all $c > 0$, for all sufficiently large n, we have $\mathsf{neg}(n) < 1/n^c$. We write $\tilde{O}(T(n))$ for $T(n) \cdot \mathbf{polylog}(n)$.

2.1   The Database Model

In the following discussion, we do not distinguish between the case of a vertically partitioned database (in which the columns are distributed among several servers) and a "whole" database (in which all the information is in one place).

We model a database as an $n \times k$ binary matrix $d = \{d_{i,j}\}$. Intuitively, the columns in d correspond to Boolean attributes $\alpha_1, \ldots, \alpha_k$, and the rows in d correspond to individuals where $d_{i,j} = 1$ iff attribute α_j holds for individual i. We sometimes refer to a row as a *record*.

Let $\mathcal{D}$ be a distribution on $\{0,1\}^k$. We say that a database $d = \{d_{i,j}\}$ is chosen according to distribution $\mathcal{D}$ if every row in d is chosen according to $\mathcal{D}$, independently of the other rows (in other words, d is chosen according to $\mathcal{D}^n$).

In our privacy analysis we relax this requirement and allow each row i to be chosen from a (possibly) different distribution $\mathcal{D}_i$. In that case we say that the database is chosen according to $\mathcal{D}_1 \times \cdots \times \mathcal{D}_n$.

Statistical Queries. A statistical query is a pair (q, g), where $q \subseteq [n]$ indicates a set of rows in d and $g : \{0, 1\}^k \to \{0, 1\}$ denotes a function on attribute values. The *exact* answer to (q, g) is the number of rows of d in the set q for which g holds (evaluates to 1):

$$a_{q,g} = \sum_{i \in q} g(d_{i,1}, \ldots, d_{i,k}) = |\{i : i \in q \text{ and } g(d_{i,1}, \ldots, d_{i,k}) \text{ holds}\}|.$$

We write (q, j) when the function g is a projection onto the jth element: $g(x_1, \ldots, x_k) = x_j$. In that case (q, j) is a query on a subset of the entries in the jth column: $a_{q,j} = \sum_{i \in q} d_{i,j}$. When we look at vertically partitioned single-attribute databases, the queries will all be of this form.

Perturbation. We allow the database algorithm to give perturbed (or "noisy") answers to queries. We say that an answer $\hat{a}_{q,j}$ is within perturbation $\mathcal{E}$ if $|\hat{a}_{q,j} - a_{q,j}| \le \mathcal{E}$. Similarly, a database algorithm $\mathcal{A}$ is within perturbation $\mathcal{E}$ if for every query (q, g)

$$\Pr[|\mathcal{A}(q, g) - a_{q,g}| \le \mathcal{E}] = 1 - \mathsf{neg}(n).$$

The probability is taken over the randomness of the database algorithm $\mathcal{A}$.

2.2 Probability Tool

Proposition 1. *Let $s_1, \ldots, s_t$ be random variables so that $|\mathbf{E}[s_i]| \le \alpha$ and $|s_i| \le \beta$ then*

$$\Pr[|\sum_{i=1}^{T} s_t| > \lambda(\alpha + \beta)\sqrt{t} + t\beta] < 2e^{-\lambda^2/2}.$$

Proof. Let $z_i' = s_i - \mathbf{E}[s_i]$, hence $|z_i'| \le \alpha + \beta$. Using Azuma's inequality[3] we get that $\Pr[\sum_{i=1}^{T} z' \ge \lambda(\alpha + \beta)\sqrt{t}] \le 2e^{-\lambda^2/2}$. As $|\sum_{i=1}^{T} s_t| = |\sum_{i=1}^{T} z' + \sum_{i=1}^{T} \mathbf{E}[s_i]| \le |\sum_{i=1}^{T} z'| + t\beta$ the proposition follows.

3 Privacy Definition

We give a privacy definition that extends the definitions in [5, 7]. Our definition is inspired by the notion of semantic security of Goldwasser and Micali [11]. We first state the formal definition and then show some of its consequences.

Let $p_0^{i,j}$ be the a priori probability that $d_{i,j} = 1$ (taking into account that we assume the adversary knows the underlying distribution $\mathcal{D}_i$ on row i. In

---

[3] Let $X_0, \ldots, X_m$ be a martingale with $|X_{i+1} - X_i| \le 1$ for all $0 \le i < m$. Let $\lambda > 0$ be arbitrary. Azuma's inequality says that then $\Pr[X_m > \lambda\sqrt{m}] < e^{\lambda^2/2}$.

general, for a Boolean function $f : \{0,1\}^k \rightarrow \{0,1\}$ we let $p_0^{i,f}$ be the a priori probability that $f(d_{i,1}, \ldots, d_{i,k}) = 1$. We analyze the a posteriori probability that $f(d_{i,1}, \ldots, d_{i,k}) = 1$ given the answers to T queries, *as well as* all the values in all the rows of d other than i: $d_{i',j}$ for all $i' \neq i$. We denote this a posteriori probability $p_T^{i,f}$.

Confidence. To simplify our calculations we follow [5] and define a monotonically-increasing 1-1 mapping conf : $(0,1) \rightarrow \mathbb{R}$ as follows:

$$\text{conf}(p) = \log \frac{p}{1-p}.$$

Note that a small additive change in conf implies a small additive change in p [4]. Let $\text{conf}_0^{i,f} = \log \frac{p_0^{i,f}}{1-p_0^{i,f}}$ and $\text{conf}_T^{i,f} = \log \frac{p_T^{i,f}}{1-p_T^{i,f}}$. We write our privacy requirements in terms of the random variables $\Delta\text{conf}^{i,f}$ defined as [5]:

$$\Delta\text{conf}^{i,f} = |\text{conf}_T^{i,f} - \text{conf}_0^{i,f}|.$$

Definition 1 ((δ,T)-Privacy). *A database access mechanism is (δ,T)-private if for every distribution $\mathcal{D}$ on $\{0,1\}^k$, for every row index i, for every function $f : \{0,1\}^k \rightarrow \{0,1\}$, and for every adversary $\mathcal{A}$ making at most T queries it holds that*

$$\Pr[\Delta\text{conf}^{i,f} > \delta] \leq \text{neg}(n).$$

The probability is taken over the choice of each row in d according to $\mathcal{D}$, and the randomness of the adversary as well as the database access mechanism.

A *target set* F is a set of k-ary Boolean functions (one can think of the functions in F as being selected by an adversary; these represent information it will try to learn about someone). A target set F is *δ-safe* if $\Delta\text{conf}^{i,f} \leq \delta$ for all $i \in [n]$ and $f \in F$. Let F be a target set. Definition 1 implies that under a (δ,T)-private database mechanism, F is δ-safe with probability $1 - \text{neg}(n)$.

Proposition 2. *Consider a (δ,T)-private database with $k = O(\log n)$ attributes. Let F be the target set containing all the 2^{2^k} Boolean functions over the k attributes. Then, $\Pr[F \text{ is } 2\delta\text{-safe}] = 1 - \text{neg}(n)$.*

Proof. Let F' be a target set containing all 2^k conjuncts of k attributes. We have that $|F'| = \text{poly}(n)$ and hence F' is δ-safe with probability $1 - \text{neg}(n)$.

To prove the proposition we show that F is safe whenever F' is. Let $f \in F$ be a Boolean function. Express f as a disjunction of conjuncts of k attributes:

---

[4] The converse does not hold – conf grows logarithmically in p for $p \approx 0$ and logarithmically in $1/(1-p)$ for $p \approx 1$.

[5] Our choice of defining privacy in terms of $\Delta\text{conf}^{i,f}$ is somewhat arbitrary, one could rewrite our definitions (and analysis) in terms of the a priori and a posteriori probabilities. Note however that limiting $\Delta\text{conf}^{i,f}$ in Definition 1 is a stronger requirement than just limiting $|p_T^{i,f} - p_0^{i,f}|$.

$f = c_1 \vee \ldots \vee c_\ell$. Similarly, express $\neg f$ as the disjunction of the remaining $2^k - \ell$ conjuncts: $\neg f = d_1 \vee \ldots \vee d_{2^k-\ell}$. (So $\{c_1, \ldots, c_\ell, d_1, \ldots, d_{2^k-\ell}\} = F$.)
We have:

$$\Delta\mathrm{conf}^{i,f} = \left| \log\left(\frac{p_T^{i,f}}{p_0^{i,f}} \cdot \frac{p_0^{i,\neg f}}{p_T^{i,\neg f}} \right) \right| = \left| \log\left(\frac{\sum p_T^{i,c_j}}{\sum p_0^{i,c_j}} \cdot \frac{\sum p_0^{i,d_j}}{\sum p_T^{i,d_j}} \right) \right|.$$

Let k maximize $|\log(p_T^{i,c_k}/p_0^{i,c_k})|$ and k' maximize $|\log(p_0^{i,d_{k'}}/p_T^{i,d_{k'}})|$. Using $|\log(\sum a_i / \sum b_i)| \le \max_i |\log(a_i/b_i)|$ we get that $\Delta\mathrm{conf}^{i,f} \le |\Delta\mathrm{conf}^{i,c_k}| + |\Delta\mathrm{conf}^{i,d_{k'}}| \le 2\delta$, where the last inequality holds as $c_k, d_{k'} \in F'$.

(δ, T)-*Privacy vs. Finding Very Heavy Sets.* Let f be a target function and $\delta = \omega(\sqrt{n})$. Our privacy requirement implies $\delta' = \delta'(\delta, \Pr[f(\alpha_1, \ldots, \alpha_k)])$ such that it is infeasible to find a "very" heavy set $q \subseteq [n]$, that is, a set for which $a_{q,f} \ge |q| (\delta' + \Pr[f(\alpha_1, \ldots, \alpha_k)])$. Such a δ'-heavy set would violate our privacy requirement as it would allow guessing $f(\alpha_1, \ldots, \alpha_k)$ for a random record in q.

Relationship to the Privacy Definition of [7]. Our privacy definition extends the definition of p_0-to-p_1 privacy breaches of [7]. Their definition is introduced with respect to a scenario in which several users send their sensitive data to a center. Each user randomizes his data prior to sending it. A p_0-to-p_1 privacy breach occurs if, with respect to some property f, the a priori probability that f holds for a user is at most p_0 whereas the a posteriori probability may grow beyond p_1 (i.e. in a worst case scenario with respect to the coins of the randomization operator).

4   Privacy of Multi-attribute SuLQ Databases

We first describe our SuLQ Database algorithm, and then prove that it preserves privacy.
Let $T(n) = O(n^c)$, $c < 1$, and define $R = (T(n)/\delta^2) \cdot \log^\mu n$ for some $\mu > 0$ (taking $\mu = 6$ will work). To simplify notation, we write d_i for $(d_{i,1}, \ldots, d_{i,k})$, $g(i)$ for $g(d_i) = g(d_{i,1}, \ldots, d_{i,k})$ (and later $f(i)$ for $f(d_i)$).

SuLQ Database Algorithm $\mathcal{A}$
Input: a query (q, g).

1. Let $a_{q,g} = \sum_{i \in q} g(i) \left(= \sum_{i \in q} g(d_{i,1}, \ldots, d_{i,k}) \right)$.
2. Generate a perturbation value: Let $(e_1, \ldots, e_R) \in_R \{0,1\}^R$ and $\mathcal{E} \leftarrow \sum_{i=1}^R e_i - R/2$.
3. Return $\hat{a}_{q,g} = a_{q,g} + \mathcal{E}$.

Note that $\mathcal{E}$ is a binomial random variable with $\mathbf{E}[\mathcal{E}] = 0$ and standard deviation $\sqrt{R}$. In our analysis we will neglect the case where $\mathcal{E}$ largely deviates from zero, as the probability of such an event is extremely small: $\Pr[\|\mathcal{E}\| > \sqrt{R}\log^2 n] = \text{neg}(n)$. In particular, this implies that our SuLQ database algorithm $\mathcal{A}$ is within $\tilde{O}(\sqrt{T(n)})$ perturbation.

We will use the following proposition.

Proposition 3. *Let B be a binomially distributed random variable with expectation 0 and standard deviation $\sqrt{R}$. Let L be the random variable that takes the value $\log\left(\frac{\Pr[B]}{\Pr[B+1]}\right)$. Then*

1. $\log\left(\frac{\Pr[B]}{\Pr[B+1]}\right) = \log\left(\frac{\Pr[-B]}{\Pr[-B-1]}\right)$. *For $0 \le B \le \sqrt{R}\log^2 n$ this value is bounded by $O(\log^2 n/\sqrt{R}))$.*
2. $\mathbf{E}[L] = O(1/R)$, *where the expectation is taken over the random choice of B.*

Proof. 1. The equality follows from the symmetry of the Binomial distribution (i.e. $\Pr[B] = \Pr[-B]$).

To prove the bound consider $\log(\Pr[B]/\Pr[B+1]) = \log(\binom{R}{R/2+B}/\binom{R}{R/2+B+1}) = \log\frac{R/2+B+1}{R/2-B-1}$. Using the limits on B and the definition of R we get that this value is bounded by $\log(1+O(\log^2 n/\sqrt{R})) = O(\log^2 n/\sqrt{R})$.

2. Using the symmetry of the Binomial distribution we get:

$$\mathbf{E}[L] = \sum_{0 \le B \le R/2} \binom{R}{R/2+B} 2^{-R}\left[\log\frac{R/2+B+1}{R/2-B} + \log\frac{R/2-B+1}{R/2+B}\right]$$

$$= \sum_{0 \le B \le \log^2 n\sqrt{R}} \binom{R}{R/2+B} 2^{-R}\log\left(1+\frac{R+1}{R^2/4-B^2}\right) + \text{neg}(n) = O(1/R)$$

Our proof of privacy is modeled on the proof in Section 4 of [5] (for single attribute databases). We extend their proof (i) to queries of the form (q, g) where g is any k-ary Boolean function, and (ii) to privacy of k-ary Boolean functions f.

Theorem 1. *Let $T(n) = O(n^c)$ and $\delta = 1/O(n^{c'})$ for $0 < c < 1$ and $0 \le c' < c/2$. Then the SuLQ algorithm $\mathcal{A}$ is $(\delta, T(n))$-private within $\tilde{O}(\sqrt{T(n)}/\delta)$ perturbation.*

Note that whenever $\sqrt{T(n)}/\delta < \sqrt{n}$ bounding the adversary's number of queries to $T(n)$ allows privacy with perturbation magnitude less than $\sqrt{n}$.

Proof. Let $T(n)$ be as in the theorem and recall $R = (T(n)/\delta^2) \cdot \log^\mu n$ for some $\mu > 0$.

Let the $T = T(n)$ queries issued by the adversary be denoted $(q_1, g_1), \ldots, (q_T, g_T)$. Let $\hat{a}_1 = \mathcal{A}(q_1, g_1), \ldots, \hat{a}_t = \mathcal{A}(q_T, g_T)$ be the perturbed answers to these queries. Let $i \in [n]$ and $f : \{0,1\}^k \to \{0,1\}$.

We analyze the *a posteriori* probability p_ℓ that $f(i) = 1$ given the answers to the first ℓ queries $(\hat{a}_1, \ldots, \hat{a}_\ell)$ and $d^{\{-i\}}$ (where $d^{\{-i\}}$ denotes the entire database except for the ith row). Let $\text{conf}_\ell = \log_2 p_\ell/(1 - p_\ell)$. Note that $\text{conf}_T = \text{conf}_T^{i,f}$ (of Section 3), and (due to the independence of rows in d) $\text{conf}_0 = \text{conf}_0^{i,f}$.

By the definition of conditional probability[6] we get

$$\frac{p_\ell}{1 - p_\ell} = \frac{\Pr[f(i) = 1|\hat{a}_1, \ldots, \hat{a}_\ell, d^{\{-i\}}]}{\Pr[f(i) = 0|\hat{a}_1, \ldots, \hat{a}_\ell, d^{\{-i\}}]} = \frac{\Pr[\hat{a}_1, \ldots, \hat{a}_\ell \wedge f(i) = 1|d^{\{-i\}}]}{\Pr[\hat{a}_1, \ldots, \hat{a}_\ell \wedge f(i) = 0|d^{\{-i\}}]} = \frac{\textbf{Num}}{\textbf{Denom}}.$$

Note that the probabilities are taken over the coin flips of the SuLQ algorithm and the choice of d. In the following we analyze the numerator (the denominator is analyzed similarly).

$$\textbf{Num} = \sum_{\sigma \in \{0,1\}^k, f(\sigma)=1} \Pr[\hat{a}_1, \ldots, \hat{a}_\ell \wedge d_i = \sigma|d^{\{-i\}}]$$

$$= \sum_{\sigma \in \{0,1\}^k, f(\sigma)=1} \Pr[\hat{a}_1, \ldots, \hat{a}_\ell|d_i = \sigma, d^{\{-i\}}] \Pr[d_i = \sigma]$$

The last equality follows as the rows in d are chosen independently of each other. Note that given both d_i and $d^{\{-i\}}$ the random variable $\hat{a}_\ell$ is independent of $\hat{a}_1, \ldots, \hat{a}_{\ell-1}$. Hence, we get:

$$\textbf{Num} = \sum_{\sigma \in \{0,1\}^k, f(\sigma)=1} \Pr[\hat{a}_1, \ldots, \hat{a}_{\ell-1}|d_i = \sigma, d^{\{-i\}}] \Pr[\hat{a}_\ell|d_i = \sigma, d^{\{-i\}}] \Pr[d_i = \sigma].$$

Next, we observe that although $\hat{a}_\ell$ depends on d_i, the dependence is weak. More formally, let $\sigma_0, \sigma_1 \in \{0,1\}^k$ be such that $f(\sigma_0) = 0$ and $f(\sigma_1) = 1$. Note that whenever $g_\ell(\sigma) = g_\ell(\sigma_1)$ we have that $\Pr[\hat{a}_\ell|d_i = \sigma, d^{\{-i\}}] = \Pr[\hat{a}_\ell|d_i = \sigma_1, d^{\{-i\}}]$. When, instead, $g_\ell(\sigma) \neq g_\ell(\sigma_1)$, we can relate $\Pr[\hat{a}_\ell|d_i = \sigma, d^{\{-i\}}]$ and $\Pr[\hat{a}_\ell|d_i = \sigma_1, d^{\{-i\}}]$ via Proposition 3:

Lemma 1. *Let σ, σ_1 be such that $g_\ell(\sigma) \neq g_\ell(\sigma_1)$. Then $\Pr[\hat{a}_\ell|d_i = \sigma, d^{\{-i\}}] = 2^\epsilon \Pr[\hat{a}_\ell|d_i = \sigma_1, d^{\{-i\}}]$ where $|\mathbf{E}[\epsilon]| = O(1/R)$ and*

$$\epsilon = \begin{cases} -(-1)^{g_\ell(\sigma_1)}O(\log^2 n/\sqrt{R}) & \text{if } \mathcal{E} \leq 0 \\ (-1)^{g_\ell(\sigma_1)}O(\log^2 n/\sqrt{R}) & \text{if } \mathcal{E} > 0 \end{cases}$$

and $\mathcal{E}$ is noise that yields $\hat{a}_\ell$ when $d_i = \sigma$.

Proof. Consider the case $g_\ell(\sigma_1) = 0$ ($g_\ell(\sigma) = 1$). Writing $\Pr[\hat{a}_\ell|d_i = \sigma, d^{\{-i\}}] = \Pr[\mathcal{E} = k]$ and $\Pr[\hat{a}_\ell|d_i = \sigma_1, d^{\{-i\}}] = \Pr[\mathcal{E} = k - 1]$ the proof follows from Proposition 3. Similarly for $g_\ell(\sigma_1) = 1$.

Note that the value of ϵ does not depend on σ.

Taking into account both cases ($g_\ell(\sigma) = g_\ell(\sigma_1)$ and $g_\ell(\sigma) \neq g_\ell(\sigma_1)$) we get

$$\textbf{Num} = \sum_{\sigma \in \{0,1\}^k, f(\sigma)=1} \Pr[\hat{a}_1, \ldots, \hat{a}_{\ell-1}|d_i = \sigma, d^{\{-i\}}] 2^\epsilon \Pr[\hat{a}_\ell|d_i = \sigma_1, d^{\{-i\}}] \Pr[d_i = \sigma].$$

---

[6] I.e. $\Pr[E_1|E_2] \cdot \Pr[E_2] = \Pr[E_1 \wedge E_2] = \Pr[E_2|E_1] \cdot \Pr[E_1]$.

Let $\hat{\gamma}$ be the probability, over d_i, that $g(\sigma) \neq g(\sigma_1)$. Letting $\gamma \geq 1$ be such that $2^{1/\gamma} = \hat{\gamma}$, we have

$$\mathbf{Num} = 2^{\epsilon/\gamma} \Pr[\hat{a}_\ell | d_i = \sigma_1, d^{\{-i\}}] \sum_{\sigma \in \{0,1\}^k, f(\sigma)=1} \Pr[\hat{a}_1, \ldots, \hat{a}_{\ell-1} | d_i = \sigma, d^{\{-i\}}] \Pr[d_i = \sigma]$$

$$= 2^{\epsilon/\gamma} \Pr[\hat{a}_\ell | d_i = \sigma_1, d^{\{-i\}}] \sum_{\sigma \in \{0,1\}^k, f(\sigma)=1} \Pr[\hat{a}_1, \ldots, \hat{a}_{\ell-1} \wedge d_i = \sigma | d^{\{-i\}}]$$

$$= 2^{\epsilon/\gamma} \Pr[\hat{a}_\ell | d_i = \sigma_1, d^{\{-i\}}] \Pr[\hat{a}_1, \ldots, \hat{a}_{\ell-1} \wedge f(i) = 1 | d^{\{-i\}}]$$

$$= 2^{\epsilon/\gamma} \Pr[\hat{a}_\ell | d_i = \sigma_1, d^{\{-i\}}] \Pr[f(i) = 1 | \hat{a}_1, \ldots, \hat{a}_{\ell-1}, d^{\{-i\}}] \Pr[\hat{a}_1, \ldots, \hat{a}_{\ell-1} | d^{\{-i\}}]$$

$$= 2^{\epsilon/\gamma} \Pr[\hat{a}_\ell | d_i = \sigma_1, d^{\{-i\}}] p_{\ell-1} \Pr[\hat{a}_1, \ldots, \hat{a}_{\ell-1} | d^{\{-i\}}]$$

and similarly

$$\mathbf{Denom} = 2^{\epsilon'/\gamma'} \Pr[\hat{a}_\ell | d_i = \sigma_0, d^{\{-i\}}](1 - p_{\ell-1}) \Pr[\hat{a}_1, \ldots, \hat{a}_{\ell-1} | d^{\{-i\}}].$$

Putting the pieces together we get that

$$\mathrm{conf}_\ell = \log_2 \frac{\mathbf{Num}}{\mathbf{Denom}} = \mathrm{conf}_{\ell-1} + (\epsilon/\gamma - \epsilon'/\gamma') + \log_2 \frac{\Pr[\hat{a}_\ell | d_i = \sigma_1, d^{\{-i\}}]}{\Pr[\hat{a}_\ell | d_i = \sigma_0, d^{\{-i\}}]}.$$

Define a random walk on the real line with $\mathrm{step}_\ell = \mathrm{conf}_\ell - \mathrm{conf}_{\ell-1}$. To conclude the proof we show that (with high probability) T steps of the random walk do not suffice to reach distance δ. From Proposition 3 and Lemma 1 we get that

$$|\mathbf{E}[\mathrm{step}_\ell]| = O(1/R) = O\left(\frac{\delta^2}{T \log^\mu n}\right)$$

and

$$|\mathrm{step}_\ell| = O(\log^2 n / \sqrt{R}) = O\left(\frac{\delta}{\sqrt{T} \log^{\mu/2-2} n}\right).$$

Using Proposition 1 with $\lambda = \log n$ we get that for all $t \leq T$,

$$\Pr[|\mathrm{conf}_t - \mathrm{conf}_0| > \delta] = \Pr[|\sum_{\ell \leq t} \mathrm{step}_\ell| > \delta] \leq \mathsf{neg}(n).$$

5   Datamining on Vertically Partitioned Databases

In this section we assume that the database is chosen according to $\mathcal{D}^n$ for some underlying distribution $\mathcal{D}$ on rows, where $\mathcal{D}$ is independent of n, the size of the database. We also assume that n, is sufficiently large that the true database statistics are representative of $\mathcal{D}$. Hence, in the sequel, when we write things like "$\Pr[\alpha]$" we mean the probability, over the entries in the database, that α holds.

Let α and β be attributes. We say that α *implies* β *in probability* if the conditional probability of β given α exceeds the unconditional probability of β. The ability to measure implication in probability is crucial to datamining. Note that since $\Pr[\beta]$ is simple to estimate well, the problem reduces to obtaining a good estimate of $\Pr[\beta|\alpha]$. Moreover, once we can estimate the $\Pr[\beta|\alpha]$, we can use Bayes' Rule and de Morgan's Laws to determine the statistics for any Boolean function of attribute values.

Our key result for vertically partitioned databases is a method, given two single-attribute SuLQ databases with attributes α and β respectively, to measure $\Pr[\beta|\alpha]$.

For more general cases of vertically partitioned data, assume a k-attribute database is partitioned into $2 \leq j \leq k$ databases, with $k_1, \ldots, k_j$ (possibly overlapping) attributes, respectively, where $\sum_i k_i \geq k$. We can use functional queries to learn the statistics on k_i-ary Boolean functions of the attributes in the ith database, and then use the results for two single-attribute SuLQ databases to learn binary Boolean functions of any two functions f_{i_1} (on attributes in database i_1) and f_{i_2} (on attributes in database i_2), where $1 \leq i_1, i_2 \leq j$.

5.1   Probabilistic Implication

In this section we construct our basic building block for mining vertically partitioned databases.

We assume two SuLQ databases d_1, d_2 of size n, with attributes α, β respectively. When α implies β in probability with a gap of Δ, we write $\alpha \overset{\Delta}{\to} \beta$, meaning that $\Pr[\beta|\alpha] = \Pr[\beta] + \Delta$. We note that $\Pr[\alpha]$ and $\Pr[\beta]$ are easily computed within error $O(1/\sqrt{n})$, simply by querying the two databases on large subsets. Our goal is to determine Δ, or equivalently, $\Pr[\beta|\alpha] - \Pr[\beta]$; the method will be to determine if, for a given Δ_1, $\Pr[\beta|\alpha] \geq \Pr[\beta] + \Delta_1$, and then to estimate Δ by binary search on Δ_1.

Notation. We let $p_\alpha = \Pr[\alpha]$, $p_\beta = \Pr[\beta]$, $p_{\beta|\alpha} = \Pr[\beta|\alpha]$ and $p_{\beta|\bar{\alpha}} = \Pr[\beta|\neg\alpha]$.

Let X be a random variable counting the number of times α holds when we take N samples from $\mathcal{D}$. Then $\mathbf{E}[X] = Np_a$ and $\mathbf{Var}[X] = Np_a(1 - p_a)$.

Let

$$p_{\beta|\alpha} = p_\beta + \Delta. \tag{1}$$

Note that $p_\beta = p_\alpha p_{\beta|\alpha} + (1 - p_\alpha)p_{\beta|\bar{\alpha}}$. Substituting $p_\beta + \Delta$ for $p_{\beta|\alpha}$ we get

$$p_{\beta|\bar{\alpha}} = p_\beta - \Delta \frac{p_\alpha}{1 - p_\alpha}, \tag{2}$$

and hence (by another application of Eq. (1))

$$p_{\beta|\alpha} - p_{\beta|\bar{\alpha}} = \frac{\Delta}{1 - p_\alpha}. \tag{3}$$

We define the following testing procedure to determine, given Δ_1, if $\Delta \geq \Delta_1$. Step 1 finds a heavy (but not very heavy) set for attribute α, that is, a set q for which the number of records satisfying α exceeds the expected number by more than a standard deviation. Note that since $T(n) = o(n)$, the noise $|\hat{a}_{q,1} - a_{q,1}|$

is $o(\sqrt{n})$, so the heavy set really has $Np_\alpha + \Omega(\sqrt{N})$ records for which α holds. Step 2 queries d_2 on this heavy set. If the incidence of β on this set sufficiently (as a function of Δ_1) exceeds the expected incidence of β, then the test returns "1" (ie, success). Otherwise it returns 0.

Test Procedure $\mathcal{T}$
Input: $p_\alpha, p_\beta, \Delta_1 > 0$.

1. Find $q \in_R [n]$ such that $a_{q,1} \geq Np_\alpha + \sigma_\alpha$ where $N = |q|$ and $\sigma_\alpha = \sqrt{Np_\alpha(1-p_\alpha)}$.
   Let $\text{bias}_\alpha = a_{q,1} - Np_\alpha$.
2. If $a_{q,2} \geq Np_\beta + \text{bias}_\alpha \frac{\Delta_1}{1-p_\alpha}$ return 1, otherwise return 0.

Theorem 2. *For the test procedure $\mathcal{T}$:*

1. *If $\Delta \geq \Delta_1$, then $\Pr[\mathcal{T}\ outputs\ 1] \geq 1/2$.*
2. *If $\Delta \leq \Delta_1 - \varepsilon$, then $\Pr[\mathcal{T}\ outputs\ 1] \leq 1/2 - \gamma$,*

where for $\varepsilon = \Theta(1)$ the advantage $\gamma = \gamma(p_\alpha, p_\beta, \varepsilon)$ is constant, and for $\varepsilon = o(1)$ the advantage $\gamma = c \cdot \varepsilon$ with constant $c = c(p_\alpha, p_\beta)$.

In the following analysis we neglect the difference between $a_{q,i}$ and $\hat{a}_{q,i}$, since, as noted above, the perturbation contributes only low order terms (we neglect some other low order terms). Note that it is possible to compute all the required constants for Theorem 2 explicitly, in polynomial time, without neglecting these low-order terms. Our analysis does not attempt to optimize constants.

Proof. Consider the random variable corresponding to $a_{q,2} = \sum_{i \in q} d_{i,2}$, given that q is biased according to Step 1 of $\mathcal{T}$. By linearity of expectation, together with the fact that the two cases below are disjoint, we get that

$$\mathbf{E}[a_{q,2}|\text{bias}_\alpha] = (Np_\alpha + \text{bias}_\alpha)p_{\beta|\alpha} + (N(1-p_\alpha) - \text{bias}_\alpha)p_{\beta|\bar{\alpha}}$$
$$= Np_\alpha p_{\beta|\alpha} + N(1-p_\alpha)p_{\beta|\bar{\alpha}} + \text{bias}_\alpha(p_{\beta|\alpha} - p_{\beta|\bar{\alpha}})$$
$$= Np_\beta + \text{bias}_\alpha \frac{\Delta}{1-p_\alpha}.$$

The last step uses Eq. (3). Since the distribution of $a_{q,2}$ is symmetric around $\mathbf{E}[a_{q,2}|\text{bias}_\alpha]$ we get that the first part of the claim, i.e. if $\Delta \geq \Delta_1$ then

$$\Pr[\mathcal{T}\ outputs\ 1] = \Pr[a_{q,2} > Np_\beta + \text{bias}_\alpha \frac{\Delta_1}{1-p_\alpha}|\text{bias}_\alpha] \geq 1/2.$$

To get the second part of the claim we use the de Moivre-Laplace theorem and approximate the binomial distribution with the normal distribution so that we can approximate the variance of the sum of two distributions (when α holds and when α does not hold) in order to obtain the variance of $a_{q,2}$ conditioned on bias_α. We get:

$$\mathbf{Var}[a_{q,2}|\text{bias}_\alpha] \approx (Np_\alpha + \text{bias}_\alpha)p_{\beta|\alpha}(1-p_{\beta|\alpha}) + (N(1-p_\alpha) - \text{bias}_\alpha)p_{\beta|\bar{\alpha}}(1-p_{\beta|\bar{\alpha}}).$$

Assuming N is large enough, we can neglect the terms involving bias$_\alpha$. Hence,

$$\mathbf{Var}[a_{q,2}|\text{bias}_\alpha] \approx N[p_\alpha p_{\beta|\alpha} + (1-p_\alpha)p_{\beta|\bar{\alpha}}] - N[p_\alpha p^2_{\beta|\alpha} + (1-p_\alpha)p^2_{\beta|\bar{\alpha}}]$$

$$\approx Np_\beta - N[p_\alpha p^2_{\beta|\alpha} + (1-p_\alpha)p^2_{\beta|\bar{\alpha}}]$$

$$= N[p_\beta - p^2_\beta] - N\Delta^2\frac{p_\alpha}{1-p_\alpha} < N[p_\beta - p^2_\beta] = \mathbf{Var}_\beta.$$

The transition from the second to third lines follows from $[p_\alpha p^2_{\beta|\alpha}+(1-p_\alpha)p^2_{\beta|\bar{\alpha}}]-p^2_\beta = \Delta^2\frac{p_\alpha}{1-p_\alpha}$ [7].

We have that the probability distribution on $a_{q,2}$ is a Gaussian with mean and variance at most $Np_\beta + \text{bias}_\alpha(\Delta_1 - \varepsilon)/(1 - p_\alpha)$ and $\mathbf{Var}_\beta$ respectively. To conclude the proof, we note that the conditional probability mass of $a_{q,2}$ exceeding its own mean by $\varepsilon \cdot \text{bias}_\alpha/(1-p_\alpha) > \varepsilon\sigma_\alpha/(1-p_\alpha)$ is at most

$$\frac{1}{2} - \gamma = \Phi\left(-\frac{\varepsilon\sigma_\alpha/(1-p_\alpha)}{\sqrt{\mathbf{Var}_\beta}}\right)$$

where Φ is the cumulative distribution function for the normal distribution. For constant ε this yields a constant advantage γ. For $\varepsilon = o(1)$, we get that $\gamma \geq \frac{\varepsilon}{2}\frac{\sigma_\alpha/(1-p_\alpha)}{\sqrt{\mathbf{Var}_\beta}\sqrt{2\pi}}$.

By taking $\varepsilon = \omega(1/\sqrt{n})$ we can run the Test procedure enough times to determine with sufficiently high confidence which "side" of the interval $[\Delta_1 - \varepsilon, \Delta_1]$ Δ is on (if it is not inside the interval). We proceed by binary search to narrow in on Δ. We get:

Theorem 3. *There exists an algorithm that invokes the test T*

$$O_{p_\alpha,p_\beta}(\log(1/\epsilon)\frac{\log(1/\delta) + \log\log(1/\epsilon)}{\epsilon^2})$$

times and outputs $\hat{\Delta}$ such that $\Pr[|\hat{\Delta} - \Delta| < \varepsilon] \geq 1 - \delta$.

6 Datamining on Published Statistics

In this section we apply our basic technique for measuring implication in probability to the real-life model in which confidential information is gathered by a trusted party, such as the census bureau, who publishes aggregate statistics. The published statistics are the results of queries to a SuLQ database. That is, the census bureau generates queries and their noisy responses, and publishes the results.

---

[7] In more detail: $[p_\alpha p^2_{\beta|\alpha} + (1 - p_\alpha)p^2_{\beta|\bar{\alpha}}] - p^2_\beta = p^2_{\beta|\alpha}p_\alpha(1 - p_\alpha) + p^2_{\beta|\bar{\alpha}}(1 - p_\alpha)p_\alpha - 2p_\alpha(1-p_\alpha)p_{\beta|\alpha}p_{\beta|\bar{\alpha}} = p_\alpha(1-p_\alpha)[p^2_{\beta|\alpha}+p^2_{\beta|\bar{\alpha}}-2p_{\beta|\alpha}p_{\beta|\bar{\alpha}}] = p_\alpha(1-p_\alpha)(p_{\beta|\alpha}-p_{\beta|\bar{\alpha}})^2 = \Delta^2\frac{p_\alpha}{1-p_\alpha}$.

Let k denote the number of attributes (columns). Let $\ell \leq k/2$ be fixed (typically, ℓ will be small; see below). For every ℓ-tuple of attributes $(\alpha_1, \alpha_2, \ldots, \alpha_\ell)$, and for each of the 2^ℓ conjunctions of literals over these ℓ attributes, $(\bar{\alpha}_1 \bar{\alpha}_2 \ldots \bar{\alpha}_\ell, \bar{\alpha}_1 \bar{\alpha}_2 \ldots \alpha_\ell$, and so on), the bureau publishes the result of some number t of queries on these conjunctions. More precisely, a query set $q \subseteq [n]$ is selected, and noisy statistics for all $\binom{k}{\ell} 2^\ell$ conjunctions of literals are published for the query. This is repeated t times.

To see how this might be used, suppose $\ell = 3$ and we wish to learn if $\alpha_1 \alpha_2 \alpha_3$ implies $\bar{\alpha}_4 \bar{\alpha}_5 \alpha_6$ in probability. We know from the results in Section 4 that we need to find a heavy set q for $\alpha_1 \alpha_2 \alpha_3$, and then to query the database on the set q with the function $\bar{\alpha}_4 \bar{\alpha}_5 \alpha_6$. Moreover, we need to do this several times (for the binary search). If t is sufficiently large, then with high probability such query sets q are among the t queries. Since we query all triples (generally, ℓ-tuples) of literals for each query set q, all the necessary information is published. The analyst need only follow the instructions for learning the strength Δ of the implication in probability $\alpha_1 \alpha_2 \alpha_3 \stackrel{\Delta}{\to} \bar{\alpha}_4 \bar{\alpha}_5 \alpha_6$, looking up the results of the queries (rather than randomly selecting the sets q and submitting the queries to the database).

As in Section 4, once we can determine implication in probability, it is easy to determine (via Bayes' rule) the statistics for the conjunction $\alpha_1 \alpha_2 \alpha_3 \bar{\alpha}_4 \bar{\alpha}_5 \alpha_6$. In other words, *we can determine the approximate statistics for any conjunction of 2ℓ literals of attribute values*. Now the procedure for arbitrary 2ℓ-ary functions is conceptually simple. Consider a function of attribute values $\beta_1 \ldots \beta_{2\ell}$. The analyst first represents the function as a truth table: for each possible 2ℓ-tuple of literals over $\beta_1 \ldots \beta_{2\ell}$ the function has value either zero or one. Since these conjunctions of literals are mutually exclusive, the probability (overall) that the function has value 1 is simply the sum of the probabilities that each of the positive (one-valued) conjunctions occurs. Since we can approximate each of these statistics, we obtain an approximation for their sum. Thus, we can approximate the statistics for each of the $\binom{k}{2\ell} 2^{2^{2\ell}}$ Boolean functions of 2ℓ attributes. It remains to analyze the quality of the approximations.

Let $T = o(n)$ be an upper bound on the number of queries permitted by the SuLQ database algorithm, e.g., $T = O(n^c), c < 1$. Let k and ℓ be as above: k is the total number of attributes, and statistics for ℓ-tuples will be published. Let ε be the (combined) additive error achieved for all $\binom{k}{2\ell} 2^{2\ell}$ conjuncts with probability $1 - \delta$.

Input: a database $d = \{d_{i,j}\}$ of dimensions $n \times k$.
Repeat t times:

1. Let $q \in_R [n]$. Output q.
2. For all selections of ℓ indices $1 \leq j_1 < j_2 < \ldots < j_\ell \leq k$, output $\hat{a}_{q,g}$ for all the 2^ℓ conjuncts g over the literals $\alpha_{j_1}, \ldots, \alpha_{j_\ell}$.

Privacy is preserved as long as $t \cdot \binom{k}{2\ell} 2^{2\ell} \leq T$ (Theorem 1). To determine utility, we need to understand the error introduced by the summation of estimates.

Let $\varepsilon' = \varepsilon/2^{2\ell}$. If our test results in a ε' additive error for each possible conjunct of 2ℓ literals, the truth table method described above allows us to compute the frequency of every function of 2ℓ literals within additive error ε (a lot better in many cases). We require that our estimate be within error ε' with probability $1 - \delta'$ where $\delta' = \delta/\binom{k}{2\ell}2^{2\ell}$. Hence, the probability that a 'bad' conjunct exists (for which the estimation error is not within ε') is bounded by δ.

Plugging δ' and ε' into Theorem 3, we get that for each conjunction of ℓ literals, the number of subsets q on which we need to make queries is

$$t = O\left(2^{4\ell}(\log(1/\epsilon) + \ell)(\log(1/\delta) + \ell \log k + \log\log(1/\epsilon))/\epsilon^2\right).$$

For each subset q we query each of the $\binom{k}{\ell}2^{\ell}$ conjuncts of ℓ attributes. Hence, the total number of queries we make is

$$t \cdot \binom{k}{\ell}2^{\ell} = O\left(k^{\ell}2^{5\ell}(\log(1/\epsilon) + \ell)(\log(1/\delta) + \ell \log k + \log\log(1/\epsilon))/\epsilon^2\right).$$

For constant ϵ, δ we get that the total number of queries is $O(2^{5\ell}k^{\ell}\ell^2 \log k)$. To see our gain, compare this with the naive publishing of statistics for all conjuncts of 2ℓ attributes, resulting in $\binom{k}{2\ell}2^{2\ell} = O(k^{2\ell}2^{2\ell})$ queries.

7   Open Problems

Datamining of 3-ary Boolean Functions. Section 5.1 shows how to use two SuLQ databases to learn that $\Pr[\beta|\alpha] = \Pr[\beta] + \Delta$. As noted, this allows estimating $\Pr[f(\alpha, \beta)]$ for any Boolean function f. Consider the case where there exist three SuLQ databases for attributes α, β, γ. In order to use our test procedure to compute $\Pr[f(\alpha, \beta, \gamma)]$, one has to either to find heavy sets for $\alpha \wedge \beta$ (having bias of order $\Omega(\sqrt{n})$), or, given a heavy set for γ, to decide whether it is also heavy w.r.t. $\alpha \wedge \beta$. It is not clear how to extend the test procedure of Section 5.1 in this direction.

Maintaining Privacy for All Possible Functions. Our privacy definition (Definition 1) requires for every function $f(\alpha_1, \ldots, \alpha_k)$ that with high probability the confidence gain is limited by some value δ. If k is small (less than $\log \log n$), then, via the union bound, we get that with high probability the confidence gain is kept small for all the 2^{2^k} possible functions.

For large k the union bound does not guarantee simultaneous privacy for all the 2^{2^k} possible functions. However, the privacy of a randomly selected function is (with high probability) preserved. It is conceivable that (e.g. using cryptographic measures) it is possible to render infeasible the task of finding a function f whose privacy was breached.

Dependency Between Database Records. We explicitly assume that the database records are chosen independently from each other, according to some underlying distribution $\mathcal{D}$. We are not aware of any work that does not make this assumption (implicitly or explicitly). An important research direction is to come up with definition and analysis that work in a more realistic model of weak dependency between database entries.

References

1. D. Agrawal and C. Aggarwal, *On the Design and Quantification of Privacy Preserving Data Mining Algorithms*, Proceedings of the 20th Symposium on Principles of Database Systems, 2001.
2. N. R. Adam and J. C. Wortmann, *Security-Control Methods for Statistical Databases: A Comparative Study*, ACM Computing Surveys 21(4): 515-556 (1989).
3. R. Agrawal and R. Srikant, *Privacy-preserving data mining*, Proc. of the ACM SIGMOD Conference on Management of Data, pp. 439–450, 2000.
4. S. Chawla, C. Dwork, F. McSherry, A. Smith, and H. Wee, *Toward Privacy in Public Databases*, submitted for publication, 2004.
5. I. Dinur and K. Nissim, *Revealing information while preserving privacy*, Proceedings of the Twenty-Second ACM SIGACT-SIGMOD-SIGART Symposium on Principles of Database Systems, pp. 202-210, 2003.
6. G. Duncan, Confidentiality and statistical disclosure limitation. In N. Smelser & P. Baltes (Eds.), International Encyclopedia of the Social and Behavioral Sciences. New York: Elsevier. 2001
7. A. V. Evfimievski, J. Gehrke and R. Srikant, *Limiting privacy breaches in privacy preserving data mining*, Proceedings of the Twenty-Second ACM SIGACT-SIGMOD-SIGART Symposium on Principles of Database Systems, pp. 211-222, 2003.
8. S. Fienberg, Confidentiality and Data Protection Through Disclosure Limitation: Evolving Principles and Technical Advances, IAOS Conference on Statistics, Development and Human Rights September, 2000, available at
   http://www.statistik.admin.ch/about/international/fienberg_final_paper.doc
9. S. Fienberg, U. Makov, and R. Steele, *Disclosure Limitation and Related Methods for Categorical Data*, Journal of Official Statistics, 14, pp. 485–502, 1998.
10. L. Franconi and G. Merola, *Implementing Statistical Disclosure Control for Aggregated Data Released Via Remote Access*, Working Paper No. 30, United Nations Statistical Commission and European Commission, joint ECE/EUROSTAT work session on statistical data confidentiality, April, 2003, available at
    http://www.unece.org/stats/documents/2003/04/confidentiality/wp.30.e.pdf
11. S. Goldwasser and S. Micali, *Probabilistic Encryption and How to Play Mental Poker Keeping Secret All Partial Information*, STOC 1982: 365-377
12. T.E. Raghunathan, J.P. Reiter, and D.B. Rubin, *Multiple Imputation for Statistical Disclosure Limitation*, Journal of Official Statistics 19(1), pp. 1 – 16, 2003
13. D.B. Rubin, *Discussion: Statistical Disclosure Limitation*, Journal of Official Statistics 9(2), pp. 461 – 469, 1993.
14. A. Shoshani, *Statistical databases: Characteristics, problems and some solutions*, Proceedings of the 8th International Conference on Very Large Data Bases (VLDB'82), pages 208–222, 1982.

Optimal Perfectly Secure Message Transmission

K. Srinathan*, Arvind Narayanan, and C. Pandu Rangan

Department of Computer Science and Engineering,
Indian Institute of Technology, Madras, Chennai-600036, India
ksrinath@cs.iitm.ernet.in, arvindn@meenakshi.cs.iitm.ernet.in,
rangan@iitm.ernet.in

Abstract. In the *perfectly secure message transmission* (PSMT) problem, two synchronized non-faulty players (or processors), the *Sender* **S** and the *Receiver* **R** are connected by n wires (each of which facilitates 2-way communication); **S** has an ℓ-bit message that he wishes to send to **R**; after exchanging messages in phases[1] **R** should *correctly* obtain **S**'s message, while an *adversary* listening on and *actively* controlling any set of t (or less) wires should have *no information* about **S**'s message.

We measure the quality of a protocol for securely transmitting an ℓ-bit message using the following parameters: the number of wires n, the number of phases r and the total number of bits transmitted b. The optima for n and r are respectively $2t+1$ and 2. We prove that any 2-phase reliable message transmission protocol, and hence any secure protocol, over n wires out of which at most t are faulty is required to transmit at least $b = \left(\frac{n\ell}{n-2t}\right)$ bits. While no known protocol is simultaneously optimal in both communication and phase complexity, we present one such optimum protocol for the case $n = 2t+1$ when the size of message is large enough, viz., $\ell = \Omega(t \log t)$ bits; that is, our optimal protocol has $n = 2t+1$, $r = 2$ and $b = O(n\ell)$ bits. Note that privacy is *for free*, if the message is large enough.

We also demonstrate how randomness can effectively improve the phase complexity. Specifically, while the (worst-case) lower bound on r is 2, we design an efficient optimally tolerant protocol for PSMT that terminates in a *single* phase with arbitrarily high probability.

Finally, we consider the case when the adversary is *mobile*, that is, he could corrupt a different set of t wires in different phases. Again, the optima for n and r are respectively $2t+1$ and 2; However we show that $b \geq \left(\frac{n\ell}{n-2t}\right)$ bits irrespective of r. We present the first protocol that is (asymptotically) optimum in b for $n = 2t+1$. Our protocol has a phase complexity of $O(t)$.

1 Introduction

Consider a synchronous network $\mathcal{N}(\mathcal{P}, \mathcal{E})$ represented by an undirected graph where $\mathcal{P} = \{P_1, P_2, \ldots, P_N\} \cup \{\mathbf{S}, \mathbf{R}\}$ denotes the set of players (nodes) in

---

* Financial support from Infosys Technologies Limited, India, is acknowledged.
[1] A phase is a send from **S** to **R** or from **R** to **S** or both simultaneously.

M. Franklin (Ed.): CRYPTO 2004, LNCS 3152, pp. 545–561, 2004.
© International Association for Cryptologic Research 2004

the network that are connected by 2-way communication links as defined by $\mathcal{E} \subset \mathcal{P} \times \mathcal{P}$. The players **S** and **R** do not trust the network connecting them. Nevertheless, the sender **S** wishes to *securely* send a message to the receiver **R** through the network. Security here means that **R** should receive exactly what **S** sent to him while other players should have no information about it, even if up to t of the players (excluding **S** and **R**) collude and behave maliciously. This problem, known as *perfectly secure message transmission* (PSMT), was first proposed and solved by Dolev *et al.*[3]. In essence, it is proved in [3] that PSMT from **S** to **R** across the network $\mathcal{N}$, tolerating a static[2] adversary that corrupts up to t players (nodes), is possible if and only if $\mathcal{N}$ is at least $(2t + 1)$-(**S**,**R**)-connected[3]. We use the approach of [3] and abstract away the network entirely and concentrate on solving the PSMT problem for a single pair of synchronized processors, the *Sender* **S** and the *Receiver* **R**, connected by some number n of wires denoted by $w_1, w_2, \ldots w_n$. We may think of these wires as a collection of vertex-disjoint paths between **S** and **R** in the underlying network[4].

The PSMT problem is important in its own right as well as a very useful primitive in various secure distributed protocols. Note that if **S** and **R** are connected directly via a private and authenticated link (like what is assumed in generic secure multiparty protocols [15, 6, 1, 12]), secure communication is trivially guaranteed. However, in reality, it is not economical to directly connect *every* two players in the network. Therefore, such a complete network can only be (virtually) realized by simulating the missing links using SMT protocols as primitives.

In this paper, we shall use the simple and standard model of a synchronous network wherein any communication protocol evolves as a series of *phases*, during which the players (**S** or **R**) send messages, receive them and perform (polynomial time) local computations according to the protocol.

There are three basic aspects contributing to the quality of an algorithm for PSMT: the maximum tolerable number t of faulty wires, the number of phases r, and the total number of bits sent b. The optima for the above quality parameters are as follows: $n = 2t + 1$ [3], $r = 2$ [13]. The lower bound for b is proved in this work to be $\left(\frac{n\ell}{n-2t}\right)$ bits when $r = 2$ for any $n \geq 2t + 1$.

In the last few years, there have been some attempts toward improving the quality of protocols. All protocols proposed so far, securely communicate an element of a finite field $\mathbb{F}$; extending this to securely communicate ℓ field elements would result in a proportional increase of communication complexity. Dolev *et al.* [3] proposed three protocols: the first one with $n = 2t + 1, r = t + 1, b = O(t^3\ell)$ field elements, the second one with $n = 2t + 1, r = 3, b = O(t^5\ell)$ field elements

---

[2] By static adversary, we mean an adversary that decides on the set of players to corrupt before the start of the protocol.

[3] We say that a network $\mathcal{N}$ is κ-(P_i, P_j)-connected if the deletion of no $(\kappa - 1)$ or less nodes from $\mathcal{N}$ disconnects P_i and P_j.

[4] The approach of abstracting the network as a collection of n wires is justified using Menger's theorem [8] which states that a graph is c-(**S**,**R**)-connected if and only if **S** and **R** are connected by at least c vertex-disjoint paths.

and the third one with $n = 2t + 1$, $r = 2$ and b not polynomial in n. This was substantially improved by the protocol of [13], that has $n = 2t + 1$, $r = 2$ and $b = O(t^3 \ell)$ field elements. The protocol of [14] has $n = 2t + 1$, $r = \log t$ and $b = O(t^2 \log t\ell)$ field elements.

However, no known protocol is simultaneously optimal in both b and r. In this paper, we present (in Section 4.1) an (asymptotically) optimal protocol to perfectly securely transmit a message consisting of ℓ field elements, viz., our protocol has $n = 2t + 1$, $r = 2$ and $b = O(t\ell)$ field elements, if $\ell = \Omega(t)$. Since we require the field size to be at least n, this means that the message size is $\Omega(t \log t)$ bits.

Unfortunately, due to the stringent requirements of privacy and reliability, (even optimal) PSMT protocols are not always as efficient as we would like them to be in practice. Therefore, one often relaxes either the reliability or the privacy requirement, or both, and tries to achieve *statistical* reliability/privacy.

Thus, we look for protocols in which the probability that **R** will receive the correct message is $1 - \delta$ and the probability that the adversary will learn the message is ϵ for arbitrarily small δ and ϵ. Of course, PSMT is the case $\epsilon = \delta = 0$; broadcast satisfies $\epsilon = 1, \delta = 0$, and so on. In [5], a $(0, \delta)$-protocol with $n = 2t+1$, $r = 3$ and $b = O(t^2)$ field elements was presented to securely communicate one field element. For an extensive discussion of (ϵ, δ)-secure protocols see [5].

In this paper we introduce a new way of relaxing the requirements. We study the *average case* efficiency of SMT protocols, rather than the worst case. We do not require that the worst case complexity be polynomially bounded, or even finite; we feel that nonterminating protocols that nevertheless complete quickly with high probability and have perfect security and reliability are very useful constructions.

In Section 5 we present an optimally fault-tolerant protocol that terminates in a *single* with high probability, and having $b = O(t)$ field elements. We note that the significance of a single phase protocol is more than merely a gain in efficiency (in terms of network latency): **S** and **R** are not required to be on the network at the same time for executing a single phase protocol, and therefore they are applicable in a much bigger set of scenarios than are multi-phase protocols.

Most of the results in the literature model the sender's distrust in the network via a centralized static adversary that can corrupt up to t of the n wires and assume the worst-case that the adversary can completely control the behavior of the corrupted wires [3, 13]. In line with this, we assume up to section 6 that the adversary is static, i.e., he (a) decides on the set of t wires to corrupt before the start of the protocol and (b) a wire once corrupted remains so subsequently.

However, in practice the bound on the number of corrupted wires may depend on the total time of the protocol execution. Thus motivated, in section 6 we model the faults via a *mobile* adversary, in line with [10]. In this model, the adversary can corrupt any set of wires in the lifetime of the protocol but is constrained to corrupt at most t wires in any single phase of the protocol.

We show that our ideas in the case of static adversaries can be extended to withstand mobile adversaries. We prove that the lower bound of $b = (\frac{n\ell}{n-2t})$

for reliable message transmission holds for mobile adversaries irrespective of the
number of rounds. We also give a bit-optimal protocol when $n = 2t + 1$.

2   Preliminaries

Notation. Throughout the paper, we use $\mathbf{M}$ to denote the message that $\mathbf{S}$ wishes
to securely communicate to $\mathbf{R}$. The message is assumed to be a sequence of ℓ
elements from the finite field $\mathbb{F}$. The only constraint on $\mathbb{F}$ is that its size must be
no less than the number of wires n. Since we measure the size of the message in
terms of the number of field elements, we must also measure the communication
complexity in units of field elements; we follow this convention in the rest of
the paper. We assume that there exists a publicly specified one-to-one mapping
$\alpha : \{1, 2, \ldots, n\} \to \mathbb{F}$. For convenience, we use α_i to denote $\alpha(i)$.

We say that a wire is *faulty* if it is controlled by the adversary; all other
wires are called *honest*. A faulty wire is *corrupted* in a specific phase if the value
sent along that wire was changed. When the context makes clear which phase
is being referred to, we simply say that a wire is *corrupted*. Observe that a wire
may be faulty but not corrupted in a particular phase.

2.1   Efficient Single Phase Reliable Communication

To *reliably* communicate a message $\mathbf{m}$, a sequence of k field elements, to $\mathbf{R}$,
one simple way is for $\mathbf{S}$ to send $\mathbf{m}$ along each wire – i.e, broadcast. However,
when $n > 2t + 1$, where t out of n wires are corrupted, broadcast, requiring
$O(nk)$ field elements, is not the most efficient method of (single phase) reliable
communication. Instead, it is possible to use an *error-correcting code* to improve
the communication complexity of reliable communication to $\frac{nk}{n-2t}$ field elements.
A *block* error correcting code encoding a *message* of k field elements to a *codeword*
of n symbols is an injective mapping $\mathcal{C} : \mathbb{F}^k \to \mathbb{F}^n, (n > k)$. The encoding function
is used in conjunction with a decoding function $\mathcal{D} : \mathbb{F}^n \to \mathbb{F}^k$ with the property
that if its input differs from a valid codeword in at most t field elements, then
$\mathcal{D}$ outputs the message corresponding to that codeword. We say that the code
corrects t errors. Clearly, such a decoding function will always exist if any two
valid codewords differ in at least $2t + 1$ symbols, that is, the distance of the code
$d \geq 2t + 1$.

The efficiency of an error correcting code is subject to the *Singleton bound*:

Lemma 1. *Let C be a block code which reliably transmits k field elements by
communicating a total of n field elements and has a distance of d. Then $n \geq
k + d - 1$.*

We observe that for a t-error correcting code, the distance d (which is the
minimum Hamming distance between any two codewords) is at least $2t + 1$. Thus
we have

Corollary 1. *Let C be a t-error correcting block code as in lemma 1. Then
$k \leq n - 2t$.*

We now consider a special class of error correcting codes called Reed-Solomon codes (RS codes).

Definition 1. *Let $\mathbb{F}$ be a finite field and $\alpha_1, \alpha_2, \ldots \alpha_n$ be a collection of distinct elements of $\mathbb{F}$. Given $k \leq n \leq |\mathbb{F}|$, and a block $\mathbf{B} = [m_0 \; m_1 \; \ldots \; m_{k-1}]$ the encoding function for the Reed-Solomon code $RS(n, k)$ is defined as $[p_\mathbf{B}(\alpha_1) \; p_\mathbf{B}(\alpha_2) \; \ldots \; p_\mathbf{B}(\alpha_n)]$ where $p_\mathbf{B}(x)$ is the polynomial $\sum_{i=0}^{k-1} m_i x^i$.*

Theorem 1 ([7]). *The Reed-Solomon code meets the Singleton bound.* $\qquad\square$

The following special property of the RS-code will be of use in our subsequent discussion:

Lemma 2. *Let $[p_\mathbf{B}(\alpha_1) \; p_\mathbf{B}(\alpha_2) \; \ldots \; p_\mathbf{B}(\alpha_n)]$ be an $RS(n, k)$-encoding of $\mathbf{B}$. Then for any $n' < n$, any subsequence of $[p_\mathbf{B}(\alpha_1) \; p_\mathbf{B}(\alpha_2) \; \ldots \; p_\mathbf{B}(\alpha_n)]$ of length n' forms a valid $RS(n', k)$-encoding of $\mathbf{B}$.*

Proof: Easy observation. $\qquad\square$

Constructing message transmission protocols using error correcting codes is a typical application, for example see [2, 11]. We now describe REL-SEND$(\mathbf{m}, k)$, a protocol for reliable communication obtained by using the corresponding Reed-Solomon code $RS(n, k)$. REL-SEND$(\mathbf{m}, k)$ will be used as a sub-protocol later on.

Protocol REL-SEND: optimal single phase reliable message transmission of $\mathbf{m}$.

- $\mathbf{S}$ breaks up $\mathbf{m}$ into blocks of k field elements.
- For each block $\mathbf{B} = [m_0 \; m_1 \; \ldots \; m_{k-1}]$:
  - $\mathbf{S}$ computes $RS(n, k)$ to obtain $[p_\mathbf{B}(\alpha_1) \; p_\mathbf{B}(\alpha_2) \; \ldots \; p_\mathbf{B}(\alpha_n)]$.
  - $\mathbf{S}$ sends $p_\mathbf{B}(\alpha_i)$ along the wire w_i.
  - $\mathbf{R}$ receives the (possibly corrupted) $p_\mathbf{B}(\alpha_i)$'s and applies the decoding algorithm and constructs $\mathbf{B}$.
- $\mathbf{R}$ concatenates the $\mathbf{B}$'s to recover the message $\mathbf{m}$.

We note that the resulting protocol is a single phase protocol. The reverse process is equally valid - given a single phase reliable communication protocol, we can convert it into a block error correcting code. Thus, the maximum attainable efficiency for single phase reliable communication is also subject to the Singleton bound.

Remark: This conversion to an error correcting code is straightforward if the messages sent along each wire in the protocol are of the same length. Suppose, however, that there is exists a protocol Π that does not have this symmetry property and beats the Singleton bound. Then consider the protocol Π' which consists of n sequential executions of protocol Π with the identities or numbers of the wires being "rotated" by a distance of i in the ith execution. Clearly, this protocol achieves the symmetry property by "spreading the load"; further its message expansion factor is equal to that of Π'. It therefore beats the Singleton bound as well, which is a contradiction.

Lemma 3. *Suppose that the receiver* **R** *knows* f *faults among the* n *wires, and* t' *be the number of faulty wires apart from those* f. *Then REL-SEND*($\mathbf{m}, k$) *works if* $n - f \geq k + 2t'$.

Proof: Since **R** knows f faults, he simply ignores those wires; and by lemma 2, this converts the code into an RS code with parameters $n - f$ and k. The result now follows from lemma 1 and theorem 1.   □

2.2 Extracting Randomness

In several of our protocols we have the following situation: **S** and **R** by some means agree on a sequence of n numbers $\mathbf{x} = [x_1, x_2, \ldots x_n] \in \mathbb{F}^n$ such that

- The adversary knows $n - f$ of the components of $\mathbf{x}$
- The adversary has no information about the other f components of $\mathbf{x}$
- **S** and **R** do not necessarily know which values are known to the adversary.

The goal is for **S** and **R** to agree on a sequence of f numbers $y_1, y_2, \ldots y_f \in \mathbb{F}$ such that the adversary has no information about $y_1, y_2, \ldots y_f$. This is achieved by the following algorithm:

Algorithm EXTRAND$_{n,f}(\mathbf{x})$. Let **V** be a $n \times f$ Vandermonde matrix with members in $\mathbb{F}$. This matrix is published as a part of the protocol specification. **S** and **R** both locally compute the product $[y_1 \; y_2 \; \ldots \; y_f] = [x_1 \; x_2 \; \ldots \; x_n]\mathbf{V}$.

Lemma 4. *The adversary has no information about* $[y_1 \; y_2 \; \ldots \; y_f]$ *in algorithm EXTRAND.*

Proof. We need to show that there is a bijective mapping between the f tuple of values that are not known to the adversary and the f tuple $y_1, y_2, \ldots y_f$. But this is a direct consequence of the fact that every f-subdeterminant in an $n \times f$ Vandermonde matrix is nonzero.

3 Lower Bound on Communication Complexity

Theorem 2. *Any 2-phase perfectly reliable message transmission (PRMT) of* ℓ *bits requires communicating* $\left(\frac{n\ell}{n-2t}\right)$ *bits.*

We first observe that a probabilistic polynomial time (PPT) protocol for PRMT with a worst-case communication complexity of b bits exists if and only if there is a deterministic protocol with the *same* communication complexity. Since perfect reliability is required, the algorithm must succeed for every possible choice of coin tosses; in particular, it must succeed when all the random bits of **S** and **R** are zeroes. Thus we convert any PPT protocol into a deterministic protocol by fixing the sequence of coin-tosses to all zeros. Hence, we assume that **S** and **R** are deterministic polynomial time algorithms.

We recall that in a phase, both **S** and **R** may simultaneously send messages to the other player. If this happens we call it a *bidirectional* phase. On the other hand, if only one of the players sends a message, we call it a *unidirectional* phase.

Without loss of generality we assume that in the first phase communication is from **R** to **S** and in the second phase it is from **S** to **R**. (Clearly there is no point in communication from **R** to **S** in the second phase; similarly if **S** sends any messages to **R** in the first phase we can consider these to be part of the second phase as well.) In the rest of the paper we assume that communication in each phase is unidirectional.

We prove the stronger statement that any 2-phase PRMT of ℓ bits requires communicating t least $(\frac{n\ell}{n-2t})$ bits even against a weaker adversary, namely, one that is passive in the first phase.

Thus, let Π be a two phase protocol in which M_1 is the totality of messages sent by **R** to **S** in phase I and M_2 the totality of messages sent by **S** to **R** in phase 2. The steps of Π are as follows:

1. **R** computes M_1 and sends it to **S**.
2. **S**, using M_1 and the message computes M_2 and sends it to **R**.
3. **R** recovers the message using M_2.

In the above protocol, we see that step 1 is "useless": consider the protocol Π' in which step 1 is replaced with the following step:

S and **R** both locally compute M_1 by simulating **R**'s execution in step 1 of Π. (Since the adversary is passive, M_1 is guaranteed to have been received by **S** in the first phase of Π.)

It is clear that Π' succeeds whenever Π succeeds.

Π' being a single phase protocol, can also be viewed as an error correcting code. That is, the concatenation of the data sent along all the wires forms the codeword. Let $\mathbb{S}_i$ be the set of possible values of the data sent along the wire w_i. Thus, each codeword is of length at least $\sum_{i=1}^{n} \log |\mathbb{S}_i|$, consisting of n elements one from each $\mathbb{S}_i$, $1 \le i \le n$. Now, the removal of any $2t$ elements from each of the codewords should result in shortened codewords that are all distinct. For if any two were identical, the original codewords could have differed only among at most $2t$ elements implying that there exist two original codewords c_1 and c_2 and an adversarial strategy such that the receiver's view is the *same* on the receipt of either c_1 or c_2. In more detail, without loss of generality assume that c_1 and c_2 differ only in their last $2t$ elements. That is, $c_1 = \alpha \circ \beta$ and $c_2 = \alpha \circ \gamma$, where $\circ$ denotes concatenation and $|\beta| = |\gamma| = 2t$ elements. Let β_1 denote the first t elements of β, while β_2 be the last t elements. That is, let $\beta = \beta_1 \circ \beta_2$, $|\beta_1| = |\beta_2| = $ telements. Similarly, let $\gamma = \gamma_1 \circ \gamma_2$, $|\gamma_1| = |\gamma_2| = t$. Now, consider the two cases: (a) c_1 is sent and the adversary corrupts it to (by corrupting the last t wires and changing β_2 to γ_2) $\alpha \circ \beta_1 \circ \gamma_2$ and (b) c_2 is sent and the adversary corrupts it to (by corrupting the penultimate set of t wires and changing γ_1 to β_1) $\alpha \circ \beta_1 \circ \gamma_2$. Thus, the receiver cannot distinguish between the receipt of c_1 and c_2, which violates the reliable communication property. Therefore, all shortened codewords are distinct and there are as many shortened codewords as original codewords. But the number of shortened codewords is at most the minimum $\prod_{j=1}^{n-2t} |\mathbb{S}_{i_j}|$ among all $(n-2t)$ sized subsets $(i_1, i_2, \ldots, i_{n-2t})$ of $1, 2, \ldots, n$. Thus we may sort the $|\mathbb{S}_i|$'s in a non-decreasing order and multiply the first $(n-2t)$ values to obtain the number of original codewords denoted by C. Thus, reliable communication of $k = \log C$ bits incurs a communication cost of at least $\sum_{i=1}^{n} \log |\mathbb{S}_i|$ bits. But $\log C = \sum_{j=1}^{n-2t} |\mathbb{S}_{i_j}|$. Thus, in the best case all the domains are of equal size and is thus subject to the Singleton bound. By corollary to lemma 1, reliable communication of $k = n - 2t$ bits incurs a communication cost of n bits. Since Π' communicates an l bit message, it follows that Π' has a communication complexity of $(\frac{nl}{n-2t})$ bits.   □

Corollary 2. *Any 2-phase perfectly reliable message transmission (PRMT) of ℓ field elements requires communicating $(\frac{nl}{n-2t})$ field elements.*

The above corollary follows from the fact that a field element can be represented as a string of $\lceil \log |\mathbb{F}| \rceil$ bits.

4   An Optimal Protocol for PSMT

4.1   The PSMT Protocol

In this section, we present our 2-phase protocol for PSMT for any message that is a sequence of ℓ field elements, with $n = 2t + 1$ and $b = O(t\ell)$ field elements, for a sufficiently large l. It turns out that we require $\ell = \Omega(t)$ bits.

Suppose that there exists a protocol that securely transmits a message consisting of $\Omega(t)$ field elements with $n = 2t+1$, $r = 2$ and $b = O(nt)$ field elements. It is evident that for any integer $j \geq 1$, tj field elements can be sent in $b = O(ntj)$ field elements whilst maintaining $n = 2t + 1$ and $r = 2$; this is because we can run the j sub-protocols in parallel. Setting $j = \lceil \frac{\ell}{t} \rceil$, we obtain a protocol that communicates ℓ field elements by sending $O(n\ell)$ field elements. Thus, our goal now reduces to the *design of a protocol that achieves secure transmission of $\Omega(t)$ field elements over a network of $n = 2t + 1$ wires, in two rounds by communicating $b = O(nt)$ field elements.* We now present one such protocol. Specifically, our protocol sends $\lfloor \frac{t}{3} \rfloor$ field elements by sending $O(nt)$ field elements.

In our protocol, the first phase is a send from **R** to **S** and the second phase is from **S** to **R**. We denote the set of n wires by $\mathcal{W} = \{w_1, w_2, \ldots, w_n\}$. We assume that **S** wishes to communicate a block, denoted by **m**, that consists of $\lfloor \frac{t}{3} \rfloor$ field elements from $\mathbb{F}$.

Phase I (R to S)

The receiver **R** selects at random n polynomials p_i, $1 \leq i \leq n$ over $\mathbb{F}$, each of degree t.

Next, through each wire w_i, **R** sends the following to **S**:

- The polynomial p_i [5].
- For each j, $1 \leq j \leq n$, the value of $p_j(\alpha_i)$ (which we denote by r_{ij}), where α_i's are arbitrary distinct publicly specified members of $\mathbb{F}$.

---

[5] We assume that the polynomial is sent by sending a $(t + 1)$-tuple of field elements.

Phase II (S to R)

S begins this phase after receiving what R has sent in the first phase. Let S receive the polynomial p_i' and the values r_{ij}' along the wire w_i. In this phase, S must locate all the corruptions that occurred in the previous phase, communicate the corruptions and send the message securely. A naive and straightforward way of doing this is as follows: communicate the list of $\Theta(n^2)$ contradictions (we say that wire i contradicts wire j if $r_{ij}' \neq p_j'(\alpha_i)$) among the wires (in the worst case). However, there are two problem with this approach: (a) the method requires communicating $\Theta(n^3)$ field elements, and (b) such an approach necessitates more than two phases.

We solve the former problem by using a two step technique to communicate the list of contradictions. In the first step we broadcast a selected set of contradictions; this will enable the players to find sufficiently many faults to facilitate sending the remaining contradictions in $O(n^2)$ field elements using the REL-SEND protocol in the second step. The second problem is solved using some new techniques described in the sequel.

S's computation.

- S initializes his *fault-list*, denoted by L_{fault}, to $\emptyset$.
- S constructs a directed graph $G = (\mathcal{W}, A)$ where arc $(w_i, w_j) \in A$ if $r_{ij}' \neq p_j'(\alpha_i)$.
- Let $H = (\mathcal{W}, E)$ be the undirected graph based on G; that is, $(w_i, w_j) \in E$ if $(w_i, w_j) \in A$ or $(w_j, w_i) \in A$.
- For each i, $1 \leq i \leq n$, such that the degree of node w_i in the graph H constructed above is greater than t (i.e., $degree(w_i) \geq t+1$), S adds w_i to L_{fault}.
- Let $H' = (\mathcal{W}', E')$ be the induced subgraph of H on the vertex set $\mathcal{W}' = (\mathcal{W} \setminus L_{fault})$.
- Next, S finds a maximum matching[6] $M \subseteq E'$ of the graph H'; this can be done efficiently using the algorithms of [4, 9].
- For each arc (w_i, w_j) in G that does not belong to M, S associates the four-tuple $\{\alpha_i, \alpha_j, r_{ij}', p_j'(\alpha_i)\}$. Let $\{a_1, a_2, \ldots, a_N\}$ be the arcs in G that are not in M. Replacing each arc with its associated 4-tuple, S gets a set of $4N$ field elements, $X = \{X_1, X_2, \ldots, X_{4N}\}$.
- Let $u = \lfloor \frac{t}{3} \rfloor$. Next, S creates the $2t + u$-degree message-carrying polynomial $s(x) = \sum_{i=0}^{2t+u} k_i x^i$ as follows: let $\mathbf{m} = [m_0 m_1 \ldots m_{u-1}]$.

$$\text{Assign } k_i = \begin{cases} m_i & \text{if } 0 \leq i < u. \\ 0 & \text{if } w_{i-u+1} \in L_{fault}. \\ p_{i-u+1}'(0) & \text{otherwise.} \end{cases}$$

---

[6] A subset M of the edges of H, is called a *matching* in H if no two of the edges in M are adjacent. A matching M is called *maximum* if H has no matching M' with a greater number of edges than M has.

- **S** initializes the set Y as follows:

$$Y = \{s(\alpha_1), s(\alpha_2), \ldots, s(\alpha_{t+1})\}$$

- Finally, the sender **S** selects at random n polynomials q_i, $1 \leq i \leq n$ over $\mathbb{F}$, each of degree t, such that the values $q_i(0)$ lie on a polynomial of degree $\lfloor \frac{4t}{3} \rfloor$.
- **S** computes $\mathbf{y} = [y_0 \; y_1 \; \cdots \; y_{u-1}] = \text{EXTRAND}_{n,u}([q_1(0) \; q_2(0) \; \cdots \; q_n(0)])$.
- For each j, $1 \leq j \leq n$, let v_{ij} denote the value of $q_j(\alpha_i)$. With each of the $N + |M|$ arcs in the graph G, **S** associates the four-tuple $\{\alpha_i, \alpha_j, v_{ij} \text{ and } q_j(\alpha_i)\}$. He initiates the set Z in similar lines as X to contain $4(N + |M|)$ field elements.

S's communication.

S sends the following to **R** through all the wires:

1. The blinded message $\mathbf{m} \oplus \mathbf{y}$.
2. The set L_{fault}.
3. For each edge $(w_i, w_j) \in M$, the following four field elements: $\{\alpha_i, \alpha_j, r'_{ij} \text{ and } p'_j(\alpha_i)\}$.

Along each wire w_i, **S** sends the following as specified by the REL-SEND$(\cdot, \cdot)$ Algorithm:

1. REL-SEND$(X, |M| + |L_{fault}| + 1)$.
2. REL-SEND$(Y, |M| + |L_{fault}| + 1)$.
3. REL-SEND$(Z, |M| + |L_{fault}| + 1)$.

Again, through each wire w_i, **S** sends the following to **R**:

- The polynomial q_i.
- For each j, $1 \leq j \leq n$, the value of $v_{ij} = q_j(\alpha_i)$.

Message recovery by R.

R receives what **S** sent in the second phase and locally deciphers the message $\mathbf{m}$ as follows:

1. **R** reliably receives L_{fault} and knows that the wires in this set are faulty. He initializes $L_{fault}^{\mathbf{R}} = L_{fault}$.
2. For each arc $(w_i, w_j) \in M$, **R** reliably receives $\{\alpha_i, \alpha_j, r'_{ij} \text{ and } p'_j(\alpha_i)\}$. He locally verifies: $r'_{ij} \stackrel{?}{=} r_{ij}$ and $p'_j(\alpha_i) \stackrel{?}{=} p_j(\alpha_i)$. If the former check fails (that is the values are unequal), then **R** adds w_i to $L_{fault}^{\mathbf{R}}$. If the latter check fails, then **R** adds w_j to $L_{fault}^{\mathbf{R}}$ (note that both w_i and w_j may be identified as faulty; in any case, at least one of them is guaranteed to be found faulty). Thus, at least $|M|$ new faults are caught in this step.

3. From Lemma 5, it is clear that **R** receives the set X reliably. Again **R** locally verifies for each arc's (say (w_i, w_j)) 4-tuple: $r'_{ij} \overset{?}{=} r_{ij}$ and $p'_j(\alpha_i) \overset{?}{=} p_j(\alpha_i)$. If the former check fails (that is the values are unequal), then **R** adds w_i to $L^{\mathbf{R}}_{fault}$. If the latter check fails, then **R** adds w_j to $L^{\mathbf{R}}_{fault}$. At the end of this step, *all* the faults that occurred during transmission in **Phase I** are guaranteed to have been identified (see Lemma 6).

4. We know from Lemma 5 that the **R** receives the set Y correctly. If the number of faults (which are not in L_{fault}) that occurred in **Phase I** was $\leq \frac{2t}{3}$, then in the polynomial $s(x)$, **R** has $\leq t$ unknowns and $t + 1$ equations (which are bound to be consistent whatever the number of unknowns may be, since all faults have been eliminated). Thus, in this case, **R** obtains the message **m**.

5. Similarly, from Lemma 5, we know that **R** receives the set Z correctly. If the number of faults (which are not in L_{fault}) that occurred in **Phase I** was $> \frac{2t}{3}$, then from Lemma 9, we know that **R** can obtain the message using the polynomials $q(\cdot)$ and Z. Thus, in this case too, **R** obtains the message **m**.

Lemma 5. **R** *is guaranteed to receive the sets X, Y and Z correctly.*

Proof: From lemma 3, the REL-SEND$(\cdot, k)$ protocol succeeds provided that $n - f \geq k + 2(t - f)$; here, $k = |M| + |L_{faults}| + 1$ and $n = 2t + 1$. Therefore, REL-SEND succeeds if $(2t + 1) - f - (|M| + |L_{faults}| + 1) \geq 2t - 2f$, or if, $f \geq |M| + |L_{faults}|$. Since, **R** is guaranteed to have identified at least $|M| + |L_{faults}|$ faulty wires at this stage, the lemma follows. □

Lemma 6. *If the set X was received correctly, then **R** can find all the corruptions that occurred during **Phase I**.*

Proof: Suppose wire w_i was corrupted in **Phase I**, i.e, $p'_i \neq p_i$. Then the two polynomials can intersect in at most t points. Since there are $t + 1$ honest wires, there is guaranteed to be at least one honest wire which contradicts w_i. Since the correct values corresponding to every contradiction have been received by **R**, **R** can find all the corruptions. □

Lemma 7. *If the number of corruptions that occurred in **Phase I** was $\leq \frac{2t}{3}$, **R** obtains the polynomial $s(x)$ correctly.*

Proof: To find the message **R** must find the polynomial $s(x)$. To find $s(x)$, **R** must find k_i for $0 \leq i \leq 2t + u$. Of these **R** does not yet know k_i for $0 \leq i < u$ and does not know $\leq \frac{2t}{3}$ of the $p'_i(0)$, a total of at most $\lfloor \frac{t}{3} \rfloor + 1 + \frac{2t}{3} = t + 1$ field elements. But the set Y gives **R** $t + 1$ values of $s(x)$, which yield $t + 1$ linear equations on the coefficients, and using these values **R** can determine all k_i. □

Lemma 8. *For some i, if the wire w_i was corrupted in **Phase I** then **R** can correct any corruption of the corresponding q_i in **Phase II**.*

Proof: Let the in-degree of w_i in G be d. Then there must have been at least $t - d + 1$ corruptions in **Phase I** (since the number of honest wires is $t + 1$). Thus in the second phase, there are at most $2t + 1 - (t + 1 - d) = t + d$ legitimate wires. The maximum number of faults that **R** needs to correct in this phase is $t - (t + 1 - d) = d - 1$. We verify that these parameters satisfy the constraint in lemma 3, and therefore **R** will be able to correct all corruptions of q_i.   □

Lemma 9. *If the set Z was received correctly and if the number of faults that occurred in* **Phase I** *was* $> \frac{2t}{3}$, *then* **R** *can obtain the message* **m**.

Proof: By lemma 8, **R** can correct all except a maximum of $\frac{t}{3}$ of the q_i's. The degree of the polynomial that they lie on is $\lfloor \frac{4t}{3} \rfloor$; since these parameters satisfy the constraints of lemma 3, it follows from the correctness of the REL-SEND protocol that **R** can obtain all the q_i's and hence **m**.   □

Theorem 3. *The protocol presented in Section 4.1 achieves perfect reliability.*

Proof: Perfect reliability is a consequence of lemmas 7 and 9.   □

Lemma 10. *For every honest wire w_i, the adversary has no information about $p_i(0)$ and $q_i(0)$.*

Proof: Obvious. p_i is a random polynomial of degree t but the adversary has seen only t points on it. The same argument holds for q_i as well.   □

Theorem 4. *The protocol presented in Section 4.1 achieves perfect security.*

Proof (sketch): First we prove that the adversary has no information about the coefficients $k_0, k_1, \ldots k_{u-1}$ of the polynomial $s(x)$. There are at least $t + 1$ values of $p_i(0)$ which are not known to the adversary. The adversary obtains $t + 1$ linear equations on the coefficients k_j by knowing the values of $s(\alpha_1), s(\alpha_2), \ldots s(\alpha_{t+1})$ which are sent reliably by **S**. Thus the adversary has $t + 1$ linear equations on $u + t + 1$ unknowns, which implies that he has no information about any u-tuple of them.

Next we observe that among the $\lfloor \frac{4t}{3} \rfloor$ values of $q_i(0)$, the adversary knows at most t, and hence by the security of the EXTRAND algorithm (lemma 4) it follows that the adversary gets no information about **m**.   □

4.2   Performance

Theorem 5. *Given an undirected graph $H = (V, E)$, with a maximum degree of Δ, and a maximum matching M, the number of edges $|E|$ is less than or equal to $(2|M|^2 + |M|\Delta)$.*

Proof: We first fix a representation of the maximum matching M as a set of *ordered* pairs of vertices as described below.

We say that a vertex i belongs to vertex-set of the matching, denoted by $Vertex(M)$, if there exists another vertex j such that the edge $(i, j) \in M$. A vertex $i \in Vertex(M)$ is called the *match-vertex* if the degree of i in the subgraph H'_i induced by H over the vertices $i \cup (V \setminus Vertex(M))$ is ≤ 1.

Given a maximum matching M, a match-vertex i may have at most one incident edge $e_i = (i, \cdot)$ in H'_i. We call the edge e_i as a *match-edge* (corresponding to the match-vertex i). We now define X to be the set of all match-edges (corresponding to each of the match-vertices in M).

Claim. Every edge $(i, j) \in M$ has at least one match-vertex.

Proof: On the contrary, if neither i nor j was a match-vertex, then, both i and j are adjacent to at least two vertices in $(V \setminus Vertex(M))$. Let i be adjacent to vertices u and v in $(V \setminus Vertex(M))$ and let j be adjacent to a vertex w ($\neq u$) in $(V \setminus Vertex(M))$. Now, removing the edge (i, j) from M and adding the edges (u, i) and (j, w) to M gives rise to a new matching in H of size $|M| + 1$ which contradicts the maximality of the matching M. Hence the claim holds.

Hereafter, we represent every edge in $M \cup X$ as (i, j) if and only if i is a match-vertex; in case of both i and j being match-vertices, i is the one with the lower number of corresponding match-edges (ties broken by random choice). We fix one such representation of the edges in $M \cup X$. To avoid confusion between the unordered pair (i, j) and the ordered pair used in the representation of an edge in $M \cup X$ hereafter, we denote the *ordered* pair as $\langle i, j \rangle$. A vertex i belonging to $Vertex(M)$ is called a *left-vertex* if, in the representation of M that we fixed earlier, there exists a vertex j such that $\langle i, j \rangle \in M$. We call all the non-left-vertices belonging to M as *right-vertices*.

Note that the number of left-vertices is equal to the number of right-vertices is equal to $|M|$. Also note that by definition, every left-vertex is a match-vertex.

Now, it is easy to place an upper bound on $|E|$ as follows: the maximum number of edges among the $2|M|$ vertices in $Vertex(M)$ is $|M|(2|M|-1)$. Again, the maximum number of edges from the left-vertices to $(V \setminus Vertex(M))$ is $|M|$, since each left vertex is a match-vertex (having at most one edge to $(V \setminus Vertex(M))$) and there are $|M|$ left vertices. Furthermore, each right vertex can have at most $\Delta - 1$ edges to $(V \setminus Vertex(M))$ (by the definition of Δ). Thus, $|E| \leq (|M|(2|M| - 1) + |M| + |M|(\Delta - 1)) \leq (2|M|^2 + |M|\Delta)$. □

Theorem 6. *The PSMT protocol presented in Section 4.1 communicates $O(t^2)$ field elements in order to securely transmit $\lfloor \frac{t}{3} \rfloor$ field elements.*

Proof: We have already proved that the protocol securely transmits $\lfloor \frac{t}{3} \rfloor$ field elements. From the description of the protocol, it is easy to verify that all steps except possibly the invocations of REL-SEND$(X, .)$ and REL-SEND$(Z, .)$ in Phase II have a communication cost of $O(t^2)$ field elements. Since the maximum degree of a node in H' is at most $t + 1$ and $|M|$ is also at most $t + 1$, from theorem 5 it follows that $|X|$ (and hence also $|Z|$) is $O((|M| + |L_{fault}|)t)$. Since the efficiency of REL-SEND$(X, |M| + |L_{fault}| + 1)$ is $O(\frac{|X|t}{|M| + |L_{fault}|})$, the theorem follows. □

The main result now follows from the discussion at the beginning of Section 4.1:

Corollary 3. *There exists a 2-round PSMT protocol that securely communicates a message consisting ℓ field elements and has a communication complexity of $O(n\ell)$ field elements when $n = 2t + 1$, if $\ell = \Omega(t)$.*

5   A Las Vegas Single Phase Protocol

In this section we present an optimally tolerant ($n = 2t + 1$) PSMT protocol which terminates in a single phase with (arbitrarily) high probability. We represent the block of field elements **m** that **S** wishes to send to **R** as $\mathbf{m} = [m_0\ m_1\ \ldots\ m_t]$.

The Single Phase Protocol

1. **S** selects at random n polynomials $p_i, 1 \le i \le n$ over $\mathbb{F}$, each of degree t.
2. For each wire w_i
   - **S** sends the polynomial p_i through w_i
   - For each j, **S** randomly selects one of the t points of intersection of p_i and p_j (denote the selected point by r_{ij}). **S** ensures that $r_{ij} \ne r_{ji}$. **S** sends r_{ij} through w_i.
3. **S** computes $\mathbf{y} = [y_0\ y_1\ \ldots\ y_t] = \text{EXTRAND}_{n,t+1}\ ([p_1(0)\ p_2(0)\ \ldots\ p_n(0)])$ and broadcasts $\mathbf{m} \oplus \mathbf{y}$.
4. Let p'_i and r'_{ij} be the values received by **R**. We say that wire i *contradicts* wire j if p'_i and p'_j do not intersect at r_{ij}.
5. **R** checks if there is a wire contradicted by at least $t+1$ wires. All such wires are removed.
6. If there is at least one contradiction among the remaining wires, **R** broadcasts "FAILURE"; **S** and **R** now execute the PSMT protocol of section 4.1.
7. If there is no contradiction, **R** corrects the polynomials $p_i(.)$ of each corrupted wire w_i (i.e, he "corrects" those wires) using the values of r_{ij} received along the uncorrupted wires. **R** now knows all the polynomials p_i.
8. **R** computes $\mathbf{y} = [y_0\ y_1\ \ldots\ y_t] = \text{EXTRAND}_{n,t+1}([p_1(0)\ p_2(0)\ \ldots\ p_n(0)])$ and recovers $\mathbf{m} = (\mathbf{m} \oplus \mathbf{y}) \oplus \mathbf{y}$.

Let ϵ be a bound on the probability that the protocol does not terminate in a single phase. We require that the size of the field $\mathbb{F}$ be $\Omega(\frac{Q(n)}{\epsilon})$, for some polynomial $Q(n)$, but this is of course acceptable since the complexity of the protocol increases logarithmically with field size. We now discuss the correctness of the protocol.

Lemma 11. R *will never output an incorrect value.*

Proof. Since any corruption involves changing the polynomial corresponding to that wire, it is clear that no corrupted wire can escape contradiction by at least one other wire. If p and p' agree on $t + 1$ points (corresponding to the $t + 1$ honest wires) then p and p' must be equal. Therefore, at the start of step 7, all the wires which were used in calculation of the output could not have corrupted their values. This guarantees that **R**'s output in step 8 is correct. □

Lemma 12. *The protocol terminates in a single phase with high probability.*

Proof. Since no uncorrupted wire changes the value sent on the wire, it follows that no honest wire can contradict another honest wire. Thus, if wire i contradicts wire j, then either wire i or wire j is faulty. From this it is easy to see that an honest wire can be contradicted by at most t other wires, and therefore any wire that is contradicted $t + 1$ or more wires has to be faulty. Hence **R** can be sure that all the wires removed by him are indeed faulty.

We need to show that if a wire is corrupted, then it will be contradicted by all the honest players with high probability. Let π_{ij} be the probability that the corrupted wire j will not be contradicted by i. This means that the adversary can ensure that $p_j(r_{ij}) = p'_j(r_{ij})$ with a probability of π_{ij}. Since there are only t points at which these two polynomials intersect, this allows the adversary to guess the value of r_{ij} with a probability of at least $\frac{\pi_{ij}}{t}$. But since r_{ij} was selected uniformly in $\mathbb{F}$, the probability of guessing it is at most $\frac{1}{|\mathbb{F}|}$. Therefore we have $\pi_{ij} \leq \frac{t}{|\mathbb{F}|}$ for each i, j. Thus the total probability that the adversary can find i, j such that corrupted wire j will not be contradicted by i is at most $\sum_{i,j} \pi_{ij} \leq \frac{n^2 t}{|\mathbb{F}|}$.

Since $\mathbb{F}$ is chosen such that $|\mathbb{F}| \geq \frac{Q(n)}{\epsilon}$, it follows that the protocol terminates in a single phase with probability $\geq 1 - \epsilon$ if we set $Q(n) = n^3$. $\square$

Lemma 13. *The adversary gains no information about the message.*

Proof: We observe that the adversary has no information about $p_i(0)$ for each honest wire w_i. This is because p_i is a random polynomial of degree t and the adversary has seen only t points on it (one corresponding to each faulty wire.) The proof now follows from lemma 4. $\square$

6   Mobile Adversaries

6.1   Lower Bound on Bit Complexity

The lower bound on the bit complexity of perfectly reliable message transmission proved in section 3 holds for mobile adversaries with no restriction on the number of phases. We give below a brief sketch of the proof.

Since the adversary can corrupt a different set of wires in each phase, the protocol cannot adapt as it finds corrupted wires; thus it can be considered to be memoryless. Therefore the total number of bits transmitted reliably is no more than the sum of the number of bits transmitted reliably in each phase; we have already shown in section 3 that the Singleton bound implies a lower bound of $\frac{nl}{n-2t}$ for a single phase; therefore this bound holds for multiple phase protocols as well.

6.2   An Optimal Protocol

The protocol with optimal fault tolerance and optimal communication complexity is presented below. Let **m** be a block of $t + 1$ field elements that **S** wishes to communicate securely to **R**.

The Optimal Protocol

1. **S** selects at random n polynomials p_i, $1 \leq i \leq n$ over $\mathbb{F}$, each of degree t and sends through each wire w_i, the polynomial p_i and the set $\{p_j(\alpha_i)\}_{1 \leq j \leq n}$.

2. After the completion of Phase I, **R** has computed the directed graph G of contradictions among the wires. The rest of the protocol involves finding all the wires that were corrupted in phase I using the graph G as follows:

   While there exists a wire w_j in G that contradicts another wire w_i
   - **R** broadcasts i, j, $p_i'(\alpha_j)$ and r_{ij}' to **S**.
   - **S** uses this information to determine which of w_i and w_j is faulty and broadcasts the identity of the faulty wire to **R**.
   - **R** removes the faulty wire from the protocol, i.e, sets G to the projection of G onto the remaining wires.

3. **R** now knows all the wires which were corrupted in phase I. Thus he can now use the r_{ij} values from the uncorrupted wires to find the correct values of the corrupted wires.

4. **S** and **R** both compute $\mathbf{y} = [y_0 \; y_1 \; \dots \; y_t] = \text{EXTRAND}_{n,t+1}([p_1(0) \; p_2(0) \; \dots \; p_n(0)])$.

5. **S** broadcasts $\mathbf{m} \oplus \mathbf{y}$ to **R**. **R** recovers $\mathbf{m} = (\mathbf{m} \oplus \mathbf{y}) \oplus \mathbf{y}$.

Theorem 7. *The adversary gains no information about the message.*

Proof: First we note that at the end of the first phase, the adversary has no information about $p_i(0)$ for each honest wire w_i. This is because p_i is a random polynomial of degree t and the adversary has seen only t points on it (one corresponding to each faulty wire.) Furthermore, the adversary gains no new information in step 2. This can be seen as follows: each phase in step 2 involves broadcast of the 4-tuple $(i, j, p_i(\alpha_j), r_{ij}')$. Since either wire w_i or wire w_j is faulty, this information is already known to the adversary. The other information that is broadcasted is which of wire w_i and wire w_j is faulty, which is also known to the adversary. The theorem follows. □

6.3 Complexity

The first phase of the protocol involves communication of $O(n^2)$ field elements. In step 2, each phase of communication results in the elimination of one wire. Therefore the number of phases in this step is $O(n)$. Since each phase involves the broadcast of $O(n)$ field elements, step 2 has a communication complexity of $O(n^2)$ field elements. The final phase involves broadcast a string of length $O(t)$ field elements. Therefore the entire protocol has communication complexity $O(n^2)$ field elements. Thus the protocol for a message $\mathbf{M}$ consisting of an arbitrary number ℓ of field elements obtained by executing this protocol in parallel $\lceil \frac{\ell}{t+1} \rceil$ times has a communication complexity of $O(\frac{n^2 \ell}{t})$ field elements, which is optimal when $n = 2t + 1$.

7 Conclusion

In this paper we have contributed significantly to the progress of the state of the art in the problem of Perfectly Secure Message Transmission. The protocol of section 4.1 constitutes a major improvement over existing protocols tolerating static adversaries; in fact we have achieved the optimal communication complexity when $n = 2t + 1$ and $r = 2$. The protocol can be extended to achieve the optimal communication complexity when $n > 2t + 1$ as well, though we have not presented it here. It would be interesting to see if the lower bound we have proved in section 3 holds when $r > 2$ as well (for PSMT); we conjecture that it does.

Perhaps our most interesting result is the average case single phase PSMT protocol. It is in fact surprising that such a protocol even exists; in addition our protocol is also very efficient in terms of communication complexity.

References

1. M. Ben-Or, S. Goldwasser, and A. Wigderson. Completeness theorems for non-cryptographic fault-tolerant distributed computation. In *20th ACM STOC*, pages 1–10, 1988.
2. Y. Desmedt and Y. Wang. Perfectly secure message transmission revisited. In *EUROCRYPT '02*, volume 2332 of *LNCS*, pages 502–517. Springer-Verlag, 2002.
3. D. Dolev, C. Dwork, O. Waarts, and M. Yung. Perfectly secure message transmission. *JACM*, 40(1):17–47, January 1993.
4. J. Edmonds. Paths, trees and flowers. *Canadian Jl. of Math.*, 17:449–467, 1965.
5. M. Franklin and R.N. Wright. Secure communication in minimal connectivity models. *Journal of Cryptology*, 13(1):9–30, 2000.
6. O. Goldreich, S. Micali, and A. Wigderson. How to play any mental game. In *19th ACM STOC*, pages 218–229, 1987.
7. F. J. MacWilliams and N. J. A. Sloane. *The Theory of Error Correcting Codes*. North Holland Publishing Company, 1978.
8. K. Menger. Zur allgemeinen kurventheorie. *Fundamenta Mathematicae*, 10:96–115, 1927.
9. S. Micali and V. Vazirani. An $O(\sqrt{|V|}\,|E|)$ algorithm for maximum matching in general graphs. In *21st IEEE FOCS*, pages 17–27, 1980.
10. R. Ostrovsky and M. Yung. How to withstand mobile virus attacks. In *10th ACM PODC*, pages 51–61, 1991.
11. M.O. Rabin. Efficient dispersal of information for security, load balancing, and fault tolerance. *JACM*, 36:335–348, 1989.
12. T. Rabin and M. Ben-Or. Verifiable secret sharing and multiparty protocols with honest majority. In *21st ACM STOC*, pages 73–85, May 1989.
13. H. Sayeed and H. Abu-Amara. Efficient perfectly secure message transmission in synchronous networks. *Information and Computation*, 126(1):53–61, 1996.
14. K. Srinathan, V. Vinod, and C. Pandu Rangan. Brief announcement: Efficient perfectly secure communication over synchronous networks. In *22nd ACM PODC*, page 252, 2003.
15. A. C. Yao. Protocols for secure computations. In *23rd IEEE FOCS*, pages 160–164, 1982.

Pseudo-signatures, Broadcast, and Multi-party Computation from Correlated Randomness

Matthias Fitzi[1], Stefan Wolf[2], and Jürg Wullschleger[2]

[1] Department of Computer Science
University of California, Davis, USA
fitzi@cs.ucdavis.edu
[2] Département d'Informatique et R.O.
Université de Montréal, Canada
{wolf,wullschj}@iro.umontreal.ca

Abstract. Unconditionally secure multi-party computations in general, and broadcast in particular, are impossible if any third of the players can be actively corrupted and if no additional information-theoretic primitive is given. In this paper, we relativize this pessimistic result by showing that such a primitive can be as simple as noisy communication channels between the players or weakly correlated pieces of information. We consider the scenario where three players have access to random variables X, Y, and Z, respectively, and give the exact condition on the joint distribution P_{XYZ} under which unconditional broadcast is possible. More precisely, we show that this condition characterizes the possibility of realizing so-called pseudo-signatures between the players. As a consequence of our results, we can give conditions for the possibility of achieving unconditional broadcast between n players and any minority of cheaters and, hence, general multi-party computation under the same condition.

Keywords: Unconditional security, pseudo-signatures, broadcast, multi-party computation, information theory.

1 Motivation and Preliminaries

1.1 Introduction

Digital signatures [11, 19] are a powerful tool not only in the context of digital contract signing, but also as a basic primitive for cryptographic protocols such as electronic voting or secure multi-party computation. Much less known are so-called *pseudo-signature schemes*, which guarantee *unconditional* security—in contrast to classical digital-signature schemes. The inherent price for their higher security, however, is the signatures' limited transferability: Whereas classical signatures can be arbitrarily transfered without losing conclusiveness, pseudo-signatures only remain secure for a fixed number λ—the *transferability*—of transfers among different parties. Since the necessary number of signature transfers in a protocol is typically bounded by the number of involved parties, pseudo-signatures are, nevertheless, useful and offer a provably higher security level than

M. Franklin (Ed.): CRYPTO 2004, LNCS 3152, pp. 562–578, 2004.

traditional signature schemes. For example, the authenticated broadcast protocol in [13] can be based on pseudo-signatures and then guarantees unconditional (instead of computational) security against any number of corrupted players [25].

A pseudo-signature scheme among a number of players can either be set up by a mutually trusted party, by a protocol among the players when given global broadcast channels, or—as we will show—by exploiting an information source that provides the players with certain correlated pieces of information—a similar model has been considered in [21] in the context of secret-key agreement.

In this paper, we consider the general case of an information source that provides a set of n players with pieces of information distributed according to some given joint probability distribution. For the case of three players, we completely characterize when such an information source allows for setting up a pseudo-signature scheme. This result can be used for deriving a complete characterization of when unconditionally secure three-party computation—or broadcast, in particular—is achievable in the presence of an actively corrupted player. Furthermore, we give, in the same model, a sufficient condition for the achievability of unconditionally secure multi-party computation for any number n of players secure against $t < n/2$ actively corrupted players.

1.2 Context and Previous Work

Pseudo-signature schemes (PSS). The first pseudo-signature-like scheme was given in form of an information-checking protocol among three players [26]. In contrast to real pseudo-signatures, however, the signer is required to commit to her input value already during the setup of the scheme.

The first PSS was introduced in [7] with the restriction to be secure only with respect to a correct signer. In [25], finally, a complete PSS was proposed for any transferability λ and any number of corrupted players.

Setting up a PSS. It was shown in [25] how to set up a PSS using global broadcast channels, where the *dining-cryptographers protocol* [5, 4] was used. Obtaining a PSS from a common random source was considered in [15, 16], but only with respect to three players and *one particular* probability distribution.

Broadcast. The *broadcast problem* was introduced in [20]. It was proven that, in the standard model with secure channels between all pairs of players, but without the use of a signature scheme, broadcast is achievable if and only if the number t of cheaters satisfies $t < n/3$. Furthermore, it was shown that when additionally a signature scheme is given among the n players, then computationally secure broadcast is achievable for any number of corrupted players. The first efficient such protocol was given in [12]. In [25], an efficient protocol was given with unconditional security based on a pseudo-signature scheme with transferability $t + 1$.

Multi-party computation (MPC). Broadcast—or the availability of signatures with sufficiently high transferability—is a limiting factor for general *multi-party computation* introduced in [27]. A complete solution with respect to computational security was given in [18]. In [2, 6], it was shown that in a model with only pairwise secure channels, MPC unconditionally secure against an active adversary is achievable if and only if $t < n/3$ players are corrupted. As shown in [1, 26], $t < n/2$ is achievable when global broadcast channels are additionally given—and this bound was shown tight. A protocol more efficient than those in [1, 26] was given in [10].

1.3   Our Results

We first consider a set of three players, connected in pairs by secure channels, where an additional information source provides the players with correlated pieces of information. We give a necessary and sufficient condition on the joint probability distribution of this side information for when a pseudo-signature scheme can be set up among the three players with a *designated* signer. Furthermore, we show that the tight condition for the achievability of broadcast or multi-party computation among three players unconditionally secure against one actively corrupted player is exactly the same as the one for a pseudo-signature scheme with respect to an *arbitrary*. The derived condition shows that pseudo-signature schemes and broadcast among three players are possible under much weaker conditions than previously known.

We further consider the general case of n players, connected in pairs by secure channels, where, again, an additional information source provides the players with side information. For this model, and under the assumption that an active adversary can corrupt up to $t < n/2$ players, we show that MPC is possible under much weaker conditions than previously known.

1.4   Model and Definitions

We consider a set $P = \{P_1, \ldots, P_n\}$ of n players that are connected by a complete, synchronous network of pairwise secure channels—in the presence of an active adversary who can select up to t players and corrupt them in an arbitrary way. Furthermore, we assume this adversary to be computationally unbounded. A player which does not get corrupted by the adversary is called *correct*.

Pseudo-signatures. We follow the definition of pseudo-signature schemes as given in [25].

Definition 1. A *pseudo-signature scheme (PSS) with transferability* λ among the players $P_1, \ldots, P_n$, where P_1 is the signer, satisfies the following properties.

Correctness. If player P_1 is correct and signs a message, then a correct player P_i accepts this message from P_1 except with small probability.
Unforgeability. A correct player P_i rejects any message that has not been signed by P_1 except with small probability.

Transferability. A message signed by the correct player P_1 can be transfered λ times, e.g., via

$$P_1 \to P_{i_1} \to \cdots \to P_{i_{\lambda+1}} ,$$

such that we have for each $j \leq \lambda$ and correct players P_{i_j} and $P_{i_{j+1}}$ that if P_{i_j} accepts a message m, then $P_{i_{j+1}}$ accepts the same message except with small probability.

If the path $i_1, \ldots, i_{\lambda+1}$ can be arbitrary, we call the scheme a *PSS with arbitrary transfer paths*, if the transfer is restricted to a specific path $i_1, \ldots, i_{\lambda+1}$, we call it a *PSS with transfer path* $i_1, \ldots, i_{\lambda+1}$.

The choice $\lambda = 1$ will be sufficient in our case since any such PSS allows for broadcast for $t < n/2$ corrupted players [17].

Broadcast and Multi-party Computation. *Broadcast* is the problem of having a (possibly corrupted) sender distribute a value to every player such that all correct players are guaranteed to receive the same value.

Definition 2. A protocol among players $P_1, \ldots, P_n$, where P_1 is the sender and holds input x_s, and where every player P_i computes an output y_i, achieves *broadcast* if it satisfies the following conditions.

Validity. If the sender P_1 is correct, then every correct player P_i computes the output $y_i = x_s$.
Consistency. All correct players P_i and P_j compute the same output value, i.e., $y_i = y_j$ holds.

Broadcast is a special case of the more general problem of *multi-party computation (MPC)*, where the players want to evaluate in a distributed way some given function of their inputs and hereby guarantee privacy of these inputs as well as correctness of the computed result. From a qualitative point of view, the security of multi-party computation is often broken down to the conditions *privacy, correctness, robustness*, and *fairness*. In [8], it was shown that all these conditions can only be satisfied simultaneously if $t < n/2$ holds—the case to which we restrict our considerations in this paper.

2   Dependent Parts and Simulation of Random Variables

In this section we introduce the notion of the dependent part of a random variable with respect to another, and a certain simulatability condition, defined for a triple of random variables. The *dependent part of X from Y* isolates the part of X that is dependent on Y. Note that we always assume that the joint distribution is known to all the players.

Definition 3. Let X and Y be two random variables, and let $f(x) = P_{Y|X=x}$. The *dependent part of X from Y* is defined as $X \searrow Y := f(X)$.

The random variable $X \searrow Y$ is a function of X and takes on the value of the conditional probability distribution $P_{Y|X=x}$.

Lemma 1. *For all X and Y, we have $X \longleftrightarrow (X \searrow Y) \longleftrightarrow Y$, i.e., the sequence $X, X \searrow Y, Y$ is a Markov chain[1].*

Proof. Let $K = f(X) = X \searrow Y$. For all $x \in \mathcal{X}$—the range of X—and $k = f(x)$, we have $P_{Y|X=x,K=k} = P_{Y|K=k}$, and, hence, $P_{Y|XK} = P_{Y|K}$.

We will now show that $K = X \searrow Y$ is the part of X that a player who knows Y can verify to be correct. Lemma 2 shows that every $\overline{K}$ a player knowing X can construct that has the same joint distribution with Y as the actual K must indeed be *identical* with K. Lemma 3 shows that from K, a random variable $\overline{X}$ can be constructed which has the same joint distribution with Y as X. Hence, K is the largest part of X that someone knowing Y can verify to be correct.

Lemma 2. *Let X, K, $\overline{K}$, and Y be random variables such that $K = X \searrow Y$, $Y \longleftrightarrow X \longleftrightarrow \overline{K}$, and $P_{KY} = P_{\overline{K}Y}$ hold. Then we have $\overline{K} = K$.*

Proof. We have $K = f(X)$, $P_{\overline{K}|XY} = P_{\overline{K}|X}$, $P_K = P_{\overline{K}}$, and $P_{Y|K} = P_{Y|\overline{K}}$. Let us have a look at a value k for which $P_{Y|K=k}$ cannot be expressed as a linear combination of $P_{Y|X=x_i}$ for $x_i \in \mathcal{X}$ with $f(x_i) \neq k$. (It is easy to see that such a k must exist.) Let S be the set of all x with $f(x) = k$. In order to achieve $P_{Y|\overline{K}=k} = P_{Y|K=k}$, no x' not in S can be mapped to k by $P_{\overline{K}|X}$. Since $P_{\overline{K}}(k) = P_K(k)$ holds, $P_{\overline{K}|X}$ must map all values from S to k.

We remove the elements of S from $\mathcal{X}$, repeat the same argument for the next k, and continue this process until $\mathcal{X}$ is empty. Hence, $P_{\overline{K}|X}$ maps all x to $f(x)$, and $\overline{K} = K$ holds.   □

Lemma 3. *Let X and Y be random variables, and let $K = X \searrow Y$. There exists a channel $P_{\overline{X}|K}$—which is equal to $P_{X|K}$—such that $P_{XY} = P_{\overline{X}Y}$ holds, where $P_{\overline{X}Y} = \sum_k P_{KY} P_{\overline{X}|K}$.*

Proof. Using Lemma 1, we get $P_{\overline{X}Y} = \sum_k P_{KY} P_{\overline{X}|K} = \sum_k P_{KY} P_{X|K} = P_{XY}$.
                                                  □

The *simulatability condition*, which allows for determining the possibility of secret-key agreement over unauthenticated channels, was defined in [22] and further analyzed in [24]. It defines whether given Z, it is possible to simulate X in such a way that someone who only knows Y cannot distinguish the simulation of X from the true X.

---

[1] A sequence of three random variables A, B, C forms a *Markov chain*, denoted by $A \longleftrightarrow B \longleftrightarrow C$, if $I(A; C|B) = 0$ holds or, equivalently, if we have $P_{C|AB}(c, a, b) = P_{C|B}(c, b)$ for all $(a, b, c) \in \mathcal{A} \times \mathcal{B} \times \mathcal{C}$.

Definition 4. Let X, Y, and Z be random variables. Then X is *simulatable by* Z *with respect to* Y, denoted by

$$\text{sim}_Y(Z \to X),$$

if there exists a conditional distribution $P_{\bar{X}|Z}$ such that $P_{\bar{X}Y} = P_{XY}$ holds, where $P_{\bar{X}Y} = \sum_z P_{YZ} P_{\bar{X}|Z}$.

Lemma 4. *For all P_{XYZ}, we have* $\text{sim}_Y(Z \to X)$ *if and only if*

$$\text{sim}_Y(Z \to (X \setminus Y)) .$$

Proof. Let $K := X \setminus Y$. K is a function of X and can be simulated whenever the same holds for X. On the other hand, let $P_{\bar{K}|Z}$ be a channel that simulates K. It follows from Lemma 3 that there exists a channel $P_{\bar{X}|\bar{K}}$—which is equal to $P_{X|K}$— such that the channel $P_{\bar{X}|Z} := \sum P_{\bar{X}|\bar{K}} P_{\bar{K}|Z}$ simulates X. □

Lemma 5. *For all P_{XYZ}, we have* $\text{sim}_Z(Y \to [X,Y])$ *if and only if* $X \longleftrightarrow Y \longleftrightarrow Z$.

Proof. Suppose first that we have $\text{sim}_Z(Y \to [X,Y])$. There must exist a channel $P_{\hat{X}\hat{Y}|Y}$ such that $P_{\hat{X}\hat{Y}Z} = P_{XYZ}$ holds, where $P_{\hat{X}\hat{Y}Z} = \sum P_{YZ} P_{\hat{X}\hat{Y}|Y}$. Let $K := Y \setminus Z$ and $\hat{K} := \hat{Y} \setminus Z$. Because of $Z \longleftrightarrow Y \longleftrightarrow \hat{Y}$ and $P_{\hat{Y}Z} = P_{YZ}$, we have $Z \longleftrightarrow Y \longleftrightarrow \hat{K}$ and $P_{\hat{K}Z} = P_{KZ}$. It follows from Lemma 2 that $\hat{K} = K$ holds. From Lemma 1 follows that $P_{Y|KZ} = P_{Y|K}$. We also have $P_{\hat{X}|\hat{K}\hat{Y}} = P_{\hat{X}|\hat{Y}}$. Now,

$$P_{\hat{X}\hat{Y}Z} = \sum_y P_{YZ} P_{\hat{X}\hat{Y}|Y} = \sum_y \sum_k P_{KZ} P_{Y|KZ} P_{\hat{X}\hat{Y}|Y}$$

$$= \sum_k P_{KZ} \sum_y P_{Y|K} P_{\hat{X}\hat{Y}|Y} = \sum_k P_{KZ} P_{\hat{X}\hat{Y}|K}$$

$$= \sum_k P_{\hat{K}Z} P_{\hat{X}\hat{Y}|\hat{K}} = \sum_k P_{\hat{K}Z} P_{\hat{Y}|\hat{K}} P_{\hat{X}|\hat{K}\hat{Y}} = P_{\hat{Y}Z} P_{\hat{X}|\hat{Y}} .$$

It follows that $P_{X|YZ} = P_{\hat{X}|\hat{Y}Z} = P_{\hat{X}|\hat{Y}} = P_{X|Y}$ holds and, hence, $X \longleftrightarrow Y \longleftrightarrow Z$.

Suppose now that we have $X \longleftrightarrow Y \longleftrightarrow Z$. It follows $P_{X|YZ} = P_{X|Y}$. Let $P_{\hat{X}\hat{Y}|Y} := P_{XY|Y}$. We get

$$P_{\hat{X}\hat{Y}Z} = \sum_y P_{YZ} P_{\hat{X}\hat{Y}|Y} = P_{YZ} P_{X|Y} = P_{YZ} P_{X|YZ} = P_{XYZ}.$$

□

3   Pseudo-signature Schemes

3.1   The Case of Three Players

We will state the exact condition under which a PSS can be set up from correlated pieces of information. We need the following lemma.

Lemma 6. *Let P_{XYZ} be the probability distribution of three random variables X, Y, and Z. Then the following three conditions are equivalent:*

1. *There exist two channels $P_{\overline{X}|X}$ and $P_{\hat{X}\hat{Y}|Y}$ such that $P_{\overline{X}Y} = P_{XY}$ and $P_{\overline{X}YZ} = P_{\hat{X}\hat{Y}Z}$ hold, where $P_{\overline{X}Y} = \sum_x P_{XY} P_{\overline{X}|X}$, $P_{\overline{X}YZ} = \sum_x P_{XYZ} P_{\overline{X}|X}$, and $P_{\hat{X}\hat{Y}Z} = \sum_y P_{YZ} P_{\hat{X}\hat{Y}|Y}$,*
2. $\mathrm{sim}_Z(Y \to [X \setminus Y, Y])$,
3. $(X \setminus Y) \longleftrightarrow Y \longleftrightarrow Z$.

Proof. Lemma 5 implies that 2. and 3. are equivalent. In the following we will prove that 1. and 2. are equivalent.

Assume that 1. is true. We have $\mathrm{sim}_Z(Y \to [\overline{X}, Y])$ for some $\overline{X}$ with $P_{XY} = P_{\overline{X}Y}$ and $Y \longleftrightarrow X \longleftrightarrow \overline{X}$. Let $K = X \setminus Y$ and $\overline{K} = \overline{X} \setminus Y$. We have $P_{KY} = P_{\overline{K}Y}$ and $Y \longleftrightarrow X \longleftrightarrow \overline{K}$. From Lemma 2, it follows $K = \overline{K}$. Since $\overline{K}$ is a function of $\overline{X}$, we get $\mathrm{sim}_Z(Y \to [X \setminus Y, Y])$.

Assume now that 2. is true. Hence, there exists a channel $P_{\overline{KY}|Y}$ such that $P_{\overline{KY}Z} = P_{KYZ}$ holds for $K := X \setminus Y$. Lemma 3 implies that there exists a channel $P_{\overline{X}|K}$—which is equal to $P_{X|K}$—such that $P_{\overline{X}Y} = P_{XY}$ holds. We set $P_{\hat{X}\hat{Y}|Y} := \sum_k P_{X|K} P_{\overline{KY}|Y}$ and $P_{\overline{X}|X} := \sum_k P_{X|K} P_{K|X}$ to get

$$P_{\overline{X}Y} = \sum_x P_{XY} P_{\overline{X}|X} = \sum_x P_{XY} \sum_k P_{\overline{X}|K} P_{K|X} = \sum_k P_{\overline{X}|K} \sum_x P_{XY} P_{K|X}$$

$$= \sum_k P_{\overline{X}|K} P_{KY} = \sum_k P_{X|K} P_{KY} = P_{XY} ,$$

$$P_{\hat{X}\hat{Y}Z} = \sum_y P_{YZ} P_{\hat{X}\hat{Y}|Y} = \sum_y P_{YZ} \sum_k P_{\hat{X}|K} P_{\overline{KY}|Y} = \sum_k P_{\hat{X}|K} \sum_y P_{YZ} P_{\overline{KY}|Y}$$

$$= \sum_k P_{\hat{X}|K} P_{\overline{KY}Z} = \sum_k P_{\hat{X}|K} P_{KYZ} = \sum_k P_{\hat{X}|K} \sum_x P_{XYZ} P_{K|X}$$

$$= \sum_x P_{XYZ} \sum_k P_{\hat{X}|K} P_{K|X} = \sum_x P_{XYZ} P_{\overline{X}|X} = P_{\overline{X}YZ} .$$

$\square$

Our pseudo-signature protocol makes use of typical sequences. Intuitively, a sequence of independent realizations of a random variable is *typical* if the actual rate of occurrences of every specific outcome symbol in the sequence is close to the probability of this symbol.

Definition 5. [3, 9] Let X be a random variable with distribution P_X and range $\mathcal{X}$, let $n > 0$ be an integer, and let $\gamma > 0$. A sequence $x^n = (x_1, \ldots, x_n) \in \mathcal{X}^n$

is called *(strongly) γ-typical* if, for all $a \in \mathcal{X}$, the actual number $N(a, x^n)$ of appearances of a in x^n satisfies

$$\left| \frac{N(a, x^n)}{n} - P_X(a) \right| \leq \frac{\gamma}{|\mathcal{X}|} .$$

It is a consequence of the law of large numbers that for every $\gamma > 0$, sufficiently long sequences of independent realizations of a random variable are γ-typical with overwhelming probability.

Theorem 1. [3, 9] *Let $X^n = X_1 \cdots X_n$ be a sequence of n independent realizations of the random variable X with distribution P_X and range $\mathcal{X}$, and let $0 < \gamma \leq 1/2$. Then*

$$\text{Prob}\left[X^n \text{ is strongly } \gamma\text{-typical} \right] = 1 - 2^{-\Omega(n\gamma^2)} .$$

The following protocol allows P_1 for signing a bit along the transfer path $P_1 \rightarrow P_2 \rightarrow P_3$.

Protocol 1 Let P_{XYZ} be such that $\text{sim}_Z(Y \rightarrow [X \searrow Y, Y])$ does not hold. Let $K := X \searrow Y$ and $L := [K, Y] \searrow Z$. Lemma 4 implies that there must exist $\delta > 0$ such that for all channels $P_{\overline{L}|Y}$, the statistical distance between the distributions P_{LZ} and $P_{\overline{L}Z}$ is at least δ.

Let n be an even integer, and let $(X_1, Y_1, Z_1), \ldots, (X_n, Y_n, Z_n)$ be n triples distributed independently according to P_{XYZ}. Let $\gamma > 0$ be a security parameter and n be large enough. Let P_1, P_2, and P_3 know $(X_1, \ldots, X_n)$, $(Y_1, \ldots, Y_n)$, and $(Z_1, \ldots, Z_n)$, respectively. Let, finally, $m \in \{0, 1\}$ be the value P_1 wants to sign.

- P_1 calculates $K_i := X_i \searrow Y_i$ and sends $(m, K_{1+(n/2)m}, \ldots, K_{n/2+(n/2)m})$ to P_2.
- P_2 checks whether the received K_i and the corresponding Y_i are a γ-typical sequence with respect to P_{KY}. If so, he accepts, calculates $L_i := [K_i, Y_i] \searrow Z_i$, and sends $(m, L_{1+(n/2)m}, \ldots, L_{n/2+(n/2)m})$ to P_3.
- P_3 checks whether the received L_i and the corresponding Z_i are a $\delta/2$-typical sequence with respect to P_{LZ}. If so, he accepts.

Theorem 2. *Let $(X_1, Y_1, Z_1), \ldots, (X_n, Y_n, Z_n)$ be n triples distributed independently according to P_{XYZ}. Let P_1, P_2, and P_3 know the values X_i, Y_i, and Z_i, respectively. Let P_1 be able to send messages to P_2, and P_2 to P_3.*

If $\text{sim}_Z(Y \rightarrow [X \searrow Y, Y])$ does not hold and n is large enough, then Protocol 1 achieves PSS for the three players with the transfer path $P_1 \rightarrow P_2 \rightarrow P_3$.

Proof. We prove that Protocol 1 implements a PSS. First of all, it follows from Theorem 1 that the value from a correct sender P_1 is accepted by P_2 except with exponentially small probability. If P_2 is correct and accepts a value and if γ is small enough, Lemma 2 implies that P_1 must indeed have sent an arbitrarily large fraction (for sufficiently large N) of correct values $K_i = X_i \searrow Y_i$ to P_2.

(Note that the knowledge of the values X_j for $j \neq i$ do not help P_1 to cheat since they are independent of X_i and Y_i.)

Therefore, also an arbitrarily large fraction of the values $L_i = [K_i, Y_i] \searrow Z_i$ are correct and—if P_3 is correct—P_3 will accept the values L_i sent to him by P_2 (except with exponentially small probability).

P_2, however, cannot (except with exponentially small probability) send any other value than the one sent by P_1. Indeed, his ability to do so would imply the existence of a channel $P_{\overline{L}|Y}$ such that $P_{\overline{L}Z}$ and P_{LZ} are identical (see the proof of Lemma 6 in [23]); such a channel, however, does not exist because of the assumption stated at the beginning of the protocol.   □

We now show that the condition of Theorem 2 for the achievability of a PSS among three players is tight, in other words, that $\mathrm{sim}_Z(Y \to [X \searrow Y, Y])$ and $\mathrm{sim}_Y(Z \to [X \searrow Z, Z])$ imply that no PSS with signer P_1 is possible. In order to demonstrate impossibility, we use a similar technique as in [14]. There, the impossibility of broadcast among three players secure against one corrupted player was shown by analyzing a related system obtained by copying some of the players and rearranging the original players together with their copies in a specific way.

Theorem 3. *Let $(X_1, Y_1, Z_1), \ldots, (X_n, Y_n, Z_n)$ be n triples distributed independently according to P_{XYZ}. Let P_1, P_2, and P_3 know the values X_i, Y_i, and Z_i, respectively. Let the players be connected by pairwise secure channels.*

If $\mathrm{sim}_Z(Y \to [X \searrow Y, Y])$ and $\mathrm{sim}_Y(Z \to [X \searrow Z, Z])$ hold, then there does not exist—for any n—a PSS for the three players with any transfer path and with P_1 as the signer.

Proof. Let us assume that there exists a protocol among the players P_1, P_2, and P_3 that achieves a PSS for the three players with transfer path $P_1 \to P_2 \to P_3$. From Lemma 6, it follows that there exist channels $P_{\overline{X}|X}$ and $P_{\hat{X}\hat{Y}|Y}$ such that $P_{XY} = P_{\overline{X}Y}$ and $P_{\overline{X}YZ} = P_{\hat{X}\hat{Y}Z}$ hold, and $P_{\overline{X}'|X}$ and $P_{\hat{X}'\hat{Z}'|Z}$ such that $P_{XZ} = P_{\overline{X}'Z}$ and $P_{\overline{X}'YZ} = P_{\hat{X}'Y\hat{Z}'}$ hold.

Let P_1' be an identical copy of P_1. We now rearrange the four players P_1, P_2, P_3, and P_1' in the following way to form a new system. The analysis of that system then reveals that no PSS among the three *original* players is possible. Note that, in the new system, no player is corrupted: It is rather the *arrangement* of this new system that simulates corruption in the original system towards the players in the new system.

- P_1 is still connected to P_2 as originally, but disconnected from P_3, i.e., all messages P_1 would send to P_3 are discarded and no message P_3 would send to P_1 is ever received by P_1.
- P_2 is still connected to P_1 and P_3 as in the original system.
- P_3 is still connected to P_2 as originally, but disconnected from P_1. Instead, P_3 is connected to P_1': All messages that P_3 would send to P_1 are delivered to P_1' instead, and all messages P_1' would send to P_3 are indeed delivered to P_3.
- P_1' is connected to P_3 as originally, but disconnected from P_2.

Furthermore, instead of X_i, let P_1 have input $\overline{X}_i$ and P_1' have input $\overline{X}_i'$. Let them execute their local programs defined by the PSS protocol, where P_1 signs the message m and P_1' signs the message m'.

- Since $P_{XY} = P_{\overline{X}Y}$ holds, the joint view among P_1 and P_2 is indistinguishable from their view in the original system where P_1 holds input m and P_3 is corrupted in the following way: P_3 cuts off communication to P_1, simulates P_1' using the channel $P_{\hat{X}'\hat{Z}'|Z}$ to produce the values $\hat{X}'$ and $\hat{Z}'$, and acts towards P_2 as if communicating with P_1' instead of P_1 (indistinguishability follows from $P_{\overline{X}'YZ} = P_{\hat{X}'Y\hat{Z}'}$). Hence, by the *correctness property*, P_2 must accept m as signed by P_1.
- The joint view of P_2 and P_3 is indistinguishable from their view in the original system where P_1 is corrupted in the following way: P_1 simulates player P_1', uses the channel $P_{\overline{X}|X}$ for his own and the channel $P_{\overline{X}'|X}$ for P_1''s input, and acts towards P_3 as P_1'. Thus, by the *transferability property*, P_3 must accept the transfered message m from P_2.
- Since $P_{XZ} = P_{\overline{X}'Z}$ holds, the joint view of P_1' and P_3 is indistinguishable from their view in the original system[2] where P_1' holds input m' and P_2 is corrupted in the following way: P_2 cuts off communication to P_1', simulates P_1 using the channel $P_{\hat{X}\hat{Y}|Y}$ to produce the values $\hat{X}$ and $\hat{Y}$, and acts towards P_3 as if communicating with P_1 instead of P_1' (indistinguishability follows from $P_{\overline{X}YZ} = P_{\hat{X}\hat{Y}Z}$). Hence, by the *unforgeability property*, P_3 must reject the signature transferred to him by P_2.

However, this is impossible since P_3 cannot accept and reject m at the same time. The proof for the transfer path $P_1 \rightarrow P_3 \rightarrow P_2$ is analogous. Hence, there does not exist a PSS for *any* transfer path. □

If the condition of Theorem 3 does not hold, then there exists a transfer path—namely either $P_1 \rightarrow P_2 \rightarrow P_3$ or $P_1 \rightarrow P_3 \rightarrow P_2$—for which Theorem 2 can be applied. Therefore, the bound of Theorem 3 is tight, and we can state the *exact* condition under which a PSS for three players and a designated signer exists.

Theorem 4. *Let* $(X_1, Y_1, Z_1), \ldots, (X_n, Y_n, Z_n)$ *be* n *triples distributed independently according to* P_{XYZ}. *Let* P_1, P_2, *and* P_3 *know the values* X_i, Y_i, *and* Z_i, *respectively. Let the players pairwisely be connected by secure channels.*

There exists a PSS for the three players with transfer path $P_1 \rightarrow P_j \rightarrow P_k$ $(j \neq k)$ *for large enough* n *if and only if either* $\mathrm{sim}_Z(Y \rightarrow [X \searrow Y, Y])$ *or* $\mathrm{sim}_Y(Z \rightarrow [X \searrow Z, Z])$ *does not hold.*

Application of Lemma 5 leads to the following corollary.

---

[2] For simplicity, we assume the original system to consist of the players $\{P_1', P_2, P_3\}$ for this case.

Corollary 1. *Let* $(X_1, Y_1, Z_1), \ldots, (X_n, Y_n, Z_n)$ *be* n *triples distributed independently according to* P_{XYZ}. *Let* P_1, P_2, *and* P_3 *know the values* X_i, Y_i, *and* Z_i, *respectively. Let the players pairwisely be connected by secure channels.*

There exists a PSS for the three players with transfer path $P_1 \to P_j \to P_k$ *(*$j \neq k$*) for large enough* n *if and only if either* $(X \searrow Y) \longleftrightarrow Y \longleftrightarrow Z$ *or* $(X \searrow Z) \longleftrightarrow Z \longleftrightarrow Y$ *does not hold.*

We will now present a special case of noisy channels among three players for which our PSS works. This special case is related to the "satellite scenario" of [21] for secret-key agreement.

Corollary 2. *Let* R *be a binary random variable and let* $X, Y,$ *and* Z *be random variables resulting from the transmission of* R *over three binary symmetric channels with error probabilities* $\varepsilon_X, \varepsilon_Y,$ *and* ε_Z, *respectively, such that* $0 \leq \varepsilon_X < 1/2$, $0 < \varepsilon_Y < 1/2$ *and* $0 < \varepsilon_Z < 1/2$ *hold. Let* $(X_1, Y_1, Z_1), \ldots, (X_n, Y_n, Z_n)$ *be* n *triples generated independently this way. Let* P_1, P_2, *and* P_3 *be three players and assume that they know* X_i, Y_i, *and* Z_i, *respectively. Let, finally, the players pairwisely be connected by secure channels. Then, for large enough* n, *there exists a PSS for the three players with arbitrary transfer path.*

Proof. We have that $X \searrow Y$, $X \searrow Z$ and X are—up to renaming—equal, and neither $X \longleftrightarrow Y \longleftrightarrow Z$ nor $X \longleftrightarrow Z \longleftrightarrow Y$ holds.   □

Corollary 3. *Let the players* P_1, P_2, *and* P_3 *be connected by a noisy broadcast channel. This is a channel for which* P_1 *has an input bit* X, *and* P_2 *and* P_3 *get output bits* Y *and* Z, *respectively, which result from sending* X *over two independent noisy channels with error probabilities* $0 < \varepsilon_Y < 1/2$ *and* $0 < \varepsilon_Z < 1/2$. *Then a PSS for the three players with arbitrary transfer path can be realized.*

Proof. Let the transfer path be $P_1 \to P_2 \to P_3$. P_1 sends n random bits over the channel. Both P_2 and P_3 check whether the received values are indeed random, that is, whether they are γ_2- and γ_3-typical. The values γ_2 and γ_3 are chosen such that even if P_1 cheats, P_2 does not accept if P_3 does not either—except with small probability. The resulting joint distribution satisfies the condition of Corollary 2.   □

3.2   The Case of More Than Three Players

Theorem 2 can be generalized to $p > 3$ players in a natural way. Assume that p players $P_1, \ldots, P_p$ want to implement a PSS along the transfer path $P_1 \to \cdots \to P_p$. Let $(X_1^1, \ldots, X_1^p), \ldots, (X_n^1, \ldots, X_n^p)$ be n lists distributed independently according to $P_{X^1 \ldots X^p}$. Let player P_j know the values X_i^j.

As in the protocol for three players, player P_1 sends m together with his signature $(m, K_{1+(n/2)m}^1, \ldots, K_{n/2+(n/2)m}^1)$, where $K_i^1 := X_i^1 \searrow X_i^2$, to P_2. P_2 is able to check whether P_1 sent the correct values K_i^1 or not, and he only accepts the signature if almost all values K_i^1 were correct.

Now we let P_2 sign the value m himself, using the random variable $[X_i^2, K_i^1]$. (Since he only received half of the values K_i^1, he is able to sign m, *but not* $1 - m$.) He sends $(m, K_{1+(n/2)m}^2, \ldots, K_{n/2+(n/2)m}^2)$, where $K_i^2 := [X_i^2, K_i^1] \searrow X_i^3$, to P_3. Now P_3 can check the signature and, if he accepts, sign the value m himself, and so forth. Note that the security parameter for every signature must be less restrictive than the previous one, because some of the received K_i^j may have been faulty. Nevertheless, the error probability remains exponentially small in n. Player P_j is not able to forge a signature if

$$\text{sim}_{X^{j+1}}(X^j \to [K^{j-1}, X^j])$$

does not hold. Hence, we get the following theorem.

Theorem 5. *Let $(X_1^1, \ldots, X_1^p), \ldots, (X_n^1, \ldots, X_n^p)$ be n lists distributed independently according to $P_{X^1 \ldots X^p}$. Let $P_1, \ldots, P_p$ be p players, and let P_j know all the X_i^j. Assume that for all i, player P_i can send messages to P_{i+1} in a secure way (where $P_{p+1} = P_1$). Let $K^1 := X^1 \searrow X^2$ and $K^j := [X^j, K^{j-1}] \searrow X^{j+1}$ for $j \in \{2, \ldots, n-1\}$.*

Then, for large enough n, there exists a PSS for p players with the transfer path $P_1 \to \cdots \to P_p$ and tolerating one corrupted player if there does not exist $j \geq 2$ with

$$\text{sim}_{X^{j+1}}(X^j \to [K^{j-1}, X^j]) .$$

4   Broadcast and Multi-party Computation

4.1   The Case of Three Players

We will now apply the results of Section 3 and state the exact condition under which broadcast is possible for three players.

Theorem 6. *Let $(X_1, Y_1, Z_1), \ldots, (X_n, Y_n, Z_n)$ be n triples distributed independently according to P_{XYZ}. Assume that P_1, P_2, and P_3 know the values X_i, Y_i, and Z_i, respectively. Let all players pairwisely be connected by secure channels.*

If n is large enough and $\text{sim}_Z(Y \to [X \searrow Y, Y])$ or $\text{sim}_Y(Z \to [X \searrow Z, Z])$ does not hold, then there exists a broadcast protocol for three players with sender P_1.

Proof. If either $\text{sim}_Z(Y \to [X \searrow Y, Y])$ or $\text{sim}_Y(Z \to [X \searrow Z, Z])$ does not hold, it is possible to set up a PSS with either the transfer path $P_1 \to P_2 \to P_3$ or $P_1 \to P_3 \to P_2$. It was shown in [16] that this is sufficient to construct a broadcast protocol for three players. □

Theorem 7. *Let $(X_1, Y_1, Z_1), \ldots, (X_n, Y_n, Z_n)$ be n triples distributed independently according to P_{XYZ}. Assume that P_1, P_2, and P_3 know the values X_i, Y_i, and Z_i, respectively. Let all players pairwisely be connected by secure channels.*

If both $\text{sim}_Z(Y \to [X \searrow Y, Y])$ and $\text{sim}_Y(Z \to [X \searrow Z, Z])$ hold, then there exists no broadcast protocol (for any n) for three players with sender P_1.

Proof. From Lemma 6, it follows that there exist channels $P_{\overline{X}|X}$ and $P_{\hat{X}\hat{Y}|Y}$ such that $P_{XY} = P_{\overline{X}Y}$ and $P_{\overline{X}YZ} = P_{\hat{X}\hat{Y}Z}$ hold, as well as $P_{\overline{X}'|X}$ and $P_{\hat{X}'\hat{Z}'|Z}$ such that $P_{XZ} = P_{\overline{X}'Z}$ and $P_{\overline{X}'YZ} = P_{\hat{X}'Y\hat{Z}'}$ hold.

As in the proof of Theorem 3, we duplicate the sender P_1 and rearrange the four resulting players in the following way: We disconnect P_1 and P_3 but connect P_3 to P_1' instead, whereas P_2 stays connected as originally.

P_1 gets input $\overline{X}_i$, constructed by applying the channel $P_{\overline{X}|X}$ on X_i. P_1' gets input $\overline{X}'$, constructed by applying the channel $P_{\overline{X}'|X}$ on X_i. P_2 gets input Y_i, and P_3 gets input Z_i.

We give P_1 and P_1' two different inputs m and m' and let them all execute the protocol; they all output a value. We now consider three scenarios of an original system involving some of the players P_1, P_1', P_2, and P_3 of the new system obtained by interconnecting all four players as described above.

- Let P_1 and P_2 be correct and P_3 be corrupted. Using his variables Z_i, P_3 can produce $\hat{X}_i'$ and $\hat{Z}_i'$ such that P_2 cannot distinguish them from $\overline{X}_i'$ and Z_i. Furthermore, P_2 cannot distinguish $\overline{X}_i$, which he receives from P_1, from X_i. P_3 simulates P_1', giving him the values $\hat{X}_i'$ as input, and using the values $\hat{Z}_i'$ himself.
- Let P_1' and P_3 be correct and P_2 be corrupted. Using his variables Y_i, P_2 can produce $\hat{X}_i$ and $\hat{Y}_i$ such that P_3 cannot distinguish them from $\overline{X}_i$ and Y_i. Furthermore, P_3 cannot distinguish $\overline{X}_i'$, which he receives from P_1, from X_i. P_2 simulates P_1, giving him the values $\hat{X}_i$ as input, and using the values $\hat{Y}_i$ himself.
- Let P_2 and P_3 be correct and P_1 be corrupted. Using his variables X_i, P_1 can produce $\overline{X}_i$ and $\overline{X}_i'$. He can simulate player P_1' with $\overline{X}_i'$ as input and use $\overline{X}_i$ for himself.

The joint view of the players P_1 and P_2 in the new system is indistinguishable from their view in the first scenario, and they must thus output m. The joint view of the players P_1' and P_3 in the new system is indistinguishable from their joint view in the second scenario, and they, therefore, output m'. But also the joint view of players P_2 and P_3 in the new system is indistinguishable from their view in the third scenario, and thus they must agree on their output value, which contradicts what we derived above. Therefore, no broadcast protocol can exist. $\square$

Using Theorems 6 and 7 we can now state the exact condition under which broadcast and MPC among three players are possible.

Theorem 8. *Let* $(X_1, Y_1, Z_1), \ldots, (X_n, Y_n, Z_n)$ *be* n *triples distributed independently according to* P_{XYZ}. *Let* P_1, P_2, *and* P_3 *know the values* X_i, Y_i, *and* Z_i, *respectively. Let all players pairwisely be connected by secure channels. Broadcast with sender* P_1 *is possible if and only if*

$$\neg \left(\mathrm{sim}_Z(Y \to [X \searrow Y, Y]) \land \mathrm{sim}_Y(Z \to [X \searrow Z, Z]) \right)$$

holds.

Corollary 4. *Let* $(X_1, Y_1, Z_1), \ldots, (X_n, Y_n, Z_n)$ *be* n *triples distributed independently according to* P_{XYZ}. *Let* P_1, P_2, *and* P_3 *know the values* X_i, Y_i, *and* Z_i, *respectively. Let all players pairwisely be connected by secure channels.*
Broadcast with sender P_1 *is possible if and only if*

$$\neg \Big((X \searrow Y) \longleftrightarrow Y \longleftrightarrow Z \;\wedge\; (X \searrow Z) \longleftrightarrow Z \longleftrightarrow Y \Big)$$

holds.

Lemma 7. *Given three players* P_1, P_2, *and* P_3, *connected pairwisely by secure channels and additionally by broadcast channels from* P_1 *to* $\{P_2, P_3\}$ *and from* P_2 *to* $\{P_1, P_3\}$ *(but no other primitive such as a PSS among the players). Then broadcast from* P_3 *to* $\{P_1, P_2\}$ *is impossible.*

Proof. This follows from a generalization of the proof in [14], where only pairwise channels are assumed. □

Theorem 9. *Let* $(X_1, Y_1, Z_1), \ldots, (X_n, Y_n, Z_n)$ *be* n *triples distributed independently according to* P_{XYZ}. *Let* P_1, P_2, *and* P_3 *know the values* X_i, Y_i, *and* Z_i, *respectively. Let all players pairwisely be connected by secure channels.*
Broadcast with arbitrary sender as well as general multi-party computation secure against one corrupted player are possible if and only if

$$\neg \Big(\mathrm{sim}_Z(Y \to [X \searrow Y, Y]) \;\wedge\; \mathrm{sim}_Y(Z \to [X \searrow Z, Z]) \Big) \;\wedge$$
$$\neg \Big(\mathrm{sim}_X(Z \to [Y \searrow Z, Z]) \;\wedge\; \mathrm{sim}_Z(X \to [Y \searrow X, X]) \Big) \;\wedge$$
$$\neg \Big(\mathrm{sim}_X(Y \to [Z \searrow Y, Y]) \;\wedge\; \mathrm{sim}_Y(X \to [Z \searrow X, X]) \Big)$$

holds.

Proof. The condition is sufficient for the possibility of broadcast because of Theorem 8 and Lemma 7. The achievability of multi-party computation then follows from [1, 26, 10]. Furthermore, since broadcast is a special case of multi-party computation, the impossibility of broadcast immediately implies the impossibility of MPC. □

Corollary 5. *Let* $(X_1, Y_1, Z_1), \ldots, (X_n, Y_n, Z_n)$ *be* n *triples distributed independently according to* P_{XYZ}. *Let* P_1, P_2, *and* P_3 *know the values* X_i, Y_i, *and* Z_i *respectively. Let all players pairwisely be connected by secure channels.*
Broadcast with arbitrary sender as well as general multi-party computation secure against one corrupted player are possible if and only if

$$\neg \Big((X \searrow Y) \longleftrightarrow Y \longleftrightarrow Z \;\wedge\; (X \searrow Z) \longleftrightarrow Z \longleftrightarrow Y \Big) \;\wedge$$
$$\neg \Big((Y \searrow Z) \longleftrightarrow Z \longleftrightarrow X \;\wedge\; (Y \searrow X) \longleftrightarrow X \longleftrightarrow Z \Big) \;\wedge$$
$$\neg \Big((Z \searrow Y) \longleftrightarrow Y \longleftrightarrow X \;\wedge\; (Z \searrow X) \longleftrightarrow X \longleftrightarrow Y \Big)$$

holds.

4.2   The Case of More than Three Players

Corollary 6. *Let $P_1, \ldots, P_n$ be n players. Let all players pairwisely be connected by secure channels. Furthermore, let every triple of players (P_i, P_j, P_k) have enough independent realizations of X^i, X^j, and X^k, respectively, such that either $\mathrm{sim}_{X^k}(X^j \to [X^i \setminus X^j, X^j])$ or $\mathrm{sim}_{X^j}(X^k \to [X^i \setminus X^k, X^k])$ does not hold. Then broadcast and multi-party computation unconditionally secure against $t < n/2$ corrupted players are achievable.*

Proof. From Theorem 9, it follows that any triple of players can execute a broadcast protocol. Using the protocol from [17], broadcast for n players tolerating $t < n/2$ corrupted players can be achieved. Using [1, 26, 10], a protocol for unconditional MPC can be constructed that can tolerate $t < n/2$ corrupted players. □

5   Concluding Remarks

In the model of unconditional security, we have completely characterized the possibility of pseudo-signatures, broadcast, and secure multi-party computation among three players having access to certain correlated pieces of information. Interestingly, this condition is closely related to a property called (non-) simulatability previously studied in an entirely different context, namely information-theoretic secret-key agreement.

As a consequence of this result, we gave a new, weaker condition for the possibility of achieving unconditional broadcast between n players and any minority of cheaters and, hence, general multi-party computation under the same conditions.

Acknowledgments

The first author was supported by the Packard Foundation and the second and the third author by Canada's NSERC.

References

1. Donald Beaver. Multiparty protocols tolerating half faulty processors. In *Advances in Cryptology: CRYPTO '89*, volume 435 of *Lecture Notes in Computer Science*, pages 560–572. Springer-Verlag, 1989.
2. Michael Ben-Or, Shafi Goldwasser, and Avi Wigderson. Completeness theorems for non-cryptographic fault-tolerant distributed computation. In *Proceedings of the 20th Annual ACM Symposium on Theory of Computing (STOC '88)*, pages 1–10. Springer-Verlag, 1988.
3. Richard E. Blahut. *Principles and practice of information theory.* Addison-Wesley, Reading, MA, 1988.
4. Jurjen Bos and Bert den Boer. Detection of disrupters in the DC protocol. In *Advances in Cryptology: EUROCRYPT '89*, volume 434 of *Lecture Notes in Computer Science*, pages 320–327. Springer-Verlag, 1990.

5. David Chaum. The Dining Cryptographers Problem: Unconditional sender and recipient untraceability. *Journal of Cryptology*, 1(1):65–75, 1988.
6. David Chaum, Claude Crépeau, and Ivan Damgård. Multiparty unconditionally secure protocols (extended abstract). In *Proceedings of the 20th Annual ACM Symposium on Theory of Computing (STOC '88)*, pages 11–19. ACM Press, 1988.
7. David Chaum and Sandra Roijakkers. Unconditionally-secure digital signatures. In *Advances in Cryptology: CRYPTO '90*, volume 537 of *Lecture Notes in Computer Science*, pages 206–214. Springer-Verlag, 1990.
8. Richard Cleve. Limits on the security of coin flips when half the processors are faulty. In *ACM Symposium on Theory of Computing (STOC '86)*, pages 364–369, ACM Press, 1986.
9. Thomas M. Cover and Joy A. Thomas. *Elements of Information Theory*. Wiley-Interscience, New York, USA, 1991.
10. Ronald Cramer, Ivan Damgård, Stefan Dziembowski, Martin Hirt, and Tal Rabin. Efficient multiparty computations secure against an adaptive adversary. In *Advances in Cryptology: EUROCRYPT '99*, volume 1592 of *Lecture Notes in Computer Science*, pages 311–326. Springer-Verlag, 1999.
11. Whitfield Diffie and Martin Hellman. New directions in cryptography. *IEEE Transactions on Information Theory*, IT-22(6):644–654, 1976.
12. Danny Dolev and H. Raymond Strong. Polynomial algorithms for multiple processor agreement. In *Proceedings of the 14th Annual ACM Symposium on Theory of Computing (STOC '82)*, pages 401–407, 1982.
13. Danny Dolev and H. Raymond Strong. Authenticated algorithms for Byzantine agreement. *SIAM Journal on Computing*, 12(4):656–666, 1983.
14. Michael J. Fischer, Nancy A. Lynch, and Michael Merritt. Easy impossibility proofs for distributed consensus problems. *Distributed Computing*, 1:26–39, 1986.
15. Matthias Fitzi, Nicolas Gisin, and Ueli Maurer. Quantum solution to the Byzantine agreement problem. *Physical Review Letters*, 87(21):7901-1–7901-4, 2001.
16. Matthias Fitzi, Nicolas Gisin, Ueli Maurer, and Oliver von Rotz. Unconditional Byzantine agreement and multi-party computation secure against dishonest minorities from scratch. In *Advances in Cryptology: EUROCRYPT '02*, volume 2332 of *Lecture Notes in Computer Science*, pages 482–501. Springer-Verlag, 2002.
17. Matthias Fitzi and Ueli Maurer. From partial consistency to global broadcast. In *32nd Annual Symposium on Theory of Computing, STOC '00*, pages 494–503. ACM, 2000.
18. Oded Goldreich, Silvio Micali, and Avi Wigderson. How to play any mental game. In *Proceedings of the 19th Annual ACM Symposium on Theory of Computing (STOC '87)*, pages 218–229. ACM Press, 1987.
19. Leslie Lamport. Constructing digital signatures from a one-way function. Technical Report SRI-CSL-98, SRI International Computer Science Laboratory, 1979.
20. Leslie Lamport, Robert Shostak, and Marshall Pease. The Byzantine generals problem. *ACM Transactions on Programming Languages and Systems*, 4(3):382–401, 1982.
21. Ueli Maurer. Secret key agreement by public discussion. *IEEE Transaction on Information Theory*, 39(3):733–742, 1993.
22. Ueli Maurer. Information-theoretically secure secret-key agreement by NOT authenticated public discussion. In *Advances in Cryptology: EUROCRYPT '97*, volume 49 of *Lecture Notes in Computer Science*, pages 209–225. Springer-Verlag, 1997.

23. Ueli M. Maurer and Stefan Wolf. Secret-key agreement over unauthenticated public channels—Part I: Definitions and a completeness result. *IEEE Transactions on Information Theory*, 49:822–831, 2003.

24. Ueli M. Maurer and Stefan Wolf. Secret-key agreement over unauthenticated public channels—Part II: The simulatability condition. *IEEE Transactions on Information Theory*, 49:832–838, 2003.

25. Birgit Pfitzmann and Michael Waidner. Information-theoretic pseudosignatures and Byzantine agreement for $t >= n/3$. Technical Report RZ 2882 (#90830), IBM Research, 1996.

26. Tal Rabin and Michael Ben-Or. Verifiable secret sharing and multiparty protocols with honest majority. In *Proceedings of the 21st Annual ACM Symposium on Theory of Computing (STOC '89)*, pages 73–85, 1989.

27. Andrew C. Yao. Protocols for secure computations. In *Proceedings of the 23rd Annual IEEE Symposium on Foundations of Computer Science (FOCS '82)*, pages 160–164, 1982.

Author Index

Lecture Notes in Computer Science

For information about Vols. 1–3047

please contact your bookseller or Springer-Verlag